ADVANCES IN CHEMICAL PHYSICS

VOLUME 139

EDITORIAL BOARD

ADVANCES IN CHEMICAL PHYSICS

VOLUME 139

Series Editor

STUART A. RICE

Department of Chemistry
and
The James Franck Institute
The University of Chicago
Chicago, Illinois

WILEY-INTERSCIENCE
A JOHN WILEY & SONS, INC. PUBLICATION

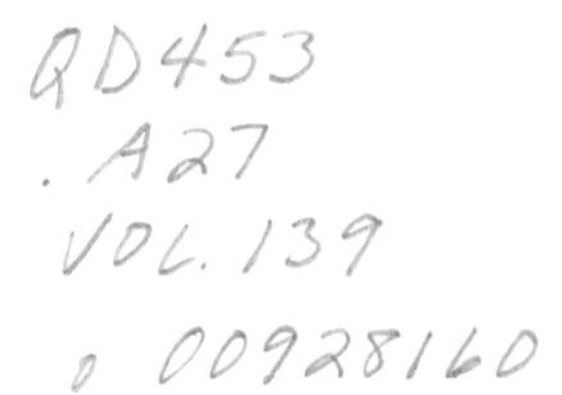

Published by John Wiley & Sons, Inc., Hoboken, New Jersey
Published simultaneously in Canada

Library of Congress Catalog Number: 58-9935

ISBN: 978-0-470-25389-2

Printed in the United States of America

10 9 8 7 6 5 4 3 2 1

CONTRIBUTORS TO VOLUME 139

PAUL BLAISE, Laboratoire de Mathématiques, Physique et Systèmes, Université de Perpignan, 66860 Perpignan Cedex, France

JEAN-MARC BOMONT, Laboratoire de Physique des Milieux Denses, Université Paul Verlaine, 57078 Metz, France

OLIVIER HENRI-ROUSSEAU, Laboratoire de Mathématiques, Physique et Systèmes, Université de Perpignan, 66860 Perpignan Cedex, France

MARK N. KOBRAK, Department of Chemistry, Brooklyn and the Graduate Center of the City University of New York, Brooklyn, NY 11210 USA

FRANÇOIS O. LAFORGE, Department of Chemistry and Biochemistry, Queens College—CUNY, Flushing, NY 11367 USA

MICHAEL V. MIRKIN, Department of Chemistry and Biochemistry, Queens College—CUNY, Flushing, NY 11367 USA

UDAYAN MOHANTY, Eugene F. Merkert Chemistry Center, Department of Chemistry, Boston College, Chestnut Hill, MA 02467-3800 USA

ALEX SPASIC, Baker Laboratory of Chemistry and Chemical Biology, Cornell University, Ithaca, NY 14853-1301 USA

PENG SUN, Department of Chemistry and Biochemistry, Queens College—CUNY, Flushing, NY 11367 USA

ALBERT STOLOW, Steacie Institute for Molecular Sciences, National Research Council Canada, Ottawa ON K1A 0R6, Canada

JONATHAN G. UNDERWOOD, Department of Physics and Astronomy, The Open University, Walton Hall, Milton Keynes, MK7 6AA, UK

INTRODUCTION

Few of us can any longer keep up with the flood of scientific literature, even in specialized subfields. Any attempt to do more and be broadly educated with respect to a large domain of science has the appearance of tilting at windmills. Yet the synthesis of ideas drawn from different subjects into new, powerful, general concepts is as valuable as ever, and the desire to remain educated persists in all scientists. This series, *Advances in Chemical Physics*, is devoted to helping the reader obtain general information about a wide variety of topics in chemical physics, a field that we interpret very broadly. Our intent is to have experts present comprehensive analyses of subjects of interest and to encourage the expression of individual points of view. We hope that this approach to the presentation of an overview of a subject will both stimulate new research and serve as a personalized learning text for beginners in a field.

STUART A. RICE

CONTENTS

RECENT ADVANCES IN THE FIELD OF INTEGRAL EQUATION THEORIES: BRIDGE FUNCTIONS AND APPLICATIONS TO CLASSICAL FLUIDS

JEAN-MARC BOMONT

Laboratoire de Physique des Milieux Denses, Université Paul Verlaine, 57078 Metz, Cedex 3, France

CONTENTS

Advances in Chemical Physics, Volume 139, edited by Stuart A. Rice

ABSTRACT

We offer a nonexhaustive presentation of the advances in the field of the integral equation theories (IETs) in terms of bridge functions, by describing their application to the determination of structural and thermodynamic properties of simple fluids. After having exposed some basic necessary definitions in the structural description of fluid systems, various IETs are first presented, by recalling their basic expressions and underlying physical assumptions. In this context, a special attention is devoted to the thermodynamic consistency concept, and to the role that this natural constraint plays with the improvement of their performances. In this framework, we shall have as a specific purpose in this presentation to investigate the r– and q–space structural predictions of the IETs together with their relationship with thermodynamics. It is already known that these older theories, that provide in principle a direct source of information, yield however only qualitative agreement for correlation functions, because of suffering of a severe thermodynamic inconsistency. Improved closure relations, the self-consistent integral equation theories (SCIETs) are then introduced. Their applications are examined and systematically compared against related computer simulation data, when available. Satisfying to a single thermodynamic consistency condition provides better correlation functions, but this criterion is not always necessarily sufficient in obtaining accurate description of bridge functions, that are shown to be sensitive to the long-range part of the potential. Introducing supplementary tools, like partitioning schemes for the interaction potential, is needed. Furthermore, the calculation of the excess chemical potential and of the related entropic quantities requires the best possible bridge function $B(r)$ and, at least, the fulfilment of a second thermodynamic consistency condition. In this framework, recent developments are presented. Among numerically solvable theories, only few of them turn out to be accurate in the reproduction of the correlation functions up to the bridge function. These SCIETs are capable of providing accurate predictions for fluid models as compared to simulation calculations. The accuracy of theoretical predictions is also discussed for real systems involved with many-body forces with the aim of systematic assessment of theories. The q–space structural predictions of the latters are of primary importance since they are measurable quantities. In order to provide a complete scenario of calculations in classical fluids, both thermodynamic and structural properties are usually presented in parallel with the available experimental measurements and with computer simulation data.

I. INTRODUCTION

The primary goal of liquid theory is to predict the macroscopic properties of classical fluids from the knowledge of the interaction potential between the constituent particles of a liquid. This area is a very challenging task, because liquids are of vital interest for technology, physics, and chemistry, and for life itself. Seventy years ago, the very existence of liquids seemed a little mysterious. Today, one can make fairly precise predictions of the microscopic and macroscopic static properties of liquids. More than a century of effort since the pioneering work of van der Waals has led to a complete basic understanding of the physicochemical properties of liquids. Advances in statistical mechanics (integral equations, perturbation theories, computer simulation), in knowledge of intermolecular forces and in experimental techniques have all contributed to this. In this presentation, we will be concerned with recent advances in the liquid theory devoted to the description of simple classical fluids properties as determined by means of integral equation theories (IETs).

The availability of a satisfactory theory for simple fluids properties means that these last can successfully be predicted and described at the microscopic statistical mechanics level. This means, once the interparticle law force for a certain fluid has been fixed, one in principle should be able to determine, by means of exact equations relating the interaction potential to some structural functions and thermodynamical quantities, the properties the system will exhibit. However, in practice, a certain number of approximations need to be done in such a theoretical approach, which can be recalled according to the following uncontournable aspects:

1. In order to study a system, one first has to assume a model interaction potential between the particles that are defined as the constituents of the fluid under investigation. Such a modelization is necessary if it is desired not to perform a quantum mechanical description of the system at the level of a first principle Hamiltonian composed of elementary forces. In the latter case, the *ab initio* molecular dynamics technique, developed by Car and Parrinello [1, 2], was revealed to be a powerful investigation tool that was adopted by many authors the last two decades.
2. The physical properties of the fluid model are then calculated through either classical computer simulation techniques, or by some adequate liquid-state theories.

On the one hand, as far as classical computer simulation is concerned, this kind of approach provides, for a given potential model, virtually exact results for structural and thermodynamic properties accompanied with a statistical error, which essentially is due to the use of a finite number of particles [3 and references cited therein]. However, the evaluation of desired quantities requires considerable computational times. Other difficulties arise in the simulation of near critical thermodynamic states, since correlations between the constituents tend to extend over "infinite" distances while the simulation cell edge is at most a few ten angströms [4]. However, despite these problems, it is observed that considerable progresses have been made recently in the simulation calculations thanks to a number of computational strategies derived by several authors together with the increasing power of computers. Nowadays, large-scale simulations [5], in which the calculation of the forces are parallel, allow us to simulate systems with an increasing number of particles in the cell and are suitable to study physical systems involved with many-body forces.

On the other hand, equilibrium statistical mechanics offers appropriate theoretical tools for a complete microscopic determination of properties under interest. In fact, basic thermodynamic quantities, such as pressure and internal energy, wherefrom most of the other thermodynamic quantities involved in the description of the fluid can be determined, are expressed in terms of a structural function that measures the degree of correlation between pairs of particles [6]. For a homogenous and isotropic system, this is the well-known "pair correlation function" (pcf) $g(r)$. Integral equation theories for the liquid structure, whose purpose is to determine this function, have developed rapidly in the late 1980s and early 1990s from atomic to more complex systems. The pcf, which describes the local arrangement of particles, is in fact related to the interparticle potential by exact equations that involve the so-called "bridge function" $B(r)$ [7], which, as will be seen further, is expressed as a density ρ infinite series weighted in terms of irreductible diagrams. Unfortunately, it is not expressed in terms of the pcf itself in a closed form. At this stage, some approximations must be introduced in order to solve the structural problem. To study this problem, an integral equation is typically generated in which the pcf $g(r)$ (or some other structural function closely related to it) is the *a priori* unknown function to be determined. It is obvious from what has been said that the IETs introduce in the description of a fluid model a certain degree of approximation with respect to the "exact" computer simulation treatment. Therefore, thermodynamic and structural properties predicted by the theories must be conveniently and systematically assessed against the corresponding results provided by molecular dynamics (MD) or Monte Carlo (MC) calculations. For a long time, these IETs have suffered from a fundamental shortcoming, the lack of an accurate closure relation [8]. Hence, the integral equation method has seemed to be a relative weak field among its sister methods.

Nevertheless, as will be seen, IETs possess their own peculiarities that make them an irreplaceable tool of investigation of the fluid state. In fact, note first that the approximations made within IET about the form of the pair correlation function, and about its relationship with the potential either amount to, or explicitly express, some simplifying representation of the full many-body structural problem [7, 9–11]. In this respect, comparison with the simulation results possibly followed by that with experimental data for some real system mimicked by the model also implies a test of the hypothesis made on structural correlations and of the physical picture adopted. Also, the solution of an IET is not in general conditioned to the use of a finite system, as it happens with simulation. This advantage is very precious. It allows us, for example, to obtain in a very short time results at the low wavelength limit that can be directly compared to Small Angle Neutron Scattering (SANS) measurements [12–15], while the numerical simulation requires the treatment of very large cells, with increasing large execution times. Quite frequently, IETs can be solved only through numerical procedures that require the use of spatial grids whose extent is by necessity finite. The grid size, however, is generally much larger than typical simulation box sizes. Moreover, such numerical solutions usually do require shorter computational times than comparably accurate simulations. Another advantage of IETs is that these theories can be inverted. Contrary to the usual scheme, where the structure is calculated from a hypothetical interaction potential, the inverted method [16] allows us to determine with a reasonable accuracy an effective potential from the experimental structure factor $S(q)$, that is the Fourier transform of $g(r)$. It is not difficult to guess that the more the IET is accurate, the more the extracted inverted potential is confident. This scheme has proved to be a precious source of information for atomic fluids, and is somehow the counterpart of the Car–Parrinello approach.

It is probably an understatement to say that the quest in integral equation studies is the search for accurate closures. Perhaps closures constitute the most obstreperous bottleneck in achieving high accuracy and furthering advances for IETs. The study of $B(r)$ and the development of new and better closure relations for this function have been the subject of increasing interest. As attested in the literature, this is over the two last decades that several attempts have been made to improve upon these closure relations and to extend the range of validity of integral equation theory. As seen in the following sections, the "bridge function" is one of the keywords in liquid theory. For example, the configurational chemical potential depends explicitly on the bridge function $B(r)$ and has been shown [8] that its calculation is mainly affected by the contribution of $B(r)$ inside the core region (98% in the case of hard-sphere fluid). That is one of the reasons why a detailled knowledge of the bridge function is crucial to perform such a calculation.

Answers to the following questions are sought. (1) Can a closure, or several closures, be found to satisfy theorems for simple liquids? (2) If such closures exist, will they be improvements over conventional ones? Do they give better thermodynamic and structural information? (3) Will such closure relations render the IE method more competitive with respect of computer simulations and with other methods of investigation?

Here, we propose to give an overview of the present status of the applications of self-consistent integral equation theories (SCIETs) aimed to predict the properties of simple fluids and of some real systems that require pair and many-body interactions. We will not therefore be concerned with a number of attempts that have been achieved by various authors to extend the IETs approach to fluids with quantum effects, either with several existing studies of specific systems, as, for example, liquid metals, whose treatment yields a modification of the IETs formalism. Our attention will be restricted to simple fluid models, whose description is, however, an essential step to be reached before investigating more complex systems.

In order to introduce basic equations and quantities, a preliminary survey is made in Section II of the statistical mechanics foundations of the structural theories of fluids. In particular, the definitions of the structural functions and their relationships with thermodynamic quantities, as the internal energy, the pressure, and the isothermal compressibility, are briefly recalled together with the exact equations that relate them to the interparticular potential. We take advantage of the survey of these quantities to introduce what is a natural constraint, namely, the thermodynamic consistency.

In Section III, a number of nonconsistent IETs is first introduced together with their numerical solution procedures or, when available, with their analytical solutions. The accuracy of each envisaged theory is also shortly summarized. Then, the problem of the thermodynamic consistency preliminarily addressed in Section II is fully developped together with the SCIETs. Reference is made to very recent works.

Specific results of the SCIETs for fluid models, including thermodynamic concepts and quantities necessary to describe phase equilibria, are reported in Section IV. In particular, a section is devoted to recent developments in the calculations of the excess chemical potential and related entropic quantities.

The extension of SCIETs to the many-body interactions is presented in Section V. Rare gases, whose constituents interact through three-body forces, are a test case to examine the validity of the SCIETs in describing real systems. Again, the problem of the thermodynamic consistency is covered in this section, since recent SANS measurements provide the structure factor $S(q)$ at very low–q and allow us to deduce the strength of the three-body interactions. A direct comparison of the theoretical results against sharp experiments is feasible. The conclusions are given in Section VI.

II. CLASSICAL STATISTICAL MECHANICS FOR LIQUID STATE

A. Pair Correlation Function and Thermodynamics Quantities

The liquid state of a material has a definite volume, but it does not have a definite shape and takes the shape of a container, unlike that of the solid state. Unlike the gas state, a liquid does not occupy the entire volume of the container if its volume is larger than the volume of the liquid. At the molecular level, the arrangement of the particles is random, unlike that of the solid state in which the molecules are regular and periodic. The molecules in the liquid state have translational motions like those in a gas state. There is short-range interparticular ordering or structure, however.

Briefly, we recall some basic definitions involving the short-order structural functions typical of the liquid state and their relationships with thermodynamic quantities. Considering a homogenous fluid of N particles, enclosed in a definite volume V at a given temperature T (canonical ensemble), the two-particles distribution function [7, 9, 17, 18] is defined as

$$g^{(N)}(\mathbf{r}_1, \mathbf{r}_2) = \frac{1}{Z^{(N)}} V^2 \int \cdots \int \exp[-\beta u(\mathbf{r}_1, \mathbf{r}_2, \ldots, \mathbf{r}_N)] d\mathbf{r}_3 \ldots d\mathbf{r}_N \tag{1}$$

where $\mathbf{r}_1,\ \mathbf{r}_2, \ldots, \mathbf{r}_N$ denote the set of $3N$ spatial coordinates of the N particles and $u(\mathbf{r}_1,\ \mathbf{r}_2, \ldots, \mathbf{r}_N)$ is the total potential energy. The partition function expressed as

$$Z^{(N)} = \int \cdots \int \exp[-\beta u(\mathbf{r}_1, \mathbf{r}_2, \ldots, \mathbf{r}_N)] d\mathbf{r}_1 \ldots d\mathbf{r}_N \tag{2}$$

and $\beta = (k_B T)^{-1}$ is the inverse temperature. According to Eq. (1), $g^{(N)}(\mathbf{r}_1,\ \mathbf{r}_2)$ measures the probability of finding the particle labeled 1 in a volume $d\mathbf{r}$ at $\mathbf{r}_1$, and the particle labeled 2 in a volume $d\mathbf{r}$ at $\mathbf{r}_2$, irrespective of the positions of the $N-2$ remaining particles. The appearance of the factor V^2 in Eq. (1) results of a normalization to $1/V^2$, which is the probability of obtaining the same configuration of particles in the absence of correlations. In this case, each particle has the probability $1/V$ of occupying any position in the volume V. In this framework, it turns out that when the distance between particles 1 and 2 tends to infinity, $g^{(N)}(\mathbf{r}_1, \mathbf{r}_2)$ tends to 1. This expresses the loss of correlation between particles at large distances. If the system is not only homogenous, but also isotropic, $g^{(N)}(\mathbf{r}_1, \mathbf{r}_2)$ depends only on the distance $|\mathbf{r}_1 - \mathbf{r}_2| = r$, that is to say that $g(\mathbf{r}_1, \mathbf{r}_2) = g(r)$. In this case, the latter structural function is known as the pair correlation function, which measures the probability that given a particle at the origin, another particle of the fluid can be found at a distance r from it (see Fig. 1). In the following, we will focus our attention on the determination of $g(r)$ as a solution of integral equations. We will restrict our attention first to systems of particles that interact through central pair forces (an extension to many-body interactions is presented in Section V) for which the total potential energy can be written as a sum of pairwise additive terms, so that

$$u(\mathbf{r}_1, \mathbf{r}_2, \ldots, \mathbf{r}_N) = \frac{1}{2} \sum_{i \neq j} u_2(\mathbf{r}_{ij}) \tag{3}$$

where $r_{ij} = |r_i - r_j|$ and $i, j = 1, \ldots, N$. In this case, it is easy to prove that the excess internal energy per particle is given by

$$\frac{E^{\text{ex}}}{N} = 2\pi\rho \int g(r)u(r)r^2 dr \tag{4}$$

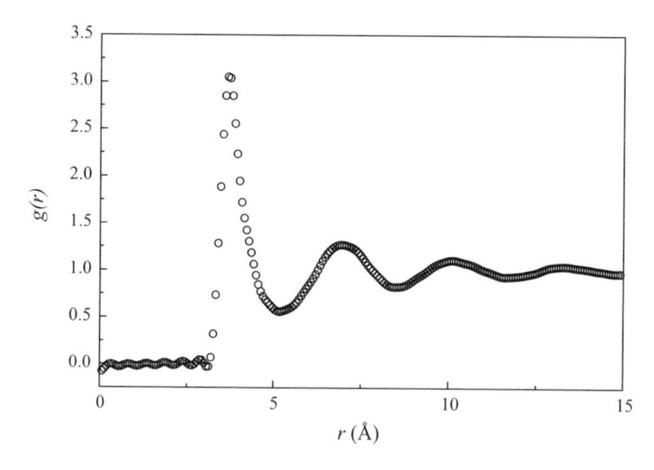

Figure 1. Neutron-scattering experiment result for the pair correlation function $g(r)$ of liquid argon at $T = 85$ K and $V = 28.26\,\text{cm}^3\,\text{mol}^{-1}$, near the triple point. Notice that the ripples at small r are artifacts of the data treatment. Taken from Ref. [19].

while the virial pressure, given by the equation of state (EOS), can be written as

$$\frac{\beta P}{\rho} = 1 - \frac{2}{3}\beta\pi\rho \int g(r) r^3 \frac{du(r)}{dr} dr \tag{5}$$

where $\rho = N/V$ is the average number density of particles in the system. Equations (4) and (5) illustrate the importance of the knowledge of the pcf in order to achieve an estimate of fundamental thermodynamic quantities [7, 9, 17, 18]. Extending Eqs. (1) and (2) to the grand-canonical ensemble case leads to a first indication, but poor one, of the link between the interactions and the structure. It can be shown easily that, in the case of a dilute gas ($\rho \to 0$), one has $g(r) \to e^{-\beta u(r)}$. In this densities regime, the pressure can be formally written

$$\frac{\beta P}{\rho} = \sum_{n=1}^{+\infty} B_{n+1}(T)\rho^n \tag{6}$$

This series is known as the virial expansion, where B_n are the virial coefficients. In principle, these last can be calculated if the potential is known. In practice, however, the calculation is feasible only for the first few coefficients. In this framework, Eq. (6) is only applicable to low density regimes. Obviouly, a complete theory is expected to confidently calculate the highest possible number of known virial coefficients. Such a calculation is one of the benchmarks of its overall accuracy.

Since it is important to judge the accuracy of the description that will be given to the fluid, a third important quantity, the isothermal compressibility χ_T of the system, can be defined via two independent routes. On the one hand, by making use of the thermodynamic fluctuation theory, one obtains

$$\rho k_B T \chi_T = 1 + 4\pi\rho \int [g(r) - 1] r^2 dr \tag{7}$$

This relationship is known as the compressibility equation. On the other hand, from Eq. (5), at a given temperature T one has

$$\chi_T = \frac{1}{\rho} \frac{\partial \rho}{\partial P}\bigg|_T \tag{8}$$

These last relations clearly establish the link between the short ordered structure and the compressibility of the fluid.

B. Pair Correlation Function and Structure Factor

As mentioned above, when the distance separating a pair of particles tends to infinity, the correlations vanish and $g(r)$ tends to 1. That means that the

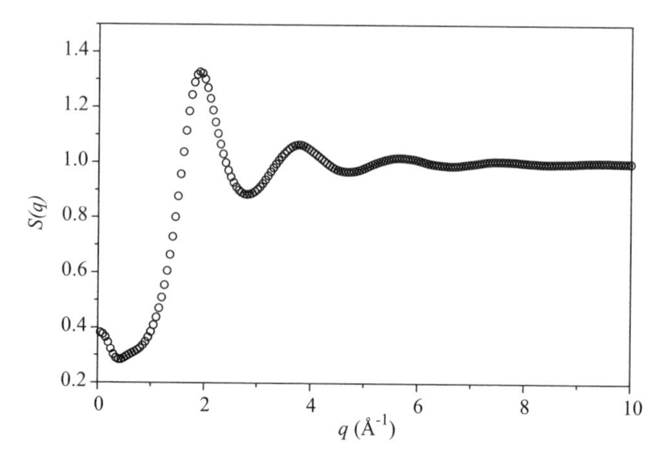

Figure 2. Neutron-scattering experiment result for the structure factor of argon at $T = 350\,\mathrm{K}$ and $\rho = 12.3$ atoms nm^{-3}. Taken from Ref. [20].

total correlation function defined as $h(r) = g(r) - 1$ tends to 0. For the theory of liquids, the Fourier transforms (FT) of $g(r)$ and $h(r)$ are of useful interest [7, 18]. Actually, they are directly related to the structure factor $S(q)$ that is experimentally measurable by X-ray diffraction or neutron scattering (see Fig. 2). One has

$$S(q) = 1 + \rho \int d\mathbf{r}[g(r) - 1] \exp(-i\mathbf{q}.\mathbf{r}) \tag{9}$$

and

$$S(q) = 1 + \rho h(q) \tag{10}$$

where $h(q)$ is the FT of the total correlation function $h(r)$, so that

$$h(q) = \int d\mathbf{r} h(r) \exp(-i\mathbf{q}.\mathbf{r}) \tag{11}$$

In other words, applying the inverse FT to the measured $S(q)$ leads to the knowledge of the pcf $g(r)$, that is

$$\rho[g(r) - 1] = \frac{1}{(2\pi)^3} \int d\mathbf{q}(S(q) - 1) \exp(i\mathbf{q}.\mathbf{r}) \tag{12}$$

An interesting feature appears from Eqs. (8) and (9). The structure in the q–space is found to be directly related to the thermodynamics since, in the $q = 0$

limit [9, 18], one obtains the striking result

$$S(q = 0) = \rho k_B T \chi_T \tag{13}$$

This relationship corresponds to what is called a *thermodynamic consistency condition*, whose fulfilment is one of the criterions of accuracy of a self-consistent IET: since $S(q = 0)$ is calculated from the structure (9) and χ_T is obtained from the pressure (8), the left-hand side (l.h.s.) and right-hand side (r.h.s.) of the last relation have to coincide with a precision of 1%. This important aspect is developed below.

C. Thermodynamic Consistency: A Natural Constraint

The thermodynamic consistency of IETs has been the subject of several investigations in the past [21]. Since the thermodynamic quantities are derived from integrals over $g(r)$ and $c(r)$, the desired goal is to obtain exact results at both structural and thermodynamic levels. But the reader has to be aware that a theory rendered self-consistent will not necessarily also become exact at each of these levels simultaneously. In principle, the following is the main adopted stategy: One essentially forces the theory to satisfy some equalities for the pressure calculated from the structure.

There are several routes to the determination of $\beta P/\rho$ from the structure. One is given by Eq. (5). A second possibility is offered by the isothermal compressibility

$$\rho \left.\frac{\partial P}{\partial \rho}\right|_T = \frac{1}{\chi_T} \tag{14}$$

In fact, the quantity $1/\chi_T$ can be integrated with repect to the density and along an isothermal path to yield the pressure. We will therefore use the notation $(\beta P/\rho)_c$ to indicate the compressibility EOS so obtained. A third route is based on the use of the excess internal energy E^{ex}, related to the excess Helmholtz free energy F^{ex} by $E^{\mathrm{ex}} = [\partial(\beta F^{\mathrm{ex}}/\partial\beta]_V$. Then, E^{ex} can be integrated with repect to the inverse temperature β along an isochore path to yield F^{ex}. The excess pressure P^{ex} is then obtained as

$$P^{\mathrm{ex}} = -\left.\frac{\partial F^{\mathrm{ex}}}{\partial V}\right|_E \tag{15}$$

The energy EOS so obtained will henceforth be indicated as $(\beta P/\rho)_E$. If the exact $g(r)$, that is if the true pcf of the system is used in Eqs. (4)–(7), then the virial, the compressibility, and the energy EOS should all take the same value.

So, an accurate predictive theory should satisfiy the following equality:

$$\left(\frac{\beta P}{\rho}\right)_v = \left(\frac{\beta P}{\rho}\right)_c = \left(\frac{\beta P}{\rho}\right)_E \tag{16}$$

The fulfilment of this condition is termed as one of the *thermodynamic consistency conditions*. But, doing so requires the use of external information to the theory that cannot be used in a predictive manner. Of course, obtaining a full thermodynamic consistency without resorting to any external data is the practical purpose of self-consistent IETs.

Integral equation theories of $g(r)$ do in general yield only an approximate estimate of this quantity, and hence they are, to more or less extent, thermodynamically inconsistent. In practice, instead of Eq. (16), one prefers to apply the pressure–compressibility $(P - \chi_T)$ condition expressed by

$$\left.\frac{\beta \partial P}{\partial \rho}\right|_T = \frac{1}{\rho k_B T \chi_T} \tag{17}$$

where P is the pressure obtained from the virial EOS and χ_T is the isothermal compressibility as obtained from the fluctuations route. The advantage of Eq. (17) with respect to Eq.(16) relies on the fact that [22] one can verify the consistency for any thermodynamic state without the necessity of performing an integral of χ_T over successive thermodynamic states, as would be the case for the estimate of $(\beta P/\rho)_c$.

Moreover, since Maxwell's equation relates the internal energy to the pressure $(E - P)$

$$\left.\frac{\partial E^{\text{ex}}}{\partial V}\right|_T = T^2 \frac{\partial}{\partial T}\left(\frac{P}{T}\right)_V \tag{18}$$

two consistency conditions in the form of integrals are obtained [23]:

$$I(\rho, T) = \int \left[c(r) - \frac{r}{6}\frac{d\beta u(r)}{dr}\left(2g(r) + \rho\frac{\partial g(r)}{\partial \rho}\right)\right] r^2 dr = 0 \tag{19}$$

and

$$J(\rho, T) = \int \left[u(r)\left(g(r) + \rho\frac{\partial g(r)}{\partial \rho}\right) + \frac{r}{3}\frac{du(r)}{dr}\left(g(r) - T\frac{\partial g(r)}{\partial T}\right)\right] r^2 dr = 0 \tag{20}$$

These obtained equations set the self-consistent relationship between the correlation functions and their derivatives. Vompe and Martynov [24] defined the total thermodynamic consistency (including the $E - P$ and $P - \chi_T$ terms) by stating that the quantity $S(\rho, T) = \omega_1 I^2(\rho, T) + \omega_2 J^2(\rho, T)$, ω_i's being weight factors, has to be as small as possible for a given thermodynamic state.

Nevertheless, the reader has to notice that the use of the latter two consistency conditions is not always sufficient in obtaining an accurate description of structural functions (e.g., the bridge function) and, consequently, thermodynamics quantities. The problem of identifying and fulfiling at least a second thermodynamic condition is under the scope in this review article. We will turn back to the crucial point of *thermodynamic consistency* in Sections III and IV when discussing the solution of integral equation theories and their application to simple liquids.

III. INTEGRAL EQUATION THEORIES

A. Fundamental Set of Equations for the Structure

1. *Correlations in a Liquid : The Ornstein–Zernike Equation*

The correlation functions play an essential role in the static description of homogenous classical liquids whose particles are taken to interact through an effective pair potential. The starting point of the liquid-state theory, in terms of correlation functions, is the well-known Ornstein–Zernike equation [25]. The total correlation function $h(r)$ defined in Section II is actually a sum of two contributions that is illustrated by the following relationship

$$h(r) = c(r) + \rho \int c(|\mathbf{r} - \mathbf{r}'|)h(\mathbf{r}')d\mathbf{r}' \tag{21}$$

The first contribution to $h(r)$ is the direct correlation function $c(r)$ that represents the correlation between a particle of a pair with its closest neighbor separated by a distance r. The second contribution is the indirect correlation function $\gamma(r)$, which represents the correlation between the selected particle of the pair with the rest of the fluid constituents. The total and direct correlation functions are amenable to an analysis in terms of configurational integrals clusters of particles, known as diagrammatic expansions. Providing a brief resume of the diagrammatic approach of the liquid state theory is beyond the scope of this chapter. The reader is invited to refer to appropriate textbooks on this approach [7, 9, 18, 26].

By FT (21), one obtains

$$h(q) = \frac{c(q)}{1 - \rho c(q)} \tag{22}$$

where $c(q)$ has a similar definition as $h(q)$. It is easy to show that

$$\rho c(0) = 1 - \frac{1}{\rho k_B T \chi_T} \tag{23}$$

So, $c(q = 0)$ is a finite quantity even when the isothermal compressibility χ_T tends to diverge in the critical region of any fluid. As will be seen, the former quantity is useful, because of being accessible from direct experiment [12] thanks to recent SANS measurements (see Section V). At variance from $g(r)$, the direct correlation function $c(r)$ in not zero inside the core region $(r < \sigma)$ and its knowledge is crucial in this range of distances, since it is directly related to χ_T so that

$$\frac{1}{\rho k_B T \chi_T} = 1 - \rho \int c(r) d\mathbf{r} \tag{24}$$

In other words, $c(r)$ can be directly obtained from the structure factor $S(q)$ by

$$\rho c(r) = \frac{1}{(2\pi)^3} \int dq \left(1 - \frac{1}{S(q)}\right) \exp(i\mathbf{q}.\mathbf{r}) \tag{25}$$

That means, as for $g(r)$, the direct correlation function $c(r)$ can be, in principle, derived from experiment. As can be seen, the OZ relation deals with *a priori* two unknown functions $h(r)$ and $c(r)$. The cluster expansion provides an important exact feature for $c(r)$ with respect to the potential interaction at large distances $(r \to +\infty)$, with the result

$$c(r) \simeq -\beta u(r) \tag{26}$$

The direct correlation function decays fairly rapidly with the distance. However, this information, which correlates the structure to the interactions in the fluid, is too poor to provide a complete and accurate theory.

In order to remedy this drawback, the OZ equation can be rewritten by the use of an iterative substitution of $h(\mathbf{r}')$ inside the kernel of (21) [7]. A series of integral terms is obtained, with the result

$$h(r) = c(r) + \rho \int c(|\mathbf{r} - \mathbf{r}'|) c(\mathbf{r}') d\mathbf{r}' + \rho^2 \int\int c(|\mathbf{r} - \mathbf{r}'|) c(|\mathbf{r}' - \mathbf{r}''|) c(\mathbf{r}'') d\mathbf{r}' d\mathbf{r}'' + \cdots \tag{27}$$

here truncated at the second iteration. The chain structure of (27) shows that the OZ equation amounts of describing the total correlation function $h(r)$ between a pair of particles as a sum of different contributions, the first being the direct correlation function $c(r)$ and the rest being the indirect correlations. This algorithm may have difficulties in converging if the input functions (or some starting solutions) one uses are not close enough to the requested solution. Furhtermore, the final solution has to fulfil of course the thermodynamic consistency condition one uses. As observed, the use of this density expansion is not tractable enough in practice to determine $h(r)$ and $c(r)$.

2. *A Necessary Closure Relation : Introduction of the Bridge Function*

From what precedes, it is obvious that in order to determine the latter two correlation functions for a given pair potential $u(r)$, Eq. (21) must be supplemented by an auxiliary closure relation [7, 17, 18, 27] derived from a cluster diagram analysis that reads

$$h(r) + 1 = \exp[-\beta u(r) + h(r) - c(r) + B(r)] \tag{28}$$

Equations (21) and (28) form the fundamental set of equations of the theory of the liquid state and Eq. (28) may be regarded as a definition for $B(r)$. The term $h(r) - c(r) + B(r)$ is the so-called negative excess potential of mean force $\omega(r)$ ([18] and references cited therein). The bridge function $B(r)$ was shown to be the sum of an infinite number of terms, each consisting of integrals whose kernel are products, of increasing order, of correlation functions. This infinite series is graphically represented by so-called "bridge" diagrams, hence the name attributed to $B(r)$. In other words, the diagrams that comprise the negative excess potential of mean force can be classified [28] according to the number of h bonds connected to particle 1,

$$\omega(r_{12}) = \sum_{n=1}^{+\infty} \omega^{(n)}(r_{12}) \tag{29}$$

The h-bond diagrams of $\omega(r_{12})$ cannot contain any articulation pairs, and so the diagrams connecting the root point 2 with the n h-bonds connected to root point 1 do not have nodal points. This definition is just of the $(n+1)$-particle direct correlation function $c(\mathbf{r}_2,\ \mathbf{r}_3, \ldots, \mathbf{r}_{n+2})$ and one has

$$\omega^{(n)}(r_{12}) = \frac{\rho^n}{n!} \int \cdots \int d\mathbf{r}_3 d\mathbf{r}_4 \ldots d\mathbf{r}_{n+2} h(r_{13}) h(r_{14}) \ldots h(r_{1,n+2}) c(\mathbf{r}_2, \mathbf{r}_3, \ldots, \mathbf{r}_{n+2}) \tag{30}$$

The reader has to notice that $\omega^{(1)}(r_{12}) = h(r_{12} - c(r_{12})$, which is the indirect correlation function $\gamma(r_{12})$. Unfortunately, even if the types of diagrams are known, the resultant series cannot be transformed to any tractable analytical formula or evaluated numerically with a good accuracy. Recently, from the bridge diagram series, simple phenomenological forms for $B(r)$ haven been proposed for the LJ fluid [29]. They present the advantage of bringing some repulsion between particles at short range, and also attraction at longer distances. However, these forms, even if simple, depend on an adjustable parameter that requires external information. Furthermore, they seem to diverge at zero separation ($r \rightarrow 0$). Similar approach has been presented for the HS fluid [30]. It is shown that the knowledge of the coefficients of the h-bond expansion of $B(r)$ up to order ρ^4 in Eq. (30) requires the evaluation of many of diagrams. For example, the evaluation of the fourth coefficient of $B(r)$ involves 1731 distinct diagrams, and, of course, investigating higher order coefficients would require the treatment of an exponentially increasing number of diagrams. It is easy to understand that the convergence is very slow (if it converges at all). Despite these drawbacks, it is clear that the knowledge of the exact bridge function would close the problem of the *thermodynamic consistency conditions*. Unfortunately, the bridge function has to be approximated in practice. Though a number of theoretical and simulation procedures have been derived in recent years in order to get an approximate estimate of $B(r)$ for different fluid models, the exact bridge function is not known for any system. As will be discuss next, a class of integral equations arises from the joint use of (28) and OZ. In fact, Eqs. (21) and (28) contain three unknown structural functions. Once some sort of approximation is introduced for $B(r)$, then Eq. (28) can be used as a closure relation to OZ. Then, $h(r)$ and $c(r)$ can be determined by solving the set of two nonlinear integral equations.

In order to be competitive with respect to molecular simulation, the purpose of the IETs is to establish a bridge in between semianalytical relations and the numerical methods. By analyzing a large amount of simulation studies, Rosenfeld and Ashcroft [27] concluded in favor of the existence of a "universality" in the short-range structure of a wide class of fluids models. On this basis, they proposed that the bridge function relative to such systems could be parametrized by the HS fluid bridge function. Then, two important thresholds have been crossed when: (1) Carnahan and Startling [31] derived an exact EOS for HS fluid, and (2) Groot et al. [32] obtained an exact construction of the bridge function for HS fluid, by a suitable fit of numerical simulation results. However, the problem has been found to be more complex, since another studies have shown that, outside the region where repulsive core effects are dominant, the bridge function exhibits a "non universal" behavior.

At this stage, undiscutable data, external of the IETs, were necessarily required to shed some light on these peculiar behaviors, which provides exact reference data for more realistic potentials. First, Nicolas et al. [33] derived an EOS for the Lennard-Jones fluid and Johnson et al. [34] provided MD results for the classical thermodynamic quantities. Notice that Heyes and Okumura [35] recently derived an EOS of the Weeks–Chandler–Andersen fluid.

A second important step has been reached by Llano-Restrepo and Chapman [36]. With the aid of the two former EOS, they performed Monte Carlo simulation calculations for this system over several thermodynamic states, to evaluate the direct and indirect correlation functions, and then to extract the bridge function $B(r)$. These results became useful in testing the accurateness of IETs for the Lennard-Jones potential. Since the thermodynamic properties involve integral over the correlation functions, comparison of just the thermodynamic properties could hide, *a priori*, inaccuracies in the theory.

These prior works offered new motivation to several authors, whose purpose was to obtain adequate theories that would be able to match, at last, with simulation results. Huge efforts were made in this direction. Thus, in order to overcome the problem of thermodynamic consistency (or inconsistency), some attempts consisted in performing judicious interpolations between already existing nonconsistent IETs, while others were devoted to improve these last by introducing parameters to render them consistent [37]. This finding is the subject of what follows.

B. Nonconsistent Integral Equations

We present a brief survey of the most employed auxiliary closure relations that have to be used in conjunction with the OZ equation (21) to determine the total and direct functions $h(r)$ and $c(r)$, or closely related functions. We emphasize that for any proposed approximation for $B(r)$ presented below (but PY and MSA, that are analytically solvable), the numerical procedure imposes the adoption of a finite grid of points in r-space over which the structural functions have to be defined. This obviously implies that also the wave vector in q-space is discretized, and the accurateness of the final solution obviously depends both on the overall extent of such grids and on their spacings (i.e., the total number of points used). The rapidity through which a convergent solution can be obtained crucially depends on the solution strategy. Great progresses have been made in this respect mostly as a result of the application of the Newton–Raphson technique [37, 38] that allows us to improve the accuracy of the successive guessed correlation functions.

1. Percus–Yevick Approximation

A very popular closure relation is the Percus–Yevick (PY) approximation [39]. For a generic potential $u(r)$, this approximation assumes that

$$c(r) = [1 - e^{\beta u(r)}] \times g(r) \tag{31}$$

In other words, the bridge function corresponding to PY relation reads

$$B(r) = \ln[1 + \gamma(r)] - \gamma(r) \tag{32}$$

The solution of the system formed by Eqs. (21) and (32) is usually obtained thanks to numerical iterative procedures. However, in the special case of hard-sphere (HS) fluid, the solution is found to be analytical. For this system, the potential reads $u(r) = +\infty$ for $r < \sigma$ and $u(r) = 0$ for $r > \sigma$, where σ is the HS diameter. So, it follows that $g(r) = 0$ for $r < \sigma$ and $c(r) = 0$ for $r > \sigma$. This result satisfies the exact asymptotic limit of $c(r)$ at large distances [see Eq.(26)]. Percus–Yevick is a typical candidate for the thermodynamic inconsistency: indeed, the virial and the compressibility EOS provide the following analytic results in terms of the packing fraction η

$$\left(\frac{\beta P}{\rho}\right)_v = \frac{1 + 2\eta + 3\eta^2}{(1-\eta)^2} \qquad \left(\frac{\beta P}{\rho}\right)_c = \frac{1 + \eta + \eta^2}{(1-\eta)^3} \tag{33}$$

where $\eta = \frac{\pi}{6}\rho\sigma^3$. It appears that the PY virial and compressibility EOS, respectively, underestimate and overestimate the HS simulation results. These last are quite accurately parametrized in terms of a heuristic EOS developed by Carnahan and Starling (CS) [31], which reads

$$\left(\frac{\beta P}{\rho}\right)_{\text{CS}} = \frac{1 + \eta + \eta^2 - \eta^3}{(1-\eta)^3} \tag{34}$$

This parametrization is usually considered as exact and is useful to test the quality of the results from IETs calculations for the HS fluid. However, another EOS proposal has been provided [40, 41].

$$\frac{\beta P}{\rho} = \frac{1 + \eta + \eta^2 - 2(\eta^3 + \eta^4)/3}{(1-\eta)^3} \tag{35}$$

whose predictions compare well with MC results [41].

2. *Mean-Spherical Approximation*

This another popular closure [43] deals with spherical particles fluids that interact through an infinite repulsive potential at short range $u(r) = +\infty$ for $r < \sigma$ and through a tail at longer distances. Similarly to PY, the mean-spherical approximation (MSA) is formulated in terms of an ansatz for the direct correlation function. In this approach, $c(r)$ is supposed to be

$$c(r) = -\beta u(r) \tag{36}$$

in a range of distances larger that the HS diameter. In the limit $r \to +\infty$, the behavior of $c(r)$ is correct. Consequently, one also has $g(r) = 0$ for $r < \sigma$. It appears that the domain of validity of Eq. (31) is extended at shorter range distances. This approach can be generalized in the presence of continuous potentials. Obviously, MSA reduces to PY for the HS system and presents an internal inconsistency as well.

3. *Hypernetted-Chain Approximation*

An analysis of clusters expansion to higher order (as compared to PY equation) leads to the hypernetted-chain (HNC) approximation [44–46]. In other words, directly solving the OZ relation in conjunction with Eq. (28) is possible, under a drastic assumption on $B(r)$. The total correlation function is given simply by

$$h(r) = \exp[-\beta u(r) + h(r) - c(r)] - 1 \tag{37}$$

implying that the bridge function $B(r)$ is identically equal to 0. In this formalism, all the information supposed to be hidden into $B(r)$ to make a bridge between IETs and the molecular dynamics approach are omitted. Contrary to PY or MSA, the HNC is not analytically solvable and requires an iterational procedure. Note that the PY closure is recovered by linearizing the HNC closure. In this framework, HNC could be expected to be more accurate than the PY. From a general point of view, HNC has been shown to be more accurate than the PY system. Also it is more accurate than the MSA for systems governed by long-range potentials. However, for short-range potentials, as in the simpliest case of HS, the accuracy of the results is less than the one of the PY approximation.

4. *Verlet Approximation*

A semiphenomenological equation for the pcf has been proposed by Verlet (V) [47] for the case of HS, on the basis of the following functional assumption for the bridge function

$$B(r) = \frac{-\gamma^2(r)}{2(1 + \alpha\gamma(r))} \tag{38}$$

where α is a suitable parameter that is usually taken to be equal to four-fifths. This approximation yields excellent results for both thermodynamic and structural properties. As seen below, it has been successfully employed with some modifications to be applied to other systems than the HS fluid.

5. *Martynov–Sarkisov Approximation*

Another approximation, which has been proposed by Martynov and Sarkisov (MS) [48] sets the bridge function as

$$B(r) = [1 + 2\gamma(r)]^{1/2} - \gamma(r) - 1 \tag{39}$$

This expression comes from the fact that the authors have truncated Eq. (29) to the second order $n = 2$. The MS approximation, which has been originally applied to the HS system, has no adjustable parameter. In the dense regime, it has been shown that MS improved the PY and HNC results, by reducing considerably the thermodynamic inconsistency of the two latters.

C. First Attempts to Impose Thermodynamic Consistency

1. *The Soft-Core MSA (SMSA) Approximation*

At variance from the HS system, it had been observed that PY is not as accurate for attractive potentials. Hence, an alternative closure has been derived and consists in a generalization of the MSA closure [49, 50]. This has been feasible by incorporating the division scheme introduced by Weeks et al. [51] for the (12-6) Lennard-Jones (LJ) fluid composed of particles interacting through the potential

$$u(r) = 4\epsilon\left[\left(\frac{\sigma}{r}\right)^{12} - \left(\frac{\sigma}{r}\right)^{6}\right] \tag{40}$$

This system is characterized in terms of the reduced temperature $T^* = (k_B T)/\epsilon$, and the reduced density usually taken as $\rho^* = \rho\sigma^3$, where σ is the node of the potential so that $u(r = \sigma) = 0$. According to Weeks et al. [51], this continuous potential $u(r)$ is separated into a short-range $u_{\mathrm{SR}}(r)$ and a long-range part $u_{\mathrm{LR}}(r)$ by setting

$$u_{\mathrm{SR}}(r) = \begin{cases} u(r) - u(r_m) & r \leq r_m \\ 0 & r > r_m \end{cases} \tag{41}$$

and

$$u_{\mathrm{LR}}(r) = \begin{cases} u(r_m) & r \leq r_m \\ u(r) & r > r_m \end{cases} \tag{42}$$

respectively, where r_m denotes the position of the minimum of the potential. The Weeks–Chandler–Andersen (WCA) potential, $u_{\mathrm{SR}}(r)$, is the LJ potential, shifted upward by $u(r_m)$ and truncated at the LJ minimum $r_m = 2^{1/6}\sigma$. Although this potential is repulsive, it is composed of repulsive and attractive components. Originally, the LJ potential was divided into a reference part entirely repulsive (the WCA potential) and a perturbative part entirely attractive without altering the potential since, *in fine*, one has $u_{\mathrm{SR}}(r) + u_{\mathrm{LR}}(r) = u(r)$. The SMSA bridge function reads as:

$$B(r) = -\gamma(r) + \beta u_{\mathrm{LR}}(r) + \ln[1 + \gamma(r) - \beta u_{\mathrm{LR}}(r)] \tag{43}$$

As compared to Eq. (32), the long-range part of the potential has been added to the indirect correlation function. For convenience, the term $\gamma(r) - \beta u_{\mathrm{LR}}(r)$ is noted $\gamma^*(r)$, which is the so-called renormalized indirect correlation function, whose importance is crucial (see the next paragraph). When $r > r_m$, this closure reduces to the MSA approximation $c(r) = -\beta u(r)$. In the case of a purely repulsive potential, one has $r_m \to +\infty$ and $u_{LR}(r) = 0$. That is to say that SMSA reduced to the PY closure. In this framework, it clearly appears that SMSA interpolates between the PY and MSA [22]. This approach is the first improvement of the IETs presented in Section, III.B since it is the first theory that suggests an interpolation between two other existing theories.

2. *Reference HNC and Modified HNC Approximations*

An extension of the HNC approach is the reference hypernetted chain (RHNC) approximation [52–54]. In this approach, studying a system preliminarily requires that we determine the properties of a reference fluid (RF). Usually, it has to be assumed that the properties of the actual system are closed to those of the RF. For example, the interaction potential of the system under study can be written as the sum of the RF potential and of a small perturbation, namely, $u(r) = u^{\mathrm{RF}}(r) + \Delta u(r)$. Correspondingly, one has $\gamma(r) = \gamma^{\mathrm{RF}}(r) + \Delta\gamma(r)$, and the direct correlation function is given by

$$c(r) = g^{\mathrm{RF}}(r) \exp[\Delta\gamma(r) - \beta\Delta u(r)] - \gamma(r) - 1 \tag{44}$$

where $g^{\mathrm{RF}}(r)$ is the pcf of the reference fluid. As attested in the references quoted before, the performances of this method are as much accurate as those of the HNC.

Having at our disposal accurate structural and thermodynamic quantities for HS fluid, the latter has been naturally considered as a RF. Although real molecules are not hard spheres, mapping their properties onto those of an equivalent HS fluid is a desirable goal and a standard procedure in the liquid-state theory, which is known as the modified hypernetted chain (MHNC) approximation. According to Rosenfeld and Ashcroft [27], it is possible to postulate that the bridge function of the actual system of density ρ reads

$$B(r) = B^{\mathrm{HS}}(r, \eta^*) \tag{45}$$

where $\eta^* = \frac{\pi}{6}\rho(\sigma^*)^3$. The hard core diameter of the RF is chosen so that consistency between virial and compressibility EOS is obtained [27]. This is feasible since $B^{\mathrm{HS}}(r, \eta^*)$ can easily be calculated for any desired density. In order to achieve thermodynamic consistency, another possibility concerns the fulfilment of an extremum condition of the Helmholtz free energy. As said earlier, $g(r)$ and $c(r)$ can be determined from experiment, but neither $B(r)$ nor $u(r)$ unfortunately. However, the MHNC approach is useful to overcome this drawback [16], since the B^{HS} can be employed to extract an inverted effective potential, so that

$$\beta u^{\mathrm{inv}}(r) = g^{\mathrm{exp}}(r) - 1 - c^{\mathrm{exp}}(r) - \ln g^{\mathrm{exp}}(r) + B^{\mathrm{HS}}(r, \eta^*) \tag{46}$$

The extraction of $u^{\text{inv}}(r)$ of course necessitates an iterative procedure and the criterion to optimize the choice of the hard-core diameter of the reference fluid HS [54] reads

$$\int dr[g(r) - g^{\text{HS}}(r)] \frac{\partial B^{\text{HS}}(r, \sigma^*)}{\partial \sigma^*} = 0 \tag{47}$$

Another possibility to select the hard-core diameter consists in equating the second density derivative of the free energy of the actual system to that of the RF [55]. The RHNC and MHNC present some disavantages. As seen, they do not allow us to study a fluid without making reference to another one and the reference fluid has to be close enough to the actual fluid under study.

Regarding what precedes, it is clear that one of the challenges of the liquid-state theory is to ascribe an effective hard sphere diameter σ_{HS} to the real molecule. As stated in the literature [56–59], a number of prescriptions for σ_{HS} exist through empirical equations. Among them, Verlet and Weiss [56] proposed

$$\sigma_{\text{HS}}/\sigma = \frac{0.3837 + 1.068\epsilon/k_{\text{B}}T}{0.4293 + \epsilon/k_{\text{B}}T} \tag{48}$$

while more recently, Ben-Amotz and Herschbach [60] proposed

$$\sigma_{\text{HS}}/\sigma = \frac{\alpha_0}{(1 + (T/T_0)^{1/2})^{1/6}} \tag{49}$$

where α_0 and T_0 can be density-dependent parameters [61]. In the case where $\alpha_0 = 2^{1/6}$ and $T_0 = 1$, the proposal of Boltzmann [62] is recovered, with the result

$$\sigma_{\text{HS}}/\sigma = \frac{2^{1/6}}{(1 + (T)^{1/2})^{1/6}} \tag{50}$$

where σ_{HS} is free from any density dependence. Note that this last expression has been found to be useful to develop an accurate EOS for the WCA fluid [35], whose expression is similar to the Carnahan and Starling hard-sphere EOS

$$\frac{\beta P}{\rho} = \frac{1 + \eta^* + a\eta^{*2} - b\eta^{*3}}{(1 - \eta^*)^3} \tag{51}$$

where η^* is the packing fraction defined by $\pi\sigma_{HS}^3 N/6V$ and a least-squares fit to the range of data provides $a \approx 3.597$ and $b \approx 5.836$.

D. Self-Consistent Integral Equation Theories (SCIETs)

As pointed out above, the lack of accurate information about the detailed form of $B(r)$, especially in the core region ($r < \sigma$), and the difficulty in solving the thermodynamic consistency conditions, have prompt the authors to formulate what has been termed as "mixed closure" approaches. The practical justifications can be seen as follows.

Since HNC and PY approximations results braket those of the exact EOS for HS, an appropriate mixing of two inadequate EOS would yield improved results as compared to reference data. Moreover, fromwhat we said, it turns out that HNC and PY have somehow complementary behaviors. The former is more accurate for long-range forces, while the latter is better for short-range forces. Another avantageous possibility is to formulate direct approximations for the bridge function based on physical arguments [37]. This part is devoted to the presentation of these improved closures.

1. *The HNC–SMSA Approximation*

The previous basic observations have suggested to adopt a closure that allows us to interpolate continuously between two existing theories. In this framework, the famous HNC–SMSA(HMSA) closure [22] has been proposed for the Lennard-Jones fluid. It interpolates between HNC and SMSA. The HMSA has strong theoretical basis since it can be derived from Percus' functional expansion formalism and its bridge function expresses as a functional of the remormalized indirect correlation function $\gamma^*(r) = \gamma(r) - \beta u_{LR}(r)$ so that

$$B(r) = -\gamma^*(r) + \ln\left[1 + \frac{\exp\{f(r)\gamma^*(r)\} - 1}{f(r)}\right] \tag{52}$$

where $f(r)$ is the "switching" function defined as:

$$f(r) = 1 - \exp(-\alpha r) \tag{53}$$

The parameter α is varied in a self-consistent manner until virial and compressibility routes for the pressure coincide. Bretonnet and Jakse [63] showed that $f(r)$ could be accurately simplified by setting $f(r) = f_0$. This last parameter is determined in the same way as $f(r)$. This is a local consistency criterion that can be adopted with confidence, since f_0 turns out to be a slightly varying function of the density. The HMSA closure reduces to HNC for $f_0 = 1$ and to SMSA when $f_0 \to 0$.

For systems dominated by short-range forces, one has to set $u_{\mathrm{LR}}(r) = 0$, so that $u_{\mathrm{SR}}(r) = u(r)$. In this case, $\gamma^*(r) = \gamma(r)$ and HMSA reduces to the famous Rogers and Young (RY) [64] approximation that reads

$$B(r) = -\gamma(r) + \ln\left[1 + \frac{\exp\{f(r)\gamma(r)\} - 1}{f(r)}\right] \tag{54}$$

It turns out that RY interpolates continuously between HNC ($f_0 = 1$) and PY ($f_0 \to 0$). This SCIET has been applied with satisfactory results for the HS system [64]. But, of course, this scheme is inapplicable to the LJ fluid study.

As seen in Section IV, HMSA failed in reproducing the bridge function calculated from simulation [36]. This is the reason several authors imagined another class of bridge functions, which are inspired from existing IETs by extending them to provide accurate results as compared to reference data.

2. *Extended Verlet Approximations*

Even if the bridge function due to Verlet [47] is one of the oldest bridge functions, recalling that it describes the structure and thermodynamics of the HS fluid quite well, it has been considered as a good starting point for improvements.

a. Duh and Haymet Approximation. Based on the inversion of simulation data for the bridge function, Duh and Haymet (DHH) [65] proposed a new approximation for $B(r)$ so that

$$B(r) = \frac{-\gamma^{*2}(r)}{2\left[1 + \left(\frac{5\gamma^*(r)+11}{7\gamma^*(r)+9}\right)\gamma^*(r)\right]} \tag{55}$$

As seen, compared to (38) the parameter α has been replaced by a more complicated term involving itself the indirect correlation function $\gamma^*(r)$. Even if reputed to be accurate, this approximation suffers from a slight thermodynamic inconsistency. As recognized by the authors, "the compressibility obtained from the virial route differs little from that obtained directly from the compressibility equation". Nevertheless, this approximation is here considered as a SCIET.

b. Lloyd L. Lee Approximation (ZSEP). In order to take into account the correct behavior of the correlation functions inside the core region ($r < \sigma$), Lee

[8] proposed a flexible approximation that reads

$$B(r) = -\frac{\zeta\gamma^{*2}(r)}{2}\left(1 - \frac{\alpha\phi\gamma^{*}(r)}{(1+\alpha\gamma^{*}(r))}\right) \tag{56}$$

This closure contains three parameters that will have to be determined thanks to consistency conditions and theorems (the Zero-Separation Theorems) connecting the properties of the bulk fluid with the correlation functions at coincidence ($r = 0$). This last relation reduces to (38) if ζ and ϕ are set to one. In general, the numerical implementation of consistency conditions with more than one parameter in the context of an iterational theory is far from being trivial.

c. Choudhury and Ghosh Approximation. More recently, these authors proposed calculating the bridge function by using the originaly form of the Verlet approximation,

$$B(r) = \frac{-\gamma^{*2}(r)}{2(1+\alpha\gamma^{*}(r))} \tag{57}$$

but assuming that the parameter α is related to the density of the fluid through an empirical relation [66].

3. *Extended Martynov–Sarkisov Approximations*

As in the case of the Verlet IE, the Martynov–Sarkisov IE has been the subject of a number of extensions too.

a. BPGG Approximation. The first improvement is due to Ballone, Pastore, Galli, and Gazzillo (BPGG) [67] that assumed

$$B(r) = [1 + s\gamma(r)]^{1/s} - \gamma(r) - 1 \tag{58}$$

For $s = 1$, the HNC is recovered, while for $s = 2$, Eq. (58) reduces to the MS closure. The parameter s has been used to fit some simulation results or to impose the thermodynamic consistency condition.

b. The Vompe–Martynov Approximation (VM). It is important to mention that the authors of the MS have themselves refined their own approximation. After a carefull analysis of diagrams expansion, it has been concluded that it was more efficient to expand Eq. (29) in terms of $[\omega(r) - \beta u_{\mathrm{LR}}(r)]$ rather than in terms of $\omega(r)$ [24]. Instead of $B(r) = -1/2\omega^2(r)$, the bridge function is approximated by $B(r) = -1/2[\omega(r) - \beta u_{\mathrm{LR}}(r)]^2$, providing an extension in terms of $\gamma^*(r)$, so that

$$B(r) = [1 + 2\gamma^*]^{1/2} - \gamma^*(r) - 1 \tag{59}$$

In this framework, inspired by the BPGG approach, a parameter α, whose role is to fulfil the thermodynamic consistency condition, has been introduced by Charpentier and Jakse [68]. In this semiempirical approach, the bridge function expresses as $B(r) = -\alpha[\omega(r) - \beta u_{\mathrm{LR}}(r)]^2$, so that

$$B(r) = \frac{1}{2\alpha}[(1 + 4\alpha\gamma^*(r))^{1/2} - 1 - 2\alpha\gamma^*(r)] \tag{60}$$

Note that the latter reduces to the MS closure relation by setting the parameter α to one-half and by removing the WCA splitting in $\gamma^*(r)$.

c. Bomont and Bretonnet Approximation. More recently, another extension of the MS approximation has been proposed [69–70]. Starting from the fact that, in order to obtain the OZ equation solution, the bridge function is expanded on chosen basis functions $\omega^{(n)}(r)$ [see Eq. (29)], the authors have found an accurate basis to expand the bridge function on. This method allows us to incorporate directly a self-consistent parameter *f*. A carefull analysis of the diagram approach makes clear that the first term of $B(r)$ must always to be of order $\omega^2(r)$ or $[\omega(r) - \beta u_{\mathrm{LR}}(r)]^2$. The selected basis respects this last prescription and provides the following form $-\frac{1}{2}[\omega(r) - \beta u_{\mathrm{LR}}(r)]^2 + \frac{f}{2}\gamma^{*2}(r)$, so that the bridge function reads

$$B(r) = [1 + 2\gamma^*(r) + f\gamma^{*2}(r)]^{1/2} - 1 - \gamma^*(r) \tag{61}$$

Note that the last closure relation interpolates between two others. By setting *f* to 1, HNC is recovered, while the Bomont and Bretonnet (BB) approximation reduces to the VM when *f* is set to 0 (and also to MS if the WCA division scheme is removed).

E. Comments on SCIETs

Given a potential of interaction in the Hamiltonian, from first principles, there corresponds uniquely a pcf $g(r)$. Then, all other correlation functions are also uniquely determined. This connection is expressed in general terms as a functional relation $B = B[g(r)]$. That means, given a $g(r)$ or some related function $\gamma(r)$, one can go through a functional to obtain the bridge function. Only under certain circumstances, the functional reduces uniquely to a function $B = B(\gamma)$. In this way, all common closure relations (PY, HNC, MS, BPGG, and V) are attempts to capture this unique function relation. This is the unique functionality condition. On the one hand, it is not clear from a theoretical aspect that such a relation exists for arbitrary pair potentials. This is the existence problem. On the other hand, if such function exists, is the expression unique? Do there exist other function forms that are equally well representative? This is the uniqueness problem.

As seen above, most of the recent SCIETs are involved with the function $\gamma^*(r)$. The use of this renormalized indirect correlation function in the diagrams expansions has a simple and practical justification: The $h(r)$ diagrams for the bridge function contain an $\exp[\omega(r) - \beta u(r)]$ factor in the integrand expressions, so they can be expanded in terms of the mean force powers $[\omega(r) - \beta u(r)]$. In order to avoid problems caused by the unlimited increasing of the pair potential inside the core, it is useful to take the expansion in powers of $[\omega(r) - \beta u_{\mathrm{LR}}(r)]$.

This phenomenological treatment first employed in the mean spherical approximation [Eq.(43)], in a form suitable for a potential made up of soft core and an attractive part [49, 50], has proven its reliability [22].

In this framework, optimizing the division scheme of the potential seems to allow us to approach such uniqueness [65]. By searching through the function space, one hopes to find a unique representation $B = B(\gamma^*)$ after renormalization. Such endeavors are beyond the scope of research in the field of SCIETs. The logic is as follows: If the closure is exact, it must necessary satisfy all known thermodynamic and structural conditions. The reverse question, the sufficiency condition, requires more care. If such a functionality provides, *a posteriori*, accurate results for all properties, though not proof of unique functionality, it is favorably disposed to it.

IV. TESTING THE SCIETS FOR FLUIDS MODELS

For a given potential, it is clear that the predictive results of SCIETs must be systematically compared against the corresponding results provided by numerical simulation. Theoretically, an accurate theory has to fulfill some criteria or theorems (not necessarily a single thermodynamic consistency condition) that we briefly describe below.

A. Several Aspects of the Thermodynamic Consistency and Theorems

As previously described, there is the partial consistency (e.g., the pressure consistency), and there are the Helmholtz free energy functionals. Unfortunately, the exact Helmholtz functional suffers from intractability. The evaluation of such functionals is equivalent to thermodynamic integration, where the integrands are not exact. New efficient methods in the integration of functionals should be developed first. Again, approximate functionals provide inconsistent thermodynamic properties.

Thus, another approach consists in selecting some boundary conditions and properties. It is obvious that all exact correlation functions must satisfy and incorporate them in the closure expressions at the outset, so that the resulting correlations and properties are consistent with these criteria. These criteria have to include the class of Zero-Separation Theorems (ZSTs) [71,72] on the cavity function $y(r)$, the indirect correlation function $\gamma(r)$ and the bridge function $B(r)$ at zero separation ($r = 0$). As will be seen, this concept is necessary to treat various problems for open systems, such as phase equilibria. For example, the calculation of the excess chemical potential $\beta\mu_{ex}$ is much more difficult to achieve than the calculation of usual thermodynamic properties since one of the constraints it has to satisfy is the Gibbs–Duhem relation

$$\beta\frac{\partial\mu_{ex}}{\partial\rho} = -\int c(r)d\mathbf{r} \tag{62}$$

This last decade, the chemical potential has been the subject of intensive efforts in the IETs field, while it can be easily obtained from simulation through the test particle insertion method [73, 74].

Here, we report some basic results that are necessary for further developments in this presentation. The merging process of a test particle is based on the concept of cavity function (first adopted to interpret the pair correlation function of a hard-sphere system [75]), and on the potential distribution theorem (PDT) used to determine the excess chemical potential of uniform and nonuniform fluids [73, 74]. The obtaining of the PDT is done with the test-particle method for nonuniform systems assuming that the presence of a test particle is equivalent to placing the fluid in an external field [36].

Consider a classical system of N particles interacting with a potential energy equal to the sum of pairwise interactions $u(r)$. If the particle labeled 1 is selected for special consideration, the total potential energy can be partitioned as follows:

$$V_N(\mathbf{r}^N) = u(r_{12}) + \sum_{j>2}^{N} u(r_{1j}) + V_{N-1}(\mathbf{r}^{N-1}) \tag{63}$$

In the rhs of Eq. (63), the first term is the interaction between particles 1 and 2, the second term represents the excess interaction due to the particle 1 on each other, except 2, and the third term corresponds to the sum of the pairwise interactions of the remaining particles.

The pair correlation function $g(r_{12})$ is then evaluated in the grand canonical ensemble with the familiar expression [7]

$$g(r_{12}) = \frac{1}{\rho^2} \frac{1}{Q_{\mu,V,T}} \sum_{N=2}^{\infty} \frac{z^N}{(N-2)!} \int \exp[-\beta V_N(\mathbf{r}^N)] d\mathbf{r}_3 \cdots d\mathbf{r}_N \tag{64}$$

where ρ is the number density, $Q_{\mu,V,T}$ is the grand canonical partition function given by

$$Q_{\mu,V,T} = \sum_{N=0}^{\infty} \frac{z^N}{N!} \int \exp[-\beta V_N(\mathbf{r}^N)] d\mathbf{r}^N$$

and z is the fugacity related to the chemical potential μ and to the de Broglie wavelength Λ by the standard relation $z = \frac{1}{\Lambda^3}\exp(\beta\mu)$. Combining Eq. (63) with Eq. (64), taking $e^{-\beta u(r_{12})}$ out of the integral, and introducing the integral over $\mathbf{r}_2$ with $\frac{1}{V}\int d\mathbf{r}_2$, yields the pair correlation function under the form

$$\begin{aligned} g(r_{12}) = & \frac{e^{-\beta u(r_{12})}}{\rho^2 V} \frac{1}{Q_{\mu,V,T}} \sum_{N=2}^{\infty} \frac{z^N}{(N-2)!} \int \exp[-\beta V_{N-1}(r^{N-1})] \\ & \times \exp[-\beta \sum_{j>2}^{N} u(r_{1j})] d\mathbf{r}_2 \cdots d\mathbf{r}_N \end{aligned} \tag{65}$$

To simplify Eq. (65), it is necessary to change the index in the summation so that the system of N particles is labeled $2, 3, \ldots, N+1$. Particle 1 becomes a fictitious test particle because it does not influence the movement of the other particles. When the rearragement of the index is carried out, $g(r_{12})$ is expressed as a function of the ensemble average $\langle N \exp(-\beta \sum_{j>2}^{N+1} u(r_{1j}))\rangle$ and the cavity distribution function, defined by $\ln y(r) = B(r) + \gamma(r)$, reduces to the form

$$y(r_{12}) = \frac{1}{\rho^2 V} \frac{\exp(\beta\mu)}{\Lambda^3} \left\langle N \exp\left(-\beta \sum_{j>2}^{N+1} u(r_{1j})\right)\right\rangle \tag{66}$$

This relation is the potential distribution theorem [73, 74], which gives a physical interpretation of the cavity function in terms of the chemical potential, and the excess interaction generated by the test particle, $\sum_{j>2}^{N+1} u(r_{1j})$, via the ensemble average of its Boltzmann factor. In numerical simulation, the use of such a test-particle insertion method is of prime importance in calculating the cavity function at small distances and particularly at zero separation. Note that if the particle labeled 1 approaches the particle labeled 2, a dumbbell particle [41] is created with a bond length $L = |\mathbf{r}_2 - \mathbf{r}_1|$ corresponding to a dimer at infinite

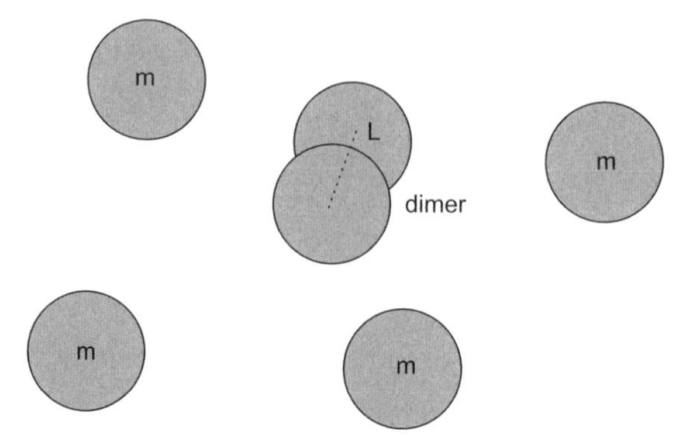

Figure 3. A dimer molecule at infinite dilution in a bath of monomers (m). The dimer is formed of two monomers fused to a bond length of L. In the limit $L \to 0$, we form a coincident dimer or a strong "monomer" with same size σ but twice energetic (2ε). Taken from Ref. [76].

dilution in the solvent (see Fig. 3). As the merging process proceeds, the bond length L of the dumbbell reduces continuously until total coincidence between particles 1 and 2. Both of them occupy the same space so that the resulting particle, referred to as a strong test particle, interacts with the other particles with the combined potentials of the two original particles. Note that the cavity function is related to the work, $-\ln y(0)$, required to insert a strong test particle, of the same size as a single particle, with twice the potential energy (2ϵ) according to the relation [70, 80]

$$-\ln y(0) = \beta\mu_2(0) - 2\beta\mu_1 \tag{67}$$

where $\mu_2(0)$ is the excess chemical potential of the strong test particle of bond length $L = 0$ and μ_1 is that of a single particle. The main consequence for our purpose is that the PDT of the strong test particle yields the zero-separation value of $y(r)$, which allows us to calculate the bridge function $B(r)$ at $r = 0$. For HS spheres, the potential energy is either $+\infty$ or 0. Doubling of $+\infty$ or 0 gives back $+\infty$ or 0, respectively. Then, $\beta\mu_2(0) = \beta\mu_1$. Thus one recovers $\ln y^{\mathrm{HS}}(0) = \beta\mu_1$, the well-known ZST for HS. Equation (67) is more general and is applicable to soft-sphere potentials. For these fluid models, one has a second theorem that reads

$$B(0) = \ln y(0) - \gamma(0) \tag{68}$$

In the case of HS fluid, the last relation reduces to $B(0) = \beta\mu_1 - \gamma(0)$ and making use of the Carnahan-Starling EOS, one obtains

$$B^{CS}(0) = -\frac{15\eta^2 - 12\eta^3 + 3\eta^4}{(1-\eta)^4} - I_C. \tag{69}$$

I_C, that is a non analytical term given by $I_C = \rho \int g(r)c(r)d\mathbf{r}$, has been found to be a small correction [8]. At very high density, I_C is only 3% of $\gamma(0)$, so that it could be neglected in Eq. (69) without any nuisance. Then, the ZSTs serve as connecting the thermodynamics properties of the bulk fluid with the correlation functions at coincidence. Some other interesting boundary conditions can be found elsewhere [78]. As attested in the literature, this is over the last decade that several attempts have been made to improve upon these closure relations and to extend the range of validity of IETs.

B. Accuracy of the Bridge Function

In order to make the SCIETs competitive with respect to numerical simulation results, an intriguing strategy has been followed by Duh and Haymet [65], while working on the bridge function for ionic fluids. By using a double dosage of (1) separating the pair potential into a reference and a perturbative part (that was termed as optimization) and (2) renormalization, say $u_{\mathrm{LR}}(r)$, they showed promise of obtaining the unique functionality condition. Then, an important point is to mention that not only the WCA potential separation [51] between a reference and a perturbative part, but other different prescriptions for the partitioning of the pair potential are available. Optimizing the perturbative part of the pair potential has been one of the keys of success in calculating the bridge function [79, 80]. We present the two major attempts of division schemes of the potential.

1. Duh and Haymet Division Scheme (DHHDS)

To address the problem of uniqueness of the closure, these authors [65] proposed a method for partitioning the potential energy, by defining the perturbation part to be density dependent (see Fig. 4), so that

$$u_{\mathrm{LR}}(r) = -4\epsilon\left(\frac{\sigma}{r}\right)^6 \exp\left[\frac{-1}{\rho^*}\left(\frac{\sigma}{r}\right)^{6\rho^*}\right] \tag{70}$$

The physical picture behind this choice is that (1) at high densities, the system is dominated by the repulsive part of the potential, while (2) at low densities, the attractive part begins to play a more important role. This division scheme has been improved [81] according to an additional criterion that reads

$$\left.\frac{\partial u_{\mathrm{LR}}(r)}{\partial r}\right|_{r=\sigma} = 0 \tag{71}$$

and according to the partitioning $u_{\mathrm{SR}}(r) + u_{\mathrm{LR}}(r) = u(r)$, one obtains

$$\left.\frac{\partial u_{\mathrm{SR}}(r)}{\partial r}\right|_{r=\sigma} = \left.\frac{\partial u(r)}{\partial r}\right|_{r=\sigma} \tag{72}$$

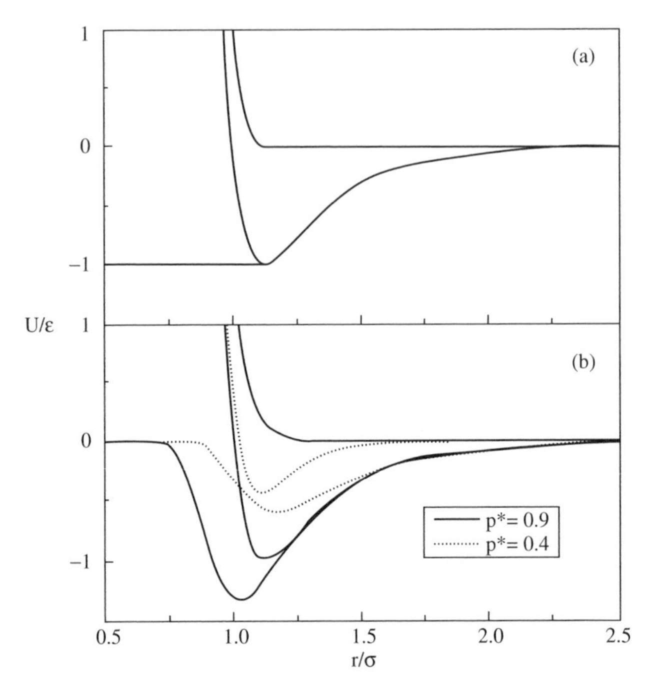

Figure 4. Partitioning of the Lennard-Jones potential into a reference part and a perturbation part by (a) the WCA scheme and (b) the DHHDS scheme. The thick line in both panels is the full Lennard-Jones potential. Taken from Ref. [65].

In this framework, the force acting on the particles at $r = \sigma$ is not altered by the introduction of the potential perturbative part.

2. Bomont and Bretonnet: Optimized Division Scheme (ODS).

These authors wanted to take advantage of the utility of the test particle insertion usually involved in molecular dynamics, by transfering and adapting this concept to the integral equation method [73, 74]. They showed that it is possible to obtain consistently $B(r)$ in good agreement with MC calculations, only if the renormalized $\gamma^*(r)$ function is conjugated with an optimized separation of the pair potential [80]. To achieve this, a new division scheme has been assumed and developed following the idea emphasized above.

In order to improve the description of the atomic arrangement, according to the authors, an optimization consists in determining an adequate short-range part, $u_{\mathrm{SR}}(r)$, and long-range part, $u_{\mathrm{LR}}(r)$, so that $u(r) = u_{\mathrm{SR}}(r) + u_{\mathrm{LR}}(r)$. To achieve this, they proposed a partition of the pair potential so-called optimized division scheme (ODS), that contrary to DHHDS is not density dependent, for which

$$u_{\mathrm{LR}}(r) = \begin{cases} p\epsilon & r \leq r_1 \\ a_1 + a_2 r + a_3 r^2 + a_4 r^3 & r_1 < r \leq r_2 \\ u(r) & r > r_2 \end{cases} \tag{73}$$

and $u_{\mathrm{SR}}(r) = u(r) - u_{\mathrm{LR}}(r)$. This partitioning (see Fig. 5) was put forward not as a final result, but as a starting point for further study. As seen, the range of distances is divided into three parts delimited by r_1 and r_2. Note that the extent between r_1 and r_2 is that of the main peak of $g(r)$, the most crucial region in which $B(r)$ has to be specified [27]. Also, the present separation of the pair potential reduces to the WCA scheme [51] when $p = 1$ and $r_1 = r_2 = r_m$. To avoid singularities in $u_{\mathrm{LR}}(r)$, the coefficients $a_1,\ a_2,\ a_3$, and a_4 of the polynomial function are determined in order to keep safe the continuity and the first derivative of the potential $u_{\mathrm{LR}}(r)$ at the positions r_1 and r_2. The analytical solution of the simultaneous equations provides the values of the a_i parameters versus $r_1,\ r_2,\ p\epsilon, u(r_2)$, and $u'(r_2)$ [$u'(r)$ being the first derivative of $u(r)$ vs. r] under the forms

$$\begin{aligned} a_1 &= \frac{r_1^3 u(r_2) - r_2 r_1^3 u'(r_2) - 3r_2 r_1^2 u(r_2) + r_1^2 r_2^2 u'(r_2) + 3p\epsilon r_1 r_2^2 - p\epsilon r_2^3}{(r_1 - r_2)^3} \\ a_2 &= -\frac{r_1(-r_1^2 u'(r_2) - r_1 r_2 u'(r_2) + 6p\epsilon r_2 - 6u(r_2) + 2r_2^2 u'(r_2))}{(r_1 - r_2)^3} \\ a_3 &= \frac{-2r_1^2 u'(r_2) + 3p\epsilon r_1 - 3r_1 u(r_2) + r_2 r_1 u'(r_2) + 3p\epsilon r_2 - 3r_2 u(r_2) + r_2^2 u'(r_2)}{(r_1 - r_2)^3} \\ a_4 &= -\frac{-r_1 u'(r_2) - 2u(r_2) + r_2 u'(r_2) + 2p\epsilon}{(r_1 - r_2)^3} \end{aligned} \tag{74}$$

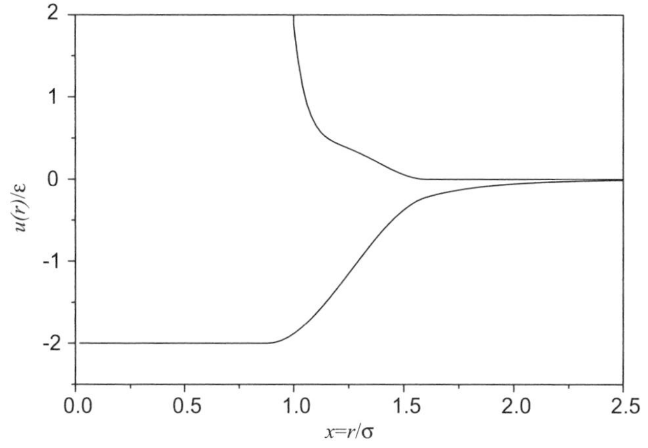

Figure 5. Optimized division scheme for the LJ potential. This partitioning is not state dependent. Taken from Ref. [80].

Specifically, the model relies on three parameters r_1, r_2, and p that are determined according to the following prescription. First, the parameter r_1 is chosen for convenience to be the distance at which $g(r)$ becomes nonzero and r_2 the first minimum of $g(r)$, roughly estimated by $r_2 = r_m + 2(r_m - r_1)$. Strictly, for the LJ potential, these values are $r_1 = 0.88\sigma$ and $r_2 = 1.6\sigma$, whatever the temperature and the density [82]. Second, the authors adopted the test-particle insertion method (strong monomer of diameter σ interacting with a doubled potential energy 2ϵ), to fix the parameter $p = 2$ so that the depth of the long-range part of the potential equals 2ϵ into the core region. Here, the fictitious character of the strong test particle is reflected by the fact that the pair potential is recovered *in fine* by adding both parts $u^{\text{SR}}(r)$ and $u^{\text{LR}}(r)$. So the presence of strong test particle does not modify the pair potential and does not affect the movement of the real particles.

C. Classical Thermodynamic and Structural Quantities for Fluid Models

1. *The Hard-Sphere Fluid*

We start with the supposed simplest fluid model. For hard spheres, the essential thermodynamic [7] properties are the pressure and the isothermal compressibility (there is no energy) that read

$$\frac{\beta P^v}{\rho} = 1 + \frac{2}{3}\pi\rho\sigma^3 g(\sigma^+) \tag{75}$$

and

$$\frac{\beta \partial P^v}{\partial \rho} = 1 + \frac{\partial}{\partial \rho}\left(\frac{2}{3}\pi\rho^2\sigma^3 g(\sigma^+)\right) \tag{76}$$

According to the Carnahan and Starling formula [31], the contact value $g(\sigma^+)$ is accurately given by

$$g(\sigma^+) = \frac{1 - 0.5\eta}{(1 - \eta)^3} \tag{77}$$

where η is the packing fraction. Table I offers a good overview of the accuracy of some IETs in predicting the classical thermodynamic properties of HS. The

TABLE I
Pair Correlation Function $g(r)$ at Contact and Thermodynamic Quantities $\beta P/\rho$ and $\beta\partial P/\partial\rho$ for HS Fluid

ρ^*	$g^a(\sigma^+)$	$g^b(\sigma^+)$	$g^{CS}(\sigma^+)$	$\beta P^a/\rho$	$\beta P^b/\rho$	$\beta P^{CS^c}/\rho$	$\beta\partial P^a/\partial\rho$	$\beta\partial P^b/\partial\rho$	$\beta\partial P^{CS}/\partial\rho$
0.1	1.14	1.15	1.14	1.24	1.24	1.24	1.51	1.52	1.51
0.3	1.53	1.56	1.54	1.96	1.98	1.97	3.40	3.48	3.39
0.5	2.15	2.20	2.16	3.26	3.31	3.26	7.61	7.83	7.59
0.7	3.20	3.30	3.21	5.70	5.84	5.71	17.43	18.05	17.54
0.9	5.11	5.29	5.17	10.64	10.98	10.74	43.18	43.66	43.55

[a]ZSEP approximation [8].
[b]BB approximation [83].
[c]CS : Carnahan–Starling EOS.

results obtained from ZSEP [8] and BB [85] are similar and are in good agreement with CS reference data given by

$$\frac{\beta P^{CS}}{\rho} = \frac{1+\eta+\eta^2-\eta^3}{(1-\eta)^3} \tag{78}$$

and

$$\frac{\beta\partial P^{CS}}{\partial\rho} - 1 = \frac{8\eta-2\eta^2}{(1-\eta)^4} \tag{79}$$

The reader has to be aware that the values of these last quantities, contrary to the chemical potential, for example, are not very sensitive to the bridge function values inside the core. Then, accurate values are obtained with others IETS of course.

But, as far as the bridge function $B(r)$ is concerned, things are different. Table II lists the bridge function's zero-separation values, compared to exact data [32] at four densities for different closures: Percus–Yevick (PY), Verlet (V), Ballone–Pastore–Galli–Gazillo (BPGG),Martynov–Sarkisov (MS), Lee (ZSEP), and Bomont–Bretonnet (BB). As seen, all closures, except ZSEP and BB, systematically overestimate the exact $B^{CS}(0)$ given by Eq. (69) whatever the density. Among the nonconsistent integral equations, only the Verlet closure is acceptable at high density ($\rho^* \sim 0.7$), while fails at lower densities ($0.3 < \rho^* < 0.5$). For example, at $\rho^* = 0.8$, near $r = 0$, VM is merely 4% off, while PY is -41% ; MS, -22% and BGPP, -18%. From this comparison, it

TABLE II
Zero-sepertion Values of $B(r)$ Calculated from Several Approximations for HS Fluid, Compared to Carnahan–Starling EOS Data

ρ^*	Exact	BB	ZSEP	VM	PY	MS	BPGG
0.3	−0.683	−0.693	−0.725	−0.995	−1.19	−1	−0.927
0.5	−2.905	−3.04	−3.04	−3.52	−4.65	−3.90	−3.68
0.7	−9.549	−9.83	−9.82	−9.84	−14	−12	−11.5
0.8	−16.52	−17.05		−15.86	−23.26	−20.22	−19.45
0.9	−29.42	−29.00	−30.30	−26.7	−40.1	−35.5	−34.3

turns out that ZSEP and BB closure relations provide a highly accurate representation of $B(r)$ at zero separation. Fig. 6 plots the bridge functions calculated with the BB approximation, versus the reduced distance, with the one based on the Groot's construction that is usually considered to be exact. An

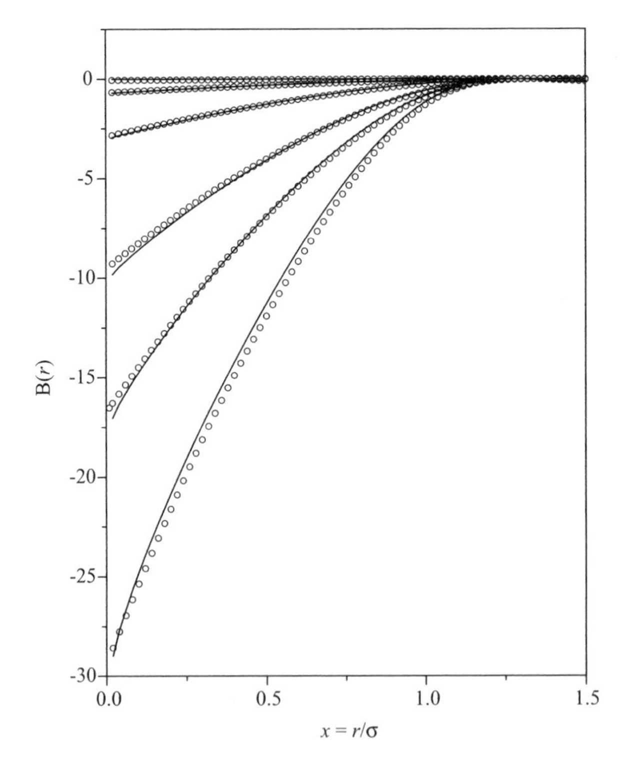

Figure 6. The bridge function $B(r)$ calculated with the BB approximation for five density states: $\rho^* = 0.1, 0.3, 0.5, 0.7, 0.8$, and 0.9 (full line, from top to bottom), compared to Groot's construction (open circles). Taken from Ref. [83].

excellent agreement is achieved over all densities. The ZSEP provides very similar results compared to Groot's construction. As seen, $B(r)$ is always negative inside the core, in agreement with most literature closures.

In order to obtain such results, the ZSEP necessitates the knowledge of three parameters (per state), determined with the aid of several consistency conditions [78], whereas the BB requires only the knowledge of a single parameter adjusted in a self-consistent manner so that Eq. (17) is satisfied within 1%. These two SCIETs yield accurate and consistent results up to high densities. This is in contrast with other theories, where results are deficient in consistency. The accuracy produced here is comparable to that of simulation data.

2. *The Lennard–Jones Fluid*

As far as the LJ fluid is concerned, things are less straightforward and require, as it has been said above, a renormalization of the indirect correlation function $\gamma(r)$ that has to be accompanied by an optimized division scheme of the potential. As will be seen, the WCA [51] splitting is not sufficient to get accurate results for the correlation functions inside the core region, while these values are crucial for phase-equilibria treatment.

Thermodynamic properties, such as the excess energy [Eq. (4)], the pressure [Eq. (5)], and the isothermal compressibility [Eq. (7)] are calculated in a consistent manner and expressed in terms of correlation functions [$g(r)$, or $c(r)$], that are themselves determined so that Eq. (17) is satisfied within 1%. It is usually believed that for the thermodynamic quantities, the values of the correlation functions [$B(r)$ and $c(r)$, e.g.] do not matter as much inside the core. This may be true for quantities dependent on $g(r)$, which is zero inside the core. But this is no longer true for at least one case: the isothermal compressibility that depends critically on the values of $c(r)$ inside the core, where major contribution to its value is derived. In addition, it should be stressed that the final $g(r)$ is slightly sensitive to the consistent isothermal compressibility.

In this framework, we present the repercussions on the physical properties of a renormalized indirect correlation function $\gamma^*(r)$ conjugated with an optimized division scheme. All the units are expressed in terms of the LJ parameters, that is, reduced temperature $T^* = k_B T/\epsilon$ and reduced density $\rho^* = \rho\sigma^3$. In order to examine the consequences of a renormalization scheme, the direct correlation function $c(r)$ calculated from ZSEP conjugated with DHH splitting is compared in Fig. 7 to those obtained with the WCA separation. For high densities, the differences arise mainly in the core region for $\gamma(r)$ and $c(r)$ [77]. These calculated quantities are in excellent agreement with simulation data. The reader has to note that similar results have been obtained with the ODS scheme (see Ref [80]). Since the acuracy of $c(r)$ can be affected by the choice of a division scheme, the isothermal compressibility is affected too, as can be seen in Table III for the $\rho k_B T \chi_T$ quantity. As compared to the values obtained with

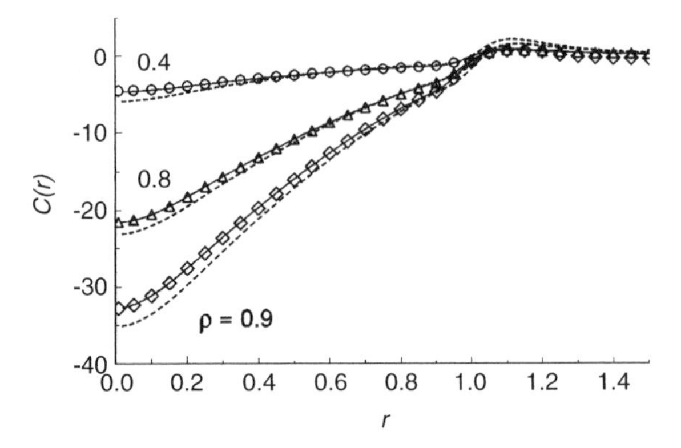

Figure 7. The direct correlation functions predicted by ZSEP at $T^* = 1.5$ using WCA scheme (dotted lines) and DHHDS partitioning (solid lines), compared to MC simulation data [36] (symbols). Taken from Ref. [77].

WCA separation, the values stemming for the ODS are smaller for each case studied. At $T^* = 1.5$, the relative difference varies between 0.5 and 9% over the range of densities. And when the temperature is decreasing, the relative difference increases until 19% at $T^* = 0.72$ and $\rho^* = 0.85$. In general, quantities, such as E^{ex}/N and $\beta P/\rho$, are well described by SCIETs compared to molecular dynamics data of Johnson [34]. This is also the case for the BB approximation conjugated with the ODS separation scheme whose results are displayed in Table IV. In passing, notice the quality of the results obtained by Sarkisov [84].

As we are rather interested in this presentation in the effects of an optimized potential separation on the correlation functions, we turn now to some results

TABLE III
Isothermal Compressibility $\rho k_B T \chi_T$ of the LJ Fluid for Several States, Calculated by Using the HMSA Approximation Conjugated with the WCA and the ODS Division Schemes

		$\rho k_B T \chi_\tau$	
T^*	ρ^*	WCA	ODS
1.5	0.4	1.389	1.384
1.5	0.6	0.246	0.231
1.5	0.7	0.122	0.112
1.5	0.8	0.066	0.061
1.5	0.9	0.038	0.035
1	0.8	0.065	0.056
0.81	0.8	0.066	0.055
0.72	0.85	0.044	0.035

TABLE IV
Thermodynamic properties $\beta E^{ex}/N$ and $\beta P/\rho$ of the LJ Fluid at Supercritical Temperatures

T^*	ρ^*	$-\beta E_{ex}/N$	$-\beta E^a_{ex}/N$	$-\beta E^b_{ex}/N$	$\beta P/\rho$	$\beta P^a/\rho$	$\beta P^b/\rho$
5	0.1	0.102	0.102	0.101	1.066	1.066	1.069
	0.3	0.298	0.299	0.299	1.318	1.321	1.327
	0.5	0.471	0.475	0.475	1.863	1.856	1.860
	0.7	0.580	0.589	0.581	2.940	2.900	2.932
	0.9	0.540	0.573	0.538	4.950	4.796	4.930
	1.1	0.226	0.318	0.225	8.446	8.011	8.418
	1.2	−0.091	0.058	−0.077	10.990	10.293	10.993
2.74	0.1	0.222	0.223	0.222	0.974	0.974	0.977
	0.2	0.439	0.440	0.439	0.990	0.987	0.993
	0.3	0.652	0.654	0.653	1.050	1.056	1.060
	0.4	0.862	0.865	0.862	1.197	1.204	1.204
	0.5	1.067	1.071	1.066	1.456	1.471	1.469
	0.6	1.261	1.266	1.257	1.893	1.912	1.914
	0.7	1.426	1.436	1.421	2.601	2.593	2.616
	0.8	1.547	1.563	1.539	3.656	3.602	3.665
	0.9	1.593	1.627	1.587	5.184	5.036	5.174
	1	1.531	1.601	1.532	7.339	7.008	7.294
1.35	0.1	0.580	0.579	0.575	0.718	0.717	0.722
	0.2	1.131	1.130	1.121	0.495	0.497	0.507
	0.35	1.822	1.836	1.839	0.298	0.307	0.327
	0.4	2.036	2.055	2.061	0.261	0.286	0.303
	0.5	2.487	2.509	2.510	0.266	0.347	0.329
	0.6	2.967	2.983	2.984	0.507	0.640	0.566
	0.7	3.440	3.451	3.461	1.194	1.311	1.202
	0.8	3.859	3.875	3.886	2.487	2.538	2.438
	0.9	4.165	4.205	4.199	4.639	4.523	4.516
	0.95	4.253	4.317	4.290	6.130	5.871	5.981

[a]Ref. [34].
[b]Ref. [84].

for $g(r)$ and $B(r)$. We start with the MSA that is known to yield acceptable $g(r)$ except at medium densities. Moreover, it greatly overestimates the magnitude of the bridge function inside the core [65]. In contrast, usual SCIETs provide accurate results for the pair correlation function as illustrated in Fig. 8, but not necessarily accurate results for $B(r)$. As seen in Fig. 9, it turns out that SCIETs involved with the WCA separation, even based on self-consistent schemes, provide poor results for $B(r)$ in the core region. This is the case for HMSA and VM or for the approximation derived by Charpentier and Jakse [68], who present a similar behavior. This trend tends to suggest that (1) the employment of a single criterion of consistency is not sufficient, (2) and (/or) that these relations are not flexible enough, and finally (3) that WCA is not an enough optimized separation scheme.

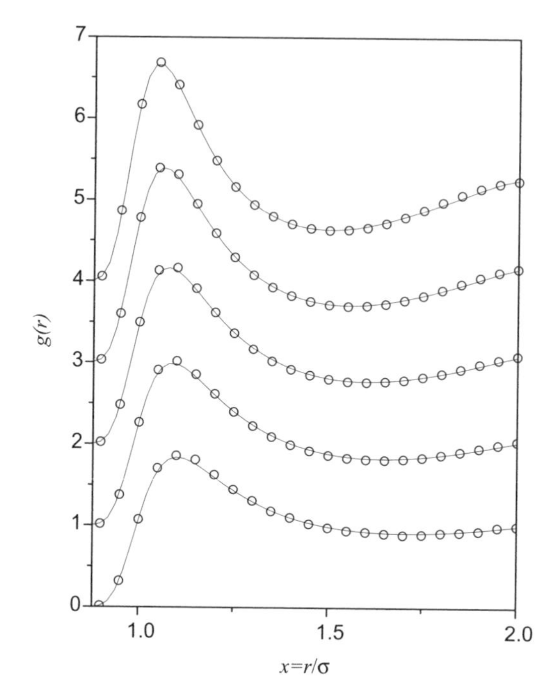

Figure 8. Pair correlation function $g(r)$ at $T^* = 1.5$, for five reduces densities $\rho^* = 0.4$, 0.6, 0.7, 0.8, and 0.9, calculated with the ODS scheme, compared to MC simulation data [36]. Taken from Ref. [80].

We would like to mention that RHNC is able to reproduce qualitatively accurate bridge functions inside the core, especially when a reference system (RS) of soft spheres is utilized rather than the conventional hard-sphere fluid. However, the prescription used to determine the RS requires *a priori* knowledge of both the RS and the system under investigation. This feature restricts the applicability of RHNC to sytems for which the EOS is available. The optimization of the RS proposed by Lado does not have such a drawback, but requires intensive computation since the optimization has to be performed at each state point. Furthermore, it was shown [85] that for the LJ potential, there exists a region around the critical point where the RHNC [54] has no solution.

Fortunately, extended approximations conjugated with judicious separation schemes yield adequate results. These approximations do not rely on the knowledge of any other system and are automonous. For example, HMSA+ODS, DHH+DHHDS (except at moderate densities, where the accuracy is less), ZSEP+DHHDS, and BB+ODS seem to reproduce the expected behavior of $B(r)$ inside the core as illustrated in Fig. 9. Contrary to what could have been thought, HMSA is seen to be flexible enough if conjugated with an appropriate division scheme. Another picture of these results at lower temperatures is given in Fig. 10.

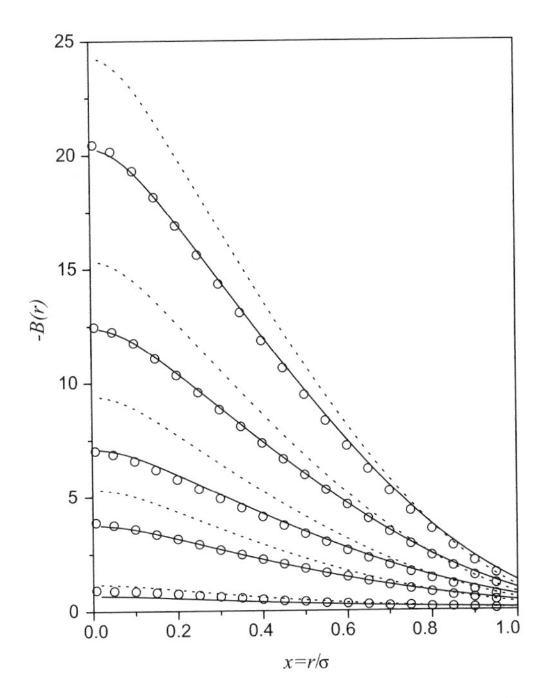

Figure 9. Bridge function $B(r)$ at $T^* = 1.5$, for five densities (see Fig. 8), calculated with the HMSA closure by using theWCA division scheme (dotted line) and ODS scheme (solid lines), compared to MC simulation data [36] (open circles). Taken from Ref. [80].

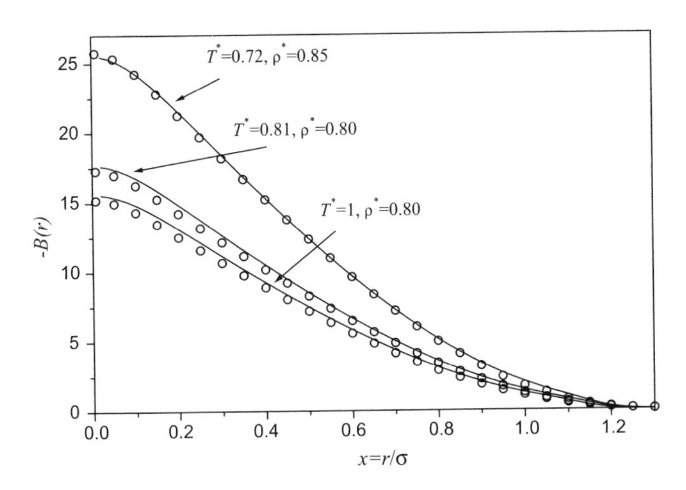

Figure 10. Bridge function $B(r)$ at $T^* = 0.72$, 0.81, and 1 (from top to buttom) calculated with the ODS scheme, compared to MC simulation data [36] (open circles). Taken from Ref. [80].

As seen, for example, BB+ODS, which requires only a single self-consistent parameter, is able to predict the expected behavior of $B(r)$ for a large range of temperatures at high densities. The three other theories, of course, provide similar results.

Thus, these last approaches seem to be promising enough to be tested on the entropic quantities calculation.

D. Excess Chemical Potential and Entropy

One of the most challenging tasks in the theory of liquids is the evaluation of the excess entropy S^{ex}, which is representative of the number of accessible configurations to a system. It is well known that related entropic quantities play a crucial role, not only in the description of phase transitions, but also in the relation between the thermodynamic properties and dynamics. In this context, the prediction of S^{ex} and related quantities, such as the residual multiparticle entropy in terms of correlation functions, free of any thermodynamic integration (means direct predictive evaluation), is of primary importance. In evaluating entropic properties, the key quantity to be determined is the excess chemical potential $\beta\mu_{ex}$. Calculation of $\beta\mu_{ex}$ is not straightforward and requires a special analysis.

After recalling the underlying physical meaning of the excess chemical potential, we provide a survey of the two common approaches in obtaining it from the correlation functions. From these last two, a recent approach is described. This one offers important information about one of the most unknown correlation functions: The so-called "one particle bridge function" $B^{(1)}(r)$, which is the key to calculate $\beta\mu_{ex}$ in a consistent manner.

1. *Kirkwood Definition*

On one hand, most existing theories are based on the Kirkwood charging formula [86]. In this framework, a rigorous way of expressing the chemical potential in excess of the ideal-gas contribution reads

$$\beta\mu_{ex} = \rho \int_0^1 d\lambda \int \frac{\partial \beta u(r,\lambda)}{\partial \lambda} g(r,\lambda) d\mathbf{r} \tag{80}$$

where $u(r)$ is the pair potential and $g(r,\lambda)$ is the pair correlation function between a partially and fully coupled particle in a system of fully coupled particles of diameter d. In order to insert a test particle in the fluid, a cavity has to be created whose diameter has to grow gradually from 0 to d, and $\beta\mu_{ex}$ is just the work required of forming a cavity of diameter d. The parameter λ is the so-called charging or coupling parameter that couples a test particle to the rest of the particles in the system. When $\lambda = 0$, the test particle is absent from the system, and when $\lambda = 1$, the test particle is fully coupled to the system, that is, it is not distinguishable from the others. In a first approximation, the growing diameter d_g of the cavity can be taken as λd, but there are, of course, other equivalent expressions [28] for d_g. The direct use of Eq. (80) calls for a numerical integration along a path of integration. This is carried out in assuming the repeated calculations of the pair correlation function $g(r, \lambda)$ for different values of λ, which requires much calculation. In order to remedy this drawback, a closed form for the chemical potential has been proposed [72, 87, 88]. It is based on the differentiation with respect to λ of the following exact closure relation [similar to Eq. (28)] for the total correlation function, $h(r, \lambda) = g(r, \lambda) - 1$, defined [7] as

$$1 + h(r, \lambda) = \exp[-\beta u(r, \lambda) + h(r, \lambda) - c(r, \lambda) + B(r, \lambda)] \tag{81}$$

Substituting the expression of the integrand $\frac{\partial \beta u(r, \lambda)}{\partial \lambda} g(r, \lambda)$ obtained from Eq. (81) into the Kirkwood formula yields

$$\beta\mu_{\text{ex}} = \rho \int \left[\frac{1}{2} h^2(r) - c(r) + B(r)\right] d\mathbf{r} + \rho \int \left[\int_0^1 h(r, \lambda) \frac{\partial}{\partial \lambda} [B(r, \lambda) - c(r, \lambda)] d\lambda\right] d\mathbf{r} \tag{82}$$

We take advantage of this presentation to make mention of the Scaled Particle Theory (SPT) [89,90] that depicts the mechanism of a test particle (or solute particle) insertion. Among the particle of diameter d, there is also a spherical cavity of radius $\lambda d/2$ that contains no particle centers. The local density of particles centers in contact with the surface of the cavity is denoted by $\rho G(\lambda, \rho)$, where ρ is the bulk density. $G(\lambda, \rho)$, whose precise form is not known, is the central function of the SPT. This function is related to the excess chemical potential by

$$\beta\mu_{ex} = 24\eta \int x^2 G(x, \eta) dx, \tag{83}$$

where $x = r/d$. This approach has been recently revisited [91]. Providing new information about the form of G, improves the accuracy of the results for $\beta\mu_{ex}$ in the case of HS fluid.

Once the excess chemical potential has been introduced, we turn now to the entropy.

The multiparticle correlation expansion of the entropy of simple classical fluids expresses as an infinite series [18]

$$S^{\text{ex}} = \sum_{n=2}^{+\infty} S_n \tag{84}$$

where S^{ex} is the excess entropy per particle and s_n values the partial entropies, obtained from the integrated contribution between n-tuples of particles. The residual multiparticle entropy is defined as $\Delta S = S^{ex} - s_2$. For a homogeneous and isotropic fluid, the pair term reads

$$s_2 = -\frac{1}{2}\rho \int [g(r) \ln g(r) - g(r) + 1] d\mathbf{r} \tag{85}$$

The residual multiparticle entropy ΔS appears to be a very sensitive indicator of structural changes that occur in the system with increasing density [92, 93]. Another exact possibility to define S^{ex} consists in using the thermodynamic equality [94].

$$S^{ex} = \beta \frac{E^{ex}}{N} + \beta P^v/\rho - 1 - \beta\mu_{ex} \tag{86}$$

This last relation involves previouly defined thermodynamic quantities. Note that, in the case of the HS system, $\beta E^{ex}/N = 0$. Once again, it is easy to guess that an accurate predictive SCIET is needed first to obtain the excess chemical potential with a good degree of confidence and then to obtain accurate results on the excess entropy. In order to be calculated in a consistent manner, the excess chemical potential has to satisfy the following condition

$$\beta \left.\frac{\partial \mu_{ex}}{\partial \rho}\right|_T = -\int c(r) d\mathbf{r} \tag{87}$$

At this stage, the analytical evaluation of the excess chemical potential requires necessary assumptions, such as choosing an integration path.

2. *Lloyd L. Lee Formula*

The first assumption involves the particular coupling parameter integration path in Eq. (82). This important concept of topological homotopy, which formalizes the notion of continuous deformation of a curve passing through two fixed points, has been addressed by Kjellander and Sarman [88] and Lee [72]. In fact, these authors proposed using a simple linear dependence on λ for the correlation functions $[h(r,\lambda) = \lambda h(r)$ and $c(r,\lambda) = \lambda c(r)]$. That is to say that they implicitly suppose d_g to be equal to λd. This is the commonly adopted rule (but not unique). Then (82) can be rewritten as:

$$\beta\mu_{ex} = \rho \int \left[\frac{1}{2}h^2(r) - c(r) + B(r) - \frac{h(r)c(r)}{2}\right] d\mathbf{r} + \rho \int \mathbf{dr}\, h(r) \int_0^1 \lambda \frac{\partial B(r,\lambda)}{\partial \lambda} d\lambda \tag{88}$$

The second assumption is the unique functionality [72] of the bridge function, meaning that $B = B[\gamma]$ is a simple function of the indirect correlation function, as expressed in the majority of the approximations for the bridge function. Changing the integration variable from λ to γ, and integrating by parts yields

$$\rho \int \mathbf{dr}\, h(r) \int_0^1 \lambda \frac{\partial B(r,\lambda)}{\partial \lambda} d\lambda = \rho \int \mathbf{dr}[h(r)B(r) - S(r)] \tag{89}$$

where

$$S(r) = \frac{h(r)}{\gamma(r)} \int_{\gamma^*(\lambda=0)}^{\gamma^*(\lambda=1)} B(\gamma\prime)d\gamma\prime \tag{90}$$

Again, a homotopy has to be constructed (the parameter λ varies from 0 to 1), so that

$$\gamma^*(r,\lambda) = \gamma_0(r) + \lambda\gamma_1(r) \tag{91}$$

Then, the calculation of the chemical potential requires the knowledge of the Star function [72] defined by $S^* = \rho \int drS(r)$. Most of the previously defined bridge functions can be easily integrated through their primitive forms. For example, from PY

$$\int_0^\gamma B(x)dx = (1+\gamma)\ln(1+\gamma) - \gamma - \gamma^2/2 \tag{92}$$

From MS,

$$\int_0^\gamma B(x)dx = \frac{1}{3}(1+2\gamma)^{3/2} - \gamma - \gamma^2/2 \tag{93}$$

A similar expression is obtained for the BPGG relation. For VM, we have

$$\int_0^\gamma B(x)dx = -(4\alpha^3)^{-1}[(1+\alpha\gamma)^2 - 4(1+\alpha\gamma) + 2\ln(1+\alpha\gamma) + 3] \tag{94}$$

From ZSEP,

$$\int_0^{\gamma} B(x)dx = -\frac{\zeta(\alpha\gamma)^3}{6\alpha^3} + \frac{\zeta\phi}{12\alpha^3}[2(1+\alpha\gamma)^3 - 9(1+\alpha\gamma)^2 + 18(1+\alpha\gamma) - 6\ln(1+\alpha\gamma) - 11] \tag{95}$$

and from DHH, one has

$$\int_0^{\gamma} B(x)dx = \frac{81}{50}\gamma - \frac{7}{20}\gamma^2 - \frac{9}{2}\ln(3+\gamma) - \frac{9}{125}\ln(3+5\gamma) \tag{96}$$

Then $S(r)$ can be calculated from Eq. (90). The validity of these expressions have to be assessed against exact machine results.

3. *Kiselyov–Martynov Approach*

On the other hand, there is a possibility to express the excess chemical potential exclusively in terms of the direct correlation function with an exact expression [95] for $\beta\mu_{ex}$ equivalent to Eq. (80), namely,

$$\beta\mu_{ex} = -\rho\int c^{(1)}(r)d\mathbf{r} \tag{97}$$

where $c^{(1)}$ is the "one-particle" direct correlation function defined as:

$$c^{(1)}(r) = h(r) - \omega(r) - \frac{1}{2}h(r)[\omega(r) + B^{(1)}(r)] \tag{98}$$

Here, $B^{(1)}(r)$ is the "one particle bridge" function [95] and $\omega(r)$, the thermal potential that equals the opposite of the excess potential of the mean force [18, 28] given by

$$\omega(r) = h(r) - c(r) + B(r) \tag{99}$$

Consequently, another formally exact definition for $\beta\mu_{ex}$ follows from gathering Eqs. (97–99).

$$\beta\mu_{ex} = \rho\int\left[\frac{1}{2}h^2(r) - c(r) + B(r)\right]d\mathbf{r} + \rho\int\left[-\frac{1}{2}h(r)c(r) + \frac{1}{2}h(r)[B(r) + B^{(1)}(r)]\right]d\mathbf{r} \tag{100}$$

In this approach, neither Kirkwood charging parameter nor thermodynamic integration is needed. But we face another problem related to the knowledge of the (*a priori* unknown) one particle bridge function $B^{(1)}(r)$. Thus, the accuracy of the excess chemical potential calculations is also conditioned to the approximation of $B^{(1)}(r)$. The one-particle bridge function is an infinite series of irreductible diagrams [28].

$$B^{(1)}(r) = \sum_{n=2}^{+\infty} \frac{1}{n+1} \omega^{(n)}(r) \tag{101}$$

and, unfortunately, cannot be summed up exactly. To remedy this drawback, since the authors have truncated Eq. (29) to the second-order $n = 2$, they approximated (even for $n > 2$) the one-particle bridge function to

$$B^{(1)}(r) \approx \frac{1}{3} B(r) \tag{102}$$

In this formalism, it turns out that $B^{(1)}(r)$ is directly related to $B(r)$. The quality of the calculated chemical potential is conditionned by the quality of $B(r)$, which means that $B(r)$ has to be as good as possible compared to simulation results.

4. *Bomont and Bretonnet Approach*

This very recent approach [96, 97] addresses the general problem of topological homotopy. By comparing Eqs. (82) and (100), it turns out that $B^{(1)}(r)$ can be unambiguously determined by the closure $B(r)$, namely,

$$-\frac{1}{2}h(r)c(r) + \frac{1}{2}h(r)[B(r) + B^{(1)}(r)] = \int_0^1 h(r,\lambda)\frac{\partial}{\partial\lambda}[B(r,\lambda) - c(r,\lambda)]d\lambda \tag{103}$$

So far, Eq. (103) is general and free from any approximation. To determine $B^{(1)}(r)$, it is necessary to integrate Eq. (103) over the charging parameter λ and to make some assumptions with respect to the λ dependence of the correlation functions. The usual rules are the linear dependence of the correlation functions on λ and the unique functionality of the bridge function already mentioned. But, in that case, there is some arbitrariness on it and, as pointed out by Lee [72], a quadratic dependence could equally well be assumed. In a rigorous way, a λ dependent correlation function, say $\Gamma(r,\lambda)$, has to express as $P(\lambda)\Gamma(r)$, with the conditions $P(\lambda) \geq P(\lambda = 0) = 0$ and $P(\lambda) \leq P(\lambda = 1) = 1$. Unfortunately, $P(\lambda)$ remains unknown whatever the correlation function under consideration. So, the way a test particle does couple with the rest of the fluid is an open question. The author has assumed that $P(\lambda) = \lambda^n$, namely $h(r,\lambda) = \lambda^n h(r)$ and $c(r,\lambda) = \lambda^n c(r)$, which corresponds to an extention of the Kjellander–Sarman

[88] and Lee [72] formulas. This assumption is valid since a cavity of diameter λd is equivalent to a solute particle of diameter $\lambda^n d$. Clearly, the authors take $d_g = \lambda^n d$ so that the cavity required to fully couple the test particle to the rest ($\lambda = 1$) has, *in fine*, the desired diameter d. Then, by substituting these last expressions in Eq. (103) and integrating by parts yields

$$B^{(1)}(r) = -2n \int_0^1 \lambda^{n-1} B(r, \lambda) d\lambda + B(r) \tag{104}$$

The authors have arrived at a closed form for the one particle bridge function $B^{(1)}(r)$ as a function of the approximate bridge function $B(r)$ pertaining to all the common used functional forms. In order to perform the integration in Eq. (104), they turned to the problem of the dependence of $B(r, \lambda)$ on the charging parameter λ, which is a problem of topological homotopy. From Eq. (104), it is obvious that assuming an identical λ^n dependence for $h(r, \lambda), c(r, \lambda)$ and $B(r, \lambda)$ yields $B^{(1)}(r) = 0$ whatever the thermodynamic state under study, whereas $B(r)$ is not necessarily a zero functional. Now, in order that $B^{(1)}(r) \neq 0$, they assumed, by analogy, the general dependence $B(r, \lambda) = \lambda^m B(r)$, where $m \neq n$. Inserting the last form of the bridge function in Eq. (104) to achieve the integration yields the convenient form for the one-particle bridge function

$$B^{(1)}(r) = \left(\frac{m-n}{m+n} \right) B(r) \tag{105}$$

In its present form, the quantity $(m-n)/(m+n)$ could also be expressed in terms of the ratio n/m only. But, it is clear that investigating the topological homotopy problem, as we do, requires to treat n and m separately. Therefore, in this framework, it would not make sense to deal only with n/m as a single parameter, since it would break down the concept of topological homotopy.

By analyzing the diagrams expansions of both bridge functions $B^{(1)}(r)$ and $B(r)$, and also by numerical verifications [83, 98], it turns out that $B^{(1)}(r)$ is found to be lower than $\frac{1}{3}B^{(2)}(r)$. Following this prescription, one obtains the following condition for m, namely, $n < m \leq 2n$. If n and m are taken for simplicity as whole numbers, this means that the values of m are degenerated values. In passing, it is noticeable that the Kiselyov and Martynov approximation [95] is recovered with the possible couple of values ($n = 1;\ m = 2$), namely, $B^{(1)}(r) = \frac{1}{3}B(r)$ for any thermodynamic state.

Finally, the substitution of Eq. (105) in Eq. (100) provides a single-state formula for the excess chemical potential

$$\beta\mu_{ex} = \rho \int \left[\frac{1}{2} h^{(2)}(r) - c(r) + B(r) - \frac{1}{2} h(r)c(r) + \frac{1}{2} h(r)B(r) \right] d\mathbf{r} \\ + \frac{m-n}{m+n} \rho \int \frac{1}{2} h(r)B(r) d\mathbf{r} \tag{106}$$

The advantage of this choice of the λ dependence for the correlation functions and the bridge function relies on the fact that the excess chemical potential, and the one-particle bridge function as well, can be determined unambiguously in terms of $B(r)$ as soon as n and m are known. To address this problem, the authors proposed to determine the couple of parameters $(n; m)$ in using the Gibbs–Duhem relation. This amounts to obtaining values of n and m from Eq. (87), which is considered as supplementary thermodynamic consistency condition that have to be fulfiled.

5. *Classical Thermodynamic Integration*

The configurational chemical potential can also be calculated in a complete different way [99], that is to say by means of a thermodynamic integration of the rhs of Eq. (87), namely,

$$\beta\mu_{ex} = -4\pi \int d\rho\prime \int c(r, \rho\prime) r^2 dr \tag{107}$$

Practically, the values of $\int c(r, \rho') r^2 dr$ are fitted in a wide range of densities by a polynomial of order 3 in density. Then, $\beta\mu_{ex}$ is obtained by analytically integrating the resulting polynomial as a function of density. It should be stressed that this method involves only the direct correlation function $c(r)$, but neither $B^{(1)}(r)$ nor $B(r)$, which are known to be the keys of the IETs. It must be stressed that such a thermodynamic integration process is performed along an isotherm T. This method is only accurate for supercritical temperatures, but is not at all for lower temperatures. Furthermore, it is not adapted to a predictive scheme.

E. Entropic Quantities Results for Fluid Models

1. *The Hard-Sphere Fluid*

Since we have at our disposal some highly accurate representations for $B(r)$ over the whole density domain (see Fig. 6), we turn now to the results of calculation of the excess chemical potential obtained with two accurate SCIETs. In Fig. 11, we show $\beta\mu_{ex}$, calculated via Lee formula (90) for the ZSEP, and Eq. (105) for the BB approximation compared to the CS reference data given by

$$\beta\mu_{ex} = \frac{8\eta - 9\eta^2 + 3\eta^3}{(1-\eta)^3} \tag{108}$$

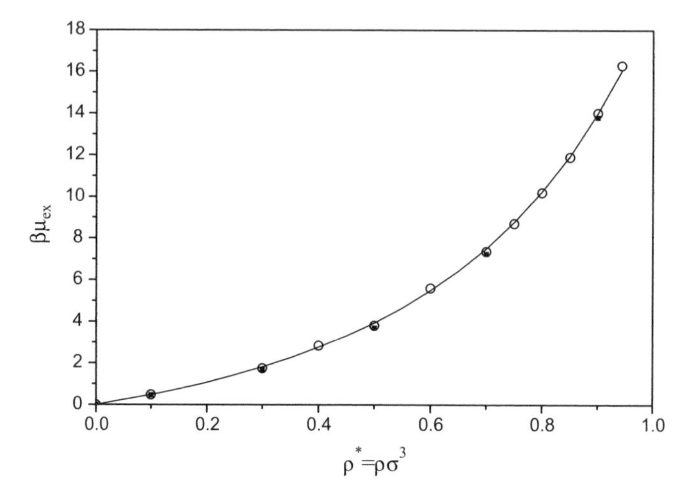

Figure 11. The HS excess chemical potential $\beta\mu_{ex}$ calculated with ZSEP closure relation (squares) and with BB approximation. Comparison with the CS equation of state results (open circles). Taken from Ref. [96].

For this system, the Lee formula requires to use the following prescription in Eq. (91), namely, $\gamma_0(r) = \rho f(r)/2$ and $\gamma_1(r) = \gamma(r)$. For the sake of comparison, the results of calculation obtained with the BB conjugated with (102) are displayed in addition. The results are in close agreement altogether from low densities up to $\rho^* = 0.9$, and both methods yield similar results, while Kiselyov–Martynov approximation for $B^{(1)}(r)$ fails at high densities. This means that the methods proposed by Lloyd L. Lee and by Bomont–Bretonnet seem to be very efficient. Nevertheless, the reader has to be aware of the fact that if the IETs provided exact values for all the correlation function (unfortunately, this is never the case), the quantity $B(0) + \gamma(0)$ should be sufficient to provide accurate values of the chemical potential. But, as attested in the literature, IETs suffer of the fact that this last quantity overestimates systematically the expected value.

We take advantage of this presentation to provide new insights for the one-particle function $B^{(1)}(r)$ derived by Bomont [96] and Bomont–Bretonnet [97]. The authors proposed to determine the couples of parameters $(n;m)$ by first considering the Gibbs–Duhem relation given by Eq. (87). Following the prescription given by $n \leq m \leq 2n$, the *a priori* possible values of the couples read

$$
\begin{array}{lllllll}
 & n & & m & & & \\
\text{step1}: & 2 & \rightarrow & 3, & 4. & & \\
\text{step2}: & 3 & \rightarrow & 4, & 5, & 6. & \\
\text{step3}: & 4 & \rightarrow & 5, & 6, & 7, & 8. \\
\ldots & \ldots & \rightarrow & \ldots & \ldots & \ldots & \ldots \quad \ldots
\end{array}
$$

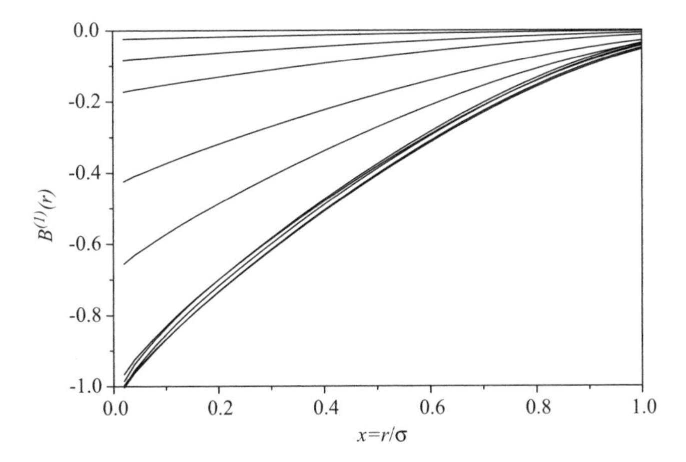

Figure 12. The "one particle" bridge $B^{(1)}(r)$ calculated for 10 density states $\rho^* = 0.3, 0.4, 0.5, 0.6, 0.7, 0.75, 0.8, 0.85, 0.9$, and 0.943 (from top to bottom) with the BB approximation. Taken from Ref. [97].

In order to find a couple (n;m) relative to each thermodynamic state, the algorithm starts at step 1 with the initial value $n = 2$. For a given value of n, the corresponding degenerated values of m are investigated. Among these last, if there exists no value of m that provides a couple satisfying the GD relation, then the parameter n is increased by one unity in the next step. The procedure is repeated until the GD is satisfied. Despite appearing to be a simple mathematical exercise, this approach leads to useful information on $B^{(1)}(r)$ in the well identified dynamical regimes of the hard sphere fluid.

The calculated one-particle bridge functions are plotted versus the reduced densities in Fig. 12. The values of the excess chemical potential $\beta\mu_{\text{ex}}$, the final couples $(n_f; m_f)$, and the quantity $B(0) + \gamma(0)$ are displayed in Table V over the entire range of densities, in which three distinct regimes in the hydrodynamic behavior have been identified [92]: (1) a dilute-gas regime (I), (2) an intermediate-density regime (II), and (3) a dense-fluid regime (III). The HS molecular dynamics calculations [100] of the mean-square displacement show that the transition between the latter two occurs at a reduced density $\rho^* \approx 0.57$. In addition, we emphasize that regime labeled III has been found to be separated into two distinct dynamical regimes (III_1 and III_2) whose threshold occurs at a reduced density of $\sim\rho^* \approx 0.74$.

In the dilute-gas regime, the fulfilment of Eq. (87) is achieved provided that $n_f = 1$ and $m_f = n_f + 1$, that is to say $B^{(1)}(r) = \frac{1}{3}B(r)$. So, values $n_f = 1$ and $m_f = 2$ are sufficient. This point is interesting since it indicates that the λ linear dependence of the correlation functions seems to be a good approximation

TABLE V
Results for HS Fluid versus the Reduced Density. Taken from Ref. [97]

Regimes	ρ^*	$(n_f; m_f)^a$	$\beta\mu_{ex}^b$	$\beta\mu_{ex}^{CS}$	$\beta\mu_{ex}^{KM}$	$-B(r=0)$	$B(0)+\gamma(0)$
I	0.1	1;2	0.48	0.46	0.48	0.001	
	0.15	1;2	0.76			0.13	
II	0.2	20;21	1.07			0.25	1.07
	0.25	18;19	1.43			0.45	
	0.3	14;15	1.82	1.75	1.85	0.69	1.82
	0.35	11;12	2.26			1.01	
	0.4	8;9	2.77	2.83	2.83	1.41	2.88
	0.45	6;7	3.32			2.08	
	0.5	7;8	3.94	3.81	4.12	3.04	3.99
	0.55	6;7	4.66			4.05	4.87
III_1	0.6	6;7	5.50	5.60	5.85	5.51	5.77
	0.65	6;7	6.43			7.37	6.88
	0.7	7;8	7.50	7.36	8.25	9.83	8.16
III_2	0.75	6;7	8.77	8.70	9.81	12.95	9.78
	0.8	8;9	10.25	10.20	11.70	17.05	11.73
	0.85	11;12	11.95	11.90	14.05	22.50	14
	0.9	14;15	14.01	14.10	16.90	29.00	17.50
	0.943	18;19	16.05	16.30	19.90	37.05	20.55

[a]Self-consistent couples $(m_f; m_f)$.
[b]Excess chemical potential $\beta\mu_{ex}$ calculated from Eq. (106) compared to exact Carnahan–Startling EOS of state results and to Kiselyov and Martynov [95] approximation results, corresponding zero exporation values of $-B(r)$ and $B(0)+\gamma(0)$.

in this range of densities and shows why the Lee formula is quite accurate for such a range of densities. So, assuming that $d_g = \lambda d$ is sufficient in regime I. This means that the fluid is so diluated that a test particle enters the fluid in a linear way.

In contrast, at higher densities (intermediate and dense-fluid regimes), this previous finding is no longer accurate. Because of the change of local density, n_f presents a nontrivial density dependence, while a common feature is found for m_f systematically equals $n_f + 1$. This indicates that, for regimes II and III, major information necessary to fulfill the constraint imposed by Eq. (87) can be simply obtained with couples $(n_f; n_f + 1)$. The authors stress that this feature is in agreement with the previous implicit statement of Kiselyov and Martynov: According to them, if $h(r, \lambda)$ and $c(r, \lambda)$ have a linear dependence on λ, then $B(r, \lambda)$ should have a quadratic dependence on λ (see Eqs. (26) and (31) in Sarkisov [84]. Provided that $B(r)$ is in close agreement with simulation data, the main result is that Eq. (105) reduces, *a posteriori* in regimes II, III and I, to

$$B^{(1)}(r) = \frac{1}{2n_f + 1} B(r), \tag{109}$$

and the excess chemical potential reads

$$\beta\mu_{ex} = \rho \int \left[\frac{1}{2}h^2(r) - c(r) + B(r) - \frac{1}{2}h(r)c(r) + \frac{1}{2}h(r)B(r) \right] d\mathbf{r} + \frac{1}{2n_f + 1}\rho \int \frac{h(r)}{2} B(r) d\mathbf{r}, \tag{110}$$

where n_f is so that $d_g = \lambda^{n_f} d$. So, it is clear that calculating the excess chemical potential by analytical formulas is highly conditioned to the way a particle inserts in the fluid. As it can be seen, this way is not limited to the supposed linear one. The results of the calculated $\beta\mu_{ex}$ are in close agreement with the CS reference data for the whole range of densities. For the sake of comparison, they also calculated, for some states, the excess chemical potential with the approximation $B_{KM}^{(1)}(r) = \frac{1}{3}B(r)$. In this case, the Gibbs–Duhem relation is satisfied only in the dilute-gas regime I, while an increasing discrepancy is observed with increasing density.

Now, by focusing on regimes II and III, a striking feature appears: It is found that $(n_f; m_f)_{\rho_0 - \Delta\rho} = (n_f; m_f)_{\rho_0 + \Delta\rho}$, where $\rho_0 = 0.6$, which is very close to the threshold ($\rho^* \approx 0.57$) between regimes II and III. Thus, from Eq. (109), this feature leads to an organic relation between the one-particle bridge function in the intermediate-fluid regime and the one in the dense-fluid regime, namely,

$$\left. \frac{B^{(1)}(r)}{B(r)} \right|_{\rho_0 - \Delta\rho} = \left. \frac{B^{(1)}(r)}{B(r)} \right|_{\rho_0 + \Delta\rho} \tag{111}$$

This feature has been interpreted as a signature of the transition between regimes II and III. So, once $B^{(1)}(r; \rho_0 + \Delta\rho)$ is known in the dense-fluid regime III, it is easy to calculate $B^{(1)}(r; \rho_0 - \Delta\rho)$ in the intermediate-density regime II. Unfortunately, since the highest liquid density is ~ 0.95 ($\Delta\rho_{max} \approx 0.35$), the authors could not provide further information for densities < 0.25. Therefore, Eq. (109) is of course useable for these diluted states.

By examining the plots of $B^{(1)}(r)$ in more detail (unfortunately, no simulation data are available to compare with), $B^{(1)}(r)$ is found to tend systematically to -1 for reduced densities > 0.75. This range of densities has been identified by Giaquinta and Giunta [92] as a distinct dynamical regime III_2 in the dense-fluid regime III. In fact, as far as the bridge function is considered as exact, meaning that $B(r)$ is in excellent agreement with the exact construction of Groot et al. [32] inside the core ($r < \sigma$), a second striking feature appears here. From Eq. (109), $B^{(1)}(r)$ equals $B(r)/(2n_f + 1)$, while in regime III_2 we note that $-B(r = 0)$ very nearly equals $2n_f + 1$ (see Table V). This suggests that the one-particle bridge function $B^{(1)}(r)$ can be expressed as the normalized

TABLE VI
Excess Entropy Calculated from Several Approximations for HS fluid, Compared to Carnahan–Starling EOS Data

ρ^*	Exact[a]	S_b^{ex}	S_c^{ex}	S_d^{ex}	S_e^{ex}
0.1	−0.22	−0.219		−0.23	−0.25
0.3	−0.78	−0.735	−0.83	−0.81	−0.94
0.5	−1.54	−1.440	−1.63	−1.58	−2.15
0.7	−2.65	−2.530	−2.66	−2.69	−4.48
0.9	−4.36	−4.163	−4.05	−4.45	−9.54

[a]Exact = Carnahan–Starling EOS.
[b] = ZSEP approximation.
[c] = BB approximation.
[d] = VM approximation.
[e] = HNC approximation [95].

bridge function $B(r)$, so that

$$B^{(1)}(r) = \frac{-B(r)}{B(r=0)} \tag{112}$$

This trend has been checked to be valid up to the solid–liquid transition. This feature has been considered as a signature of the transition between regimes III_1 and III_2. Furthermore, since $B(r=0) \equiv B^{CS}(r=0)$, that is given by Eq. (69), this suggests that $B^{(1)}(r)$ can be also expressed in terms of η and $B(r)$, so that

$$B^{(1)}(r) = \frac{(1-\eta)^4}{15\eta^2 - 12\eta^3 + 3\eta^4} B(r). \tag{113}$$

The present relations differ from the KM approximation since the factor 3 is replaced by the bridge function at zero separation. This feature does not seem to be unreasonable because, from diagrammatic expansions, $B^{(1)}(r) = B(r)/3$ is supposed to be accurate only at very low densities. Eq. (112) presents two advantages at high density: *i*) it provides a closed-form expression for $B^{(1)}(r)$ that could be used for other fluids than the HS model and *ii*) it allows to ensure a consistent calculation of the excess chemical potential by requiring only the use of the pressure consistency condition (the Gibbs-Duhem constraint, no longer required, is nevertheless implicitly satisfied within 1%).

Once the thermodynamic properties and the excess chemical potential have been calculated in a consistent manner, the entropy of the system can be derived. As seen in Table VI, a close agreement for the previous thermodynamic properties is found against CS data at all densities, while the HNC approximation fails at higher densities.

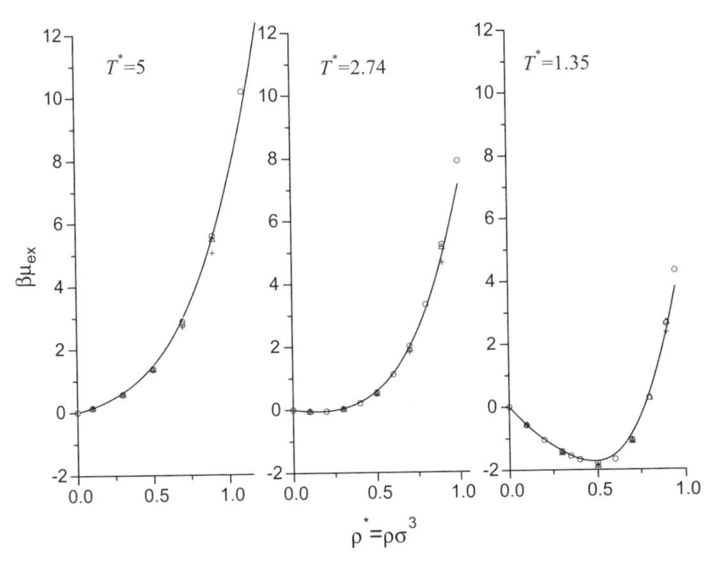

Figure 13. Plot of the excess chemical potential versus the reduced density for three different temperatures. The solid line represents the results from BB approximation conjugated with the direct formula of Lee, while the open circles correspond to the Johnson et al. [34] data. Triangles and crosses are for the results taken from Sarkisov [84] and from Choudhury and Ghosh [66] for comparison, respectively. Taken from Refs. [69,70].

2. *The Lennard-Jones Fluid*

For this fluid model, the chemical potential is one of the most difficult thermodynamic quantities to calculate. Only few SCIETs approximations succeed in obtaining direct accurate results (ZSEP, DHH, BB, ...). The reader has to be informed that the method developed by Bomont and Bretonnet [96, 97] has not yet been applied to the LJ fluid, but works are in progress along these lines.

Figure 13 showns the values of $\beta\mu_{ex}$ calculated with the BB conjugated with the Lee formula [with $\gamma_0(r) = -\beta u_{LR}(r)$ and $\gamma_1(r) = \gamma(r)$ that have to be used in Eq. (91)], along the three supercritical isotherms $T^* = 1.35, 2.74$, and 5. They are compared to the Johnson et al. [34] data, and also to recent results obtained by Sarkisov [84] and Choundhury–Ghosh [66] because the excess chemical potential is a very sensitive quantity to defects in a theory. For all the states under study, they are in close agreement from low densities up to $\rho^* = 0.9$. The high density results are also surprisingly well predicted taking into account that small inaccuracies in the bridge function are apt to affect significantly the chemical potential calculations at higher densities.

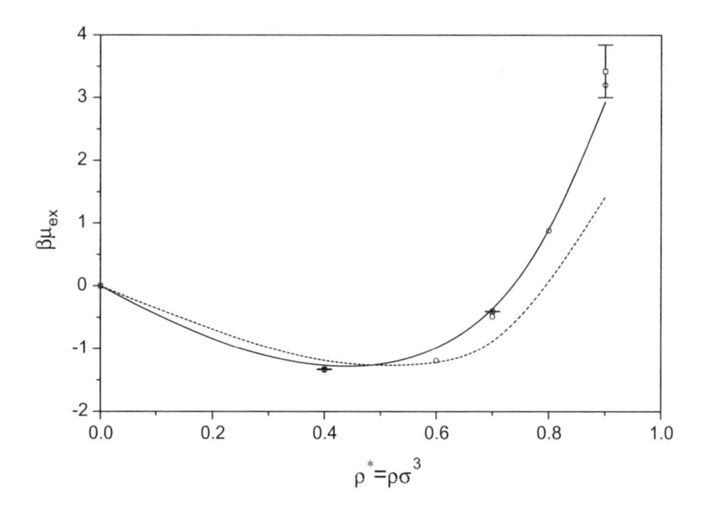

Figure 14. Plot of the excess chemical potential vs the reduced density at $T^* = 1.5$. The solid line represents the results from BB approximation conjugated with the direct formula of Lee, while the symbols (open circles and squares) correspond, respectively, to MD simulation of Johnson et al. [34] and to MC simulation of Lee et al. [76] and Lee [77] results. For comparison, the results (dash line) obtained with Charpentier and Jakse [68] bridge function. Taken from Refs. [69,70] are shown.

In order to extend the presentation, the results along the $T^* = 1.5$ isotherm are presented in Fig. 14. Included for comparison are those from MD simulation [34] and from MC simulation [76, 77], as well as those obtained with VM bridge function by Charpentier and Jakse [68]. The results obtained from BB conjugated with the Lee formula still remain in quite good agreement with MC and MD simulation data over a wide range of densities. The worst concordance is for the highest density $\rho^* = 0.9$, where the calculated excess chemical potential taking the value 2.97 is almost within the error bar of the $\beta\mu_{ex}$ from the MC data (3.42 ± 0.42). This value is hardly different from that of 3.05 obtained by Lee [77] from its zero-separation approach. In contrast, the value of $\beta\mu_{ex} = 1.45$ for the same state condition, obtained by Charpentier and Jakse [68] using the VM bridge function with the optimized value of $\alpha = 0.475$, remains far from the expected data.

Further attempts have been made this last decade to obtain competitive results for $\beta\mu_{ex}$ as compared to simulation data. Recently, Bomont proposed the approximation $B^{(1)}(r) = \alpha(T, \rho)B(r)$ [98]. Once the correlation functions, the excess internal energy, the pressure, and the isothermal compressibility are calculated with respect to the *first thermodynamic consistency condition*, the parameter $\alpha(T, \rho)$ is iterated until $\beta\partial\mu_{ex}/\partial\rho$ satisfies the *second thermodynamic consistency condition* within 1% [Eq. (87)]. At the end of the iteration cycle

TABLE VII
Excess Chemical Potential $\beta\mu_{ex}$ of the Lennard-Jones Fluid at Supercritical Temperatures

	BB approximation[a]		Simulation[b]	Theories[c]	
ρ^*	$\alpha^{coh}(T,\rho)$	$\beta\mu_{ex}$	$\beta\mu_{ex}$	$\beta\mu_{ex}$	
			$T^* = 5$		
0.1	0	0.125	0.129	0.124	0.124
0.3	0.1705	0.551	0.562	0.542	0.542
0.5	0.1275	1.356	1.371	1.315	1.334
0.7	0.1845	2.851	2.883	2.699	2.807
0.9	0.1740	5.450	5.582	5.044	5.473
			$T^* = 2.74$		
0.1	0	−0.060	−0.056	−0.062	−0.062
0.3	0.0030	0.009	0.022	0.001	−0.002
0.5	0.1140	0.523	0.536	0.479	0.488
0.7	0.1570	1.971	1.996	1.801	1.898
0.9	0.1715	5.155	5.229	4.651	5.130
			$T^* = 1.35$		
0.1	0	−0.576	−0.571	−0.578	−0.579
0.35	0	−1.568	−1.557	−1.456	−1.468
0.5	0	−1.818	−1.802	−1.858	−1.882
0.7	0.111	−0.928	−1.060	−1.103	−1.107
0.9	0.147	2.928	2.633	2.362	2.670

[a]Results taken from Ref. [98].
[b]Taken from Ref. [34].
[c]Taken from Ref. [66, 84].

$\alpha(T,\rho) = \alpha^{coh}(T,\rho)$ and the excess chemical potential reads

$$\beta\mu_{ex}^{coh} = \rho \int \left[-h(r) + \gamma(r) + B(r) + \frac{h(r)}{2}[\gamma(r) + (1 + \alpha^{coh}(T,\rho))B(r)] \right] d\mathbf{r} \tag{114}$$

It is a matter of a fact that the proposed method does not require the knowledge of the primitive of $B(r)$ with respect to $\gamma^*(r)$ as is the case with the direct formula of Lee. The results, displayed in Table VII are all in close agreement from low density up to $\rho^* = 0.9$. It is obvious that this method improves by several percent the previous results that we obtained by using the same bridge function (BB) in conjunction with the direct formula of Lee. In Fig. 14, along the isotherm $T^* = 1.5$, a close agreement is found with those of MC [76] and do compare very well to the equation of state of Nicolas et al. [33]. At $\rho^* = 0.9$, the calculated excess chemical potential takes the value 3.17 that is in very good agreement with the ZSEP integral equation [77] result that is 3.05. Yet the results are rather

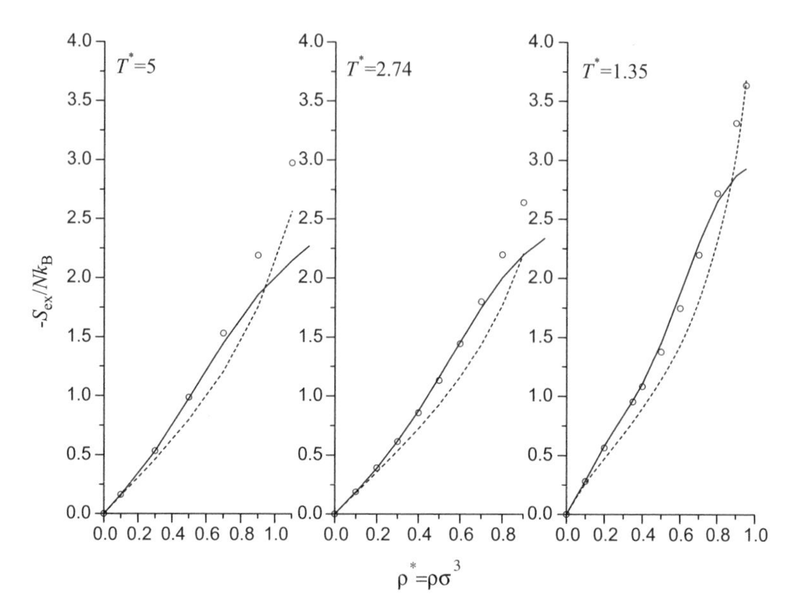

Figure 15. Plot of the excess entropy versus the reduced density for three different temperatures. The solid line represents the results from the BB approximation conjugated with the direct formula of Lee, while the open circles correspond to the Johnson et al. [34] data. The dash line stands for the pair contribution $S^{(2)}$ to the excess entropy defined by Eq. (85). Taken from Refs. [69,70].

sensitive to the approximation of $B^{(1)}(r)$. The state-dependent parameter $\alpha^{coh}(T, \rho)$ presents a nontrivial density dependence due to the changing local liquid structure. Nevertheless, it is noticeable that it takes zero value at low density, which is not surprising, since for this range of densities, $B(r)$ is almost zero. This approach presents several advantages: (1) it allows us to satisfy the Gibbs–Duhem relation in a simple way, (2) it is more tractable than the direct formula proposed by Lee [77], and (3) it can be used in conjunction with any consistent (or not) integral equation.

Now, we turn to the entropy. In Fig. 15, the excess calculated entropy is plotted in comparison with the reference data drawn from MD simulation [34] owing to Eq. (86). For the three supercritical temperatures investigated, the curves follow the reference data extremely well from low densities up to $\rho^* \sim 0.7$. Beyond this limit, the accordance becomes less and less satisfactory. This is not surprising since the excess chemical potential (see Fig. 13) is itself less accurately predicted for high densities. The bump observed on the curve at $T^* = 1.35$ is also not surprising as it lies around the critical point located, with different theoretical and simulation approaches, at temperature between 1.30 and 1.35 and density between 0.30 and 0.35.

Also displayed in Fig. 15 is the pair contribution $S^{(2)}$ to the excess entropy. It is seen quite clearly that the three- and more-body terms to the excess entropy cannot be neglected to reproduce the reference values since their contributions augment as the density increases, until $\sim\rho^* \sim 0.7$. As expected, at low densities the differences between the calculated excess entropy S_{ex} and pair contribution $S^{(2)}$ are small. For the three temperatures considered, the largest difference appears at $\rho^* \sim 0.7$, for which the two-body entropy contributes $\sim 80\%$ of the total excess entropy, while the difference cancels at $\sim\rho^* \sim 0.9$. Note that the crossing of the curves associated with S_{ex} and $S^{(2)}$, respectively, also was observed by Schmidt [94]. It is worth noting that Baranyai and Evans [101] calculated the contributions $S^{(2)}$ and $S^{(3)}$, directly from MD simulation, for the Lennard-Jones fluid at $T^* = 1.15$ and 1.5. Their calculations show unambiguously that the excess entropy cannot be satisfactory estimated without the knowledge of $S^{(3)}$.

Mention has to be made to the inevitable crossing between the curves representative of $S^{(2)}$ and to the reference excess entropy, respectively, at high density. In fact, it has been observed that the residual multiparticle entropy $(\Delta S = S^{(3)} + S^{(4)} + S^{(5)} + \cdots)$ does undergo a change of sign, in the supercritical regime, for a unique value of the density corresponding quite remarkably to the freezing point [92]. Although no clear reason why ΔS changes sign at a thermodynamic phase boundary, this observation can be viewed as the signature of the germination in the fluid phase. Specifically, the MD freezing line of the Lennard-Jones fluid has been well reproduced owing to this criterion [93]. Then the intersection point between the $S^{(2)}$ term and the reference data of S_{ex}, observed in Fig. 15 for $T^* = 1.35$, is located at $\rho^* \sim 0.95$, in good agreement with the available data in the literature.

We cannot conclude this section without presenting some SCIETs predictions for the phase diagrams. As summarized, the accuracy of the chemical potential is crucial for phase equilibria. For a one-component system, the conditions of coexistence of the gaseous (g) and the liquid (l) phases in contact with each other at a given temperature T are

$$\begin{cases} P(T, \rho_g) = P(T, \rho_l) \\ \mu_{\mathrm{ex}}^{g}(T, P) = \mu_{\mathrm{ex}}^{l}(T, P) \end{cases} \tag{115}$$

The condition of phase stability for such a system is closely related to the behavior of the Helmholtz free energy, by stating that the isothermal compressibility $\chi_T > 0$. The positiveness of χ_T expresses the condition of the mechanical stability of the system. The binodal line at each temperature and densities of coexisting liquid and gas determined by equating the chemical potential of the two phases. The conditions expressed by Eq. (115) simply say that the gas–liquid phase transition occurs when the $P - \mu_{\mathrm{ex}}$ surface from the gas branch intersects the $P - \mu_{ex}$ surface from the liquid branch along the same isotherm. It is clear that a drastic difference between two $P - \mu_{ex}$ surfaces emphasizes the important role of the bridge function inside the core. Although it has a limited effect on the structure, for example, $g(r)$ outside the core, it greatly affects the overall thermodynamic quantities.

Figure 16 shows the gas–liquid phase diagrams calculated with DHH+DHHDS [65, 81], compared to the ones calculated from MSA and from BB+ODS. Also include the data of Lotfi et al. [102] and those obtained from the Gibbs ensemble method by Panagiotopoulos [103]. Note that the reported values obtained from DHH+DHHDS are only estimates, because as said above, this approximation, even if accurate, exhibits little thermodynamic inconsistencies. As

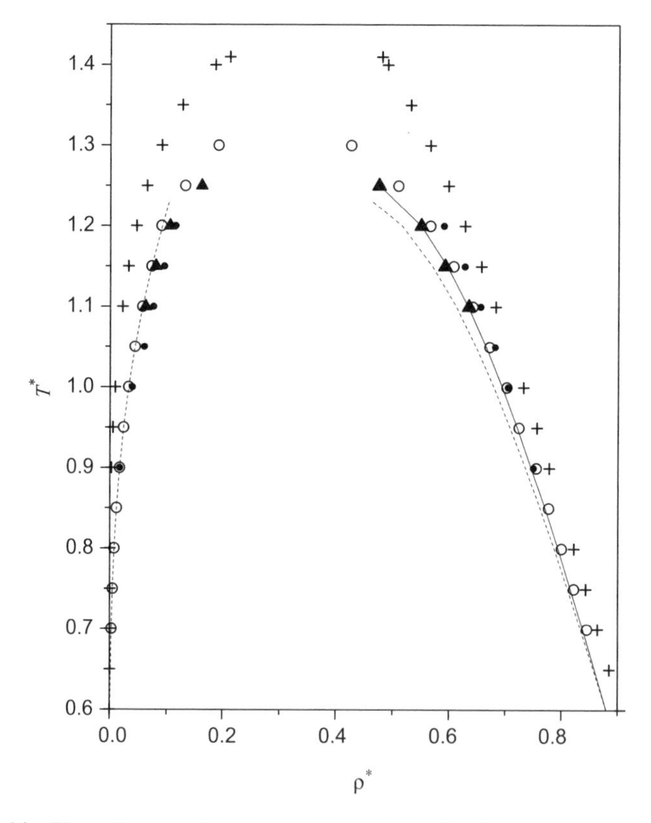

Figure 16. Phase diagram of the Lennard-Jones fluid with different approximations: MSA (crosses) and DHH (dot line) (taken from Ref. [65]), DHH+DHHDS (solid line) (see Ref. [81]), BB approximation (black circles) (courtesy of the author), the chemical potential had been calculated by using Lee formula). Simulation data are from Lotfi et al. [102] (open circles) and from Panagiotopoulos [103] (solid triangles).

emphasized previouly, every approximate integral equation has difficulty near the critical point, and the SCIETs presented here are no exception. However, the good agreement with subcritical phase-coexistence data is encouraging. The discrepancy between the MSA results and those of the SCIETs is a measure of the progresses performed in the knowledge of the bridge function these last 10 years.

For other fluid models, the reader is encouraged to refer to Caccamo [99] and references cited therein for detailed discussions.

V. TESTING THE SCIETs FOR RARE GASES FLUIDS

A. Three-Body Forces in Rare Gase Fluids

In 1980s, it was concluded from neutron diffraction data on krypton that the structure of this sytem could not be described adequately with a single two-boby potential. A supplementary contribution to the energy had to be taken into account, which is the triple–dipole three-body correction [104].

Rare gase fluids are a test case to assess the validity of the SCIETs in describing real systems. As can be guessed, the r– and q–space predictions of structural and thermodynamic properties of dense fluids require an accurate knowledge of the interatomic potential $u(r)$. The most direct probe of a realistic potential is in the experimental observation of the static structure factor $S(q)$, closely related to density fluctuations and containing useful information about short- and long-range parts of the pair and higher- order potentials. Clearly, the short-range part of $u(r)$ is better understood than the long-range part, and the outcome of the structural properties cannot be fully attributed to the effect of the two-body interactions alone. The importance of a quantitative investigation of the long-range part of $u(r)$, including two- and three-body dispersion forces, have been emphasized for long time. As an example, Barker and Henderson [105] and Aziz and Slaman [106, 107] derived empirical pair potentials for noble gases, using a large diversity of experimental data as sources of information and disregarding many-body interactions. These works were useful because the two-body interactions have to be known perfectly before including any higher order contribution. But such a description becomes not strictly valid as the interaction between two particles is disturbed by the presence of a close third particle [105]. Therefore, a correction with the pair potential, established at an electronic level and suspected to be the triple–dipole dispersion interaction [108], has to be included. In this framework, the total potential energy, previouly defined by Eq.(3), has to be rewritten

$$u(\mathbf{r}_1, \mathbf{r}_2, \ldots, \mathbf{r}_N) = \frac{1}{2}\sum_{1\neq j} u_2(\mathbf{r}_{ij}) + \frac{1}{6}\sum_{i\neq j\neq k} u_3(\mathbf{r}_{ij}, \mathbf{r}_{jk}, \mathbf{r}_{ik}) \qquad (116)$$

Neglecting terms of higher order than third-order appears to be a reasonable approximation for the physical properties presented in this section. Among the main factors inhibiting progress in the important role of the triple–dipole interactions, we can mention (1) the lack of accurate experimental structure factor $S(q)$ and (2) the difficulty of implementing a reliable theory that would permit a fine analysis of various contributions of $u(r)$. Since the small-q values of $S(q)$ are related to two- and three-body effects of the long-range potential, precise measurements of $S(q)$ at small angle neutron scattering (SANS) have been fortunately performed the last 10 years on fluids like Ar [109], Kr [12, 14, 110] and Xe [13]. Progress has also been made to overcome the second obstacle allowing the determination of the strength of the long-range potential as well as the undiscutable detection of the three-body effects, whose magnitude can be directly obtained in combining theory and experiment [111, 112]. Therefore the experimental structure factors of xenon and krypton provide an important test for theoretical models when triple–dipole dispersion interaction has to be taken into account.

This section reports the results obtained from two self-consistent schemes that have been extended to deal with the triple–dipole interaction, which reduces to a state-dependent effective pair potential. The pair potential of Aziz and Slaman [106] (AS) is combined with the Axilrod and Teller [108] (AT) triple–dipole potential to describe the interactions in Xe and Kr fluids. Note that other investigations have been performed with *ab initio* potentials [113, 114].

As u_2, the AS pair potential is selected for this presentation. It reads

$$u_2(x_{ij}) = A\ \exp(-\alpha x_{ij} + \beta x_{ij}^2) - F(x_{ij}) \sum_{j=0}^{2} \frac{C_{2j+6}}{x_{ij}^{2j+6}} \tag{117}$$

where $x_{ij} = r_{ij}/\sigma$ is the reduced distance and σ the position of the node of the potential. According to the authors, the repulsive and attractive parts have to be matched with the switching function

$$F(x_{ij}) = \begin{cases} \exp\left[-\left(\frac{D}{x_{ij}} - 1\right)^2\right] & \text{if} \quad x_{ij} < D \\ 1 & \text{if} \quad x_{ij} \geq D \end{cases} \tag{118}$$

The relevant parameters in Eqs. (117) and (118) are listed in the paper of Aziz and Slaman [106].

As u_3, the usual expression derived by Axilrod and Teller (AT), is adopted for our purpose,

$$u_3(\mathbf{r}_i, \mathbf{r}_j, \mathbf{r}_k) = \nu \frac{1 + 3\cos\theta_i \cos\theta_j \cos\theta_k}{r_{ij}^3 r_{ik}^3 r_{jk}^3} \tag{119}$$

which corresponds to an irreducible triple–dipole potential between closed-shell atoms. The parameter ν stands for the strength and θ_i, θ_j, and θ_k denote, respectively, the angles at vertex i, j, and k of the triangle $(i,\ j,\ k)$ with sides $r_{ij} = |\mathbf{r}_j - \mathbf{r}_i|$, $\mathbf{r}_{ik} = |\mathbf{r}_k - \mathbf{r}_i|$, and $r_{jk} = |\mathbf{r}_k - \mathbf{r}_j|$.

As will be seen, a direct comparison with MD calculations, as well as recent SANS measurements [12–14] and available thermodynamic properties [115, 116], attests the efficiency of this approach. The principal motivation for this application is to validate the accuracy of some SCIETs on a concrete problem in investigating the effects of the triple–dipole contribution on structural and thermodynamic properties of Xe and Kr.

B. Three-Body Interactions Treatment

1. IETs Framework

Contrary to numerical simulation, taking the three-body contribution into account in IETs requires that we define an effective pair potential. Making use of the two- and three-body contributions allows one to write the state-dependent effective potential $u_{\text{eff}}(r)$ under the standard form [10, 112, 117].

$$\beta u_{\text{eff}}(r_{12}) = \beta u_2(r_{12}) - \rho \int g(r_{13}) g(r_{23}) [\exp\{-\beta u_3(r_{12}, r_{13}, r_{23})\} - 1] d\mathbf{r}_3 \tag{120}$$

where ρ is the number density, $\beta = (k_B T)^{-1}$ is the inverse temperature, $u_2(r_{12})$ is the (AS) two-body potential, and $u_3(r_{12}, r_{13}, r_{23})$ stands for the (AT) form for the triple–dipole interaction. It is assumed that the other three-body dispersion and exchange overlap interactions have little influence on Xe [118] and that the triple–dipole dispersion interaction is the only effect still dominant at low densities.

The integral equation theory consists in obtaining the pair correlation function $g(r)$ by solving the set of equations formed by (1) the Ornstein–Zernike equation (OZ) (21) and (2) a closure relation [76, 80] that involves the effective pair potential $u_{\text{eff}}(r)$. Once the pair correlation function is obtained, some thermodynamic properties then may be calculated. When the three-body forces are explicitly taken into account, the excess internal energy and the virial pressure, previously defined by Eqs. (4) and (5) have to be, extended respectively [112, 119] so that

$$E^{\text{ex}} = \frac{\rho^2}{2!} \int d\mathbf{r}_1 d\mathbf{r}_2 g(r_{12}) u_2(r_{12}) + E^{(3)} \tag{121}$$

$$P = \frac{\rho}{\beta} - \frac{\rho^2}{2!3\,V} \int d\mathbf{r}_1 d\mathbf{r}_2 g(r_{12}) r_{12} \frac{\partial u_2(r_{12})}{\partial r_{12}} + P^{(3)} \tag{122}$$

where

$$P^{(3)} = \frac{3}{V}E^{(3)} = \frac{3}{V}\frac{\rho^3}{3!}\int d\mathbf{r}_1 d\mathbf{r}_2 d\mathbf{r}_3 g^{(3)}(r_{12}, r_{13}, r_{23})u_3(r_{12}, r_{13}, r_{23}) \tag{123}$$

These expressions are formally exact and the first equality in Eq. (123) comes from Euler's theorem stating that the AT potential $u_3(r_{12}, r_{13}, r_{23})$ is a homogeneous function of order -9 of the variables r_{12}, r_{13}, and r_{23}. Note that Eq. (123) is very convenient to realize the thermodynamic consistency of the integral equation, which is based on the equality between both expressions of the isothermal compressibility stemmed, respectively, from the virial pressure, $\chi_T = \frac{1}{\rho}(\partial\rho/\partial P)_T$, and from the long-wavelength limit $S(0)$ of the structure factor, $\chi_T = \beta[S(0)/\rho]$. The integral in Eq. (123) explicitly contains the triple–dipole interaction and the triplet correlation function $g^{(3)}(r_{12}, r_{13}, r_{23})$ that is unknown and, according to Kirkwood [86], has to be approximated by the superposition approximation, with the result

$$g^{(3)}(r_{12}, r_{13}, r_{23}) = \exp[-\beta u_3(r_{12}, r_{13}, r_{23})]g(r_{12})g(r_{13})g(r_{23}) \tag{124}$$

2. *Molecular Dynamics Framework*

In performing molecular dynamics calculations, it is necessary to consider the force acting on a particle i from a particle j, which is derived from the AS potential under the form

$$\mathbf{F}_{ij}(x_{ij}) = \Bigg\{A\exp(-\alpha x_{ij} + \beta x_{ij}^2)(-\alpha + 2\beta x_{ij}) - F(x_{ij})\left[\sum_{j=0}^{2}\frac{(2j+6)C_{2j+6}}{x_{ij}^{2j+7}} - \frac{2D(D - x_{ij})}{x_{ij}^3}\sum_{j=0}^{2}\frac{C_{2j+6}}{x_{ij}^{2j+6}}\right]\Bigg\}\mathrm{e}_{ij} \tag{125}$$

where e_{ij} is a unit vector in the $\mathbf{r}_{ij}$–direction.

For the AT potential, the force acting on the particle i from the particles j and k is given by

$$\mathbf{F}_{i,jk}(r_{ij}, r_{jk}, r_{ik}) = \frac{\partial u_3}{\partial r_{ij}}\mathbf{e}_{ij} + \frac{\partial u_3}{\partial r_{ik}}\mathbf{e}_{ik} \tag{126}$$

while the forces acting on j and k are, given respectively, by

$$\begin{cases} \mathbf{F}_{j,ik}(r_{ij}, r_{jk}, r_{ik}) = -\frac{\partial u_3}{\partial r_{ij}}\mathbf{e}_{ij} + \frac{\partial u_3}{\partial r_{jk}}\mathbf{e}_{jk} \\ \mathbf{F}_{k,ij}(r_{ij}, r_{jk}, r_{ik}) = -\frac{\partial u_3}{\partial r_{ik}}\mathbf{e}_{ik} - \frac{\partial u_3}{\partial r_{jk}}\mathbf{e}_{jk} \end{cases} \tag{127}$$

The standard expressions of the partial derivatives of u_3 can be found in the paper of Hoheisel [120]. At variance from SCIETs framework, the two- and three-body forces are explicitly calculated.

C. Confronting Theoretical Results Against Experiment

1. *Details of Calculation*

The first step of this presentation consists in analyzing the structure of fluid Xe and Kr for some thermodynamic states recently investigated by experiment. For Xe, we refer to measurements performed by Formisano et al. [13] along the isotherm $T = 297.6\,K$ at densities $\rho = 0.95,\ 1.37,\ 1.69$, and $1.93\,\mathrm{nm}^{-3}$. The SCIET whose results are presented is the HMSA approximation [22] conjugated with the ODS division scheme [80]. For Kr, the calculation of the structure at $T = 297\,\mathrm{K}$ for different densities in the range between 1.52 and $4.277\,\mathrm{nm}^{-3}$, correspond to the thermodynamic states studied by Formisano et al. [12] and by Guarini et al. [14]. For this system, the theoretical results are from the SCIET conjugated with the ODS scheme [11]. As far as we know, these are the most recent approaches that have been devoted to the study of rare gases.

In order to eliminate any ambiguity arising from the self-consistent schemes, calculation have been performed by MD in the microcanonical ensemble. For Xe, a number $N = 256$ atoms are set in a cubic box subject to the usual periodic boundary conditions, whose volume V is fixed to reach the desired number density ρ. The atoms are initially displayed in a face-centered cubic (fcc) lattice and their velocities are randomly attributed according to the Maxwell–Boltzmann distribution. The AT potential is used in MD simulation under its original form with the standard expressions of the three-body forces drawn from the paper of Hoheisel [120]. The equations of motion of these 256 pointlike particles are integrated in a discrete form by means of the finite difference method [121], using Verlet's algorithm under the velocity form [3, 122]. After an equilibration of 10^4 steps with a time step of $\Delta t = 10^{-15}s$, $g(r)$ is extracted of a sample of 4×10^4 time-independent configurations every $10\Delta t$. Taking these typical values of time and system size, the numerical evaluation of $g(r)$ is not prohibitively expensive in terms of computer time.

At variance from Xe, the presented properties for Kr require more computional efforts. In order to reach the small-q range of $S(q)$, large-scale molecular dynamics have been carried out in the microcanonical ensemble (NVE) with the usual periodic boundary conditions. The equations of motion are integrated in the same discrete form as for Xe. The time step Δt is the same as for Xe and $g(r)$ is extracted over a sample of 8000 time-independent configurations every $10\Delta t$.

The calculation of the forces at each time step is one of the most demanding task. Since $\mathrm{F}_{ij} = -\mathrm{F}_{ji}$ for the two-body forces, and $\mathrm{F}_{i,jk} = -\mathrm{F}_{j,ik} - \mathrm{F}_{k,ij}$ for the three-body forces, the forces can be calculated once only, leading to a computation time reduced roughly by a factor 2. In addition, taking advantage of the short-ranged potentials that allow the use of cut-off radii r_c, the linked-cell list [3, 122] was used in order to reduce the complexity of computations to $O(N)$, because only pairs and triplets of particles within the cut-off radius are taken into account [123, 124]. In this presentation, $r_c = 2.5\sigma$ for two- and three-body forces.

Nevertheless, for large-scale simulations, typically involving $N > 10^4$ particles, the computational execution time is still large. Thus, algorithms suitable for parallel computers are nowadays commonly used [125], and applied to liquid-state studies [5, 126]. Usually, an algorithm is based on a spatial decomposition (SD) method [127] that balances the computation among the processors of the parallel machine. This SD method consists of dividing the simulation box into P regions. Each of them is assigned to a processor that performs the calculations for the particles situated in it and communicates the data to the other processors. Such strategy, in which the calculations of the forces and the pcf $g(r)$ are parallelized, reduces the execution time by a factor RP, where $R = t_1/(P.t_P)$ is the speed-up, t_1 and t_P being, respectively, the execution time with 1 and P processors. For this purpose, the plate decomposition associated with a torus communication scheme was used [127] with $N = 6912$ particles and $P = 6$ processors. It should be stressed that $N = 16,384$ with $P = 8$ processors had to be used in order to show that the results became insensible to system sizes beyond a certain value. In both cases, R takes values ~ 0.95.

a. The Structure in the r–Space. The pair correlation functions $g(r)$ of Xe and Kr calculated with SCIETs applied to the two- and three-body forces are displayed in Figs. 17 and 18, respectively. All the curves of $g(r)$ show similar trends with a rise governed by the repulsive part of $u_{\text{eff}}(r)$ and a first maximum located at the position of the potential well. The difference between the two-body interaction and the two- plus three-body interactions is maximal on the first peak of $g(r)$ and increases with density. For Xe, it turns out that the AT triple–dipole potential affects nearly the whole of $g(r)$ in reducing its magnitude outside the core contrarily to that is observed [11, 128] for Kr. While in the present case, $\Delta g(r)$ at the first peak position is 0.14 for the highest density ($\rho \approx 1.93$ atoms/nm^3), in the case of Kr the change in $g(r)$ is much smaller [11], reaching a maximum of 0.04 for the reduced density $\rho \approx 5.15$ atoms/nm^3). The reason invoked to explain this change in the main peak of $g(r)$, for Kr, was a partial compensation between the AT triple–dipole interaction and the three-body exchange interaction since both contributions are opposite in sign. But concerning Xe, no compensation of the three-body exchange effect is expected since the densities of Xe investigated here are very low. Thus, the deviation observed outside the core can be the direct consequence of the strength of the AT potential that is over three times as much

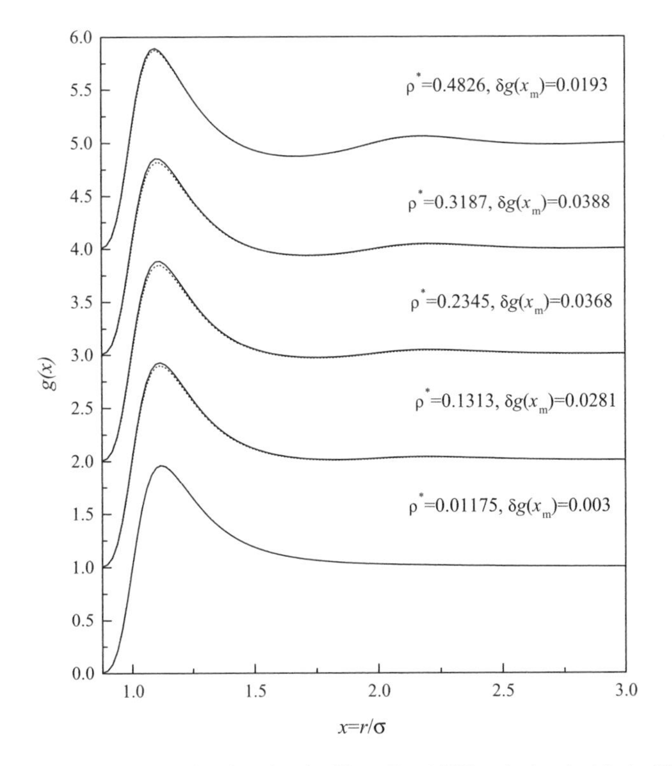

Figure 17. Radial distribution function for Kr at $T = 297\,\mathrm{K}$, calculated with the HMSA: the dotted lines correspond to the two-body interaction and the full lines correspond to the two- plus three-body interaction. The variation $\delta g(x_m)$, induced by the inclusion of the three-body interaction at the first maximum x_m of $g(r)$ is indicated for each density. Taken from Ref. [11].

larger for Xe than for Kr. The next part of this presentation deals with the q–space predictions of the SCIETs.

b. The Structure in the q–Space. Semianalytical results of $S(q)$ for Xe are compared with the experiments of Formisano et al. [13] for the same four previous thermodynamic states. The effects of the two- and three-body interactions on $S(q)$ are individually confronted with experimental data. As seen in Fig. 19, the influence of the AT potential becomes very significant in the small-q region. Compared to the experiments, the two-body interactions greatly overestimate the magnitude of $S(q)$, whereas the addition of the triple–dipole potential shows an appreciable improvement for the highest density. Specifically, at small angle scattering, $S(q)$ very closely approaches the experimental data of Formisano et al. [13] with a qualitatively correct curvature and a value of the long-wavelength limit nearby that is obtained from the thermodynamic

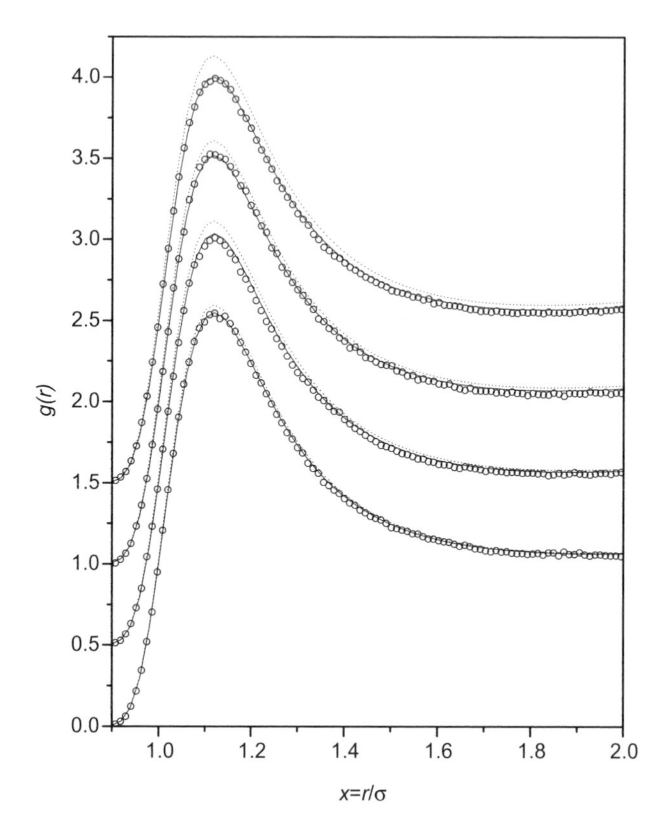

Figure 18. Pair correlation function $g(r)$, at $T = 297.6\,\mathrm{K}$, for $\rho = 0.95$, 1.37, 1.69, and $1.93\,\mathrm{nm}^{-3}$ (from the bottom to the top), calculated by the ODS integral equation with both the two-body interactions (dotted lines), and the two- plus three-body interactions (solid lines). The curves for $\rho = 1.37$, 1.69, and $1.93\,\mathrm{nm}^{-3}$ are shifted upward by 0.5, 1, and 1.5, respectively. The comparison is made with molecular dynamics simulation (open circles). Taken from Ref. [129].

properties data [115]. The values of $S(0)$ calculated from the isothermal compressibility are reported in Table VIII. Clearly, more than for Kr, the triple–dipole contributions cannot be ignored for Xe.

Since experiments for Kr have been performed at small angle neutron scattering for some low density states, we present the results of the Fourier transform of the direct correlation function, $c(q) = (S(q) - 1)/\rho S(q)$, rather than those of the structure factor $S(q)$. Figure 20 shows the curves of $c(q)$. As it can be seen, the theoretical results, obtained by HMSA+WCA and MD with the AS plus AT potentials, are in excellent agreement with the experimental data [12]. While the AT contribution is included by means of an effective pair potential in the SCIET, it is used under its original form owing to Eqs.(119) and

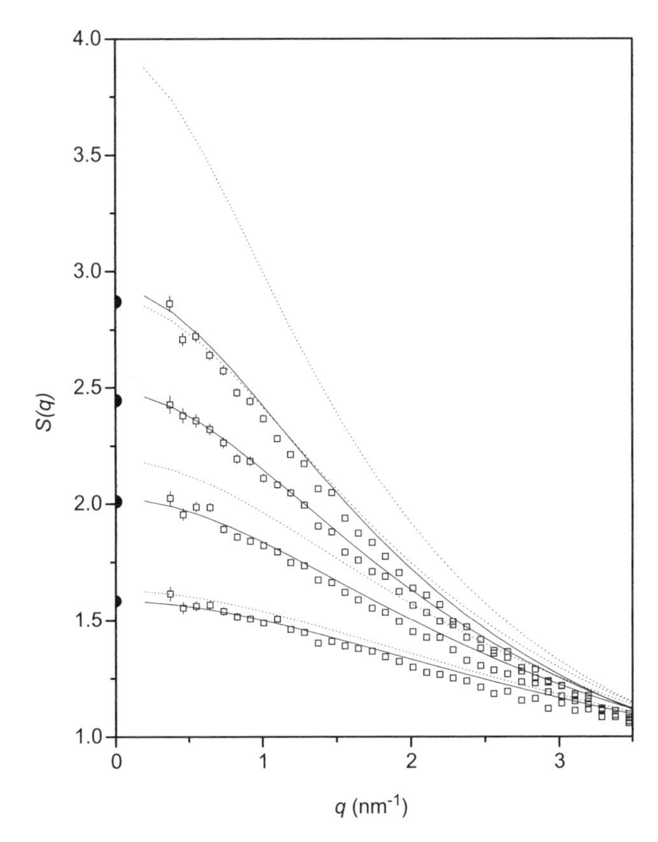

Figure 19. Static structure factor $S(q)$ in the small-q region, at $T = 297.6\,\mathrm{K}$, for $\rho = 0.95$, 1.37, 1.69, and $1.93\,\mathrm{nm}^{-3}$ (from the bottom to the top), calculated with the ODS integral equation. The dotted lines correspond to the calculations with the twobody potential alone and the solid lines correspond to those that combine the two- and three-body interactions. The squares are for the experimental data of Formisano et al. [13] and the filled circles, at zero-q value, for the PVT data of Michels et al. [115]. Taken from Ref. [129].

(126) in MD. This demonstrates that the treatment of the three-body potential in the SCIET is valid in this range of densities. The effect of the three-body contribution, which is only visible in the range of $q < 5\,\mathrm{nm}^{-1}$, is to lower the values of $c(q)$, that is to say to reduce the density fluctuations in the gas.

Since MD results compare favorably to experiment on the small–q part of $c(q)$, it is possible to affirm without ambiguity that the interaction scheme, which consists of combining the AS two-body potential with the AT three-body contribution, is suitable for studying rare gases fluids.

On the other hand, the good agreement found between HMSA and the MD attests that the self-consistent integral equation, and its extension to the

TABLE VIII
Static Structure Factor $S(q)$ at $q = 0$, Calculated with the HMSA+ODS Integral Equation Scheme by the Using Two-Body Potential Alone and the Two- plus Three Body Potentials[a]

$\rho(\mathrm{nm}^{-3})$	$S^{\mathrm{MD}}(0)^{(a)}$	$S^{\mathrm{AS}}(0)^{(b)}$	$S^{\mathrm{AS+AT}}(0)^{(b)}$
0.95	1.584	1.629	1.584
1.37	2.010	2.165	2.006
1.69	2.445	2.878	2.450
1.93	2.860	3.856	2.880

[a]PVT data are those of Michels et al. [116].
[b]Taken from Ref. [129].

three-body potential (120), is very convenient. This finding shows once again the efficiency of the self-consistent predictions at small q and reinforces previous conclusions.

c. Low Scattering Angle Behavior of $c(q)$. Since the pioneering work of Johnson and March [130], it has been recognized that important features can be extracted from the measured structure factor. However, it has emerged that the effective potentials obtained are very sensitive to (1) the accuracy of the experimental data of $S(q)$, in the small-q region, and (2) the particular liquid-state theory invoked. Consequently, it can be unclear which underlying features are truly physical in origin. Formisano et al. [12] attempted to account for their

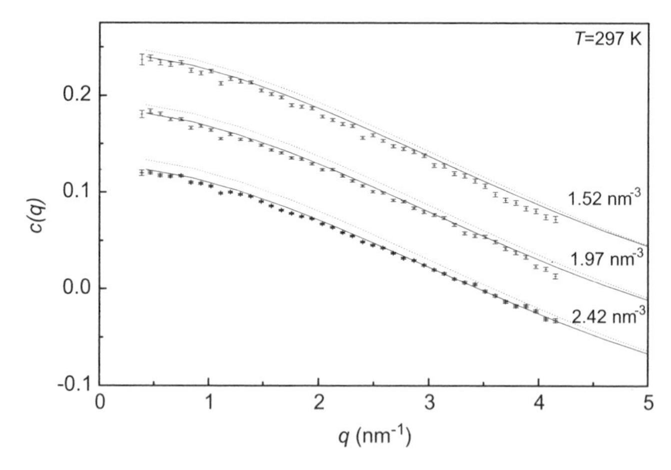

Figure 20. Fourier transform of the direct correlation function $c(r)$ at $T = 297\,\mathrm{K}$ for $\rho = 1.52$, 1.97, and $2.42\,\mathrm{nm}^{-3}$ calculated with HMSA. The dashed lines correspond to the two-body interaction and the full lines to the two- plus three-body interaction. Open squares are for the experimental data of Formisano et al. [12]. The values of $c(0)$ are taken from the PVT data of Trappeniers et al. [116]. Taken from Ref. [11].

observed variations of $c(q)$ to extract useful information on the effective potential for Kr. In contrast, the present predictive scheme is intending to deduce the strength of the AT potential by inverting the IET structure factor that is seen to fit very well the experimental data. This procedure has the merit of yielding convenient analytical expressions for certain Fourier coefficients of the small–q expansion of $c(q)$, and provides a useful test of the asymptotic form of the SCIET, for the specific case of the dispersion forces including the three-body AT interactions.

If the structure is decided by an effective potential $u_{\text{eff}}(r)$, it was demonstrated in the mean spherical approximation (MSA) that the direct correlation function $c(r)$ should rapidly approach $-\beta u_{\text{eff}}(r)$ for large r (see Section III). According to Reatto and Tau [131], this relationship, which is asymptotically exact for large distance and low density, holds quite well when the long-range dispersion term of the AS potential, $-C_6/r^6$, and the AT triple–dipole potential, $< u_3(r) >\sim (8\pi/3)\nu\rho/r^6$, are considered, so that the direct correlation function reads

$$c(r) \sim \beta[C_6 - (8\pi/3)\nu\rho]/r^3 \quad \text{as}\, r \to \infty \tag{128}$$

and can be predicted directly from given C_6 and ν. When using Eq. (124) and the Fourier asymptotic analysis, it is readily shown [131] that $c(q)$, at small–q, expresses as:

$$c(q) = c(0) + c_2 q^2 + c_3|q^3| + c_4 q^4 + O(q^5) \tag{129}$$

without limitation to the SCIET used. In that expansion, c_2 has no tractable expression, while c_3 takes the suitable form

$$c_3 = \beta(\pi^2/12)[C_6 - (8\pi/3)\nu\rho] \tag{130}$$

Thus, from the theoretical aspect, it follows that c_3 depends linearly on the density when the AT interaction model is involved, whereas it is a constant in its absence. It is possible to extract the coefficients c_2, c_3, and c_4 by considering the function

$$\lambda(q) = \frac{c(q) - c(0)}{q^2} \tag{131}$$

which can be extracted directly from the SCIET results, as well as the experimental data. Figure 21 compares the theoretical curves of $\lambda(q)$, including the AT contribution, to the experimental data. The agreement with experiment is very good for the three densities involved. By rewritting the last term under the form

$$\lambda(q) = c_2 + c_3|q| + c_4 q^2 \tag{132}$$

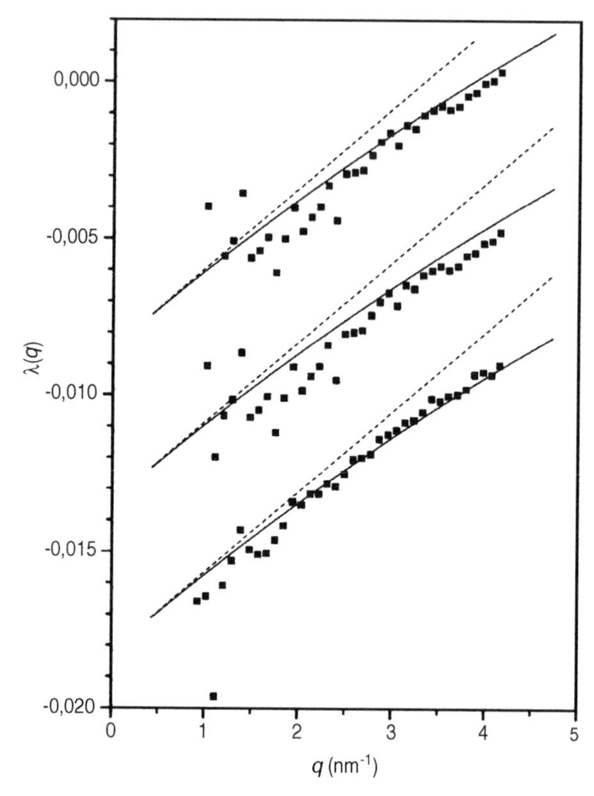

Figure 21. Function $\lambda(q)$ for $T = 297\,\text{K}$ at $\rho = 1.52$, 1.97 and $2.42\,\text{nm}^{-3}$ from top to the buttom calculated directly from the small-q expansion [Eq. (132)] (solid lines), and also from Eq. (128) in which the q^2 term was removed (dashed lines). The symbols represent the experimental data from Formisano et al. [12]. Taken from Ref. [128].

it is shown that the expansion up to q^2 term is sufficient. The curvature of $\lambda(q)$ at the upper q values is well reproduced, revealing the presence of at least a q^4 term, while its finite limit, when q tends to zero, corroborates the absence of a linear term in the low–q expansion of $c(q)$. In order to test the linearity of the IET at low q, the term q^2 is removed. The merit of this expression is to exhibit the q^3 dependence of $c(q)$ and to reveal the range in which the q^4 term can be neglected. This result is particularly important from an experimental point of view since c_3 allows the determination of the C_6 coefficient of the London dispersion forces and the stength ν of the AT three-body potential. The curves show that the linearity of the low–q expansion of $c(q)$ is limited to a range $1 < q < 3\,\text{nm}^{-1}$, which is slightly lower than the one estimated by Reatto and Tau [117, 131].

2. *The Thermodynamic Properties*

a. Internal Energy and Pressure. The satisfying approach of the self-consistent schemes used for describing the structural properties of Xe and Kr should also be able to reproduce the well-established trends present in the thermodynamic properties. Figures 22 and 23 display results of the excess internal energy and the virial pressure as a function of the density. Since the integral equation has no physical solution as the spinodal line is approached, the calculations are performed over a wide range of densities along

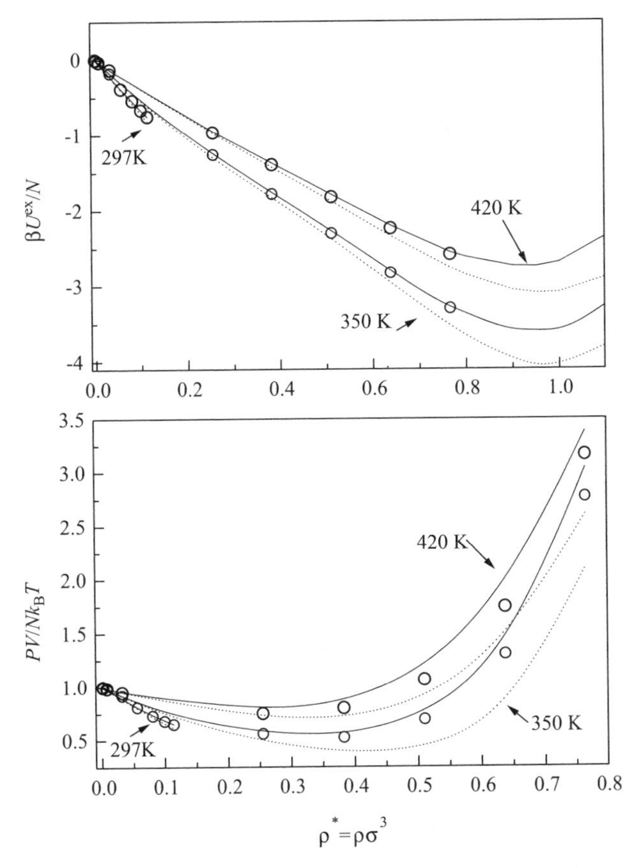

Figure 22. Excess internal energy, $\beta E^{ex}/N$, and virial pressure, $\beta P/\rho$, calculated with the ODS integral equation versus the reduced densities $\rho^* = \rho\sigma^3$, along the isotherms $T = 297.6$, 350 and 420 K (from bottom to top), by using the two-body potential alone (dotted lines) and the two- plus three-body potentials (solid lines). The experimental data (open circles) are those of Michels et al. [115]. Taken from Ref. [129].

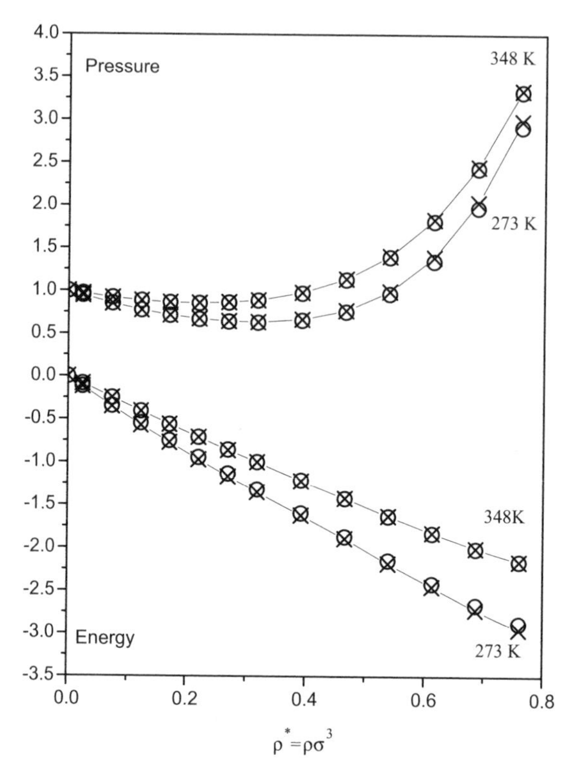

Figure 23. Excess internal energy and EOS of Kr for several reduced densities along the isotherms $T = 273$ and 348 K. AS only (dotted lines), AS+AT contribution (solid lines), symbols are from Trappeniers et al. [116]. Courtesy of the author.

supercritical isotherms above the critical temperature, namely $T = 350$ and 420 K (for Xe), and $T = 273$ and 348 K (for Kr), for which the experimental data are available [115, 116]. As expected, the general trend of the internal energy is to diminish monotonously with the density in the region of existence of the fluid phase, whereas the pressure increases after a slow initial decrease underneath the ideal gas value. For all thermodynamic states investigated, the improvement is manifest for Xe and Kr when the triple–dipole contribution is included in the effective pair potential, though the most significant improvement is for the excess energy. The predicted values of the excess energy are seen to be in perfect concordance with the experimental values, while those of the virial pressure are somewhat larger at medium densities. Note that the use of the two-body potential alone systematically underestimates both the excess energy and the virial pressure. In investigating the role of the three-body dispersion interaction in liquid Xe near its triple point,

Levesque et al. [132] also arrived at the conclusion that the AT potential increases the pressure. In addition, they shown that the local structure is not reinforced by the AT potential and that the collective dynamical properties are affected only in a minor way.

b. Isothermal Compressibility. The isothermal compressibility factor $\rho k_B T \chi_T$ are shown in Figs. 24 and 25, respectively, for Xe and Kr, as a function of the reduced density for several thermodynamic states. For Kr, the calculations have been achieved over the same range of densities for four different temperatures $T = 273.15$, 297.15, 348.15, and 423.15 K. The results reproduce the general features, with a maximum close to the critical density as precursor of the infinite compressibility that takes place at the approach of the spinodal line, and compare favorably with tabulated data [133]. It is at once apparent that the triple–dipole contribution is more influent when the temperature is decreasing. In addition, at a given temperature, the isothermal compressibility is in better agreement with the experiment when the triple–dipole interaction is involved. This fact is fully consistent with the low-q behavior of the structure factor (Fig. 19), for which the fluctuations in density are expected to be smaller when the AT potential is included in the calculations.

At a given temperature, it is shown that the rares gases described with AS+AT potential are much less compressible than that modeled with AS interaction only. These plots show that the difference between the heights of the maxima is increasing when the temperature is becoming closer and closer to the

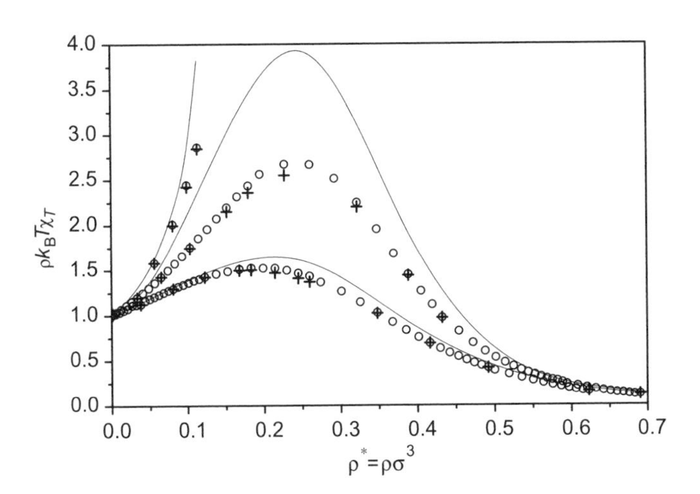

Figure 24. Isothermal compressibility for Xe at $T = 297$, 350, and 420 K (from top to buttom) calculated with the ODS scheme. Two-body potential only (solid lines), and two- plus three-body contribution (crosses), compared to data of Michels et al. [115]. Taken from [129].

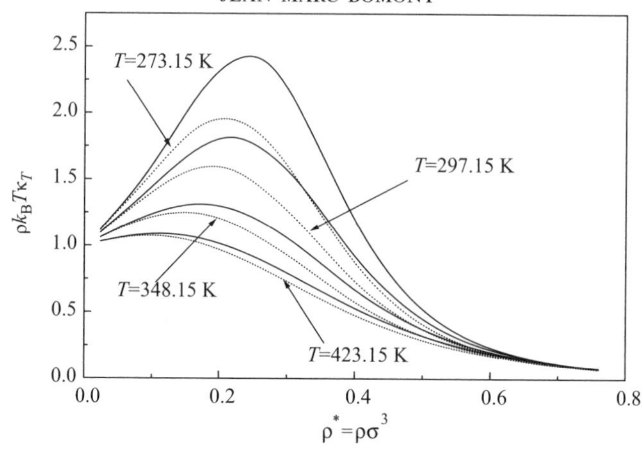

Figure 25. Isothermal compressibility for Kr at four temperatures calculated with HMSA by using AS potential (dotted lines) only and by including the AT contribution (solid lines). Taken from Jakse Ref. [128].

experimental critical temperature (for Kr, $T_c = 209\,\mathrm{K}$). It has been demonstrated [134] that T_c changes by 10% due to the presence of the three-body forces, and the curves indicate that the critical temperature of the model is substantially larger than the experimental one if only two body forces are taken into account. Therefore, the AT potential is a necessary ingredient and the critical region might be well better described in this case.

Thus, compared to the two-body potential, the addition of the triple–dipole dispersion interaction decreases the isothermal compressibility and increases the excess internal energy and the virial pressure all over the region of the densities investigated. The interpretation can be partially sustained by physical reasons in considering the asymptotically exact expression of the effective potential given by Eq. (120), that is, $u_{\mathrm{eff}}(r) \sim [-C_6 + (8\pi/3)\nu\rho]r^{-6}$, where the first term of the right-hand side represents the two-body dispersion interaction and the second term corresponds to the triple–dipole dispersion interaction. The qualitative effect of this resulting attractive potential is a tendency for the system to form a bound state. Consequently, if the attraction is sufficiently strong no external wall is required to contain the system because the emphasis of the attraction produces a decrease in the pressure that the fluid exerts on an external wall. In return, the presence of the triple–dipole interaction causes a reduction in the strength of the attraction and a small increase of the pressure. As far as the results can be compared with the experiments, it is remarkable that the excess energy is predicted with a very good degree of confidence when the AT potential is taken into account. In contrast, the effect of the AT potential is to yield a small excess of virial pressure. Presumably, additional corrections for higher order many-body interactions could be requisite to get an excellent agreement for the pressure, but the average accuracy of the SCIE schemes adopted here in different thermodynamic regions (from the critical to the triple point) remains good.

To summarize, the inclusion of the AT triple–dipole potential yields the following features for the thermodynamic properties, that is, the increasing of the internal energy and the pressure and the decreasing of the isothermal compressibility, compared to the pure AS potential. Given that the three-body AT potential is positive for most configurations of three atoms, it does not stand to reason. The AT potential gives rise to a repulsive two-body contribution that has the same behavior as the AS potential at greater distance [see Eq. (129)] and that reduces the strength of the attractive tail. Therefore, the treatment of the effective potential has to be performed in its entirety to justify the preceding conclusion on the isothermal compressibility. We also have to point out that the influence of the three-body AT interaction on the internal energy and the pressure becomes more significant as the density increases, while, for the isothermal compressibility, it is just important at intermediate densities where the critical point is located.

VI. CONCLUSION

We have offered a nonexhaustive presentation of the advances in the field of the IETs in terms of bridge function $B(r)$, by describing their application to the determination of structural and thermodynamic properties of simple fluids. First, various theories were presented, by recalling their basic expressions and underlying physical assumptions. Their performances were briefly discussed in Sections II and III. In this context, special attention was devoted to the thermodynamic consistency of the various theories, and to the role that this natural constraint plays in the improvement of their performances. Since earlier studies in the statistical mechanics of fluids, it was already known that these older theories (which in principle provide a direct source of information) yield only qualitative agreement for correlation functions, because of being inconsistent. Improved closure relations (SCIETs) were then introduced. Their applications were examined and systematically compared against related computer simulation data (Section IV). It is obvious that the fulfilment of the first thermodynamic consistency condition provides better correlation functions, but this criterion is not always sufficient to obtain accurate bridge functions. As was seen, the calculation of the excess chemical potential and the related entropic quantities requires the best possible bridge function and, at least, the fulfilment of a second thermodynamic consistency condition. Furthermore, a supplementary tool is to propose division schemes for the interaction potential (with strong physical basis). In this framework, recent developments have been presented. Among numerically solvable theories, only a few of them turn out to be accurate in the reproduction of the correlation functions up to the bridge function. As was observed, SCIETs are capable of providing accurate predictions. They are often a complementary source of information with respect to simulation, or might be able to explore the behavior of some correlation functions (e.g., rare gases) without requiring huge computational times.

In order to test their applicability to real systems, one of these last has been selected for the purpose of this presentation. The predicted results have been compared to direct experimental measurements. Section V was concerned with the application to rare gases. To perform this study, interatomic interactions modeled by the two-body Aziz–Slaman potential and the three-body Axilrod–Teller potential have been selected. Then, both the SCIET conjugated with an effective potential that combines the two- and three-body potentials, as well as the molecular dynamics simulation have been used. The selected self-consistent integral equation extended to three-body forces correctly reproduces the MD and experimental results of the pair correlation function. A comparative study of the small-q behavior of $S(q)$ has also been performed for Xe and Kr at room temperature and subcritical densities. It comes out that a very good agreement is obtained between the semianalytical results and the experimental data of Formisano et al. [12, 13] and Guarini et al. [14] when the AT potential with the standard value of ν is employed. To appraise the role of the three- and more-body potentials, there are three regions of the phase diagram to be investigated by integral equation theory, namely, that (1) at high temperatures over a wide range of densities, (2) close to the critical point, and (3) at low temperatures and high densities in the liquid phase. For high temperatures, it was shown that the AT potential has a large effect when the density increases so that its presence is essential to get the correct thermodynamic properties. Furthermore, in the liquid phase, accurate experimental data of the low–q behavior of $S(q)$ are needed in order to confirm the validity of this theoretical approach. The potential energy function (Aziz–Slaman conjugated with Axilrod–Teller) predicts the essential characteristics of structural properties in liquid Kr, even though it is believed that an accurate description of $S(q)$ at low q requires the inclusion of other three-body contributions, such as dipole–dipole–quadrupole, or exchange overlap potentials. The reader is invited to refer to Jakse et al. [135] for further details.

In addition, mention is made of some recent measurements of small angle neutron scattering for biological systems [136–138], like protein solutions. This technique is of course useful for the understanding of such macromolecules, because a full comprehension of effective protein interactions is a challenge [139–141]. The obtained data suggest that the effective interactions should have more features in addition to the well-known short-range attraction and electrostatic repulsion. The further could be interpreted in terms of the presence of a week long-range attraction between protein molecules. So, such systems are good canditates to extend the validity of IETs to the study of complex fluids. For instance, preliminary low–q results for $S(q)$ calculated with the Bomont-Bretonnet (BB) integral equation (among some others) are very encouraging. Works are on these lines.

To conclude, it is probably an understatement to say that, even if the Integral Equation Theories provide, for a given potential, results much faster than simulation methods, the advances in this field progress, however, more slowly than these of its sister method. But, where there is a will, there is a way. Our basic understanding of the liquid state is now at least comparable with our understanding of the physics of solids.

Acknowledgments

It has been a great pleasure to acknowledge with thanks the discussions and interest of Jean-Louis Bretonnet, to whom I am indebted.

The author would like to thank sincerely Lloyd L. Lee, Jean-Louis Rivail, Claude Millot, Hong Xu, Jean-François Wax, Jean-Georges Gasser and other direct collaborators along the last ten years. I am thankful to Professor J. P. Hansen for his kind interest. The "Laboratoire de la Théorie de la Matière Condensée, Université de Metz," the "UMR-CNRS 7565, Université Nancy I" and the "Laboratoire de Physique des Milieux Denses, Université de Metz" are gratefully acknowledged for their hospitality along these years. Finally, I thank respective publishers for permission to reproduce figures from Journal of Chemical Physics and Physical Review.

References

1. R. Car and M. Parinello, *Phys. Rev. Lett.* **55**, 2471 (1985).
2. G. Galli and M. Parrinello, in *Computer Simulation in Materials Science*, M. Meyer and M. Pontikis, eds., Kluwer, Dordrecht, The Netherlands, 1991.
3. M. P. Allen and D. J. Tildesley, *Computer Simulation of Liquids,* Clarendon Press, New York, (1989).
4. D. Frenkel, *Molecular Dynamics Simulation of Statitical Mechanics Systems*, G. Ciccotti and W. G. Hoover, eds., North-Holland, Amsterdam, The Netherlands, 1986.
5. N. Jakse and I. Charpentier, *Mol. Sim.* **13**, 293 (2000).
6. L. D. Landau, and E. M. Lifshitz, *Statistical Physics*, Pergamon, London-New York, 1958.
7. J. P. Hansen, and I. R. McDonald, *Theory of Simple Liquids*, Academic Press, London, 2006.
8. L. L. Lee, *J. Chem. Phys.* **103**, 9388 (1995).
9. N. H. March and M. P. Tosi, *Atomic Dynamics in Liquids*, Macmillan, London, 1976.
10. G. Casanova, R. J. Dulla, D. A. Johan, J. S. Rowlinson, and G. Savile, *Mol. Phys.* **18**, 589 (1970).
11. J. M. Bomont, N. Jakse, and J. L. Bretonnet, *Phys. Rev. B* **57**, 10217 (1998).
12. F. Formisano, C. J. Benmore, U. Bafile, F. Barocchi, P. A. Egelstaff, R. Magli, and P. Verkerk, *Phys. Rev. Lett.* **79**, 221 (1997).
13. F. Formisano, F. Barocchi, and R. Magli, *Phys. Rev. E* **58**, 2648 (1998).
14. E. Guarini, G. Casanova, U. Bafile, and F. Barocchi, *Phys. Rev. E* **60**, 6682 (1999).
15. E. Guarini, R. Magli, M. Tau, F. Barocchi, G. Casanova, and L. Reatto, *Phys. Rev. E* **63**, 052201 (2001).

16. Y. Rosenfeld and J. Kahl, *Phys.: Condens. Matter* **9**, 89 (1997).
17. L. L. Lee, *Molecular Thermodynamics of Nonideal Fluids*, Butterworths, Boston, 1988.
18. G. A. Martynov, *Fundamental Theory of Liquids : Methods of Distribution Functions*, Adam Hilger, ed., 1992.
19. J. L. Yarnell, M. J. Katz, R. G. Wengel, and S. K. Koenig, *Phys. Rev. A* **7**, 2130 (1973).
20. T. Pfleiderer, I. Waldner, H. Bertagnolli, K. Tödheide B. Kirchner, H. Huber, H. E. Fischer, *J. Chem. Phys.* **111**, 2641 (1999).
21. G. Stell, *Phys. Rev.* **184**, 135 (1969).
22. G. Zerah and J. P. Hansen, *J. Chem. Phys.* **84**, 2336 (1986).
23. G. M. Martynov, G. N. Sarkisov, and A. G. Vompe, *J. Chem. Phys.* **110**, 3961 (1999).
24. A. G. Vompe and G. A. Martynov, *J. Chem. Phys.* **100**, 5249 (1994).
25. L. S. Ornstein and F. Zernike, *Proc. Acad. Sci. (Amsterdam)* **17**, 793 (1914).
26. R. Balescu, *Equilibrium and Nonequilibrium Statistical Mechanics*, John Wiley & Sons, Inc., New York, (1975).
27. Y. Rosenfeld and N. W. Ashcroft, *Phys. Rev. A* **20**, 1208 (1979).
28. P. Attard, *J. Chem. Phys.* **98**, 2225 (1993).
29. J. S. Perkyns, K. M. Dyer, and B. M. Pettitt, *J. Chem. Phys.* **116**, 9404 (2002).
30. S. K. Kwak and D. A. Kofke, *J. Chem. Phys.* **122**, 104508 (2005).
31. N. F. Carnahan and K. E. Starling, *J. Chem. Phys.* **51**, 635 (1969).
32. R. D. Groot, J. P. van der Eerden, and N. M. Faber, *J. Chem. Phys.* **87**, 2263 (1987).
33. J. J. Nicolas, K. E. Gubbins, W. B. Streett, and D. J. Tildesley, *Mol. Phys.* **37**, 1429 (1979).
34. J. K. Johnson, J. A. Zollweg, and K. E. Gubbins, *Mol. Phys.* **78**, 591 (1993).
35. D. M. Heyes and H. Okumura, *J. Chem. Phys.* **124**, 164507 (2006).
36. M. Llano-Restrepo and W. G. Chapman, *J. Chem. Phys.* **97**, 2046 (1992).
37. S. Labik, A. Malijevski, and P. Vonka, *Mol. Phys.* **56**, 709 (1985).
38. M. J. Gillan, *Mol. Phys.* **41**, 75 (1980).
39. J. K. Percus and G. J. Yevick, *Phys. Rev.* **110**, 1 (1958).
40. J. Kolafa, unpublished, see Ref. 41.
41. T. Boublik, *Mol. Phys.* **59**, 371 (1986).
42. J. J. Erpenbeck and W. W. Wood, *J. Stat. Phys.* **35**, 321 (1984).
43. J. L. Lebowitz and J. K. Percus, *Phys. Rev.* **144**, 251 (1966).
44. J. M. J. Van Leeuwen, J. Groenveld, and J. De Boer, *Physica* **25**, 792 (1959).
45. T. Morita and K. Hiroike, *Progr. Theor. Phys.* **23**, 1003 (1960).
46. L. Verlet and D. Levesque, *Physica (Utrecht)* **28**, 1124 (1960).
47. L. Verlet, *Mol. Phys.* **41**, 183 (1980).
48. G. A. Martynov and G. N. Sarkisov, *Mol. Phys.* **49**, 1495 (1983).
49. J. Chihara, *Prog. Theor. Phys.* **50**, 409 (1973).
50. W. G. Madden and S. A. Rice, *J. Chem. Phys.* **72**, 4208 (1980).
51. J. D. Weeks, D. Chandler, and H. C. Andersen, *J. Chem. Phys.* **54**, 4931 (1970).
52. F. Lado, *Phys. Rev. A* **135**, 1013 (1964).
53. G. Stell, *Phase Transitions and Critical Phenomena*, C. Domb and M. S. Green eds., Academic Press, New York, Vol. 5b, Ch. 3, 1976.

54. F. Lado, M. S. Foiles, and N. W. Ashcroft, *Phys. Rev. A* **28**, 2374 (1983).
55. Y. Rosenfeld and L. Blum, *J. Chem. Phys.* **85**, 2197 (1986).
56. L. Verlet and J. J. Weis, *Phys. Rev. A* **5**, 939 (1972).
57. L. E. S. de Sousa and Ben-Amotz, *Mol. Phys.* **78**, 137 (1993).
58. J. Kolafa and I. Nezbeda, *Fluid Phase Equilib.* **100**, 1 (1994).
59. Y. Tang and J. Wu, *J. Chem. Phys.* **119**, 7388 (2003).
60. D. Ben-Amotz and D. R. Herschbach, *J. Phys. Chem.* **94**, 1038 (1990).
61. D. Ben-Amotz and G. Stell, *J. Phys. Chem. B* **108**, 6877 (2004).
62. L. Boltzmann, *Lectures on Gas Theory*, translated by S. Bush, University of California Press, Berkeley, p. 169 (1964).
63. J. L. Bretonnet and N. Jakse, *Phys. Rev. B* **46**, 5717 (1992).
64. F. J. Rogers and D. A. Young, *Phys. Rev. A* **30**, 999 (1984).
65. D. M. Duh and A. D. J. Haymet, *J. Chem. Phys.* **103**, 2625 (1995).
66. N. Choudhury and S. K. Ghosh, *J. Chem. Phys.* **116**, 8517 (2002).
67. P. Ballone, G. Pastore, G. Galli, and D. Gazzillo, *Mol. Phys.* **59**, 275 (1986).
68. I. Charpentier and N. Jakse, *J. Chem. Phys.* **114**, 2284 (2001).
69. J. M. Bomont and J. L. Bretonnet, *J. Chem. Phys.* **119**, 2188 (2003a).
70. J. M. Bomont and J. L. Bretonnet, *Mol. Phys.* **101**, 3249 (2003b).
71. L. L. Lee, K. S. Shing, *J. Chem. Phys.* **91**, 477 (1989).
72. L. L. Lee, *J. Chem. Phys.* **97**, 8606 (1992).
73. B. J. Widom, *Chem. Phys.* **39**, 2808 (1963).
74. B. J. Widom, *J. Stat. Phys.* **19**, 563 (1978).
75. E. Meeron and A. J. F. Siegert, *J. Chem. Phys.* **48**, 3139 (1968).
76. L. L. Lee, D. Ghonasgi, and E. Lomba, *J. Chem. Phys.* **104**, 8058 (1996).
77. L. L. Lee, *J. Chem. Phys.* **107**, 7360 (1997).
78. L. L. Lee, *J. Chem. Phys.* **110**, 7589 (1999).
79. J. M. Bomont, N. Jakse, and J. L. Bretonnet, *J. Chem. Phys.* **107**, 8030 (1997).
80. J. M. Bomont and J. L. Bretonnet, *J. Chem. Phys.* **114**, 4141 (2001).
81. D. M. Duh and D. Henderson, *J. Chem. Phys.* **104**, 6742 (1996).
82. L. Verlet, *Phys. Rev.* **165**, 201 (1968).
83. J. M. Bomont and J. L. Bretonnet, *J. Chem. Phys.* **121**, 1548 (2004).
84. G. N. Sarkisov, *J. Chem. Phys.* **114**, 9496 (2001).
85. E. Lomba, *Mol. Phys.* **68**, 87 (1989).
86. J. G. Kirkwood, *J. Chem. Phys.* **3**, 300 (1935).
87. T. Morita, *Progr. Theor. Phys.* **20**, 920 (1958).
88. R. Kjellander and S. Sarman, *J. Chem. Phys.* **90**, 2768 (1989).
89. H. Reiss, H. L. Frisch and J. L. Lebowitz, *J. Chem. Phys.* **31**, 369 (1959)
90. M. J. Nandell and H. Reiss, *J. Stat. Phys.* **13**, 113 (1975)
91. M. Heying and D. S. Corti, *J. Phys. Chem.* **B108**, 19576 (2004)
92. P. V. Giaquinta and G. Giunta, *Physica A* **187**, 145 (1992a).
93. P. V. Giaquinta, G. Giunta, and S. Prestipino Giarritta, *Phys. Rev. A* **45**, 6966 (1992b).
94. A. B. Schmidt, *J. Chem. Phys.* **99**, 4225 (1993).

95. O. E. Kiselyov and G. A. Martynov, *J. Chem. Phys.* **93**, 1942 (1990).
96. J. M. Bomont, *J. Chem. Phys.* **124**, 206101 (2006).
97. J. M. Bomont and J. L. Bretonnet, *J. Chem. Phys.* **126** 214504 (2007).
98. J. M. Bomont, *J. Chem. Phys.* **119**, 11484 (2003).
99. C. Caccamo, *Integral Equation Theory : Description of Phase Equilibria in Classical Fluids*, Elsevier, Science, New York, 1996.
100. J. J. Erpenbeck and W. W. Wood, *Phys. Rev. A* **43**, 4254 (1991).
101. A. Baranyai and D. J. Evans, *Phys. Rev. A: At., Med., Opt. Phys.* **40**, 3817 (1989).
102. A. Lotfi, J. Vrabec, and J. Fischer, *Mol. Phys.* **76**, 1319 (1992).
103. A. Z. Panagiotopoulos, *Int. J. Thermophys.* **15**, 1057 (1994).
104. W. Schommers, *Phys. Rev. A* **22**, 2855 (1980).
105. J. A. Barker and D. Henderson, *Rev. Mod. Phys.* **48**, 589 (1976).
106. R. A. Aziz and M. J. Slaman, *Mol. Phys.* **57**, 827 (1986).
107. R. A. Aziz and M. J. Slaman, *Chem. Phys.* **130**, 187 (1989).
108. B. M. Axilrod and E. Teller, *J. Chem. Phys.* **11**, 299 (1943).
109. R. Magli, F. Barocchi, P. Chieux, and R. Fontana, *Phys. Rev. Lett.* **77**, 846 (1996).
110. C. J. Benmore, F. Formisano, R. Magli, U. Bafile, P. Verkerk, P. A. Egelstaff, and F. Barocchi, *J. Phys. : Condens. Matter* **11**, 3091 (1999).
111. M. Tau, L. Reatto, R. Magli, P. A. Egelstaff, and F. Barocchi, *J. Phys.: Condens. Matter* **1**, 7131 (1989).
112. P. Attard, *Phys. Rev. A: At., Mol., Opt. Phys.* **45**, 3659 (1992).
113. J. M. Bomont, J. L. Bretonnet, T. Pfleiderer, and H. Bertagnolli, *J. Chem. Phys.* **113**, 6815 (2000).
114. J. M. Bomont, J. L. Bretonnet, and M. A. van der Hoef, *J. Chem. Phys.* **114**, 5674 (2001).
115. A. Michels, T. Wassena, and P. Louwerse, *Physica (Utrecht)* **20**, 99 (1954).
116. N. J. Trappeniers, T. Wassenaar, and G. J. Wolker, *Physica* **24**, 1503 (1966).
117. L. Reatto and M. Tau, *J. Chem. Phys.* **86**, 6474 (1987).
118. J. A. Barker, *Phys. Rev. Lett.* **57**, 230 (1986).
119. J. A. Anta, E. Lomba, and M. Lombardero, *Phys. Rev. E: Stat. Phys., Plasmas* **55**, 2707 (1997).
120. C. Hoheisel, *Phys. Rev. A* **23**, 1998 (1981).
121. J. M. Haile, *Molecular Dynamics Simulation : Elementary Methods*, John Wiley and Sons Inc., New York, 1992.
122. F. Yonezawa, *Molecular Dynamics Simulation*, Springer-Verlag, New York, 1992.
123. K. Esselink and P. A. J. Hilbers, *J. Comput. Phys.* **106**, 108 (1993a).
124. K. Esselink, B. Smit, and P. A. J. Hilbers, *J. Comput. Phys.* **106**, 101 (1993b).
125. S. J. Plimpton, *Comput. Phys.* **117**, 1 (1995).
126. S. Munejiri, F. Shimojo, and K. Hoshino, *J. Phys.: Condens. Matter* **10**, 4963 (1998).
127. I. Charpentier and N. Jakse, *Proc. SPIE* **3345**, 266 (1998).
128. N. Jakse, J. M. Bomont, I. Charpentier, and J. L. Bretonnet, *Phys. Rev. E* **62**, 3671 (2000).
129. J. M. Bomont and J. L. Bretonnet, *Phys. Rev.* **B 65**, 224203 (2002).
130. M. D. Johnson and N. H. March, *Phys. Lett.* **3**, 313 (1963).
131. L. Reatto and M. Tau, *J. Phys.: Condens. Matter* **4**, 1 (1992).
132. D. Levesque, J. J. Weis, and J. Vermesse, *Phys. Rev. A* **37**, 918 (1988).

133. V. A. Rabinovich, A. A. Vasserman, V. I. Nedostup, and I. S. Veksler, Thermodynamic Properties of Neon, Argon, Krypton and Xenon, Springer, Berlin, 1988.

134. S. Celi, L. Reatto, and M. Tau, *Phys. Rev. A* **39**, 1566 (1989).

135. N. Jakse, J. M. Bomont, and J. L. Bretonnet, *J. Chem. Phys.* **116**, 8504 (2002).

136. S. Finet, F. Bonnete, J. Frouin, K. Provost and A. Tardieu, *Eur. Biophys. J.* **27**, 263 (1998)

137. A. Stradner et al., Nature (London) **432**, 492 (2004)

138. F. Sciortino et al., *Phys. Rev. Lett.* **93**, 055701 (2004)

139. Y. Liu et al., *J. Chem. Phys.* **122**, 044507 (2005a)

140. Y. Liu et al., *Phys. Rev. Lett.* **95**,118102 (2005b)

141. M. Broccio, D. Costa, Y. Liu and S. H. Chen, *J. Chem. Phys.* **124**, 084501 (2006)

THE CHEMICAL ENVIRONMENT OF IONIC LIQUIDS: LINKS BETWEEN LIQUID STRUCTURE, DYNAMICS, AND SOLVATION

MARK N. KOBRAK

Department of Chemistry, Brooklyn College and the Graduate Center of the City University of New York, Brooklyn, NY 11210

CONTENTS

Advances in Chemical Physics, Volume 139, edited by Stuart A. Rice

I. INTRODUCTION

The past decade has seen an explosion of research devoted to the study of organic salts that are molten at low temperatures. These systems, commonly referred to as room temperature ionic liquids (ILs), are normally defined to include salts with a melting point $<100°C$. An enormous number of species are known, and while their properties vary significantly, as a class they possess low flammabilities, low volatilities, and display a wide liquidus temperature range [1–4]. They have been proven capable of acting as solvents for many molecular and ionic solutes [5–7], and have generally been observed to possess solvation properties comparable to moderately polar solvents, such as short-chain alcohols [8–10]. This solvation behavior has made it possible to use these media in an enormous range of synthetic [3, 11–17], separatory [18–23], and electrochemical [24–30] processes at the laboratory scale, and has led to their adoption in industrial processes [31, 32]. It was found that ions of many different chemical structures can be incorporated in ILs, and by one estimate, it may be possible to "mix and match" existing cations and anions to create as many as 10^{18} different liquids [33]. The availability of so many different species offers the possibility of using ILs as "designer solvents", since research has shown that variation of ionic structure may optimize ILs to specific tasks and achieve performance characteristics unavailable in conventional media [34, 35].

However, despite a substantial number of theoretical and experimental physical studies of ionic liquids, the physical principles underlying their function as reaction media remain poorly understood. Ionic liquids represent an exotic chemical environment for chemical reactions. Prior to their emergence, relatively few studies considered the influence of an ionic solvent on a chemical reaction, and those that did focused on a handful of industrial processes, such as aluminum refining [36] and heavy metals processing [37] in high temperature molten salts. The solvent systems of interest were primarily composed of monatomic or small inorganic molecular ions, and the solutes were invariably ionic species or coordination complexes. More recent studies of ILs consider a broader range of solutes, including organic and inorganic molecules and ions. Understanding the influence of ILs on the reactions of these species necessitates consideration of a host of chemical concepts, such as solvent polarity, that were not explicitly considered in studies of high temperature systems. A clear theoretical description connecting the chemical structure of an ion with its activity as a solvent is necessary if researchers are to design novel liquids with desired properties, or to predict how changes in the composition of a solvent will affect reaction outcomes. The derivation of such a framework requires a careful redevelopment of theories for solvation and kinetics in these novel environments.

The goal of this chapter is to compile existing knowledge on the behavior of ionic liquids and their influence on solvation and chemical reactivity. The intent is not to list reactions and their outcomes, but rather to review the results of studies that offer physical insight into the microscopic environment of ILs and their interaction with solute species. While many excellent reviews of ILs have been written [1, 4, 23, 30, 38–40], this chapter is distinct in its attempt to identify the basic physical principles relevant to solvation in ILs.

To this end, we review the physics of high temperature fused salts and draw on observations made in these systems to understand the microscopic structure of ionic liquids. We also review some physics of glass-forming liquids, focusing on concepts necessary to understand structural and dynamic inhomogeneity in ILs. We provide a broad review of attempts to characterize ILs empirically, and discuss those results with reference to simulation and theoretical studies. The overall objective of this study is to develop a conceptual toolbox that can be used to interpret experimental results in ILs and help identify useful new questions for the field. To this end, we present a series of principles describing the nature of solvation in ionic liquids at the conclusion of this chapter.

The balance of this Introduction will be committed to an overview of the chemical structures and macroscopic properties of ionic liquid systems. Section II provides a brief overview of the properties of high temperature molten salts, to provide a reference against which room temperature species may be compared. Section III considers the liquid structure and dynamics of neat ILs, and Sections IV and V discuss their operation as solvents at the microscopic level.

A. Structures of Ionic Liquids

There are in fact many classes of ILs, and we must begin by limiting the scope of this chapter. First, we consider only pure compounds that are molten below 100°C, eliminating "deep eutectic" systems formed by mixtures of organic salts and molecular species [41]. Though these systems show potential as solvents, their microenvironment is highly complex and more than can be adequately addressed in this chapter. We also eliminate ionic systems composed of ions known to undergo covalent rearrangements in melts, such as those based on the tetrachloroaluminate anion, which creates an equilibrium with a dialuminum complex

$$2AlCl_4^- \rightleftharpoons Al_2Cl_7^- + Cl^- \tag{1}$$

This decision excludes a large literature [26, 42, 43], but simplifies questions of the solvent environment, as indicated above. We place DIMCARB [44] and similar ionic liquids in this category as well. These salts are formed by the equilibration of carbon dioxide and protic alkylammonium species at high pressure, and their structure leaves open the possibility of covalent rearrangement in the liquid state.

Given the desire for covalent stability noted above, we must also be wary of "protic" ionic liquids in general. These species can be written as the product of the reaction of an Arrhenius acid and a base

$$HA(l) + B(l) \rightarrow [HB^+][A^-](l) \tag{2}$$

and while this is not always the synthetic route taken in practice [3], it leaves open the possibility of an equilibrium between molecular and ionic forms in the liquid. However, while experiments have shown that the existence of the molecular form can become significant at high temperatures [45], proton transfer is generally not significant at ambient temperatures for the extremely weakly basic anions employed in most IL studies. Thus, we will assume these species represent a purely ionic liquid unless noted otherwise.

Finally, while the majority of physical studies on ionic liquids concern monovalent ILs, many species containing di- or trivalent cations and/or anions are known [46, 47]. While there have been some attempts to characterize their physical properties [48], these species have not received the breadth and depth of attention that monovalent species have, so we will not consider them in this chapter.

Despite these restrictions, we are left with a daunting range of structural types. Some representative species of cations and anions are given in Fig. 1. Cations fall loosely into two structural types: Relatively small, rigid ions, such as those based on imidazolium or other cyclic structures, and large, highly flexible species, such as alkylammonium and alkylphosphonium species. Typically, such cations form low temperature molten salts in combination with smaller inorganic anions, though examples of ILs incorporating comparably sized organic anions are known [49].

Perhaps the common theme in understanding IL chemical structures is simply that they are prototypical "bad" crystals. A great deal is known about the lattice energies of ionic systems [50], and while the lattice energy is not

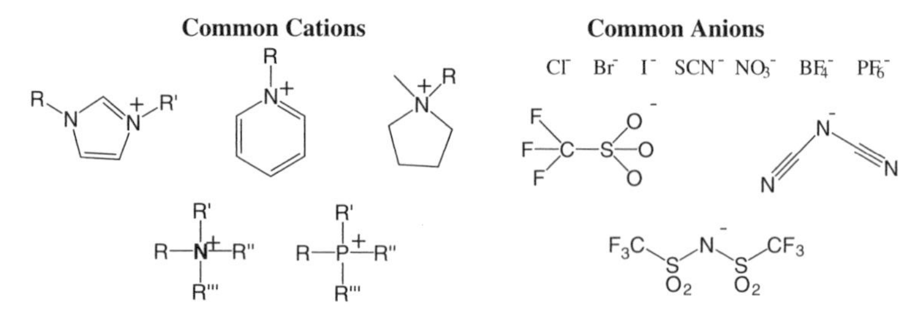

Figure 1. Structures of some common ions incorporated in ionic liquids.

directly correlated with melting point (it is an energy for a solid–vapor transition, not solid–liquid), one may expect the same general principles to apply. In particular, a large mismatch in ionic size and symmetry is known to reduce lattice energies, and this is a feature typically observed in ILs. High electronic polarizabilities for ions are also correlated with reduced lattice energies [51], an important observation because the components ions of ILs are generally polarizable. This polarizability extends not only to their electron clouds, but also to the presence of flexible aliphatic side chains. The existence of these substituents makes it possible for an ion to redistribute mass about charged moieties (e.g., the nitrogen center in a tetraalkylammonium ion), effectively changing the distribution of charge within the volume defined by the dispersion interactions of the ion. We will discuss the nature of interion forces in greater detail in Section III.

B. Physical Properties of Ionic Liquids

The goal of this chapter is to understand the behavior of ionic liquids as solvents and their influence on reaction based on their chemical structure and microscopic environment. We will therefore provide only a basic overview of their macroscopic physical properties. An online database, compiled by a research team operating under the auspices of the International Union of Pure and Applied Chemists (IUPAC), is now available detailing the physical properties of many known IL species [52].

As indicated above, the working definition of an IL is a material with a melting point $<100°C$, perhaps a rather optimistic definition of "room temperature". However, many of these materials are molten at far lower temperatures; some species of ILs based on imidazolium [53] and pyridinium [54] cations are liquid to temperatures of roughly $-80°C$, though most are significantly higher. The ILs based on alkylammonium cations typically melt at somewhat higher temperatures [55], and tremendous variation with anion is observed as well. Ionic liquids are not observed to boil at ambient pressures, though recent studies have revealed that some species possess significant vapor pressures and may be distilled at elevated temperatures or reduced pressures [56, 57]. However, the vapor pressures of ILs are generally sufficiently low that researchers may work with them under vacuum without measurable loss, a property that has facilitated some unique reactor designs [19, 58].

At sufficiently elevated temperatures, ILs decompose chemically. Thermogravimetric analysis has given some insight on the decomposition process for some species [55, 59], though it has not been well characterized for many classes of ILs. Temperatures for thermal decomposition vary widely between species, but can be as high as 450°C [54, 55, 60].

Ionic liquids are typically dense fluids, with specific gravities ranging from 1.0 to 1.6 [61]. In pure form, most are colorless to the naked eye, though some aromatic species show significant optical activity [62]. Under ambient conditions, ILs generally possess high viscosities, ranging from 30 to $>10,000$ cP. [54, 61]. Viscosity generally increases with increasing formula mass in a fashion similar to that of molecular liquids, though ionic structure also plays an important role [49, 61]. Species based on tetraalkylammonium and tetraalkylphosphonium cations, for example, often show viscosities much higher than comparably massive systems based on imidazolium or pyrrolidinium cations when combined with the same anion. Likewise, the incorporation of anions of low symmetry (e.g., trifluoromethylsulfonate) generally leads to lower viscosities than that observed for anions of higher symmetry. Section III provides discussion of the relationship between chemical structure and viscosity.

Ionic liquids show interesting properties with respect to their miscibility with other liquids. They can be broadly divided into "hydrophilic" and "hydrophobic" species, which are either miscible or largely immiscible with water. The aqueous miscibility of ionic liquids can be understood by applying the well-known Hofmeister [63] scale of hydrophobicity to their component ions [64]. The aqueous miscibility of ionic liquids also depends strongly on temperature, a fact that has been exploited in novel chemical processes [65]. Mixtures of ILs and organic phases also show interesting properties [66, 67], though we will not discuss them in detail in this chapter.

The designation of certain IL species as "hydrophobic" is misleading when considering the possibility of aqueous contamination. All known ILs are hygroscopic to some degree [68], and even hydrophobic species take up water in significant proportion. Because of the relatively high molecular masses of common ILs, even relatively low water content by weight implies significant contamination based on mole fraction, and physical properties, such as viscosity, are known to be extremely sensitive to the presence of water [68–70]. Where pure liquids are desired, it is necessary to dry ILs in a vacuum oven and store them under anhydrous conditions. Measurement of their water content via Karl–Fischer titration when reporting physical properties is essential. The presence of undetermined water and halide impurities has been noted as a cause for concern in some of the values for physical properties reported in the literature [64, 68].

We turn now to understanding the connection between the structure of ILs and their properties as solvents. We begin by considering the nature of high temperature fused salts, and consider ILs further in subsequent sections.

II. HIGH TEMPERATURE FUSED SALTS

Most chemists have considerable experience with molecular liquids, and possess a working knowledge of the principles underlying their structure–property relationships. However, the properties of high temperature fused salts, such as alkali halides, are not as well known, and a review of their properties is necessary to frame the current discussion of room temperature ionic liquids. These high temperature fused salts are often referred to as "ionic liquids" in the literature [71], but in the present discussion we will reserve this term for room temperature species and refer to high temperature species as "fused salts".

A. Liquid Structure of a Fused Salt

The literature on high temperature fused salts is extensive [72–74], and we make only the most cursory review here. The first and most obvious statement about fused salts is that they are fundamentally different than molecular liquids, in that they retain a substantial degree of order on melting. The strength of interion Coulomb interactions mandates that ions be surrounded by counterions, and so maintain the most uniform possible charge distribution throughout the liquid. This expectation is born out by X-ray [75, 76] and neutron diffraction [76, 77] experiments, which indicate that molten salts retain much of their solid-state structure in the liquid state; a representative radial distribution function for a molten salt is given in Fig. 2.

The melting behavior of fused salts is consistent with the idea that charge ordering is preserved. For example, the heats of fusion for alkali halide systems are usually $<5\%$ of the lattice energy, indicating that most of the lattice structure is preserved [78]. It has also been observed that nearest-neighbor distances decrease on melting for many molten salts, while ion

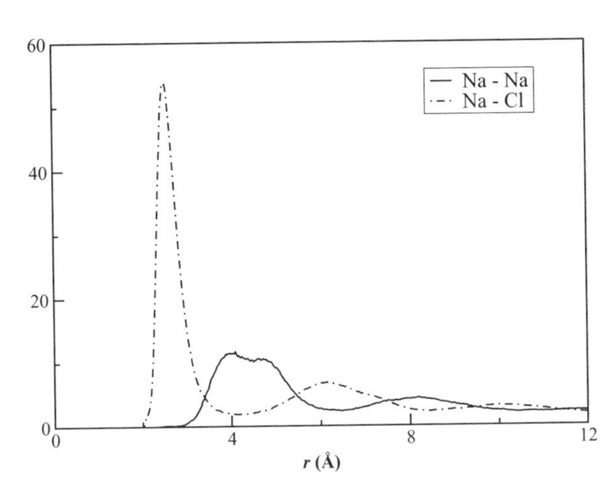

Figure 2. Cation–cation and cation–anion radial distribution functions for molten NaCl, obtained from simulation (force field given in Ref. [285]). Charge ordering is clearly visible, as described in the text.

coordination numbers decrease. Cation–anion distances in NaCl, for example, go from 2.95 to 2.78 Å on melting, but the coordination number of the ions drops from 6 to ~ 4 [72]. The overall picture of the liquid is one in which the solid-state lattice is largely preserved, but is disrupted by a high population of voids and defects.

This principle serves as the basis for a number of models of fused salt systems. Perhaps the best known of these is the Temkin model, which uses the properties of an ordered lattice to predict thermodynamic quantities for the liquid state [79]. However, certain other models that have been less successful in making quantitative predictions for fused salts may be of interest for their conceptual value in understanding room temperature ionic liquids. The interested reader can find a discussion of the early application of these models in a review by Bloom and Bockris [71], though we caution that with the exception of hole theory (discussed in Section II.C) these models are not currently in widespread use. The development of a general theoretical model accurately describing the full range of phenomena associated with molten salts remains a challenge for the field.

B. Distribution of Charge Density

The structure of a molten salt also allows characterization of the charge density about an ion. Stillinger et al. [80] have shown analytically that the charge distribution surrounding a given ion is best described by a spherical charge density wave, with alternating positively and negatively charged layers; an example of the radial distribution of charge about an ion is given in Fig. 3. At large distances, the distribution cancels out and electroneutrality is maintained. Such charge ordering may be viewed as a manifestation of electrostatic screening [81]. Fused

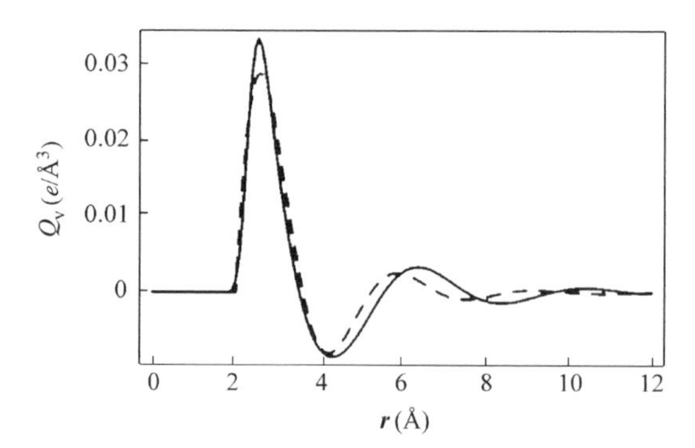

Figure 3. Charge density about a chloride anion in molten NaCl (simulation as per Fig. 2), taken from Ref. [239]. **Solid:** Results of simulation data. **Dashed:** Best-fit based on Eq. 3. Cation–cation and cation–anion radial distribution functions for molten NaCl, obtained from simulation (force field given in Ref. [285]).

salts are conductive media, and an applied electric field must be screened at the conductor's surface to maintain a uniform potential within the bulk. If one takes the view that an ion in the liquid represents a charge inside a cavity within a conductive medium, its electric field must be screened by the distribution of charge around it. The granular nature of the charge carriers in a salt system implies there should be a characteristic length scale associated with this screening process. Keblinski et al. [82] showed in simulation that the charge density about an ion in fused sodium chloride may be fitted to the equation

$$Q_v(r) = \frac{A}{r}\exp\left[-r/\lambda_d\right]\sin\left[2\pi r/d + \phi\right] \tag{3}$$

where r is the distance from the ion, d is the period of oscillations (i.e., the distance between like ions in the IL), ϕ is a phase factor, and λ_d is the Debye screening length of the liquid. This equation is normally used to apply Debye–Hückel theory in concentrated electrolytes, but it provides a good description of the charge density of fused salts as well. A sample fit to simulation data is given in Fig. 3; the fit does not appear to be very good due to a poor description of the innermost layer of charge density, but the qualitative features of the distribution are well reproduced.

The existence of charge ordering has a dramatic effect on liquid properties, perhaps best illustrated in the behavior of metal–molten salt solutions. It is well known that neutral metal atoms introduced to a solid-state crystalline ionic lattice will spontaneously ionize, creating a metal cation and an excess electron [83]. The solute metal will then occupy a cationic site in the crystal lattice, while the excess (solvated) electron will occupy hydrogen-like states centered on an anionic vacancy (i.e., a crystalline defect corresponding to a missing anion). The site occupied by the excess electron is referred to as an "F center", and spectroscopic evidence indicates that the electron occupies hydrogen-like states [83]. The optical spectral features associated with such hydrogen-like states persist on melting [84], and nuclear magnetic resonance (NMR) spectroscopy confirms the presence of the solvated electron [85]. This is not the result of thermal excitation of the electron, as the phenomenon is observed even at thermal energies well below the vapor-state ionization energy of the metal center. Rather, it is due to the favorable energetic interactions associated with the charge-separated state. In essence, the fused salt imposes its order on the solute, a phenomenon that is relevant to our discussion of solvation in room temperature ionic liquids in Section IV.

The magnitude of such electrostatic effects makes the question of ionic polarizability particularly important. Simulations have shown that while simulations employing nonpolarizable models adequately reproduce equilibrium properties in fused salts [86], ionic polarizability may have important dynamic effects. Simulations of phenomena, such as diffusion [87, 88] and dynamic light scattering [89, 90], require careful treatment of ionic polarizability to reproduce experimental observations. Madden and co-workers [91–94] developed highly advanced methodologies for incorporating the effect of ionic polarizability in simulation, not only with regard to the distribution of charge within an ion, but also asymmetric distortions of its dispersive interaction (i.e., changes in ionic shape). Their models have proven capable of reproducing purely dynamic phenomena, and have also shown that certain transient complexes known to exist in fused salts (and previously ascribed to covalent effects) can be reproduced by highly accurate polarizable models [95, 96]. The sensitivity of liquid dynamics to ionic shape may be relevant in understanding room temperature ILs based on molecular ions.

It would be helpful to be able to make analogies between solvation phenomena in high temperature molten salts and solvation in room temperature ionic liquids, but the comparisons are hard to draw. Most of the information on the former topic concerns mixtures of different fused salts [78, 97], where it is observed that solute ions intercalate into the structure of the liquid in a manner that minimizes disruption of the existing charge distribution. A particularly interesting class of such experiments of this type is the solvation of an optically active ion, which permits study of its coordination with counterions via its spectral shift [98, 99]; studies of the solvated electron, such as those noted above, can provide similar insight. However, knowledge of the coordination structure about a single ion does not provide direct insight into questions regarding solvation of a neutral solute species, and these are questions of great current interest in understanding reactions in room temperature ionic liquids. It would be informative, for example, to conduct photophysical experiments, such as time-resolved fluorescence Stokes shift studies, on a neutral solute in a fused salt. Such studies have been performed in a wide range of chemical environments [100–104], and characterization of fused salts in this way could help interpret observations in room temperature ILs [105–113]. We are not aware of any suitable molecular probe for such experiments, which would have to be photoactive and thermally stable in these extreme environments, but we raise the possibility of such studies in the hope that others will consider the issue.

While it is important to understand the behavior of high temperature fused salts, some caution is in order before applying these principles to room temperature ionic liquids. Studies of high temperature molten salts typically involve either single atom or very small ($\leq$4 atom) molecular ions, leading to both very compact distributions of charge and little if any structural flexibility. The molecular ions composing room temperature ionic liquids typically involve a more diffuse spatial distribution of charge and a larger number of atoms, which may be expected to change their interactions significantly. However, the radial distribution functions of ions in ILs are observed to show the same charge ordering as their high temperature brethren [114–116], so the principles governing liquid structure and charge ordering must apply.

C. Transport Properties

Because of their relevance to electrochemistry, the transport properties of fused salts have been the subject of considerable study [117, 118]. We do not attempt a comprehensive review of observed phenomena, but instead attempt to focus on those aspects we believe most relevant to understanding chemistry in room temperature ILs. The first and most general observation concerning ion self-diffusion in fused salts is that the motion of cations and anions appears strongly coupled. In NaCl, for example, the diffusion constants for cations and anions are the same to within 25% [72], despite a substantially larger difference in their volume, implying that the dynamic motions underlying ionic diffusion are coupled. Other experimental [119] and theoretical [120, 121] studies have indicated a high degree of cooperativity in ionic diffusion. While in some cases this cooperativity can be attributed to the formation of complex ions, it is observed even in noncoordinating salts, such as NaCl noted above. The existence of these correlations is unsurprising, as strong Coulomb forces prevent independent motion of cations and anions, and so make diffusion a collective process.

One model that has proven successful at describing this process is hole theory [71], which uses knowledge of the macroscopic liquid surface tension to predict an energy for cavity creation within the liquid. In this approach, the diffusion process is treated as the creation of a hole in that liquid followed by movement of a neighboring ion into that space. The model accurately predicts ionic diffusion constants, but the insight obtained from the model is limited. The approach makes use of the macroscopic surface tension of the liquid, and so does not provide insight on how the structure of an individual ion influences the creation of this volume, or diffusion. Thus, while its success suggests the model has certain correct qualitative features, it does not provide direct insight on the microscopic environment.

A more mechanistic approach, Instantaneous Normal Mode (INM) theory [122], can be used to characterize the collective modes of a liquid. Ribeiro and Madden [123] applied this theory to a series of fused salts, including both noncoordinating and coordinating species. They found that the INM analysis provided a good estimate of the diffusion constants for noncoordinating fused salts. For coordinating ions, however, the situation was complicated by the existence of transient, quasimolecular species. While a more detailed analysis is possible [124], the spectrum becomes sufficiently complicated that it would be difficult to characterize specific motions in the system.

The collective motions discussed above have been studied extensively via light-scattering experiments on fused salts [125–127]. In general, these spectra are dominated by a single low frequency peak referred to as the "boson peak", which represents charge and mass transport modes within the fused salt. Many fused salts show higher frequency shoulders to the boson peak, which are remnants of solid-state phonon modes that persist on melting. The survival of such modes is likely a consequence of charge ordering in the molten state, which makes it possible for the melt to sustain long-lived collective motion. Indeed, a phonon mode description can be used to interpret the light scattering spectra of fused salts [125, 128]. This is an interpretation that may also be relevant in understanding room temperature ILs; Ribeiro et al. [129] already observed the boson peak in a room temperature species, suggesting qualitatively similar collective motion to that observed in high temperature fused salts. In addition Urahata and Ribeiro [130] studied the density of states for the IL 1-butyl-3-methylimidazolium chloride in simulation, and have observed that the motions responsible for transport indeed correspond to a single low frequency peak, directly analogous to the boson peak of high temperature fused salts.

This discussion of the structure and dynamics of fused salts provides only the briefest overview of their properties. However, even this minimal background will prove a useful reference point as we turn our attention to room temperature ionic liquids.

III. MICROSCOPIC ENVIRONMENT OF NEAT IONIC LIQUIDS

Two distinct, but closely related, challenges emerge when attempting to characterize room temperature ionic liquids: How should their interionic structure and dynamics be described, and how does their chemical structure influence their properties? The former question ties very closely to studies of high temperature fused salts, as one expects interion Coulomb interactions to be a dominant force in both types of liquids. But because most studies of high temperature melts focus on monatomic or very small molecular ionic systems, details of their electrostatic structure can largely be neglected except as determinants of a total ionic volume and polarizability. In contrast, the chemical structure of the ions in ILs can be quite complex, and some thought must be given to structure–property relationships.

We will first consider the issue of liquid structure and dynamics, and then offer some insight on how the chemical structure of an ion influences liquid properties.

A. Liquid Structure and Dynamics

A number of simulation [114, 115, 131–133] and experimental [116, 134] studies have explored the structure of ILs. The radial distribution functions obtained from these studies are similar to those of high temperature fused salts, in that they show definite association between oppositely charged ions. Close examination reveals that specific interactions, such as interion hydrogen-bonding, influence the details of the distribution [114, 132, 135, 136], but the overall picture of local ion–ion interactions is largely the same for high temperature fused salts and room temperature ionic liquids.

Before going further, we must address a concept that is often invoked to explain the properties of fused salts: Ion pairing. There is strong evidence, which we will discuss below, that ions in ILs do not move independently, and that the distribution of ionic mobilities is quite complex. This represents inhomogeneous dynamics in the liquid, and suggests some form of ionic aggregation. However, we are not aware of any experiment or simulation that directly identifies these aggregates as being composed of a single pair. Below we discuss some evidence that in fact the relevant structures are substantially larger. Further, we question whether the concept of an ion pair can be productively applied in ionic liquids. In a fused salt or an ionic liquid, each ion is associated with multiple counterions at comparable distances. How can one identify which ions are "paired"? While in some cases theoretical electronic structure studies of isolated ion pairs can provide valuable insight on specific and nonspecific energetic interactions between ions [137], there is no reason to expect an identifiable pair of ions should be a relevant construct in the liquid.

There are in fact several forms of structural inhomogeneity known in ILs. X-ray and Raman spectroscopic structures of imidazolium-based ILs indicate that aliphatic substituent groups associate with each other, suggesting some degree of segregation between ionic and aliphatic regions in the liquid [138]. Simulation work by Canongia-Lopes and Pádua [139] and by Wang and Voth [140, 141] has provided a detailed description of this segregation, which amounts to the creation of two chemically distinct regions within ionic liquids. The first of these is formed by the association of the charged components of the ions (e.g., the ring of an alkylimidazolium ion) and the second by the association of neutral substituents, such as aliphatic chains. Within the charged domains, charge ordering is maintained as described above, producing a continuous network of ionic interactions that is interspersed with the domains of the neutral substituents. The effect is most pronounced in species including long-chain neutral substituents, but is observed even for systems incorporating relatively small substituents. These domains have been observed to remain identifiable on time scales of hundreds of picoseconds. The results are qualitatively consistent with the observed X-ray structures of the solid state of similar ionic compounds [142], which show distinct ionic and aliphatic domains.

This spatial inhomogeneity may also help explain observed fluorescence phenomena. Mandal et al. [143] reported evidence of a red edge effect (REE) [144] in the fluorescence spectra of chromophores solvated in a series of imidazolium-based ILs. In this phenomenon, the emission spectrum is observed to be dependent on excitation frequency, in violation of Kasha's rule. Two criteria must be met to observe the REE in experiment: First, the chromophore must exist in multiple chemical environments (as occurs in inhomogeneous media), and second, the fluorescence lifetime of the chromophore must be short compared to the time required for thermal sampling of the environment. Mandal et al. observed the existence of this effect only for 2-amino-7-nitrofluorene (ANF), a probe observed to have a lifetime of $\sim$100 ps, substantially shorter than that of other common probes in ILs. We may infer from this that the structure of the IL is longer lived than this timescale, and that the chromophore does not diffuse sufficiently quickly to repartition itself between solvent environments on this time scale. Hu and Margulis [145, 146] reproduced the red edge effect in simulation, and showed that different regions of the emission spectrum do indeed appear to arise from distinct distributions of solvent around the chromophore. We will consider other implications of the spatial inhomogeneity on solvation in ILs in Section IV.

While the segregation of charged and aliphatic groups discussed above explains some observations, experiments suggest there may be other types of structural inhomogeneity present. Coherent anti-Stokes Raman experiments reveal evidence for nanoscale domains in room temperature ILs, with sizes on the order of tens of nanometers [147]. This is far larger than the aliphatic–ionic domains observed in simulation, and was observed for alkylimidazolium structures of with relatively short alkyl substituents ($n = 4$–8). Such chains are not nearly long enough to create structures on this length scale. Xiao et al. [148] report that the Kerr effect spectra of binary mixtures of two different ILs are additive, and can be described as a sum of the spectra for the individual liquids. They interpret this result as evidence of nanoscale organization, on the grounds that the distinctive motions associated with the spectra of each liquid are preserved by creating microscopic domains that are "pure" in one compound or the other. This does not appear to be adequately explained by the presence of charged and aliphatic domains, since these could be formed by uniform distributions of unlike ions rather than requiring the existence of different regions of each type.

To understand these properties, it is necessary to recognize that ionic liquids are glass formers [149], and form amorphous rather than crystalline solids on cooling. Recent work on the theory of glass-forming liquids has led to the development of the "inherent structures" model, in which glass-forming liquids near their melting point form localized, metastable configurations that evolve in time [150]. Thermodynamically, the liquid is viewed as occupying local, but not global, minima on an energetic landscape, with dynamics described as a series of activated transitions between these minima [151]. The activation process limits the rate of these transitions, making local structures persistent features of the medium. These structures may be the nanoscale objects inferred from experiments noted above. Simulations of a model molecular liquid suggest that the existence of this structural inhomogeneity would promote dynamic inhomogeneity by creating regions of greater and lesser energetic stability [152].

Such dynamical inhomogeneity is well established in ILs. Neutron scattering experiments [153] show non-Debye–Waller relaxation dynamics characteristic of dynamical inhomogeneity. Optical Kerr effect experiments by Hyun and co-workers [154] and by Cang, Li, and Fayer [155] indicate a bimodal distribution of intermolecular dynamics, consistent with dynamical inhomogeneity. Other Kerr effect experiments by Giraud and co-workers [156] also indicate the presence of dynamic inhomogeneity, but the authors do not explore the question in detail. Investigations using broadband terahertz spectroscopy are also enlightening, and several recent dielectric spectroscopic studies [157–159] find evidence of glassy dynamics. The observed spectra are consistent with the forms of earlier work [160, 161], though the authors do not directly consider the question of glassy dynamics in these studies.

Shirota and Castner [162] performed an interesting Kerr effect study comparing the spectrum of an ionic liquid to that of an equimolar mixture of two molecular liquids representing an isoelectronic neutral pair (i.e., a pair of neutral molecules that were isoelectronic with the ion pair). The IL displayed a significantly broader spectrum than the molecular mixture, indicative of glassy or supercooled dynamics. It was observed that the slowest dynamics were roughly two orders of magnitude slower for the IL than for the neutral pair. Thus, there is no doubt that the dynamical inhomogeneity of ILs arises solely from the presence of Coulomb interactions.

Many simulation studies of ILs [130, 131, 163] have noted distinct time scales for diffusion, with rapid diffusion over very small distances followed by slower diffusion on longer time scales. Other studies [164] have found that cation and anion diffusion constants may vary between different ILs, but are always nearly identical within the same liquid. These observations are generally taken to imply the existence of "cages" of counterions from which diffusing ions must escape, a dynamic phenomenon characteristic of glasses and supercooled liquids. Hu and Margulis [145] studied the van Hoff correlation functions for ion self-diffusion and found evidence of distinct populations of "fast" and "slow" groups of ions in the liquid. They further demonstrated the existence of spatial correlations within the two groups, implying distinct coordinate domains. These scenarios are consistent with the inherent structures model and its dynamical inhomogeneity.

The existence of dynamical inhomogeneity also explains certain measurements for self-diffusion in ILs that have previously been attributed to ion pairing. Watanabe and co-workers [165–170] conducted studies on a range of ILs, comparing the molar conductivities calculated via PGSE–NMR (nuclear magnetic resonance) measurement against those obtained via electrochemical impedance measurements. The former estimate assumes all ions in the sample are mobile and are diffusing at the rate determined by spectroscopic measurements, and so may be viewed as a theoretical maximum. The latter estimate is based on direct measurement of the conductivity, and so reflects real ion mobility in the sample. The ratio of the two estimates, $\Lambda_{imp}/\Lambda_{NMR}$ is therefore a proxy for the degree of ion association in the sample. Values for this ratio fall between 0.4 and 0.8 for the ILs studied, suggesting a significant fraction of the ions move at reduced rates relative to the value that would be expected based on direct measurement of diffusion. This seems consistent with the idea of inherent structures, where ionic populations in different microenvironments would possess different mobilities. Mobility would then be a function of both ion size and mass, and a collective property of the medium (related to the properties of the inherent structural domains).

Tsuzuki et al. [165] report a relationship between the dynamics of the ionic liquid and the interactions between ions in an isolated pair. The authors perform electronic structure calculations on an isolated ion pair, and both identify the most stable configuration for the ions and calculate their binding energy. The authors observe that the degree of association inferred from the experimental $\Lambda_{imp}/\Lambda_{NMR}$ ratio increases with the calculated binding energy of the ions, suggesting that strong interactions lead to greater aggregation. They also observe a decrease in ion mobility associated with increased specificity of ionic orientation in the electronic structure of an isolated ion pair. In other words, ions that form isolated ion pairs possessing well-defined steric interactions form liquids of reduced ion mobility. The origin of this effect is unclear, but may relate to changes in librational dynamics that hinder ionic translation. This suggests a close relationship between structural and dynamic inhomogeneity and ionic structure.

Another interesting possibility in the application of the inherent structure model to ILs is the existence of polymorphism in IL compounds. Several studies have observed crystalline polymorphism in the crystallization of 1-butyl-3-methylimidazolium halide ILs [171–174], and the authors of these studies have quite reasonably suggested this dimorphism may in some way frustrate crystallization, keeping these materials liquid. Whether this is true or not, dimorphism could certainly contribute to the existence of inherent structures. The similarity between solid and molten states for ionic compounds noted in Section II.A suggests that liquid domains of distinct morphology could exist, maintaining structural and dynamic inhomogeneity. It is impossible at this point to assess how broadly this phenomenon might apply in different classes of ILs, but we note the issue.

The phenomenology of glass formers has been extensively studied, and existing analytical frameworks are readily extended to ionic liquids. Xu and co-workers [149] studied a number of aspects of ILs from this perspective, most notably the variation of viscosity with temperature. In simple liquids, fluidity (the inverse of viscosity) is observed to follow an Arrhenius equation, from which an activation energy for viscous flow can be inferred. For glassy systems, however, this is generally not a good description, and the relationship is more often described by the Vogel–Fulcher–Tammann (VFT) equation [175, 176]

$$\eta = \eta_0 \exp\left[\frac{DT_0}{T - T_0}\right] \tag{4}$$

where T is the temperature, T_0, D, and η_0 are fitted parameters for a given material, and η is the fluidity. Related quantities, such as the self-diffusion constant and (for ionic systems) the molar conductivity, can also be modeled using the VFT approach.

Experiments on ILs have generally shown a highly non-Arrhenius behavior that is well described by the VFT equation. Xu et al. [149] report the temperature-dependent viscosity of a series of covalently stable ILs, and note that the VFT equation fits the temperature dependence of the fluidity quite well. A series of studies by Watanabe and co-workers [167–169] on a range of different ILs shows that the VFT provides a good fit to diffusion constants, molar conductivity and viscosity.

Another conceptual tool used in the study of glassy systems is the Walden plot [177]. Its use is based on the observation that if ions in an ionic system move independently, then the product of their viscosity and conductivity should be a constant

$$C = \eta\gamma \tag{5}$$

where η is the viscosity, γ is the equivalent conductivity, and C is a constant for a given material. In the limit where ions can migrate freely at all temperatures, the value of this constant should be independent of temperature, as changes in viscosity are balanced by a reduction in liquid density (and thus a reduction in the density of charge carriers). However, if ions form aggregate structures, C is no longer temperature independent as the distribution of free and bound ions may change with temperature.

Xu et al. [149] construct Walden plots for a series of covalent ILs of the type of interest in this chapter, and observe behaviors ranging from "good" [Eq. (5) obeyed] to "poor", with the latter indicative of ionic aggregation. The authors infer that some ILs do not form aggregate structures, while others do. The details of aggregation inferred from the data suggest the process is very sensitive to chemical structure, with even minor changes in ionic structure producing significant differences in aggregation. These results suggest the formation of aggregate domains may not be a universal feature of ILs, as particularly nonpolarizable ions (e.g., tetrafluoroborate) seem to act as "good" ILs. Yet the same systems display some degree of non-Arrhenius dependence of viscosity on temperature, suggestive of aggregation. We cannot explain these seemingly contradictory observations, and interpreting the temperature dependence of viscosity and conductivity in ILs will require detailed knowledge of the underlying structure and dynamical inhomogeneity in these systems.

Sidestepping the question of liquid structure with an empirical approach, Abbott has had considerable success in modeling IL viscosity [178] and conductivity [179] using hole theory, as described in Section II.A. While the model is a useful tool, as noted previously, it is based on the macroscopic surface tension and does not provide direct insight on many of the underlying properties of the liquid. Nevertheless, any framework that is developed to interpret the behavior of ILs must also explain the success of hole theory.

B. Electrostatic Description of Molecular Ions

The issues described above, while important, largely sidestep a critical question: How does the chemical structure of a component ion affect the properties an IL? We have touched on this in the Introduction in noting the features of ions that make them good components for ILs. Large mismatches in ionic size favor a low melting point, as do ions of high polarizability and low symmetry. But we have not considered in detail how the distribution of charge within an individual ion should influence the properties of a liquid, or its interaction with a solute. The dipole moment, the lowest order description of the charge distribution of a molecule, serves as the basis for the discussion of this issue in molecular liquids. But this is not a well-defined physical quantity for an ion [180], and we must therefore consider in detail how an ion should be described.

We start by considering the origin of the dipole moment, which represents the lowest order nonzero term in a Taylor series expansion of the electrostatic potential arising from a neutral body (i.e., a molecule). For an assembly of n discrete charges, the electrostatic potential at a coordinate $\mathbf{r}$ may be written

$$\phi(\mathbf{r}) = \sum_{i=1}^{n} \frac{q_i}{R_i} \tag{6}$$

where $R_i = |\mathbf{r} - \mathbf{r_i}|$, and q_i and $\mathbf{r_i}$ represent the charge and coordinate of the ith site, respectively. Assume for the moment that the coordinate origin is located within the assembly of points defining the molecule (the scheme for the "best" placement is explained below). If the distances between molecular sites are small compared to the distance between the molecule and the point at which the electrostatic potential is measured (i.e., if the observer is distant), it is productive to expand in a Taylor series in $|\mathbf{r_i}|/|\mathbf{r}|$ [180]. Performing this expansion, and grouping terms of like order in $r = |\mathbf{r}|$, yields

$$\phi(\mathbf{r}) = \frac{1}{r}\sum_{i=1}^{n} q_i + \frac{1}{r^2}\sum_{i=1}^{n} q_i r_i \cos[\theta_i] + \frac{1}{r^3}\sum_{i=1}^{n} \frac{q_i r_i^2}{2}(3\cos^2[\theta_i] - 1) + \cdots \quad (7)$$

where θ_i is the angle between $\mathbf{r}$ and $\mathbf{r_i}$. The terms on the right-hand side of Eq. (7) represent the monopole, dipole, and quadrupole interactions, respectively. For a neutral molecule, the sum over charges $\sum_{i=1}^{n} q_i = 0$, and the monopole moment is zero. The dipole moment is then the lowest order nonzero term in the expansion. It is easy to show [181] that for a molecule possessing a nonzero dipole moment, the quadrupole moment is coordinate dependent, and the origin should be chosen such that the quadrupole moment is zero. This corresponds to describing the molecule as a point dipole centered at the origin, and is corrected by higher order terms that vary as $1/r^{\geq 4}$. This is the underlying reason for the utility of the dipole moment: It is the optimal mathematical description of the electrostatic distribution of the molecule.

The same approach may be applied to ionic systems using what is known as the "center of charge" framework [181]. In this case, the first (Coulomb) term in Eq. (7) is nonzero and dominates electrostatic interactions. It is the dipole term that is coordinate dependent and is eliminated by the choice of origin. This is accomplished by placing the origin $\mathbf{R_{cq}}$

$$\mathbf{R_{cq}} = \sum_{i=1}^{n} \frac{q_i \mathbf{r_i}}{Q} \quad (8)$$

where $Q = \sum_{i=1}^{n} q_i$. This models the ion as a point charge localized at the origin, and the description is corrected only by quadrupole or higher order terms.

The elimination of the dipole moment is not mathematical chicanery, but stems from the need to describe different physics for ions and polar molecules. The electrostatic interactions of a polar molecule are dominated by the asymmetry of its charge distribution, which is described in a series expansion as a gradient. By contrast, an ion represents a maximum or minimum (depending on its charge) in the electrostatic potential. For an optimal description, the gradient (dipole) term should be eliminated.

In this framework, polarization of an ion leads to movement of the center of charge as charge density is redistributed. In some applications, it may be advantageous to work in an ion-fixed coordinate system and track the dipole moment as an indicator of the extent of polarization, as has been done [182]. But an increase in the dipole moment in such a fixed coordinate system does not make the ion more "polar", it is simply an artifact of the choice of coordinate.

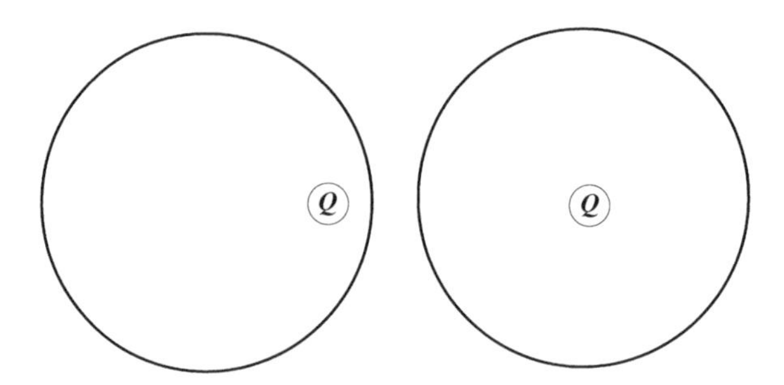

Figure 4. Two spherical ions, with equivalent charges at different centers. In a purely center of charge description the two ions are equivalent, but their interactions will be governed by a different balance of dispersive and Coulombic interactions.

The charge of the ion is unchanged, and the formal description of the polarization process is the translation of the center of charge.

In comparing the center of charge framework for an ion to the dipole moment of a molecule, it is important to note that where the dipole moment is a continuous quantity, the charge of an ion is quantized. Charge in chemical systems is distributed only in increments of e, and so two ions of the same charge are, at this level of description, indistinguishable. Of course, this does not take into account the distribution of mass (and, by implication, dispersive interactions) within the ion, which could profoundly affect interactions with neighboring species. To see this, consider two spherical ions containing a single point charge, one in which the charge is placed at the center of the sphere and the other in which it is placed near the surface, as shown in Fig. 4. Their interactions with neighboring ions or molecules will be substantially different, and thus the important factor in understanding structure–property relationships for ions is not the location of the center of charge per se, but rather the distribution of charge in relation to the dispersive interactions of the ion. This has not to our knowledge been explored.

An issue that has been explored is how the relative distribution of charge and mass affect the viscosity of an ionic liquid. Kobrak and Sandalow [183] pointed out that ionic dynamics are sensitive to the distance between the centers of charge and mass. Where these centers are separated, ionic rotation is coupled to Coulomb interactions with neighboring ions; where the centers of charge and mass are the same, rotational motion is, in the lowest order description, decoupled from an applied electric field. This is significant, because the Kerr effect experiments and simulation studies noted in Section III.A imply a separation of time scales for ionic libration (fast) and translation (slow) in ILs. Ions in which charge and mass centers are displaced can respond rapidly to an applied electric field via libration. Time-dependent electric fields are generated by the motion of ions in the liquid

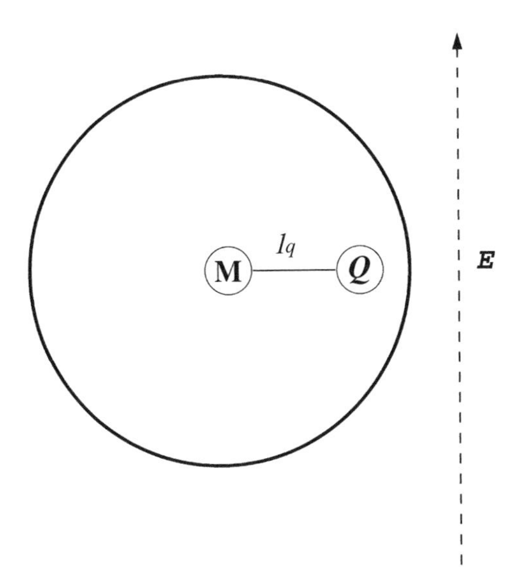

Figure 5. Schematic representation of an ion with charge and mass centers displaced. An applied electric field E couples to both translational and rotational motion in the ion.

(as in viscous flow or electrical conduction), and so ILs composed of asymmetric ions should possess lower viscosities and higher conductivities.

To see this, consider a spherical ion in which the center of mass resides at the center and the center of charge is displaced from it by some distance l_q, as shown in Fig. 5. Assuming translation of the ion is slow, the angular acceleration of the ion experienced due to an applied electric field $\mathcal{E}$ is given by

$$\ddot{\alpha} = -\frac{\mathcal{E} l_q Q \sin\alpha}{I} \tag{9}$$

where I is the moment of inertia for the ion, and α is the angle between the applied electric field and the vector connecting the centers of charge and mass. Kobrak and Sandalow point out that the quantity $L_c = l_q|Q|$ is a property of the ion reflective of the strength of the coupling of Coulomb interactions to rotation, and label this quantity the "charge arm". They show that for rigid ions of roughly equivalent mass, the viscosity of an ionic liquid does indeed decrease with increasing charge arm. Shirota and Castner [162] invoked the charge arm to explain observed viscosities in silyl-containing ILs.

The arguments presented above do not directly consider the association of the ion with neighboring ions, which may be very important. Given the evidence of collective transport modes for cationic and anionic species, one might be inclined to wonder whether the rotational motion of an individual ion is a relevant quantity. But the real question is whether that rotational motion can reduce the energy associated with transport processes. The existence of a significant charge arm in ionic species creates additional degrees of freedom within ionic assemblies, facilitating their response to an electric field and reducing the activation energy for transport processes.

While the charge arm is a useful quantity, it is not a proxy for the dipole moment. The dipole moment is a direct determinant of the energetics of molecule–molecule interactions in a liquid, and so is relevant to both thermodynamic and dynamic phenomena. By contrast, the derivation of the charge arm considers only dynamic phenomena, and are based on the separation of time scales for ionic rotation and diffusion. The charge arm is thus not directly relevant to describing thermodynamic phenomena, such as the melting points or solvent polarities. It is worth considering, however, that the coupling of rotational motion to Coulomb interactions will affect the density of states of a liquid, which must in turn affect its thermodynamic properties. This may relate to some of the structural features of ions common to ILs discussed above, and may be worthy of further study.

The issue of the relationship between rotational dynamics and the distribution of charge and mass in a molecular ion has seen some attention in studies of high temperature fused salts, most notably cyanide-containing species. Experimental [184] and theoretical [185] studies note complex rotational relaxation dynamics for these systems. These phenomena have been interpreted with the framework of rotational–translational coupling [186], a more detailed but less transparent description than the charge arm framework described above. This may be a useful approach to understanding IL dynamics, but has not to our knowledge been applied to these systems.

Another important quantity in understanding the electrostatic interactions in an IL is the polarizability of the ions. *Ab initio* molecular dynamics studies by Prado et al. [182] find that electronic polarization redistributes charge within the ion to roughly the same degree as intraionic vibrational motion. In studies using a polarizable force field, Yan and co-workers [187] showed that the exclusion of electronic polarizability in simulation reduces the rate of diffusion for ions by a factor of 3. This is cause for concern where dynamic properties are of interest, though their results also indicate that thermodynamic properties do not appear to be as sensitive to the description of polarization. Adams and Hills [86] made a similar observation on the properties of high temperature fused salts. The reason for the insensitivity of such time-independent properties may be that significant as polarization is, it only leads to a small displacement of the center of charge. So long as this displacement is small relative to interion differences, its net effect on energetic interactions is minimal. However, for slow processes with high activation barriers, even a small energetic change may significantly affect dynamics, making polarizability a crucial parameter in these phenomena.

IV. IONIC LIQUID SOLUTIONS

Questions surrounding solubility and the influence of a solvent on the properties and reactivity of a solute are central to the practice of chemistry. While the details of solute–solvent interactions and the study of solvation dynamics remain active areas of research, the general level of understanding in the chemical community is sufficient to guide researchers working with conventional liquids. This understanding is based on certain simple ideas concerning molecular size and electronic polarizability, the potential for nonspecific dipole–dipole interactions between solute and solvent, and the capacity for specific interactions, such as hydrogen-bonding or π–π interactions.

As discussed below, ionic liquids often behave comparably to conventional polar organic solvents [6, 8, 10]. But the physics underlying solvation are entirely different. As noted above, ILs are characterized by considerable structural and dynamic inhomogeneity, and even simple concepts, such as the dipole moment, cannot be productively applied. We are therefore in the unusual position of needing to explain how an exotic microscopic environment produces conventional macroscopic behavior. To this end, we will review empirical characterizations of the ionic liquid environment, and then turn our attention to the underlying physics of solute–solvent interactions.

A. Empirical Characterization of Solvent Polarity

Solvent polarity is the property most often invoked to explain the solubility of a species and to interpret the influence of a solvent on a reaction. As such, it is the most widespread expression of solvation thermodynamics in chemistry. But there is no definition of polarity that can be constructed from first principles. The characterization of solvent polarity is always empirical, with approaches ranging from simple single-parameter models, such as the dielectric constant of the medium, to multiparameter fits based on a linear free energy relationship formalism [188, 189]. We loosely divide schemes for the estimation of solvent polarity into three categories. The first includes nonreactive methods that characterize the solvent environment by measuring the partitioning of solute probes between the phase of interest and a reference phase (e.g., an aqueous layer). The second category includes spectroscopic characterization of the medium, either by the direct interaction of the medium with electromagnetic radiation (as in dielectric spectroscopies) or by the study of the influence of the environment on the absorption or emission spectrum of a probe solute. The third category consists of reactive methods, which use the outcome of a chemical reaction to draw inferences about the environment. The outcomes of these experiments are generally interpreted with reference to two solvent properties: The electrostatic polarizability of the liquid (which for molecular liquids includes both permanent and induced dipoles) and its capacity for specific interactions.

We review the results of experimental measurements of polarity in ILs and discuss how solute–solvent interactions should be viewed in ionic liquids. We focus primarily on the solvation of molecular species, though we include some discussion of ionic solvation at the end of the section. In the interest of brevity, we avoid discussion of mixtures and focus on dilute solutions.

1. *Partitioning Methods*

A common nonreactive approach to the characterization of solvent polarity is the measurement of the partitioning of a series of probe molecules between the phase of interest and a reference phase. Two sets of systematic studies have been conducted on ionic liquids using this approach. The first is due to Rogers and co-workers [5, 190, 191], who studied the partitioning of organic molecules between IL phases and a reference liquid (water). The second series of studies is due to Armstrong and co-workers [6, 192, 193], and involves the use of ionic liquids as stationary phases in gas chromatography (GC). In this case, the reference phase is the vapor state, and the partitioning is inferred from chromatographic retention times. The two methods are complementary in that while the chromatographic approach restricts studies to the use of volatile probes, the use of a vapor-state reference phase eliminates the possibility of contamination of the ionic liquid by the reference liquid.

Both of these studies were ultimately parameterized as linear free energy relationships (LFERs) of a form due to Abraham and co-workers [194, 195]. The core of this LFER is an equation of the form

$$\log[k] = c + rR_2 + s\pi_2^H + a\alpha_2^H + b\beta_2^H + l\log[L] \tag{10}$$

where k represents the equilibrium constant for partitioning into the phase of interest (or the adjusted relative retention time for chromatographic experiments), and c is a constant of the solvent. In each other term on the right-hand side of Eq. (10), the first variable represents a property of the solvent and the second variable represents a property of the solute. Since the latter terms have already been parameterized for a series of probe molecules, the former may be inferred by fitting the data. Specifically, α_2^H and β_2^H represent hydrogen-bond acidity and basicity of the solute, respectively, and π_2^H represents the solute dipolarity–polarizability. The R_2 term represents the excess molar refraction calculated from the solute's refractive index. It is also included (see below) in π_2^H, and its explicit inclusion in the equation allows separate analysis of contributions from the solvent ϵ_∞ and ϵ_0 (electronic and nuclear polarizabilities, respectively). The parameter L is the solute gas–hexadecane partition coefficient, included as an estimator of dispersive interactions. The values of a and b represent solvent hydrogen-bond basicity and acidity, respectively, and r and s are measures of solvent dipolarity–polarizability; l is a related measure of dispersion forces.

The results of the chromatographic characterization by Anderson et al. [192, 193] are in many ways consistent with expectations. The hydrogen-bond basicity of the solvent is controlled largely by the anion. The acidity of the cation is controlled by both the acidity of the cation and the basicity of the anion, suggesting anion–solute competition for hydrogen-bond donation from the cation. Species possessing an aromatic anion evince a significant capacity for π–π bonding (through the *s* parameter), though aromatic cations rarely show the same effect unless the cation ring includes an electron-donating substituent that increases its polarizability. These features appear to be readily explained by conventional chemical principles, and lead to solute–solvent interactions comparable to those observed in moderately polar organic solvents, such as short-chain alcohols.

The only parameter that differs significantly from expectations for comparable molecular liquids is the value of r, indicating the existence of anomalously strong dispersive interactions. This is interesting, as previous work by the same group [6] provides evidence of a "dual nature" for ILs. When ILs are used as stationary phases, nonpolar analytes show retention times comparable to those observed in nonpolar stationary phases, while polar analytes show retention times characteristic of polar phases. Given the discussion of structural inhomogeneity in Section III, it is tempting to propose that this may correspond to the partitioning of solutes in different regimes. But it may also be true that the anomalously large dispersive interactions arise from some other feature of the IL, and make for more favorable solute–solvent interactions than would be expected for nonpolar solutes in other polar liquids.

Rogers and co-workers [190, 191] studied the partitioning of probe molecules between an IL and an aqueous layer, and have fitted their data to an equation similar to Equation (10). Their results with respect to hydrogen bonding are generally consistent with those inferred from chromatographic studies, in that the resultant partitioning is comparable to that observed in short-chain alcohols. In these studies, however, the results seem to indicate an anomalously large dipolarity–polarizability rather than anomalously large dispersive interactions. This may be due to the presence of water in the IL phase; the authors attempt a meta-analysis of their fitted coefficients using principal component analysis, and conclude that the presence of water is likely at least partly responsible. Nevertheless, these anomalous values of dipolarity–polarizability are intriguing given the differences between the electrostatic structure of ionic and molecular solvents.

We note that Lee [196] recently challenged the forms of the LFERs used in the studies above, claiming that ILs can only be adequately described by an equation that includes additional information on molar internal volume. Subsequent work by Acree and Abraham [197] provides evidence that the LFERs used in the studies described above are adequate. While it is possible that this debate may lead to some modification of the preferred form of LFER for ILs, the central physical observations of the studies above do not appear to be in jeopardy.

2. *Spectroscopic Methods and the Dielectric Constant*

Another nonreactive route to the characterization of solvent polarity is the study of the optical absorption and emission spectra of chromophores [188]. These spectra are sensitive to the molecular environment, and because different solvatochromic probes may have different capacities for specific interactions, it is possible to characterize the solvent environment in detail and to construct LFERs analogous to those described above. Studies of the spectra of solvatochromic probes in ionic liquids have in general been consistent with the results of partitioning studies described above [8–10, 69, 70, 198–200], though we will discuss one observed anomaly below [198].

The most commonly used probes for such studies are of the betaine dyes. Reichardt has recently made an excellent review of the application of these experiments [201], and so we will only briefly summarize those results here. The solvatochromic shift of betaine-30 is observed to depend on the identity of the cation rather than the anion [8, 9], perhaps due to a strong cation association with the pendant oxygen as observed in simulation [202]. While these experimental studies also show a relatively weak dependence on the chain length of alkyl substituents, Dzubya and Bartsch [199] showed that greater variation of the solvent polarity is possible through derivatization of the alkyl substituent. They observe a particularly large solvatochromic shift in betaine-30 when a hydroxy substituent is added to the alkyl substituent of the cation, but this shift was not observed in a second (less hydrogen-bond labile) chromophore. This is consistent with the known sensitivity of the betaine-30 spectrum to hydrogen bonding [188]. Studies with Nile Red [10], a species that might be expected to interact more strongly with hydrogen-bond accepting anions, do indeed show a greater sensitivity to the anion and relatively weak dependence on the cation. Thus, once the preferential association of the solute with specific ions is taken into account, standard chemical logic appears to explain solvatochromism in ILs.

The Kamlet–Taft scheme [189, 203, 204] represents a LFER for the analysis of a solvent environment via solvatochromic effects. The Kamlet–Taft model is similar to the form of Eq. (10), but is simplified

$$\nu = \nu_0 + s\pi^* + a\alpha + b\beta \tag{11}$$

where ν and ν_0 are optical absorbance maxima, s, a, and b are parameters of a given dye, and π^*, α, and β are parameters describing the dipolarity–polarizability, hydrogen-bond donor character, and hydrogen-bond acceptor character, respectively [this notation is standard, but assigns Roman characters to solute properties and Greek to solvent properties in reversal of those for Eq. (10)]. Reports of Kamlet–Taft parameters for ILs [69, 205, 206] give relatively high values for dipolarity–polarizability, in line with the partitioning studies of Rogers and co-workers noted above.

Crowhurst et al. [205] find evidence that where hydrogen bonding between cation and anion can compete with solute–ion hydrogen bonding, the solvent's net capacity for hydrogen bonding is reduced; this phenomenon has been noted in simulation as well [202]. Here again, while there appear to be some differences between ILs and molecular solvents, established chemical logic appears capable of explaining the discrepancy.

While the study of solvatochromic dyes is well established as a means of probing solvent polarity, these are not the only solutes that can be used in this fashion. A more exotic solvatochromic probe is an excess electron in solution. Optical absorption studies of the thermalized (solvated) electron generated in the pulse radiolysis of a series of ILs show a strong dependence on cation character, with a relatively low frequency for tetraalkylammonium systems and a higher frequency for cyclic (pyrrolidinium-based) cations [48, 207]. The solvated electron spectrum is often interpreted in a "particle-in-a-box" framework, which would imply that the cyclic cations (which possess smaller ionic volumes) simply coordinate more closely with the electron and so create a smaller domain in which the electron must localize. The breadth of the absorptions and their maximum fall within the range of values expected for moderately polar organic solvents.

Novel approaches using infrared (IR) vibrational spectroscopy to determine the polarity of the solvent environment have also been employed. Tao and co-workers [208] used the vibrational dynamics of small molecular probes, such as acetone to infer polarities of common ILs, and Dahl et al. [209] used the C—N stretch in cyano-containing ILs to determine polarity without the need for a solute probe. The results of both studies indicate polar, aprotic solvents of comparable polarity to those determined above.

While most of the empirical studies noted above generally give consistent descriptions of the solvent properties of ILs, we make note of one observed anomaly. The fluorescence spectrum of pyrenecarboxaldehyde in IL solution is weak and highly structured in a fashion more consistent with a nonpolar molecular solvent [198]. This contrasts sharply with a study of pyrene by the same authors, which indicates a polar environment. The difference may arise from the distinct fluorescence lifetimes of the two species. As noted in the discussion of the red edge effect in Section III, ILs are extremely slow to equilibrate and the observed spectrum does not always represent that of an ergodically sampled excited state. The authors report lifetimes for pyrene in an IL on the order of 140–300 ns, depending on the presence of oxygen, which is adequate to sample excited-state dynamics of the liquid. While they do not report fluorescence lifetimes for pyrenecarboxaldehyde, an earlier study [210] reported the lifetime of this species to be on the order of 2–5 ns in a series short-chain alcohols and moderately polar solutions. If this is the lifetime in the IL, it may be too short a time for adequate sampling of excited-state dynamics, and so the observed fluorescence spectrum does not represent relaxation from the equilibrated environment. This would make the solvent appear less polar, as observed in experiment.

Thus far, we have avoided reference to perhaps the most commonly used estimate for the polarity: The dielectric constant of the pure liquid. The dielectric constant can be determined from a range of experiments, including capacitance [211] and dielectric reflectance spectroscopy [212]. While the dielectric constant does not characterize the solvent environment as fully as multiparameter LFER approaches, it is simple to measure and accurate values are available for an enormous number of solvents [213].

A number of studies have attempted to characterize ionic liquids through their dielectric constant, and all have observed inconsistencies between the measured dielectric constant and the solvation properties of the liquid. Recent experiments making use of dielectric reflectance spectroscopy [214] indicate dielectric constants in the range of 10–15 for a series of imidazolium-based ILs, substantially lower than those for molecular solvents observed to possess comparable polarities as estimated by solvatochromism. Weingärtner [215] has recently published a series of static dielectric constants obtained from dielectric reflectance spectroscopy, and compared them with those of common molecular liquids. The analysis includes comparison with the Kamlet–Taft π^* parameter for the liquids from Eq. (11); we have prepared a plot of π^* versus dielectric constant in Fig. 6. The relationship between π^* and ϵ for molecular liquids

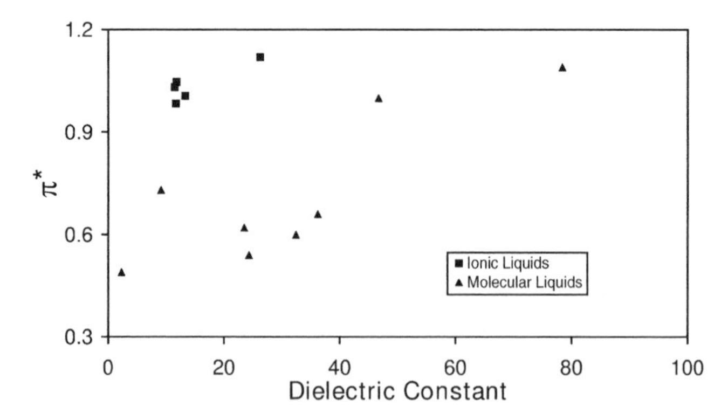

Figure 6. Kamlet–Taft π^* parameter versus dielectric constant for a series of ionic and molecular liquids. Data are taken from Ref. [215]

shows a clear if somewhat scattered trend, while that for the ionic liquids is murky. With the exception of a single outlying point (corresponding to ethylammonium nitrate, to our knowledge the IL possessing the lowest possible formula mass), almost no variation is seen in the measured IL dielectric constants, and little variation in the value of π^* is seen for this limited sample of liquids.

While the magnitude of the variation in ϵ and π^* may simply reflect an unfortunate choice of ILs for the study, it is significant that the points are clustered well away from the line corresponding to the behavior of molecular liquids. Dielectric continuum theory is a useful description of molecular liquids because the response of the medium to a microscopic (solute) electric field is similar to its response to a macroscopic field. The fact that the relationship is different in ionic liquids suggests that the polarization response may be qualitatively different on microscopic and macroscopic lengthscales. This likely reflects the fact that, as noted in Section III, ionic liquids are highly structured media. The distribution of charge in the liquid oscillates on the lengthscale of interion distances, which are typically similar to or larger than the length scale of a solute molecular dipole. The solvent response to a molecular dipole is therefore likely quite different than it is for a macroscopic electric field.

One can address this issue more rigorously by considering the work of Rovere and Tosi [73]. The authors show that, in the limit of a salt composed of monovalent, nonpolarizable ions of uniform size, linear response theory predicts the dielectric constant $\epsilon(k)$ should be

$$\frac{1}{\epsilon(k)} = 1 - \frac{4\pi\rho e^2}{k^2 k_B T} S_{\alpha\alpha}(k) \tag{12}$$

where e is the unit charge on an ion, k is the wavelength of the applied field, ρ is the number density of the liquid, and $S_{\alpha\alpha}(k)$ is a normalized form of the concentration–concentration structure factor. The latter is defined as

$$S_{\alpha\alpha}(k) = S_{++}(k) + S_{--}(k) - 2S_{+-}(k) \tag{13}$$

where $S_{++}(k), S_{--}(k)$ and $S_{+-}(k)$ denote the cation–cation, anion–anion, and cation–anion structure factors, respectively. While the structural complexity of molecular ions in ILs suggests their dielectric behavior may not be well described by this simple model, it is clear that the length scale for charge distribution within the solvent plays a key role in the dielectric behavior of the solvent. Dielectric continuum models must therefore be modified to account for this k-dependence, a formidable challenge. We are not aware of any theoretical studies addressing this issue for solvation in either high or low temperature molten salts.

3. *Reactive Methods*

Another approach used in the empirical characterization of liquid polarity is the study of the outcome of a chemical reaction. Earle et al. [216] report a preliminary study of the keto–enol tautomerization of pentane-2,4-dione, and create an empirical correlation between the degree of tautomerization and the dielectric constant of molecular liquids. They then predict dielectric constants for a series of ILs based on the observed keto–enol equilibrium; the values range from 40 to 50, slightly higher than those of short-chain alcohols. A more detailed study by Angelini et al. [217] considers the tautomerization of a nitroketone complex in a series of five imidazolium-based ILs. The results, parameterized as a linear free energy analysis of the behavior of the equilibrium constant, indicates an overall polarity comparable to that of acetonitrile, consistent with the partitioning and spectroscopic studies referenced above.

A great number of reactions have been studied in ILs, and many have been interpreted as evidence of particular properties of the solvent. An overview of the insights gained from such reactions would be a review in itself, and indeed, the interested reader should consult any of a number of reviews on the subject [3, 14, 218]. We note the above reactions because their analysis is specifically geared toward detailed parameterization of polarity, and do not consider cases where the analysis of solvent effects is less detailed.

B. Ionic Solutes in Ionic Liquids

The above characterizations primarily concern the interactions between molecular solutes and ILs. However, ILs are also good solvents for ionic compounds, and have been studied extensively as media for transition metal catalysis [4, 38, 219] and for the extraction of heavy metals [23]. ILs are capable of solvating even simple salts, such as NaCl, to some degree [219], and in fact the removal of halide impurities resulting from synthesis can be a considerable challenge [68]. However, ionic complexes are generally far more soluble than simple salts [220], and we focus our attention on these systems as they have received greater study and are more relevant to the processes noted above.

In general, the solubility of metal ions in ILs can be increased through complexation with organic ligands, as is commonly seen in organic solvents. A number of studies using crown ethers [5, 21, 22] and other agents [222] have noted such increases. An early study by Dai et al. [21] on the crown ether extraction of aqueous strontium nitrate notes that the coefficient for partitioning into the IL layer is roughly 0.3–0.7, indicating that the ion is somewhat less soluble in an IL phase than in an aqueous one (though still far more soluble in the IL than in a molecular solvent of comparable polarity). The addition of a crown ether to the IL phase increases the partition coefficient by as much as three orders of magnitude, depending on the IL.

There is also evidence that a novel extraction mechanism may be at work in ILs. Electroneutrality must be maintained in any extraction process, and in extractions involving two molecular liquids this is achieved through the formation of neutral complexes and/or coextraction of counterions. In ILs, however, charge balance between a molecular and an ionic liquid can be maintained by the partitioning of IL solvent ions in the molecular phase (i.e., an ion exchange process). Jensen et al. [223] point this out for the strontium reaction discussed above, and note that unlike extractions to molecular phases, the strontium–crown ether complex formed in the IL carries a net charge, suggestive of an exchange reaction. Intriguingly, the same study found that when a pure sample of the solid strontium nitrate–crown ether complex is added to the IL, it forms a neutral complex and does not appear to coordinate with solvent ions. Thus, it is not that the complex is unstable, but rather that the thermodynamics of ion exchange drive an alternative reaction for complex formation. Additional studies provide other examples of an ion-exchange mechanism underlying extraction [224, 225], though others [222] have found that for at least some ions, the coordination environment of the IL phase matches the neutral coordination shell observed in molecular solvents. A particularly thorough study by Gaillard and co-workers [226] shows that alteration of the anion of an IL can change the coordination of a dissolved cation from a charged to a neutral complex; the authors note that this can be understood by considering the nucleophilicity of the anions involved, though many details remain unclear. Studies of the coordination of actinides in ILs have been reviewed by Cocalia et al. [39].

There have been a number of molecular dynamics simulation studies of the coordination of metal ions in ILs [227–231], which offer some insight. Metal ions in IL solution are observed to coordinate strongly with anions [227, 228], and display a greater solubility in IL phases when complexed with an organic ligand [230]. A study of the coordination of Eu(III) in an IL reveals an $[EuF_6]^{3-}$ complex that is found to be thermodynamically unstable in isolation, and the authors characterize the solute–solvent interactions that stabilize the species in solution. A similar study of uranyl complexes [229] including complexation with water and anionic substituents offers insight on the process. The novelty of the compounds generated in all of these studies is not the structure of the complexes themselves, but the predominance of cations in the medium surrounding them. Until the structure of this counterion solvation shell is understood, it will be impossible to predict the structures of coordination complexes in ILs.

Given the association of solvent ions with charged solutes, one would anticipate that the solvation properties of ions should be very sensitive to the chemical structure of the IL. This fact has been exploited in the creation of "task specific" ILs, in which relatively small modifications are made to known ILs to alter their coordination properties with a given solute ion. Visser et al.

[34, 35] observed that the addition of electron-rich functional groups to a long ($n \geq 4$) aliphatic substituent of a cation can increase coordination to metal centers, dramatically increasing the partitioning of these species into the IL phase.

These empirical and simulation studies of ionic and molecular solvation in ILs are extremely valuable, as the insight they provide can guide researchers seeking to exploit the novel properties of these systems. However, further advancement of the field requires a deeper understanding of solute–solvent interactions in ILs, and that is the subject of the next section.

C. Solute–Solvent Interactions in Ionic Liquids

Typically, solute–solvent interactions are divided into two broad categories: Specific and nonspecific interactions. Specific interactions include phenomena such as hydrogen bonding and π–π interactions, which depend on the presence of particular functional groups or steric structures. They are short ranged, and are "specific" in the sense that they involve individual solvent species within the first solvation shell of the liquid. In contrast, nonspecific interactions represent interactions that are not associated with the presence of individual functional groups. In molecular liquids, these include dispersion and electrostatic interactions, such as dipole–dipole forces. We will discuss the nature of each type of interaction in ionic liquids in the sections that follow.

1. Specific Interactions in Ionic Liquids

In general, most experimental and theoretical studies of ionic liquids display specific interactions that can be explained in conventional ways, without the need to invoke novel concepts for ionic liquids. As indicated in Section IV.A, capacities for solute–solvent hydrogen bonding of a given liquid are usually consistent with expectations based on chemical structures. Simulations involving hydrogen bonding also generally conform to expectations [202, 232, 233], though competition between solute–ion and counterion–ion hydrogen bonding can affect interactions.

A potentially unique form of specific solute–solvent interaction has been proposed for ionic liquids. Blanchard and Brennecke [234] note that the solubilities of aromatic species are anomalously high in an imidazolium-based IL when compared to solutes of comparable molecular weight and dipole moment. This cannot be explained purely by π–π interactions, because while π–π interaction energies can be significant [235], the solubilization of a pure aromatic liquid must disrupt at least as many π–π contacts as it creates. However, work by Holbrey and co-workers [171] characterizes a cocrystalline "clathrate" form of an imidazolium-based IL with benzene, which shows distinctive π–π stacking.

Hanke et al. [236] use molecular dynamics simulation to study the solvation of aromatics in imidazolium-based ILs, and find that the quadrupole moment of an aromatic species leads to stronger solute–solvent interactions than are present in an isostructural species species of zero quadrupole moment. They attribute this effect to ion–quadrupole interactions, which manifest in π–π stacking of the aromatic cation and the neutral solute. While this is certainly related to π–π interactions in neutral systems, an analogy might also be drawn to cation–π interactions for biological systems [237], where it is known that a cationic species, such as a potassium ion, may coordinate strongly with an aromatic ring. If so, it suggests such interactions may exist in ILs based on nonaromatic cations, though the absence of π–π interactions and/or the presence of steric effects may limit their significance. We are not aware of any experimental studies addressing this possibility.

2. *Nonspecific Interactions*

The highly structured distribution of charge in an ionic liquid has significant implications for its interaction with a solute. The spontaneous ionization of metal atoms in high temperature fused salts noted in Section II.B suggests the charge-ordered nature of an ionic solvent can drive reactions leading to the separation of charges, though the more diffuse charge distribution of molecular ions may make this effect less dramatic in ILs. We must also consider the nature of nonspecific solute–solvent interactions for stable neutral solutes, and attempt to construct a framework analogous to the polarization of a molecular liquid noted above. These are the objectives for this section.

First, we note that the charge ordering of the solvent can impose itself on the distribution of products in reaction. Chiappe and Pieraccini [40] report that in their study of electron transfer between Micheler's ketone and tetracyanoethene, they observed that the formation of a radical ion pair to be preferred over formation of a single, neutral complex. Such a preference is only observed for the most highly polar molecular liquids, and is analogous to the spontaneous ionization of metal atoms in fused salts noted above. This represents a novel phenomenon for moderately polar solvents, though its generality is unclear at this time.

A more subtle, but closely related phenomenon, is the effect of the solvent on the formation of coordination complexes. As discussed in Section IV.B, coordination complexes in ILs often take on charged configurations even when neutral complexes are the norm in molecular liquids. Some of this preference for charged structures may reflect stabilization of ionic complexes by ion–ion interactions, analogous to the forces driving the separation of charges noted above. But in some cases, the formation of charged processes is driven by the thermodynamics of ion exchange with an aqueous layer. As noted in Section IV.C.I, a different coordination complex may form when a compound is extracted from an aqueous layer than when it is prepared by solvation of the pure solid. This is a novel nonspecific interaction in ILs, in that complex formation is thermodynamically driven at least in part by the free energy of solvation of a solvent ion in an aqueous layer. We are not aware of any analogous process in molecular liquids.

Now, we turn to the electrostatic interactions of a polar molecule with the solvent. As discussed in Section III, ILs show structural inhomogeneity in at least two ways: By the segregation of charged and neutral substituents in solution, and by the formation of nanoscale domains. Solute species presumably partition within these phases to minimize their free energies, so immediately the discussion of nonspecific interactions begs the question of whether we are truly discussing a single environment. With regard to segregation between charged and neutral substituent domains, simulation studies have begun to explore this issue [238]. Though preliminary, the evidence suggests that more polar molecular solutes do partition preferentially in contact with charged substituents, while nonpolar solutes associate with neutral substituents.

This association is likely quite complex. First, for relatively small alkyl chains ($n \leq 4$), the substituent domain must be relatively small. How does solute size affect its partitioning? And can it really be said to be partitioned if its size is on the same order of magnitude as the domain itself? Could the presence of the solute disrupt microphase separation, making the question moot?

With regard to nanoscale domains, a solute species could preferentially partition within particular environments. The relevance of this effect would depend on the solvation character of the domain, as well as its size and lifetime. We are not aware of any studies that could provide insight on this possibility or how it might affect the solvation process.

Even if structural inhomogeneity were not present, however, one would still be left with the question of how the electrostatic interaction of a neutral solute and a fused salt should be described. As noted in Section II.B, this is a relatively new problem, as neutral solutes have not been studied in high temperature fused salts. As discussed in Section IV.A, the use of a dielectric continuum approximation is unlikely to succeed unless detailed knowledge of its wavelength dependence is available on molecular length scales. Kobrak [239] has recently side stepped this latter issue by modeling the distribution of charge about the solute, and has proposed a theory explaining how a neutral solute should interact with an ionic solvent. While the theory neglects the possibility of structural inhomogeneity, it may serve as a useful starting point for understanding neutral solute–ionic solvent interactions.

The theory is based on two observations. First, solute–solvent interactions are characterized by dipole–ion interactions, and so are much weaker than the ion–ion interactions between solvent species. Thus, the presence of the solute dipole should not greatly perturb the liquid from the electrostatic structure of the neat liquid. Second, because the ionic liquid is a conductor, the electric field of the solute must be screened by the solvent. This observation has been confirmed by simulation of charged surfaces in fused salts [240] and ionic liquids [241], and through simulation of a polar solute in an ionic liquid [202]. Taken together, these facts imply that the solute must sample solvent configurations corresponding to configurations of the neat liquid that, on average, screen the charge distribution of the solute.

This suggests a framework for interpretation of the distribution of charge about the solute. The solute dipole may be replaced by two point charges of unit magnitude and opposite sign, placed a distance apart that reproduces the magnitude of the solute dipole. One may then make the ansatz that the distribution of charge density about each of these point charges has the form of the distribution of charge density about an ion in a concentrated electrolyte, as given in Eq. (3). The total charge density about the solute is then simply the sum of these two distributions. Kobrak has studied the charge distribution about a dipolar solute in a fused salt for a model system in simulation, and found that the observed charge density is well described by such a summation.

One implication of this framework is that the electrostatic component of solute–solvent interactions should correlate strongly with the charge density of the liquid. This result is confirmed by study of the variation of the Kamlet–Taft π^* parameter with the number density of an ionic liquid. The result is shown in Fig. 7 (taken from [239]). This figure shows a clear relationship between π^* and the number density of the IL, where no such relationship exists in molecular liquids. It

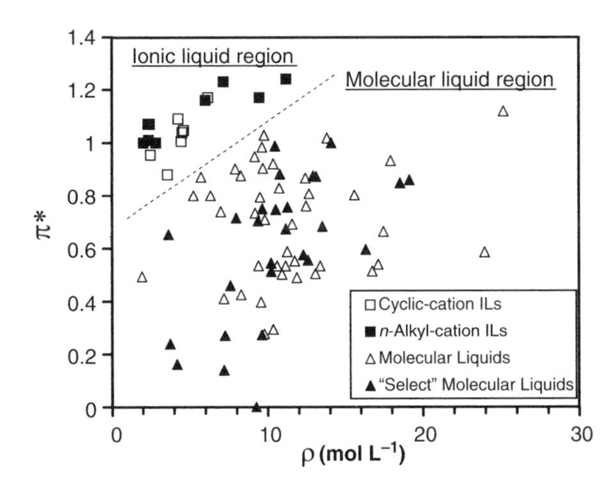

Figure 7. The Kamlet–Taft π^* parameter *vs.* solvent number density. "Select" molecular solvents possess little or no capacity for hydrogen bonding, and their interactions are controlled by electrostatics. Ionic liquids are categorized based on whether they possess a cyclic cation or an alkylammonium cation. A wide range of anions are employed in both categories, and no attempt is made to restrict specific interactions. Taken from Ref. [239]

is noteworthy that all of the ionic liquids, including both cyclic and acyclic cations in combination with an enormous range of anions, fall on the same trendline. This implies that the chemical structure of the ion is of secondary importance to the distribution of the density of charge in the liquid. Thus, other than total charge and ionic volume, there is no structural feature of an ion that significantly affects its electrostatic interactions with the solute. This is an entirely different situation from molecular liquids, where the dipole moment of a solvent molecule is known to be the greatest determinant of electrostatic interactions [242].

A problem with this interpretation relates to electrostriction, a process in which the density of the solvent changes about a solute. Shim et al. [243] noted evidence of electrostriction in molecular dynamics simulations of a model chromophore in an IL, and the degree of electrostriction was sensitive to the charge distribution of the solute. This observation does not necessarily contradict the framework above, as some local disruption of solvent structure due to dispersive interactions is inevitable. However, it is desirable to obtain a clearer understanding of the competition between these local interactions and the need to maintain a uniform charge distribution in the liquid.

The subtlety of these interactions, and the multiple forms of inhomogeneity present in ILs, dictate that an understanding of nonspecific solute–solvent interactions in ILs will be exceptionally complex. This is an area worthy of further attention.

V. REACTION DYNAMICS IN IONIC LIQUIDS

Thus far, we have focused primarily on understanding the properties of the neat liquid, and the nature of solute–solvent energetic interactions. We must now turn our attention to the question of how these factors influence dynamic processes.

A solvent can influence reaction kinetics in a number of ways. Most obviously, liquid viscosity affects the diffusion of reactants, controlling the rate of formation of a reactive complex. Another mechanism is the association of solvent species with reactive solutes, which can block the formation of a reactive complex or prevent its dissociation on completion of the reaction. The extent of this effect will depend largely on the nature of the solute–solvent interactions that stabilize reactive configurations. Nonspecific interactions can also influence reaction dynamics. The best known example of this is Marcus theory for charge transfer [244, 245], in which the kinetics of the process are dominated by fluctuations of the solvent configuration that bring reactant and product electronic states into degeneracy. However, this phenomenon can influence kinetics in any process where reactant and intermediate states possess significantly different charge distributions that the solvent must accommodate. We will review what is known about each of these phenomena in ILs.

A. Diffusion

Bimolecular reactions are frequently described as "diffusion-limited", meaning that the limiting step in reaction is the rate of formation of a reactive complex. This situation often applies in ILs, as they are highly viscous media. A number of studies [246, 247] have found activation energies for reactions in ILs that are equivalent to the activation energies for viscous flow in the liquid, a feature of diffusion-limited kinetics.

Yet the relationship between solute chemical structure and diffusion is not always simple. Werner et al. [248] conducted fluorescence correlation spectroscopic studies of three fluorescent probes in 1-butyl-3-methylimidazolium hexafluorophosphate. The probes were chosen to be of comparable molecular structure, but possessed positive, negative, and neutral charges. The authors found that while the neutral probe diffused more rapidly than the cationic probe, the anionic probe diffused the most quickly.

This stands in opposition to conventional expectations, where the presence of the charge should lead to association with solvent species (either molecules or ions), increasing the effective hydrodynamic radius of the solute and decreasing its diffusion constant. This phenomenon is known to be important in high temperature fused salts, especially for highly polarizable ions [249], and there is no clear explanation as to why it should not apply in ILs. The authors note that the probe, fluorescein, possesses a highly diffuse distribution of charge, which may weaken cation–solute interactions, perhaps preventing the stable coordination of cations and lowering the activation energy for transport. One could easily make the opposite interpretation, that the anionic solute coordinates strongly with the cation, and diffuses through the liquid as a neutral body. However, this should increase the species' hydrodynamic radius, slowing its diffusion relative to the neutral dye. Thus, neither explanation addresses why the anionic solute should diffuse significantly faster than the neutral solute.

Electron spin resonance (ESR) studies of radical probe species also suggest complexity. Evans et al. [250] study the temperature dependence of IL viscosity and the diffusion of probe molecules in a series of dissimilar IL solvents. The results indicate that, at least over the temperature range studied, the activation energy for viscous flow of the liquid correlates well with the activation energies for both translational and rotational diffusion, indicative of Stoke–Einstein and Debye–Stokes–Einstein diffusion, respectively. Where exceptions to these trends are noted, they appear to be associated with structural inhomogeneity in the solvent. However, Strehmel and co-workers [251] take a different approach, and use ESR to study the behavior of spin probes in a homologous series of ILs. In these studies, comparisons of viscosity and probe dynamics across different (but structurally similar) ILs do not lead to a Stokes–Einstein correlation between viscosity and solute diffusion. Since the capacities for specific interactions are held relatively constant between these ILs, this suggests that differences in solvent ion volume may be important determinants of probe dynamics. Thus, while diffusive dynamics appear to correlate significantly with macroscopic diffusion in ILs, their microscopic ordering can lead to novel diffusive behaviors.

Solute size may also contribute to unconventional diffusive behaviors. A number of studies [252–254] noted that ILs typically possess high gas solubilities, at least for gaseous species capable of significant dispersive interactions, and that the solubilization of a gas does not lead to an appreciable change in volume [255]. This has been interpreted as evidence that gases partition into preexisting cavities in the neat liquid. The proposal is supported by simulation results [256] and is consistent with the idea that fused salts possess a high population of voids and defects as noted in Section II.A. Recent experiments by Morgan et al. [257] note that while gaseous diffusion constants in ILs are lower than those for molecular liquids, they depend relatively weakly on IL viscosity. The authors also observe that the diffusion constants depend more strongly on solute size, and taken together, these observations strongly support the theory given above. Thus, the diffusion of small, neutral solute species in ILs may be qualitatively different from that of larger solutes.

Small, charged solutes, however, do not possess the same mobility. Voltammetric studies by Buzzeo et al. [258] indicate that the rate of diffusion of molecular oxygen in an IL is 30 times faster than that of the superoxide radical $O_2^{-\cdot}$. Similarly, pulse radiolysis experiments [259] indicate that the diffusion of a neutral hydrogen atom in an IL is roughly an order of magnitude faster than that of an unbound electron. These charged species are clearly slowed by their interaction with the charge distribution of the medium.

An interesting series of molecular dynamics simulations by Shim et al. [260] characterizes inhomogeneous dynamics in the rotation of a model solute in an IL. For small, nonpolar solutes, the authors observe a rapid initial decay in the solute orientational correlation function, followed by a slower, inhomogeneous response. As the solute size increases, however, the dynamics become better fit by a single exponential, suggestive of Stokes–Einstein–Debye behavior. A similar trend toward dynamic homogeneity is observed with increasing solute dipole moment. This dynamical inhomogeneity, and its possible relationship to solute size and liquid structural inhomogeneity, will require further study.

The rate of diffusion in ILs appears to vary inversely with macroscopic viscosity, as expected, but there do appear to be significant effects arising from the chemical structure of solute and solvent species.

B. Desolvation Effects

In reactions involving the association of reactants and/or the dissociation of products, solute–solvent interactions can affect kinetics. Solvent species must sometimes dissociate from solute species to permit formation of a complex, or associate with product species to stabilize them. Both of these processes involve transitions between "bound" and "free" solvent states, and understanding this process is important to kinetics.

The most obvious phenomenon in which solvent coordination can affect kinetics is in catalytic processes, where the substrate must compete with the solvent for access to catalytic sites. We have already discussed this issue to some degree in Section IV.A. Here, we simply note that ILs composed of relatively noncoordinating anions (e.g., tetrafluorborate, hexafluorophosphate, or trifluormethylsulfonate species) tend to have good physical properties, such as low viscosities and stability against reactions with water. Consequently, anionic coordination in commonly used ILs is often weak enough that it does not interfere with processes, such as transition metal catalysis in ILs [1], and so the question of desolvation in an IL has not received a great deal of attention. A flash photolysis study [261] of the interaction of 1-butyl-3-methylimidazolium hexafluorophosphate with a neutral chromium complex showed not only that the binding of the hexafluorophosphate ion is extremely weak, its rate of exchange with the medium is two orders of magnitude faster than that of dichloroethane, a comparably polar molecular liquid. Thus, ILs' reputation as noncoordinating solvents appears to be well deserved.

The reorganization of the solute about reacting species is not limited to catalytic complexes, however. Studies of nucleophilicity of reactant species in ILs [262, 263] found that in some cases the nucleophilicity of solute species depends strongly on the character of the cation, though in others the association appears to be relatively weak [264]. In some cases, desolvation of solvent ions is limited by unfavorable entropic effects rather than enthalpic ones [262]. This likely relates to the highly structured nature of the solvent, which can run counter to intuitive relationships between the entropy of "free" and "bound" states. Harper and Kobrak [14] reviewed this literature in detail, and we will not discuss it further here.

It is also possible that the solvent may form structures about a bimolecular complex, inhibiting dissociation and thereby affecting reaction kinetics. Gordon and McLean [265] observed in their studies of bimolecular electron transfer that the IL did indeed appear to slow the rate of dissociation and increase the lifetime of the complex. They note the possibility that the positively charged reactive complex may have been stabilized by interaction with solvent anions. The generality of this phenomenon is not known at this time.

C. Solvation Dynamics

As noted above, the polarization of a solvent can profoundly affect a chemical reaction, and the rate at which polarization takes place is a critical parameter in understanding the kinetics of many different reactions. Dynamic studies of this polarization process are generally built around time-resolved photophysical experiments, such as observation of the time-resolved fluorescence spectrum of probe molecules in the solvent of interest. ILs have received considerable attention in this regard; for recent reviews, see Samanta [266] or Mandal et al. [267]. To our knowledge, the earliest studies of this type in ILs were due to Huppert and co-workers [268, 269]. A large number of studies have been published in the area more recently [105, 107–112, 270–274] on a wide range of chromophores and ionic liquids. The observed dynamics are quite complex, and there is no consensus as to how they should be interpreted.

However, certain general trends are clear. First, ILs composed of small, rigid ions tend to show significant subpicosecond dynamics [107], similar in time scale (but not necessarily microscopic character) to the ballistic response observed in molecular liquids. Depending on the system, this rapid response can account for 20–50% of the observed dynamic Stokes shift. In most cases, this value is inferred from a "time zero" shift [106, 107] as the dynamics are faster than the instrumental response. A recent study by Lang et al. [272] suggests that the magnitude of this initial response may be exaggerated by systematic error, at least where coumarin-153 is used as a probe [272]. However, Arzhantsev and co-workers [108] recently made a direct observation of the subpicosecond response of 4-dimethylamino-4'-cyanostilbene (DCS) in a series of ILs, and the extent of the initial dynamics is observed to fall within the expected range of magnitudes. Thus, whether the details of the coumarin studies are questionable or not, the subpicosecond response is a significant fraction of the total fluorescence response in some ILs.

Intriguingly, this subpicosecond response is absent in studies of ILs based on large, highly flexible ions, such as tetraalkylammonium and tetraalkylphosphonium species [107, 268, 269]. Given our earlier discussion relating viscosity to charge arm, it is tempting to infer that the rapid response of rigid ions stems from rotational motion, but theoretical analysis indicates that the response is primarily translational [275, 276] and involves both cation and anion motion even where the anion possesses zero charge arm [275–277]. The presence of ions of significant charge arm may act to facilitate these collective motions, but rotational motion does not make a large, direct contribution to the subpicosecond relaxation process. Analysis of the solvation dynamics using the Steele method [278, 279] indicates that the subpicosecond response of ILs is highly collective [276]. The dynamics thus appear to be dominated by high frequency, dipole-active interionic vibrational modes [276]. The rapidity of the response is due to the presence of thermal energy in these collective modes at the time of excitation of the chromophore, and the subpicosecond dynamics represent inertial motion in these modes rather than a response to a change in interaction with the chromophore. While the subpicosecond response of molecular solvents is also dominated by such inertial dynamics, the motion is best described as the uncorrelated rotation of individual molecules [278]. The predominance of collective translational motion in the subpicosecond response is unique to ILs, and reflects the importance of interion Coulomb interactions.

This characterization of the response is, however, not universally accepted. Petrich and co-workers [105, 274, 280] noted the strong dependence of the subpicosecond response on cationic structure, and have studied the ultrafast photophysical response of ILs in comparison with the response of molecular liquids that are structurally analogous to the cation. The results show a high degree of similarity in the short-time response, which the authors interpret as evidence that the response is dominated by the cation. This would imply a significant contribution from the rotational response of the ions, as is known to be responsible for the subpicosecond response of molecular liquids [278]. In light of the charge arm discussion given in Section III.B and the identification of collective translational motion in simulation referenced above, we believe the similarity in the form of the responses of molecular and ionic liquids is not reflective of substantial mechanistic similarity between the media. However, we note the debate, and encourage the interested reader to review the studies referenced above.

Another consistent observation is the wide range of time scales observed. Relaxation responses typically involve motion on time scales ranging from picosecond or subpicosecond to nanosecond time scales. Indeed, the time scales are so long that the fluorescence lifetimes of some commonly used dyes are not adequate to sample the dynamics. To our knowledge, the only study that addresses the full range of time scales for relaxation dynamics is the study of DCS due to Arzhantsev et al. noted above [108], though other studies span picosecond to nanosecond scales and so observe most of the dynamics [270].

While the observed temporal responses are qualitatively similar between the studies, different authors disagree as to the best functional form for their description. Some groups claim a biexponential function yields a good fit [111, 112, 270], while others argue in favor of a stretched exponential [107, 109, 110]. In their study of the subpicosecond to nanosecond response of DCS, Arzhantsev and co-workers [108] found the response best described by the sum of a single exponential and a stretched exponential, and the fitted parameters suggest that it is the short-time response described by the single exponential. Thus, the debate centers on the form of the slow dynamics of the system.

The difference is more than simple mathematics. At issue is whether the time scales can be clearly resolved into two distinct regimes, or whether the dynamics vary continuously across all time scales involved. In the former case, one could hope to characterize the motions responsible for each time scale and so gain a deeper understanding of the dynamics, while in the latter, the idea of characterizing distinct responses must be abandoned in favor of a search for a single theory governing the full dynamics. The issue is complicated by the fact that a number of different systems have been studied by the different groups involved, and the underlying dynamics may be genuinely different between the systems.

These time scales are not readily accessible by simulation, though the spatial and dynamical heterogeneity noted for the neat liquid in Section II.A may well be at the heart of the issue. Kobrak [276] showed that the subpicosecond response in ILs is dominated by the motions of nearby ions, and longer ranged solute–solvent interactions can affect the fluorescence signal on longer time scales. One may infer from this a separation of length scales, with long-ranged interactions evolving more slowly than short-ranged ones. If this is the case, the relaxation process is likely heterogeneous itself (the most common interpretation of the stretched exponential), and will not be elucidated until the nature of the inhomogeneities in ILs is clearer.

An innovative approach due to Halder et al. [113] may help to sidestep the challenges involved in explicit molecular dynamics simulation and obtain information on these slow dynamics. The authors use the results of dielectric reflectance spectroscopy to model the IL as a dielectric continuum, and study the solvation response of the IL in this framework. The calculated response is not a good description of the subpicosecond dynamics, a problem the authors ascribe to limited data on the high frequency dielectric response, but may be qualitatively correct at longer times. We have already expressed concern regarding the use of the dielectric continuum model for ILs in Section IV.A, but believe that if the wavelength dependence of the dielectric constant can be adequately modeled, this approach may be the most productive theoretical analysis of these slow dynamics.

Another class of photophysical experiment involves the time-resolved absorption spectroscopy of the excess electron generated in pulse radiolysis. Electrons generated in these experiments typically possess a high kinetic energy, and interactions with the solvent eventually bring them to thermal equilibrium. These preequilibrium and equilibrium states are referred to as "presolvated" and "solvated", respectively. An excess electron is an excellent probe, as the solvation process involves the localization of the electron from an initially delocalized state, and localization leads to a significant blue shift in the spectrum. Wishart and Neta [48] used the picosecond time-resolved absorption spectroscopy and have inferred that the presolvated electron has a lifetime of roughly 4 ns in methyltributylammonium bis(trifluoromethylsulfonyl)imide. While electrons in low temperature solids can remain in nonequilibrium states apparently indefinitely [281–283], no other known liquid system requires such a long equilibration time for the presolvated electron; time scales in molecular liquids are more typically on the order of tens of picoseconds or less [284]. The nanosecond time scale for solvation is comparable to the slow solvation response for the fluorescence studies noted above, despite the fact that those studies employed a neutral probe and the electron is charged. Certain features of the solvent response therefore appear to be very general.

As noted in Section II.C, the dynamics of high temperature fused salts are dominated by collective motion, and this is likely the case for the slow dynamics of ILs as well. The presence of a charged or neutral solute adds an additional layer of complexity to this motion, however, as a given solute will interact only with those modes that are susceptible to certain interactions (e.g., the solvation dynamics of a neutral dipolar molecule will be determined by dipole-active collective modes). The issue is further complicated by the relative weakness of neutral solute–solvent interactions, since as noted in Section IV.C these are not sufficiently strong to drive solvent motion. Instead, one must view the relaxation process as a weakly biased random walk, in which thermal motion of the solvent gradually brings the system to equilibrium. Additional layers of complication are posed by the structural and dynamic inhomogeneity of ILs, as different solutes may well sample different chemical environments in the same liquid. Until these issues are better understood, the debate surrounding these photophysical experiments cannot be resolved.

VI. CONCLUSIONS

Ionic liquids are an intriguing class of solvents. Their macroscopic solvation properties make it possible to recreate many conventional processes in these novel materials, but their convenience in this regard should not lull researchers into complacency. As salts, they represent fundamentally different media than molecular liquids and conventional chemical logic must be modified to account for these differences in their underlying physics.

The goal of this chapter has been to compile the results of theories and experiments relevant to this understanding, and to create a conceptual framework for further analysis of the area. While the problem is extremely subtle, we summarize here the most significant principles necessary to understand solvation phenomena in ionic liquids.

- Coulomb effects dominate liquid structure. The local structure of ions in the neat liquid is controlled primarily by Coulomb interactions. In solution, ion–ion interactions are far larger than ion–dipole interactions, and the presence of a neutral solute represents a local disruption of the neat liquid structure but is otherwise a relatively weak perturbation on the charge distribution of the neat liquid.
- Motion in ionic liquids is predominantly collective. The strength of interion Coulomb interactions makes independent motion of ions impossible, and transport modes are collective. Further, solvation dynamics involve

interaction with collective, dipole-active interionic vibrational modes of the solvent. When seeking to understand these phenomena, researchers must consider how the properties of an individual ion affect these collective modes rather than considering the interaction of an individual ion with a solute.

- Structural and dynamic inhomogeneity are fundamental to understanding ionic liquids. Multiple forms of structural inhomogeneity exist in ionic liquids. The simplest is the segregation of ionic liquids into regions of charged and neutral substituent groups, which has implications for solute–solvent interactions. Additionally, ions form aggregates on larger scales. This aggregation is a general feature of ILs and creates structural and dynamic inhomogeneity that affect transport and solvation phenomena in complex ways. While this inhomogeneity must not become a "bogey man" used to explain every observed anomaly, researchers must keep these phenomena in mind when interpreting experiments.
- Ions must be described by the center of charge. The dipole moment of an ion is an artifact of the choice of coordinate, and possesses no physical significance. The correct, lowest order electrostatic description of an ion is as a point charge localized at the point that best approximates the total distribution of charge. Polarization of the ion represents motion of the center of charge. An important implication of this observation is that, at this level of description, all ions of a given charge possess identical electrostatic properties. This explains why ILs of radically different structure follow the same trend in π^* versus ρ shown in Fig. 7; future work may uncover other structure-independent behaviors.
- Unique chemical interactions are possible in ionic liquids. Many observations in ionic liquids can be explained using conventional chemical phenomena, such as hydrogen bonding. However, novel charge–quadrupole interactions have been characterized in ionic liquids, and the coordination chemistry of solutes in ILs can be influenced by ion exchange with a neighboring phase. The latter creates a link between coordination chemistry and the thermodynamics of solvation for an ion in an alternative medium that has no direct analogue in molecular liquids. Researchers must be alert to these and other as yet undiscovered interactions in ILs.

Given the current state of knowledge in the field, we make no claim that the above list is either comprehensive or immutable. We have throughout this work pointed out areas in which existing knowledge is incomplete, and as these gaps are filled the list of principles must be extended and rewritten. Nevertheless, we hope that the principles presented here will be of use to researchers both as a reference with which to interpret their results and as a springboard to further discussion of the chemical physics underlying solvation in ionic liquids.

Acknowledgments

We gratefully acknowledge Prof. Austen Angell, Prof. Edward Castner, Dr. Jason Harper, Prof. Mark Maroncelli, Prof. Edward Quitevis, and Dr. James Wishart for helpful discussions of ionic liquids.

References

1. T. Welton, *Chem. Rev.* **99**, 2071–2083 (1999).
2. K. R. Seddon, *J. Chem. Tech. Biotechnol.*, **68**, 351–356 (1997).
3. P. Wasserscheid and T. Welton, *Ionic Liquids in Synthesis*, Wiley–VCH, Mörlenbach, Germany, 2002.
4. J. Dupont, R. F. de Souza, and P. A. Z. Suarez, *Chem. Rev.* **102**, 3667–3692 (2002).
5. A. E. Visser, R. P. Swatloski, and R. D. Rogers, *Green Chem.* **2**, 1–5 (2000).
6. D. W. Armstrong, L. He, and Y.-S. Liu, *Anal. Chem.* **71**, 3873–3876 (1999).
7. A. E. Visser and R. D. Rogers, *J. Solid State Chem.* **171**, 109–113 (2003).
8. M. J. Muldoon, C. M. Gordon, and I. R. Dunkin, *Perkin Trans.* **2**, 433–435 (2001).
9. S. N. V. K. Aki, J. F. Brennecke, and A. Samanta, *Chem. Commun.* 413–414 (2001).
10. A. J. Carmichael and K. R. Seddon, *J. Phys. Org. Chem.* **13**, 591–595 (2000).
11. K. Jin, X. Huang, L. Pang, J. Li, A. Appel, and S. Wherland, *Chem. Commun.* 2872–2873 (2002).
12. H. Zhao and S. V. Malhotra, *Aldrichim. Acta* **35**, 75–83 (2002).
13. M. J. Earle, P. B. McCormac, and K. R. Seddon, *Chem. Commun.* 2245–2246 (1998).
14. J. B. Harper and M. N. Kobrak, *Mini-Rev. Org. Chem.* **3**, 253–269 (2006).
15. T. Welton, *Coord. Chem. Rev.* **248**, 2459–2477 (2004).
16. N. E. Leadbeater, H. M. Torenius, and H. Tye, *Comb. Chem. High Throughput Screen* **7**, 511–528 (2004).
17. M. J. Earle and K. R. Seddon, *Proc. Electrochem. Soc.* **2002-19**, 177–189 (2002).
18. R. P. Swatloski, S. K. Spear, J. D. Holbrey, and R. D. Rogers, *J. Am. Chem. Soc.* **124**, 4974–4975 (2002).
19. A. G. Fadeev and M. M. Meagher, *Chem. Comm.* 295–296 (2001).
20. L. A. Blanchard, D. Hancu, E. J. Beckman, and J. F. Brennecke, *Nature* (*London*) **399**, 28–29 (1999).
21. S. Dai, Y. H. Ju, and C. E. Barnes, *Dalton Trans.* 1201–1202 (1999).
22. J. L. Anderson, D. W. Armstrong, and G. Tzo, *Anal. Chem.* **78**, 2892–2902 (2006).
23. M. L. Dietz, *Sep. Sci. Tech.* **41**, 2047–2063 (2006).
24. W. Lu, A. G. Fadeev, B. Qi, E. Smela, B. R. Mattes, J. Ding, G. M. Spinks, J. Mazurkiewicz, D. Zhou, G. G. Wallace, D. R. MacFarlane, S. A. Forsyth, and M. Forsyth, *Science* **297**, 983–987 (2002).
25. Y. Ito and T. Nohira, *Electrochim. Acta* **45**, 2611–2622 (2000).
26. J. S. Wilkes, J. A. Levisky, R. A. Wilson, and C. L. Hussey, *Inorg. Chem.* **21**, 1263–1264 (1982).
27. H. L. Chum, V. R. Koch, L. L. Miller, and R. A. Osteryoung, *J. Am. Chem. Soc.* **97**, 3264–3265 (1975).
28. M. Galinski, A. Lewandowski, and I. Stepniak, *Electrochim. Acta* **51**, 5567–5580 (2006).

29. A. P. Abbott and K. J. McKenzie, *Phys. Chem. Chem. Phys.* **8**, 4265–4279 (2006).
30. M. C. Buzzeo, R. G. Evans, and R. G. Compton, *Chem. Phys. Chem.* **5**, 1107–1120 (2004).
31. M. Freemantle, *Chem. Eng. News* **March 31**, 9 (2003).
32. P. L. Short, *Chem. Eng. News* **April 24**, 15–21 (2006).
33. J. D. Holbrey and K. R. Seddon, *Clean Prod. Proc.* **1**, 223–236 (1999).
34. A. E. Visser, R. P. Swatloski, W. M. Reichert, R. Mayton, S. Sheff, A. J. Wierzbicki, J. H. Davis, and R. D. Rogers, *Environ. Sci. Technol.* **36**, 2523–2529 (2002).
35. A. E. Visser, R. P. Swatloski, W. M. Reichert, R. Mayton, S. Sheff, A. Wierzbicki, J. H. Davis, and R. D. Rogers, *Chem. Commun.* 135–136 (2001).
36. T. A. Utigard, R. R. Roy, and K. Friesen, *Canad. Metal. Quart.* **40**, 327–334 (2001).
37. E. C. Gay, W. E. Miller, and J. J. Laidler, *Proc. Electrochem. Soc.* **94**, 721–727 (1994).
38. P. Wasserscheid and W. Keim, *Angew. Chem. Int. Ed. Engl.* **39**, 3772–3789 (2000).
39. V. A. Cocalia, K. E. Gutowski, and R. D. Rogers, *Coord. Chem. Rev.* **250**, 755–764 (2005).
40. C. Chiappe and D. Pieraccini, *J. Phys. Chem. A* **110**, 4937–4941 (2006).
41. A. P. Abbott, G. Capper, D. L. Davies, R. K. Rasheed, and V. Tambyrajah, *Chem. Commun.* 70–71 (2003).
42. H. Olivier-Bourbigou and L. Magna, *J. Mol. Catal. A: Chemical* **182–183**, 419–437 (2002).
43. Y. Chauvin and H. Olivier-Bourbigou, *Chemtech* **25**, 26–30 (1995).
44. U. P. Krecher, A. E. Rosamilia, C. L. Raston, J. L. Scott, and C. R. Strauss, *Molecules* **9**, 387–393 (2004).
45. M. Yoshizawa, W. Xu, and C. A. Angell, *J. Am. Chem. Soc.* **125**, 15411–15419 (2003).
46. R. Engel, J. I. Cohen, and S. I. Lall, *Phosphorus, Sulfur Silicon* **177**, 1441–1445 (2002).
47. R. Engel and J. I. Cohen, *Curr. Org. Chem.* **6**, 1453–1467 (2002).
48. J. F. Wishart and P. Neta, *J. Phys. Chem. B* **107**, 7261–7267 (2003).
49. O. O. Okoturo and T. J. VanderNoot, *J. Electroanal. Chem.* **568**, 167–181 (2004).
50. H. D. B. Jenkins, H. K. Roobottom, J. Passmore, and L. Glasser, *Inorg. Chem.* **38**, 3609–3620 (1999).
51. V. Kumar, G. M. Prasad, A. R. Chetal, and D. Chandra, *Cryst. Res. Tech.* **31**, K16–K19 (1996).
52. *http:ilthermo.boulder.nist.gov.*
53. J. D. Holbrey and K. R. Seddon, *J. Chem. Soc. Dalton* **8**, 2133–2139 (1999).
54. J. M. Crosthwaite, M. J. Muldoon, J. K. Dixon, J. L. Anderson, and J. F. Brennecke, *J. Chem. Thermo.* **37**, 559–568 (2005).
55. H. L. Ngo, K. LeCompte, L. Hargens, and A. B. McEwen, *Thermochim. Acta* **357–358**, 97–102 (2000).
56. M. J. Earle, J. M. S. S. Esperança, M. A. Gilea, J. N. Canongia-Lopes, L. P. N. Rebelo, J. W. Magee, K. R. Seddon, and J. A. Widegren, *Nature (London)* **439**, 831–834 (2006).
57. L. P. N. Rebelo, J. N. Canongia-Lopes, J. M. S. S. Esperança, and E. Filipe, *J. Phys. Chem. B* **109**, 6040–6043 (2005).
58. T. Schäfer, C. M. Rodrigues, C. A. M. Afonso, and J. G. Crespo, *Chem. Commun.* 1622–1623 (2001).
59. T. J. Wooster, K. M. Johanson, K. J. Fraser, D. R. MacFarlane, and J. L. Scott, *Green Chem.* **8**, 691–696 (2006).
60. D. M. Fox, J. W. Gilman, H. C. DeLong, and P. C. Trulove, *J. Chem. Thermo.* **37**, 900–905 (2005).

61. R. A. Mantz and P. C. Trulove, in P. Wasserscheid, and T. Welton (eds.) *Ionic Liquids in Synthesis*; Wiley–VCH, Weinheim, Germany, 2002.
62. A. Paul, P. K. Prasun, and A. Samanta, *J. Phys. Chem. B* **109**, 9148–9153 (2005).
63. K. D. Collins and M. W. Washabaugh, *Q. Rev. Biophys.* **1985**, 323–422 (1985).
64. J. G. Huddleston, A. E. Visser, W. M. Reichert, H. D. Willauer, G. A. Broker, and R. D. Rogers, *Green Chem.* **3**, 156–164 (2001).
65. P. J. Dyson, D. J. Ellis, and T. Welton, *Can. J. Chem.* **79**, 705–708 (2001).
66. S. N. V. K. Aki, A. M. Scurto, and J. F. Brennecke, *Ind. Eng. Chem. Res.* **45**, 5574–5585 (2006).
67. H. Tokuda, S.-J. Baek, and M. Watanabe, *Electrochemistry* **73**, 620–622 (2005).
68. K. R. Seddon, A. Stark, and M.-J. Torres, *Pure Appl. Chem.* **72**, 2275–2287 (2000).
69. S. N. Baker, G. A. Baker, and F. V. Bright, *Green Chem.* **4**, 165–169 (2002).
70. K. A. Fletcher and S. Pandey, *Appl. Spectrosc.* **56**, 266–271 (2002).
71. H. Bloom and J. O. Bockris, in B. R. Sundheim, (ed.) *Fused Salts* McGraw-Hill, New York, 1964.
72. M. P. Tosi, D. L. Price and M.-L. Saboungi, *Ann. Rev. Phys. Chem.* **44**, 173–211 (1993).
73. M. Rovere and M. P. Tosi, *Rep. Prog. Phys.* **49**, 1001–1081 (1986).
74. K. S. Pitzer, *J. Phys. Chem.* **88**, 2689–2697 (1984).
75. J. Zarzycki, *Faraday Soc. Discuss* **32**, 38–48 (1961).
76. A. K. Adya, *J. Ind. Chem. Soc.* **82**, 1197–1225 (2005).
77. H. A. Levy and M. D. Danford, Diffraction Studies of Molten Salts. in M. Blander, (ed.), *Molten Salt Chemistry* Interscience Publishers, New York, 1964.
78. G. E. Blomgren, *J. Phys. Chem.* **66**, 1500–1508 (1962).
79. M. Blander, in M. Blander, (ed.), *Molten Salt Chemistry*, Interscience Publishers, New York, 1964.
80. F. H. Stillinger, J. G. Kirkwood, and P. J. Wojtowicz, *J. Chem. Phys.* **32**, 1837–1845 (1960).
81. J. P. Hansen and I. R. McDonald, *Theory of Simple Liquids*, Academic Press, New York, 1976.
82. P. Keblinski, J. Eggebrecht, D. Wolf, and S. R. Phillpot, *J. Chem. Phys.* **113**, 282–291 (2000).
83. W. B. Fowler, *The Physics of Color Centers*, Academic Press, New York, 1968.
84. W. Freyland, K. Garbade, and E. Pfeiffer, *Phys. Rev. Lett.* **51**, 1304–1306 (1983).
85. W. W. Warren, S. Sotier, and G. F. Brennert, *Phys. Rev. B* **30**, 65–77 (1984).
86. D. Adams and G. Hills, in D. Inman and D. G. Lovering, (eds.) *Ionic Liquids*, Plenum Press, New York, 1981.
87. A. Baranyai, I. Ruff, and R. L. McGreevy, *J. Phys. C: Solid State* **19**, 453–465 (1986).
88. G. Jacucci, I. R. McDonald, and A. Rahman, *Phys. Rev. A* **13**, 1581–1592 (1976).
89. K. F. O'Sullivan and P. A. Madden, *J. Phys. Condens. Matter* **3**, 8751–8756 (1991).
90. P. A. Madden, M. Wilson, and F. Hutchinson, *J. Chem. Phys.* **120**, 6609–6620 (2004).
91. M. Wilson and P. A. Madden, *J. Phys. Condens. Matter* **5**, 2687–2706 (1993).
92. A. Aguado and P. A. Madden, *J. Chem. Phys.* **119**, 7471–7483 (2003).
93. M. Wilson, P. A. Madden, N. C. Pyper, and J. H. Harding, *J. Chem. Phys.* **104**, 8068–8081 (1996).
94. A. J. Rowley, P. Jemmer, M. Wilson, and P. A. Madden, *J. Chem. Phys.* **108**, 10209–10219 (1998).
95. P. A. Madden and M. Wilson, *J. Phys. Condens. Matter* **12**, A95–A108 (2000).

96. F. Hutchinson, M. K. Walters, A. J. Rowley, and P. A. Madden, *J. Chem. Phys.* **110**, 5821–5830 (1999).

97. O. J. Kleppa, in R. C. Newton, A. Navrotsky, B. J. Wood, (eds.) *Thermodynamics of Minerals and Melts*, Springer-Verlag, New York, 1981.

98. G. P. Smith, in M. Blander, (ed.) *Molten Salt Chemistry*, Interscience Publishers, New York, 1964.

99. I. B. Polovov, V. A. Volkovich, S. A. Shipulin, S. V. Maslov, A. Khokhryakov, B. D. Vasin, T. R. Griffiths, and R. C. Thied, *J. Mol. Liq.* **103–104**, 387–394 (2003).

100. P. Hazra, D. Chakrabarty, and N. Sarkar, *Langmuir* **18**, 7872–7879 (2002).

101. L. R. Martins and M. S. Skaf, *Chem. Phys. Lett.* **370**, 683–689 (2003).

102. B. M. Luther, J. R. Kimmel, and N. E. Levinger, *J. Chem. Phys.* **116**, 3370–3377 (2002).

103. N. Kometani, S. Arzhantsev, and M. Maroncelli, *J. Phys. Chem. A* **110**, 3405–3413 (2006).

104. S. J. Rosenthal, X. Xie, M. Du, and G. R. Fleming, *J. Chem. Phys.* **95**, 4715–4718 (1991).

105. P. K. Chowdhury, M. Halder, L. Sanders, T. Calhoun, J. L. Anderson, D. W. Armstrong, X. Song, and J. W. Petrich, *J. Phys. Chem. B* **108** (2004).

106. N. Ito, S. Arzhantsev, and M. Maroncelli, *Chem. Phys. Lett.* **396**, 83–91 (2004).

107. S. Arzhantsev, N. Ito, M. Heitz, and M. Maroncelli, *Chem. Phys. Lett.* **381**, 278–286 (2003).

108. S. Arzhantsev, H. Jin, N. Ito, and M. Maroncelli, *Chem. Phys. Lett.* **417**, 524–529 (2006).

109. J. A. Ingram, R. S. Moog, N. Ito, R. Biswas, and M. Maroncelli, *J. Phys. Chem. B* **107**, 5926–5932 (2003).

110. N. Ito, S. Arzhantsev, M. Heitz, and M. Maroncelli, *J. Phys. Chem. B* **108**, 5771–5777 (2004).

111. R. Karmakar and A. Samanta, *J. Phys. Chem. A* **106**, 4447–4452 (2002).

112. R. Karmakar and A. Samanta, *J. Phys. Chem. A* **107**, 7340–7346 (2003).

113. M. Halder, L. S. Headley, P. Mukherjee, X. Song, and J. W. Pet rich, *J. Phys. Chem. A* **110**, 8623–8626 (2006).

114. C. G. Hanke, S. L. Price, and R. M. Lynden-Bell, *Mol. Phys.* **99**, 801–809 (2001).

115. J. K. Shah, J. F. Brennecke, and E. J. Maginn, *Green Chem.* **4**, 112–118 (2002).

116. M. Deetlefs, C. Hardacre, M. Nieuwenhuyzen, A. A. H. Padua, O. Sheppard, and A. K. Soper, *J. Phys. Chem. B* **110**, 12055–12061 (2006).

117. I. Okada, in J. O. Bockris, B. E. Conway, R. E. White, (eds.) *Modern Aspects of Electrochemistry No. 34*; Kluwer Academic/Plenum Publishers, New York, 2001.

118. G. J. Janz and R. D. Reeves, in C. W. Tobias, (ed.), Interscience, New York, 1967.

119. H. Tatlipinar, M. Amoruso, and M. P. Tosi, *Phys. B* **275**, 281–284 (2000).

120. J. Trullás and J. A. Padró, *Phys. Rev. B* **55**, 12210–12217 (1997).

121. A. Verdaguer and J. A. Padró, *J. Chem. Phys.* **114**, 2738–2744 (2001).

122. R. M. Stratt and M. Cho, *J. Chem. Phys.* **100**, 6700–6708 (1994).

123. M. C. C. Ribeiro and P. A. Madden, *J. Chem. Phys.* **106**, 8616–8619 (1997).

124. M. C. C. Ribeiro, M. Wilson, and P. A. Madden, *J. Chem. Phys.* **109**, 9859–9869 (1998).

125. J. Giergiel, K. R. Subbaswamy, and P. C. Eklund, *Phys. Rev. B* **29**, 3490–3499 (1984).

126. R. L. McGreevy, *Faraday Trans. 2* **83**, 1875–1889 (1987).

127. C. Raptis and E. W. J. Mitchell, *J. Phys. C* **20**, 4513–4528 (1987).

128. R. A. J. Bunten, R. L. McGreevy, E. W. J. Mitchell, and C. Raptis, *J. Phys. C* **17**, 4705–4724 (1984).

129. M. C. C. Ribeiro, L. F. C. de Oliveira, N. S. Gonçalves, *Phys. Rev. B* **63**, 104303/1–8 (2001).
130. S. M. Urahata and M. C. C. Ribeiro, *J. Chem. Phys.* **122**, 024511/1-9 (2005).
131. C. J. Margulis, H. A. Stern, and B. J. Berne, *J. Phys. Chem. B* **106**, 12017–12021 (2002).
132. M. G. Del Pópolo and G. A. Voth, *J. Phys. Chem. B* **108**, 1744–1752 (2003).
133. S. M. Urahata and M. C. C. Ribeiro, *J. Chem. Phys.* **120**, 1855–1863 (2004).
134. A. Mele, G. Romanò, M. Giannone, E. Ragg, G. Fronza, G. Raos, and V. Marcon, *Angew. Chem. Int. Ed. Engl.* **45**, 1123–1126 (2006).
135. M. G. Del Pópolo, R. M. Lynden-Bell, and J. Kohanoff, *J. Phys. Chem. B* **109**, 5895–5902 (2005).
136. M. Bühl, A. Chaumont, R. Schurhammer, and G. Wipff, *J. Phys. Chem. B* **109**, 18591–18599 (2005).
137. P. A. Hunt, B. Kirchner, and T. Welton, *Chem. Eur. J.* **12**, 6762–6775 (2006).
138. H. Katayanagi, S. Hayashi, H. Hamaguchi, and K. Nishikawa, *Chem. Phys. Lett.* **392**, 460–464 (2004).
139. J. N. A. Canongia-Lopes and A. A. H. Pádua, *J. Phys. Chem. B* **110**, 3330–3335 (2006).
140. Y. Wang and G. A. Voth, *J. Phys. Chem. B* **110**, 18601–18608 (2006).
141. Y. Wang and G. A. Voth, *J. Am. Chem. Soc.* **127**, 12192–12193 (2005).
142. C. M. Gordon, J. D. Holbrey, A. R. Kennedy, and K. R. Seddon, *J. Mat. Chem.* **8**, 2627–2636 (1998).
143. P. K. Mandal, M. Sarkar, and A. Samanta, *J. Phys. Chem. A* **108**, 9048–9053 (2004).
144. A. P. Demchenko, *Luminescence* **17**, 19–42 (2002).
145. Z. Hu and C. J. Margulis, *Proc. Natl. Acad. Sci. U.S.A.* **103**, 831–836 (2006).
146. Z. Hu and C. J. Margulis, *J. Phys. Chem. B* **110**, 11025–11028 (2006).
147. S. Shigeto and H. Hamaguchi, *Chem. Phys. Lett.* **427**, 329–332 (2006).
148. D. Xiao, J. R. Rajian, S. Li, R. A. Bartsch, and E. L. Quitevis, *J. Phys. Chem. B* **110**, 16174–16178 (2006).
149. W. Xu, E. I. Cooper, and C. A. Angell, *J. Phys. Chem. B* **107**, 6170–6178 (2003).
150. F. H. Stillinger and T. A. Weber, *Science* **225**, 983–989 (1984).
151. S. Sastry, P. G. Debenedetti, and F. H. Stillinger, *Nature (London)* **393**, 554–557 (1998).
152. M. S. Shell, P. G. Debenedetti, and F. H. Stillinger, *J. Phys. Chem. B* **108**, 6772–6777 (2004).
153. A. Triolo, O. Russina, C. Hardacre, M. Nieuwenhuyzen, M. Angel, and H. Grimm, *J. Phys. Chem. B* **109**, 22061–22066 (2005).
154. B.-R. Hyun, S. V. Dzyuba, R. A. Bartsch, and E. L. Quitevis, *J. Phys. Chem. A* **106**, 7579–7585 (2002).
155. H. Cang, J. Li, and M. D. Fayer, *J. Chem. Phys.* **119**, 13017–13022 (2003).
156. G. Giraud, C. M. Gordon, I. R. Dunkin, and K. Wynne, *J. Chem. Phys.* **119**, 464–477 (2003).
157. A. Rivera and E. A. Rössler, *Phys. Rev. B* **73**, 212201/1-4 (2006).
158. S. Schroedle, G. Annat, D. R. MacFarlane, M. Forsyth, R. Buchner, and G. Hefter, *Chem. Commun.* 1748–1750 (2006).
159. C. Daguenet, P. J. Dyson, I. Krossing, A. Oleinikova, J. Slattery, C. Wakai, and H. Weingärtner, *J. Phys. Chem. B* **110**, 12682–12688 (2006).
160. M. L. T. Asaki, A. Redondo, T. A. Zawodinski, and A. J. Taylor, *J. Chem. Phys.* **116**, 10377–10385 (2002).

161. H. Weingärtner, A. Knocks, W. Schrader, and U. Kaatze, *J. Phys. Chem. A* **105**, 8646–8650 (2001).

162. H. Shirota and E. W. Castner, *J. Phys. Chem. B* **109**, 21576–21585 (2005).

163. T. I. Morrow and E. J. Maginn, *J. Phys. Chem. B* **106**, 12807–12813 (2002).

164. C. Cadena, Q. Zhao, R. Q. Snurr, and E. J. Maginn, *J. Phys. Chem. B* **110**, 2821–2832 (2006).

165. S. Tsuzuki, H. Tokuda, K. Hayamizu, and M. Watanabe, *J. Phys. Chem. B* **109**, 16474–16481 (2005).

166. H. Tokuda, S. Tsuzuki, M. A. B. H. Susan, K. Hayamizu, and M. Watanabe, *J. Phys. Chem. B* **110**, 19593–19600 (2006).

167. H. Tokuda, K. Hayamizu, K. Ishii, M. A. B. H. Susan, and M. Watanabe, *J. Phys. Chem. B* **108**, 16593–16600 (2004).

168. H. Tokuda, K. Hayamizu, K. Ishii, M. A. B. H. Susan, and M. Watanabe, *J. Phys. Chem. B* **109**, 6103–6110 (2005).

169. H. Tokuda, K. Ishii, M. A. B. H. Susan, S. Tsuzuki, K. Hayamizu, and M. Watanabe, *J. Phys. Chem. B* **110**, 2833–2839 (2006).

170. A. Noda, K. Hayamizu, and M. Watanabe, *J. Phys. Chem. B* **105**, 4603–4610 (2001).

171. J. D. Holbrey, W. M. Reichert, M. Nieuwenhuyzen, S. Johnston, K. R. Seddon, and R. D. Rogers, *Chem. Commun.* 1636–1637 (2003).

172. S. Hayashi, R. Ozawa, and H. Hamaguchi, *Chem. Lett.* **32**, 498–499 (2003).

173. H. Hamaguchi and R. Ozawa, *Adv. Chem. Phys.* **131**, 85–104 (2005).

174. A. Downard, M. J. Earle, C. Hardacre, S. E. J. McMath, M. Nieuwenhuyzen, and S. J. Teat, *Chem. Mater.* **16**, 43–48 (2004).

175. C. A. Angell, Strong and Fragile Liquids. in K. L. Ngai and G. B. Wright, (eds.) *Relaxation in Complex Systems*, Naval Research Laboratory, Washington, DC, 1984.

176. C. A. Angell, *J. Non-Cryst. Solids* **131–133**, 13–31 (1991).

177. J. O. Bockris and A. K. N. Reddy, *Modern Electrochemistry*, Plenum Press, New York, 1970.

178. A. P. Abbott, *Chem. Phys. Chem.* **5**, 1242–1246 (2004).

179. A. P. Abbott, *Chem. Phys. Chem.* **6**, 2502–2505 (2005).

180. R. K. Wangsness, *Electromagnetic Fields*, John Wiley & Sons, Inc., New York, 2nd. ed., 1986.

181. A. J. Stone, *The Theory of Intermolecular Forces*, Clarendon Press, Oxford, 1997.

182. C. E. R. Prado, M. G. Del Pópolo, T. G. A. Youngs, J. Kohanoff, and R. M. Lynden-Bell, *Mol. Phys.* **104**, 2477–2483 (2006).

183. M. N. Kobrak and N. Sandalow, in R. A. Mantz, (ed.), *Molten Salts XIV* The Electrochemical Society, Pennington, NJ, 2006.

184. J. M. Rowe and S. Susman, *Phys. Rev. B: Cond. Matter* **29**, 4727–4732 (1984).

185. S. Miller and J. H. R. Clarke, *Faraday Trans. 2* **74**, 160–173 (1978).

186. R. Chelli, G. Cardini, and S. Califano, *J. Chem. Phys.* **107**, 8041–8050 (1997).

187. T. Yan, C. J. Burnham, M. G. Del Pópolo, and G. A. Voth, *J. Phys. Chem. B* **108**, 11877–11881 (2004).

188. C. Reichardt, *Chem. Rev.* **94**, 2319–2358 (1994).

189. M. J. Kamlet and R. W. Taft, *J. Am. Chem. Soc.* **98**, 377–383 (1976).

190. M. H. Abraham, A. M. Zissimos, J. G. Huddleston, H. D. Willauer, R. D. Rogers, and W. E. J. Acree, *Ind. Eng. Chem. Res.* **42**, 413–418 (2003).

191. J. G. Huddleston, H. D. Willauer, R. P. Swatloski, A. E. Visser,, and R. D. Rogers, *Chem. Commun.* 1765–1766 (1998).

192. J. L. Anderson, J. Ding, T. Welton, and D. W. Armstrong, *J. Am. Chem. Soc.* **124**, 14247–14254 (2002).

193. J. L. Anderson and D. W. Armstrong, *Anal. Chem.* **75**, 4851–4858 (2003).

194. M. H. Abraham, *Chem. Soc. Rev.* **22**, 73–83 (1993).

195. M. H. Abraham, G. S. Whiting, R. M. Doherty, and W. J. Shuely, *J. Chromatogr.* **587**, 213–228 (1991).

196. S. B. Lee, *J. Chem. Tech. Biotech.* **80**, 133–137 (2005).

197. W. E. Acree and M. H. Abraham, *J. Chem. Tech. Biotech.* **81**, 1441–1446 (2006).

198. P. Bonhôte, A.-P. Dias, N. Papageorgiou, K. Kalyanasundaram, and Grätzel, M. *Inorg. Chem.* **35**, 1168–1178 (1996).

199. S. V. Dzyuba and R. A. Bartsch, *Tetrahedron Lett.* **43**, 4657–4659 (2002).

200. P. Wasserscheid, C. M. Gordon, C. Hilgers, M. J. Muldoon, and I. R. Dunkin, *Chem. Commun.* 1186–1187 (2001).

201. C. Reichardt, *Green Chem.* **7**, 339–351 (2005).

202. V. I. Znamenskiy and M. N. Kobrak, *J. Phys. Chem. B* **108**, 1072–1079 (2004).

203. T. Yokoyama, R. W. Taft, and M. J. Kamlet, *J. Am. Chem. Soc.* **98**, 3233–3237 (1976).

204. R. W. Taft and M. J. Kamlet, *J. Am. Chem. Soc.* **98**, 2886–2894 (1976).

205. L. Crowhurst, P. Mawdsley, J. Perez-Arlandis, P. Salter, and T. Welton, *Phys. Chem. Chem. Phys.* **5**, 2790–2794 (2003).

206. A. Oehlke, K. Hofmann, and S. Spange, *New J. Chem.* **30**, 533–536 (2006).

207. A. M. Funston and J. F. Wishart, *ACS Symp. Ser.* **901**, 102–116 (2005).

208. G.-H. Tao, M. Zou, X.-H. Wang, Z. Chen, D. G. Evans, and Y. Kou, *Aust. J. Chem.* **58**, 327–331 (2005).

209. K. Dahl, G. M. Sando, D. M. Fox, T. E. Sutto, and J. C. Owrutsky, *J. Chem. Phys.* **123**, 084504/1-11 (2005).

210. K. Kalyanasaundaram and J. K. Thomas, *J. Phys. Chem.* **81**, 2176–2180 (1977).

211. C. W. Garland, J. W. Nibler, and D. P. Shoemaker, *Experiments in Physical Chemistry*, McGraw-Hill, Boston, 7th ed., 2003.

212. U. Kaatze and K. Giese, *J. Phys. E* **13**, 133–141 (1980).

213. R. C. Weast, ed., *CRC Handbook of Chemistry and Physics*, CRC Press, Inc., Boca Raton, FL, 64th ed., 1983.

214. C. Wakai, A. Oleinikova, M. Ott, and H. Weingärtner, *J. Phys. Chem. B* **109**, 17028–17030 (2005).

215. H. Z. Weingärtner, *Z. Physik. Chem.* **220**, 1395–1405 (2006).

216. M. J. Earle, B. S. Engel, and K. R. Seddon, *Aust. J. Chem.* **57**, 149–150 (2004).

217. G. Angelini, C. Chiappe, P. D. Maria, A. Fontana, F. Gasparrini, D. Pieraccini, M. Pierini, and G. Siani, *J. Org. Chem.* **70**, 8193–8196 (2005).

218. C. Chiappe and D. Pieraccini, *J. Phys. Org. Chem.* **18**, 275–297 (2005).

219. P. J. Dyson, *Trans. Met. Chem. (N.Y.)* **27**, 353–358 (2002).

220. J. D. Holbrey, A. E. Visser, and R. D. Rogers, P. Wasserscheid, and T. Welton,(eds.) in *Ionic Liquids in Synthesis*; Wiley–VCH, Mörlenbach, Germany, 2002.

221. A. E. Visser, R. P. Swatloski, W. M. Reichert, S. T. Griffin, and R. D. Rogers, *Ind. Eng. Chem. Res.* **39**, 3596–3604 (2000).

222. V. A. Cocalia, M. P. Jensen, J. D. Holbrey, S. K. Spear, D. C. Stepinski, and R. D. Rogers, *Dalton Trans.* 1966–1971 (2005).
223. M. P. Jensen, J. A. Dzielawa, P. Rickert, and M. L. Dietz, *J. Am. Chem. Soc.* **124**, 10664–10665 (2002).
224. M. L. Dietz and D. C. Stepinski, *Green Chem.* **7**, 747–750 (2005).
225. M. P. Jensen, J. Neuefeind, J. V. Beitz, S. Skanthakumar, and L. Soderholm, *J. Am. Chem. Soc.* **125**, 15466–15473 (2003).
226. C. Gaillard, I. Billard, A. Chaumont, S. Mekki, A. Ouadi, M. A. Denecke, G. Moutiers, and G. Wipff, *Inorg. Chem.* **44**, 8355–8367 (2005).
227. A. Chaumont, E. Engler, and G. Wipff, *Inorg. Chem.* **42**, 5348–5356 (2003).
228. A. Chaumont and G. Wipff, *Phys. Chem. Chem. Phys.* **5**, 3481–3488 (2003).
229. A. Chaumont and G. Wipff, *Phys. Chem. Chem. Phys.* **8**, 494–502 (2006).
230. N. Sieffert and G. Wipff, *J. Phys. Chem. B* **110**, 19497–19506 (2006).
231. A. Chaumont and G. Wipff, *Phys. Chem. Chem. Phys.* **7**, 1926–1932 (2005).
232. C. G. Hanke, N. A. Atamas, and R. M. Lynden-Bell, *Green Chem.* **4**, 107–111 (2002).
233. R. M. Lynden-Bell, N. A. Atamas, A. Vasilyuk, and C. G. Hanke, *Mol. Phys.* **100**, 3225–3229 (2002).
234. L. A. Blanchard and J. F. Brennecke, *Ind. Eng. Chem. Res.* **40**, 287–292 (2001).
235. S. Tsuzuki, K. Honda, T. Uchimaru, M. Mikami, and K. Tanabe, *J. Am. Chem. Soc.* **124**, 104–112 (2002).
236. C. Hanke, A. Johansson, J. Harper, and R. Lynden-Bell, *Chem. Phys. Lett.* **374**, 85–90 (2003).
237. D. A. Dougherty, *Science* **271**, 163–168 (1996).
238. J. N. Canongia-Lopes, M. F. C. Gomes, and A. A. H. Pádua, *J. Phys. Chem. B* **110**, 16816–16818 (2006).
239. M. N. Kobrak, *J. Phys. Chem.* B **111**, 4755–4762 (2007).
240. O. J. Lanning, P. A. Madden, *J. Phys. Chem. B* **108**, 11069–11072 (2004).
241. C. Pinilla, M. G. Del Pópolo, R. M. Lynden-Bell, and J. Kohanoff, *J. Phys. Chem. B* **109**, 17922–17927 (2005).
242. M. J. Kamlet, J. L. Abboud, and R. W. Taft, *J. Am. Chem. Soc.* **99**, 6027–6038 (1977).
243. Y. Shim, M. Y. Choi, and H. J. Kim, *J. Chem. Phys.* **122**, 044510 (2005).
244. R. A. Marcus, *Ann. Rev. Phys. Chem.* **15**, 155–196 (1964).
245. W. J. Albery, *Ann. Rev. Phys. Chem.* **31**, 227–63 (1980).
246. C. M. Gordon, A. J. McLean, and M. J. Muldoon, *ACS Symp. Ser.* **856**, 357–369 (2003).
247. A. J. McClean, M. J. Muldoon, C. M. Gordon, and I. R. Dunkin, *Chem. Commun.* 1880–1881 (2002).
248. J. H. Werner, S. N. Baker, and G. A. Baker, *The Analyst* **128**, 786–789 (2003).
249. R. Brookes, A. Davies, G. Ketwaroo, and P. A. Madden, *J. Phys. Chem. B* **109**, 6485–6490 (2005).
250. R. G. Evans, A. J. Wain, C. Hardacre, and R. G. Compton, *Chem. Phys. Chem.* **6**, 1035–1039 (2005).
251. V. Strehmel, A. Laschewsky, R. Stoesser, A. Zehl, and W. Herrmann, *J. Phys. Org. Chem.* **19**, 318–325 (2006).
252. J. Jacquemin, P. Husson, V. Majer, and M. F. G. Gomes, *Fluid Phase Equil.* **240**, 87–95 (2006).
253. J. L. Anderson, J. K. Dixon, E. J. Maginn, and J. F. Brennecke, *J. Phys. Chem. B* **110**, 15059–15062 (2006).

254. J. L. Anthony, J. L. Anderson, E. J. Maginn, and J. F. Brennecke, *J. Phys. Chem. B* **109**, 6366–6374 (2005).
255. L. A. Blanchard, Z. Y. Gu, and J. F. Brennecke, *J. Phys. Chem. B* **105**, 2437–2444 (2001).
256. X. Huang, C. J. Margulis, Y. Li, and B. J. Berne, *J. Am. Chem. Soc.* **127**, 17842–17851 (2005).
257. D. Morgan, L. Ferguson, and P. Scovazzo, *Ind. Eng. Chem. Res.* **44**, 4815–4823 (2005).
258. M. C. Buzzeo, O. V. Klymenko, J. D. Wadhawan, C. Hardacre, K. R. Seddon, and R. G. Compton, *J. Phys. Chem. A* **107**, 8872–8878 (2003).
259. J. Grodkowski, P. Neta, and J. F. Wishart, *J. Phys. Chem. A* **107**, 9794–9799 (2003).
260. Y. Shim, D. Jeong, M. Y. Choi, and H. J. Kim, *J. Chem. Phys.* **125**, 061102/1-4 (2006).
261. K. Swiderski, A. McClean, C. M. Gordon, and D. H. Vaughan, *Chem. Commun.* 590–591 (2004).
262. L. Crowhurst, N. L. Lancaster, J. M. P. Arlandis, and T. Welton, *J. Am. Chem. Soc.* **126**, 11549–11555 (2004).
263. D. Landini and A. Maia, *Tetrahedron Lett.* **46**, 3961–3963 (2005).
264. C. Chiappe, D. Pieraccini, and P. Saullo, *J. Org. Chem.* **68**, 6710–6715 (2003).
265. C. M. Gordon and A. J. McLean, *Chem. Commun.* **15**, 1395–1396 (2000).
266. A. Samanta, *J. Phys. Chem. B* **110**, 13704–13716 (2006).
267. P. K. Mandal, S. Saha, R. Karmakar, and A. Samanta, *Curr. Sci.* **90**, 301–310 (2006).
268. E. Bart, A. Meltsin, and D. Huppert, *J. Phys. Chem.* **98**, 3295–3299 (1994).
269. E. Bart, A. Meltsin, and D. Huppert, *J. Phys. Chem.* **98**, 10819–10823 (1994).
270. S. Saha, P. K. Mandal, and A. Samanta, *Phys. Chem. Chem. Phys.* **6**, 3106–3110 (2004).
271. H. Shirota, A. M. Funston, J. F. Wishart, and E. W. Castner, *J. Chem. Phys.* **122**, 1–12 (2005).
272. B. Lang, G. Angulo, and E. Vauthey, *J. Phys. Chem. A* **110**, 7028–7034 (2006).
273. D. Chakrabarty, P. Hazra, A. Chakraborty, D. Seth, and N. Sarkar, *Chem. Phys. Lett.* **381**, 697–704 (2003).
274. L. S. Headley, P. Mukherjee, J. L. Anderson, R. Ding, M. Halder, D. W. Armstrong, X. Song, and J. W. Petrich, *J. Phys. Chem. A* **110**, 9549–9554 (2006).
275. Y. Shim, M. Y. Choi, and H. J. Kim, *J. Chem. Phys.* **122**, 044511/1-12 (2005).
276. M. N. Kobrak, *J. Chem. Phys.* **125**, 064502 (2006).
277. M. N. Kobrak, V. Znamenskiy, *Chem. Phys. Lett.* **395**, 127–132 (2004).
278. B. M. Ladanyi and M. Maroncelli, *J. Chem. Phys.* **109**, 3204–3221 (1998).
279. W. A. Steele, *Mol. Phys.* **61**, 1031–1043 (1987).
280. P. Mukherjee, J. A. Crank, M. Halder, D. W. Armstrong, and J. W. Petrich, *J. Phys. Chem. A* **110**, 10725–10730 (2006).
281. J. P. Suwalski and J. Kroh, *Radiat. Phys. Chem.* **64**, 197–201 (2002).
282. H. A. Gillis and D. C. Walker, *J. Chem. Phys.* **65**, 4590–4595 (1976).
283. K. Kawabata, H. Horii, and S. Okabe, *Chem. Phys. Lett.* **14**, 223–225 (1972).
284. X. Zhang, Y. Lin, and C. D. Jonah, *Radiat. Phys. Chem.* **54**, 433–440 (1999).
285. M. P. Tosi and F. G. Fumi, *J. Phys. Chem. Sol.* **25**, 45–52 (1964).

COUNTERION CONDENSATION IN NUCLEIC ACID

ALEX SPASIC* AND UDAYAN MOHANTY

Eugene F. Merkert Chemistry Center, Department of Chemistry, Boston College, Chestnut Hill, Massachusetts 02467

CONTENTS

*Current address: Baker Laboratory of Chemistry and Chemical Biology, Cornell University, Ithaca, New York 14853–1301.

Advances in Chemical Physics, Volume 139, edited by Stuart A. Rice

I. INTRODUCTION

Due to the polyelectrolyte nature of nucleic acids, the concentration, the valence, and the type of counterions exquisitely influence the structure, the stability, and the biological activity of these macromolecules in aqueous solution [1, 2]. Divalent metal ions, such as magnesium, induce folding of hairpin ribozyme [3]. Substantial conformation fluctuations occur in a three-way junction from eubacterial 16S ribosomal ribonucleic acid (rRNA) in the presence of magnesium cations. These conformation fluctuations enhance the association rate constant of binding of ribosomal protein S15 to the folded junction [4]. In nucleic acids, metal ions play a significant role in phosphodiester cleavage, as well as ligation reactions [5, 6].

The interaction of various charged groups in a nucleic acid with each other and with the cations and the co-ions in an aqueous solution dictate the characteristics of electrostatic components of the free energy. Van der Waals and hydrophobic interactions, for example, contribute to the nonelectrostatic part of the free energy [7, 8, 9]. The electrostatic interactions are destabilizing and can be offset by association of cations with the nucleic acid [9].

When a nucleic acid interacts with cations in aqueous solution, bound states form in the interfacial region of the macromolecule [10]. These states are classified either as site bound or diffusely bound [9, 11–14]. If the translational energy of the ions is unable to overcome the electrostatic forces between the ions and the nucleic acid, then the ions are viewed as trapped in a volume encasing the surface of the nucleic acid. These ions are site bound [9, 11, 13, 14].

The site bound ions accounts for its hydration state and are grouped either as "outer–sphere" or as "inner-sphere" [9, 11, 14]. In the later case, it is assumed that the water molecules in the hydration shell do not participate. The ions directly interact with the phosphate charges and anionic ligands [9, 11, 14]. Since both outer- and inner-sphere interactions lead to formation of ion pairs, site bound ions are describable in terms of an association constant satisfying the law of mass action [9, 11, 14–16].

In the case of diffusive binding, the counterions form a delocalized cloud around the nucleic acid [9, 11, 17]. The interactions between the counterions and the nucleic acid are weak, long-range Coulombic, and nonspecific [9, 11, 12, 14, 17]. In diffusive binding, specific binding sites are not involved, and the interactions are not expressible in terms of the law of mass action [9, 11, 14, 16, 17]. Quantitative differences between site and diffuse bindings can be only resolved though interpretation of the energetics of the nucleic acid in solution [9, 11, 14].

Specific metal ion binding sites are directly observed in the crystal structures of hammerhead ribozymes [18], P4–P6 domain of *Tetrahymena* group I intron [19], transfer (tRNA) [20], GAAA tetraloop receptor [21], sarcin–ricin loop [22], and MMTV pseudoknots [23], for example. High resolution crystal structures of DNA oligonucleotides reveal localization of monovalent and divalent cations [6, 24–27]. DNA sequences and characteristics of crystal packing dictate metal ion bonding, and the bound cations modulate the conformation of the nucleic acid [6, 24–28]. However, which of the bound metal ions are an essential part of a DNA duplex or RNA motif that allows folding are complicated to resolve experimentally [6, 24–28]. Even though crystal structures of the loop fragment of 5S rRNA reveal five magnesium ions [26], only one of them is important for biological function [6, 27].

A four-way junction from hairpin ribozyme exhibits two types of ion binding [3]. At low concentrations of monovalent ions, binding of magnesium to the junction is diffuse [3]. In contrast, at high concentrations of monovalent ions, site binding of magnesium occurs at low magnesium concentration [3]. Other nucleic acids exhibit similar characteristics [6].

A number of methodologies have been developed and generalized in recent years to quantitatively describe the ion atmosphere around nucleic acids [11, 12, 17, 28, 29]. These include models based on Poisson–Boltzmann equation [11, 12], counterion condensation [17], and simulation methods, such as Monte Carlo, molecular dynamics, and Brownian dynamics [28, 29].

The physical ideas behind Manning's counterion condensation follow from thermodynamic arguments [17, 31]. The DNA is taken to be a linear array of equally spaced phosphate charges. The DNA is immersed in an aqueous solution. The total Gibbs free energy of the polyion with its ion atmosphere is a sum of two terms. One of the terms is a result of transfer of some of the counterions [17, 31] from the bulk solution to a small region near the surface of the polyion. The remaining term is the electrostatic free energy of assembling the charges into a conformation that resembles a linear array [17, 31]. In the model, each charge is reduced by theta: the total number of counterions that condense per polyion charge. The equilibrium state is obtained by minimizing the total Gibbs free energy with respect to theta [17, 31].

In the limit of low salt concentration, Manning predicted that a fraction of the counterions remains associated with the DNA. These counterions are viewed as condensed [17, 31]. The condensed fraction is found to be independent of the salt concentration up to 0.1 M [17, 31]. The uncondensed counterions form a Debye–Hückel cloud. The fraction of condensed counterions per polyion charge charge is 76 and 88% for monovalent and divalent cations, respectively, for B-DNA [17, 31]. Thus, according to the counterion condensation model, the renormalized charge on a phosphate in B-DNA is 0.12 and 0.24 for magnesium and sodium counterions, respectively [17, 31]. The predictions of colligative properties, transport properties, binding equilibrium and melting temperatures of DNA, RNA, and various charged polymers, based on a counterion condensation model are in qualitative, and in some cases, quantitative agreement with experimental data [17, 30, 31].

In this chapter, some recent advances in counterion condensation are surveyed. We also discuss how some of these advances have been applied to elucidate the characteristics of nucleic acids. Due to the scope of this chapter, a variety of topics on the ion atmosphere of nucleic acid has been excluded.

II. COUNTERION CONDENSATION

A. Manning Model

The phosphate groups in DNA are ionized at physiological pH [31]. Consequently, the macromolecule is a polyanion [31]. In solution the polyion is endowed with an equal number of counterions due to electroneutrality. However, if the solution contains excess salt, then these ions together with the mobile counterions form the ion atmosphere [31].

Now, consider B-DNA molecules in an aqueous solution. Manning showed that the electrostatic stability of the DNA is controlled by the dimensionless linear charge density parameter ξ [17, 31]. The ratio of the Bjerrum length l_B to the charge spacing b of the phosphates defines the linear charge density [17, 31]

$$\xi = \frac{l_B}{b} \tag{1}$$

The distance at which the repulsive Coulomb energy between two unit charges balances energy due to thermal fluctuations, at a specific temperature, is called the Bjerrum length l_B [17, 31]. In water, and at room temperature, the Bjerrum length is 7.1 Å. For B-DNA, the average spacing b of the charge is 1.7 Å, where we have assumed the phosphate charges are projected onto the axis of symmetry of the DNA [17, 31].

Observe that the linear charge density of B-DNA is greater than one. If this constraint is satisfied, Manning and Onsager argued that the system is electrostatically unstable [17, 31]. To lower the total energy of the system, counterions from the bulk condense on the backbone of DNA to reduce the bare phosphate charges [17, 31].

In the counterion condensation model, the total free energy G of the system consists of two parts. First, there is an electrostatic free energy G_1 of assembling the phosphate charges. The second contribution to the free energy, G_2, arises from the entropy of mixing of the solvent molecules and the bound and the free counterions [17, 31].

The Coulombic repulsion between closely spaced phosphate charges is considerable. Consequently, work is required to overcome the electrostatic repulsion if the phosphate charges are assembled into a linear configuration, with constant spacing b, from charges that are infinitely far apart in aqueous solution. Thus, G_1 is the reversible work to assemble a linear configuration of charges [31, 32]

$$\frac{G_1}{k_B T} = \frac{2q_{\text{net}}^2}{\varepsilon k_B T b} \sum_{i=1, i \neq j}^{P} \sum_{j=1}^{P} \frac{e^{-\kappa |i-j| b}}{|i-j|} \tag{2}$$

Here, T is the absolute temperature, ε is the bulk dielectric constant of the solvent, P is the number of phosphate charges, κ is the inverse of the Debye screening length, k_B is the Boltzmann constant, q_{net} is the renormalized charge, the interaction between the charges is screened Debye–Hückel potential, and $|i-j|b$ is the distance between a pair of charges labeled i and j.

Due to condensed counterions of unsigned valence z, the bare charge of each phosphate group q is renormalized by $q(1-z\theta)$, where θ is the total number of counterions per charge of the polyion [31, 32]. The double sums in Eq. (2) can be explicitly calculated. If the length of the chain $L = Pb \gg \kappa^{-1}$, the Gibbs free energy per charge reduces to [31, 32]

$$\frac{G_1}{k_B T} = -\xi (1 - z\theta)^2 \ln(1 - e^{-\kappa b}) \tag{3}$$

where ξ is the linear charge density parameter. The Debye screening parameter is governed by bulk salt concentration c_s [31]

$$\kappa^2 = 8\pi \times 10^{-3} N_A l_B c_s \tag{4}$$

Here, N_A is the Avogadro's number, the salt is monovalent, and its concentration is expressed in molarity, while κ has units of reciprocal angstroms Å^{-1}.

The free energy G_2 per charge arises from moving the ions from the bulk aqueous solution to the region around the DNA that defines the condensed layer [31, 32]

$$\frac{G_2}{k_B T} = \theta \ln\left(\frac{10^3 \theta}{c_s \gamma Z Q}\right) \tag{5}$$

where Z is the number of counterions in the salt, and γ is an activity coefficient that describes nonideality of the solvent. The parameter Q is viewed as the partition function of the condensed layer, and is a measure of an effective volume around the DNA within which the condensed counterions reside. The partition function Q in Eq. (5) is expressed in units of cubic centimeters per mole of phosphate charge ($\text{cm}^3\,\text{mol}^{-1}$) [31].

To obtain an equilibrium state, the total free energy must be a minimum with respect to variations in θ. Imposing this constraint leads to a nonlinear equation for θ [31, 32]

$$
\begin{aligned}
&2z(1 - z\theta)\xi \ln\left(\frac{1 - e^{-\kappa b}}{\kappa b}\right) + \ln(10^3\theta c_s \gamma Z Q) + 1 \\
&+ 2z(1 - z\theta)\xi\left[\ln[b(8\pi 10^{-3} N_A l_B)^{1/2}] + \frac{1}{2}\ln c_s\right] = 0
\end{aligned}
\tag{6}
$$

We regroup those terms in Eq. (6) that are associated with $\ln c_s$. This is so because $\ln c_s$ diverges as c_s approaches zero. Hence, the coefficient of $\ln c_s$ must vanish. This leads to an explicit connection between θ and the linear charge density [17, 31]

$$
z\theta = 1 - \frac{1}{z\xi} \tag{7}
$$

The non-$\ln c_s$ terms in Eq. (6) must also vanish identically. This yields an expression for the condensed-layer partition function [31, 32]

$$
Q = \frac{4\pi e N_A b^3 z'}{\gamma Z}(Z + Z')\left(\xi - \frac{1}{z}\right)\left(\frac{1 - e^{-\kappa b}}{\kappa b}\right)^2 \tag{8}
$$

Substituting the expression for $z\theta$ and Q in Eqs. (3) and (4) lead to the desired expression for the total electrostatic free energy of DNA per unit charge [31, 32]

$$
\frac{G}{N k_B T} = \frac{1}{z}\left(\frac{1}{z\xi} - 2\right)\ln(1 - e^{-\kappa b}) - \frac{1}{z} + \frac{1}{z^2\xi} \tag{9}
$$

Equations (7)–(9) are the basic results of the counterion condensation model.

B. Poisson–Boltzmann

Despite the inherent weaknesses of the Poission–Boltzmann (PB) equation that includes neglect of polyion–polyion interactions and the size of mobile ions, it has nevertheless played an important role in elucidating the characteristics of electrostatic interactions and ion atmopshere in nucleic acids [12]. Two salient results have emerged from such an analysis. First, for cylindrical-shaped molecule of infinite length, there is a critical linear charge density beyond which counterion condensation takes place [33, 34]. Second, the phenomenon of counterion condensation takes place in one and two dimensions only [33, 35]. An implication of this result is that the counterion condensation does not occur for a charged sphere of fixed size [33].

In an important work, Ramanathan analyzed the colloidal and the micellar limits of a charged sphere [36]. In the latter case, the Debye screening length is larger than the radius of the sphere. In the colloidal limit, the reverse is true, namely, the radius of the sphere is larger than the Debye screening length.

The PB equation for a sphere of radius a is [36]

$$\Delta u = u'' + \frac{2}{r}u' = \varepsilon^2 \sinh u \tag{10}$$

where $r = x/a$, x is the distance from the center of the sphere, ϕ is the electrostatic potential, u is the reduced electrostatic potential, $e\phi/k_{\mathrm{B}}T$, k_{B} is the Boltzmann constant, T is the temperature, e is the electron charge, κ^{-1} is the Debye length, and $\hat{\varepsilon} = \kappa a$. The charge of the sphere is $-Z_0 e$, where Z_0 is the absolute value of the valence of the sphere. The boundary conditions on the reduced electrostatic potential are [36]

$$u'(1) = -Z_0\delta < 0, \quad u(\infty) = 0 \tag{11}$$

where $\delta = e^2 z/\varepsilon kTa$, and z is the valence of the small ions, ε is the dielectric constant. Thus, $\hat{\varepsilon} \gg 1$ and $\hat{\varepsilon} \ll 1$ correspond, respectively, to the colloidal and the micellar limits of the sphere.

The PB analysis of spherically charged macroion leads to new insights into the phenomenon of counterion condensation. First, assume that the surface charge density of the sphere is held fixed while increasing its size. Then, if the size of the sphere gets larger than a threshold value, the counterions will condense [36]. To put it differently, if the valence $Z_0 \geq -2al_{\mathrm{B}}^{-1}\ln(\kappa a)$, and $\hat{\varepsilon} \ll 1$ (micellar limit), then counterion condensation occurs and a sheath of counterions encase the macroion [36]. Second, the effective charge of the sphere is $-2al_{\mathrm{B}}^{-1}\ln(\kappa a)$ [36]. Finally, if the radius of the sphere is such that $\hat{\varepsilon} \gg 1$, then the condensed counterions will neutralize the charge of the sphere [36]. In this case, the counterions are localized in a shell whose thickness is much less than the radius of the polyion [36].

C. Ionic Oligomers

The ion atmosphere of a polyion of finite length L in an aqueous solution has been studied using the BBGY hierarchy. The basic idea is to introduce scaled variable $\tilde{\varepsilon} = \kappa d$, where d is the distance of closest approach of a mobile ion to a polyion, as a measure of a small parameter. The small parameter allows one to separate the insignificant from the significant terms in the hierachy of BBGY correlation functions [33].

Ramanathan–Woodbury carried out such an analysis and showed that if $L \gg d$, but in which $\kappa L = 0(1)$ [33], then there is critical value of a dimensionless charge density, defined as $\xi_o = \log(\kappa a)^{-1}/\log(L/d)$, beyond which counterions will condense. Furthermore, if the length of the polyion is of the order of the Debye length, then the charge density is independent of the Debye screening parameter [33]. These predictions are in accord with experimental results [37–39].

Extended Debye–Huckel and density function approaches have also been developed by Manning and Mohanty [40] to provide insights into Ramanathan–Woodbury results [33]. In the extended Debye–Huckel approach, the polyion is taken to be linear, with an array of P charged sites, and in which the spacing between the charges is b [40]. Each site has a charge $q(1-\theta)$, where θ accounts for the reduction of the bare charge q due to counterion condensation. The interaction between a pair of charges is governed by screened Debye–Huckel potential. The total work w is the sum of (a) the work w_P in assembling the charges on the polyion from infinity, and (b) the work of transferring θP counterions, w_{tr}, from bulk to a region near the polyion [40]. The equilibrium state of the polyion is that state for which the total work is a minimum with respect to θ.

A noteworthy conclusion is that for a polyion in aqueous solution, and in which κL is held fixed, the equilibrium state of the macromolecule is stable provided $\theta = 1 - \xi^{-1}$ [40]. Furthermore, the number of condensed counterions, as well as the threshold value for linear charge density, is unchanged for a polyion whose length is of the order of κ^{-1} [40].

The situation is complicated for a short polyion. In short, we mean a polyion whose length is much smaller than the Debye screening length, $L \ll \kappa^{-1}$. More precisely, under the conditions $\kappa \to 0$, $L \to \infty$, $L = o(\kappa^{-1})$, the total work in assembling the polyion charge and the ion atmosphere is [40]

$$w/k_{\mathrm{B}}T \approx (1-\theta)^2 \xi P \ln(L/b) + \theta P \ln[\theta/c_s Q(\theta)] \tag{12}$$

where c_s is the salt concentration and Q is the partition function of the adsorbed or the condensed layer.

The equilibrium value of θ is obtained by minimizing Eq. (12); one finds for $\xi > \xi_{\mathrm{crit}}$ [40]

$$\theta = 1 - \frac{\xi_{\mathrm{crit}}}{\xi} \tag{13}$$

The threshold value ξ_{crit} for counterion condensation is given by [40]

$$\xi_{\mathrm{crit}} = \frac{\ln(\kappa b)^{-1}}{\ln(L/b)} \tag{14}$$

Several conclusions follow from the Manning–Mohanty model. First, for ionic oligomers whose length $L \ll \kappa^{-1}$, counterion condensation occurs [40]. Second, the threshold linear charge density, beyond which counterions condense, increases with a decrease of salt concentration [40]. Third, in agreement with Ramanathan–Woodbury assessment [33], the number of condensed counterions that are in the condensed layer is identical whether one considers a long polyelectrolyte chain or an ionic oligomer of the order of one Debye length [40]. However, the free energy of the condensed layer is distinct in the two cases. Finally, a surprising result is that the number of condensed counterions next to the ends of a long polyelectrolyte chain is equal to the number of counterions condensed on a short oligomer [40].

D. Helical Model

A counterion condensation model has been generalized to include the helical structure of DNA. A single helix is defined by the position vector $\vec{r} = (a\cos\phi, a\sin\phi, h\phi)$, where the helical rise and radius are h and a, respectively, and ϕ is the rotational angle. The length between a pair of charges along the helix is given by [41, 42]

$$r_n = nb(1 + \left(\frac{\sqrt{2}a}{nb}\right)^2 (1 - \cos(nb/h))^{1/2} \qquad (15)$$

Introduce the sum Ξ defined as [41, 42]

$$\Xi = \sum_{n=1}^{\infty} \frac{e^{-\kappa r_n}}{r_n} \qquad (16)$$

The free energy per unit charge associated with assembling the helical charges is

$$\frac{G_1}{k_B T} = (1 - z\theta)^2 l_B \Xi \qquad (17)$$

There is also an additional free energy G_2 resulting from transferring counterions from the bulk solution into a region near the backbone of the chain [41, 42].

As before, the total free energy is minimized with respect to θ. Minimization leads to the following results [41]. (a) The minimum value of θ is identical with Eq. (7) from the line model. This implies that the number of condensed ions is unaffected by the helix and is governed only by the linear charge-density parameter ξ. (b) The partition function Q_{helix} is influenced by the helical structure of the phosphate charges

$$Q_{\text{helix}} = Qe^{-2\mathbb{R}} \qquad (18)$$

where $\mathbb{R}$ is defined as

$$\mathbb{R} = \sum_{n=1}^{\infty} \left(\frac{e^{-\kappa r_n}}{r_n/b} - \frac{e^{-\kappa nb}}{n} \right) \tag{19}$$

(c) The total free energy of the line charge model [see Eq. (9)] is larger than that of the helical array, the later is given by [41, 42]

$$\frac{G}{k_B T} = \frac{1}{z}\left(\frac{1}{z\xi} - 2\right)[\mathbb{R} - \ln(1 - e^{-\kappa b})] - \frac{1}{z} + \frac{1}{z^2 \xi} \tag{20}$$

The arguments given above have been generalized to the DNA double helix [42]. Each helical strand consists of P phosphate charges that are equally spaced. The total number of charges is given by $N = 2P$. The linear charge density of one helix is $\xi_1 = l_B/b_1$, where b_1 is the spacing between the charges projected onto the symmetry axis of the helix. Similarly, let us denote the linear charge density for a line charge consisting of $N = 2P$ charges by $\xi_2 = l_B/b_2$. Since $b_2 = b_{1/2}$ and $\xi_2 = 2\xi_1$, this suggests that ξ_2 is an average liner charge density [41, 42].

To determine θ, the number of condensed charges per polyion charge of the double helix, one adds the free energy of assembling the phosphate charges onto a double-helical configuration to the transfer free energy in Eq. (5), minimize the total free energy with respect to θ, and then analyze the $\ln c_s$ singularity. One finds [41, 42]

$$z\theta = \left(1 - \frac{1}{2z\xi_1}\right) = \left(1 - \frac{1}{z\xi_2}\right) \tag{21}$$

This result is the same as the line charge formula [see Eq. (7)] due to Manning.

Another prediction of the DNA double helix model is the partition function of the condensed layer [41, 42]

$$Q_{\text{double helix}} = Q(b = b_2)\left(\frac{1 - e^{-\kappa b_1}}{1 - e^{-\kappa b_2}}\right)^2 e^{-2(\mathbb{R})}\text{double helix} \tag{22}$$

$Q(b = b_2)$ is the partition function of the condensed layer for the line charge, but with $b = b_2$.

The total electrostatic free energy of the DNA double helix is [41, 42]

$$\frac{G}{2Pk_B T} = \left(\frac{G}{2Pk_B T}\right)_{\text{line}} - \frac{1}{z}\left(\frac{1}{z\xi_2} - 2\right)(\mathbb{R})_{\text{double helix}} - \frac{1}{z}\left(2 - \frac{1}{z\xi_2}\right)\ln\left(\frac{1 - e^{-\kappa b_1}}{1 - e^{-\kappa b_2}}\right) \tag{23}$$

where $(G/2Pk_BT)_{\text{line}}$ is the total Gibbs free energy for line charge with the number of charges equal to $2P$ and with linear charge density ξ_2. The factor $(\mathbb{R})_{\text{double helix}}$ is analogous to Eq. (19) and is a function of parameters that define the DNA double helix. The total electrostatic free energy of double helix is lower than that of the single helix at high salt [41, 42]. At low salt concentrations, the reverse is true.

E. Counterion and Co-ion Distribution Functions

Consider a counterion of absolute valence Z_0 that is held fixed at a distance r from the polyion. The polyion is modeled within the framework of the Manning line model [31] discussed in Section Chapter II.A. The idea here is that the reduction of a polyion charge is dictated by how far the counterion is from that charge. Hence, the renormalized charge due to counterion condensation is written as $q_{\text{net}} = (1 - \theta(r))q$, where q is the bare charge [43].

To assemble N polyion charges, with the constraint that the counterion is held fixed at a distance r, requires work w_1. There is an addition work, w_2, due to moving θN counterions from the bulk solution to the region that defines the layer of condensed ions [43]. Finally, there is the work w_{1c} due to interaction between the counterion at r and the charges on the polymer [43]

$$w_{1c} = -2z_0\xi(1 - \theta)K_o(r) \tag{24}$$

where $K_o(r)$ is the Bessel function of zeroth order. The total work is $w(r) = w_1(r) + w_2(r) + w_{1c}(r)$. To obtain the optimized counterion–polyion potential, one minimizes $w(r)$ with respect to variable $\theta(r)$ to obtain the equilibrium number of condensed counterions per polyion charge and the partition function $Q(r)$ of the condensed layer [43].

In order to discuss the predictions of the model, it is fruitful to partition the space external to the line charge into three regions. These regions are asymptotically defined with respect to $\kappa \to 0$ limit [43]: (a) far region, (b) intermediate region, and (a) near region. The distance r that defines these regions are, respectively, of $O(1/\kappa)$, $O(1)$, and $O(1/\kappa^x)$, where x is an exponent, whose value ranges between zero and unity [43].

There are several characteristics of the near region. First, if a counterion of valence Z is brought from a distance far away from the polyion to the near region, then Z univalent counterions are released into the bulk aqueous solution [43]. Thus, the near region is associated with the condensed layer. Second, the interaction potential between counterion–polyion is unscreened, attractive, and is proportional to $2Z_0\xi \ln(r/2b)$ [43].

Let a counterion of charge Z, whose valence is the negative of the polyelectrolyte, approach the intermediate region from far away. Analysis reveals that the number of counterions that are released increases from zero to Z

as the distance between the counterion and the intermediate region decreases [43]. In contrast, in the far away region $\theta(r) = 1 - \xi^{-1}$ even though the partition function of the condensed layer is distance dependent $Q(r) = Q(\infty)e^{[(2Z_0\xi/N)K_o(\kappa r)]}$ [43]. Another intriguing result in this region is that the polymer–counterion potential is exactly Debye–Hückel interaction, that is, $w(r) = -2Z_0\xi K_o(\kappa r)$, where ξ is the unrenormalized linear charge density [43].

The radial distribution function of polymer–counterion as a function of distance from the polyion surface exhibits two peaks. These peaks are interpreted as reflecting two populations of counterions: a diffuse Debye–Hückel cloud at a distance characteristic of the Debye screening length, and the condensed layer near the polymer surface where the condensed counterions reside [43].

The polyion–co-ion distribution displays distinctive characteristics. (1) In the far away region, the polymer–co-ion potential of mean force is the negative of the near-region polymer-counterion potential [43]. (2) In the intermediate region, the co-ion charge is to some extent incorporated with the polyelectrolyte charge. The radial distribution function as a function of distance indicates a single peak outside the condensed layer, but inside the Debye–Hückel cloud [43]. This peak is due to accumulation of co-ions. (3) When a co-ion is located in the near region, the number of condensed ions is increased by one and the polymer–co-ion potential of mean force is repulsive [43].

III. TOPICS IN COUNTERION CONDENSATION

A. Absorption Excess

By viewing the polyion and the bound ions as a "dressed" macromolecule, a deep connection was unraveled between Manning's θ and adsorption excess per charged monomer [44]. To be precise, the dressed polyion picture emerges from the PB equation through a statistical mechanical quantity called the absorption excess per charged monomer

$$\Gamma = -\frac{\beta\kappa}{2}\frac{\partial g_{\mathrm{elec}}}{\partial\kappa} \tag{25}$$

where $\beta = 1/k_{\mathrm{B}}T$, g_{elec} is the Gibbs free energy associated with charging the polyion, and $1/\kappa$ is the Debye screening length [44].

Assume that the polyion is a charged cylinder of contour length L and radius R. On evaluating the surface charge density σ of the cylinder by a self-consistent methodology, the electrostatic free energy βg_{elec} was expressible as [44]

$$\beta g_{\mathrm{elec}} = \frac{e\phi(x_o)}{k_{\mathrm{B}}T} - \frac{2}{\sigma x_o}\int_{x_o}^{w} [1 + (1 + w^2)^{1/2}]\frac{dw}{w} \tag{26}$$

In Eq. (26), ϕ is the electrostatic potential, $x_o = \kappa R$, $w = x_o(1+y)/2$, and $y = \cosh\left(\frac{e\phi(x_o)}{2k_BT}\right)$

The integral in Eq. (26) can be explicitly evaluated to yield an expression for the absorption excess per charged monomer [44]

$$\Gamma = \frac{2}{x_o\sigma}[(1+w^2)^{1/2} - (1+x_o^2)^{1/2}] \tag{27}$$

A key observation is that Eq. (27) reduces, under the constraint $\kappa L \gg 1$ and for large y, to that predicted by the counterion condensation model, namely, $\Gamma \approx \theta = 1 - 1/\xi$ [44].

B. Cell Model

In this section, we discuss a recent advance in the cell model of polyelectrolytes. In the cell model, a cylindrically shaped polyion with all its counterions reside in a cell. The cell is assumed to be cylindrical. The total charge in each cell is zero. Within the framework of cell model, thermodynamic quantities including counterion and co-ion distributions have been studied and compared with the counterion condensation model [45, 46].

A generalization of the cell model to the case of dilute solution of polyions was recently formulated by Deshkovski et al. [47]. For dilute solutions of polyions, it is fruitful to partition the volume around each polyion into two regions. One of the regions (outer) has spherical symmetry, while the other (inner) region has cylindrical symmetry [47].

A cylindrical volume is constructed around each polyion of length L, radius r_o, and charge (negative) Q_o. The radius of the cylindrical region is smaller than the length of the cylinder [47]. A second region is constructed by drawing a sphere, about each polyion, whose radius R is the average distance between the polyions [47].

In the model, univalent counterions are present both in the outer and in the inner regions. The dielectric constant of the solvent in cylindrical region is ε. It is entropically less than favorable to have all counterions in the inner region [47]. Consequently, the inner region is not electroneutral [47]. Due to the presence of counterions, the total charge of the inner region is nonzero. Let this charge be denoted by Q_R. Hence, two linear charge densities are introduced [47]

$$\gamma_o = -\frac{Q_0}{e}\frac{l_B}{L} \tag{28}$$

$$\gamma_R = -\frac{Q_R}{e}\frac{l_B}{L} \tag{29}$$

where l_B is the Bjerrum length and e is the unit charge. Thus, γ_R can be interpreted as the total charge per Bjerrum length along the symmetry axis of the inner region.

The solution of the PB equation for the electrostatic potential with appropriate inner and outer boundary conditions is [47]

$$\varphi(r) = \frac{kT}{e} \ln\left(\frac{r}{\nu l_B}\left(\left(\frac{r}{\Upsilon}\right)^{\nu} - \left(\frac{r}{\Upsilon}\right)^{-\nu}\right)^2\right) + \varphi_o \tag{30}$$

where r is the distance from the axis of the polyion. Both Υ and ν are unknown parameters and φ_o is a constant related to the zero of the potential. The distribution of counterions is given by [47]

$$c(r) = \frac{2}{\pi} \frac{\nu^2}{r^2 l_B} \left[\left(\frac{r}{\Upsilon}\right)^{\nu} - \left(\frac{r}{\Upsilon}\right)^{-\nu}\right]^{-2} \tag{31}$$

The two unknown parameters in Eq. (30) are obtained from the boundary conditions imposed at the inner and at the outer regions [47]

$$\zeta^{2\nu} = r_o^{2\nu}\left(\frac{\gamma_o - 1 - \nu}{\gamma_o - 1 + \nu}\right) \tag{32}$$

$$\zeta^{2\nu} = R^{2\nu}\left(\frac{\gamma_R - 1 - \nu}{\gamma_R - 1 + \nu}\right) \tag{33}$$

For fixed values of the linear charge densities, the counterion distribution can be studied by numerically solving Eqs. (31)–(33). Such an analysis reveals three distinct phases of counterion distributions labeled I, II, and III.

Phase I corresponds to polyions that have small values of linear charge density γ_o [47]. The electrostatic attractions in this region are unable to localize the counterions close to the polyion surface [47].

Phase II describes "saturated" counterion condensation [47]. The polyelectrolyte charge is nearly compensated by the counterions [47]. The uncondensed counterions are dispersed in a self-similar fashion throughout the cylindrical region [47]. In fact, the number of counterions bounded between radius r and, say, $2r$ is independent of r [47].

Phase III depicts "unsaturated" counterion condensation [47]. In this phase, counterion condensation occurs, but there are few counterions in the inner-cylindrical region [47].

C. Flexibile Polyelectrolytes

The nature of counterions in the vicinity of a flexible polyelectrolyte chain recently was investigated by taking into accounting the various competing factors between the free energy associated with electrostatic interactions between the polymer chain and the counterions, and the free energy changes due to entropy loss of the condensed counterions [48]. Another basic ingredient in the model is a self-consistent assessment of the degree of adsorption of the counterions and the size of the chain measured, for example, by the radius of gyration [48].

The polyelectrolyte chain consists of N monovalent charged monomers. The chain is immersed in an aqueous solution. The volume of the system is V. The solution is electroneutral. If the number of counterions adsorbed on the surface of the polyion is M, then the degree of ionization is given by $f = 1 - M/N$ [48]. The salt added to the solution is completely dissociated into n_- co-ions and n_+ counterions. The salt concentration is, therefore, $c_s = n_+/V$.

In the model, the total free energy of the system is decomposed into six terms. The first term accounts for the entropy of the condensed counterions [48]

$$\frac{F_1}{Nk_\mathrm{B}T} = f\log f + (1-f)\log(1-f) \tag{34}$$

The next contribution to free energy comes from the translational entropy of both co-ions and uncondensed counterions and is given by [48]

$$\frac{F_2}{Vk_\mathrm{B}T} = (f\rho + c_s)\log(f\rho + c_s) + c_s\log c_s - (f\rho + 2c_s) \tag{35}$$

where $\rho = N/V$ is the monomer density.

Another contribution to the free energy is due to fluctuations due to interactions between dissociated ions [48]

$$\frac{F_3}{Vk_\mathrm{B}T} = -\frac{1}{3}\sqrt{4\pi} l_\mathrm{B}^{3/2}(f\rho + 2c_s)^{3/2} \tag{36}$$

where l_B is the Bjerrum length.

The fourth contribution to the free energy reflects the increase in entropy as a result of ion pairs being formed from the adsorbed counterions [48]

$$\frac{F_4}{Nk_\mathrm{B}T} = -(1-f)\delta l_\mathrm{B}/l \tag{37}$$

Here, $\delta = (\varepsilon/\varepsilon_l)(l/d)$, ε_l is the local dielectric constant, ε is the bulk dielectric constant, l is the distance between two charges on the polylectrolyte, and d is the length of the dipole that is formed between one condensed couterion and the closest charged monomer [48].

The fifth term describes the free energy of a polyelectrolyte chain associated with M dipoles and N–M charges. The chain is in aqueous solution of N–M counterions and salt ions [48]. This free energy is obtained from the Edwards Hamiltonian [49]

$$\frac{F_5}{k_B T} = \frac{3}{2}(\tilde{l}_1 - 1 - \log\tilde{l}_1) + \frac{4}{3}\left(\frac{3}{2\pi}\right)^{3/2} w\sqrt{N}\frac{1}{\tilde{l}_1^{3/2}} + 2\sqrt{\frac{6}{\pi}}f^2\tilde{l}_B\frac{N^{3/2}}{\tilde{l}_1^{1/2}}\Theta_0(a) \quad (38)$$

where $\Theta_0(a)$ is defined by

$$\Theta_0(a) = \frac{\sqrt{\pi}}{2}\left(\frac{2}{a^{5/2}} - \frac{1}{a^{3/2}}\right)\exp(a)\mathrm{erfc}(\sqrt{a}) + \frac{1}{3a} + \frac{2}{a^2} - \frac{\sqrt{\pi}}{a^{5/2}} - \frac{\sqrt{\pi}}{2a^{3/2}} \quad (39)$$

w is a parameter that is a measure of short-ranged interactions that includes excluded volume and hydrophobic effects. Finally, the scaled Debye screening parameter, the number density of monomers and the salt concentration are defined [48], respectively, by $\tilde{\kappa}^2 = 4\pi\tilde{l}_B(f\tilde{\rho} + 2\tilde{c}_s)$, $\tilde{\rho} = \rho l^3$, and $\tilde{c}_s = c_s l^3$, where $\tilde{\kappa} = \kappa l$. The parameter $\tilde{l}_1 = l_1/l$ is the scaled expansion factor l_1 that defines the mean-square end-to-end distance

$$\langle R^2 \rangle = Nll_1 \equiv Nl^2\tilde{l}_1 \quad (40)$$

and $a = \tilde{\kappa}^2 N\tilde{l}_1/6$.

The last term in the free energy describes the correlation between the ion pairs on the polyelectrolyte chain. One term in this free energy, $F_{6,1}$, depicts correlations between the ion pairs, while the remaining term, $F_{6,2}$, accounts for the interaction between an ion-pair and a charged monomer and monomer charge and ion pair, respectively [48]

$$\frac{F_{6,1}}{k_B T} = \frac{4}{3}\left(\frac{3}{2\pi}\right)^{3/2} w_1\delta^2\tilde{l}_B^2\tilde{d}^6(1-f)^2\sqrt{N}\frac{1}{\tilde{l}_1^{3/2}} \quad (41)$$

$$\frac{F_{6,2}}{k_B T} = \frac{4}{3}\left(\frac{3}{2\pi}\right)^{3/2} w_2\delta^2\tilde{l}_B^2\tilde{d}^4 f(1-f)\sqrt{N}\frac{1}{\tilde{l}_1^{3/2}} \quad (42)$$

The parameters w_1 and w_2 reflect the change in the excluded volume parameter due to attractive interaction between the ion pair and the monomer and monomer charge and ion pair, respectively [48].

The degree of ionization f and the scaled expansion parameter l_1 as a function of the various parameters that enter the model are obtained by minimizing the total free energy. Analysis revels that the degree of ionization decreases if the chain flexibility or the salt concentration increases [48]. This decease is also observed if the chain length or the dielectric constant is decreased [48].

In Manning model, the slope of f versus the Coulomb strength or the linear charge density parameter $\tilde{l}_{\mathrm{B}} = l_{\mathrm{B}}/l$ is discontinuous at $\tilde{l}_{\mathrm{B}} = 1$ [31, 48]. This is a consequence of the fact, that for all $\tilde{l}_{\mathrm{B}}$ less than or equal to unity, f is linear function of $1/\varepsilon$ and saturates at unity [31, 35, 48]. The present model indicates a sigmoidal variation of f with $1/\varepsilon$ and saturates as $\tilde{l}_{\mathrm{B}} \rightarrow 0$ [48].

Another prediction of the model is the decrease in the size of the polymer as the salt concentration is increased [48]. If counterion adsorption is dominant, the interactions between the ion pairs are attractive in nature, and leads to the radius of gyration being less than that of a Gaussian chain [48].

D. Fluctuations

The Manning counterion condensation line model provides insights into the mean-square number fluctuations of the condensed counterions, $\langle(\Delta\theta)^2\rangle$, where θ is the number of condensed counterions per polyion charge [50]. Denote by θ_o the value of θ for which the total polyelectrolyte free energy G per charge is a minimum, that is, $(\partial G/\partial\theta)_{\theta=\theta_o}$ vanishes. Expanding G in powers of $\theta - \theta_o$ to quadratic order leads to [50]

$$\Delta G = G - G(\theta_o) \approx \frac{1}{2}\left(\frac{\partial^2 G}{\partial\theta^2}\right)_{\theta=\theta_0} \tag{43}$$

On evaluating the second derivative on the right-hand side of Eq. (43), one obtains [50]

$$\Delta G = \frac{k_{\mathrm{B}}T}{2}\frac{(\Delta\theta)^2}{\theta_o}[1 - 2z^2\xi\theta_o \ln(\kappa b)] \tag{44}$$

Now, let us consider the fluctuations of a segment of M charges. The segment is located at least one Debye length away from the ends of the polyion. If Δg denotes the deviation of the free energy of the segment from its minimum value, then $\Delta g = M\Delta G$. But, the thermal average of ΔG is, by definition, $\langle\Delta G\rangle = k_{\mathrm{B}}T/2$. Consequently, the mean-square number fluctuations of condensed counterions, $\langle(\Delta\theta)^2\rangle$, is [50]

$$\langle(\Delta\theta)^2 > = \frac{\theta_o}{[1 - 2z^2\xi\theta_o \ln(\kappa b)]M} \tag{45}$$

The contribution of the term in the square brackets is greater than unity and is due to electrostatic interactions between the polyion and the condensed counterions, as well as between the condensed counterions. Consequently, the mean-square number fluctuations are smaller in the presence of interactions [50].

Another model in which fluctuation effects of the condensed counterions has been explored is based on a planar geometry [51]. Consider a single plate of charge density $\sigma_o\delta(z)$, together with its counterions, each of charge $-Z_o$e, such that the system is neutral overall. It is fruitful to partition the counterions as condensed and as free fraction [51].

The free counterions, of average concentration $c = n_R/Z_o\lambda_R$, have three-dimensional (3D) characteristics [51]. These counterions are treated as an ideal gas and are confined to a slab, whose thickness is the Gouy–Chapman screening length $\lambda_R = 1/\pi l_B Z_o n_R$. The free energy of the free counterions is approximated as [51]

$$F_{3D}/k_B T \approx c\lambda_R\{\ln(ca^3) - 1\} - \frac{\kappa^3}{12\pi}\lambda_R \tag{46}$$

where a is a cutoff distance of the order of size of a counterion. The first term in Eq. (46) is the entropy of a 3D ideal gas, while the second term depicts the fluctuation corrections to it, at the level of Debye–Hückel theory [51, 52], and κ as usual is the Debye screening parameter.

The condensed fraction, of density n_c, is localized on a two-dimensional (2D) planar surface [51]. The fraction of condensed counterions in 2D is $\tau = Z_o e n_c/\sigma_o$, and the reduced surface charge density is $en_R = \sigma_o - Z_o e n_c$. The free energy of the condensed counterions per unit area is [51]

$$F_{2D}/k_B T \approx n_c\{\ln(n_c a^2) - 1\} + \frac{1}{2}\int \frac{d^2 q}{(2\pi)^2}[\ln(1 + \frac{1}{q\lambda_D}) - \frac{1}{q\lambda_D}] \tag{47}$$

where $\lambda_D = 1/2\pi l_B Z_o^2 n_c$ is the 2D Gouy–Chapman screening length.

Minimizing the total free energy $F_{2D} + F_{3D}$, with respect to the fraction τ of condensed counterions, leads to a nonlinear equation [51]

$$1 + \tau g \ln(\frac{\pi}{\tau g \vartheta}) - \ln(\frac{\tau}{(1-\tau)^2 g \vartheta}) - \frac{4}{3}g(1-\tau) = 0 \tag{48}$$

Here, $\vartheta = a/Z_o^2 l_B$, $g = Z_o^2 l_B/\lambda$, and where λ is the unrenormalized Gouy–Chapman length and l_B is the Bjerrum length.

If the coupling is weak, that is, for $g \ll 1$, there is no condensation of counterions [51]. If density of surface charges is high, and for ϑ less than a critical value ϑ_c, analysis reveals that τ as a function of g exhibits a jump at $g_o(\vartheta)$. This feature is characteristic of a first-order phase transition [51]. However, for ϑ larger than a critical value ϑ_c, there is no phase change even though counterion condensation occurs. This phase change is analogous to what occurs when a line of first-order phase transitions end at the critical point [51]. Taking $a \sim 1$ Å and Bejrrum length for water at 25°C, the counterion valency Z_c corresponding to the critical value ϑ_c is found to be ~ 1.6. Thus, for $Z_o < Z_c$, there is no first-order phase condensation of the counterions [51].

The theory breaks down for large values of g. To deal with the strong coupling regime, the condensed counterions are assumed to behave more as a strongly correlated 2D liquid [51, 53]. Some of the features predicted by the 2D correlated liquid model are the existence of renormalized Gouy–Chapman length, as well as reduced surface charge density [51, 53].

A field theoretical description of counterion fluctuations has also been formulated. In this approach, fluctuating electrostatic potential is introduced in which the surface charge density is treated as a variational parameter [54]. The methodology captures the nonlinearity of the counterion distributions of highly charged systems.

For a charged cylinder, the various parameters that appear in the above model are [54]: the radius R, the linear charge density υ, the surface charge density $\sigma = \upsilon/2\pi R$, the Debye screening length, the Gouy–Chapman length $\lambda = 1/2\pi Z\sigma$, where Z is the valence of the positive and negative ions, the Bjerrum length, and the uniform charge renormalization factor η_o. As the salt concentration decreases, the model predicts [54] that the charge renormalization factor η_o is governed by the Manning limit: $\eta_o = 1/\bar{R} = 1/Z\upsilon l_B$, where $\bar{R} = R/\lambda = Z\upsilon l_B$ is the Manning parameter [31].

For $\kappa R \ll 1$, the charge renormalization is (54)

$$\eta_o = \frac{1}{\bar{R}}\left(1 + \frac{\ln \bar{R}}{2\ln(1/\kappa R)}\right) \tag{49}$$

This is in agreement with the Manning–Mohanty prediction [40, 54]. At high salt concentration, asymptotic analysis leads to charge renormalization that vanishes as [54]

$$\eta_o = 1 - \frac{1}{24(\kappa\lambda)^2} \tag{50}$$

These results have been generalized to planar and spherical surfaces [54]. For the later case, at high salt concentration, the charge renormalization is also given by Eq. (49) [54]. For low salt, however, and for $\kappa R \ll 1$, one obtains a logarithmic dependence of η_o on the Manning parameter [31, 54]

$$\eta_o = \frac{1}{2\bar{R}}\ln\left(\frac{1}{\bar{R}(\kappa\lambda)^2}\right) \tag{51}$$

The charge renormalization factor corresponds to an effective charge on the sphere [54]

$$z_{\text{eff}} = z\eta_o = \frac{R}{Zl_B}\ln\left(\frac{Zl_B z}{R^3(\kappa)^2}\right) \tag{52}$$

In Eqs. (51)–(52) z and R are the charge and the radius of the sphere, respectively.

E. Binding Isotherm Model

In the binding isotherm model [55], a polyion labeled P has N fixed charges. The polyion is divided into sections [55]. There is no section of the polyion that has more than one binding site. Let the counterion valence be z_c. Let J denote a specific configuration of the n_c counterions. In this configuration, the counterions bind to the polyion and form the complex $D_J(n_c)$, and are described by the equilibrium [55]

$$P + n_c C \rightleftarrows D_J(n_c) \tag{53}$$

In terms of chemical potential μ of the various species, the equilibrium condition for the above reaction is [55]

$$\mu_P + \mu_C n_c \rightleftarrows \mu_{D_J(n_c)} \tag{54}$$

If the standard state chemical potential of species i is denoted by μ_i^o, then the chemical potentials of the species i satisfy [55]

$$\mu_i = \mu_i^o + k_B T \ln(m_i/m_o) \tag{55}$$

where m_o and m_i are, respectively, the molarity of water and molar concentration of species labeled i.

Let ΔA_{loc}, $A_{\mathrm{elec}}(0)$, $A_{\mathrm{elec}}(J, n_c)$ denote the free energy associated with moving a counterion to the binding site under standard conditions, the free energy due to electrostatic interactions of the bare polyion, and the free energy of the polyion in the complex $D_J(n_c)$, respectively [55]. Making use of Eqs. (52)–(54), one can express the total concentration of the complexes m_D as [55]

$$\frac{m_D}{m_P} = \sum_{n_c=1}^{N} \left(\frac{\tilde{\beta} m_C}{m_o}\right)^{n_c} \sum_J e^{-(A_{elec}(J,n_c) - A_{elec}(0))/k_B T} \tag{56}$$

Here, $\tilde{\beta} = e^{-\Delta A_{\mathrm{loc}}/k_B T}$.

The right-hand side of Eq. (56) can be simplified under the following assumptions [55]. First, it is assumed that a single configuration, say $\bar{J}$, dominates the sum over J, and that this configuration corresponds to uniformly distributed bound counterions. Second, uniform configurations have the same electrostatic energy. Third, the number of uniform configurations is $(N/n_c)^{n_c}$. Fourth, one can approximate $\ln(m_D/m_P)$ by its maximum term in the sum over n_c. Carrying out the maximum term method leads to [55]

$$r = m_c \left(\frac{\tilde{\beta}}{e m_o}\right) |z_c| e^{-\frac{\partial A_{elec}/k_B T}{\partial \bar{n}_c}} \tag{57}$$

where $r = \bar{n}_c |z_c| / N$ is the fraction of condensed counterions, $\bar{n}_c$ is the maximum value of n_c, and here $e = 2.718$.

Schurr and Fujimoto [55] evaluated the fraction of condensed counterions, and the local binding constant $\tilde{\beta}$ for various geometries, such as two parallel lines, double helix, and single helix. Several comments are in order. First, as $m_c \to 0$, Manning's prediction for the number of condensed counterions per polyion charge is recovered [55, 31]. Second, the binding constant $\tilde{\beta}$ is a function of salt concentration and the actual geometry of the charged sites. Third, for a single-line charge, the value for $\tilde{\beta}$ is the same as that obtained by Manning [55, 31]. Finally, as m_c increases, the fraction of condensed counterions drops below the Manning limiting value [55].

F. Other Models

There has also been a recent attempt to generalize the cell model to two dimensions. The Hamiltonian for such a system is [56]

$$\beta H = 2\xi \sum_{i=1}^{n} \log(r_i/a) - \frac{\xi}{n} \sum_{i \neq j} \log\left|\vec{r}_i - \vec{r}_j\right| \tag{58}$$

In the model, there are n point ions, each of charge q', interacting with a disk of radius a located at the origin. The charge on the disk is q. The Manning charge-density parameter ξ is $\beta q q'$, where $\beta = 1/k_B T$. The ions are restricted to the region $a \leq r \leq R$ [56]. Observe that the polyion–ion and ion–ion interactions are logarithmic.

The Hamiltonian is transformed exactly to one having planar symmetry, and defined on the (u, ϕ) strip, where $u = \log(r/a)$ and varies between zero and $L = \log(R/a)$, and ϕ is between zero and 2π [56]. The canonical partition function is explicitly evaluated in the strip. In the mean-field approximation, the threshold for countercondensation is the same as that predicted by Manning [56]. As the Manning parameter ξ is increased, transitions that reflect condensation due to a single ion is observed [56]. The unique feature has been verified by Monte Carlo simulations [57].

A connection between Manning counterion condensation [31] and Kosterlitz–Thouless transition [58a] was conjectured by Mohanty, and supported by phenomenological arguments [58b]. An explicit calculation using nonlinear PB confirmed the hypothesis [59].

In another study, the nonlinear PB equation with added salt has been approximately solved for a cylindrical polyion of radius a, by matching the near- and far-field solutions in an asymptotic expansion in terms of a small parameter, $\varepsilon = 1/\ln(1/\kappa a)$, and where κ^{-1} is the Debye screening length [60].

Let $\tilde{q}$ be the sum of the polymer charge and the charge of the ions, that are enclosed within a distance r from the surface of the polyion, per Bjerrum length [60]. The PB equation is rewritten in terms of the variable $\tilde{q}$. Let us consider the far-field solution of this PB equation for a thin rod infinitely long whose charge is renormalized by the condensed counterions [60]. Analysis reveals that the mass of the condensed fraction satisfies the Manning condition [31, 60]

$$\tilde{q}(R_{\rm M}) = 1 \tag{59}$$

where $R_{\rm M}$ is the thickness of the condensed layer.

The inner solutions are constructed from the fact that near the polyion, the ions are mainly those with charges opposite to that of the macromolecule [60]. The technical aspect of the analysis involves finding the region where the inner and the outer solutions overlap, and are accurate as $\varepsilon \to 0$ [60]. Such an analysis reveals that the thickness of the condensed layer is [60]

$$R_{\rm M} = (a/\kappa)^{1/2} e^{-[(\xi-1)^{-1}+C_o]/2} \tag{60}$$

Here, a is the polyion radius, $C_o = \gamma - \ln 8$ and in which the Euler constant is denoted by γ.

The results for the thickness of the condensed layer agree with matched asymptotic expansions by Ramanathan and co-workers [33, 34, 60]. In comparison, numerical solutions to PB indicate that the thickness of the condensed layer goes as $R_{\rm M} = (a/\kappa)^{1/2}$ [60, 61]. In summary, the analysis provides [60] (a) an analytic solution that is globally accurate, (b) information about the structure of the condensed fraction, and (c) the density profile of the uncondensed ions.

In another investigation of the ion atmosphere of DNA, a coarse-grained model was developed [62]. In this model, the DNA is taken to be a double helix, and in which a charged sphere is used to represent a phosphate group. A neutral sphere, located in between the polyion and the phosphate, and a sphere of charge e placed near a phosphate group describes a nucleotide [62]. The coordinates of the phosphates are input to the model, and are obtained from the X-ray fiber diffraction data [63]. The bound counterions are assumed to be either diffuse or tightly bound. The region of space around the DNA defined as tightly bound is a cell that extends from the surface of the grooves to the phosphates [62]. Thus, an ion that is tightly bound exists in the grooves or is bound to the phosphates.

The diffuse ions are handled within the framework of the PB equation [62]. Fluctuations of the tightly bound ions are taken into account by introducing binding modes within a statistical mechanical framework [62]. For fixed occupancy of ions that are tightly bound, the free energy of binding is expressed in terms of the potential of the mean force for intra-cell and inter-cell interactions [62].

A key assumption in the model is that the ion distribution of the diffuse layer is dictated by the preequilibrated distribution of the ionic charges that are tightly bound [62]. The assumption is valid at low salt. At high salt, however, the free energy of the diffuse layer is expected to depend on the characteristics of the charge distribution, and hence the mode of binding of the tightly bound ions [62].

Quantities such as the mean thickness of the tightly bound region, the electrostatic free energy, the radial distribution of the total bound charge, and the net charge of tight bound counterions, are approximately evaluated and compared with MC simulations [62].

In another generalization of counterion condensation model, the polyelectrolyte was viewed as an infinitely long rod, of radius a, and charge $\xi e/l_B$ per length, where ξ is the linear charge density [64]. The polyelectrolyte is immersed in a salt solution. The Debye–Huckel cloud neutralizes part of the charge in the ion atmosphere [64]. The volume between the surface of the rod, and the condensed volume, at $r = c$, is occupied by the condensed counterions [64]. The electrostatic potential is split into two parts: $\psi_1(r) + \psi_2(r)$, where $\psi_1(r)$ is the Debye–Huckel potential that is valid for $r > a$ [64]. The potential $\psi_2(r)$ is due to the condensed and all remaining charges in the region $a < r < c$. The potential $\psi_2(r)$ is governed by the Poisson equation [64]. From Gauss's law, the boundary condition is that the slope of $\psi_2(r)$ vanishes at $r = c$. The total electrostatic energy E_{elec} of the rod is proportional to an integral over the total volume of the product of the potential $\psi(r)$ at a distance r from the rod, and the charge density $\rho(r)$ at r in the ion atmosphere [64]. The entropy S_{el} of the ionic double layer is also obtained form the electrostatic potential and ion potentials in the various regions [64]. The total free energy of the polyelectrolyte is constructed via the thermodynamic relation $F_{el} = E_{\text{el}} - TS_{\text{el}}$.

IV. TOPICS IN COUNTERION CONDENSATION: NUCLEIC ACID

A. Counterion Release

An experimental measurable quantity that is of fundamental importance in nucleic acid is the preferential interaction coefficient Γ. The parameter Γ is a statistical thermodynamic quantity that probes nonideality effects of nucleic acids in aqueous solution. This quantity also provides information on the characteristics of co-ions and counterions distributions in the vicinity of polyions [65–76].

Consider an equilibrium reaction between a polyion D and a charged ligand X of valence $+z$. The binding equilibrium for such a process is [65, 66, 71–73]

$$\mathrm{D} + \mathrm{X} \underset{k_r}{\overset{k_f}{\rightleftarrows}} \mathrm{XD} \tag{61}$$

k_f and k_r are the forward and the reverse rate constants. If K_{eq} denotes the equilibrium constant for the binding equilibrium, we have $K_{eq} = K_{\text{obs}} K_{\gamma}$, where K_{γ} is the activity coefficients of the products divided by that of the reactants, and the intrinsic binding constant, K_{obs}, is the corresponding concentration of the products over that of the reactants [65, 66, 71–74].

Model based on counterion condensation predicts that the variation of the intrinsic binding constant with salt concentration is given by [31]

$$\frac{\partial K_{\text{obs}}}{\partial \log[\text{M}^+]} = -z \tag{62}$$

In Eq. (62), the concentration of cations in aqueous solution is symbolized by $[\text{M}^+]$. An assumption, explicit in Eq. (62), is that the binding equilibrium does not lead to changes in the polyion conformation [31, 65–73]. The physical content of Eq. (62) is simple, namely, that if a ligand of charge z neutralizes z charges on the polyion, then this process leads to the release of z counterions into the bulk solution [31, 65–73].

Now, consider a three-component system. The three components are water, polyion and an electrolyte, the latter two are tagged as D and 3, respectively. Explicit thermodynamic arguments show that the variation of K_{obs}, with respect to salt concentration under constant temperature T and pressure P, is related to the polyion–ion preferential interaction coefficient [65–73]

$$-\left(\frac{\partial \ln K_{\text{obs}}}{\partial \ln c_3}\right)_{T,P;c_{\text{D}}\to 0} = (1 + 2\Gamma_{3\text{D}}) \tag{63}$$

Here, c_3 and c_{D} are the concentrations of salt and polyion, respectively. The difference in association of counterions between the products and the reactants is given by $(1 + 2\Gamma_{3\text{D}})$ [65–73]. Consequently, counterions associated with the polyion will be released into the bulk solution if the $(1 + 2\Gamma_{3\text{D}})$ is negative. The release of counterions leads to density changes in the condensed layer and provides an additional driving force, except that it is entropic in nature [65–73].

An alternate way to define preferential interaction coefficient is to consider the ionic density $n_i(\rho)$ of species i at a radial distance ρ from the polyion. The parameter Γ_i can then be defined, per polyion charge and per unit length, as an integral over the excess local density [68, 71]

$$\Gamma_i = \phi b \int_a^{R_o} (n_i(\rho) - n^{\text{bulk}})\rho d\rho \tag{64}$$

The quantity $\phi = 2\pi N_{\text{A}} 10^{-27}$, N_{A} is the Avogardo number, n^{bulk} is the bulk density and where b is the axial spacing of the charges. The polyion is taken to be infinitely long, but has a radius a. The entire system is enclosed in a cylindrical volume of radius R_o.

It is insightful to view counterion association through the Donnan equilibrium experiment. Let c_{D} be the concentration of polyion, and let c_3 be the salt concentration in the sample chamber [65–70, 74]. In the presence of nucleic acid, the distribution of salt is perturbed. However, there are counterions in solution that are not associated with the polyion [65–70, 74]. To reduce the free counterion concentration, some of these counterions must move across the membrane. As a result of electroneutrality, each counterion that is not associated has with it a co-ion [65–69, 74]. At equilibrium, the counterion and the co-ion concentrations are balanced in the two chambers [65–70, 74].

The partitioning of the salt ions across the membrane leads to another interpretation of the preferential interaction coefficient [65–69, 74]

$$\Gamma = \lim_{c_D \to 0} \left(\frac{c_3 - c_3'}{c_D} \right)_{T,P} \tag{65}$$

In Eq. (65), the prime indicates the salt concentration in the reference chamber of the Donnan dialysis experiment with T and P held fixed.

A more fundamental way to describe the preferential interaction coefficient follows by a variational formulation of the PB equation. The variational technique applied to the PB equation allows the electrostatic free energy of charges in solution to be expressed as [75]

$$\Delta G_{\text{elec}} = \int (k_B T \omega^D u - \varepsilon E^2/8\pi) dV - \Delta\Pi \tag{66}$$

where $\Delta\Pi$ is

$$\Delta\Pi = k_B T \sum_i \int n_i (e^{-z_i u} - 1) dV \equiv \sum_i \Delta\Pi_i \tag{67}$$

The dimensionless electrostatic potential u satisfies the PB equation, and ω^D is the charge distribution of the polyion. In Eq. (66), E is the electric field, ε is the dielectric, n_i is the concentration of ionic species i, integration is over entire volume of the aqueous solution, and total excess of ith ion is denoted by Π_i [75].

Let the salt be of the form $A_p B_q$. Denote the corresponding bulk concentrations of ions to be c_A and c_B. Then, the change in free energy with respect to salt is linked to the dearth or the excess number of ions from the salt [75]

$$\begin{aligned} \frac{\partial \Delta G_{\text{elec}}}{\partial \ln c_{AB}} &= -k_B T p c_{AB} \int (e^{-z_A u} - 1) dV - k_B T q c_{AB} \int (e^{-z_B u} - 1) dV \\ &= -(\Delta\Pi_p + \Delta\Pi_q) \end{aligned} \tag{68}$$

In obtaining Eq. (68), from Eqs. (67) and (66), one has used the fact that $c_A = p c_{AB}$ and $c_B = q c_{AB}$ [75]. Note that if the source of nonideality is due to electrostatic interactions, and the salt is 1-1, such as NaCl, then the activity coefficient of the polyion is related to the electrostatic free energy $\Delta G_{\text{elec}}/k_{\text{B}}T = \ln \gamma_{\text{D}}$ [75].

The salt dependent of the polylectrolyte activity coefficient varies as [75]

$$\begin{aligned} -\frac{\partial \ln \gamma_{\text{D}}}{\partial \ln c_s} &= z_{\text{D}}(2\Gamma_{\text{D}} + 1) \\ &= (\Delta\Pi_+ + \Delta\Pi_-) \end{aligned} \tag{69}$$

z_D is the charge of the polyion. Equations (68, 69) provide a fundamental statistical mechanical relation for the preferential interaction coefficient.

B. Preferential Interaction Coefficient

We now discuss an asymptotic methodology of calculating the preferential interaction coefficient of nucleic acids. Assume that the electric potential of a cylindrical polyion satisfies the nonlinear PB equation. The polyion is immersed in an aqueous solution containing symmetric electrolytes. If $y \equiv ze\psi/k_{\text{B}}T$ denotes the reduced potential, where ψ is the actual potential, then the preferential interaction coefficient can be expressed in terms of the surface potential y_a[65–67, 77, 78]

$$\Gamma_u^{\text{coul}} = -\frac{1}{2} + \frac{\xi}{4} - \frac{1}{16\varepsilon^2\xi}(\exp(y_a) - 2 + \exp(-y_a)) \tag{70}$$

In Eq. (70), ξ is the axial charge density of the cylinder and is related to the surface charge density $\sigma = e/(2\pi ab)$, where b is the axial separation of the charges and a is the radius of the polyion [31, 77, 78].

Recently, the PB equation was solved as an asymptotic series in terms of the reduced potential $y = y_o + O(f(\varepsilon))$, where ε^{-1}, the ratio of the radius a of the polyion to the Debye screening length κ^{-1}, is a small parameter [77]. In the limit $f(\varepsilon) \to 0$, the remainder $O(f(\varepsilon))$ in the expansion is of order $f(\varepsilon)$ [77].

The zeroth-order term, y_o, is [77]

$$y_o = 2K_o(x\varepsilon^{-1}) + 2\gamma - 2\ln(\beta^{-1}S(\beta K_o(x\varepsilon^{-1}) + \beta C)) \tag{71}$$

where $K_o(\kappa r) = K_o(x\varepsilon^{-1})$ is a modified Bessel function, $\gamma = 0.5772$, $x = r/a$ is reduced radial distance and $S\alpha = \sinh\alpha$ or $\sin\alpha$ as $\varepsilon^{-1} \to 0$, depending on whether linear charge density is less than or larger than unity, respectively. From electroneutrality, and the boundary conditions, one obtains the constants β and C in Eq. (71) [77]

$$C = e^{\gamma} + O(\beta^2) \tag{72}$$

$$\beta = \frac{\pi}{\ln\varepsilon + \alpha + (\xi - 1)^{-1}} + O(\beta^4) \tag{73}$$

with $\alpha = e^{\gamma} + \ln 2 - \gamma = 1.897$.

The potential at the surface of the polyion is obtained by imposing the boundary condition at the surface [77]

$$y_{o,a} = y_o|_{x=1} = \ln[4\varepsilon^2((\xi - 1)^2 \pm \beta^2)] \tag{74}$$

where "−" and "+" denote the solution for the cases $S(\nu) = \sinh \nu$ and $S(\nu) = \sin \nu$, respectively. Substituting the expression for the surface potential in Eq. (70) leads to the desired expressions for the preferential interaction coefficient [77]

$$\begin{aligned} \Gamma^{\text{coul}}_{u,\text{LS}} &= -\frac{1}{4\xi} - \frac{1}{4\xi}\left(\frac{\pi}{\ln \varepsilon + \alpha + (\xi - 1)^{-1}}\right)^2 & \xi > 1 \\ \Gamma^{\text{coul}}_{u,\text{LS}} &= -\frac{1}{4\xi} - \frac{1}{4\xi}\left(\frac{\pi}{2(\ln \varepsilon + \alpha)}\right)^2 & \xi \cong 1 \\ \Gamma^{\text{coul}}_{u,\text{LS}} &= -\frac{1}{2} + \frac{\xi}{4} - \frac{(1-\xi)^2 e^{-2(1-\xi)(\ln 2 + C - \gamma)}}{\xi \varepsilon^{2(1-\xi)}} & \xi < 1 \end{aligned} \tag{75}$$

where $-1/4\xi$ is the limiting law value of Γ that follows from Manning counterion condensation formulation [17, 31]. The symbol LS denotes a low salt value of Γ [31, 77].

If an asymptotic expansion is carried out in the limit $\varepsilon \to 0$, then one obtains the preferential interaction coefficient at high salt (HS) [77, 78]

$$\Gamma^{\text{coul}}_{u,\text{HS}} = -\frac{1}{2} - \frac{1 - \bar{q}}{2p} + \varepsilon\frac{(1 - \bar{q})^2}{4p\bar{q}} \tag{76}$$

where $\bar{q} = \sqrt{1 + p^2(1 + \varepsilon/2)^{-2}}$ and $p = \varepsilon\xi$.

The above results can be compared with low salt solutions obtained from numerical solutions of the nonlinear PB equation [77, 79, 80]. The preferential interaction coefficient is found to be of the form [77–79]

$$\Gamma^{\text{coul}}_{u} = \Gamma^{\text{coul}}_{u,\text{LL}}(1 + \sigma_\Gamma) \tag{77}$$

where the term σ_Γ provides correction to the LL value. In fact, at very low salt, σ_Γ decreases as $(\ln \varepsilon)^{-2}$ (77, 79).

Since low salt calculations are based on an asymptotic expansion in terms of ε^{-1}, one condition for the validity of Eq. (71) is that $\varepsilon \gg 1$ [77]. The accuracy of the solutions obtained by asymptotic expansions declines in the boundary layer [77] that separates $\varepsilon < 1$ (HS) and $\varepsilon > 1$ (LS).

Taubes et al. developed rigorous bounds on preferential interaction coefficient [72]. Their staring point is the reduced electrostatic potential u of a polyion of radius a and length L in an ionic solution containing monovalent and divalent counterions. The reduced potential $u(\rho)$ satisfies the nonlinear PB equation [72, 81]

$$\Delta u = \sigma^2(\alpha e^u + (1-\alpha)e^{2u} - e^{-u}) \tag{78}$$

where the Laplacian is denoted by Δ, $\alpha = 1/(1 + n_d/2n_m)$, n_d, and n_m are the concentrations of divalent and monovalent ions, respectively, $\sigma^2 = \varepsilon^2/(2+\alpha)$ and $\varepsilon = \kappa a$.

Since $R \gg a$, the preferential interaction coefficient simplifies to [72]

$$\Gamma = -\frac{(\kappa a)^2}{4\xi} I - \frac{(\kappa a)^2}{8\xi} \tag{79}$$

where $I = \int_1^\infty (1 - e^{-u(\rho)})\rho d\rho$. To find tight bounds on Γ, Taubes and coworkers employed the "maximum principle" [72]. The basic idea here is that if the Laplacian of a function is positive, then that function cannot have a local maximum. Alternatively, if the Laplacian of the function is negative, the function cannot have a local minimum [72].

To see how this works, consider the $\alpha = 0$ limit and case $\sigma \geq 1$. This corresponds to the case $n_d/2n_m \ll 1$. Let $w(\rho)$ satisfy the equation [72]

$$w_{\rho\rho} = \sigma^2(e^w - e^{-w}) \tag{80}$$

with the boundary conditions $w \to 0$ as $\rho \to \infty$ and $w_\rho|_{\rho=1} = -2\xi$. Equation (80) can be exactly solved [72]

$$e^{-w/2} = (1 - \bar{b}e^{-2^{1/2}\sigma(\rho-1)})/(1 + \bar{b}e^{-2^{1/2}\sigma(\rho-1)}) \tag{81}$$

$\bar{b}$ is a constant deduced from boundary conditions. From the solution, we observe that for all ρ, $w_\rho < 0$. Consequently, $w_{\rho\rho} + w_\rho/\rho \leq \sigma^2(e^w - e^{-w})$ [72]. This means that $u \leq w$ by the maximum principle, and an upper bound for the integral that defines I is evaluated in a straightforward fashion [72]

$$\int (1 - e^{-w(\rho)})\rho d\rho = 4((b/2^{1/2}(1+b)\sigma) + \frac{1}{2\sigma^2}\ln(1+b) \tag{82}$$

The strategy for obtaining a lower bound to Eq. (80) is to calculate the constant b from the boundary condition $w_\rho - w/2 = -2\xi$ at ξ equal to unity [72]. Let $z = w\rho^{1/2}$, so that z satisfies $z_{\rho\rho} + z_\rho/\rho \geq \sigma^2(e^z - e^{-z})$. Another boundary condition is $z_\rho|_{\rho=1} = -2\xi$ [72]. Consequently, for all $\rho \geq 1$, $u \geq z$. Note that from Eq. (80), $w_\rho = -2^{1/2}(e^w + e^{-w} - 2)\sigma$. Let c_* be the derivative, with respect to ρ, of z, evaluated at unity. Analysis reveals that c_* satisfies the nonlinear relation [72]

$$2\xi = c_*/2 + \sigma 2^{1/2}(e^{c_*/2} - e^{-c_*/2}) \tag{83}$$

The value of the integral for I obtained by evaluating $\int(1 - e^{-z})\rho d\rho$, constitutes, a lower bound for Γ [72].

Similar strategy can be exploited for other values of α, corresponding to mixed salts, and for a polyion of finite length [72]. For B-DNA, at 51.0 mM, the bound on preferential interaction coefficient is $-0.095 < \Gamma < -0.057$, while at 0.255 M, the bound is $-0.281 < \Gamma < -0.199$ (72).

C. Branched Nucleic Acids

The density of phosphate charges [74–83] in tertiary structures, such as pseudoknots, RNA, inter-helix junctions, Holliday and branched recombination junctions, exceeds that in linear DNA molecules [74–93]. In fact, the degree of counterion association has been linked with the overall stability of these types of structures [46, 82–84, 95, 96].

Since the preferential interaction coefficient Γ can be interpreted in terms of Donnan equilibrium [66, 74, 96, 97], a grand canonical Monte Carlo (GCMC) simulations could be used to determine it, from a knowledge of the slope of salt concentration c_3 as a function of the polyion concentration c_D [68, 73, 74]. Such an analysis was carried out by Olmsted and Hagerman for a tetrahedral four-arm DNA junction, based on the so-called "primitive model" of the electrolyte [74].

A Metropolis method with umbrella sampling was employed [74, 98–102]. For transition between states i and j, the acceptance ratio for moves is $F_{ij} = \exp(-(E_j - E_i)/k_B T)$, where E_i is the energy of configuration i, k_B is the Boltzmann constant, and T is the absolute temperature. The energy of conformation i is obtained by summing the Coulombic interactions over all charged species in a cell or its adjacent image cell [74, 101]. If $\bar{n}$ is the number of ion pairs that are deleted or inserted, then the acceptance ratio for insertions is

$$F_{ij} = B_{ij}\exp(\ln(a_3(N_A)^2)^{\bar{n}})\exp(-(E_j - E_i)/k_B T) \tag{84}$$

where N_A is Avogadro's number and a_3 is the activity of the electrolyte component [74, 101–104]. The number of ways of selecting a given number of ions for insertion into the simulation cell is different than the number of ways that this group of ions would be deleted from the cell [74, 101–104]. The coefficients B_{ij} takes this fact into account. A similar expression holds for the acceptance ratio for deletions.

In the simulation, the junction and the ion pairs were placed in the simulation cell and several types of transitions were implemented [74, 103, 104]: (1) number of ions was randomly selected and moved in random directions by small distances; (2) a number of ion pairs was randomly deleted; (3) insertion of several ion pairs at random positions. Transition involving multiple ion pairs brought the acceptance to rejection ratio near unity [74, 103, 104)] Furthermore, configurations in which ions overlap with the junction were rejected [74].

The concentration of the phosphate charges in molarity is $c_D = n_D/VN_A$, where V is the volume of the unit cell in the simulation. At the end of the simulation, the number of ion pairs N_3 will be determined. The salt concentration follows from the relation $c_3 = N_3/VN_A$. Consequently, Γ was calculated from the slope of c_3 versus the polyion concentration c_D, the later was be varied in the range around 1 mM [74].

The number of counterions associated with the tetrahedral junction was less than sensitive not only to parameters that define the junction region, but also to the interbranch angle between 90° and 109.5° [74] The number of associated counterions is substantially larger in the four-way junction than other junction geometries and constructs studied [74]. As salt concentration is increased, the stability of the junction is enhanced over a linear polyelectrolyte molecule of identical length as the junction [74]. For junctions with symmetrical branches, the counterions associated with the junction in excess of that of a linear construct increases with the length of the branches and then saturates [74].

If charges are neutralized at the center of a junction containing short branches, analysis reveals that counterions are released not near sites of neutralization, but from the entire molecule [74]. Furthermore, counterion release is enhanced for junctions with larger branches [74]. A basic conclusion of this study is that a protein recognizes a tertiary structure in part through an increase of the associated counterions [74]

The grand canonical Monte Carlo (GCMC) simulations complement a model for excess binding of counterions to various branched and kinked DNA structures [83]. In this approach, the branched structures are modeled as lines of phosphate charges that emerge from the center of the junction or the kink region. The stability for the structures is determined by a counterion condensation model [31, 83]. For four- and three-way junctions with arms lengths $>$ 50-bp/turn and in low salt concentrations, the number of excess counterions is less than fifteen if the counterions are magnesium, but larger than twenty if the counterions are sodium [84]. For branched structures, such as X-shape, T-shape, pyramid, tetrahedral, kink, or Y-shape, the excess number of counterions either does not change as the chain length increases or is found to increase with increasing chain length and then reaching a plateau value [84]. A compact junction, such as X-shaped, in ionic solution is less stable that open configuration junctions [83]. However, with increasing salt concentrations, the ionic stabilities of the compact forms increase [83].

Kim et al. [105] have employed single molecule fluorescence resonance energy-transfer spectroscopy (smFRET) to study the folding and the opening rates of a three-way junction from 16S rRNA [106] as a function of sodium and magnesium ion concentrations. By using a generalization of the counterion condensation model that is applicable to symmetric Y-shaped and asymmetric *y*-shaped junctions, Mohanty et al. [107] determined how the condensed counterions, as well as the fraction of "screening counterions" per phosphate charge vary with salt concentration. These authors predict the variation of $\log(k_f/k_o)$, where k_o and k_f are the opening and folding rates, with $\log c_s$ where c_s is the concentration of sodium ion or magnesium ion [107]. The results are in good agreement with smFRET data [105, 107].

D. *Tetrahymena* Ribozyme

Experimental and theoretical studies [108–110] have been carried out to assess how condensed counterions affect the stability and folding of *Tetrahymena* ribozyme [111, 112]. Substrate cleavage and gel elecrophoretic assays indicate that both divalent and monovalent counterions are capable of folding the *Tetrahymena* ribozyme [108, 109]. The collapse of the unfolded RNA toward the native state occurs through a set of metastable, but compact intermediates. These intermediates have the characteristics of both nonnative, as well as native conformations [108, 109]. The intermediates fold slowly to the native state, and the rate of folding is explicitly dependent on the counterion charge and valence [108, 109]. Although various divalent ions are found to partially stabilize the tertiary structure, magnesium ions are, however, required for the catalytic activity of the ribozyme [108–113]. The properties that govern the transition state ensemble between the unfolded and the folded states are dictated by the charge density of the counterions [109b].

Compact structures are also induced by size, shape, and charge of multivalent counterions, such as spermidine^{3+} and $[Co(NH_3)_6]^{3+}$ [31, 108, 109]. The concentrations of trivalent and divalent ions that are necessary to reach the intermediates states are much lower than the corresponding concentrations of monovalent ions [108, 109]. To put it differently, the number of multivalent condensed cations that reduce the net charge of RNA is smaller than the number of condensed monovalent ions needed to reach the same charge reduction [31, 108, 109].

At higher temperatures, higher concentrations of counterions are needed to induce folding in nucleic acids. This is so because thermal fluctuations resist the propensity of counterions to condense in the vicinity of the polyelectrolyte [108, 109]. This is in agreement with experiments [31, 108–110]. In fact, the midpoints of the folding transitions are monovalent ion dependent, and decreases with increasing temperatures [108, 109].

To explain the folding characteristics of *Tetrahymena* ribozyme, a simple generalization of counterion condensation model was proposed [108, 109]. In this two state model, equilibrium is considered between condensed and free counterions [108]. The chemical potential for each phase is approximately calculated as follows. If the volume fraction occupied by the counterions is denoted by ϕ, then the chemical potential of the free counterions is [108]

$$\mu_{\mathrm{free}}/k_{\mathrm{B}}T = \ln \phi \tag{85}$$

By definition, $\phi = \varpi v N_{\mathrm{A}}$ where N_{A} is Avogadro's number, v is the volume per counterion, and ϖ is the counterion concentration.

The energy due to coulomb interaction of the condensed counterions with the RNA is [108]

$$\mu_{\mathrm{cond}}/k_{\mathrm{B}}T = m\vartheta Z l_{\mathrm{B}}/R_g \tag{86}$$

The counterion valence is denoted by Z, the parameter R_g is the radius of gyration of RNA, ϑ is the average charge per nucleotide after counterion condensation, and m is the number of nucleotides.

From the chemical potential of the free and the condensed counterions and simple thermodynamic arguments one obtains [108]

$$m\vartheta = -\frac{R_g \ln \phi}{Z l_{\mathrm{B}}} \tag{87}$$

Several conclusions could be drawn from the two-state model. First, the function $\ln \phi$ does not change much with ϕ. If the conformation of the molecule is rodlike, the condition for counterion condensation is the same as that due to Manning [31, 108]. Second, if the shape of the folded *Tetrahymena* ribozyme, after counterion condensation, is assumed to be a globule, then the radius of gyration scales as $R_g = l_P m^{1/3}$, where l_p is the persistence length [108]. Since the experimental value of R_g for the intermediate states of *Tetrahymena* ribosyme is around 51 Å [113], the persistence length l_p is found to be 7 Å [108]. This value is much smaller than the l_p value for double-stranded RNA [108, 114]. Third, for multivalent counterions, there is an additional attractive force arising from Coulomb interactions between the phosphate charges and the effective charge of the counterions [108]. For multivalent cations, the dipole–dipole interactions between ion pairs vary as $1/r$, not $1/r^6$. Consequently, attractive dipole-induced dipole interactions could permit interactions between several phosphate charges. This would add appreciably to the compaction of the RNA [108]. Note that the R_g varies inversely with counterion valency [108]. Finally, since the buffer contains sodium ions, the multivalent cations would displace the sodium ions that are in the condensed volume into the bulk aqueous solution [108]. The entropy gained by displacement of the sodium ions would contribute to the equilibrium constant that defines the folding process [108].

E. Other Nucleic Acids

Recently, Schultes Spasic, Mohanty and Bartel studied in exquisite detail the effects of monovalent and divalent cations on the conformation order of random RNA sequences [115, 116]. These authors investigated the following questions: Can arbitrary RNA sequences fold into a unique structure? Is this is an evolutionary property of RNA sequences [115, 116]? Schultes et al. utilized biochemical tools, such as lead ion induced cleavage, ultracentrifugation, and gel electrophoretic mobility, to probe the structure of evolved and random RNA sequences [115, 116].

Analysis of gel electrophoretic mobility data is based on the basic idea is that ordered sequences are compact. Compact molecules migrate faster in the gel matrix [115, 116]. But random RNA sequences have a disposition to acquire multiple conformations. Only some of these conformations may be compact [115, 116]. Furthermore, when the RNA constructs are run on gels in the presence of magnesium ions, some of the conformations undergo nonspecific collapse due to condensed counterions [115, 116].

To interpret the gel mobility experiments, a quantitative predictive model was developed that describes the compactness of the RNA constructs [115, 116]. The model takes into account polyelectrolyte effects, salt concentration, pH of the buffer, screening of the hydrodynamic interactions, flexibility of the molecule, and concentration of the gel.

The RNA oligomers studied by these authors [115, 116] are several evolved sequences (hepatitis delta virus), synthetic polymers (poly-U), and various random RNA sequences. The random sequences are obtained from the HDV sequence by a permutation of the nucleotides in which base composition is fixed [115, 116]. The lengths of all RNA oligonucleotide fragments are 85 nucleotides [115, 116].

The study by Schultes et al. demonstrated for the first time that random RNA oligonucleotide sequences achieve conformations that are as compact as evolved sequences [115, 116]. Compact folds are an inherent characteristic of oligonucleotide RNA, and are not a result of constraints imposed by evolution [115, 116].

Murthy and Rose have carried out a detailed study of the packing of secondary structures observed in crystal structures of RNA, DNA, and RNA/DNA complexs [117]. Their work indicates that almost 80% of secondary structure elements pack via parallel packing of helices that are side-by-side, ridge into groove, or groove into groove motifs [118]. Parallel packing motifs, although disfavored by arguments based on repulsions among the phosphate groups, are nevertheless often observed in crystal structures of RNA molecules [117–120].

To explain the observed packing arrangements of RNA in crystals, a simple model for RNA folding [117] was introduced based on the concept of counterion condensation and its generalization to ionic oligomers [32]. When counterions condense on the RNA double helix, these counterions form a diffuse cloud that shield the phosphate charges [31, 32, 117]. The entropy that is gained from transferring some of the counterions from the bulk solution to the region of the condensed volume is sufficient to counteract the charge repulsion among the two parallel helices [31, 32, 117]. In monovalent salt, the Gibbs free energy associated with attraction between two short parallel double helices, of length 12 nucleotides, is crudely estimated to be $\Delta G \leq -2.1$-kcal mol^{-1} pair of helices [117]. For divalent salts, this value is $\Delta G \leq -1.5$-kcal mol^{-1} pair of helices [117].

What happens if the helices are not parallel, but instead are oriented at an angle relative to each other? Various studies have indicated that counterions near junction sites are generally confined, due to inhomogeneity in the electrostatic field [43, 71, 74, 117]. Furthermore, there is an orientation dependence to the distribution of the delocalized counterions [43, 74, 117, 118]. These ideas were exploited by Murthy and Rose to rationalize the observed crossing angles of various helices in crystal structures of nucleic acids [117].

Acknowledgments

This work was supported, in part, by the National Science Foundation. Over the years, the author has benefited from discussions on nucleic acids and biopolymers with Steve Chu, David Bartel, E. Schultes, James Maher, Jr., Nancy Stellwagen, Tali Haran, Donald Crothers, Gerald Manning, Wilma Olson, Larry McLaughlin, and Mary Roberts.

References

1. (a) R. Hanna and J. A. Doudna, *Curr. Opin. Chem. Biol.* **4**, 166 (2000). (b) T. C. Kuo, O. W. Odom and D. L. Herrin, *FEBS* **273**, 2631 (2006). (c) H. Ashara and O. C. Uhlenbeck, *Proc. Natl. Acad. Sci. USA* **99**, 3499 (2002). (d) W. Yang, J. Y. Lee and M. Nowotny, *Mol. Cell.* **22**, 5 (2006).
2. (a) C. L. Shenvi, K. C. Dong, E. M. Friedman, J. A. Hanson and J. H. Cate, *RNA* **11**,1898 (2005). (b) N. Korolev, A. P. Lyubarstev, L. Nordenskiold, *Biophys. J.* **90**, 4305 (2006). (c) D. Thompson and T. Simonson, *J. Biol. Chem.* **281**, 23792 (2006).
3. T. J. Wilson and D. M. J. Lilley, *RNA* **8**, 587 (2002).
4. R. T. Batey and J. R. Williamson, *RNA* **4**, 984 (1988).
5. M. J. Fedor, *Curr. Opion. Struct. Biol.* **12**, 289 (2002).
6. M. Egli and V. Tereshko, in (eds.), N.C. Stellwagen and U. Mohanty, *Nucleic Acids; Curvature and Deformation*, ACS Symposium Series vol. 884, 2004, p. 87.
7. K. A. Dill, *Biochemistry* **29**, 7133 (1990).
8. B. Honig, K. Sharp and A. Yang, *J. Phys. Chem.* **97**, 1101 (1993).
9. (a) Y. V. Bukhman and D. E. Draper, *J. Mol. Biol.* **273**, 1020 (1997). (b) V. Misra and D. E. Draper, *Biopolymers* **48**, 113 (1998). See also refs. 71 and 72.

10. D. Kremp and W. Beskrownij, *J. Chem. Phys.* **104**, 2010 (1996).
11. L. G. Laing, T. C. Gluick and D. E. Draper, *J. Mol. Biol.* **237**, 577 (1994).
12. C. F. Anderson and M. T. Record, *Annu. Rev. Phys. Chem.* **46**, 657 (1995).
13. T. L. Hill, *J. Chem. Phys.* **23**, 623 (1955).
14. (a) D. E. Draper, *RNA* **10**, 335 (2004). (b) D. E. Draper, D. Grilley and A. M. Soto, *Ann, Rev. Biophys. Bioml. Struct.* **34**, 221 (2005). (c) Y. V. Bukhman and D. E. Draper, *J. Mol. Biol.* **274**, 101 (1997). (d) See also Ref. 71.
15. J. Wyman and S. J. Gill, *Binding and Linkage. Functional Chemistry of Biological Macromolecules.* University Science Books, Mill Valley, CA 1990.
16. J. O. M. Bockris and A. K. N. Reddy, *Modern Electrochemistry.* I, Plenum Press, New York, 1970; Vol I.
17. G. S. Manning, *J.Chem.Phys.* **51**, 924 (1969). See also Ref. 31 and 32.
18. Y. W. Pley, K. M. Flaherty, and D. B. Mckay, *Nature (London)* **372**, 68 (1994).
19. J. H. Cate and J. A. Doudna, *Structure* **4**, 1221 (1996).
20. A. Jack, J. E. Ladner, D. Rhodes, R. S. Brown and A. Klug, *J. Mol. Biol.* **111**, 315 (1977).
21. J. H. Cate, et al. *Science* **273**, 1696 (1996).
22. L. X. Shen and I. J. Tinoco, *Mol. Biol.* **247**, 963 (1995).
23. A. A. Szewczak, P. B. Moore, Y. Chan and I. G. Wool, *Proc. Natl. Acad. USA* **90**, 9581 (1993).
24. T. K. Chiu and R. E. Dickerson, *J. Mol. Biol.* **301**, 915 (2000).
25. X. Shui, L. McFail-Isom, G. G. Hu and L. D. Williams, *Biochemistry* **37**, 8341 (1998).
26. C. C. Correll, B. Freeborn, P. B. Moore and T. A. Steitz, *Cell* **91**, 705 (1977).
27. M. J. Serra, J. D. Baird, T. Dale, B. L. Fey, K. Retatagos and E. Westhof, *RNA* **8,** 307 (2002).
28. (a) D. L. Beveridge and K. J. McConnell, *Curr. Opin. Struct. Biol.* **10**, 182 (2000). (b) M. A. Young, B. Jayaram and D. L. Beveridge, *J. Am. Chem. Soc.* **119**, 59 (1997). (c) D. L. Beveridge, S. B. Dixit, K. S. Byun, G. Barreiro, K. M. Thayer and S. Ponomarev, in N. C. Stellwagen, and U. Mohanty (eds.), *Nucleic Acids; Curvature and Deformation*, ACS Symposium Series vol. 884, 2004, p. 13.
29. (a) E. Giudice and R. Lavery, *Acc. Chem. Res.* **35**, 350 (2002). (b) T. E. Cheatham and P. A. Kollman, *Annu. Rev. Phys. Chem.* **51**, 435 (2000).
30. (a) M. L. Bleam, C. F. Anderson and M. T. Record, *Biochemistry* **22**, 5418, **1983**. (b) W. H. Braunlin, and V. A. Bloomfield, *Biochemistry* **30**, 754, 1991. (c) G. S. Manning, *Acc. Chem. Res.* **12**, 443, 1979. (d) F. Gago and W. G. Richards, *Mol. Pharmacol.* **37**, 341, 1990.
31. G. S. Manning, *Quart. Rev. Bioph.* **111**, 179 (1978). See also refs. 32, 40–43, and 83 cited below.
32. G. S. Manning and U. Mohanty, *Physica A.* **247,** 196 (1997).
33. G. V. Ramanathan and C. P. Woodbury, *J. Chem. Phys.* **77**, 4133 (1982).
34. G. V. Ramanathan, *J. Chem. Phys.* **78**, 3223 (1983).
35. B. H. Zimm and M. Le Bret, *J. Biomol. Struct. Dynam.* **1**, 461 (1983).
36. G. V. Ramanathan, *J. Chem. Phys.* **88,** 3887 (1988).
37. M. Satoh, J. Komiyama, and T. Iijima, *Macromolecules* **18**, 1195 (1985).
38. J. S. Moore and S. I. Stupp, *Macromolecules* **19**, 1815 (1986).
39. C. Wandrey, D. Hunkeler, U. Wendler and W. Jaeger, *Macromolecules* **33**, 7143 (2000).
40. G. S. Manning and U. Mohanty, *Physica A* **241**, 196 (1997).

41. G. S. Manning, *Macromolecules* **34**, 4650 (2001).
42. G. S. Manning, *Biophys. Chem.* **101–102**, 461 (2002).
43. (a) J. Ray and G. S. Manning, *Langmuir* **10**, 2450 (1994). (b) J. Ray and G. S. Manning, *Macromolecules* **32**, 4588 (1999).
44. U. Mohanty, B. Ninham and I. Oppenheim, *Proc. Natl. Acad. USA* **93**, 4342 (1966).
45. S. Lifson and A. Katchalski, *J. Poly. Sci.* **13**, 43 (1954).
46. C. F. Anderson and M. T. Record, Jr., *Ann. Rev. Phys. Chem.* **33**, 191 1982.
47. A. Deshkovski, S. Obukhov, M. Rubinstein, *Phys. Rev. Lett.* **86**, 11, 2341 (2001).
48. M. Muthukumar, *J. Chem. Phys.* **120**, 9343 (2004).
49. M. Muthukumar, *J. Chem. Phys.* **86**, 7230 (1987).
50. G. S. Manning, *Langmuir* **10**, 962 (1994).
51. A. W. C. Lau, D. B. Lukatsky, P. Pincus and S. A. Safran, *Phys. Rev. E* **65**, 051502 (2002).
52. L. D. Landau and E. M. Lifshitz, *Statistical Physics*, 3rd ed, Pergamon, New York, 1980.
53. (a) B. I. Shklovskii, *Phys. Rev. E* **60**, 5802 (1999). (b) I. Rouzina and V. A. Bloomfield, *J. Phys. Chem. B* **100**, 9977 (1996).
54. R. R. Netz and H. Orland, *Eur. Phys. J. E.* **11**, 301 (2003).
55. J. M. Schurr and B. S. Fujimoto, *Biophys. Chem.* **101–102**, 425 (2002). See also ref. 71.
56. Y. Burak and H. Orland, *Phys. Rev. E.* **73**, 010501 (2006).
57. A. Naji and R. R. Netz, *Phys. Rev. Lett.* **95**, 185703 (2004).
58. (a) J. Kosterlitz and D. Thouless, *J. Phys. Solid Sate Phys.* **6**, 1181 (1973). (b) U. Mohanty, *Biopolymers* **38**, 377–388 (1996) and references cited therein.
59. A. L. Kholodenko and A. L. Beyerlein, *Phys. Rev. Lett.* **74**, 4679 (1995).
60. B. O. Shaughnessy and Q. Yang, *Phys. Rev. Lett.* **94**, 048302 (2005).
61. M. Gueron and G. Weisbuch, *Biopolymers* **19**, 353 (1980).
62. Z. J. Tan and S. J. Chen, *J. Chem. Phys.* **122**, 044903 (2005).
63. S. Arnot and D. W. L. Hukins, *Biochem. Biophys. Res. Commun.* **47**, 1504 (1972).
64. D. Stigter, *Biophys. J.* **69**, 380 (1995).
65. C. F. Anderson and M. T. Record, *Ann. Rev. Phys. Chem.* **46**, 657–700 (1995).
66. C. F. Anderson and M. T. Record, *Ann. Rev. Biophys. Biophys. Chem.* **19**, 423 (1990).
67. M. T. Record, W. Zhang and C. F. Anderson, *Adv. Prot. Chem.* **51**, 281 (1998).
68. H. Ni, C. F. Anderson, M. T. Record, *J. Phys. Chem. B* **103**, 3489 (1999).
69. H. Eisenberg, *Biological Macromolecules and Polyelectrolytes in Solution*, Claredon, Oxford, 1976.
70. S. N. Tamasheff, *Ann. Rev. Biophys. Biomol. Struct.* **22**, 67 (1993).
71. U. Mohanty, A. Spasic and S. Chu, *J. Phys. Chem. B.* **109**, 21369 (2005).
72. C. H. Taubes, U. Mohanty and S. Chu, *J. Phys. Chem. B.* **109**, 21267 (2005).
73. C. F. Anderson and M. T. Record, *J. Phys. B. Chem*, **97**, 7116 (1993).
74. M. Olmsted and P. J. Hagerman, *J. Mol. Biol.* **243**, 919 (1994).
75. K. A. Sharp, *Biopolymers* **36**, 227 (1995).
76. P. Sens and J. F. Joanny, *Phys. Rev. Lett.* **84**, 4862 (20000.
77. I. A. Shkel, O. V. Tsodikov and M. T. Record, *Proc. Natl. Acad. Sci.* USA **99**, 2597 (2002).
78. I. A. Shkel, O. V. Tsodikov and M. T. Record, *J. Phys. Chem. B* **104**, 5161 (2000).

79. C. F. Anderson and M. T. Record, *Biophys. Chem.* **11**, 353 (1980).
80. L. M. Gross and U. P. Strauss, in B. F. Conway, and R. G. Barradas, (eds.), *Chemical Physics of Ionic Solutions*, John Wiley & Sons, Inc. New York, 1966 pp. 361–389.
81. Z. Alexandrowicz, *J., Polymer Sci.* **56**, 97 (1962).
82. W. H. Braunlin, *Adv. Biophys. Chem.* **5**, 89 (1995).
83. M. O. Fenley, G. S. Manning, N. L. Marky and W. K. Olson, *Biophys. Chem.* **74**, 135 (1998).
84. R. Holliday, *Genet. Res.* **5**, 282 (1964).
85. N. Sigal and B. Alberts, *J. Mol. Biol.* 71, 789 (1972).
86. J. W. Szostack, T. L. Orr-Weaver, R. J. Rothstein and F. W. Stahl, *Cell* **33**, 25 (1983).
87. D. M. Lilley and B. Kemper, *Cell* **36**, 413 (1984).
88. I. G. Panyutin and P. Hsieh, *Proc. Natl. Acad. Sci. U. S. A.* **91**, 2021 (1994).
89. D. R. Duckett, A. I. H. Murchie, R. M. Clegg, A. Zechel, E. von Kitzing, S. Dieckmann and D. M. J. Lilly, *Structure & Methods.Vol I: Human Genome Initiative Recombination*, R. H. Sarma, M. H. Sarma, (eds.), Adenine Press, 1990, 157.
90. M. J. Lilley and M. R. Clegg, *Ann. Rev. Biophys. Biomol. Struc.* **22**, 299 (1993).
91. M. Petrillo, C. Neton, R. Cunningham, R. Ma, N. Lakkenback and N. Seeman, *Biopolymers* **27**, 1337 (1989).
92. R. M. Clegg, A. I. H. Murchie and D. M. J. Lilley, *Biophys. J.* **66**, 99 (1994).
93. R. M. Clegg, A. I. H. Murchie, A. Zechel, C. Carlberg, S. Diekmann and D. M. Lilley, *Biochemistry* **31**, 4846 (1992).
94. M. T. Record, T. M. Lohman and P. D. Haseth, *J. Mol. Biol.* **107**, 145 (1975).
95. M. T. Record and T. M. Lohman, *Biopolymers* **17**, 159 (1978).
96. S. N. Timasheff, *Biochemistry* **31**, 9857 (1992).
97. S. N. Timasheff, *Biochemistry* **41**, 13473 (2002).
98. N. Metropolis, A. W. Rosenbluth, M. N. Rosenbluth, A. H. Teller and E. Teller, *J. Chem. Phys.* **21**, 1087 (1953).
99. D. Frenkel and B. Smit, 2nd ed., Academic Press, New York, 2002.
100. A. R. Leach, *Molecular modeling: principles and applications.* 2nd ed., Prentice Hall New York, 2000.
101. P. A. Mills, C. F. Anderson and M. T. Record, Jr. *J. Phys. Chem.* **90**, 6541 (1986).
102. B. Jayaram and D. L. Beveridge, *J. Phys. Chem.* **95**, 2506 (1991).
103. M. C. Olmsted, C. F. Anderson and M. T. Jr., Record, *Biopolymers* **31**, 1593 (1991).
104. B. Jayaram, S. Swaminathan, D. L. Beveridge, K. Sharp and B. Honig, *Macromolecules* **23**, 3156 (1990).
105. H. D. Kim, G. U. Nienhaus, T. Ha, J. W. Orr, J. R. Williamson and S. Chus. *Proc. Natl. Acad. Sci.* USA **99**, 4284 (2002).
106. R. T. Batey and J. R. Williams, *J. Mol. Biol.* **261**, 550 (1996).
107. U. Mohanty, A. Spasic, H. D. Kim and S. Chu, *J. Phys. Chem. B.* **109**, 21369 (2005).
108. S. L. Heilman-Miller, D. Thirumalai and S. A. Woodson, *J. Mol. Biol.* **306**, 1157 (2001).
109. (a) S. L. Heilman-Miller, J. Pan, D. Thirumalai and S. A. Woodson, *J. Mol. Biol.* **309**, 57 (2001). (b) E. Koculi, D. Thirumalai and S. A. Woodson, *J. Mol. Biol.* **359**. 446 (2006).
110. S. A. Woodson, *Nature (London)*, **438**, 566 (2005).

111. T. R. Cech and B. L. Golden, in R. F., Cech, T. R. and J. F. Atkins, (eds.), *The RNA World* Gesteland, 2nd ed., Cold Spring Harbor Laboratory Press, Cold Spring Harbor, NY, 1999, p. 321.
112. T. R. Cech, *Annu. Rev. Biochem.* **59**, 543 (1990).
113. R. Russell, I. S. Millett, S. Doniach and D. Herschlag, *Nature Struct. Biol.* **7**, 367 (2000).
114. P. J. Hagerman, *Annu. Rev. Biophys. Biomol. Struct.* **26**, 139 (1997).
115. E. Schultes, A. Spasic, U. Mohanty and D. Bartel, *Nature Struct. Biol.* **12**, 1130 (2005).
116. A. Spasic, Ph.D. Thesis, Boston College, 2006.
117. V. L. Murthy and G. D. Rose, *Biochemistry* **39**, 14365 (2000). See also references cited therein.
118. J. H. Cate, A. R. Goodwing, E. Podell, K. Zhou, B. L. Golden, C. E. Kundrot, T. R. Cech and J. A. Doudna, *Science* **273**, 1678 (1996).
119. B. T. Wimberly, R. Guymon, J. P. McCutcheon, S. W. White and V. Ramakrishnan, *Cell* **97**, 491 (1999).
120. G. L. Conn, D. E. Draper, E. E. Lattman and A. G. Gittis, *Science* **284**, 1171 (1999).

PHYSICOCHEMICAL APPLICATIONS OF SCANNING ELECTROCHEMICAL MICROSCOPY

FRANÇOIS O. LAFORGE, PENG SUN, AND MICHAEL V. MIRKIN

Department of Chemistry and Biochemistry, Queens College–CUNY, Flushing, NY 11367, USA

CONTENTS

Advances in Chemical Physics, Volume 139, edited by Stuart A. Rice

I. INTRODUCTION

Scanning electrochemical microscopy (SECM; the same abbreviation is also used for the device, i.e., the microscope) is often compared (and sometimes confused) with scanning tunneling microscopy (STM), which was pioneered by Binning and Rohrer in the early 1980s [1]. While both techniques make use of a mobile conductive microprobe, their principles and capabilities are totally different. The most widely used SECM probes are micrometer-sized amperometric ultramicroelectrodes (UMEs), which were introduced by Wightman and co-workers ~1980 [2]. They are suitable for quantitative electrochemical experiments, and the well-developed theory is available for data analysis. Several groups employed small and mobile electrochemical probes to make measurements within the diffusion layer [3], to examine and modify electrode surfaces [4, 5]. However, the SECM technique, as we know it, only became possible after the introduction of the feedback concept [6, 7].

SECM is a powerful tool for studying structures and heterogeneous processes on the micrometer and nanometer scale [8]. It can probe electron, ion, and molecule transfers, and other reactions at solid–liquid, liquid–liquid, and liquid–air interfaces [9]. This versatility allows for the investigation of a wide variety of processes, from metal corrosion to adsorption to membrane transport, as discussed below. Other physicochemical applications of this method include measurements of fast homogeneous kinetics in solution and electrocatalytic processes, and characterization of redox processes in biological cells.

SECM employs an UME probe (tip) to induce chemical changes and collect electrochemical information while approaching or scanning the surface of interest (substrate). The substrate may also be biased and serve as the second working electrode. The nature of the tip and the way it interacts with the substrate determine what information can be obtained in an SECM experiment. Many different types of UMEs have been fabricated, for example, microband electrodes, cylindrical electrodes, microrings, disk-shaped, and hemispherical electrodes [10, 11]. For reasons discussed below, the disk geometry is preferred for SECM tips, though other shapes may be suitable for specific experiments [12–19]. The UMEs offer an important advantages for electroanalytical applications including a greatly diminished ohmic potential drop in solution and double-layer charging current, the ability to reach a steady state in seconds or milliseconds, and a small size allowing one to do experiments in microscopic domains.

The precise positioning capabilities, which make high spatial resolution possible, give the SECM an important edge over other electrochemical techniques employing UMEs [20]. For example, the SECM can pattern the substrate surface, visualize its topography, and probe chemical reactivity on the micrometer or nanometer scale. Here, we briefly survey the fundamentals of various modes of the SECM operation and then focus on more recent advances in SECM theory and applications.

II. PRINCIPLES OF SECM OPERATION

Figure 1 shows a schematic diagram of the basic SECM instrument employing an amperometric microprobe. An UME tip is attached to a three-dimensional (3D) piezo positioner controlled by a computer, which is also used for data acquisition. A bipotentiostat (i.e., a four-electrode potentiostat) controls the potentials of the tip and/or the substrate versus the reference electrode and

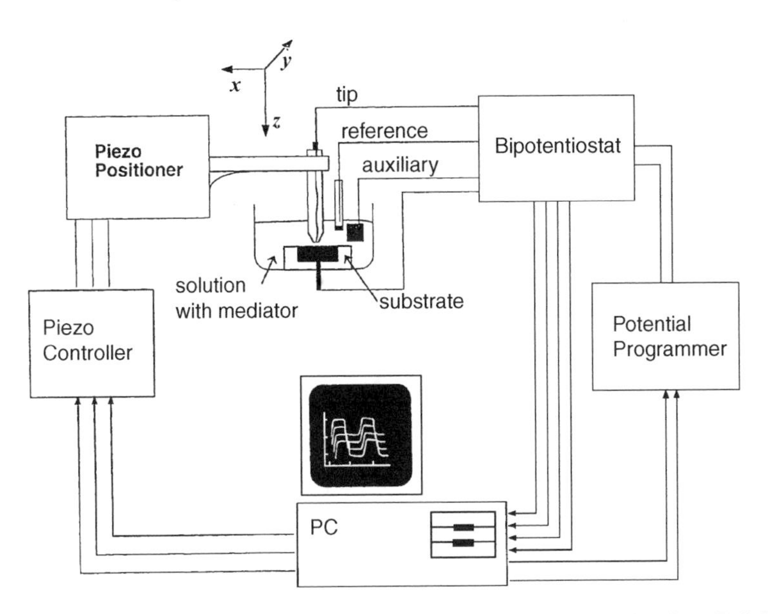

Figure 1. Block diagram of the SECM apparatus. Taken with permission from Ref. [59]. Copyright © 1991, American Association for the Advancement of Science.

measures the tip and substrate currents. The SECM instrument is often mounted on a vibration-free optical table inside a Faraday cage to isolate it from environmental vibrations and electromagnetic noise. With essentially the same setup, several SECM modes of operation can be realized including feedback mode, tip generation/substrate collection (TG/SC) mode, substrate generation/tip collection (SG/TC) mode, penetration mode and ion-transfer feedback mode.

A. Feedback Mode

In a feedback mode experiment, the tip is immersed in a solution containing a redox mediator (e.g., an oxidizable species, R). When a sufficiently positive potential is applied to the tip, the oxidation of R occurs via the reaction

$$\mathrm{R} - ne^{-} \rightarrow \mathrm{O} \tag{1}$$

at a rate governed by diffusion of R to the UME surface. If the tip is far (i.e., greater than several tip diameters) from the substrate (Fig. 2a) the steady-state current, $i_{\mathrm{T},\infty}$, is given by

$$i_{\mathrm{T},\infty} = 4nFDca \tag{2}$$

where F is the Faraday constant, n is the number of electrons transferred in reaction (1), D and c are the diffusion coefficient and the bulk concentration of R, respectively, and a is the tip radius.

When the tip is brought to within a few tip radii of a conductive substrate surface (Fig. 2b), the O species formed in the reaction (1) diffuses to the

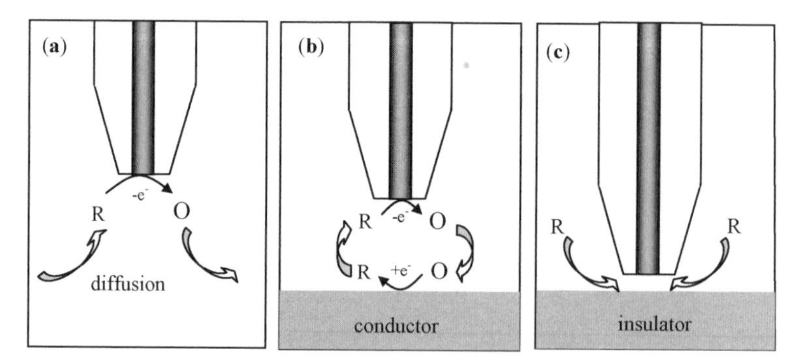

Figure 2. Feedback mode of the SECM operation. (a) The UME tip is far from the substrate. (b) Positive feedback: species R is regenerated at the substrate. (c) Negative feedback: Diffusion of R to the tip is hindered by the substrate.

substrate where it can be reduced back to R

$$O + ne^- \rightarrow R \tag{3}$$

This process produces an additional flux of R to the tip, and hence "positive feedback", that is, the increase in tip current ($i_T > i_{T,\infty}$). The shorter the tip–substrate separation distance (d), the larger the tip current. When reaction (3) is rapid, $i_T \rightarrow \infty$ as $d \rightarrow 0$.

If the substrate is an inert electrical insulator, the tip–generated species, O, cannot react at its surface. At small d, $i_T < i_{T,\infty}$ because the insulator blocks the diffusion of species R to the tip surface (negative feedback; Fig. 2c). The closer the tip to the insulating substrate, the smaller the i_T, with $i_T \rightarrow 0$ as $d \rightarrow 0$. Overall, the rate of the mediator regeneration at the substrate determines the magnitude of the tip current, and conversely the measured i_T versus d dependence (approach curve) provides information on the kinetics of the process at the substrate.

If the substrate is a conductor, the rate of reaction (3) can be controlled by applying a suitable potential to it by a potentiostat. Alternatively, the potential of a conductive substrate (E_S) may be determined by concentrations of redox species in solution without an external bias. For example, if the solution contains only the reduced form of the redox species, most of the substrate surface, which is usually much larger than that of the tip, is exposed to solution of R. The parameter E_s can be evaluated from the Nernst equation

$$E_s = E^\circ + \frac{RT}{nF} \ln \frac{c_O}{c_R} \tag{4}$$

In this case, $c_O \sim 0$, and $E_s - E^\circ \ll 0$, where E° is the standard potential of the mediator, and all oxidized species reaching the substrate get reduced at its surface.

B. Tip Generation/Substrate Collection

In the TG/SC mode experiment, the tip generates an electroactive species that diffuses across the tip/substrate gap to react at the substrate surface (Fig. 3a). A TG/SC experiment includes simultaneous measurements of both tip and substrate currents (i_T and i_S). For a one-step heterogeneous electron transfer (ET) at steady state, these quantities are almost identical if d is not very large, and consequently the collection efficiency, $i_S/i_T > 0.99$ at $d \leq 2a$. Under these conditions, the tip/generated species, R, predominantly diffuse to the large substrate, rather than escape from the tip–substrate gap. For a process with a coupled homogeneous chemical reaction, there may be large differences between i_S and i_T, and both quantities can provide important kinetic information. For example, species R may undergo a first-order irreversible

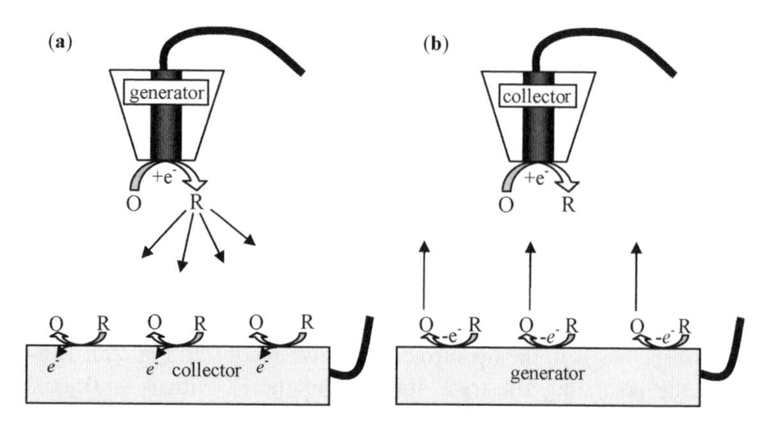

Figure 3. Scheme of TG/SC mode (a) and SG/TC mode (b). (a) The tip generates species R by reduction of O in solution; R diffuses toward the substrate and is reoxidized to O. (b) O is electrogenerated at the substrate surface and collected at the tip. The tip and substrate currents are recorded in both cases.

reaction whose product is electrochemically inactive. If the reaction is slow, the process is diffusion controlled, the i_T versus d curves follow the positive feedback theory, and $i_S/i_T \rightarrow 1$ at short separation distances. In contrast, if homogeneous reaction is very fast, most species R get converted to the electroinactive product before reaching the substrate. Hence, the very low substrate current, and $i_S/i_T \rightarrow 0$. Between these two extreme cases, the homogeneous kinetics can be determined by measuring the collection efficiency as a function of d [21–25].

C. Substrate Generation/Tip Collection Mode

Historically, the first SECM-type experiments were carried out to measure concentration profiles in the diffusion layer generated by a macroscopic substrate [3, 26]. This type of measurement represents substrate generation/tip collection (SG/TC) mode. When the tip is moved through the thick diffusion layer produced by the substrate, the changes in i_T reflect local variations of concentrations of redox species (Fig. 3b). Ideally, the tip should not perturb the diffusion layer at the substrate. This is easier to achieve with a potentiometric tip, which is a passive sensor and does not change concentration profiles of electroactive species.

The collection efficiency in SG/TC mode is much lower than that in the TG/SC mode, and the true steady state can be achieved only by using a micrometer-sized substrate. Other disadvantages of this mode are the high sensitivity to noise and the difficulty in controlling the tip–substrate separation distance.

Theoretical treatments for SG/TC mode have been reported for a spherical cap or an embedded microdisk-shaped substrate that generates stable species and for a smaller tip collecting them [27–32]. This mode can be used for monitoring corrosion, enzymatic reactions, and other heterogeneous processes at the substrate surface.

D. Penetration Experiments

In this mode, a small SECM tip is used to penetrate a microstructure, for example, a submicrometer-thick polymer film containing fixed redox centers or loaded with a redox mediator, and extract spatially resolved information (i.e., a depth profile) about concentrations, kinetic- and mass-transport parameters [33, 34]. With a tip inside the film, relatively far from the underlying conductor or insulator, solid-state voltammetry, at the tip can be carried out similarly to conventional voltammetric experiments in solution. At smaller distances, the tip current either increases or decreases depending on the rate of the mediator regeneration at the substrate. If the film is homogeneous and not very resistive, the current–distance curves are similar to those obtained in solution.

More recent penetration experiments were carried out in biological systems, that is, large intact nuclei [35], giant liposomes [36], and mammalian cells [37]. Such experiments can provide information about the distribution of electroactive species inside the cell, potentials, and ion transfers across biological membranes (see Section V.F).

E. Electron and Ion-Transfer Feedback Experiments at the Liquid–Liquid Interface

Heterogeneous charge-transfer reactions at the interface between two immiscible electrolyte solutions (ITIES) and liquid–membrane interfaces are of fundamental importance for many biological and technological systems. When SECM is used to probe ET at the ITIES, an UME tip is placed in the upper liquid phase (e.g., organic solvent) containing one form of the redox species (e.g., the reduced form, R_1). With the tip held at a sufficiently positive potential, R_1 reacts at the tip surface to produce the oxidized form of the species, O_1. When the tip approaches the ITIES, the mediator can be regenerated at the interface via the bimolecular redox reaction between O_1 in the organic phase (o) and R_2 in the aqueous phase (w)

$$O_1(o) + R_2(w) \rightarrow R_1(o) + O_2(w) \tag{5}$$

and i_T increases with the decrease in d (positive feedback).

While conventional studies of the ITIES have been carried out at externally biased polarizable ITIES, in SECM measurements, a nonpolarizable ITIES is poised by the concentrations of the potential-determining ions providing a controllable driving force for the ET process [38, 39].

In the SECM measurements shown schematically in Fig. 4a, four stages of the overall process may influence the tip current: organic mediator diffusion

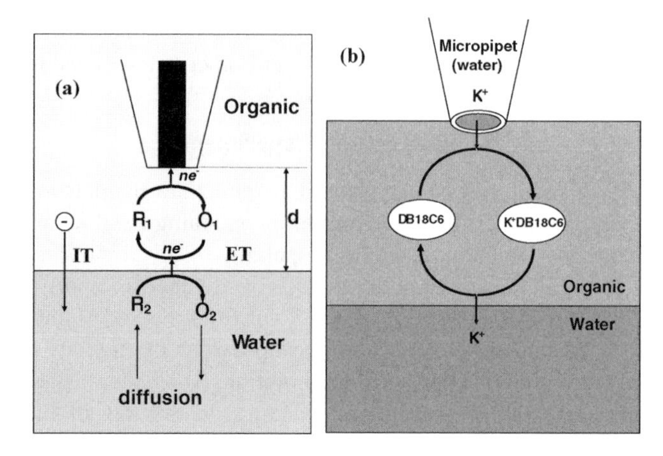

Figure 4. Charge-transfer processes at the liquid–liquid interface. (a) Probing ET at the liquid–liquid interface with the SECM. The kinetics of ET between two redox couples confined to different immiscible liquid phases can be measured with the SECM operating in the conventional feedback mode. Electroneutrality is maintained by transfer of the common ion (shown as an anion) across the interface (IT). Adapted with permission from Ref. [38]. Copyright © 1995, American Chemical Society. (b) Schematic diagram of facilitated ion transfer reaction studied by SECM.

between the tip and the ITIES, interfacial reaction (Eq. 5), diffusion of R_2 in water, and charge compensation by interfacial ion transfer (IT). The electrical current across the ITIES (i_S) caused by this multistage serial process can be expressed as [38],

$$1/i_S = 1/i_T^c + 1/i_{ET} + 1/i_d + 1/i_{IT} \tag{6}$$

where i_T^c, i_{ET}, i_d, and i_{IT} are the characteristic limiting currents for the above four stages, respectively. Any of these stages can be rate limiting, but the concentration of R_2 is usually made sufficiently high to exclude the possibility of diffusion limitations in the bottom phase. If the common ion (e.g., ClO_4^-) concentration in both phases is sufficiently high, IT is fast, and the last term in Eq. (6) is negligible. Under these conditions, the effective heterogeneous rate constant can be obtained by fitting experimental current–distance curves to Eq. (23) in Section IV.A.2.

The ion-transfer feedback mode can also be used to probe the transfers of electroinactive ionic species, (e.g., ClO_4^-), alkali metal cations, and tetra-alkylammonium cations across the ITIES. A micrometer- or nanometer-sized pipet can be filled with a solvent immiscible with the outer solution and used as a tip to approach a macroscopic ITIES (Fig. 4b). In an SECM study of facilitated ion transfer [40], the transfer of K^+ from aqueous filling solution to the external 1,2-dichloroethane (DCE) solution was assisted by dibenzo-18-crown-6 (DB18C6)

$$K^+_{(w)} + DB18C6_{(DCE)} \rightarrow [KDB18C6]^+_{(DCE)} \qquad \text{(at the pipet tip)} \qquad (7a)$$

When the tip approached the ITIES, the regeneration of DB18C6 occurred via an interfacial dissociation mechanism

$$[KDB18C6]^+_{(DCE)} \rightarrow K^+_{(w)+} DB18C6_{(DCE)} \qquad \text{(at the ITIES)} \qquad (7b)$$

and a positive feedback current was observed. Positive feedback can also be produced by a simple ion-transfer reaction [41].

In most SECM experiments, a nonpolarizable ITIES was poised by the concentrations of the potential-determining ion providing a constant driving force for either ET or IT process. Alternatively, a polarizable ITIES can be externally biased [42]. In this case, the applied interfacial voltage should be within the polarization window, where almost no current flows across a macroscopic liquid–liquid interface. The tip reaction changes the concentrations of redox species in a thin layer near the phase boundary and induces a small current across a micrometer-sized portion of the ITIES. An advantage of such an approach is in a potentially wider range of interfacial voltages available for the ET study (as compared to those at a nonpolarizable interface).

F. Impedance-Based Measurements

In steady-state experiments, the source of the tip current is a Faradaic process, that is, a redox reaction occurring at the tip. In some cases, it can be advantageous to measure non-Faradaic tip current. When a small amplitude (several tens of millivolts) ac voltage is applied to the tip, the charging current flows through the double layer capacitance, the solution resistance, and the double-layer capacitance of the counterelectrode. A small amplitude of the ac voltage does not induce any redox reaction at the tip (alternatively, a dc bias can be added to the ac voltage to purposely induce a redox reaction). Because the counterelectrode is incomparably larger than the tip, its capacitance acts as a shunt, and the solution resistance drops mostly within a few radii distance from the tip surface. Alternating current impedance measurements can therefore give information about the double-layer capacitance and the solution resistance at the tip. These properties, in turn, are governed by the nature of the local environment near the tip (e.g., local concentration of ionic species and the conductivity of a substrate). Horrocks et al. [43] demonstrated that, under certain experimental conditions, the dependence of the solution resistance versus tip–substrate distance, is mathematically equivalent to the current–distance dependence in the amperometric SECM feedback mode (see Section IV.A.1). More recently Kurulugama et al. [44] used impedance control to maintain a constant distance between the tip and the immobilized living cell during constant–distance imaging. White et al. [45] were able to discern single pores openings in a high pore density membrane by scanning a 1-μm radius electrode over the membrane in the impedance mode.

III. INSTRUMENTATION

A. Tip

A typical SECM probe is a micrometer-sized UME sharpened to allow closer approach to the substrate surface [46]. Such probes are easy to fabricate from commercially available microwires and polish, thus ensuring accuracy and reproducibility of their electrochemical responses. An established procedure involves heat sealing of a microwire or a carbon fiber in a glass capillary under vacuum and connecting it with silver epoxy to a larger copper wire on the back side [6]. The sealed side of the probe is polished down to 50-nm alumina and then sharpened to form a tip using coarse sandpaper. The important parameters of the tip geometry are the radius of the conductive core, a, and the total tip radius (i.e., a plus the insulating sheath thickness), r_g. The dimensionless parameter $RG = r_g/a$ is normally ≤ 10. This fabrication method does not require any expensive equipment, and it is widely used for the production of micron-sized tips.

Nanometer- and submicrometer-sized tips have been fabricated differently. A simple way to produce nanometer-sized conical electrodes consists in etching a metal wire and then coating it with an insulator while leaving the apex exposed. For example, a 250-μm diameter Pt—Ir wire was etched in a solution containing 3 M NaCN and 1 M NaOH by applying a 20-V ac between the wire and the solution [47]. Then, it was insulated by dipping the metal tip into molten Apiezon wax, and finally its very end was exposed by using an STM. Many other types of coating have been tried, such as varnish, molten paraffin, silica coating, poly(α-methylstyrene), polyimide [48], electropolymerized phenol, and electrophoretic paint [49]. Submicrometer-sized conical carbon tips were prepared by Zhan and Bard [36] using flame etching.

Most nanotips prepared by etching are conical and not polishable [12, 47–50]. Such tips are suitable for high resolution imaging and penetration experiments, but they are less useful for quantitative kinetic measurements and especially for feedback mode experiments. Disk-type tips can be prepared using a micropipet puller [51–53]. Shao et al. [52] used a laser puller to prepare glass-sealed Pt electrodes with effective radii ranging from 2 to 500 nm. They attempted to polish larger (> 100 nm) electrodes and to characterize the tip geometry by combination of SEM, steady-state voltammetry and SECM; but the quality of the obtained approach curves was not high. A somewhat different approach to polishing of pulled submicrometer electrodes was proposed by Kateman and Schuhmann [53]. Recently, it was shown that a tip as small as ~10-nm radius can be polished on a lapping tape under the video-microscopic control [51]. The polished flat nanotips yield more reliable and reproducible data, and can be used for fast kinetic measurements.

The addition of an electromagnetic shield to the body of the tip can greatly reduce the stray capacitance. This is particularly useful for fast-scan voltammetry [54] and high frequency impedance experiments [55].

B. Positioning

Similarly to other scanning probe microscopies, an SECM probe has to be moved in x, y, and z directions with the nanometer-scale precision. However, the required travel distance can be as long as a few hundred microns. Most SPM positioning devices are not suitable for such long scans. Also, a reasonable accuracy in the x and y directions has to be combined with a higher accuracy in z direction for the measurement of i_T versus d curves. One possibility is to use inchworm motors for x and y axes traveling and for relatively coarse z scans, while a vertical PZT piezo pusher is used for finer z control over short distances. Inchworm motors use a clamping mechanism that produces slight lateral and possibly axial discontinuities of the displacement every 2 μm of travel when the axis is clamped/unclamped. The lateral motion discontinuity can be eliminated by using the motor to drive a ball bearing linear stage to which the probe is attached, however, the axial discontinuity can be fatal for fragile nanotips, as a crash could occur if the tip is close to the substrate during clamping/unclamping. Other combinations are possible, for example, the CHI900B SECM (CH Instruments Inc.) uses stepper motors for coarse positioning and a XYZ piezoblock for finer displacements that gives an overall resolution of 1.6 nm. Close-loop positioners with position sensors, such as optical encoders, capacitive sensors, or strain gauge sensors, offer the possibility to precisely reposition the tip. An interesting advantage of the repositioning capability is the possibility of mapping and recording the topography of the surface and subsequently scanning the same area at a constant distance from the surface to deconvolute the contributions of surface topography and reactivity to the tip current.

Unlike STM and AFM tips, which are sharp cones, a typical SECM tip is a conductive disk surrounded by the flat ring of insulating glass whose thickness is equivalent to several disk radii. Thus, a proper alignment of the tip with respect to the substrate surface is crucial (see also Section IV.B.2). Unless the tip surface is flat and strictly parallel to the substrate plane, the insulator touches the substrate first and prevents the conductive disk from coming close to its surface. To facilitate the alignment, special attention must be paid to the attachment of the tip to the translation stage and to substrate mounting.

C. Potentiostat

Many SECM experiments require biasing the substrate. A bipotentiostat in Fig. 1 is used to control both the tip and substrate potentials. Unless transient measurements are made, the response of the bipotentiostat does not have to be fast. More importantly, it should be capable of measuring a broad range of current responses: a picoamp scale (or even sub-pA) tip current and a much higher current at a macroscopic substrate. For this reason, it is convenient to have several choices of preamplifiers/current-to-voltage transducers.

IV. THEORY

The quantitative SECM theory has been developed for various heterogeneous and homogeneous processes and for different tip and substrate geometries [56–59]. Here, we survey the theory pertinent to an inlaid disk electrode (Fig. 5) approaching a flat substrate, which can be considered infinitely large as compared to the tip size. The case of finite substrate size was treated by Bard et al. [60]. The theory for nondisk tips (e.g., shaped as a cone or a spherical cap) is discussed in Refs. [12–14] and [58], and in Section IV.B.2.

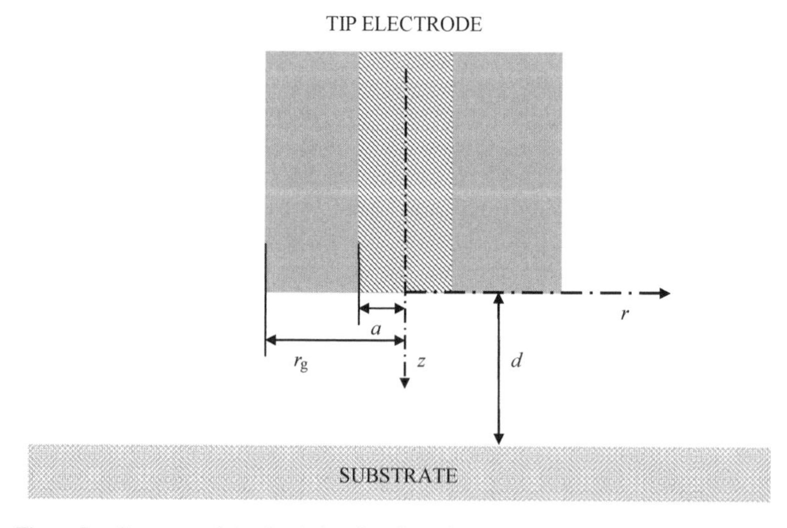

Figure 5. Geometry of the simulation domain and parameters defining the diffusion problem for SECM. Adapted with permission from Ref. [41]. Copyright © 1998, American Chemical Society.

To obtain the dependences of the tip current (the measured signal in amperometric mode) on time and position, one needs to solve the diffusion problem with the boundary conditions specific to the system geometry and chemical reactivity.

The diffusion equations in cylindrical coordinates for the redox species O and R, with no chemical reactions occurring in solution are given below:

$$\frac{\partial c_{\mathrm{O}}}{\partial t} = D_{\mathrm{O}}\left[\frac{\partial^2 c_{\mathrm{O}}}{\partial z^2} + \frac{\partial^2 c_{\mathrm{O}}}{\partial r^2} + \frac{1}{r}\frac{\partial c_{\mathrm{O}}}{\partial r}\right] \tag{8a}$$

$$\frac{\partial c_{\mathrm{R}}}{\partial t} = D_{\mathrm{R}}\left[\frac{\partial^2 c_{\mathrm{R}}}{\partial z^2} + \frac{\partial^2 c_{\mathrm{R}}}{\partial r^2} + \frac{1}{r}\frac{\partial c_{\mathrm{R}}}{\partial r}\right] \tag{8b}$$

where D_{O} and D_{R} are the diffusion coefficients of O and R, respectively.

In general, theoretical SECM dependences can be generated by numerically solving Eqs. (8a,b). Dimensionless variables are better suited for numerical simulations. The variables can be normalized as follows, assuming that only R species is initially present in the solution:

$$\mathrm{R} = \frac{r}{a} \tag{9a}$$

$$Z = \frac{z}{a} \tag{9b}$$

$$C_i = \frac{c_i}{c_{\mathrm{R}}^0} \tag{9c}$$

$$L = \frac{d}{a} \tag{9d}$$

$$T = \frac{tD_{\mathrm{R}}}{a^2} \tag{9e}$$

$$K_{\mathrm{f/b,S}} = \frac{k_{\mathrm{f/b,S}}a}{D_{\mathrm{R}}} \tag{9f}$$

$$K_{\mathrm{f/b,T}} = \frac{k_{\mathrm{f/b,T}}a}{D_{\mathrm{R}}} \tag{9g}$$

$$\mathrm{RG} = \frac{r_{\mathrm{g}}}{a} \tag{9h}$$

$$I_{\mathrm{T/S}} = \frac{i_{\mathrm{T/S}}}{i_{\mathrm{T},\infty}} = \frac{i_{\mathrm{T/S}}}{4nFD_{\mathrm{R}}c_{\mathrm{R}}^0 a} \tag{9i}$$

The assumption of equal diffusion coefficients ($D_O = D_R = D$) allows the problem to be described in terms of a single species (R). The boundary conditions are of the form:
$0 < T, 0 \leq \mathrm{R} < 1, Z = 0$ (tip electrode surface):

$$\left[\frac{\partial C_R}{\partial Z}\right] = K_{b,T}(1 - C_R) - K_{f,T}C_R = J_T \tag{10}$$

$0 < T, 1 \leq \mathrm{R} \leq \mathrm{RG}, Z = 0$ (glass insulating sheath):

$$\left[\frac{\partial C_R}{\partial Z}\right] = 0 \tag{11}$$

$0 < T, 0 \leq \mathrm{R} \leq h, Z = \mathrm{L}$ (substrate surface):

$$\left[\frac{\partial C_R}{\partial Z}\right] = K_{f,S}(1 - C_R) - K_{b,S}C_R = J_S \tag{12}$$

$$\mathrm{R} > \mathrm{RG}, 0 \leq Z \leq \mathrm{L}; \quad C_R = 1 \tag{13}$$

where J_T and J_S are the normalized fluxes of species at the tip and the substrate, and h is the ratio of the substrate radius a_S to the tip radius. The initial condition, completing the definition of the problem, is

$$T = 0, 0 \leq \mathrm{R}, 0 \leq Z \leq \mathrm{L}; \quad C_R = 1 \tag{14}$$

If the substrate is an insulator $J_S \equiv 0$. If both tip and substrate reactions are electrochemical processes, the rate constants for oxidation (k_f) and reduction (k_b) are given by the Butler–Volmer relations

$$k_f = k^0 e^{(1-\alpha)nf(E-E^{0'})} \tag{15}$$

$$k_b = k^0 e^{-\alpha nf(E-E^{0'})} \tag{16}$$

where k^0 is the standard rate constant, E is the electrode potential, $E^{\circ\prime}$ is the formal potential, α is the transfer coefficient, n is the number of electrons transferred per redox event, and $f = F/RT$ (here F is the Faraday constant, R is the gas constant, and T is the temperature). The solution of the problem can be obtained in terms of the dimensionless current versus time dependencies, $I_T(T)$ and $I_S(T)$

$$I_T(T) = -\frac{\pi}{2}\int_0^1 J_T(T,R)RdR \qquad I_S(T) = \frac{\pi}{2}\int_0^h J_S(T,R)RdR \tag{17}$$

The above formulation is somewhat overly general, since it includes the possibility of mixed diffusion–kinetic control of both the tip and substrate processes. In practice, at least one of those electrodes is held under diffusion control, so either Eq. (18a)

$$0 < T, 0 \leq R < 1, Z = 0; \quad C_R = 0 \tag{18a}$$

can be used instead of Eq. (10), or Eq. (12) can be replaced with a much simpler boundary condition

$$0 < T, 0 \leq R \leq h, Z = L; \quad C_R = 1 \tag{18b}$$

The time-dependent SECM problem was solved semianalytically in terms of 2D integral equations [60, 61] and numerically by using Krylov integrator [62] and the alternating-direction implicit finite difference (ADI) method [21, 60]. Potentiostatic transients were computed for two limiting cases: a diffusion-controlled process and totally irreversible kinetics [21, 60, 62]. The analysis of the simulation results [62] revealed several time regions typical for SECM transients (Fig. 6). In the short-time region ($T \leq 0.001$) an SECM tip current follows closely the microdisk transient, then it starts to deviate and finally levels at a constant value of i_T. For a conductive substrate $i_T \geq i_{T,\infty}$, and it is always larger than the value calculated for the same time from thin-layer cell (TLC) theory. For an insulating substrate $i_T < i_{T,\infty}$. The smaller the tip–substrate separation the earlier the deviations occur between the SECM and microdisk transients. The time when the SECM undergoes a transition from the microdisk regime to the TLC regime is related to the time needed by the species to diffuse across the gap between the tip and the substrate. This time can be determined experimentally and can be used to evaluate the diffusion coefficient [62]. The transients computed for different rates of an irreversible heterogeneous reaction [21] also showed a microdisk-type behavior at short times. The current magnitude at longer times and its eventual steady-state value are determined by the value of the dimensionless heterogeneous rate constant, $K_{b,S}$ (Fig. 7). The substrate behaves as a conductor as $K_{b,S} \to \infty$, and as an insulator as $K_{b,S} \to 0$. A few transients and non-steady-state CVs for a quasireversible process were computed for both insulating and conducting substrates [61].

Martin and Unwin simulated chronoamperometric feedback allowing for unequal diffusion coefficients of the oxidized and reduced forms of the redox mediator [63]. Unlike steady-state SECM response, the shape of the tip current transients is sensitive to the ratio of the diffusion coefficients, $\gamma = D_O/D_R$ (Fig. 8). When $D_O = D_R$, the tip current attains a steady-state value much faster

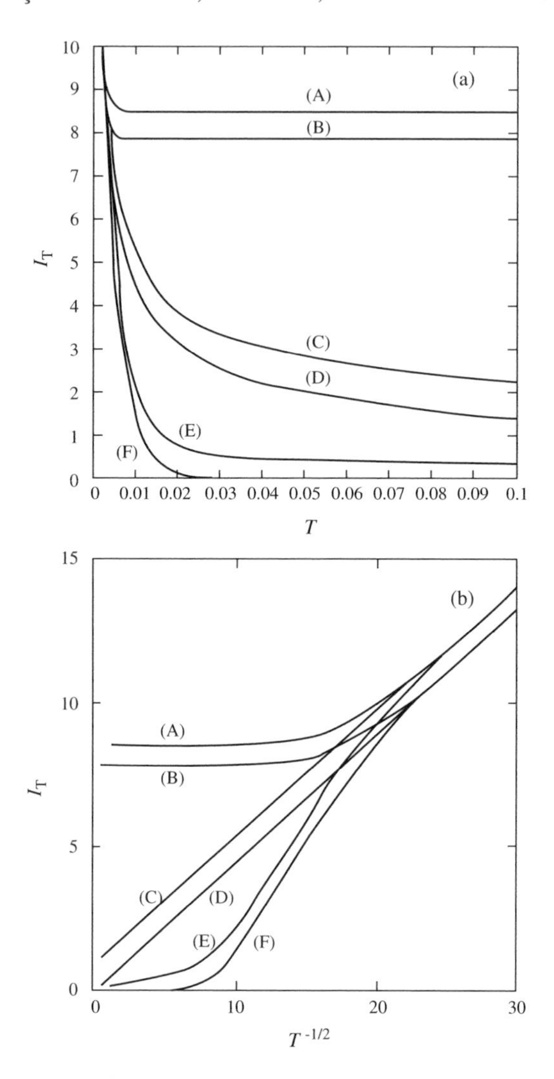

Figure 6. Comparison of simulated SECM transients with transients corresponding to different electrode geometries (all processes are diffusion controlled). (A) the SECM transient for a conductive substrate; (B) two-electrode thin-layer cell; (C) microdisk; (D) planar electrode; (E) SECM with an insulating substrate; (F) one-electrode thin-layer cell. Curves A, B, E, and F were computed with $L = d/a = 0.1$. Adapted with permission from Ref. [62]. Copyright © 1991, American Chemical Society.

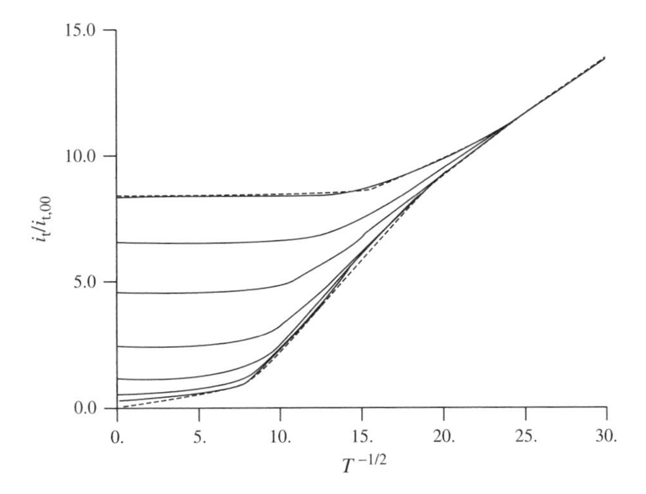

Figure 7. Feedback transients simulated for various rate constants of an irreversible heterogeneous process at the substrate. The upper and lower dashed curves correspond to the limits $K_{b,S} \to \infty$ and $K_{b,S} = 0$, respectively. The solid curves (from top to bottom): log $K_{b,S} = 3.0$, 1.5, 1.0, 0.5, 0, −0.5, and −1.0. RG = 10. Taken with permission from Ref. [60]. Copyright © 1992, American Chemical Society.

than for any $\gamma \neq 1$. At $\gamma < 1$, a characteristic minimum appears in the short-time region, which is quite unusual for potentiostatic transients.

Note that Eq. (13) implies that concentration of the mediator is equal to its bulk value everywhere beyond the limits of the tip insulating sheath (R $\geq$ RG). This approximation is applicable only if RG $\gg$ 1, and even for RG = 10 it leads to ~2% error, which is present in most published SECM simulations [41].

A. Analytical Approximations for Steady-State Responses

The solution of the above problem applies to both transient and steady-state feedback experiments. Since transient SECM measurements are somewhat less accurate and harder to perform, most quantitative studies were carried out under steady-state conditions. The non-steady-state SECM response depends on too many parameters to allow presentation of a complete set of working curves, which would cover all experimental possibilities. The steady-state theory is simpler and often can be expressed in the form of dimensionless working curves or analytical approximations.

1. *Diffusion-Controlled Heterogeneous Reactions*

The dimensionless steady-state current–distance curves were calculated numerically by Kwak and Bard [7] for both pure positive and negative feedback

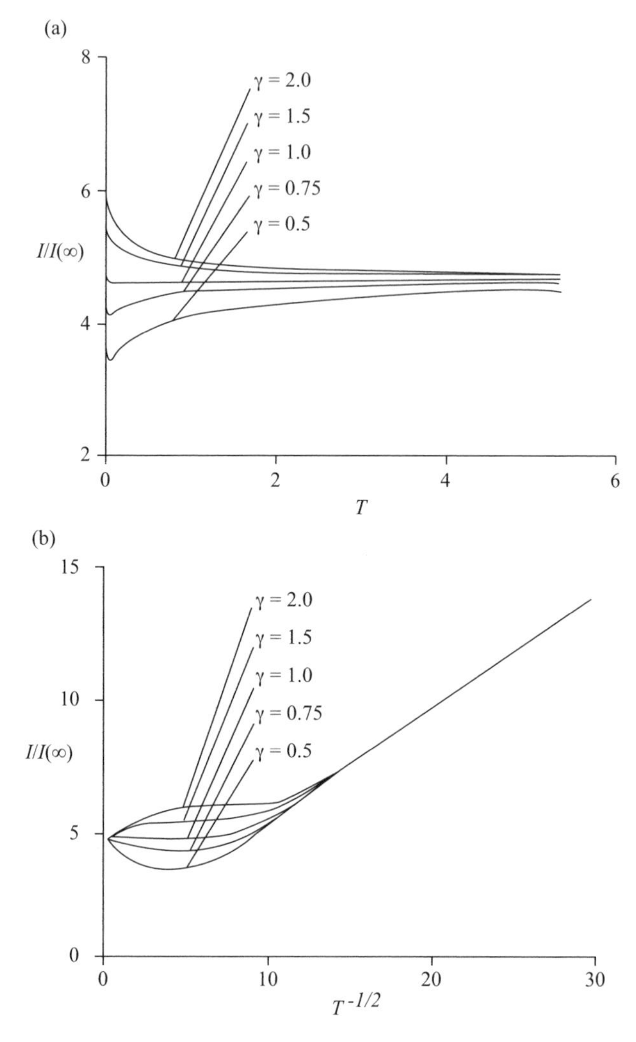

Figure 8. The effect of γ on the chronoamperometric characteristics for the positive feedback process at a tip–substrate separation, $L = 0.2$. The reduced form initially present in solution is oxidized at the tip and regenerated at the substrate. Normalized tip current is plotted as a function of dimensionless time, $\tau = tD_R/a^2$. Taken with permission from Ref. [63]. Copyright © 1997, Elsevier Science S.A.

conditions assuming a diffusion-controlled mediator turnover, equal diffusion coefficients, and an infinitely large substrate. Several previously published analytical expressions for the current versus distance curve for pure positive feedback and different RG values were recently discussed by Lefrou [64].

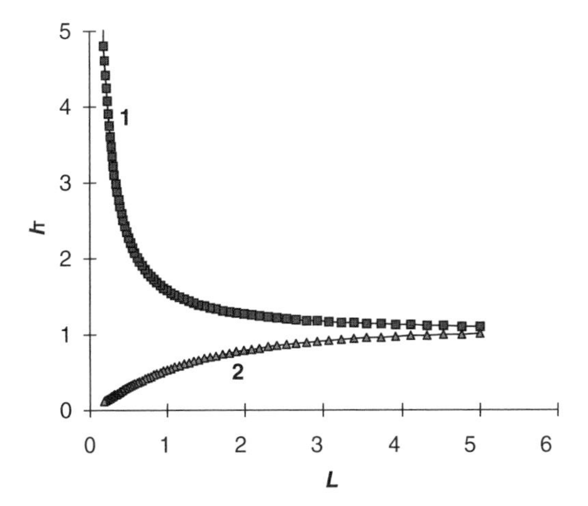

Figure 9. Theoretical approach curves for a tip electrode over a conductive (1) and insulating (2) substrate. Solid lines are computed for RG = 10 from (1) Eq. (19) and (2) Eq. (20). Symbols are from simulations in Ref. [41].

For RG = 10, Eq. (19) [12]

$$I_T^C(L) = \frac{i_\mathrm{T}}{i_{\mathrm{T},\infty}} = 0.78377/L + 0.3315\exp(-1.0672/L) + 0.68 \qquad (19)$$

where $L = d/a$ is the normalized tip–substrate distance, fits the I_T versus L curve over the L interval from 0.05 to 20 to within 0.7% (Fig. 9, curve 1). Since the shape of the approach curve at a conductive substrate does not depend strongly on RG, Eq. (19) is sufficiently accurate when the insulating sheath of the tip is not very thin (e.g., RG $\geq$ 3). For very sharp tips, equations similar to Eq. (19) with tabulated parameters are available for several RG values [41, 65]. Alternatively, one can use a more complicated analytical approximation from Ref. [64], which is reasonably accurate (within 2%) for all RG and L.

In the case of an insulating substrate, the shape of the dimensionless current–distance curve is much more sensitive to the RG value. Figure 9 shows an approach curve simulated for RG = 10 (curve 2) and a corresponding curve calculated from Eq. (20) [41]

$$I_T^{\mathrm{ins}}(L) = \frac{i_\mathrm{T}}{i_{\mathrm{T},\infty}} = \frac{1}{A + B/L + C\exp(D/L)} + \frac{EL}{F + L} \qquad (20)$$

TABLE 1.
Parameter Values for Eq. (20).

RG	A	B	C	D	E	F
1.1	1.1675164	1.0309985	0.3800855	−1.701797	0.3463761	0.0367416
1.5	1.0035959	0.9294275	0.4022603	−1.788572	0.2832628	0.1401598
2.0	0.7838573	0.877792	0.4248416	−1.743799	0.1638432	0.1993907
10	0.4571825	1.4604238	0.4312735	−2.350667	−0.145437	5.5768952

Equation (20) with the parameter values listed in Table 1 and Eq. (21) [65]

$$I_T^{\mathrm{ins}}(L) = 1/[k_1 + k_2/L + k_3 \exp(k_4/L)] \qquad (21)$$

with the parameter values listed in Table 2 cover reasonably well the entire range of RG from 1.1 to 1000. The accuracy of Eq. (20) is ~1%, which is smaller than typical experimental uncertainty. Equation (21) is not valid for small tip–substrate distances. It is suitable for $L \geq 2$ when RG = 1.11 and $L \geq 0.2$ for any other RG in the table.

By fitting an experimental current–distance curve to the theory [Eqs. (19–21)], one can determine the zero separation point ($L = 0$), which in turn allows one to establish the distance scale essential for any quantitative SECM experiment.

2. *Finite Kinetics at the Tip or Substrate*

For the finite heterogeneous kinetics at the tip and diffusion-controlled mediator regeneration at the substrate, an approximate equation (22) was recently obtained for I_T as a function of tip potential, E, and L [51]

$$I_T(E, L) = \frac{0.78377}{L(\theta + 1/\kappa)} + \frac{0.68 + 0.3315 \exp(-1.0672/L)}{\theta\left[1 + \dfrac{\pi}{\kappa\theta}\dfrac{2\kappa\theta + 3\pi}{4\kappa\theta + 3\pi^2}\right]} \qquad (22)$$

where $\kappa = \pi\lambda \exp[-\alpha F(E - E^{\circ})/RT]/(4I_T^c)$, $\theta = 1 + \exp[F(E - E^{\circ\prime})/RT]D_O/D_R$, $E^{\circ\prime}$ is the formal potential, α is the transfer coefficient, $\lambda = k^{\circ}a/D$, and $I_T^c(L)$ is the normalized tip current for the same L and the diffusion-controlled positive feedback at a conductive substrate, as defined by Eq. (19).

At constant L, Eq. (22) describes a quasireversible steady-state tip voltammogram [if kinetics is fast, $\kappa \to \infty$, and Eq. (22) reduces to one for a Nernstian tip voltammogram]. Such a curve can be obtained by scanning the potential of the tip while the substrate potential is held constant. Finite element simulations in Ref. [51] showed that Eq. (22) is more accurate than two somewhat similar expressions derived earlier [66].

TABLE 2.
Parameter Values for Eq. (21)

RG	k_1	k_2	k_3	k_4	% error	Validity range
1002	0.13219	3.37167	0.8218	−2.34719	<1%	0.3–20
100	0.27997	3.05419	0.68612	−2.7596	<1%	0.4–20
50.9	0.30512	2.6208	0.66724	−2.6698	<1%	0.4–20
20.1	0.35541	2.0259	0.62832	−2.55622	<1%	0.4–20
15.2	0.37377	1.85113	0.61385	−2.49554	<1%	0.4–20
10.2	0.40472	1.60185	0.58819	−2.37294	<1%	0.4–20
8.13	0.42676	1.46081	0.56874	−2.28548	<1%	0.4–20
5.09	0.48678	1.17706	0.51241	−2.07873	<1%	0.2–20
3.04	0.60478	0.86083	0.39569	−1.89455	<0.2%	0.2–20
2.03	0.76179	0.60983	0.23866	−2.03267	<0.15%	0.2–20
1.51	0.90404	0.42761	0.09743	−3.23064	<0.7%	0.2–20
1.11	−1.46539	0.27293	2.45648	8.995E-7	<1%	2–20

The current–distance curves for an irreversible heterogeneous reaction occurring at the substrate while the tip process is diffusion-controlled can be calculated from Eq. (23) [38]:

$$I_T(L) = I_S\left(1 - \frac{I_T^{\text{ins}}}{I_{\text{T}}^{c}}\right) + I_T^{\text{ins}} \tag{23a}$$

$$I_S = \frac{0.78377}{L(1 + 1/\Lambda)} + \frac{0.68 + 0.3315\exp(-1.0672/L)}{1 + F(L,\Lambda)} \tag{23b}$$

where I_T^C and I_T^{ins} are given by Eqs. (19) and (20), respectively, and I_S is the kinetically controlled substrate current; $\Lambda = k_{\text{f}}d/D$, k_{f} is the heterogeneous rate constant (cm^{-1} s), and $\text{F}(L,\Lambda) = (11/\Lambda + 7.3)/(110 - 40L)$.

Although Eq. (23) was derived for a one-step heterogeneous ET reaction, it was shown to be applicable to more complicated substrate kinetics (e.g., liquid–liquid interfacial charge transfer [38, 39, 67], ET through self-assembled monolayers [68, 69], and mediated ET in living cells [70–73]). The effective heterogeneous rate constant obtained by fitting experimental approach curves to Eq. (23) can be related to various parameters, which determine the rates of those processes, as discussed in the referred publications.

The radius of the portion of the substrate surface participating in the SECM feedback loop can be evaluated as $r \cong a + 1.5d$ [60]. Thus, at small tip–substrate distances (e.g., $L \leq 2$), a large substrate behaves as a virtual UME of a size comparable with that of the tip electrode. The SECM allows probing local kinetics at a small portion of the macroscopic substrate with all of the advantages of microelectrode measurements [10, 11]. The effective mass-transfer coefficient for SECM is

$$m_o = 4D_o \frac{0.68 + 0.78377/L + 0.3315\exp(-1.0672/L)}{\pi a} = \frac{I_T^C(L)}{\pi a^2 nFc} \quad (24)$$

One can see from Eq. (24) that at $L \gg 1$, $m_O \sim D/a$ (as for a microdisk electrode alone), but at $L \ll 1$, $m_O \sim D/d$, which is indicative of the TLC type behavior. By decreasing d, the mass-transport rate can be increased sufficiently for quantitative characterization of fast ET kinetics, while preserving the advantages of steady-state methods, that is, the absence of problems associated with ohmic drop, adsorption, and charging current.

3. *SG/TC Mode*

Unlike feedback mode of the SECM operation, where the overall redox process is essentially confined to the thin layer between the tip and the substrate, in SG/TC experiments the tip travels within a thick diffusion layer produced by the large substrate. The system reaches a true steady state if the substrate is an ultramicroelectrode (e.g., a microdisk or a spherical cap) that generates or consumes the species of interest. The concentration of such species can be measured by an ion-selective (potentiometric) microprobe as a function of the tip position. The concentration at any point can be related to that at the source surface. For a microdisk substrate the dimensionless expression is [74, 75]

$$c(R,Z)/c(0,L) = \frac{2}{\pi}\tan^{-1}\left[\frac{2}{R^2 + Z^2 - 1 + \sqrt{(R^2 + Z^2 - 1)^2 + 4Z^2}}\right]^{1/2} \quad (25)$$

A somewhat more complicated expression was derived to relate the concentration distribution in the diffusion layer to the flux at the source or sink surface [28].

One can also evaluate the relative change in the rate of a heterogeneous reaction at the substrate by measuring the concentration of the reaction product at the tip. In this setup, the tip is positioned at a fixed distance from the substrate, and the time dependence of concentration is measured. This simpler approach is based on the proportionality between the heterogeneous reaction rate and the product concentration. It is most useful when the substrate flux cannot be measured directly (e.g., the substrate reaction is not an electrochemical process) [76–78].

4. *Lateral Mass/Charge Transfer*

The SECM can be used in the feedback mode to probe lateral mass–charge transfer [79–83]. The theory of SECM feedback surveyed in Section IV.A.2 assumes that the substrate surface is uniformly reactive. When lateral mass and/or charge transfer occurs on the substrate surface, or within a thin film, the surface reactivity of the substrate becomes non-uniform and the SECM feedback theory must be modified. Unwin and Bard [79] developed the theory for adsorption–desorption of a redox species at the substrate that allowed for surface diffusion of the adsorbate. They introduced a new approach, the scanning electrochemical microscope induced desorption (SECMID), as a way to probe surface diffusion. The set of differential equations for the diffusion problem comprise Eqs. (8a,b), and Eq. (26), which relates the redox concentration at the substrate surface and the surface coverage by adsorbed species

$$\gamma \frac{\partial \theta}{\partial T} = D_R \gamma \left(\frac{\partial^2 \theta}{\partial R^2} + \frac{1}{R} \frac{\partial \theta}{\partial R} \right) - K_d \theta + K_a C_{Z=L}(1 - \theta) \tag{26}$$

where K_a and K_d are the dimensionless adsorption and desorption rate constants, respectively, N is the maximum number of adsorption sites on the metal oxide, $\gamma = N/c^{\text{bulk}}a, D_R$ is the ratio of the surface diffusion coefficient to the solution diffusion coefficient, c^{bulk} is the bulk concentration of the redox species, $C_{Z=L}$ is the dimensionless concentration of the solution redox species at the surface of the substrate, θ is the fractional surface coverage. Assuming Langmuirian characteristics, the initial condition at the substrate was

$$t = 0; \quad \theta = \frac{1}{1 + K_a/K_d} \tag{27}$$

By fitting experimental curves to the numerically generated working curves, Unwin and Bard were able to measure the rate of adsorption of protons on the (001) surface of rutile (TiO_2) and (010) surface of albite. They showed that surface diffusion in these systems was too slow in comparison with the solution diffusion to be measurable.

The electronic conductivity of the substrate can also be probed by SECM. In the case of finite conductivity, a non-uniform surface potential profile develops at the surface. Liljeroth et al. [80] developed the theory for probing the film resistivity with the SECM in a feedback mode. The electrochemical mass-transport equation in this case was

$$\frac{\partial^2 \tilde{\mu}}{\partial R^2} + \frac{1}{R} \frac{\partial \tilde{\mu}}{\partial R} - \frac{K^0}{\Sigma} ((1 - C_{Z=L}) e^{\tilde{\mu}/2} - C_{Z=L} e^{-\tilde{\mu}/2}) = 0 \tag{28}$$

where $\tilde{\mu} = (\mu - \mu^0)/kT$ is the dimensionless electrochemical potential of the electrons in the conductive film, K^0 is the normalized standard rate constant of the electron transfer between the solution redox species and the film, $C_{Z=L}$ is the dimensionless concentration of the solution redox species at the film surface, and Σ is the dimensionless conductivity of the film given by Eq. (29)

$$\Sigma = \frac{\sigma k T \Delta z}{e^2 a D c^{\text{bulk}} N_\text{A}} \tag{29}$$

where σ is the conductivity, k is the Boltzmann constant, T is the absolute temperature, Δz is the thickness of the film, e is the elementary charge, D and c^{bulk} are the diffusion coefficient and the bulk concentration of the redox species, and N_A is the Avogadro constant.

5. *Homogeneous Reactions*

If the product of the tip or substrate ET reaction [Eqs. (1) and (3)] participates in a homogeneous reaction within the tip–substrate gap, the feedback response is altered. In this case, the shape of the i_T versus d curve depends on the rate of the homogeneous chemical reaction [84]. If the tip and the substrate are biased at extreme potentials, so that reactions (1) and (3) are rapid, the shape of the SECM current–distance curve for a relatively simple mechanism is a function of a single kinetic parameter, $K = \text{const} \times k_c/D$, where k_c is the rate constant of the irreversible homogeneous reaction.

The SECM theory has been developed for four mechanisms involving homogeneous chemical reactions coupled with ET, that is, a first-order irreversible following reaction (E_rC_i mechanism) [21], a second-order irreversible dimerization (E_rC_{2i} mechanism) [22], ECE, and DISP1 reactions [85]. (The solution obtained for a E_qC_r mechanism in terms of multidimensional integral equations [61] has not been utilized in any calculations). Three approaches to kinetic analysis were proposed: (1) steady-state measurements in a feedback mode, (2) generation–collection experiments, and (3) analysis of the chronoamperometric SECM response. Unlike the feedback mode, generation–collection measurements include simultaneous analysis of both I_T-L and I_S-L curves or the use of the collection efficiency parameter (I_S/I_T when the tip is a generator and the substrate is a collector). The chronoamperometric measurements were found less reliable [21], so only steady-state theory will be discussed here. While for the E_rC_i and E_rC_{2i} mechanisms analytical approximations are available [24], only numerical solutions have been reported for more complicated ECE and DISP1 reactions [85].

a. First-Order Following Reaction. For the E_rC_i mechanism, the tip and substrate reactions are given by Eqs. (1) and (3) and the reaction occurring in the solution gap between the tip and the substrate is given by Eq. (30)

$$\mathrm{O} \xrightarrow{k_c} \text{products} \qquad \text{(gap)} \tag{30}$$

the SECM diffusion problem is represented by the system of differential equations

$$\frac{\partial c_{\mathrm{O}}}{\partial t} = D_{\mathrm{O}}\left[\frac{\partial^2 c_{\mathrm{O}}}{\partial z^2} + \frac{\partial^2 c_{\mathrm{O}}}{\partial r^2} + \frac{1}{r}\frac{\partial c_{\mathrm{O}}}{\partial r}\right] - k_c c_{\mathrm{O}} \tag{31a}$$

$$\frac{\partial c_{\mathrm{R}}}{\partial t} = D_{\mathrm{R}}\left[\frac{\partial^2 c_{\mathrm{R}}}{\partial z^2} + \frac{\partial^2 c_{\mathrm{R}}}{\partial r^2} + \frac{1}{r}\frac{\partial c_{\mathrm{R}}}{\partial r}\right] \tag{31b}$$

with corresponding initial and boundary conditions [i.e., Eqs. (18a, 18b) for the tip and conductive substrate]. It was solved numerically using the alternating-direction implicit (ADI) finite difference method [21]. The steady-state results were obtained as a long-time limit and presented in the form of two-parameter families of working curves [21]. These represent steady-state tip current or collection efficiency as functions of $K = ak_c/D$ and L.

Later [24], it was shown that the theory for the $\mathrm{E_rC_i}$ process under SECM conditions can be reduced to a single working curve. To understand this approach, it is useful first to consider a positive feedback situation with a simple redox mediator (i.e., without homogeneous chemistry involved) and with both tip and substrate processes under diffusion control. The normalized steady-state tip current can be presented as the sum of two terms

$$I_{\mathrm{T}} = I_f + I_{\mathrm{T}}^{\mathrm{ins}} \tag{32}$$

where I_f is the feedback current coming from the substrate and $I_{\mathrm{T}}^{\mathrm{ins}}$ is the current due to the hindered diffusion of the electroactive species to the tip from the bulk of solution given by Eq. (20); all variables are normalized by $i_{\mathrm{T},\infty}$. The substrate current is

$$I_{\mathrm{S}} = I_f + I_d \tag{33}$$

where I_f is the same quantity as in Eq. (32), representing the reduced species that eventually arrive at the tip as a feedback current, and I_d is the dissipation current, that is, the flux of species not reaching the tip. It was shown in Ref. [22] that I_S/I_T is > 0.99 at $0 < L \leq 2$, (i.e., for any L within this interval the tip and the substrate currents are essentially equal to each other). Thus from Eqs. (32) and (33)

$$I_d = I_{\mathrm{T}}^{\mathrm{ins}} \tag{34}$$

or

$$I_d/I_S = I_T^{ins}/I_T = f(L) \quad (35)$$

where $f(L)$ can be computed for any L as the ratio of the right-hand side of Eq. (20) to that of Eq. (19) assuming RG = 10.

Analogously, for an electrochemical process followed by an irreversible homogeneous reaction of any order, one can write

$$I'_T = I'_f + I_T^{ins} \quad (36)$$

$$I'_S = I'_f + I'_d \quad (37)$$

where the variables labeled with the prime are analogous to unlabeled variables in Eqs. (32) and (33), and I_T^{ins} is unaffected by the occurrence of the homogeneous reaction in Eq. (30). Since the species R are stable, the fraction of these species arriving at the tip from the substrate is also unaffected by the reaction in Eq. (30), that is, the relation $I'_d/I'_S = f(L)$ holds true [24]. Consequently,

$$\begin{aligned} I'_T &= I_T^{ins} + I'_S - I'_d \\ &= I_T^{ins} + I'_S(1 - f(L)) \end{aligned} \quad (38)$$

that is, for an SECM process with a following homogeneous chemical reaction of any order, the dependence I'_T versus I'_S at any given L is linear with a slope equal to 1-$f(L)$ and an intercept equal to $I_T^{ins}(L)$. Thus, the generation–collection mode of the SECM (with the tip electrode serving as a generator) for these mechanisms is equivalent to the feedback mode, and any quantity, I'_T, I'_S, or I'_S/I'_T, can be calculated from Eq. (38) for a given L, if any other of these quantities is known.

For mechanisms with the following irreversible reactions, one can expect the collection efficiency, I'_S/I'_T, to be a function a single kinetic parameter κ. If this parameter is known, the SECM theory for this mechanism can be reduced to a single working curve. After the function $\kappa = F(I'_S/I'_T)$ is specified, one can immediately evaluate the rate constant from I'_S/I'_T versus L or I'_T versus L experimental curves. If only tip current has been measured, the collection efficiency can be calculated as

$$\frac{I'_S}{I'_T} = \frac{1 - \frac{I_T^{ins}}{I'_T}}{1 - f(L)} \quad (39)$$

For the E_rC_i mechanism $\kappa = k_c d^2/D$. Figure 10a represents the working curve, κ versus I'_S/I'_T, along with the simulated data from Ref. [21]. The numerical results fit the analytical approximation

$$\kappa = F(x) = 5.608 + 9.347\exp(-7.527x) - 7.616\exp(-0.307/x) \qquad (40)$$

(solid curve in Fig. 10a), where $x = I'_S/I'_T$, within $\sim$1%.

The latest contribution to the theory of the EC processes in SECM was the modeling of the SG/TC situation by Martin and Unwin [86]. Both the tip and substrate chronoamperometric responses to the potential step applied to the substrate were calculated. From the tip current transient one can extract the value of the first-order homogeneous rate constant and (if necessary) determine the tip–substrate distance. However, according to the authors, this technique is unlikely to match the TG/SC mode with its high collection efficiency under steady-state conditions.

b. Second-Order Following Reaction. The detailed TG/SC theory was developed for an electrode process with a following dimerization reaction (E_rC_{2i} mechanism) [22]:

$$2\text{O} \xrightarrow{k_c} \text{products} \qquad \text{(gap)} \qquad (41)$$

This diffusion problem is similar to the one for E_rC_i mechanism [Eqs. (1), (3), and (30)] except for the different Fick's equation for c_O:

$$\frac{\partial c_O}{\partial t} = D_O\left[\frac{\partial^2 c_O}{\partial z^2} + \frac{\partial^2 c_O}{\partial r^2} + \frac{1}{r}\frac{\partial c_O}{\partial r}\right] - k_c(c_O)^2 \qquad (42)$$

Both chronoamperometric and steady-state responses were calculated by solving the related equations numerically. An analytical approach discussed in the previous section is equally applicable to the E_rC_{2i} mechanism under steady-state conditions. Equation (38) was verified using the data simulated from Ref. [22], and I'_T versus I'_S dependencies were plotted for different values of L [22]. Although for the E_rC_{2i} mechanism, the choice of κ is less straightforward than for E_rC_i, an acceptable fit for all the data points computed in Ref. [22] was obtained using $\kappa' = c^0 k'_c d^3/aD$ (Fig. 10b). This data was fit to Eq. (43):

$$\kappa' = 104.87 - 9.948x - 185.89/\sqrt{x} + 90.199/x + 0.389/x^2 \qquad (43)$$

Although this approximation is less accurate than Eq. (40), its use would not lead to an error of more than $\sim$ 5–10%, which is within the usual range of

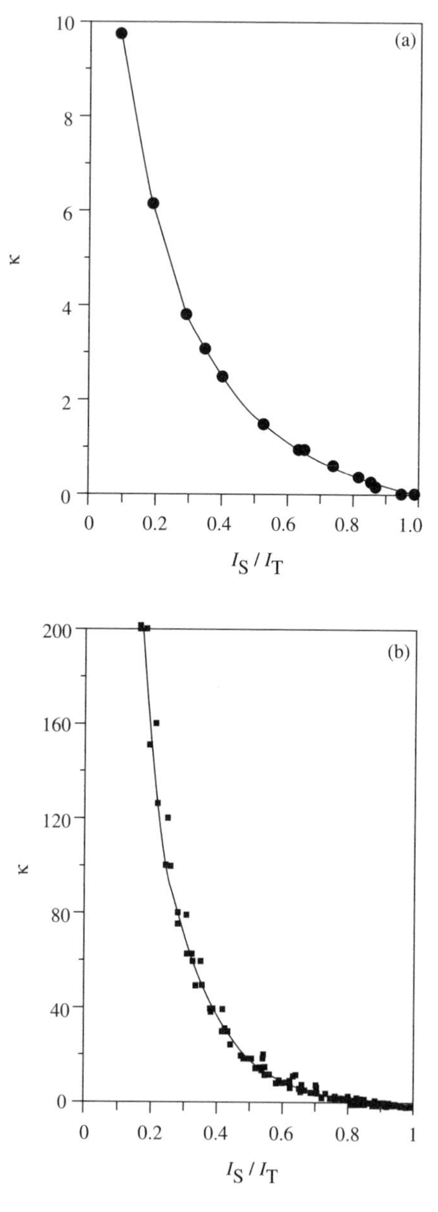

Figure 10. Kinetic parameter κ as a function of the collection efficiency (I'_S/I'_T). (a) the E_rC_i mechanism; $\kappa = k_c d^2/D$, solid line was computed from Eq. (40), triangles are simulated data taken from ref. 21. (b) the E_rC_{2i} mechanism; $\kappa' = c^\circ k_c d^3/aD$, solid line was computed from Eq. (43), squares are simulated data taken from Ref. [22]. Taken with permission from Ref. [24]. Copyright © 1994, American Chemical Society.

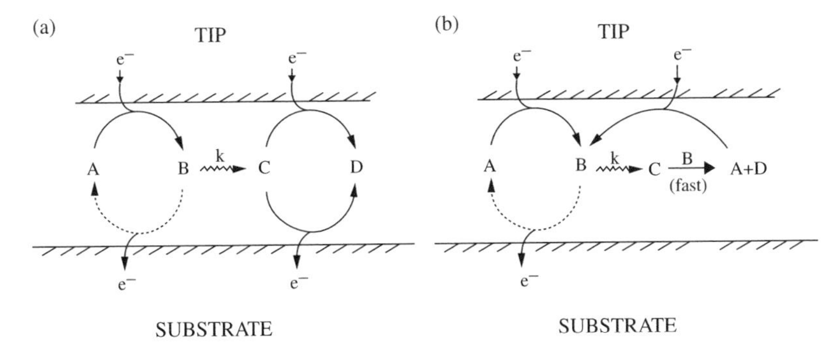

Figure 11. Diffusion and chemical processes occurring within the tip–substrate gap in the case of (a) an ECE pathway and (b) a DISP1 pathway. Taken with permission from Ref. [85]. Copyright © 1996, American Chemical Society.

experimental error. The invariability of k_c computed from different experimental points would assure the validity of the results.

c. ECE and DISP1 Mechanisms. These reactions involving two ET steps are shown schematically in Fig. 11 [85]. The product of homogeneous chemical reaction (C) often is easier to reduce than A. This second reduction can either occur at the tip electrode (ECE mechanism in Fig. 11a) or via the disproportionation (DISP1 mechanism in Fig. 11b)

$$\mathrm{C} + \mathrm{B} \rightarrow \mathrm{A} + \mathrm{D} \tag{44}$$

With the substrate biased at a potential slightly more positive than E° of A/B couple, B is oxidized to form A for both DISP1 and ECE mechanisms. However, in the latter case the reduction of C also occurs at the substrate. The numerical solution of corresponding diffusion problems (see Ref. [85] for problem formulations) yielded several families of working curves shown in Fig. 12 (DISP1 pathway) and Fig. 13 (ECE pathway). In both cases, the tip and the substrate currents are functions of the dimensionless kinetic parameter, $\kappa = ka^2/D$.

The normalization of the i_T and i_S for two-electron processes is somewhat problematic. In Ref. [85], both quantities are normalized with respect to the one-electrode steady-state current, which flows at infinite tip–substrate separation $(i_{\mathrm{T},1e,\infty} = 4FDac_a^0)$. However, this value is not equal to experimentally measured tip current at $d \rightarrow \infty$, which also includes the contribution from the second ET. Nevertheless, by comparing experimental current–distance curves to the theory one can distinguish between DISP1 and ECE pathways and evaluate the k value [85].

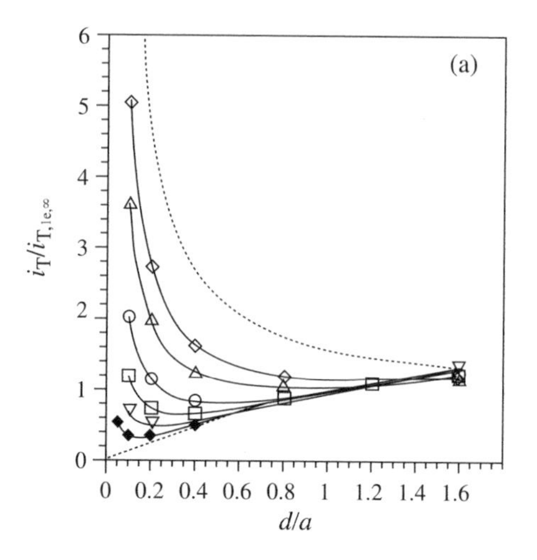

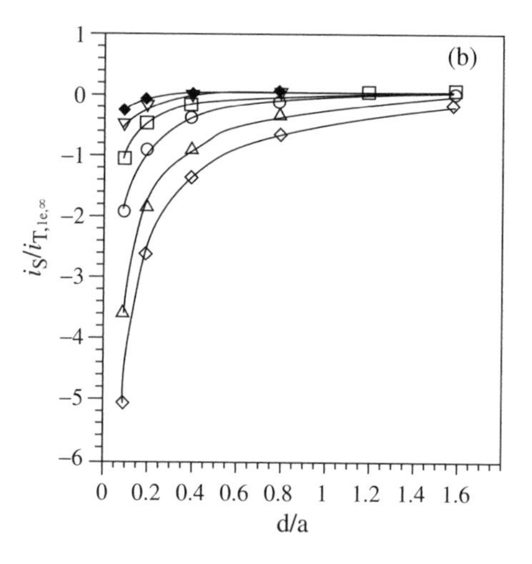

Figure 12. The DISP1 pathway. Theoretical current–distance curves for several values of $K = ka^2/D$: 1(◇), 2 (△), 5 (o), 10 (□), 20 (▽), and 50 (◆). The upper dashed line in (a) represents the one-electron pure positive feedback ($K = 0$). The lower dashed line is the two-electron pure negative feedback ($K = \infty$). The dashed line in (b) represents the one-electron pure positive feedback. Taken with permission from Ref. [85]. Copyright © 1996, American Chemical Society.

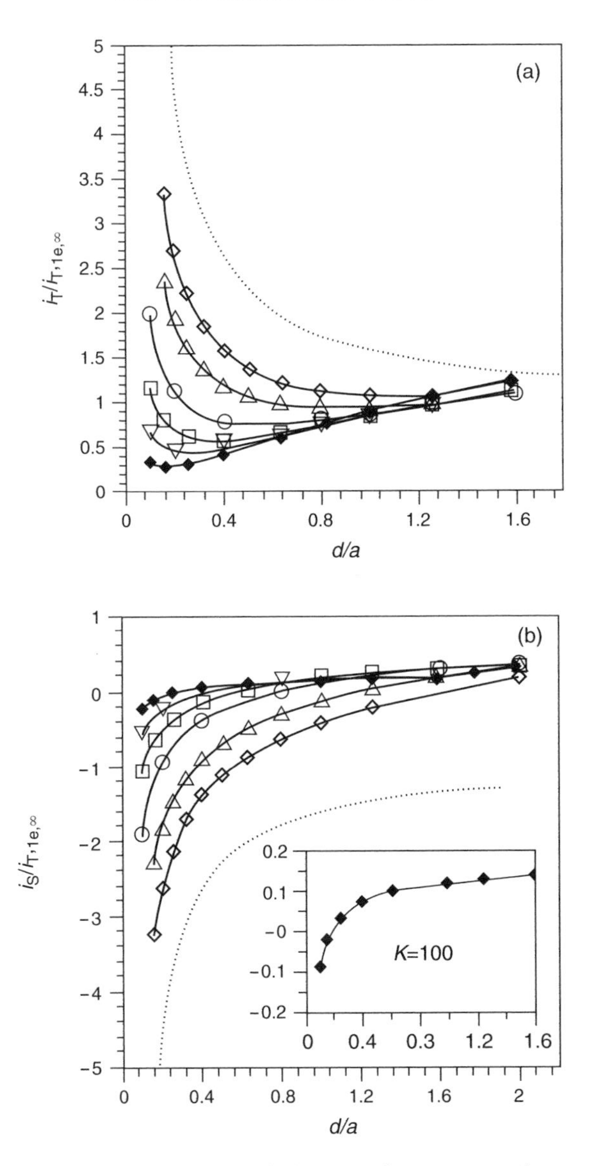

Figure 13. The ECE pathway. Theoretical current–distance curves for several values of $K = ka^2/D$: 1(◇), 2 (△), 5 (o), 10 (□), 20 (▽), and 50 (◆). The upper dashed line in (a) represents the one-electron pure positive feedback ($K = 0$). The dashed line in (b) represents the one-electron pure positive feedback. The inset shows the substrate current for $K = 100$. Taken with permission from Ref. [85]. Copyright © 1996, American Chemical Society.

B. Numerical Solutions of the SECM Diffusion Problems

The analytical approximations presented above are best fits to numerical simulations of the diffusion problems for relatively simple and well-defined electrochemical systems, for example, an inlaid disk electrode approaching a flat, infinite, and uniformly reactive substrate surface. In most quantitative SECM experiments, the use of such approximations could be justified. However, no analytical approximations are available for more complicated processes and system geometries, and so one has to resort to computer simulations.

1. Nondisk Tip Geometries

Although disk-shaped tips are typically most useful for SECM experiments, it is not always possible to produce such tips, especially when they have to be nanometer sized. For some special applications (e.g., penetration experiments), one may want to purposely fabricate tips with different geometries. To characterize nondisk shaped tips, experimental approach curves were obtained and then compared to simulated ones [12]. A number of UME tip geometries including hemispheres [14, 15], spheres [16], rings [17], ring-disks [18], and etched electrodes [19, 33] have been characterized in this way.

A relatively simple technique to produce nanoelectrodes of conical shape is electrochemical etching of partially insulated wires [13, 19]. The current–distance curves for a conical electrode approaching either a conductive or an insulating substrate were generated by an approximate procedure [12] and later simulated more accurately [13, 87]. Figure 14 shows the calculated current–distance curves for tips of different conical geometries approaching either a perfect conductor (A) or a perfect insulator (B). It can be seen that the magnitude of either positive or negative feedback decreases with increasing ratio of the cone height to its base radius (*H*). For $H > 5$, the maximum I_T^C is only $\sim$2, and the minimum $I_T^{\text{ins}} \cong 0.8$. These working curves can be used to determine the geometry of the electrode for *H* up to $\sim$3, but because of the dramatically lower feedback at $H > 0.5$, conical tips are not very useful for feedback mode experiments.

Another type of nondisk-shaped SECM tips are UMEs shaped as spherical caps. They can be obtained, for example, by reducing mercuric ions on an inlaid Pt disk electrode or simply by dipping a Pt UME into mercury [15]. An approximate procedure developed for conical geometry was also used to model spherical cap tips [12]. Selzer and Mandler performed accurate simulations of hemispherical tips using the alternative direction implicit final difference method to obtain steady-state approach curves and current transients [14]. As with conical electrodes, the feedback magnitude deceases with increasing height of the spherical cap, and it is much lower for a hemispherical tip than for the one shaped as a disk.

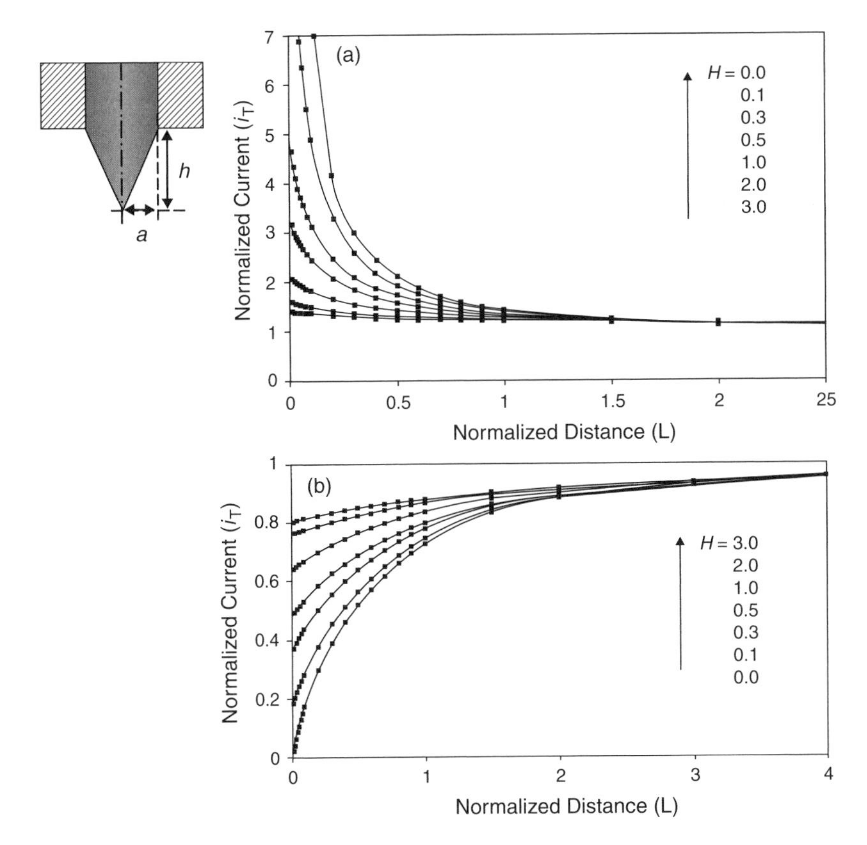

Figure 14. The SECM approach curves for a conical tip geometry (a) positive feedback, (b) negative feedback. H is the ratio of the height to the base radius, $H = h/a$. Adapted with permission from Ref. [13]. Copyright © 2004, American Chemical Society.

2. Three-Dimensional Simulations and Nonideal Conditions

The above SECM theory was developed by solving 2D axisymmetric diffusion problems. Even for an idealized situation (i.e., a flat, planar substrate, strictly perpendicular to the axis of the well-shaped disk tip) only numerical solutions could be obtained. When deviations from this ideal case occur in a real experiment, one has to evaluate the effects of such deviations and check if the available theory can still be used for data analysis. Similar questions arise when the substrate's topography is complicated and/or its surface reactivity is highly non-uniform.

Fulian et al. [88] introduced the boundary element method (BEM) (a powerful numerical method previously employed in engineering computations) for the numerical solution of SECM diffusion problems. The BEM is more suitable for problems with regions of complex or rapidly changing geometries than the finite difference methods employed in earlier SECM simulations. Fulian et al. used the BEM to simulate the current responses for different SECM situations, such as a nondisk tip approaching a flat substrate; a flat disk tip over a hemispherical or a spherical cap-shaped substrate, or a tilted substrate; a lateral scan of a flat-disk tip over an insulating–conductive boundary.

Sklyar and Wittstock [89] used the exterior Laplace formulation combined with the BEM to recalculate the steady-state feedback responses for systems treated in Ref. [88]. The exterior Laplace formulation assigns no geometrical boundaries to the solution, and therefore it does not require any assumption about the extension of the diffusion layer at long times. This approach results in a significantly better estimate of the diffusion flux from the bulk solution to the UME. The same method was also applied to numerically simulate other nonaxisymmetric experimental situations, such as a lateral scan over a band-shaped substrate. Another simulation addressed the shear-force distance control mode of the SECM. The authors concluded that for an RG of 5.1 the tilt of the probe of $\leq 8^\circ$ does not significantly affect the shapes of the i_T-d curves when d is measured from the center of the electroactive area. However, if d is assumed to be the distance between the substrate and the part of the probe closest to it (i.e., the glass edge), the shapes of i_T-d curves change dramatically, and the attainable maximum positive–negative feedback is much lower. This observation becomes significant when one uses the shear-force mode to control the tip–substrate separation distance because in such experiments the distance is measured between the glass edge and the substrate. Quantitative analysis of SECM data in the shear-force mode is therefore more sensitive to the precision of the tip–substrate alignment.

In a later publication [90], BEM was used to calculate the steady-state response of an integrated AFM–SECM probe consisting of a nonconducting AFM tip surrounded by the electroactive gold square frame. The feedback current and the substrate topography were mapped simultaneously when such a tip was scanned over the substrate. The data from the topographic image (Fig. 15a) were used to calculate the simulated current response. A comparison of the real SECM image (Fig. 15b) and the simulated one (Fig. 15d) shows the accuracy of the simulation and its applicability to the study of systems with complex geometry by SECM.

Recently, Sklyar et al. [91] developed a simulation method for diffusion problems involving two different redox species. To avoid solving complicated diffusion problems involving more than one chemical species, most previously reported experiments were carried out under special conditions or with simplifying assumptions, so that the diffusion of only one redox species needed

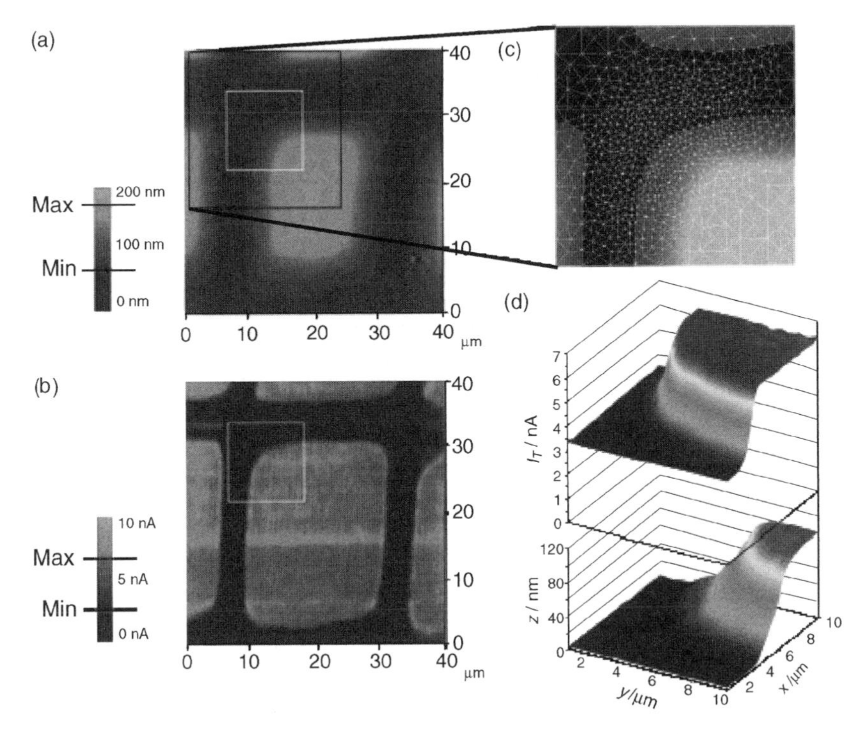

Figure 15. (a) Experimental AFM contact mode (topography) and (b) SECM feedback images obtained simultaneously with an integrated SECM-AFM probe. (c) Simulation mesh for the area is represented by a black rectangle. (d) Simulated SECM feedback current (upper graph) and real topography (lower graph) of the white rectangle area. Solution contained 50 m*M* $[Fe(CN)_6]^{4-}$ and 0.5 *M* KCl. The tip was biased at 600 mV versus Ag QRE. Taken with permission from Ref. [90]. Copyright © 2005, American Chemical Society.

to be considered. The theory developed in Ref. [91] could be useful for experiments with biological cells and other systems in which such simplifications are not appropriate.

V. SELECTED APPLICATIONS

A. Charge-Transfer Processes at Solid–Liquid and Liquid–Liquid Interfaces

The feedback mode of SECM operation is most suited for probing heterogeneous charge-transfer reactions. Electron transfer at the metal–solution interface was the first chemical reaction probed by SECM. An important advantage of this technique for studies of charge transfers at "buried" interfaces (e.g., the liquid–liquid and metal–conductive polymer interfaces) and at semiconductor electrodes is that the substrate surface does not have to be externally biased.

1. Electron-Transfer Reactions at Metal and Semiconductor Electrodes

Kinetics of ET is of primary importance for most electrochemical applications ranging from fuel cells and batteries to biosensors to solar cells to molecular electronics. To measure the fast ET kinetics under steady-state conditions, one needs a technique with the sufficiently high mass transfer rate and negligibly small uncompensated resistive potential drop in solution (IR-drop). The feedback mode of SECM meets both requirements.

The SECM can be used to measure the ET kinetics either at the tip or at the substrate electrode. In the former case, the tip is positioned in a close proximity of a conductive substrate ($d \leq a$). The substrate potential is kept at a constant and sufficiently positive (or negative) value to ensure the diffusion-controlled regeneration of the mediator at its surface. The tip potential is swept linearly to obtain a steady-state voltammogram. The kinetic parameters (k°, α) and the formal potential value can be obtained by fitting such a voltammogram to the theory [Eq. (22)]. A high value of the mass transfer coefficient (m) is achieved under steady-state conditions when $d \ll a$ [see Eq. (24) and related discussion in Section IV.A.2]. The ET rates, which are too fast to be measured when the tip is far from the substrate, can be determined at sufficiently small d. In this way, several fast rate constants ($k^\circ > 1\,\text{cm}^{-1}\,\text{s}$) were measured with micrometer-sized SECM tips [92–94].

Conceptually similar experiments employed nanometer-sized electrodes to study the kinetics of fast ET reactions [51]. Nanoelectrodes offer very high mass transfer rates in combination with practically negligible effects of the IR drop and double-layer charging current. To use a nanoelectrode for kinetic measurements, one has to determine its size and shape. Ideally, the electrode surface should be polishable to ensure the reproducibility of the data obtained with the same tip. While the electrode radius can be roughly evaluated from the steady-state diffusion limiting current, this method does not provide information about electrode geometry (inlaid disk, conical, recessed disk, etc.) From a high quality current–distance curve one can determine the tip radius, a, the RG value, and also check if the electrode surface is flat and not recessed into or protruding out of the insulator. An SECM approach curve in Fig. 16a and a steady-state voltammogram (inset) were obtained with the same polished Pt nanoelectrode, and the same radius value, $a = 46$ nm, was obtained from the limiting voltammetric current and from fitting the i_T-d curve to the theory. This radius value is highly reliable because of the very high positive feedback current (up to 8). Figure 16b shows voltammograms of 1 mM ferrocenemethanol obtained at different separation

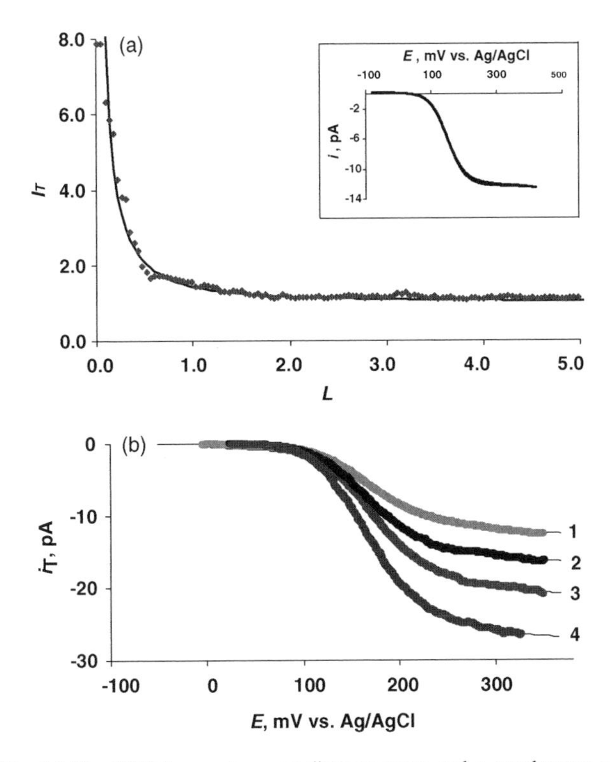

Figure 16. (a) The SECM current versus distance curve and a steady-state voltammogram (inset) obtained with a 46-nm radius polished Pt electrode. Aqueous solution contained 1 m*M* $FcCH_2OH$ and 0.2 *M* NaCl. (a) Theoretical approach curve (solid line) for diffusion-controlled positive feedback was calculated from Eq. (19). Symbols are experimental data. The tip approached the unbiased Au film substrate with a 5-nm s^{-1} speed. (b) Experimental (symbols) and theoretical (sold lines) steady-state voltammograms of 1 m*M* ferrocenemethanol obtained at different separation distances between a 36-nm Pt tip and a Au substrate. $d = \infty$ (1), 54 nm (2), 29 nm (3), and 18 nm (4). $v = 50\,mV\,s^{-1}$. Theoretical curves were calculated from Eq. (22). Adapted with permission from Ref. [51]. Copyright © 2006, American Chemical Society. See color insert.

distances between a 36-nm radius polished Pt tip and the evaporated gold substrate. The kinetic parameter values and the formal potential of the ET reaction extracted from those voltammograms were essentially independent of *d*. In this way, the kinetic parameters were obtained for several rapid ET reactions, of which the fastest was $[Ru(NH_3)_6]^{3+}$ reduction in KCl ($k^\circ = 17.0 \pm 0.9\,cm^{-1}\,s$) [51]. The self-consistent kinetic parameter values with the uncertainty margin of ~10% were obtained for electrodes of different radii and for a wide range of the tip–substrate separation distances.

To probe ET kinetics at the conductive substrate surface, the tip is held at a potential where the reaction is diffusion controlled, and the approach curves are recorded for different substrate potentials. The first experiment of this kind was reported by Wipf and Bard who measured the rate of irreversible oxidation of Fe^{2+} at the glassy carbon (GC) electrode [95]. In the feedback mode, Fe^{3+}, was reduced at the carbon fiber tip in a 1 M H_2SO_4 solution, and Fe^{2+} was oxidized at the GC substrate

$$Fe^{3+} + 1e^- \rightarrow Fe^{2+} \quad \text{(tip electrode)} \tag{45a}$$

$$Fe^{2+} - 1e^- \xrightarrow{k_f} Fe^{3+} \quad \text{(substrate electrode)} \tag{45b}$$

The tip was biased at the potential at which the reduction of Fe(III) was diffusion controlled. By changing the substrate potential over a wide range, the oxidation rate constant, k_f could be varied by several orders of magnitude. The current–distance curves obtained at different substrate potentials reflected the gradual change in substrate behavior between "perfect conductor", at which all Fe^{2+} species coming from the tip are rapidly reoxidized, and insulator, at which the reaction rate is immeasurably slow. The whole family of I_T-L curves was fitted to the theory (Fig. 17a) and yielded the Butler–Volmer-type potential dependence of the rate constant (Fig. 17b) with the kinetic parameter values, $k° = 2 \times 10^{-5}\ cm^{-1}\ s$ and $\alpha = 0.69$, in good agreement with literature data [60].

The kinetics of several heterogeneous ET reactions at semiconductor electrodes (WSe_2 and n-doped silicon) in the dark were measured by Horrocks et al. [96]. An important advantage of SECM in studies of ET at semiconductor electrodes is its relative insensitivity to parallel reactions (e.g., corrosion), which can be separated from the redox reaction of interest. Additionally, one can find smooth low defect areas on the surface most suitable for kinetic experiments. The kinetic parameters of ET in the dark to various outer-sphere redox couples (e.g., $[Ru(NH_3)_6]^{3+/2+}$, $Fc^{+/0}$, and N,N,N',N'-tetramethyl-1,4-phenylenediamine (TMPD)) at different semiconductor electrodes (i.e., n- and p-WSe_2 in aqueous electrolytes and n-Si in acetonitrile and methanol) were extracted from tip current–substrate potential (i_T-E_S) voltammograms. In these experiments, the redox reaction of interest, for example, oxidation of ferrocene, was driven at a diffusion-controlled rate at the tip. The rate of reaction at the semiconductor substrate was probed by measuring the feedback current as a function of substrate potential. In this way, different kinetics of oxidation of $[Ru(NH_3)_6]^{2+}$ were measured on the van der Waals surface (apparent transfer coefficient, $\alpha = 1$ in agreement with theories that assume an ideal semiconductor–solution interface) and at step edges.

Haram and Bard employed SECM steady-state measurements to obtain ET kinetic information about processes at an illuminated CdS surface [97]. In their

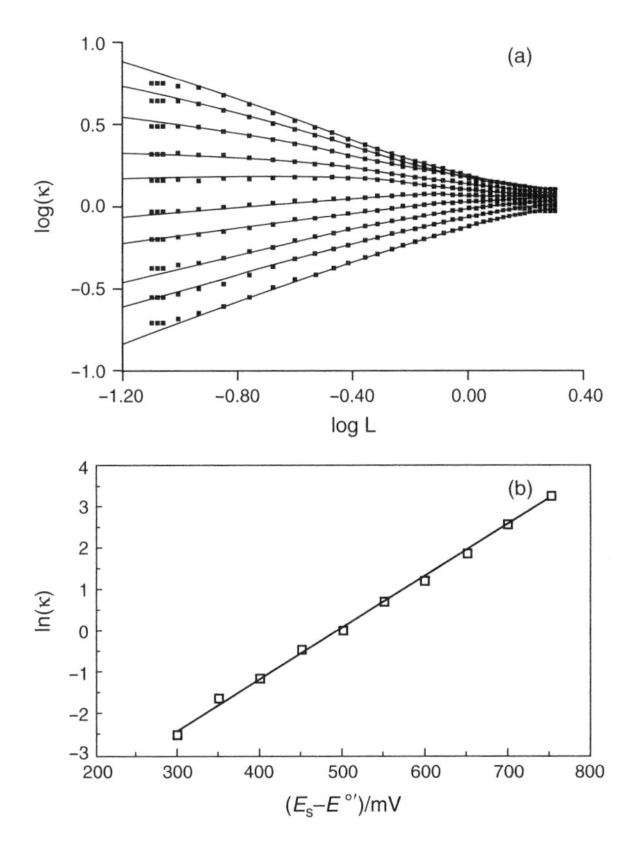

Figure 17. (a) The SECM log (current – distance) curves for the $Fe^{(III)}/Fe^{(II)}$ system (×). The tip electrode (5.5-μm radius carbon fiber) was held at a potential of −600 mV, while the GC substrate electrode was held at various potentials, from 300 to 750 mV positive of the formal potential (50-mV increments). The corresponding best theoretical fit to the data (solid lines) was obtained with various normalized rate constant values indicated in B. (b) Plot of normalized rate constant κ versus substrate overpotential $(E_s - E^{o\prime})$. Taken with permission from Ref. [60]. Copyright © 1992, American Chemical Society.

experiment, CdS thin film was used as an unbiased substrate and immersed in solution containing $MV^{+\bullet}$ and triethanolamine (TEOA). The SECM response showed negative feedback in the dark, as the substrate could not reduce tip-generated MV^{2+}, and, thus, behaved as an insulator. However, under illumination, MV^{2+} was reduced to $MV^{+\bullet}$ at the CdS–solution interface, and positive feedback could be observed. By fitting experimental approach curves to theory, apparent pseudo-first-order rate constants (k_{eff}) were estimated. The dependence of k_{eff} on light intensity and the concentration of $MV^{+\bullet}$ and TEOA mediators were determined, and the quantum efficiency was estimated as a function of light intensity.

Unwin et al. [98] described the use of transient techniques to investigate chlorophenol decomposition kinetics on illuminated TiO_2 particulate films. The same group employed the SECM in a quantitative study of the photodegradation kinetics of 4-chlorophenol (4-CP) at the TiO_2 substrate [99, 100]. In this process, the photogenerated holes mineralize the organic molecules via the rapid formation of hydroxyl radicals, and the photogenerated electrons are captured by O_2. Amperometric measurements of O_2 at the TiO_2 surface were used to study the associated kinetics. A theoretical model, which employs a Langmuir–Hinshelwood-type kinetic equation, has been developed to describe the kinetics of the photodegradation process and determine the related quantum efficiency. More recently, the ET reactions at other semiconductive surfaces including Ta_2O_5 [101], ITO [102], and boron-doped diamond electrodes [103] were probed by SECM. In the latter system, the SECM imaging revealed spatially heterogeneous ET rates that were linked to variations in intrinsic conductivity of individual grains [104]. Ghilane et al. [105] studied the ET between p-type Si coated with an alkyl monolayer and various anion radicals. The radicals were electrogenerated *in situ* and their ET kinetics were probed using the feedback mode of SECM.

2. *Electron and Ion Transfer Across the Liquid–Liquid Interface*

The SECM has been extensively used to study charge transfers at the ITIES. Several reviews of this subject are available [9, 106]. Tsionsky et al. [39, 67] studied the ET between zinc porphyrin dissolved in the organic phase and different aqueous redox species (e.g., $[Ru(CN)_6]^{4-}$, $[Fe(CN)_6]^{4-}$, or V^{2+}) at the water–benzene interface. The results were in agreement with the main predictions of existing ET theory. The effective heterogeneous rate constant of ET reaction between $ZnPor^+$ and $[Ru(CN)_6]^{4-}$ evaluated from the $i_T - d$ curves was directly proportional to the concentration of $[Ru(CN)_6]^{4-}$ in aqueous solution, and the $\log(k_f)$ versus the interfacial voltage dependence (Tafel plot) was linear with a transfer coefficient, $\alpha = 0.5$. However, the rate of the reverse reaction between ZnPor and $[Ru(CN)_6]^{3-}$ was found to be essentially potential independent at the interfaces between water and three organic solvents of different polarities [107]. Similar approaches were used to probe other ET reactions whose rates exhibited exponential driving force dependences in agreement with the Butler–Volmer model [9,108–110].

Zhang et al. [42] carried out SECM studies of the externally biased polarized ITIES. They obtained Tafel plots for different concentrations of the redox couple $[Fe(CN)_6]^{4-/3-}$, with same Fe(II)/Fe(III) concentration ratio of 1:1. Essentially the same transfer coefficient (0.38) was found from the entire set of experimental data, indicating that for a wide range of redox concentrations, the

ET reaction at the water–1,2-dichloethane interface behaved similarly to that at the metal–solution interface.

Recently [111], the ET was probed at the interface between water and a hydrophobic ionic liquid (IL), 1-octyl-3-methylimidazolium bis(trifluoromethylsulfonyl)imide. Ferrocene was dissolved in an ionic liquid, and ferrocyanide—in the aqueous phase. The tip was immersed in the aqueous phase. Ferricyanide, electrogenerated at the tip, diffused toward the interface where it was reduced by ferrocene

$$\mathrm{Fc(IL)} + [\mathrm{Fe(CN)_6}]^{3-}\mathrm{(aq)} \xrightarrow{k_f} \mathrm{Fc^+(IL)} + [\mathrm{Fe(CN)_6}]^{4-}\mathrm{(aq)} \qquad \mathrm{(ITIES)} \quad (46)$$

The interfacial potential drop at the nonpolarizable ITIES was controlled by varying the concentration of either the cation or the anion of the ionic liquid in the aqueous phase. The kinetics of interfacial ET followed the Butler–Volmer equation, and the measured bimolecular rate constant was much larger than that obtained at the water–1,2-dichloroethane interface. In the second publication, Laforge et al. [112] developed a new method for separating the contributions from the interfacial ET reaction and solute partitioning to the SECM feedback.

Interfacial ET reactions between monolayer protected gold clusters (MPCs) dissolved in 1,2-dichloroethane and aqueous redox species were also studied by SECM [113, 114]. The heterogeneous rate constant was measured for ET between organic soluble Au_{38} clusters and an aqueous $[IrCl_6]^{2-}$ oxidant [114]. The Au cluster cores, protected with phenylethylthiolate ligands, were sufficiently small ($d \sim 1.1\,\mathrm{nm}$) to exhibit molecule-like redox activity. Accordingly, the evaluation of the rate constant at the nonpolarizable liquid–liquid interface was accomplished using the approach discussed above, which was originally developed for bimolecular ET reactions between conventional redox entities.

Tsionsky et al. [67] used the SECM to test two predictions of the Marcus theory at the liquid–liquid interface, that is, the exponential distance dependence of the ET rate and the inverted region behavior. The rate of ET decreased with the number of C atoms in the hydrocarbon chain of adsorbed lipid, as predicted by ET theory. This effect was sufficiently strong to allow the driving force dependence of the ET rate to be probed for a number of aqueous redox species over a potential range of $\sim$ 1.5 V. The dependence of $\ln k_f$ on the driving force for ET between $[Fe(CN)_6]^{4-}$ and $ZnPor^+$ was linear with $\alpha \cong 0.5$ at lower overvoltages and leveled off at more negative interfacial voltages. For much more negative aqueous redox mediators, the ET rates decreased, that is, these were in the inverted Marcus region.

Barker et al. [115] developed a theoretical treatment of the ET at a nonpolarizable ITIES, in which the finite diffusion rate of the reactant in the bottom phases was taken into account. Using this approach, Ding et al. [108] extracted much faster ET rate constants from their current-distance curves and observed the inverted region behavior at the neat ITIES. Sun et al. [116] used an externally biased polarizable ITIES to control the ET driving force and also observed the inverted Marcus behavior for two different combinations of aqueous and organic redox mediators, that is, ZnPor/$[Fe(CN)_6]^{4-}$ and tetra-cyanoquinodimethane/$[Fe(CN)_6]^{3-}$.

The ion-transfer (IT) mode extends the applicability of SECM to many processes, which could not be studied at conventional metal electrodes. In Ref. [40], the current at a micropipet tip was produced by facilitated transfer of K^+ from the aqueous solution inside the pipete into DCE assisted by DB18C6 (Eq. 7a). With the concentration of K^+ inside a pipet at least 50 times higher than the concentration of DB18C6 in DCE, the tip current was limited by diffusion of DB18C6 to the pipet orifice. When the tip approached the aqueous layer, the regeneration of DB18C6 occurred via an interfacial dissociation mechanism (Eq. 7b), and a positive feedback current was observed when the tip approached the water–DCE interface. Negative feedback was observed when the tip approached a glass insulator. More recently, SECM at an externally biased nonpolarizable ITIES was used to measure the kinetics of reaction (7a) [117]. The results were in good agreement with those obtained previously by nanopipet voltammetry at a polarizable interface.

Similar SECM experiments can be performed using a simple (unassisted) IT process [41]. In this case, both the top and the bottom phases contain the same ion at equilibrium. The micropipet tip is used to deplete concentration of this common ion in the top solvent near the ITIES. The depletion results in the IT across the ITIES, which produces positive feedback. Any solid surface (or a liquid phase containing no specific ion) acts as an insulator in this experiment. The mass transfer rate for IT measurements by SECM is similar to that for heterogeneous ET measurements, and the standard rate constants in excess of $1\,cm\,s^{-1}$ should be measurable.

3. Electron-transfer Across Molecular Monolayers

Forouzan et al. [118] studied the formation of hexadecanethiol self-assembled monolayers (SAMs) on gold by SECM. With increasing duration of adsorption, the initially observed positive feedback turned negative. From the apparent ET rates, the authors were able to obtain quantitative data on the adsorption kinetics and fractional coverage. In a recent study of ET at SAMs, the SECM was used to measure independently the rates of ET mediated by monolayer-attached redox moieties and direct ET through the film, as well as the rate of the bimolecular ET reaction between the attached and dissolved redox species [68]. The SAMs were assembled onto the evaporated gold electrodes from solution containing a mixture of *n*-alkylthiol and ferrocenyl-alkanethiol. Several different situations were

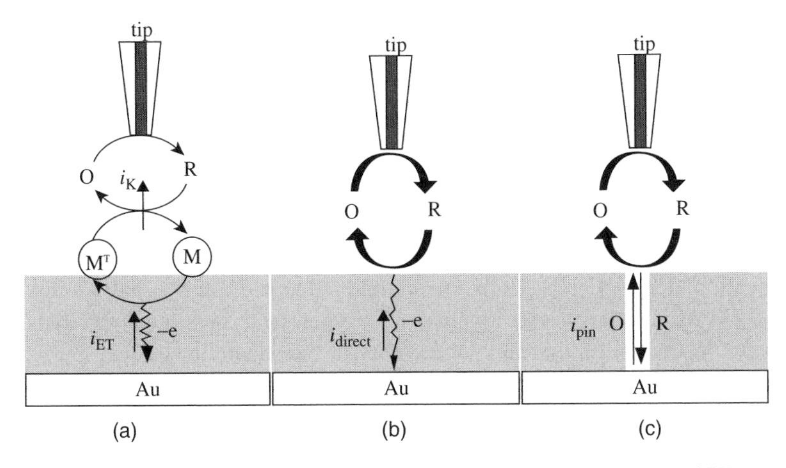

Figure 18. Schematic view of the processes involved in SECM measurements of ET across an electroactive SAM. (a) Mediated ET; (b) direct electron tunneling through the monolayer; and (c) ET through pinholes. Taken with permission from Ref. [68]. Copyright © 2004, American Chemical Society.

considered (Fig. 18): the monolayer either contained redox centers (Fig. 18a) or simply acted as a blocking layer (Fig. 18b). In Fig. 18a, ET occurred via a bimolecular reaction between dissolved redox species generated at the tip (R) and redox centers attached to the SAM (M^+), which was followed by the electron tunneling; while in Fig. 18b, ET occurred by direct tunneling between the dissolved species, R, and the electrode. Finally, the charge transfer may occur through pinhole defects in the film (Fig. 18c). The possibility of measuring the rates of all these processes and analyzing the combinations of different competing pathways for the long-distance ET was demonstrated. The upper limits for the electron tunneling and bimolecular rate constants measurable by the developed technique were given as $\sim 10^8\ s^{-1}$ and $\sim 5 \times 10^{11}\ mol^{-1}\ cm^3\ s^{-1}$, respectively. The value of the bimolecular rate constant for the ET between $[IrCl_6]^{2-}$ and SAM-bound ferrocene was found to be $1.6 \times 10^{10}\ mol^{-1}\ cm^3\ s^{-1}$ while only a lower limit of $4.5 \times 10^{10}\ mol^{-1}\ cm^3\ s^{-1}$ could be given for the ET between ferrocene and $[Ru(NH_3)_6]^{3+}$. The electron tunneling rate constants measured for alkanethiols with different chain lengths obtained at the same substrate potential decreased exponentially with the number of methylene groups in the spacer molecule. The tunneling decay constant, β, obtained from the slope was 1.0 per methylene group, a value close to those reported in previous studies.

The above approach was also used to measure the ET kinetics between ferricyanide and cytochrome *c* (immobilized on a SAM supported by a gold electrode), as well as the tunneling ET rate constant between the protein and the underlying gold electrode [119]. The value of the measured bimolecular and tunneling rate constants were $k_{BI} = 2 \times 10^8\ mol^{-1}\ cm^3\ s^{-1}$ and $k^0 = 15\ s^{-1}$, respectively.

4. *Electrocatalysis and Redox Enzymes*

The capability of SECM to detect and to image regions with different catalytic activities is well known [120–122]. So far, this technique has been applied to studies of mainly two electrocatalytic reactions, the hydrogen oxidation reaction (HOR) and the oxygen reduction reaction (ORR), which have important implications for fuel cells. Unlike reversible redox mediators usually employed in SECM experiments, the kinetics of oxygen and hydrogen reactions are strongly dependent on the catalytic activity of the substrate surface.

Kucernak et al. studied the hydrogen evolution reaction on platinum black electrocatalysts dispersed onto a flat highly oriented pyrolytic graphite (HOPG) electrode [123]. The activity of individual catalyst particles was studied as a function of substrate potential. An SECM tip was rastered over the HOPG–Pt substrate to map the concentration of molecular hydrogen and the local rate of hydrogen evolution. Zhou et al. [124] used the feedback mode of SECM to study HOR at noble metal substrates (Pt, Au). The feedback loop was initiated by proton reduction at the tip in a 0.01 *M* $HClO_4$ solution. The H_2 oxidation rate was determined quantitatively at different substrate potentials from SECM approach curves. They also investigated the effects of surface oxide and anion adsorption on HOR.

The ORR cannot be studied by feedback mode in neutral or acidic buffers. The tip-generated ORR product, that is, hydroxide ions, immediately react with aqueous protons to form water and do not have time to diffuse toward the substrate. A similar situation is encountered when the HOR is probed in neutral or alkaline media. Fernández et al. [125] employed the TG/SC mode to study the ORR on the substrate in acidic solutions. The tip was placed close to the substrate and biased at a potential at which water was oxidized to oxygen. The substrate potential was fixed at a value corresponding to oxygen reduction to water. Under these conditions, the oxygen reduction rate at the substrate surface reflected its electrocatalytic activity.

The SECM capacity for rapid screening of an array of catalyst spots makes it a valuable tool for studies of electrocatalysts. This technique was used to screen the arrays of bimetallic or trimetallic catalyst spots with different compositions on a GC support in search of inexpensive and efficient electrocatalytic materials for polymer electrolyte membrane fuel cells (PEMFC) [126]. Each spot contained some binary or ternary combination of Pd, Au, Ag, and Co deposited on a glassy carbon substrate. The electrocatalytic activity of these materials for the ORR in acidic media (0.5 *M* H_2SO_4) was examined using SECM in a rapid-imaging mode. The SECM tip was scanned in the *x*–*y* plane over the substrate surface while electrogenerating O_2 from H_2O at constant current. By scanning at step intervals of 50 μm every 0.2 s, the area as large as $7 \times 7\ mm^2$ could be screened in $\sim$ 5 h. Using this combinatorial approach, Fernández et al. proposed guidelines for the design of improved bimetallic electrocatalysts for ORR in acidic media. Most recently, the same group found two new catalysts, Pd–Co–Au (70:20:10 atom %) and Pd–Ti (50:50 atom %), that show essentially equal or slightly better performance than the more expensive Pt currently used for the ORR in PEMFC [127].

In another combinatorial study, Weng et al. [128], used poly-L-histidine (poly-his) as a matrix and ligand to complex Cu^{2+} to mimic the active sites of oxygen reductases. The electrocatalytic activity for oxygen reduction was evaluated on an array of Cu^{2+}-poly-his spots of different compositions deposited on a glassy carbon electrode. They found the highest efficiency and stability of the complex for oxygen reduction reaction at a Cu^{2+} mole fraction in the range of 0.17–0.35. Although the electrocatalytic activity of this complex is still poorer lower than that of laccase, the authors mentioned that different polypeptides and their mixtures and different preparation conditions may produce better results.

In an effort to discover and characterize new catalyst formulations for hydrogen oxidation reaction, Jayaraman and Hillier deposited a layer of Pt with a coverage gradient on a catalytically inactive indium-tin oxide substrate [129]. The reactivity gradient of this catalyst was measured directly as a function of spatial position using a SECM in the feedback mode. It was found that the non-uniform platinum coverage generates a variation in the hydrogen oxidation rate constant. The local reaction rate was proportional to the local platinum surface coverage, as determined by electron microscopy. In another report from the same group [130], the activity of Pt_xRu_y and $Pt_xRu_yMo_z$ catalysts for oxidation of hydrogen was studied as a function of composition and electrode potential.

Another type of catalytic systems on surfaces studied by SECM is immobilized redox enzymes. Two SECM-based approaches to probing redox enzymes (oxidoreductases)—generation–collection and feedback mode measurements—are illustrated in Fig. 19. The feedback mode (Fig. 19a) is more appropriate for high activity enzymes with a high surface coverage. The GC mode (Fig. 19b) is more sensitive and can be employed when enzyme kinetics are too slow for feedback measurements. The earliest example of the feedback mode application is the study of catalytic oxidation of β-D-glucose to D-glucono-δ-lactone inside a micrometer-thick layer of immobilized glucose oxidase (GO) by Pierce et al. [131]. The oxidized form of the mediator produced by the diffusion-controlled tip reaction was reduced on the substrate surface by GO. Zero-order enzyme-mediator kinetics was established, and similar apparent heterogeneous rate constants were measured for several mediators indicating saturation with respect to both β-D-glucose and mediator. A similar approach was used to observe the localized reaction of GO in the pores of track-etched polycarbonate

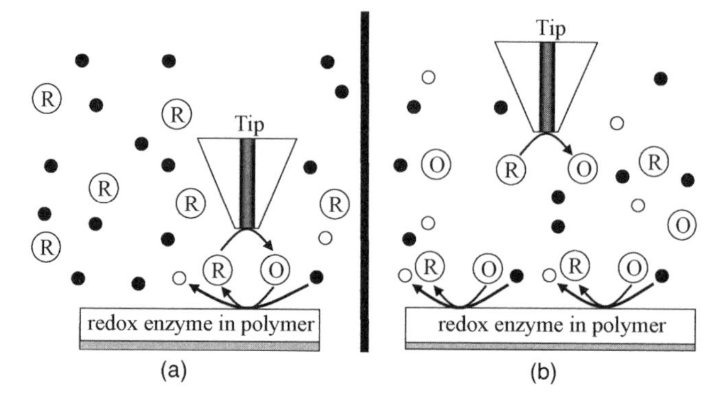

Figure 19. Feedback and GC mode SECM measurements of enzyme kinetics. (a) Feedback mode: locally electrogenerated mediator is reduced by glucose (•) at the substrate surface; the reaction is catalyzed by glucose oxidase. (b) the GC mode: the reduced form of the mediator is continuously produced by the enzyme-catalyzed reaction at the substrate and collected at the tip. The tip probes the concentration profile of reduced mediator species.

membranes and membrane-bound NADH-cytochrome *c* reductase in rat liver mitochondria [132].

The activity of immobilized alcohol dehydrogenase was probed and modified by changing the local pH [133]. The tip was positioned close to the enzyme film and used to increase the local pH by reducing water and producing hydroxide ions. A significant increase in enzyme activity was observed at higher pH. The opposite effect, that is, local inactivation of the immobilized enzyme (diaphorase) by chlorine or bromine species electrogenerated at the tip, also has been reported [134].

In a more recent study of the catalytic behavior of horseradish peroxidase (HRP), Zhou et al. [135] immobilized HRP onto glass using two methods. In the first method, the enzyme was immobilized by cross-linking with a copolymer on a glass slide. When placed in the buffer solution, the film swelled to form a hydrogel. In the second method, the same copolymer and avidin were coimmobilized on a glass slide, and biotin-labeled HRP was conjugated to the avidin in the film. The experiment was done in solution containing H_2O_2 and benzoquinone mediator. Hydroquinone was electrogenerated at the tip and catalytically reoxidized to benzoquinone by HRP in the presence of H_2O_2. This process produced positive SECM feedback. By varying the avidin loading of the film, the maximum feedback current was showed to be a linear function of the HRP surface concentration, and the detection limit of 7×10^5 HRP molecules within a 7-μm diameter area was demonstrated.

The SECM was also used to fabricate and analyze micrometer-sized enzymatic structures on surfaces, which are potentially useful in the design of miniaturized biosensors. Wittstock and Schuhmann described the preparation and imaging of micrometer-sized spots of GO on Au [136]. First, the SECM tip was used to produce μm-sized defects in a self-assembled monolayer of dodecylthiolate on Au. Then, cystaminium dihydrochloride was adsorbed on the bare Au spots and used to covalently anchor periodate oxidized GO via Schiff base chemistry. Hydrogen peroxide (H_2O_2) generated at the substrate was collected by the Pt tip in the SG/TC mode. Another approach to enzyme micropatterning was developed by Shiku et al. [137] who scanned a micropipet dispenser above the surface to create an array of 20-μm sized solution droplets on a glass slide, from which micrometer-sized spots of active HRP were formed. A very recent application of the SECM to localized enzyme (GO) immobilization involved direct microspotting of polypyrrole–biotin films [138]. Ciobanu et al. [139] used a SAM film on a micropatterned gold substrate to adsorb photosystem I (PSI). The selective adsorption of PSI onto hydroxyl-terminated thioalkane SAM ($HSC_{11}OH$) rather than the methyl-terminated SAM ($HSC_{11}CH_3$) was confirmed by SECM imaging.

B. Dissolution of Ionic Crystals

When an ionic single crystal is immersed in solution, the surrounding solution becomes saturated with respect to the substrate ions, so, initially the system is at equilibrium and there is no net dissolution or growth. With the UME positioned close to the substrate, the tip potential is stepped from a value where no electrochemical reactions occur to one where the electrolysis of one type of the lattice ion occurs at a diffusion controlled rate. This process creates a local undersaturation at the crystal–solution interface, perturbs the interfacial equilibrium, and provides the driving force for the dissolution reaction. The perturbation mode can be employed to initiate, and quantitatively monitor, dissolution reactions, providing unequivocal information on the kinetics and mechanism of the process.

In three papers [140], Macpherson et al. studied the dissolution from various regions of the (100) face of copper sulfate pentahydrate single crystals grown from the aqueous solution. The dissolution was induced by reduction of Cu^{2+} to Cu at a 25-μm diameter tip UME when the later was close to the crystal surface. In regions where the density of dislocations was high, and their spacing was considerably less than the size of the tip [e.g., in the center of the (100) face], the experimental steady-state approach curves could be fitted by the theoretical model. It was shown that the dissolution occurred at the dislocations. However, when the average interdislocation spacing was much greater than the diameter of the UME probe [e.g., near the edge of the (100) faces], SECM-induced dissolution occurred through an oscillatory rate process. After an initial period corresponding to dissolution from the growth steps present on the crystal surface, the dissolution rate rapidly decreased until the solution adjacent to the crystal surface became sufficiently undersaturated to induce the reaction again. The dissolution kinetics of the (100) surface of copper sulfate pentahydrate in highly concentrated aqueous sulfuric acid solutions (2.8–10.2 *M*) obeyed the following rate law:

$$j_{\mathrm{Cu}}^{2+} = (2.0[\pm 0.2] \times 10^{-6}\,\mathrm{mol\,cm^{-2}\,s^{-1}})\sigma \tag{47a}$$

$$\sigma = 1 - \left[\frac{(a_{Cu^{2+}} a_{SO_4^{2-}})}{(a_{Cu^{2+}}^{sat} a_{SO_4^{2-}}^{sat})}\right]^{1/2} \tag{47b}$$

where σ is the undersaturation at the crystal–solution interface, a_i denotes the activity of species i and the superscript "sat" refers to saturated solution. Other systems, like the (010) face of monoclinic potassium ferrocyanide trihydrate [141] and pressed pellets of silver chloride [142], were also studied.

In order to increase the spatial resolution of electrochemically induced dissolution imaging, Macpherson et al. [143a] introduced an integrated electrochemical-atomic force microscopic probe. A Pt coated AFM tip was used to image the topography of a dissolving crystal surface while simultaneously inducing the dissolution process electrochemically under conditions similar to those of SECM kinetic measurements. The spatial distribution of dissolution activity was also imaged by scanning the SECM tip over the surface. Macpherson and Unwin [141] mapped the dissolution rate around a single pit on the surface of potassium ferrocyanide trihydrate. An increase in tip current near the edge of the pit indicated that the local dissolution rate there was more rapid than on the planar surface. A higher spatial resolution, sufficient to observe the unwinding of screw dislocations directly on the surface, was achieved using a combined AFM–SECM probe, where the dissolution was initiated on the SECM scale by the Faradic tip current with simultaneous AFM imaging of the KBr crystal surface [143]. The same group obtained detailed information on the surface processes involved in dissolution of calcite by using combined SECM–AFM technique [144].

C. Lateral Mass/Charge Transfer

The first use of the SECM in a study of lateral mass transfer was reported by Unwin and Bard [79]. Several years later, the Unwin group studied the lateral mass transfer of a surfactant at the air–water (A–W) interface by analyzing transient current behavior [81]. The electroactive surfactant, *N*-octadecylferrocenecarboxamide ($C_{18}Fc^0$) was mixed with 1-octadecanol in a 1:1 ratio and spread onto water surface to form a Langmuir monolayer. A 25-μm diameter submarine tip (i.e., a Pt UME sealed in a U-shaped glass capillary with the conductive surface pointing upward) was placed 1–2 μm away from the A–W interface. The first "bleaching" period involved stepping the tip potential to a value at which $[Ru(bpy)_3]^{2+}$ (bpy = 2,2′-bipyridine) (the mediator form initially present in the solution) was oxidized to $[Ru(bpy)_3]^{3+}$. The oxidant diffused toward the A–W interface and reacted with $C_{18}Fc^0$ thus effectively "bleaching" the monolayer locally. During the second period, the potential step was reversed to convert the electrogenerated species to its initial form. This allowed the monolayer to recover by 2D diffusion of $C_{18}Fc^0$ and $C_{18}Fc^+$ in/out of the bleached area. The tip current–time transient was recorded during a third step when the tip potential was returned to the value of the first step. The shape of the corresponding tip current–time transient depended strongly on the duration of the second period. This dependence was used to determine the lateral diffusion coefficient of the amphiphile. The rate constant of ET between the solution mediator and the surface-confined species was also determined from the current–time transients. Computer simulations were used to fit the data, and the values of $D_{surf} = (1.0 \pm 0.2) \times 10^{-6}\,cm^2\,s^{-1}$ and $k = 0.035\,cm\,s^{-1}$ were found for a surface coverage of $\Gamma = 1.66 \times 10^{-10}\,mol\,cm^{-2}$.

A similar methodology was employed to study the lateral charge propagation in $Os(bpy)_2(PVP)_nCl]Cl$ thin films with $n = 5$ and 10 [82]. An UME tip was positioned about one radius away from the spin-coated metallopolymer. A three-step experiment was conducted using $[Ru(CN)_6]^{4-}$ as a redox mediator. In the first step, Os(II) moieties in metallopolymer were oxidized to Os(III). During the second step, the concentration of Os(II) under the tip was recovered by electron hopping in the monolayer [via self-exchange reaction between Os(III) and Os(II) moieties]. The rate of lateral charge transport was determined by fitting the current–time transients obtained in the third, final step to theoretical curves, which were generated using FEMLAB simulation package. The effective diffusion coefficient for the electron hopping as low as $10^{-10}\,cm^2\,s^{-1}$ could be measured. The rate of the ET between $[Ru(CN)_6]^{3-}$ and Os(II) at the film–solution interface was diffusion controlled.

Liljeroth et al. [80] used SECM in the feedback mode to study the electronic conductivity of a film of gold nanoparticles deposited at various pressures on a nonconductive substrate. They were able to observe an insulator-to-metal transition associated with a change in surface pressure. Unwin Whitworth et al. [83] have also developed a method to determine the electronic conductivity of ultrathin films using SECM under steady-state conditions. They obtained analytical approximations for the fitting of approach curves. The usefulness of their approach was demonstrated by investigating the effect of surface pressure on conductivity of a polyaniline monolayer at the water–air interface.

D. Adsorption–Desorption Processes

Unwin and Bard [79] studied adsorption of H^+ on metal oxide surfaces. Since the surface diffusion of electroactive species was negligibly slow in comparison with

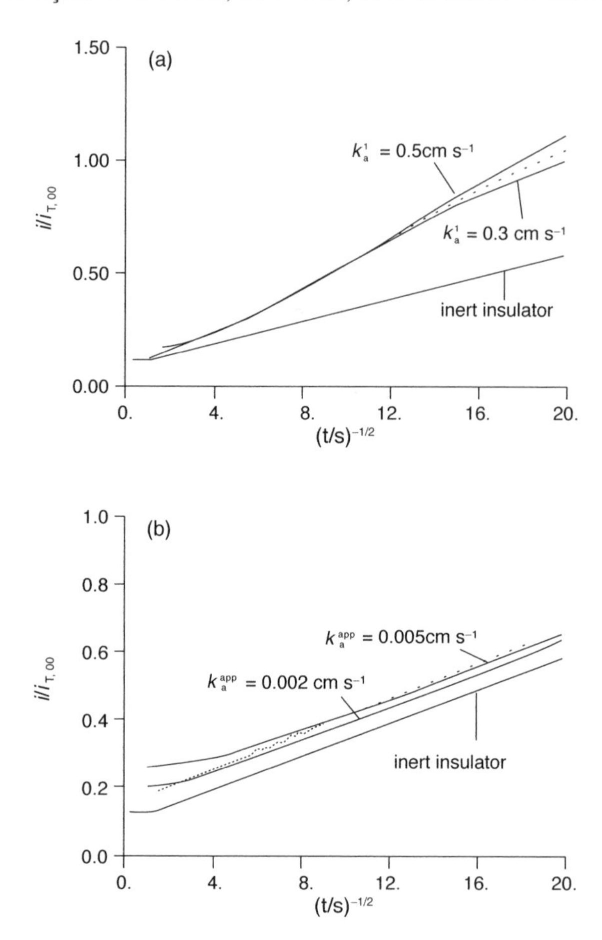

Figure 20. The SECM-induced desorption experiment. Chronoamperometric characteristics for the reduction of H^+ on (a) a rutile (001) surface with $d = 2.6\,\mu m$, and (b) an albite (010) surface with $d = 2.8\,\mu m$. In each case, solid curves from bottom to top represent the theoretical behaviors for an inert substrate and for specified adsorption rate constant values. Adapted with permission from Ref. [79]. Copyright © 1992, American Chemical Society.

the bulk diffusion, the steady-state $i_T - d$ curves were similar to those obtained at an electroinactive substrate. Chronoamperometric transients had to be used to measure adsorption/desorption rates. The tip was positioned at $L = 0.22$ from the substrate, and a potential step was applied to the tip such that H^+ was reduced at the diffusion-controlled rate. The induced desorption of H^+ from the metal oxide surface resulted in a higher transient current (Fig. 20). Fitting the current–time responses enabled the investigators to measure the rates of adsorption–desorption of H^+ on rutile (Fig. 20a, $k_a = 0.4 \pm 0.1\ \mathrm{cm\,s^{-1}}$ and $k_d = 5.5 \pm 1.4 \times 10^{-8}\ \mathrm{mol\,cm^{-2}\,s^{-1}}$), while for albite the rates were much lower (Fig. 20b, $k_a = 3 \pm 1 \times 10^{-3}\,\mathrm{cm\,s^{-1}}$ and $k_d = 8.4 \pm 3.7 \times 10^{-10}\,\mathrm{mol\,cm^{-2}\,s^{-1}}$). In general, the SECM-induced desorption effect is stronger when the tip is close to the substrate and when the adsorption/desorption kinetics is fast (cf. Fig. 20a,b).

Yang and Denuault [145–147] investigated the adsorption and desorption of hydrogen on platinum electrodes using the SG/TC and chronoamperometric modes of the SECM. The pH-dependent adsorption processes at Pt substrate were probed using tip–substrate cyclic voltammetry (T–S CV). The Pt tip potential was poised so that a pH-dependent reaction, for example, Pt oxidation, occurred at the tip. On cycling the substrate potential, the tip current provided a sensitive indication of the local pH. Large pH changes were observed during hydrogen adsorption/desorption, and the transient response was analyzed to study the kinetics of this process. Although T–S CV experiments are not straightforward to simulate quantitatively, much qualitative information is obtained from the shape $i_T - E_S$ voltammograms. A potentiometric tip could also be used in a similar manner. However, when the substrate current is high or the solution is resistive, the ohmic potential drop induced between tip and reference electrode by the substrate current must be considered.

Another example of an adsorption study employing the SECM was reported by Mansikkamäki et al. [148]. They investigated the adsorption of benzothiazole (a passivating agent for the oxidation of copper) on the surface of copper alloys using the steady-state feedback approach.

Adsorption phenomena were also studied at the liquid–liquid and liquid–air interfaces. Tsionsky et al. investigated the kinetics of adsorption of several phospholipids at the benzene–water interface and measured the related isotherm [67]. The binding of silver ions at Langmuir phospholipid monolayers formed at the water–air interface was investigated by Burt et al. [149]. A Ag^+ concentration gradient was generated close to a dipalmitoyl phosphatidic acid (DPPA) monolayer by electrooxidation of a 25-μm-diameter Ag UME positioned a few microns away from the interface. The transient current response was sensitive to local variations in Ag^+ concentration, which reflected the interaction of Ag^+ with the Langmuir monolayer. The density of adsorption sites and the capacity for metal ion adsorption at the monolayer increased at a higher surface pressure.

E. Imaging Surface Reactivity

The SECM, which does not provide atomic level spatial resolution, cannot compete with STM or AFM as a tool for topographic imaging. However, SECM is well suited for high resolution mapping of surface reactivity. This can be done in either feedback or collection mode. The former can provide a spatial distribution of the rate of a redox reaction responsible for mediator regeneration at the substrate. By proper choice of solution components to control the tip reaction and the chemistry at the substrate–solution interface, differential reaction rates at different sorts of surfaces can be probed.

The first example of surface activity mapping by SECM was reported by Engstrom et al. [3], who recorded the tip response to the substrate potential steps in a GC mode as a function of the tip position over an array of 100-μm diameter inlaid Pt microelectrodes. The changes in the tip current allow one to distinguish between the conductive Pt surface and the surrounding insulator. The steady-state feedback mode is more suitable for imaging, and it is possible not only to distinguish between conductors and insulators, but also to detect the differences in reactivities of two conductors, such as gold and glassy carbon [150], or even different portions of a polyelectrolyte coating loaded with a redox mediator [151]. The presented data suggest the possibility of distinguishing between substrate components with a ratio of rate constants as low as 2:1, when L is sufficiently small.

By mapping surface reactivity with the SECM, the precursor sites for the pit formation on oxide-covered titanium foil were detected by Casillas et al. [152, 153]. The feedback mode was used to visualize a few microscopic domains of intense faradaic activity. The tip process was the reduction of Br_2 to Br^-, and Br_2 was regenerated via electrooxidation of bromide on the TiO_2 surface. After determining the positions of the active sites, the pits on the surface were nucleated by application of a more positive potential to the substrate. For the first time, it was found that the pit nucleation occurs preferentially at surface sites of high electrochemical activity. The determined values of a typical active site diameter ($\sim$ 10–50 μm) and density ($\sim$ 20–30 sites cm^{-2}) showed that active sites represent only a small fraction ($\sim$0.04%) of the TiO_2 surface. Higher resolution maps and more information about conductivity of the oxide film on titanium were obtained later using the combination of SECM with conducting AFM [154].

Recently, Colley et al. [104] studied the distribution of electrochemical activity in microarray electrodes. The array contained $\sim$50-μm diameter boron-doped regions spaced $\sim$250 μm apart in an intrinsic diamond disk. Reaction rate imaging was done in the substrate generation/tip collection mode. The electroactive boron doped regions were biased at a suitable potential to reduce the mediator, $[Ru(NH_3)_6]^{3+}$, and the product of the reduction reaction was collected at the tip. Two-dimensional scans over different regions (Fig. 21) revealed wide variations in local electroactivity.

F. Homogeneous Reactions in Solutions

A model EC_i process used in the first study of homogeneous kinetics by SECM was the oxidation of *N,N*,-dimethyl-*p*-phenylenediamine (DMPPD), which involved a rapid two-electron oxidation followed by the pH-dependent

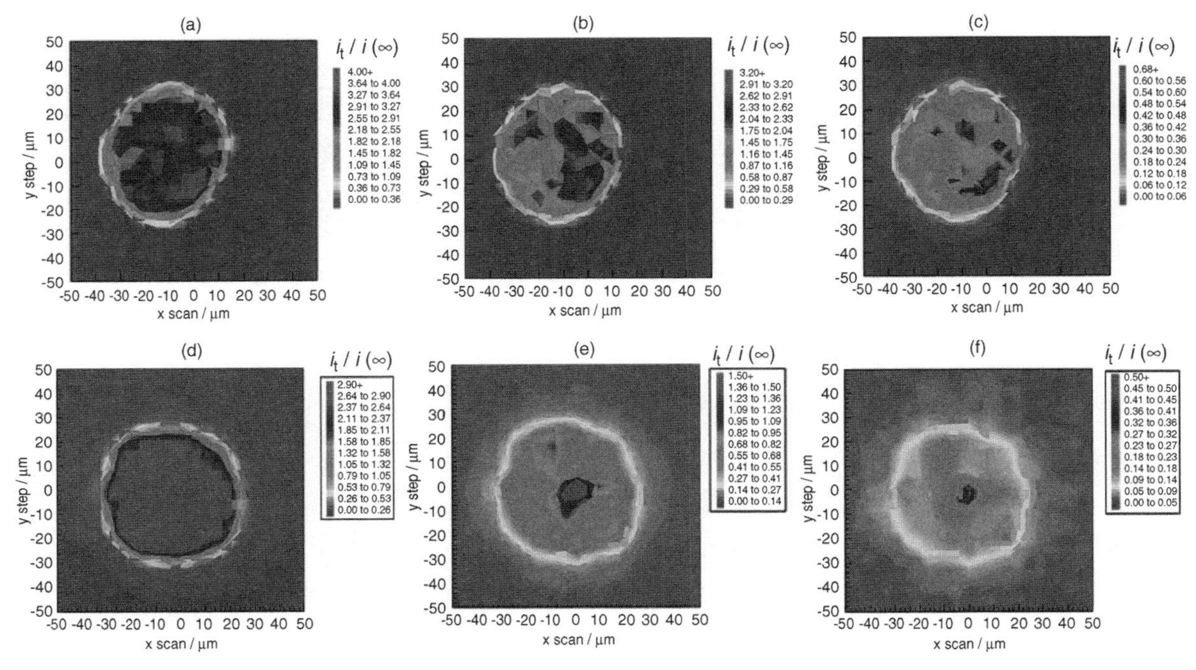

Figure 21. The SG/TC 100 × 100-μm images of two different boron doped regions (a, b, and c - region 1; d, e and f - region 2) in an intrinsic diamond disk. The substrate was biased at (a, d) −0.4 V, (b, e) – 0.3 V and (c, f) −0.2 V. The 5-μm radius Pt UME tip was kept 1 μm above region 1 and 0.6 μm above region 2. The substrate potential was 0 V versus Ag/AgCl. Adapted with permission from Ref. [104]. Copyright © 2006, American Chemical Society. See color insert.

deamination [21, 86]. The oxidation (reduction) of DMPPD at the tip (substrate) followed Eq. 48a.

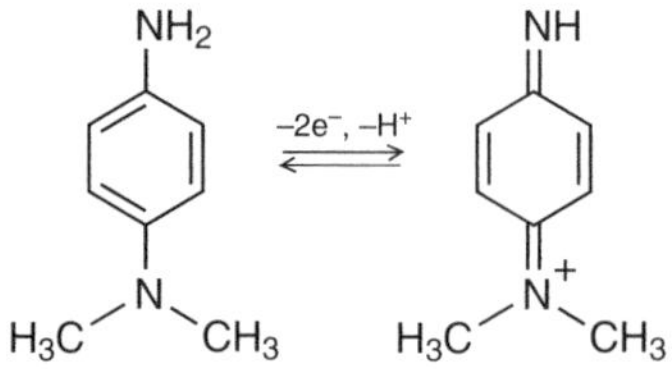

(48a)

while the decomposition of the oxidized product was via homogeneous reaction (48b)

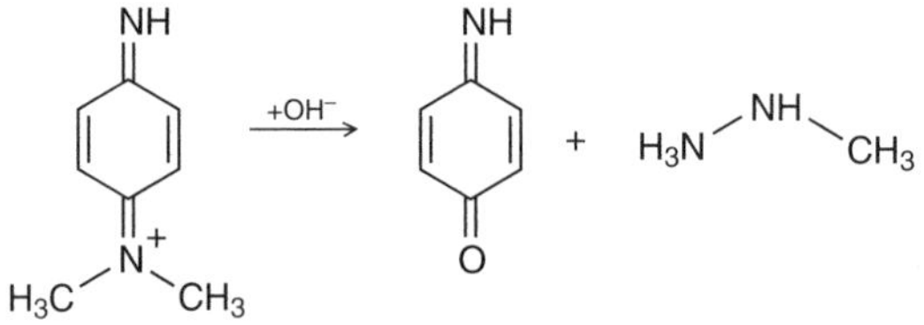

(48b)

Reaction (Eq. 48b) could be treated as a pseudo-first-order reaction as long as the pH remained constant in the tip–substrate gap. The increase in pH resulted in a higher deamination rate, and the SECM response gradually changed from positive to negative feedback. The curve fitting produced a series of homogeneous pseudo-first-order rate constants. From the rate dependence on $[OH^-]$, a bimolecular rate constant of $\sim 9.5 \times 10^3$ $mol^{-1}\,dm^3\,s^{-1}$ was obtained for the decomposition reaction (Eq. 48b). Chronoamperometric experiments at a fixed tip–substrate separation distance of 6.5 μm were also carried out to determine the kinetics of this reaction over the range of pH from 7.80 to 12.42 and produced essentially the same bimolecular rate constant value.

The SG/TC mode of SECM was also applied by Martin et al. [86] to study the oxidation of DMPPD. The generator was a 2-mm^2 substrate electrode, and the collector was a 25-μm diameter Pt disk electrode. The substrate potential was stepped from 0 V versus Ag quasi reference electrode, where no Faradic process took place, to +500 mV, where the oxidation of DMPPD was diffusion controlled. The tip potential was held at 0 V, at which the oxidized form of DMPPD could be reduced at a diffusion controlled rate. After the tip–substrate separation was found from the positive feedback current–distance curve, the rate constant was obtained from the current transient at the tip. The feedback and SG/TC modes were also used to study the reduction of

$[Cp^*Re(CO)_2\ (p\text{-}N_2C_6H_4OCH_3)][BF_4]$ in acetonitrile [155] and the oxidation of borohydride at a gold electrode [156].

In an EC_{2i} process, the initial ET step is followed by a second-order irreversible homogeneous reaction. For example, the feedback mode of SECM was employed to study the reductive hydrodimerization of the dimethyl fumarate (DF) radical anion [22]. The experiments were carried out in solutions containing either 5.15 or 11.5 m*M* DF and 0.1 *M* tetrabutylammonium tetrafluoroborate in *N,N*,-dimethylformamide (DMF). The increase in the feedback current with increasing concentration of DF indicated that the homogeneous step involved in this process is not a first-order reaction. The analysis of the data based on the EC_{2i} theory yielded the k_2 values of $180\,M^{-1}\,s^{-1}$ and $160\,M^{-1}\,s^{-1}$ for two different concentrations. Another second order reaction studied by the TG/SC mode was oxidative dimerization of 4-nitrophenolate (ArO^-) in acetonitrile [23]. In this experiment, the tip was placed at a fixed distance from the substrate. The d value was determined from the positive feedback current of benzoquinone, which did not interfere with the reaction of interest. The dimerization rate constant of $(1.2 \pm 0.3) \times 10^8\,M^{-1}\,s^{-1}$ was obtained for different concentrations of ArO^-.

The application of SECM to the study of ECE/DISP1 processes was showed in Ref. [84]. In those experiments, anthracene (AC) was reduced in DMF in the presence of phenol (PhOH). The process steps were

$$AC + e^- \rightarrow AC^{\bullet -} \tag{49a}$$

$$AC^{\bullet -} + PhOH \xrightarrow{k_1} ACH^{\bullet} + PhO^- \qquad (Ph = C_6H_5) \tag{49b}$$

$$AC^{\bullet -} + ACH^{\bullet} \xrightarrow{k_d} ACH^- + AC \tag{49c}$$

$$ACH^- + PhOH \xrightarrow{\text{fast}} ACH_2 + PhO^- \tag{49d}$$

Reaction (49b) was the rate-limiting step that could be treated as a pseudo-first-order process in the presence of excess PhOH (0.1–0.43 *M*). The tip electrode (a 7-μm C fiber) and the substrate (a 60-μm Au electrode) were placed at a fixed separation distance, which was evaluated from the positive feedback current of decamethylferrocene. A series of current–distance curves for a range of PhOH concentrations showed the decrease in feedback with increasing [PhOH]. This is because the consumption of $AC^{\bullet -}$ in the gap caused a diminution of positive feedback for $AC/AC^{\bullet -}$ couple. Fitting of the approach curves confirmed a DISP1 mechanism for the reduction of anthracene. In the presence of phenol. The results yielded a psudo-first-order rate constant for reaction (49b), k_1, from which the second-order rate constant, $k_2 = k_1/[PhOH] = 4.4 \pm 0.4\times\ 10^3 M^{-1}\,s^{-1}$ was obtained.

G. Membrane Transport and Charge-Transfer Processes in Living Cells

The transport of molecules across biological cell membranes and biomimetic membranes, including planar bilayer lipid membranes (BLMs) and giant liposomes, has been studied by SECM. The approaches used in those studies are conceptually similar to generation–collection and feedback SECM experiments. In the former mode, an amperometric tip is used to measure concentration profiles and monitor fluxes of molecules crossing the membrane. In a feedback-type experiment, the tip process depletes the concentration of the transferred species on one side of the membrane and in this way induces its transfer across the membrane.

Tsionsky et al. studied charge transport through the horizontally oriented BLM that separated the upper and lower compartments containing the same aqueous solution [157]. The bilayer was impermeable to hydrophilic ions like $[Ru(NH_3)_6]^{3+}$ and $[Fe(CN)_6]^{4-}$, and the transmembrane ET between such species was completely blocked. However, the transmembrane ET current was observed when the BLM was doped by iodine. Using a similar experimental setup, Amemiya and Bard probed the transport of K^+ through gramicidin channels formed in horizontal BLMs [158]. The amperometric ion-selective micropipet electrodes used in those experiments were filled with either a 10 m*M* valinomycin solution in 1,2-dichloroethane or a 10 mM ETH 500 solution. Both feedback and generation-collection modes were employed to investigate the transfer of K^+. In a somewhat similar manner, the transfer of hydrophilic ions through voltage-gated channels formed in a BLM by alamethicin was studied by Wilburn et al. [159]. The pores formed by several alamethicin molecules were sufficiently large and hydrophilic to transport redox species, such as $[Ru(NH_3)_6]^{3+}$ and $[Fe(CN)_6]^{3-}$, which could be detected at the Pt tip.

Two interesting examples of SECM studies of molecular transport across biological membranes were reported by Mauzeroll et al. [160, 161]. They measured the uptake of menadione by yeast cells and monitored its intracellular reaction with glutathione [160]. The export of the product of that reaction (thiodione) by adenosine triphosphate (ATP)-dependent GS–X pumps is the way for a yeast cell to get rid of toxic menadione (Fig. 22). Because yeast cells are much smaller than either mammalian cells or purple bacteria used in other SECM studies, it was not possible to probe them individually. The average flux of ~30,000 thiodione molecules per s per cell was measured by a 1-μm diameter tip, and numerical simulations were used to show that the overall process rate is limited by the uptake of menadione. The efflux through the GS–X pump was found to be at least an order of magnitude faster. In the second article [161], the thiodione efflux from individual cells and groups of highly confluent hepatoblastoma (Hep G2) cells was probed by SG/TC experiments.

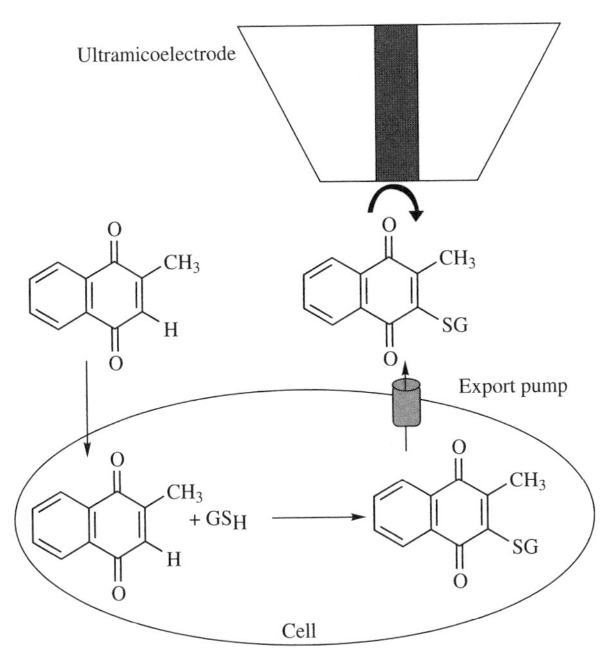

Figure 22. Oxidative stress produced by menadione in yeast cells leads to the formation and excretion of thiodione into the extracellular media, where it can be detected by SECM. Taken with permission from Ref. [160]. Copyright © 2000, National Academy of Sciences USA.

The time evolution of concentration profiles was analyzed to determine the flux of 6.0×10^6 molecules per s per cell.

A novel application of SECM was to study transport processes across the nuclear envelope. Guo and Amemiya investigated the molecular transport facilitated by nuclear pore complexes (NPCs) in a large intact nucleus of *Xenopus laevis* oocytes [35]. The NPC is a 60–120-MDa complex made of 30 or more distinct proteins. Small molecules can passively diffuse through it. The partitioning of redox species X across the nuclear envelope was characterized by obtaining approach curves and chronoamperograms. From such experiments performed with different mediators, the authors concluded that the nuclear envelope permeability is very high, and most NPCs on the nucleus surface are open. Estimates were obtained for the single-channel flux and the NPC diameter.

Most recently, Zhan and Bard used submicrometer-sized conical carbon fiber tips to approach, image, and puncture individual giant liposomes containing $[Ru(bpy)_3]^{2+}$ [36]. The leakage of $[Ru(bpy)_3]^{2+}$ through the lipid membrane was observed. A higher stability of liposomes as compared to lipid bilayers allows one to perform measurements over a more extended period of time. Such "artificial cells" can be useful for studies of molecular transport and redox regulation of cellular processes.

Liu et al. applied the SECM feedback mode to noninvasively probe the redox activity of individual mammalian cells [70, 71]. In order to probe the redox activity of a mammalian cell, both oxidized and reduced forms of the redox mediator must be capable of crossing the cell membrane and shuttling the charge between the tip electrode and the intracellular redox centers (Fig. 23a). Only hydrophobic redox mediators (e.g., menadione and 1,2-naphthoquinone) could be used in SECM experiments with mammalian cells [71]. The redox reactions at the tip and inside the cell can be presented as follows:

$$\mathrm{O} + n\mathrm{e} = \mathrm{R} \qquad \text{(at the tip electrode)} \tag{50a}$$

$$\mathrm{R} + \mathrm{O_{cell}} \xrightarrow{k} \mathrm{O} + \mathrm{R_{cell}} \qquad \text{(inside the cell)} \tag{50b}$$

The i_T versus d curves were obtained by moving the tip (negatively biased, so that reaction (50a) was diffusion controlled) toward the cell membrane and fitted to the theory to extract the value of the effective heterogeneous rate constant (k). Reaction (50b) is a complicated process involving at least three steps: (1) generation of redox centers (O_{cell}) inside the cell, (2) transport of mediator species across the cell membrane, and (3) bimolecular ET between the mediator species and intracellular redox centers. Mechanistic analysis showed that, depending on the properties of the mediator (e.g., formal potential, ionic charge, and hydrophobicity), the main factor limiting the overall charge-transfer rate can be either the membrane transport, or the availability of intracellular redox agents, or the driving force for the ET reaction, that is, the difference between the intracellular mixed-redox potential and the formal potential of the

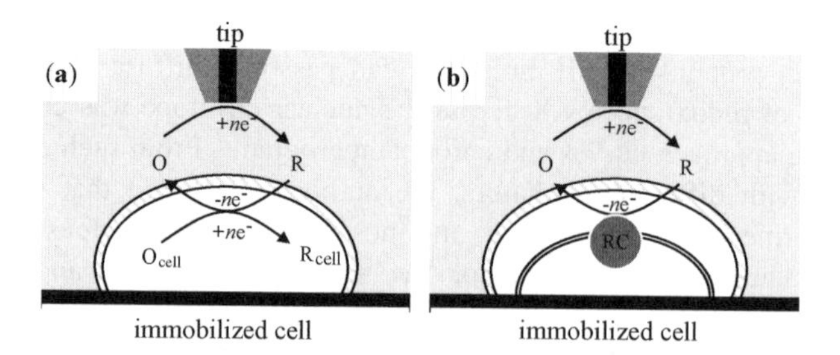

Figure 23. Schematic diagrams of the SECM experiments with cellular organisms. (a) Bimolecular ET between a hydrophobic redox mediator (O/R) and intracellular redox species (O_{cell}/R_{cell}) in a mammalian cell. (b) the ET between a redox mediator and redox centers (RC) inside a prokaryotic cell (e.g., of *Rb. Sphaeroides*).

mediator [71]. Kinetic parameters were determined for different steps of the charge-transfer process.

By using this approach, significant differences were detected in the redox responses given by three types of human breast cells with different levels of protein kinase C_α (PKC_α, an enzyme that has been linked with motility and metastasis of various cell types): non-metastatic MCF-10A cells (human breast epithelial cells), 11α cells (breast cells with engineered overexpression of PKC_α), and MDA-MB-231 cells (highly metastatic breast cancel cells expressing a high level of PKC_α). Approach curves obtained with these cells and several redox mediators demonstrated that their cellular redox activities are in the order of MCF-10A > 11α > MDA-MB-231. This result indicated that metastatic human breast cells can be electrochemically distinguished from nontransformed breast cells. The obtained kinetic data were used to identify the experimental conditions, such as the nature and concentration of the redox mediator, which would maximize the detection of metastatic cells in a field of normal breast cells and in tissue samples [73].

Purple bacteria (*Rb. sphaeroides*) were also investigated by SECM [72, 162]. Unlike mammalian cells, *Rb. Spheroides* have two membranes (outer and cytoplasmic) and no nucleus (Fig. 23b). The outer membrane is permeable to both hydrophilic and hydrophobic redox species, while the cytoplasmic membrane is impermeable to hydrophilic species and contains redox centers. Because of this difference, the intracellular potential encountered by hydrophobic and hydrophilic redox mediator were significantly different, that is, $-110\,\mathrm{mV}$ versus Ag/AgCl was found from regeneration rates of hydrophobic redox mediators, and 180 mV, for hydrophilic redox species [72].

Longobardi et al. [162] used chromatophores (specialized pigment-bearing structures obtained by mechanical rupture of the *Rb. Sphaeroides*) and liposomes in order to investigate the role played by the cytoplasmic bacterium membrane. They found that the SECM feedback process is mediated by membrane-bound redox species (Fig. 23b) and that the oxidant species inside *Rb. Sphaeroides* is most probably ubiquinone that resides in the cytoplasmic membrane pool.

The above methods for cell investigation are essentially noninvasive. However, to localize the redox activity in different cell compartments (e.g., mitochondria) the tip must penetrate the cell. The most recent effort focused on using nanometer-sized tips and nanopipet-based probes for high resolution studies of mammalian cells [37]. It was shown that amperometric nanoprobes allow quantitative spatially resolved SECM measurements inside living cells. The use of a tip ~1000 smaller than the cell greatly minimizes the damage to the cell membrane and may facilitate subcellular level studies of biologically relevant charge-transfer processes.

H. Electronically Conductive Polymer Films

Some features of conducting polymers, like the variation of conductivity and the ion movement related to doping/undoping processes can be studied by SECM. Kwak et al. [163] investigated the changes in conductivity of polypyrrole films on a Pt substrate as a function of its potential. Undoped films revealed sluggish heterogeneous kinetics in the feedback mode, whereas in the oxidized state, fast kinetics were observed. The Bard group also studied the expulsion–incorporation of redox active ions like bromide, ferrocyanide, and cationic metal complexes caused by oxidation–reduction of a polypyrrole film [164].

Troise-Frank and Denuault studied the reaction pathways of charging and discharging of polyaniline films as a function of proton concentration and ionic strength by measuring the current of proton reduction [76]. Additionally, they monitored the flux of chloride anions *in situ* by potentiometry during cyclic voltammetric experiments [165]. Heinze and co-workers presented an example of a "microwriting" and "reading" process of an undoped polyaniline film deposited on poly(ethylene terephthalate)–glycol substrate [166].

More recent efforts focused on surface modification of conductive polymers by the SECM, fabrication, and characterization of microstructures. Mandler et al. developed an approach for the formation of a 2D conducting polymer on top of an insulating layer. This approach, based on electrostatically binding a monomer (anilinium ions) to a negatively charged self-assembled monolayer of ω-mercaptodecanesulfonate [MDS, $HS(CH_2)_{10}SO_3^-$] followed by its electrochemical polymerization. The polyanion monolayer exhibited the properties similar to those of a thin polymer film [167].

Kapui et al. prepared a novel type of polypyrrole films [168]. The film was impregnated by spherical styrene–methacrylic acid block copolymer micelles with a hydrophobic core of 18 nm and a hydrophilic corona of 100 nm. The properties of the micelle-doped polypyrrole films were investigated by cyclic voltammetry and SECM. It was found that the self-assembled block copolymer micelles in polypyrrole behave as polyanions and the charge compensation by cations has been identified during electrochemical switching of the polymer films.

An *in situ* formation of conductive domains by electroreduction and local metalization of polytetrafluoroethylene (PTFE) was discussed in a series of reports by Combellas et al. [169]. The authors carried out both steady-state and transient SECM experiments to investigate the kinetics of PTFE carbonization and concluded that the propagation of the conductive PTFE zone is governed by an in-depth diffusive process.

I. AFM–SECM and Other Hybrid Techniques

The combination of spatially resolved electrochemical experiments with different types of measurements made at the same time and (if possible) at the same location is a powerful approach to studies of surface structures and dynamics. A number of analytical techniques have been combined with SECM including near-field scanning optical microscopy (NSOM), surface plasmon resonance (SPR), electrochemical quartz crystal microbalance (EQCM), fluorescence spectroscopy (FS), electrogenerated chemiluminescence (ECL), and atomic force microscopy (AFM). The corresponding hyphenated methods are NSOM-SECM [170], EQCM-SECM [171–173], SPR–SECM [174], FS-SECM [175], ECL-SECM [176], and AFM-SECM [177].

Among these hybrid methods, AFM-SECM is most widely used. One of the strength of the AFM–SECM technique is in its positioning capabilities. By using the AFM mode, one can position a submicrometer-sized tip near the substrate surface and keep a constant tip–substrate distance during the surface raster. Maintaining a constant distance between the tip and the substrate is essential for avoiding tip crashes and the damage to the sample during the imaging of a substrate with non-uniform surface reactivity and/or complex topography.

Several groups employing the SECM–AFM technique have developed different approaches to the tip fabrication [19, 178–180]. These included hand-fabricated probes, made from an etched, insulated wire with the exposed end [19, 178, 179]; AFM tips with a frame electrode fabricated by coating a commercial silicon nitride probe with a metal and an insulator and using focused ion beam technology to reshape the probe and expose the tip [180, 181]; nanoelectrode probes based on the use of single-walled carbon nanotube bundles in AFM tips [182, 183]; and batch microfabricated probes [184, 185]. The batch microfabrication was carried out on silicon wafers, yielding 60 probes in each run. The process yields sharp AFM tips, incorporating a triangular-shaped electrode at the apex. Figure 24 shows two different kinds of SECM–AFM probes. The first type (Fig. 24a), called conducting–AFM probe (C–AFM), can be used to acquire information about the electrochemical activity of a nonconductive substrate with electroactive species attached to its surface (e.g., oxidoreductase adsorbed on a glass slide). The probe in Fig. 24b is suitable for conductive substrates. The insulation of the AFM tip prevents direct electrical contact between its conductive core and the substrate. Such probes were used to study nonconductive crystal dissolution [143, 186]. The probe was first used amperometrically to induce crystal dissolution, and the resulting changes in topography were imaged in the AFM regime. Other applications like, probing diffusion through single nanoscale pores [187], the studies of electrochemical properties of a boron-doped diamond electrode [188] and a Ti/TiO_2/Pt electrode [189], have made use of C–AFM tips.

When the tip is scanned laterally above the substrate, the obtained image reflects both the surface topography and the distribution of its chemical reactivity. This makes the data interpretation difficult if no *a priori* information

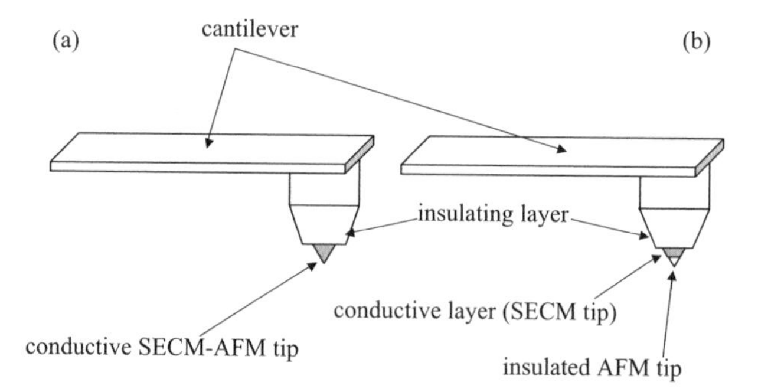

Figure 24. (a) Conducting and (b) nonconducting SECM–AFM tips. (a) The SECM and AFM tips made from the same conducting material; (b) the AFM tip is insulated to prevent direct electrical contact with the conductive substrate surface.

on the substrate topography is available. One way to avoid this problem is to maintain a constant tip–substrate distance during the raster of the substrate surface. The AFM–SECM mode achieves this by controlling the tip substrate distance with the AFM sensor. Thus, a raster in the AFM–SECM mode simultaneously produces a surface topographical map and a surface reactivity map (see, e.g., Figs. 15a,b). This makes the SECM–AFM combination very useful for imaging substrates of complex topography and variable reactivity. Davoodi et al. used AFM–SECM to study the corrosion of Al in NaCl solution [190]. The complimentary topographical and electrochemical activity maps of the same surface area have been obtained with a micrometer lateral resolution. Kranz et al. [191] integrated a submicrometer-size electrode into an AFM tip at the exactly defined distance from the apex using focused ions beam technology. They used such probes to study micropatterned samples with immobilized enzyme spots. The AFM contact mode provided the topography of the sample. Since the SECM can be used to study the enzyme activity in generation/collection mode, AFM–SECM could simultaneously probe the topography of the enzyme pattern and its local activity. A somewhat similar approach was also used in Ref. [192].

Other recent applications of AFM–SECM included the study of the iontophoretic transport of $[Fe(CN)_6]^{4-}$ across a synthetic track-etched polyethylene terephthalate membrane by Gardner et al. [193]. They made the structure and flux measurements at the single pore level and found that only a fraction of candidate pore sites are active in transport. Demaille et al. used AFM–SECM technique in aqueous solutions to determine both the static and dynamical properties of nanometer-thick monolayers of poly(ethylene glycol) (PEG) chains end-grafted to a gold substrate surface [180].

The development of comprehensive theory for SECM–AFM method is challenging because of complicated geometries of the employed tips. Holder et al. [194] reported quantitative modeling of mass transport at metal-coated AFM probes that can be used for flux-generation in SECM–AFM experiments. The theory was also developed for a frame electrode integrated with an AFM probe [90,195].

Acknowledgments

The support of our research in SECM by grants from NSF and the Petroleum Research Fund administrated by the American Chemical Society is gratefully acknowledged.

References

1. G. Binnig and H. Rohrer, *Helv. Phys. Acta* **55**, 726 (1982).
2. M. A. Dayton, J. C. Brown, K. J. Stutts, and R. M. Wightman, *Anal. Chem.* **52**, 946 (1980).
3. R. C. Engstrom, M. Weber, D. J. Wunder, R. Burgess, and S. Winguist, *Anal. Chem.* **58**, 844 (1986).
4. C. W. Lin, F.-R. F. Fan, and A. J. Bard, *J. Electrochem. Soc.* **134**, 1038 (1987).
5. M. J. Heben, M. M. Dovek, N. S. Lewis, R. M. Penner, and C. F. Quate, *J. Microscopy* **152**, 651 (1988).
6. A. J. Bard, F.-R. F. Fan, J. Kwak, and O. Lev, *Anal. Chem.* **61**, 132 (1989).
7. J. Kwak and A. J. Bard, *Anal. Chem.* **61**, 1221 (1989).
8. A. J. Bard and M. V. Mirkin, (eds.), *Scanning Electrochemical Microscopy,* Marcel Dekker, New York, 2001.
9. A. L. Barker, M. Gonsalves, J. V. Macpherson, C. J. Slevin, and P. R. Unwin, *Anal. Chim. Acta* **385**, 223 (1999).
10. R. M. Wightman and D. O. Wipf, in *Electroanalytical Chemistry,* Vol. 15, A. J. Bard, (ed.), Marcel Dekker, New York, 1988, p. 267.
11. C. Amatore, in I. Rubinstein, (ed.), *Physical Electrochemistry: Principles, Methods, and Application*, Marcel Dekker, New York, 1995, p. 131.
12. M. V. Mirkin, F.-R. F. Fan, and A. J. Bard *J. Electroanal. Chem.* **328**, 47 (1992).
13. C. G. Zoski, B. Liu, and A. J. Bard, *Anal. Chem.* **76**, 3646 (2004).
14. Y. Selzer and D. Mandler, *Anal. Chem.* **72**, 2383 (2000).
15. J. Mauzeroll, E. A. Hueske, and A. J. Bard, *Anal. Chem.* **75**, 3880 (2003).
16. C. Demaille, M. B. Brust, M. Tsionsky, and A. J. Bard, *Anal. Chem.* **69**, 2323 (1997).
17. Y. Lee, S. Amemiya, and A. J. Bard, *Anal. Chem.* **73**, 226 (2001).
18. P. Liljeroth, C. Johans, C. J. Slevin, B. M. Quinn, and K. Kontturi, *Anal. Chem.* **74**, 1972 (2002).
19. J. V. Macpherson and P. R. Unwin, *Anal. Chem.* **72**, 276 (2000).
20. A. J. Bard and L. R. Faulkner, *Electrochemical Methods, 2nd ed.*, John Wiley & Sons, Inc., New York, 2001.
21. P. R. Unwin and A. J. Bard, *J. Phys. Chem.* **95**, 7814 (1991).
22. F. Zhou, P. R. Unwin, and A. J. Bard, *J. Phys. Chem.* **96**, 4917 (1992).
23. F. Zhou and A. J. Bard, *J. Am. Chem. Soc.* **116**, 393 (1994).
24. D. A. Treichel, M. V. Mirkin, and A. J. Bard, *J. Phys. Chem.* **98**, 5751 (1994).
25. C. Demaille, P. R. Unwin, and A. J. Bard *J. Phys. Chem.* **100**, 14137 (1996).
26. R. C. Engstrom, T. Meaney, R. Tople, and R. M. Wightman, *Anal. Chem.* **59**, 2005 (1987).

27. G. Denuault, M. H. Troise-Frank, and L. M. Peter, *Faraday Discuss. Chem. Soc.* **94**, 23 (1992).
28. B. R. Horrocks, M. V. Mirkin, D. T. Pierce, A. J. Bard, G. Nagy, and K. Toth, *Anal. Chem.* **65**, 1213 (1993).
29. C. Wei, A. J. Bard, G. Nagy, and K. Toth, *Anal. Chem.* **67**, 1346 (1995).
30. C. Wei, A. J. Bard, I. Kapui, G. Nagy, and K. Toth, *Anal. Chem.* **68**, 2651 (1996).
31. E. Klusmann and J. W. Schultze, *Electrochim. Acta* **42**, 3123 (1997).
32. R. D. Martin and P. R. Unwin, *Anal. Chem.* **70**, 276 (1998).
33. M. V. Mirkin, F.-R. F.; Fan, and A. J. Bard, *Science* **257**, 364 (1992).
34. F.-R. F. Fan, M. V. Mirkin, and A. J. Bard, *J. Phys. Chem.* **98**, 1475 (1994).
35. J. Guo and S. Amemiya, *Anal. Chem.* **77**, 2147 (2005).
36. W. Zhan and A.J. Bard, *Anal. Chem.* **78**, 726 (2006).
37. P. Sun, F. O. Laforge, T. P. Abeyweera, S. A. Rotenberg, J. Carpino, and M. V. Mirkin, *Proc. Natl. Acad. Sci.* USA. **105**, 443 (2008).
38. C. Wei, A. J. Bard, and M. V. Mirkin, *J. Phys. Chem.* **99**, 16033 (1995).
39. M. Tsionsky, A. J. Bard, and M. V. Mirkin, *J. Phys. Chem.* **100**, 17881 (1996).
40. Y. H. Shao and M. V. Mirkin, *J. Electroanal. Chem.* **439**, 137 (1997).
41. Y. H. Shao and M. V. Mirkin, *J. Phys. Chem. B*, **102**, 9915 (1998).
42. Z. Zhang, Y. Yuan, P. Sun, B. Su, J. Guo, Y. Shao, and H. H. Girault, *J. Phys. Chem. B* **106**, 6713 (2002).
43. B. R. Horrocks, D. Schmidtke, A. Heller, and A. J. Bard, *Anal. Chem.* **65**, 3065 (1993).
44. R. T. Kurulugama, D. O. Wipf, S. A. Takacs, S. Pongmayteegul, P. A. Garris, and J. E. Baur, *Anal. Chem.* **77**, 1111 (2005).
45. E. Ervin, H. S. White, and L. A. Barker, *Anal. Chem.* **77**, 5564 (2005).
46. F.-R. F. Fan and C. Demaille, in A. J. Bard and M. V. Mirkin, (eds.), *Scanning Electrochemical Microscopy*, Marcel Dekker, New York, 2001, p. 75.
47. L. A. Nagahara, T. Tundat, and S. M. Lindsay, *Rev. Sci. Instrum.* **60**, 3128 (1989).
48. P. Sun, Z. Zhang, J. Guo, and Y. Shao, *Anal. Chem.* **73**, 5346 (2001).
49. C. J. Slevin, N. J. Gray, J. V. Macpherson, M. A. Webb, and P. R. Unwin, *Electrochem. Commun.* **1**, 282 (1999).
50. C. Kranz, G. Friedbacher, B. Mizaikoff, A. Lugstein, J. Smoliner, and E. Bertagnolli, *Anal. Chem.* **73**, 2491 (2001).
51. P. Sun and M. V. Mirkin, *Anal. Chem.* **78**, 6526 (2006).
52. Y. Shao, M. V. Mirkin, G. Fish, S. Kokotov, D. Palanker, and A. Lewis, *Anal. Chem.* **69**, 1627 (1997).
53. B. B. Katemann and W. Schuhmann, *Electroanalysis* **14**, 22 (2002).
54. D. O. Wipf, A. C. Michael, and R. M. Wightman, *J. Electroanal. Chem.* **269**, 15 (1989).
55. M. A. Alpuche-Aviles and D. O. Wipf, *Anal. Chem.* **73**, 4873 (2001).
56. M. V. Mirkin, in A. J. Bard and M. V. Mirkin, (eds.), *Scanning Electrochemical Microscopy*, Marcel Dekker, New York, 2001, p. 145.
57. M. Arca, A. J. Bard, B. R. Horrocks, T. C. Richards, and D. A. Treichel, *Analyst* **119**, 719 (1994).
58. A. J. Bard, F.-R. F. Fan, and M. V. Mirkin, in Vol. 18; A. J. Bard, (ed.), *Electroanalytical Chemistry*, Marcel Dekker, New York, 1993, p. 243.
59. A. J. Bard, F.-R. F. Fan, D. T. Pierce, P. R. Unwin, D. O. Wipf, and F. Zhou, *Science* **254**, 68 (1991).

60. A. J. Bard, M. V. Mirkin, P. R. Unwin, and D. O. Wipf, *J. Phys. Chem.* **96**, 1861 (1992).
61. M. V. Mirkin and A. J. Bard, *J. Electroanal. Chem.* **323**, 29 (1992).
62. A. J. Bard, G. Denuault, R. A. Freisner B. C. Dornblaser, and L. S. Tuckerman *Anal. Chem.* **63**, 1282 (1991).
63. R. D. Martin and P. R Unwin *J. Electroanal. Chem.* **439**, 123 (1997).
64. C. Lefrou, *J. Electroanal. Chem.* **592**, 103 (2006).
65. J. L. Amphlett and G. Denuault, *J. Phys. Chem. B* **102**, 9946 (1998).
66. M. V. Mirkin, T. C. Richards, and A. J. Bard, *J. Phys. Chem.* **97**, 7672 (1993).
67. M. Tsionsky, A. J. Bard, and M. V. Mirkin, *J. Am. Chem. Soc.* **119**, 10785 (1997).
68. B. Liu, A. J. Bard, M. V. Mirkin, and S. E. Creager, *J. Am. Chem. Soc.* **126**, 1485 (2004).
69. H. Yamada, M. Ogata, and T. Koike, *Langmuir* **22**, 7923 (2006).
70. B. Liu, S. A. Rotenberg, and M. V. Mirkin, *Proc. Natl. Acad. Sci. U.S.A.* **97**, 9855 (2000).
71. B. Liu, S. A. Rotenberg, and M. V. Mirkin, *Anal. Chem.* **74**, 6340 (2002).
72. C. Cai, B. Liu, M. V. Mirkin, H. A. Frank, and J. F. Rusling, *Anal. Chem.* **74**, 114 (2002).
73. W. Feng, S. A. Rotenberg, and M. V. Mirkin, *Anal. Chem.* **75**, 4148 (2003).
74. E. R. Scott and H. S. White, *Anal. Chem.* **65**, 1537 (1993).
75. B. D. Bath, R. D. Lee, H. S. White, and E. R. Scott, *Anal. Chem.* **70**, 1047 (1998).
76. M. H. Troise-Frank and G. Denuault, *J. Electroanal. Chem.* **354**, 331 (1993).
77. M. H. Troise-Frank and G. Denuault, *J. Electroanal. Chem.* **379**, 405 (1994).
78. B. R. Horrocks and M. V. Mirkin, *J. Chem. Soc. Faraday Trans.* **94**, 1115 (1998).
79. P. R. Unwin and A. J. Bard, *J. Phys. Chem.* **96**, 5035 (1992).
80. P. Liljeroth, D. Vanmaekelbergh, V. Ruiz, K. Kontturi, H. Jiang, E. Kauppinen, and B. Quinn, *J. Am. Chem. Soc.* **126**, 7126 (2004).
81. J. Zhang, C. J. Slevin, C. Morton, P. Scott, D. J. Walton, and P. R. Unwin, *J. Phys. Chem. B* **105**, 11120 (2001).
82. A. P. O'Mullane, J. V. Macpherson, J. Cervera-Montesinos, J. A. Manzanares, F. Frehill, J. G. Vos, and P. R. Unwin, *J. Phys. Chem. B* **108**, 7219 (2004).
83. A. L. Whitworth, D. Mandler, and P. R. Unwin, *Phys. Chem. Chem. Phys.* **7**, 356 (2005).
84. P. R. Unwin, in A. J. Bard and M. V. Mirkin, (eds.), *Scanning Electrochemical Microscopy*, Marcel Dekker, New York, 2001, p. 241.
85. C. Demaille, P. R. Unwin, and A. J. Bard, *J. Phys. Chem.* **100**, 14137 (1996).
86. R. D. Martin and P. R. Unwin, *J. Chem. Soc. Faraday Trans.* **94**, 753 (1998).
87. H. Xiong, J. Guo, K. Kurihara, and S. Amemiya, *Electrochem. Commun.* **6**, 615 (2004).
88. Q. Fulian, A. C. Fisher, and G. Denuault, *J. Phys. Chem. B* **103**, 4387 (1999).
89. O. Sklyar and G. Wittstock, *J. Phys. Chem. B* **106**, 7499 (2002).
90. O. Sklyar, A. Kueng, C. Kranz, B. Mizaikoff, A. Lugstein, E. Bertagnolli, and G. Wittstock, *Anal. Chem.* **77**, 764 (2005).
91. O. Sklyar, M. Träuble, C. Zhao, and G. Wittstock, *J. Phys. Chem. B* **110**, 15869 (2006).
92. M. V. Mirkin, T. C. Richards, and A. J. Bard, *J. Phys. Chem.* **97**, 7672 (1993).
93. M. V. Mirkin, L. O. S. Bulhões, and A. J. Bard, *J. Am. Chem. Soc.* **115**, 201 (1993).
94. W. J. Miao, Z. F. Ding, and A. J. Bard, *J. Phys. Chem. B* **106**, 1392 (2002).
95. D. O. Wipf and A. J. Bard, *J. Electrochem. Soc.* **138**, 469 (1991).

96. B. R. Horrocks, M. V. Mirkin, and A. J. Bard. *J. Phys. Chem.* **98**, 9106 (1994).
97. S. K. Haram and A. J. Bard *J. Phys. Chem. B* **105**, 8192 (2001).
98. T. J. Kemp, P. R. Unwin, and L. Vincze, *J. Chem. Soc., Faraday Trans.* **91**, 3893 (1995).
99. S. M. Fonseca, A. L. Barker, S. Ahmed, T. J. Kemp, and P. R. Unwin, *Phys. Chem. Chem. Phys.* **6**, 5218 (2004).
100. S. M. Fonseca, A. L. Barker, S. Ahmed, T. J. Kemp, and P. R. Unwin, *Chem. Commun.* **8**, 1002 (2003).
101. S. B. Basame and H. S. White, *Anal. Chem.* **71**, 3166 (1999).
102. A. K. Neufeld, A. P. O'Mullane. *J. Solid State Electrochem.* **10**, 808 (2006).
103. K. B. Holt, A. J. Bard, Y. Show, and G. M. Swain, *J. Phys. Chem. B* **108**, 15117 (2004).
104. A. L. Colley, C. G. Williams, U. D'Haenens Johansson, M. E. Newton, P. R. Unwin, N. R. Wilson, and J. V. Macpherson, *Anal. Chem.* **78**, 2539 (2006).
105. J. Ghilane, F. Hauquier, B. Fabre, and P. Hapiot *Anal. Chem.* **78**, 6019 (2006).
106. M. V. Mirkin and M. Tsionsky in A. J. Bard and M. V. Mirkin (eds.), *Scanning Electrochemical Microscopy*, Marcel Dekker, New York, 2001, p. 299.
107. B. Liu and M. V. Mirkin, *J. Am. Chem. Soc.* **121**, 8352 (1999).
108. Z. Ding, B. M. Quinn, and A. J. Bard, *J. Phys. Chem. B* **105**, 6367 (2001).
109. J. Zhang and P. R. Unwin *Langmuir* **18**, 1218 (2002).
110. S. Cannan, J. Zhang, F. Grunfeld, and P. R. Unwin *Langmuir* **20**, 701 (2004).
111. F. O. Laforge, T. Kakiuchi, F. Shigematsu, and M. V. Mirkin, *J. Am. Chem. Soc.* **126**, 15380 (2004).
112. F. O. Laforge, T. Kakiuchi, F. Shigematsu, and M. V. Mirkin, *Langmuir* 2006, in press.
113. B. M. Quinn, P. Liljeroth, V. Ruiz, T. Laaksonen, and K. Kontturi, *J. Am. Chem. Soc.* **125**, 6644 (2003).
114. D. G. Georganopoulou, M. V. Mirkin, and R. W. Murray, *Nano Lett.* **4**, 1763 (2004).
115. A. L. Barker, P. R. Unwin, S. Amemiya, J. Zhou, and A. J. Bard, *J. Phys. Chem. B* **103**, 7260 (1999).
116. P. Sun, F. Li, Y. Chen, M. Q. Zhang, Z. Gao, and Y. H. Shao, *J. Am. Chem. Soc.* **125**, 9600 (2003).
117. P. Sun, Z. Zhang, Z. Gao, and Y. Shao, *Angew. Chem. Int. Ed. Engl.* **41**, 3445 (2002).
118. F. Forouzan, A. J. Bard, and M. V. Mirkin, *Isr. J. Chem.* **37**, 155 (1997).
119. K. B. Holt, *Langmuir* **22**, 4298 (2006).
120. Y. Selzer, I. Turyan, and D. Mandler, *J. Phys. Chem. B* **103**, 1509 (1999).
121. S. Jayaraman, and A. C. Hillier, *J. Combinat. Chem.* **6**, 27 (2004).
122. M. Black, J. Cooper, and P. McGinn, *Meas. Sci. Technol.* **16**, 174 (2005).
123. A. R. Kucernak, P. B. Chowdhury, C. P. Wilde, G. H. Kelsall, Y. Y. Zhu, and D. E. Williams, *Electrochim. Acta* **45**, 4483 (2000).
124. J. Zhou, Y. Zu, and A.J. Bard, *J. Electroanal. Chem.* **491**, 22 (2000).
125. J. Fernández and A. J. Bard, *Anal. Chem.* **75**, 2967 (2003).
126. J. Fernández, D. A. Walsh, and A. J. Bard, *J. Am. Chem. Soc.* **127**, 357 (2005).
127. J. Fernández, V. Raghuveer, A. Manthiram, and A. J. Bard, *J. Am. Chem. Soc.* **127**, 13100 (2005).
128. Y. C. Weng, F.-R. F. Fan, and A. J. Bard, *J. Am. Chem. Soc.* **127**, 17576 (2005).

129. S. Jayaraman and A. C. Hillier, *Langmuir* **17**, 7857 (2001).
130. S. Jayaraman and A. C. Hillier, *J. Phys. Chem. B* **107**, 5221 (2003).
131. D. T. Pierce, P. R. Unwin, and A. J. Bard, *Anal. Chem.* **64**, 1795 (1992).
132. D. T. Pierce and A. J. Bard, *Anal. Chem.* **65**, 3598 (1993).
133. J. C. O'Brien, J. Shumaker-Parry, and R. C. Engstrom, *Anal. Chem.* **70**, 1307 (1998).
134. H. Shiku, T. Takeda, H. Yamada, T. Matsue, and I. Uchida, *Anal. Chem.* **67**, 312 (1995).
135. J. Zhou, C. Campbell, A. Heller, and A. J. Bard, *Anal. Chem.* **74**, 4007 (2002).
136. G. Wittstock and W. Schuhmann, *Anal. Chem.* **69**, 5059 (1997).
137. H. Shiku, T. Matsue, and I. Uchida, *Anal. Chem.* **68**, 1276 (1996).
138. S. A. G. Evans, K. Brakha, M. Billon, P. Mailley, and G. Denuault, *Electrochem. Commun.* **7**, 135 (2005).
139. M. Ciobanu, H. A. Kincaid, G. K. Jennings, and D. E. Cliffel, *Langmuir* **21**, 692 (2005).
140. (a) J. V. Macpherson and P. R. Unwin, *J. Chem. Soc. Faraday Trans.* **89**, 1883 (1993). (b) J. V. Macpherson and P. R. Unwin, *J. Phys. Chem.* **98**, 1704 (1994). (c) J. V. Macpherson and P. R. Unwin, *J. Phys. Chem.* **98**, 11764 (1994).
141. J. V. Macpherson and P. R. Unwin, *J. Phys. Chem.* **99**, 3338 (1995).
142. J. V. Macpherson and P. R. Unwin, *J. Phys. Chem.* **99**, 14824 (1995).
143. (a) J. V. Macpherson, P. R. Unwin, A. C. Hiller, and A. J. Bard, *J. Am. Chem. Soc.* **118**, 6445 (1996). (b)C. E. Jones, J. V. Macpherson, and P. R. Unwin, *J. Phys. Chem. B* **104**, 2351 (2000).
144. C. E. Jones, P. R. Unwin, and J. V. Macpherson, *Chemphyschem.* **4**, 139 (2003).
145. Y. F. Yang and G. Denuault, *J. Chem. Soc. Faraday Trans.* **92**, 3791 (1996).
146. Y. F. Yang and G. Denuault, *J. Electroanal. Chem.* **418**, 99 (1996).
147. Y. F. Yang and G. Denuault, *J. Electroanal. Chem.* **443**, 273 (1998).
148. K. Mansikkamäki, U. Haapanen, C. Johans, K. Kontturi, and M. Valden, *J. Electrochem. Soc.* **153**, B311 (2006).
149. D. P. Burt, J. Cervera, D. Mandler, J. V. Macpherson, J. A. Manzanares, and P. R. Unwin, *Phys. Chem. Chem. Phys.* **7**, 2955 (2005).
150. D. O. Wipf and A. J. Bard, *J. Electrochem. Soc.* **138**, L4 (1991).
151. C. Lee and F. C. Anson, *Anal. Chem.* **64**, 528 (1992).
152. N. Casillas, S. Charlebois, W. H. Smyrl, and H. S. White, *J. Electrochem. Soc.* **140**, L142 (1993).
153. N. Casillas, S. Charlebois, W. H. Smyrl, and H. S. White, *J. Electrochem. Soc.* **141**, 636 (1994).
154. C. J. Boxley, H. S. White, C. E. Gardner, and J. V. Macpherson, *J. Phys. Chem. B* **107**, 9677 (2003).
155. T. C. Richards, A. J. Bard, A. Cusanelli, and D. Sutton. *Organometallics* **13**, 757 (1994).
156. M. V. Mirkin, H. Yang, and A. J. Bard. *J. Electrochem. Soc.* **139**, 2212 (1992).
157. M. Tsionsky, J. Zhou, S. Amemiya, F.-R.F. Fan, A. J. Bard, and R. A. W. Dryfe, *Anal. Chem.* **71**, 4300 (1999).
158. S. Amemiya and A. J. Bard, *Anal. Chem.* **72**, 4940 (2000).
159. J. P. Wilburn, D. W. Wright and D. E. Cliffel, *Analyst* **131**, 311 (2006).
160. J. Mauzeroll and A. J. Bard, *Proc. Natl. Acad. Sci. U.S.A.* **101**, 7862 (2004).
161. J. Mauzeroll, A. J. Bard, O. Owhadian and T. J. Monks, *Proc. Natl. Acad. Sci. U.S.A.* **101**, 17582 (2004).

162. F. Longobardi, P. Cosma, F. Milano, A. Agostiano, J. Mauzeroll, and A. J. Bard, *Anal. Chem.* **78**, 5046 (2006).
163. J. Kwak, C. Lee, and A. J. Bard, *J. Electrochem. Soc.* **137**, 1481 (1990).
164. M. Arca, M. V. Mirkin, and A. J. Bard, *J. Phys. Chem.* **99**, 5040 (1995).
165. G. Denuault, M. H. Troise-Frank, and L. M. Peter, *Faraday Discuss.* **94**, 23 (1992).
166. K. Borgwarth, C. Ricken, D. G. Ebling, and J. Heinze, *Ber. Bunsenges Phys. Chem.* **99**, 1421 (1995).
167. I. Turyan and D. Mandler, *J. Am. Chem. Soc.* **120**, 10733 (1998).
168. I. Kapui, R. E. Gyurcsányi, G. Nagy, K. Tóth, M. Arca, and E. Arca, *J. Phys. Chem. B* **102**, 9934 (1998).
169. C. Combellas and F. Kanoufi *J. Electroanal. Chem.* **589**, 243 (2006). and references cited therein
170. G. Shi, L. F. Garfias-Mesias, and W. H. Smyrl, *J. Electrochem. Soc.* **145**, 3011 (1998).
171. D. E. Cliffel and A. J. Bard, *Anal. Chem.* **70**, 1993 (1998).
172. B. Gollas, P. N. Bartlett, and G. Denuault, *Anal. Chem.* **72**, 349 (2000).
173. C. Xiang, Q. Xie, J. Hu, and S. Yao, *J. Electroanal. Chem.* **584**, 201 (2005).
174. E. Fortin, Y. Defontaine, P. Mailley, T. Livache, and S. Szunerits, *Electroanalysis* **17**, 5 (2005).
175. F-M. Boldt, J. Heinze, M. Diez, J. Petersen, and M. Borsch, *Anal. Chem.* **76**, 3473 (2004).
176. F.-R. F. Fan, D. Cliffel and A. J. Bard, *Anal. Chem.* **70**, 2941 (1998).
177. C. E. Gardner and J. V. Macpherson, *Anal. Chem.* **74**, 576A (2002).
178. J. V. Macpherson and P. R. Unwin, *Anal. Chem.* **73**, 550 (2001).
179. J. Abbou, C. Demaille, M. Dret, and J. Moiroux, *Anal. Chem.* **74**, 6355 (2002).
180. J. Abbou, A. Anne, and C. Demaille, *J. Am. Chem. Soc.* **126**, 10095 (2004).
181. A. Lugstein, E. Bertagnolli, C. Kranz, A. Kueng, and B. Mizaikoff, *Appl. Phys. Lett.* **81**, 349 (2002).
182. D. P. Burt, N. R. Wilson, P. S. Dobson, J. M. R. Weaver, and J. V. Macpherson, *Nano Lett.* **5**, 639 (2005).
183. N. R. Wilson, D. H. Cobden, and J. V. Macpherson, *J. Phys. Chem. B* **106**, 13102 (2002).
184. R. J. Fasching, Y. Tao, and F. B. Prinz, *Sensors Actuators B* **108**, 964 (2005).
185. P. S. Dobson, J. M. R. Weaver, M. N. Holder, P. R. Unwin, and J. V. Macpherson, *Anal. Chem.* **77**, 424 (2005).
186. J. V. Macpherson, P. R. Unwin, A. C. Hillier, and A. J. Bard, *J. Am. Chem. Soc.* **118**, 6445 (1996).
187. J. V. Macpherson, C. E. Jones, A. L. Barker, and P. R. Unwin, *Anal. Chem.* **74**, 1841 (2002).
188. K. B. Holt, A. J. Bard, Y. Show, and G. M. Swain, *J. Phys. Chem. B* **108**, 15117 (2004).
189. J. V. Macpherson and J. L. Delplancke, *J. Electrochem. Soc.* **149**, B306 (2002).
190. A. Davoodi, J. Pan, C. Leygraf, and S. Norgren, *Electrochem. Solid State Lett.* **8**, B21 (2005).
191. C. Kranz, A. Kueng, A. Lugstein, E. Bertagnolli, and B. Mizaikoff, *Ultramicroscopy* **100**, 127 (2004).
192. Y. Hirata, S. Yabuki, and F. Mizutani, *Bioelectrochemistry* **63**, 217 (2004).
193. C. E. Gardner, P. R. Unwin, and J. V. Macpherson, *Electrochem. Commun.* **7**, 612 (2005).
194. M. N. Holder, C. E. Gardner, J. V. Macpherson, and P. R. Unwin, *J. Electroanal. Chem.* **585**, 8 (2005).
195. P. A. Kottke and A. G. Fedorov, *J. Electroanal. Chem.* **583**, 221 (2005).

THE ν_{X-H} LINE SHAPES OF CENTROSYMMETRIC CYCLIC DIMERS INVOLVING WEAK HYDROGEN BONDS

OLIVIER HENRI-ROUSSEAU AND PAUL BLAISE

Laboratoire de Mathématiques, Physique et Systèmes (LAMPS), Université de Perpignan, 66860 Perpignan Cedex, France

CONTENTS

Advances in Chemical Physics, Volume 139, edited by Stuart A. Rice

DEDICATION

This contribution is dedicated to Professor Andrej Witkowski of the Jagellion University of Cracow (Poland) who is at the origin of our works on the subject.

I. INTRODUCTION

If ultrafast nonlinear vibrational spectroscopy [1–3] has recently developed into an important tool providing original informations on the dynamics of weak hydrogen bonds (H-bonds), the simpler linear infrared (IR) ν_s(X—H) absorption spectroscopy spectra remains, however, to be an important method for the understanding of this dynamics. Considerable experimental and theoretical works have been done in this last field [4–17].

TABLE I
Classification of H-Bonds

Strength	Symmetry	Length (Å)	Bonding Enthalpy (kcal mol^{-1})
Weak	Nonsymmetrical	Long, ≈2.90	5–6
Medium	Nonsymmetrical	Short, ≈2.75	6–12
Strong	Nonsymmetrical	Very short, ≈2.50	12–15
Super strong	≈Symmetrical	Very short, ≈2.30	≈20

The H-bonded compounds are usually classified according to the strength of their H-bond. There are weak, medium, strong, and very strong H-bonded systems. Table I shows the main characteristics of these bonds, particularly the length and the bonding enthalpy of some H-bond bridges.

It is also usual to classify the H-bonds according to the property of the one dimensional (1D) proton potential function, although there are no clear discontinuities in the evolution of the properties with the increase of bond enthalpy. According to this finding, H-bonds can be classified into:

1. Those with a single, asymmetric potential minimum roughly corresponding to *weak* and *medium* H-bonds for which the tunneling effect may be ignored.
2. Those with low or without potential barrier, corresponding to *strong* or *super strong* strengths for which tunneling effects must be taken into account.

Besides, IR spectra contain interesting informations on the electronic and nuclear dynamics of H-bond species. From a spectroscopic view point, the main vibrational modes of a H-bonded system are the ν_s(X—H) high frequency stretching mode, the in-plane δ(X—H$\cdots$$Y$) bending modes, the γ(X—H$\cdots$$Y$) out-of-plane bending modes, and the ν_s(X—H$\cdots$Y) stretching low frequency mode. The most evident effects of H-bonding on the spectral density of the ν_s(X—H) stretching mode, are reported in Table II:

TABLE II
Features of the IR ν_s (X—H) Band Shapes of H-Bonded Species

Intensities strongly enhanced
Frequencies strongly shifted toward lower values (red-shift).
Band shapes extensively broadened with often peculiar subbands.
Frequencies, half-widths, and intensities involving large isotopic effects
Band shapes often keeping the same structure in gas and condensed phases

TABLE III
The ν_s (X—H) Band Half-Width

H-Bond Strength	Strength (cm^{-1})
Very weak	100–300
Weak	300–1000
Medium	1000–1500
Strong and superstrong	$\sim$ 2000

More precisely, experiments show that

1. The intensity of the ν_s(X—H) stretching mode increases up to 1000-fold.
2. The low frequency shifts may be as large as $\sim$1000 cm^{-1}.
3. The band half-width increases from $\sim$100 to 1000–2000 cm^{-1}, according to the strength of the H-bond (see Table III).
4. There are band asymmetries and the appearance of subsidiary absorption maxima and minima, such as *windows*.
5. The substitution in the X—H bond of hydrogen by deuterium lowers the angular frequency and the spectrum half-width by a factor that is sometimes $\sim\sqrt{2}$.
6. There are well-known temperature effects, particularly dealing with the two first moments of the spectra that evoke those of the thermal average appearing in the statistical mechanics of quantum harmonic oscillator coordinates.
7. There are several relationships [4, 5] between physical characteristics:

a. One between the H-bond bridge frequency shift and the X···Y H-bond bridge length.
b. One between the X···Y H-bond bridge length and the X—H equilibrium distance.
c. One between the H-bond bridge frequency shift and the X—H equilibrium distance.

The specificity of the ν_{XH} IR band shape of H bonded species, that is, the large low frequency shift, the very large half-width and the subtle features of the line shape together with the known temperature and isotopic effects are a theoretical challenge: a theory that is susceptible to cover the above phenomenology and to take into account the dramatic changes in the IR spectrum induced by the formation of the H-bond bridge, may be considered to give some understanding of the dynamics of the H-bond. Moreover, the theory has to be quantitative, by allowing the reconstruction of spectra from some physical basic accessible parameters.

The physical model that is at the basis of the majority of the theoretical approaches of the IR ν_s(X—H) line shape, is the *strong anharmonic coupling theory* according to which there is an anharmonic coupling between the X—H high frequency mode and the slow X···Y mode corresponding to the H-bond bridge [6, 18, 19]. This is generally expressed by assuming a linear dependence of the angular frequency ω of the high frequency mode with respect to the H-bond bridge elongation Q. This mechanism induces two anharmonic couplings, one linear with respect to the H-bond bridge coordinate and quadratic with respect to the fast one and the other quadratic with respect to these coordinates. Note that the strong anharmonic coupling theory is not without connection to the temporal anharmonic theory of Witkowski involving retarded effects [20].

Next, for weak H-bonded species and within the strong anharmonic coupling theory, it is possible to perform an *adiabatic separation* between the motion of the high and slow frequency modes. The reason is because the ratio of their angular frequencies is ~20. Such a separation leads to effective Hamiltonians describing the H-bond bridge, which are driven and depending on the degree of excitation of the fast mode, the driving term increasing linearly with the excitation degree. Since a driven harmonic oscillator may be viewed as an undriven but displaced harmonic oscillator, the above description in terms of effective Hamiltonians, implies that the potential of the H-bond bridge is displaced when the fast mode is passing from its ground to its first excited state. The consequence is that the ν_s(X—H) IR transition of the fast mode is accompanied by Franck–Condon transitions between the energy levels of the two displaced harmonic oscillators.

Besides, the influence of the medium on the IR ν_s(X—H) line shape may be either direct via its action on the high frequency mode, or indirect via that on the H-bond bridge. The *direct relaxation*, which is corresponding to the first mechanism, simply affects the lifetime of the first excited state of the high frequency mode, through an imaginary part of the corresponding energy level. On the other hand, the *indirect relaxation* mechanism is more complex: Within the description of the H-bond bridge in terms of effective Hamiltonians, this mechanism affects a driven quantum harmonic oscillator. The theory of the driven damped quantum harmonic oscillator has been performed by Feynmann and Vernon [21] and later by Louisell and Walker [22].

Moreover, together with the strong anharmonic coupling between the high and low frequency modes, there is also the possibility of other anharmonicities via *Fermi resonances* [23–25] between the first excited state of the fast mode and harmonics or combination bands of some bending modes. It has been shown that these possibilities of Fermi resonances are strongly assisted by the anharmonic coupling between the high and low frequency modes [26, 27].

At last, many H-bonded species may form cyclic dimers. Then, in such situations, there is the possibility of resonance between the two X—H stretching oscillators belonging to each moiety of the dimer. This leads to *Davydov coupling* that may affect the IR ν_s(X—H) line shape [18].

Since the ν_s(X—H) IR profiles of such cyclic dimers are more complex that those of single H-bonded species, they are considered as good experimental benches for testing theories. The purpose of this chapter is to give the state of the art in the quantum theory dealing with the profiles of such weak or medium H-bonded species.

The presentation will be working within the linear response theory [28] which has been early used in spectroscopy by Bratos et al. [29]. It will not consider (1) predissociation since theoretical works [30–35], showed that it is negligible; (2) tunneling effect [36–41] since it is dealing with weak or intermediate H-bonds where this mechanism cannot play a sensitive role; and (3) rotational structure [42–44].

A. The Spectral Density Within the Linear Response Theory

Within the linear response theory [28] the spectral density $I(\omega)$ is given by

$$I(\omega) = \omega \int_{-\infty}^{\infty} G(t) e^{-i\omega t} dt \tag{1}$$

Here, $G(t)$ is the quantum autocorrelation function (ACF) of the dipole moment operator responsible for the dipolar absorption transition, whereas ω is the angular frequency and t is the time. Equation (1) has been used, for example, by Bratos [45] and Robertson and Yarwood [46] in their semiclassical studies of H-bonded species within the linear response theory.

In quantum approaches the ACF $G_{Qu}(t)$ of the dipole moment operator may be given by the following trace, tr, to be performed within the base spanning the operators describing the system.

$$G_{Qu}(t) = \frac{1}{\beta} \mathrm{tr}\left\{ \rho_{\mathrm{Tot}} \int_{0}^{\beta} \{\widehat{\mu}(0)\}\{\widehat{\mu}(t + i\lambda\hbar)\} d\lambda \right\} \tag{2}$$

Here, ρ_{Tot} is the Boltzmann density operator, $\widehat{\mu}$, the dipole moment operator, λ, a variable having the dimension of β, which is given by

$$\beta = \frac{1}{k_B T}$$

where T is the absolute temperature and k is the Boltzmann constant. Besides, $\widehat{\mu}(0)$ is the dipole moment operator at initial time, whereas $\widehat{\mu}(t+i\lambda\hbar)$ is this same operator at complex time $t+i\lambda\hbar$ (where $\hbar$ is the Planck constant and $i^2=-1$) given by the following Heisenberg-like transformation:

$$\widehat{\mu}(t+i\lambda\hbar) = (e^{i(t+i\lambda\hbar)\mathrm{H}_{\mathrm{Tot}}/\hbar})\ \widehat{\mu}(0)\ (e^{-i(t+i\lambda\hbar)\mathrm{H}_{\mathrm{Tot}}/\hbar}) \tag{3}$$

Here, $\mathrm{H}_{\mathrm{Tot}}$ is the total Hamiltonian of the system. The Boltzmann density operator ρ_{Tot} is

$$\rho_{\mathrm{Tot}} = \frac{1}{Z_{\mathrm{Tot}}}(e^{-\beta\mathrm{H}_{\mathrm{Tot}}}) \quad \text{with} \quad \mathrm{tr}\,[\rho_{\mathrm{Tot}}] = 1 \quad \text{and} \quad Z_{\mathrm{Tot}} = \mathrm{tr}[(e^{-\beta\mathrm{H}_{\mathrm{Tot}}})]$$

where Z_{Tot} is the partition function.

Appendix K shows that, in the absence of damping, the IR line shape of the $\nu_{X\text{–}H}$ of weak H-bonded species calculated with the aid of Eqs. (1)–(3) is equivalent to that in which the line shape is computed with the aid of the following equations:

$$\mathrm{I}(\omega) = \int_{-\infty}^{\infty} \mathrm{G}(t)e^{-i\omega t}dt \tag{4}$$

$$\mathrm{G}(t) = \mathrm{tr}\{\rho_{\mathrm{Tot}}\widehat{\mu}(0)\ \widehat{\mu}(t)\} \tag{5}$$

Here, the dipole moment operator at time t is related to that at the initial time through the usual Heisenberg transformation:

$$\widehat{\mu}(t) = (e^{i\mathrm{H}_{\mathrm{Tot}}t/\hbar})\ \widehat{\mu}(0)\ \left(e^{-i\mathrm{H}_{\mathrm{Tot}}t/\hbar}\right) \tag{6}$$

Besides, the density operator is

$$\rho_{\mathrm{Tot}} = \frac{1}{Z_{\mathrm{Tot}}}(e^{-\beta\mathrm{H}_{\mathrm{Tot}}})$$

The linear response equations (4–6) are proved in Appendix A.

The basic quantum theories [47–50], dealing with the IR line shape of the $\nu_{X\text{–}H}$ of weak H-bonded species working within the linear response theory have been performed with the aid of Eq. (4) in place of (1) used by Bratos [45] and Robertson and Yarwood [46].

Owing to the equivalence between the two approaches in the absence of damping, we study here the IR line shape of the ν_{X-H} of weak H-bonded species, within the linear response theory, with the aid of the simple method using Eqs. (4–6), even in the presence of damping.

Note, that for the Fourier transform of the complex ACF, the following equalities hold:

$$\mathrm{I}(\omega) = \int_{-\infty}^{\infty} \mathrm{G}(t)e^{-i\omega t}dt = 2\,\mathrm{Re}\int_{0}^{\infty} \mathrm{G}(t)e^{-i\omega t}dt = 2\,\mathrm{Re}\int_{0}^{\infty} \mathrm{G}(t)^{*}e^{i\omega t}dt \qquad (7)$$

B. The IR Non-Hermitean Transition Operators

The dipole moment operator $\widehat{\mu}(0)$ of a vibrational mode depends on the coordinate q.e $\widehat{\mu}(\mathrm{q},0)$. The dipole moment operator at the initial time may be expanded, up to first order with respect to the equilibrium position $\mathrm{q}=0$. This leads to

$$\widehat{\mu}(\mathrm{q},0)=\mu(0,0)+\left[\frac{\partial\widehat{\mu}}{\partial\mathrm{q}}\right]_{\mathrm{q}=0}\mathrm{q}$$

In this expression, according to the theory of the quantum harmonic oscillator, the operator q appearing on the right-hand side, may couple two successive eigenstates $|\{k\}\rangle$ of the Hamiltonian of the harmonic oscillator. Consequently, by ignoring the scalar term $\mu(0,0)$, which does not couple these states, we may write the dipole moment operator according to

$$\widehat{\mu}(\mathrm{q},0)=\sqrt{\frac{\hbar}{2m\omega^{\circ}}}\left[\frac{\partial\widehat{\mu}}{\partial\mathrm{q}}\right]_{\mathrm{q}=0}\sum\left[\sqrt{k+1}|\{k\}\rangle\langle\{k+1\}|+\sqrt{k}|\{k\}\rangle\langle\{k-1\}|\right]$$

If one restricts operations to the ground and the first excited states, this equation reduces to

$$\widehat{\mu}(\mathrm{q},0)=\sqrt{\frac{\hbar}{2m\omega^{\circ}}}\left[\frac{\partial\widehat{\mu}}{\partial\mathrm{q}}\right]_{\mathrm{q}=0}\left[|\{0\}\rangle\langle\{1\}|+|\{1\}\rangle\langle\{0\}|\right]$$

Next, the proportionality constant may be ignored, because it will not play any role in the following. Thus, this equation may be expressed simply as:

$$\widehat{\mu}(\mathrm{q},0)=\mu_0\left[|\{0\}\rangle\langle\{1\}|+|\{1\}\rangle\langle\{0\}|\right]$$

with

$$\mu_0 = \sqrt{\frac{\hbar}{2m\omega^{\circ}}}\left[\frac{\partial\widehat{\mu}}{\partial q}\right]_{q=0} \tag{8}$$

Expressed in this way, the dipole moment operator is the sum of two operators, the first of which is the Hermitean conjugate of the other:

$$\widehat{\mu}(q,0) = \mu(0) + \mu(0)^{\dagger}$$

One corresponds to the absorption and the other to the emission. These transition operators are, respectively, given by

$$\mu(0) = \mu_0|\{1\}\rangle\langle\{0\}| \qquad \text{and} \qquad \mu(0)^{\dagger} = \mu_0|\{0\}\rangle\langle\{1\}| \tag{9}$$

In the Heisenberg picture, the operator at time t is, owing to Eq. (6):

$$\mu(t) = \mu_0(e^{+i\mathrm{H}_{\mathrm{Tot}}t/\hbar})|\{1\}\rangle\langle\{0\}|(e^{-i\mathrm{H}_{\mathrm{Tot}}t/\hbar})$$

Note that in this Heisenberg picture the term $e^{+i\mathrm{H}_{\mathrm{Tot}}t/\hbar}$ is acting on the ket at the opposite of the Shrödinger picture of the diagonal elements of a density operator where it is $e^{-i\mathrm{H}_{\mathrm{Tot}}t/\hbar}$.

Note also that, owing to Eqs. (5), and (9) the ACF is given, respectively, according to the fact that one considers the absorption or the emission, by

$$\begin{aligned} \mathrm{G}_{\mathrm{Abs}}(t) &= \mathrm{tr}\{\rho_{\mathrm{Tot}}\ \mu(0)^{\dagger}\ \mu(t)\} \\ \mathrm{G}_{\mathrm{Emis}}(t) &= \mathrm{tr}\{\rho_{\mathrm{Tot}}\ \mu(0)\ \mu(t)^{\dagger}\ \} \end{aligned} \tag{10}$$

Of course, these two ACFs are interrelated through

$$\mathrm{G}_{\mathrm{Abs}}(t) = \mathrm{G}_{\mathrm{Emis}}(t)^{*}$$

Therefore, owing to Eq. (7), the SD is given by

$$\mathrm{I}(\omega) = \int_{-\infty}^{\infty} \mathrm{G}_{\mathrm{Abs}}(t)e^{-i\omega t}dt = 2\,\mathrm{Re}\int_{0}^{\infty} \mathrm{G}_{\mathrm{Abs}}(t)e^{-i\omega t}dt \tag{11}$$

II. BARE HYDROGEN-BONDS WITHOUT DAMPING

A. The Hamiltonian in the Absence of Damping

1. *The Model of a Single Bare H-Bonded Species*

Consider a single H-bond, for example, that is involved in a carboxylic acid–ether complex depicted in Fig. 1. It may be modelized by two quantum oscillators that are, respectively, the ν_{X-H} high frequency stretching mode and the low frequency mode corresponding to the H-bond bridge.

Then, let us define the different physical terms dealing with the high and low frequency modes. This is performed in Table IV.

Within the linear response theory, in order to know the dipole moment operator at time t, it is necessary to know the Hamiltonian involved in the Heisenberg transformation (6). In a first place, we will only consider the situation where the damping of the H-bond bridge is ignored. In order to distinguish the situations with and without damping, we assign the "°" symbol to situation without damping.

2. *The Full Hamiltonian Within the Strong Anharmonic Coupling Theory*

Now, look at the Hamiltonian of the system within the strong anharmonic coupling theory [6, 18, 51]. The bare H-bond bridge potential is not only intrinsically anharmonic, because of its Morse-like form, but also anharmonically coupled with the high frequency mode. Later, it will appear that it is possible, for weak H-bonds, to separate adiabatically the motion of the H-bond bridge from that of the high frequency mode, and also to approximate the Morse-like potential of the bridge by a harmonic one. But, for the present time, let us look at the general situation that is working beyond these adiabatic and harmonic approximations.

The full Hamiltonian H_{Tot} of the H-bonded species in the absence of damping, may be written

$$H_{Tot} = H_{Fast} + H_{Slow} \tag{12}$$

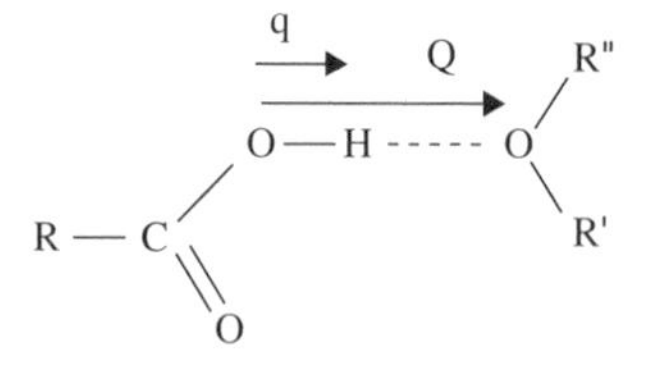

Figure 1. Single H-bond and its coordinates.

TABLE IV
Physical Parameters of Bare H-Bonded Species

H-Bond Bridge	High Frequency Mode
Coordinate position = Q	Coordinate position operator = q
Conjugate momentum = P	Conjugate momentum operator = p
$[Q, P] = i\hbar$	$[q, p] = i\hbar$
Angular frequency = Ω	Angular frequency = ω°, when Q = 0
Reduced mass = M	Reduced mass = m

Here, H_{Fast} is the Hamiltonian of the high frequency mode and H_{Slow} is that of the H-bond bridge.

The Hamiltonian of the high frequency mode may be considered to be harmonic-like with respect to its space coordinate q, but with a dependence of its angular frequency $\omega(Q)$ on the coordinate Q of the H-bond bridge:

$$H_{Fast} = \frac{p^2}{2m} + \frac{1}{2}m[\omega(Q)]^2 q^2 \tag{13}$$

On the other hand, the Hamiltonian of the H-bond bridge for which the anharmonicity of the potential may be taken into account via a Morse potential, may be given by

$$H_{Slow} = \frac{P^2}{2M} + D_e\left[1 - e^{-\Omega\sqrt{\frac{M}{2D_e}}Q}\right]^2 \tag{14}$$

where D_e is the dissociation energy involved in the H-bond bridge Morse potential.

In the following equations it will appear to be suitable to split the H-bond bridge Hamiltonian (14) into a harmonic part H° and an anharmonic perturbation V_{Anh}, according to

$$H_{Slow} = H^\circ + V_{Anh}$$

with, respectively,

$$H^\circ = \frac{P^2}{2M} + \frac{1}{2}M\Omega^2 Q^2 \tag{15}$$

and:

$$V_{Anh} = D_e\left[1 - e^{-\Omega\sqrt{\frac{M}{2D_e}}Q}\right]^2 - \frac{1}{2}M\Omega^2 Q^2 \tag{16}$$

Now, look at the high frequency mode Hamiltonian (13). If the involved angular frequency of the fast mode $\omega(Q)$ is expanded up to first order, one may write

$$\omega(Q) = \omega^{\circ} + bQ \quad \text{with Q near equilibrium H-bond bridge length} \tag{17}$$

Of course, when the H-bond bridge length $Q \rightarrow \infty$, that is, when the H-bond bridge disappears, Eq. (17) is not valid since it is necessary that $\omega(Q) \rightarrow \omega_{\text{Free}}$, where ω_{Free} is the angular frequency of the high frequency fast mode when the H-bond is missing. Now, with the aid of Eqs. (13) and (17), the following expression for the Hamiltonian of the fast mode may be obtained

$$H_{\text{Fast}} = \frac{p^2}{2m} + \frac{1}{2} m\omega^{\circ 2} q^2 + m\omega^{\circ} b q^2 Q + \frac{1}{2} m b^2 q^2 Q^2 \tag{18}$$

Note that beyond the harmonic approximation for the H-bond bridge and according to Eqs. (14) and (18), the full Hamiltonian (12) is

$$H_{\text{Tot}} = \frac{P^2}{2M} + D_e \left[1 - e^{-\Omega \sqrt{\frac{M}{2D_e}} Q} \right]^2 + \frac{p^2}{2m} + \frac{1}{2} m\omega^{\circ 2} q^2 + m\omega^{\circ} b q^2 Q + \frac{1}{2} m b^2 q^2 Q^2 \tag{19}$$

Of course, in the harmonic approximation this Hamiltonian reduces to

$$H_{\text{Tot}} = \frac{P^2}{2M} + \frac{1}{2} M\Omega^2 Q^2 + \frac{p^2}{2m} + \frac{1}{2} m\omega^{\circ 2} q^2 + m\omega^{\circ} b q^2 Q + \frac{1}{2} m b^2 q^2 Q^2 \tag{20}$$

3. *A Partition of the Hamiltonian*

Next, it will be suitable to split the Hamiltonian (18) into two parts:

$$H_{\text{Fast}} = H_{\text{Free}} + H_{\text{Int}}$$

Here, H_{Free} is the Hamiltonian of the free harmonic high frequency oscillator and H_{Int} is that giving the anharmonic interaction between the slow and fast modes, which are, respectively, given by

$$H_{\text{Free}} = \frac{p^2}{2m} + \frac{1}{2} m\omega^{\circ 2} q^2 \tag{21}$$

$$H_{\text{Int}} = m\omega^{\circ} b q^2 Q + \frac{1}{2} m b^2 q^2 Q^2 \tag{22}$$

The eigenvalue equations of the quantum harmonic oscillators Hamiltonians H_{Free} and H° given by Eqs. (21) and (15) are, respectively,

$$H^\circ|(n)\rangle = \hbar\Omega\left(n + \frac{1}{2}\right)|(n)\rangle \tag{23}$$

$$H_{Free}|\{k\}\rangle = \hbar\omega^\circ\left(k + \frac{1}{2}\right)|\{k\}\rangle \tag{24}$$

Here, $|\{k\}\rangle$ are the eigenkets of the fast mode harmonic Hamiltonian while $|(m)\rangle$ are those of the harmonic slow mode, whereas $\{k\}$ and (m) are the corresponding quantum number. These kets form two bases $\{|(n)\rangle\}$ and $\{|\{k\}\rangle\}$ that lead us to write, respectively,

$$\langle(m) \mid (n)\rangle = \delta_{mn} \qquad \text{and} \qquad \sum_n |(n)\rangle\langle(n)| = 1$$

$$\langle\{l\} \mid \{k\}\rangle = \delta_{lk} \qquad \text{and} \qquad \sum_k |\{k\}\rangle\langle\{k\}| = 1$$

The wave functions corresponding to the kets appearing in these equations are, respectively,

$$\langle \mathrm{Q}\ |(n)\rangle = \chi_n(\mathrm{Q}) \qquad \text{and} \qquad \langle \mathrm{q}\ |\{k\}\rangle = \Phi_k(\mathrm{q}) \tag{25}$$

with, of course,

$$\mathrm{q}|\mathrm{q}\rangle = q|\mathrm{q}\rangle \qquad \text{and} \qquad \mathrm{Q}\,|\mathrm{Q}\rangle = Q\,|\mathrm{Q}\rangle$$

Besides, owing to Eq. (21), the eigenvalue equation (24) is

$$\left[\frac{\mathrm{p}^2}{2m} + \frac{1}{2}m\omega^{\circ 2}\mathrm{q}^2\right]|\{k\}\rangle = \hbar\omega^\circ\left(k + \frac{1}{2}\right)|\{k\}\rangle \tag{26}$$

Again, with the help of the eigenstates of the eigenvalue equations (23) and (24), we may built up the tensorial basis:

$$\{|\{k\}(m)\rangle\} = \{|\{k\}\rangle \otimes\ |(m)\rangle\} \tag{27}$$

Note that all the eigenvalue equations of the Hamiltonian H_{Fast} given by Eq. (13), which depend on Q via ω, may be written for any value of Q according to

$$\left[\frac{\mathrm{p}^2}{2m} + \frac{1}{2}m[\omega(\mathrm{Q})]^2\,\mathrm{q}^2\right]|\Phi_k(\mathrm{Q})\rangle = \hbar\omega(\mathrm{Q})\left(k + \frac{1}{2}\right)|\Phi_k(\mathrm{Q})\rangle \tag{28}$$

It must be understood that there are as many eigenvalue equations for this Hamiltonian as there are values of Q for the H-bond bridge coordinate. Thus, the meaning of the notation $\Phi_k(Q)$ in the ket $|\Phi_k(Q)\rangle$, is that this ket is parametrically dependent on the coordinate Q. Of course, when the H-bond bridge is at equilibrium, that is, $Q = 0$, the Hamiltonian involved in Eq. (28) reduces to the Hamiltonian (21). This leads us to write the following equivalence between the ket notations met, respectively, in Eqs. (24) and (28):

$$|\Phi_k(0)\rangle \equiv |\{k\}\rangle \tag{29}$$

B. The ACF and the Corresponding SD

When the damping of the slow and high frequency mode are ignored, according to Eq. (10), the ACF of the absorption transition moment operator of the high frequency mode of the H-bonded species monomer is the following trace $\mathrm{tr}_{\mathrm{Tot}}$ to be performed over the space spanned by the total Hamiltonian $\mathrm{H}_{\mathrm{Tot}}$:

$$[\mathrm{G}^{\circ}_{\mathrm{Mono}}(t)]^{\circ} = \mathrm{tr}_{\mathrm{Tot}}\,\{\rho_{\mathrm{Tot}}\;\mu(0)^{\dagger}\;\mu(t)\}$$

Here, ρ_{Tot} is the Boltzmann density operator:

$$\rho_{\mathrm{Tot}} = \frac{1}{Z_{\mathrm{Tot}}}(e^{-\mathrm{H}_{\mathrm{Tot}}/k_{\mathrm{B}}T}) \qquad \text{with} \qquad Z_{\mathrm{Tot}} = \mathrm{tr}_{\mathrm{Tot}}[(e^{-\mathrm{H}_{\mathrm{Tot}}/k_B T})]$$

Besides, $\mu(0)$, the IR absorption transition operator at time $t = 0$ and $\mu(t)$ is this same operator at time t. Then, beyond the harmonic approximation and in view of Eq. (9) giving the expression of the transition moment operator, the ACF of the bare weak H-bond, may be written

$$\begin{aligned}[\mathrm{G}^{\circ}_{\mathrm{Mono}}(t)]^{\circ} = \left(\frac{\mu_0^2}{Z_{\mathrm{Tot}}}\right)&\mathrm{tr}_{\mathrm{Tot}}[(e^{-\mathrm{H}_{\mathrm{Tot}}/k_{\mathrm{B}}T})\;|\{0\}\rangle\langle\{1\}| \\ &\times [e^{i\mathrm{H}_{\mathrm{Tot}}t/\hbar}]|1\}\rangle\langle\;\{0\}|[e^{-i\mathrm{H}_{\mathrm{Tot}}t/\hbar}]]\end{aligned} \tag{30}$$

Here, the Hamiltonian $\mathrm{H}_{\mathrm{Tot}}$ is given by Eq. (19) or (20). Next, it is suitable to introduce twice a closeness relation using the eigenstates of the Hamiltonians (19) or (20), which are both obeying the eigenvalue equation:

$$\mathrm{H}_{\mathrm{Tot}}|\Theta_{\mu}\rangle = \hbar\omega^{\circ}_{\mu}|\Theta_{\mu}\rangle \qquad \text{with} \qquad \langle\Theta_{\mu}\mid\Theta_{\nu}\rangle = \delta_{\mu\upsilon} \tag{31}$$

This closeness relation is

$$\sum|\Theta_{\mu}\rangle\langle\Theta_{\mu}| = 1 \tag{32}$$

Moreover, with the tensorial product basis $\{|\{k\}(m)\rangle\}$ built up from the eigenvectors of the Hamiltonians of the slow and fast modes appearing in Eqs. (23) and (24), the eigenvectors involved in Eq. (31) are given by the expansion:

$$|\Theta_\mu\rangle = \sum_k \sum_m C_{\mu m}^{\{k\}} \; |\{k\}(m)\rangle \tag{33}$$

Then, inserting twice the closeness relation (32) and after a circular permutation within the trace, the ACF (30) becomes

$$[G^\circ_{\text{Mono}}(t)]^\circ = \left(\frac{\mu_0^2}{Z_{\text{Tot}}}\right) \text{tr}_{\text{Tot}} \left[|\{0\}\rangle\langle \{1\}| [e^{i\text{H}_{\text{Tot}}t/\hbar}] \left\{ \sum_\mu |\Theta_\mu\rangle\langle\Theta_\mu| \right\} \right.$$
$$\left. \times \{1\}\rangle\langle \{0\}| . \left\{ \sum_\nu |\Theta_\nu\rangle\langle\Theta_\nu| \right\} [e^{-i\text{H}_{\text{Tot}}t/\hbar}](e^{-\text{H}_{\text{Tot}}/k_B T}) \right]$$

Next, in view of Eq. (31) and again using the invariance of the trace with respect to a circular permutation, the ACF transforms into

$$[G^\circ_{\text{Mono}}(t)]^\circ = \left(\frac{\mu_0^2}{Z_{\text{Tot}}}\right)$$
$$\text{tr}_{\text{Tot}} \left[\langle\{1\}\rangle \sum_\mu \{[e^{i\omega^\circ_\mu t}] |\Theta_\mu\rangle\langle\Theta_\mu|\} |\{1\}\rangle\langle\{0\}| \right.$$
$$\left. \times \left\{ \sum_\nu |\Theta_\nu\rangle\langle\Theta_\nu| [e^{-\omega^\circ_\nu t}](e^{-\hbar\omega^\circ_\nu/k_B T}) \right\} |\{0\}\rangle \right]$$

Again, insert a closeness relation on the eigenstates of the quantum harmonic oscillator involved in Eq. (23) and then, perform the trace over the basis involving these eigenstates. That gives

$$[G^\circ_{\text{Mono}}(t)]^\circ = \left(\frac{\mu_0^2}{Z_{\text{Tot}}}\right) \sum_\nu \sum_m \sum_\mu \sum_n \langle (n)| \; \langle \; \{1\}| (e^{i\omega^\circ_\mu t}) |\Theta_\mu\rangle\langle\Theta_\mu|$$
$$|\{1\}\rangle \; | \; (m)\rangle\langle(m)| \langle \; \{0\}| |\Theta_\nu\rangle\langle\Theta_\nu| (e^{-i\omega^\circ_\nu t})(e^{-\hbar\omega^\circ_\nu/k_B T}) \; | \; \{0\}\rangle \; | \; (n)\rangle$$

Besides, this result also may be written

$$[G^\circ_{\text{Mono}}(t)]^\circ = \left(\frac{\mu_0^2}{Z_{\text{Tot}}}\right) \sum_\nu \sum_m \sum_\mu \sum_n (e^{-\hbar\omega^\circ_\nu/k_B T}) [e^{i\omega^\circ_\mu t}][e^{-i\omega^\circ_\nu t}]$$
$$\langle (n)\{1\}| |\Theta_\mu\rangle\langle\Theta_\mu| |\{1\}(m)\rangle\langle(m)\{0\}| |\Theta_\nu\rangle\langle\Theta_\nu| |\{0\}(n)\rangle$$

At last, using Eq. (33), the ACF takes the final expression:

$$[G^{\circ}_{\text{Mono}}(t)]^{\circ} = \left(\frac{\mu_0^2}{Z_{\text{Tot}}}\right) \sum_{\nu} \sum_{m} \sum_{\mu} \sum_{n} (e^{-\beta\hbar\omega^{\circ}_{\nu}}) \times (e^{-i(\omega^{\circ}_{\nu}-\omega^{\circ}_{\mu})t})[C^{\{0\}}_{\nu\, m}][C^{\{1\}}_{m\, \mu}][C^{\{1\}}_{\mu\, n}][C^{\{0\}}_{n\, \nu}]$$

Now, the direct damping of the high frequency mode may be incorporated in this ACF by multiplying it by a damping exponential operator of the form $\exp\{-\gamma^{\circ}t\}$. As a consequence, the SD, which is according to Eq. (4) the Fourier transform of the damped ACF, takes the form:

$$[I^{\circ}_{\text{Mono}}(t)] = \left(\frac{\mu_0^2}{Z_{\text{Tot}}}\right) \sum_{\nu} \sum_{m} \sum_{\mu} \sum_{n} (e^{-\beta\hbar\omega^{\circ}_{\nu}}) \; 2\,\text{Re} \int_{o}^{\infty} dt\, e^{-i\omega t}\, e^{-\gamma^{\circ}t} (e^{-i(\omega^{\circ}_{\nu}-\omega^{\circ}_{\mu})t})[C^{\{0\}}_{\nu m}][C^{\{1\}}_{m\mu}][C^{\{1\}}_{\mu n}][C^{\{0\}}_{n\nu}] \tag{34}$$

C. The Adiabatic Representation

Now, we may observe that for weak H-bonds, it is possible to perform an adiabatic separation between the motions of the high and low frequency stretching modes that will allow us to treat many important situations that cannot be handled beyond this approximation, within the straightforward approach of Section II.B.

1. Adiabatic Representation $\{I\}$

It is experimentally known that for weak H-bonds, the angular frequency of the high frequency mode is $\sim$3000 cm^{-1}, whereas that of the H-bond bridge is much lower, $\sim$100 cm^{-1}. Thus, since for weak H-bonds, the frequency of the fast mode is at least 10 times greater than that of the slow mode, it is possible to perform the adiabatic approximation allowing to separate the motions of the slow and high frequency modes [51–53]. Besides, for crude approaches, it may be suitable to neglect the intrinsic anharmonicity of the H-bond bridge, that is, to approximate the Morse potential of this bridge by a harmonic one.

Let us perform the harmonic approximation on the H-bond bridge. Then, the total Hamiltonian (12) of the system formed by the high frequency mode coupled to the H-bond bridge takes the form:

$$[H_{\text{Tot}}] = H_{\text{Free}} + H^{\circ} + H_{\text{Anh}}$$

with

$$H_{Anh} = H_{Int} + V_{Anh}$$

Because of Eqs.(12–15) and to Eqs. (21) and (22), it takes the explicit form:

$$[H_{Tot}] = \frac{P^2}{2M} + \frac{1}{2}M\Omega^2 Q^2 + \frac{p^2}{2m} + \frac{1}{2}m\omega^{\circ 2}q^2 + m\omega^{\circ}bq^2Q + \frac{1}{2}mb^2\,q^2Q^2 \quad (35)$$

Note that this equation, which is the result via Eq. (17) of the expansion of $\omega(Q)$ up to first order in Q, comes from

$$[H_{Tot}] = \frac{P^2}{2M} + \frac{1}{2}M\Omega^2 Q^2 + \frac{p^2}{2m} + \frac{1}{2}m\omega(Q)^2q^2 \quad (36)$$

Next, according to Eqs. (B.35), (B.39), and (B.40) dealing with the adiabatic approximation, one may write the total Hamiltonian according to the equations:

$$[H_{Tot}] \approx [H_{Adiab}] \quad (37)$$

$$[H_{Adiab}] = \sum [H_I^{\{k\}}]|\{k\}\rangle\,\langle\,\{k\}| \quad (38)$$

In this last equation, we see effective Hamiltonians describing the H-bond bridge that depend on the excitation degree $\{k\}$ of the fast mode. They are given by

$$[H_I^{\{k\}}] = \frac{P^2}{2M} + \frac{1}{2}M\Omega^2 Q^2 + \left(k + \frac{1}{2}\right)\hbar b Q + \left(k + \frac{1}{2}\right)\hbar\omega^{\circ} \quad (39)$$

Now, we may go to creation and annihilation operators $a^{\dagger}$ and a, defined in Table V.

Then, the effective Hamiltonians (39) becomes

$$[H_I^{\{k\}}] = \left(a^{\dagger}a + \frac{1}{2}\right)\hbar\Omega + \left(k + \frac{1}{2}\right)[\alpha^{\circ}(a^{\dagger} + a)]\hbar\Omega + \left(k + \frac{1}{2}\right)\hbar\omega^{\circ} \quad (40)$$

with

$$\alpha^{\circ} = \frac{b}{\Omega}\sqrt{\frac{\hbar}{2M\Omega}} \quad (41)$$

TABLE V
Creation and Anihilation Operators for the H-Bond Bridge, the Fast Mode, the Thermal Bath, and the Bending Modes

Slow mode	$Q = \sqrt{\frac{\hbar}{2M\Omega}}[a^\dagger + a]$	$P = i\sqrt{\frac{M\hbar\Omega}{2}}[a^\dagger - a]$	$[a, a^\dagger] = 1$
Fast mode	$q = \sqrt{\frac{\hbar}{2m\omega^\circ}}[b^\dagger + b]$	$p = i\sqrt{\frac{m_i\hbar\omega_i}{2}}[b^\dagger - b]$	$[b, b^\dagger] = 1$
Bending modes	$q_i^\delta = \sqrt{\frac{\hbar}{2m_i^\delta\omega_i^\delta}}[b_i^{\delta\dagger} + b_i^\delta]$	$p_i^\delta = i\sqrt{\frac{m_i^\delta\hbar\omega_i^\delta}{2}}[b_i^{\delta\dagger} - b_i^\delta]$	$[b_i^\delta, b_i^{\delta\dagger} = \delta_{i,j}$
Bath modes	$\tilde{q}_i = \sqrt{\frac{\hbar}{2\tilde{m}_i\tilde{\omega}_i}}[\tilde{b}_i^\dagger + \tilde{b}_i]$	$\tilde{p}_i = \sqrt{\frac{\tilde{m}_i\hbar\tilde{\omega}_i}{2}}[\tilde{b}_i^\dagger - \tilde{b}_i]$	$[\tilde{b}_i, \tilde{b}_j^\dagger] = \delta_{i,j}$

2. *From Representation $\{I\}$ to Representation $\{II\}$*

It may be of interest to remove the driven term $\alpha^\circ(a^\dagger + a)/2$, which is common to all the effective Hamiltonians (40). For this purpose, we perform the following canonical transformation:

$$[\breve{H}_{II}^{\circ\{k\}}] = [A(\alpha^\circ/2)][H_I^{\{k\}}][A(\alpha^\circ/2)]^{-1} \tag{42}$$

with

$$[A(\alpha^\circ/2)] = \left(e^{\frac{\alpha^\circ}{2}[a^\dagger - a]}\right)$$

By using the above equations, the transformed Hamiltonians take the form:

$$[\breve{H}_{II}^{\circ\{k\}}] = \left(e^{\frac{\alpha^\circ}{2}[a^\dagger - a]}\right)\left[\left(a^\dagger a + \frac{1}{2}\right)\hbar\Omega + \left(k + \frac{1}{2}\right)[\alpha^\circ(a^\dagger + a)]\hbar\Omega \right.$$
$$\left. + \left(k + \frac{1}{2}\right)\hbar\omega^\circ\right]\left(e^{-\frac{\alpha^\circ}{2}[a^\dagger - a]}\right)$$

Now, in order to perform the canonical transformation, we may use the following theorem [54]:

$$(e^{\alpha a^\dagger - \alpha a})\{f(a^\dagger, a)\}(e^{-\alpha a^\dagger + \alpha a}) = \{f(a^\dagger - \alpha, a - \alpha)\}$$

Then, by using this theorem with $\alpha = \alpha^\circ/2$, the result of the canonical transformation (43) appears to be

$$\left[\breve{H}_{II}{}^{\circ\{0\}}\right] = \left(a^\dagger a + \frac{1}{2}\right)\hbar\Omega - \frac{1}{4}\alpha^{\circ 2}\hbar\Omega + \frac{1}{2}\hbar\omega^\circ \tag{44}$$

$$[\breve{H}_{II}^{\circ\{1\}}] = \left(a^\dagger a + \frac{1}{2}\right)\hbar\Omega + \left[\alpha^\circ(a^\dagger + a) - \frac{5}{4}\alpha^{\circ 2}\right]\hbar\Omega + \left(1 + \frac{1}{2}\right)\hbar\omega^\circ \tag{45}$$

Now, since in the following we will only be interested in the differences in the eigenvalues of these two Hamiltonians, it may be desirable to simplify the equation to perform the following shift:

$$[H_{II}^{\{k\}}] = [\breve{H}_{II}^{\{k\}}] + \frac{1}{4}\alpha^{\circ 2}\hbar\Omega - \frac{1}{2}\hbar\omega^\circ$$

Then, Eqs. (44) and (45) lead, respectively, to

$$[H_{II}^{\{0\}}] = \left(a^\dagger a + \frac{1}{2}\right)\hbar\Omega = H^\circ \tag{46}$$

$$[H_{II}^{\{1\}}] = \left(a^\dagger a + \frac{1}{2}\right)\hbar\Omega + [\alpha^\circ(a^\dagger + a) - \alpha^{\circ 2}]\hbar\Omega + \hbar\omega^\circ \tag{47}$$

Within this new representation where the zero-point energy of the high frequency mode has been removed, the total Hamiltonian remains to be given by Eq. (37), whereas the adiabatic Hamiltonian is now given by

$$[H_{\mathrm{Adiab}}] = \sum [H_{II}^{\{k\}}] |\{k\}\rangle\ \langle\ \{k\}| \tag{48}$$

After coming back to the Q and P coordinates, which will be useful later, the effective Hamiltonians (46) and (47) take the form:

$$[H_{II}^{\{0\}}] = \frac{P^2}{2M} + \frac{1}{2}M\Omega^2 Q^2 = H^\circ \tag{49}$$

$$[H_{II}^{\{1\}}] = \frac{P^2}{2M} + \frac{1}{2}M\Omega^2 Q^2 + \hbar b Q + \hbar\omega^\circ - \alpha^{\circ 2}\hbar\Omega \tag{50}$$

3. *From Representation {II} to Representation {III}*

In order to remove the driven term in the effective Hamiltonian $H_{II}^{\{1\}}$ of the slow mode given by Eq. (47), without affecting the diagonal effective Hamiltonian $H_{II}^{\{0\}}$

given by Eq. (46), we will now perform a new general canonical transformation $\{III\}$ involving selective unitary transformations on the different operators $B_{II}^{\{k\}}$ that are a function of the effective Hamiltonians $H_{II}^{\{k\}}$ characterized by $\{k\} = 0, 1$ that is,

$$[B_{III}^{\{k\}}] = [A(k\alpha^{\circ})][B_{II}^{\{k\}}][A(k\alpha^{\circ})]^{-1} \tag{51}$$

These transformations are realized, using the translation operators (see Appendix E):

$$[A(k\alpha^{\circ})] = (e^{k\alpha^{\circ}[a^{\dagger} - a]}) \tag{52}$$

In this new representation $\{III\}$ given by Eq. (C.1), the effective Hamiltonians of the slow mode (46) corresponding to the ground state $|\{0\}\rangle$ of the fast mode is unmodified. On the other hand, the slow mode effective Hamiltonian (47), related to the situation where the fast mode has jumped into its first excited state $|\{1\}\rangle$, becomes diagonal. The $\{II\}$ and $\{III\}$ representations of these Hamiltonians are given in Table VI and the details of the calculations are reported in Appendix C.

TABLE VI
The H-Bond Bridge Operators Within Representations $\{II\}$ and $\{III\}$

Quantum Representation $\{II\}$	Transformation $\{II\} \rightarrow \{III\}$	Quantum Representation $\{III\}$
$H_{II}^{\{0\}} = (a^{\dagger}a + \frac{1}{2})\hbar\Omega$	$H_{III}^{\{0\}} = 1H_{II}^{\{0\}}1$	$H_{III}^{\{0\}} = (a^{\dagger}a + \frac{1}{2})\hbar\Omega$
$H_{II}^{\{1\}} = \hbar\omega^{\circ} +$ $[a^{\dagger}a + \frac{1}{2} + \alpha^{\circ}(a^{\dagger} + a) - \alpha^{\circ}2]\hbar\Omega$	$H_{III}^{\{1\}} = e^{a^{\circ}(a^{\dagger} - a)} H_{II}^{\{0\}} e^{-a^{\circ}(a^{\dagger} - a)}$	$H_{III}^{\{1\}} = [a^{\dagger}a + \frac{1}{2} - 2\alpha^{\circ}2]\hbar\Omega$
$\rho_{II}^{\{0\}} = \varepsilon e^{-\tilde{\lambda}a^{\dagger}a}$	$\rho_{III}^{\{0\}} = 1\,\rho_{II}^{\{0\}}1$	$\rho_{III}^{\{0\}} = \varepsilon e^{-\tilde{\lambda}a^{\dagger}a}$
$\rho_{II}^{\{1\}} = \varepsilon e^{-\tilde{\lambda}a^{\dagger}a}$	$\rho_{III}^{\{1\}} = e^{\alpha^{\circ}(a^{\dagger} - a)} \rho_{II}^{\{1\}} e^{-\alpha^{\circ}(a^{\dagger} - a)}$	$\rho_{III}^{\{1\}} = \varepsilon e^{-\tilde{\lambda}(a^{\dagger} - \alpha^{\circ})(a - \alpha^{\circ})}$
$\langle Q \vert (n) \rangle = \chi_n(Q)$	$\vert (n)_{III}^{\{0\}} \rangle = 1 \vert (n) \rangle$	$\langle Q \vert (n)_{III}^{\{0\}} \rangle = \chi_n(Q)$
$\langle Q \vert (n) \rangle = \chi_n(Q)$	$\vert (n)_{III}^{\{1\}} \rangle = e^{\alpha^{\circ}(a^{\dagger} - a)} \vert (n) \rangle$	$\langle Q \vert (n)_{III}^{\{1\}} \rangle = \chi_n(Q - \alpha^{\circ}Q^{\circ\circ})$
$\langle Q \vert (0) \rangle = \chi_0(Q)$	$e^{\alpha^{\circ}(a^{\dagger} - a)} \vert (0) \rangle = \vert \alpha^{\circ} \rangle$	$\langle Q \vert (0)_{III}^{\{1\}} \rangle = e^{-\frac{\alpha^{\circ 2}}{2}} \sum \frac{\alpha^{\circ}n}{n!} \chi_n(Q)$

D. Representations $\{II\}$ and $\{III\}$

The passage from representation $\{II\}$ to $\{III\}$ does not affect the eigenstates of the slow mode harmonic Hamiltonian when the fast mode is in its ground state $|\{0\}\rangle$, but affects them when this mode has jumped on its first excited state $|\{1\}\rangle$. In this last situation there is

$$|(n)_{III}^{\{k\}}\rangle = \mathrm{A}(k\alpha^{\circ})|(n)\rangle \tag{53}$$

Then, it is shown in Appendix C that the wave function corresponding to the slow mode harmonic states are, respectively, given according to the fact that the fast mode is in its ground or in its first excited state, by Eqs. (C.12) and (C.13), that is,

$$\langle \mathrm{Q} \mid (n)_{III}^{\{1\}} \rangle = \chi_n(\mathrm{Q}-\alpha^{\circ}Q^{\circ\circ}) \qquad \langle \mathrm{Q} \mid (n)_{III}^{\{0\}} \rangle = \chi_n(\mathrm{Q}) \tag{54}$$

with

$$Q^{\circ\circ} = \sqrt{\frac{\hbar}{2M\Omega}}$$

Examination of Eqs. (54) shows that, within the quantum representation $\{III\}$, the excitation of the fast mode displaces the origin of the slow mode wave functions toward shorter lengths. That may be viewed as a translation of the slow mode potential, that is induced by the excitation of the fast mode. In order to visualize this potential displacement, it is suitable to consider the potential as Morse-like. That is depicted on the right-hand side of Fig. 2. Here, the left-hand side is devoted to the quantum representation $\{II\}$, where there is no potential

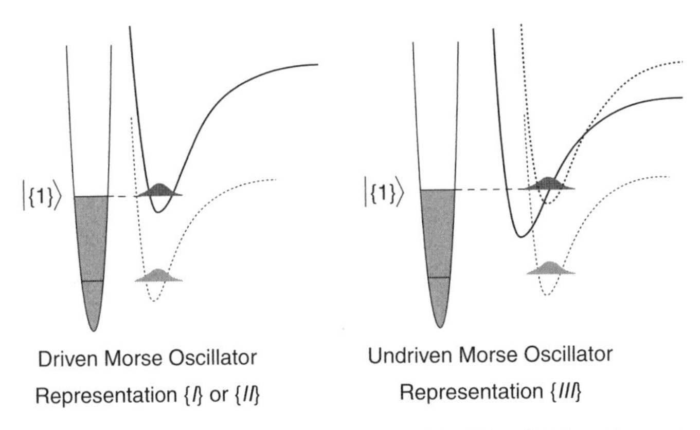

Figure 2. The equivalent quantum representations of the H-bond bridge after excitation of the fast mode.

displacement after excitation of the fast mode, and where the H-bond bridge Hamiltonian becomes driven

As a consequence of the translation of the origin of the slow mode potential induced by the excitation of the fast mode, there are in representation $\{III\}$, overlaps between the wave functions of the H-bond bridge corresponding, respectively, to the ground state of the fast mode and to its first excited state, that is,

$$\langle (m)_{III}^{\{0\}} \mid (n)_{III}^{\{1\}} \rangle = \int_{-\infty}^{\infty} \chi_m(\mathrm{Q})\chi_n(\mathrm{Q}-\alpha^\circ Q^{\circ\circ})d\mathrm{Q} = \{A_{mn}(\alpha^\circ)\}$$

These overlaps, which are matrix elements of the translation operator, are the well-known Franck–Condon factors:

$$\{A_{mn}(\alpha^\circ)\} = \langle (m)|[e^{\alpha^\circ[\mathrm{a}^\dagger - \mathrm{a}]}]|(n)\rangle \tag{55}$$

Appendix N shows that within quantum representation $\{III\}$ and according to Eq. (N12), when the fast mode is in its first excited state, the ground state of the slow mode is a coherent state, that is, an eigenstate of the lowering operator:

$$|(0)_{III}^{\{1\}}\rangle = |\alpha^\circ\rangle \qquad \text{with} \qquad \mathrm{a}|\alpha^\circ\rangle = \alpha^\circ|\alpha^\circ\rangle \tag{56}$$

Appendix N also shows that the expansion of a coherent state on the eigenstates of the number occupation operator is given by Eq. (N11), so that Eq. (56) leads to

$$|(0)_{III}^{\{1\}}\rangle = \left(e^{-\frac{\alpha^{\circ 2}}{2}}\right) \sum \frac{\alpha^{\circ n}}{\sqrt{n!}} |(n)\rangle \tag{57}$$

where keep in mind that, because of Eq. (53), there is the equivalence

$$|(n)\rangle \equiv |(n)_{III}^{\{0\}}\rangle$$

The advantage of passing from quantum representations $\{II\}$ to $\{III\}$ is that the Hamiltonian of the H-bond bridge, which is driven in representation $\{II\}$ when the fast mode is in the state $|\{1\}\rangle$, looses its driven property when passing in representation $\{II\}$. According to Eq. (57), the ground state of the H-bond bridge corresponding to the ground-state situation of the fast mode $|\{0\}\rangle$ becomes in representation $\{III\}$ a coherent state, after excitation of the high frequency mode to the state $|\{1\}\rangle$. That is illustrated in Fig. 3.

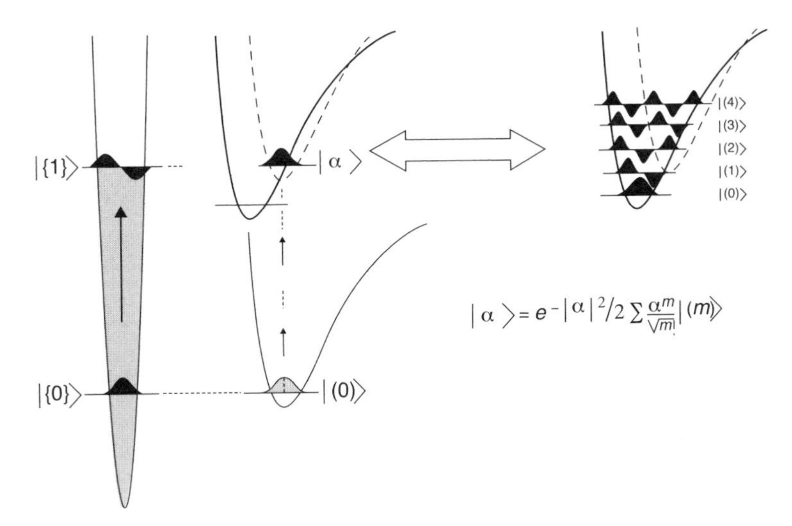

Figure 3. In representation $\{III\}$, the ground state of the H-bond bridge corresponding to the ground-state situation of the fast mode $|\{0\}\rangle$ becomes a coherent state after excitation to $|\{1\}\rangle$ of the high frequency mode.

In the same way, when the fast mode is in its ground state, the Boltzmann density operator corresponding to the H-bond bridge, viewed as a quantum harmonic oscillator, is the same in both quantum representations $\{II\}$ and $\{III\}$ (see right- and left-hand side columns of Table VI). On the other hand, when passing from representation $\{II\}$ to representation $\{III\}$, the density operator of the slow mode corresponding to the situation where the fast mode is in its excited state, $|\{1\}\rangle$ is changed according to the canonical transformation (C.1), which leads to the expression given on the right-hand side column of Table VI.

E. The Adiabatic ACF and the Corresponding SD

1. *The ACF and the Corresponding SD Within Representation* $\{II\}$

According to Eq. (10), the ACF of the non-Hermitean absorption transition moment operator of the high frequency mode is

$$[G^{\circ}_{\rm Mono}(t)]^{\circ} = {\rm tr}\{\rho_{\rm Tot}\ \mu(0)^{\dagger}\ \mu(t)\} \tag{58}$$

The IR transition non-Hermitean operator at initial time is

$$\mu(0) = \mu_0|\{1\}\rangle\langle\{0\}|$$

Again, within the adiabatic approximation, ρ_{Tot} appearing in this ACF is

$$\rho_{Tot} \sim \rho_{Adiab} = \frac{1}{Z_{Adiab}}(e^{-\beta H_{Adiab}})$$

where Z_{Adiab} is the partition function given by the following trace over the base spanned by the adiabatic Hamiltonian H_{Adiab}:

$$Z_{Adiab} = \text{tr}_{Adiab}[(e^{-\beta H_{Adiab}})]$$

and

$$\beta = 1/k_B T$$

Within the same approximation, the transition moment operator at time t is

$$\mu(t) = \mu_0[e^{iH_{Adiab}t/\hbar}]|\{1\}\rangle\langle\{0\}|\ [e^{-iH_{Adiab}t/\hbar}]$$

As a consequence, the ACF (58) takes the form:

$$[G^{\circ}_{Adiab}(t)]^{\circ} = \left(\frac{\mu_0^2}{Z_{Adiab}}\right)\text{tr}_{Adiab}[(e^{-\beta H_{Adiab}})|\{0\}\rangle\langle\{1\}|[e^{iH_{Adiab}t/\hbar}]|\{1\}\rangle\langle\ \{0\}|\ [e^{-iH_{Adiab}t/\hbar}]] \quad (59)$$

where tr holds for a trace to be performed *a priori* on both the fast and slow modes subspaces. Moreover, within representation $\{II\}$ and owing to Eq. (48), this last equation takes the form:

$$[G^{\circ}_{II}(t)]^{\circ} = \left(\frac{\mu_0^2}{Z_{II}^{\{0\}}}\right)\text{tr}_{Slow}[(e^{-\beta H_{II}^{\{0\}}})|\{0\}\rangle\langle\ \{1\}|[\ e^{iH_{II}^{\{1\}}t/\hbar}]|\{1\}\rangle\langle\ \{0\}|\ [e^{-iH_{II}^{\{0\}}t/\hbar}]]$$

where tr_{Slow} is a trace to be performed only on the slow mode subspace, because of the structure of the right-hand side term, whereas $Z_{II}^{\{0\}}$ is the following partition function:

$$Z_{II}^{\{0\}} = \text{tr}_{Slow}[(e^{-\beta H_{II}^{\{0\}}})]$$

Besides, since the Hamiltonians $H_{II}^{\{k\}}$ do not work on the space to which the kets $|\{k\}\rangle$ belong, they commute with them so that this last equation reduces, after using the orthonormality properties, to

$$[G^{\circ}_{II}(t)]^{\circ} = \left(\frac{\mu_0^2}{Z_{II}^{\{0\}}}\right)\text{tr}_{Slow}[(e^{-\beta H_{II}^{\{0\}}})[e^{iH_{II}^{\{1\}}t/\hbar}][e^{-iH_{II}^{\{0\}}t/\hbar}]] \quad (60)$$

Then, in view of Eqs. (49) and (50), Eq. (60) leads to the equations

$$[G^{\circ}_{II}(t)]^{\circ} = e^{i\omega^{\circ} t}(e^{-i\alpha^{\circ 2}\Omega t})\left(\frac{\mu_0^2}{Z_{II}^{\{0\}}}\right)\mathrm{tr}_{\mathrm{Slow}}[(e^{-\beta H_{II}^{\{0\}}})\ [U^{\circ}{}_{II}^{\{1\}}(t)]^{-1}[U^{\circ}{}_{II}^{\{0\}}(t)]] \quad (61)$$

$$[U^{\circ}{}_{II}^{\{1\}}(t)] = \left(e^{-i\left(\frac{P^2}{2M}+\frac{1}{2}M\Omega^2Q^2+\hbar bQ\right)t/\hbar}\right) \quad \text{and} \quad [U^{\circ}{}_{II}^{\{0\}}(t)] = \left(e^{-i\left(\frac{P^2}{2M}+\frac{1}{2}M\Omega^2Q^2\right)t/\hbar}\right)$$

Observe that Eq. (61) may be written within the interaction picture (IP) procedure according to

$$[G^{\circ}_{II}(t)]^{\circ} = e^{i\omega^{\circ} t}(e^{-i\alpha^{\circ 2}\Omega t})\left(\frac{\mu_0^2}{Z_{II}^{\{0\}}}\right)\mathrm{tr}_{\mathrm{Slow}}[(e^{-\beta H_{II}^{\{0\}}})[U^{\circ}{}_{II}^{\{1\}}(t)^{\mathrm{IP}}]^{-1}] \quad (62)$$

$$[U^{\circ}{}_{II}^{\{1\}}(t)] = [U^{\circ}{}_{II}^{\{0\}}(t)][U^{\circ}{}_{II}^{\{1\}}(t)^{\mathrm{IP}}]$$

In Appendix G, it is shown that the product $[U^{\circ}{}_{II}^{\{1\}}(t)]^{-1}[U^{\circ}{}_{II}^{\{0\}}(t)]$ involved in Eq. (61) transforms into Eq. (G.11), so that ACF (61) reduces to

$$[G^{\circ}{}_{II}(t)]^{\circ} = e^{i\omega^{\circ} t}(e^{-i\alpha^{\circ 2}\Omega t})\left(\frac{\mu_0^2}{Z_{II}^{\{0\}}}\right)\mathrm{tr}_{\mathrm{Slow}}\left[(e^{-\beta H_{II}^{\{0\}}})\widehat{P}\left[e^{ib\int_0^t Q(t')^{\mathrm{IP}}dt'}\right]\right] \quad (63)$$

Here, $\widehat{P}$ is the Dyson time-ordering operator and $Q(t)^{\mathrm{IP}}$ is given by the Heisenberg transformation Eq. (G.6), that is,

$$Q(t)^{\mathrm{IP}} = [U^{\circ}{}_{II}^{\{0\}}(t)]\ Q\ [U^{\circ}{}_{II}^{\{0\}}(t)]^{-1}$$

Note that Eq. (63) may be written formally as:

$$[G^{\circ}{}_{II}(t)]^{\circ} = (\mu_0^2)\ e^{i\omega^{\circ} t}(e^{-i\alpha^{\circ 2}\Omega t})\left\langle \widehat{P}\left[e^{ib\int_0^t Q(t')^{\mathrm{IP}}dt'}\right]\right\rangle_{\mathrm{Slow}} \quad (64)$$

where the meaning of the notation $\langle\cdots\rangle_{\mathrm{Slow}}$ is that of a thermal average over the Boltzmann density operator. This last expression (64) evokes the starting ACF used by Robertson and Yarwood in their semiclassical approach of the line shapes of H-bonded species [46].

Now, to go further, we will pass to the Boson representation given in Table V. Within this representation, the IP H-bond bridge coordinate appearing in Eq. (63) takes the form:

$$\mathrm{Q}(t)^{\mathrm{IP}} = Q^{\circ\circ}(e^{i\mathrm{a}^{\dagger}\mathrm{a}\Omega t})[\mathrm{a}^{\dagger} + \mathrm{a}](e^{-i\mathrm{a}^{\dagger}\mathrm{a}\Omega t})$$

where

$$Q^{\circ\circ} = \sqrt{\frac{\hbar}{2M\Omega}}$$

After performing the canonical transformation, this expression becomes

$$\mathrm{Q}(t)^{\mathrm{IP}} = Q^{\circ\circ}[\mathrm{a}^{\dagger}e^{i\Omega t} + \mathrm{a}e^{-i\Omega t}]$$

Besides, within the Bosons representation, the Boltzmann density operator appearing in Eq. (63), is given in Table VI.

Consequently, within the Bosons representation, ACF (63) appears to be given by

$$[\mathrm{G}^{\circ}{}_{II}(t)]^{\circ} = (\mu_0^2)e^{i\omega^{\circ}t}(e^{-i\alpha^{\circ 2}\Omega t})\tilde{\varepsilon}\mathrm{tr}_{\mathrm{Slow}}\left[(e^{-\tilde{\lambda}[\mathrm{a}^{\dagger}\mathrm{a}+1/2]})\widehat{\mathrm{P}}\left[e^{i\alpha^{\circ}\int_0^t[\mathrm{a}^{\dagger}e^{i\Omega t'}+\mathrm{a}e^{-i\Omega t'}]dt'}\right]\right] \quad (65)$$

with α° given by Eq. (41) and $\tilde{\lambda}$ given by

$$\tilde{\lambda} = \frac{\hbar\Omega}{k_{\mathrm{B}}T}$$

where $\tilde{\varepsilon}$ is the inverse of the partition function, that is,

$$\frac{1}{\tilde{\varepsilon}} = \mathrm{tr}_{\mathrm{Slow}}[e^{-\tilde{\lambda}[\mathrm{a}^{\dagger}\mathrm{a}+1/2]}]$$

or

$$\tilde{\varepsilon} = e^{\tilde{\lambda}/2} - e^{-\tilde{\lambda}/2}$$

It is shown in Appendix D, that the inverse IP time evolution operator appearing in Eqs. (62) and (65) and governing the dynamics of the driven harmonic oscillator is given by Eq. (D.25), that is,

$$\widehat{\mathrm{P}}\left[e^{i\alpha^{\circ}\int_0^t[\mathrm{a}^{\dagger}e^{i\Omega t'}+\mathrm{a}e^{-i\Omega t'}]dt'}\right] = (e^{-i\alpha^{\circ 2}\Omega t})\left(e^{i\alpha^{\circ 2}\sin\Omega t}\right)\left(e^{\Phi^{\circ}_{II}(t)^{*}\mathrm{a}^{\dagger}-\Phi^{\circ}_{II}(t)\mathrm{a}}\right) \quad (66)$$

Here, $\Phi^{\circ}_{II}(t)$ is a time-dependent complex scalar that is given by

$$\Phi^{\circ}_{II}(t) = \alpha^{\circ}[e^{-i\Omega t} - 1] \tag{67}$$

Thus, owing to Eq. (66) and after neglecting the zero-point energy involved in the Boltzmann operator, which is here unphysical, the ACF (65) takes the form:

$$[\mathrm{G}^{\circ}{}_{II}(t)]^{\circ} = (\mu_0^2)e^{i\omega^{\circ}t}(e^{-i\alpha^{\circ 2}\Omega t})(e^{-i\alpha^{\circ 2}\Omega t})(e^{i\alpha^{\circ 2}\sin\Omega t})\varepsilon\,\mathrm{tr}_{\mathrm{Slow}}[(e^{-\tilde{\lambda}\mathrm{a}^{\dagger}\mathrm{a}})[e^{\Phi^{\circ}_{II}(t)^{*}\mathrm{a}^{\dagger}-\Phi^{\circ}_{II}(t)\mathrm{a}}]] \tag{68}$$

$$\frac{1}{\varepsilon} = \mathrm{tr}_{\mathrm{Slow}}[e^{-\tilde{\lambda}\mathrm{a}^{\dagger}\mathrm{a}}] = (1 - e^{-\tilde{\lambda}})^{-1}$$

Again, performing the trace involved in the ACF (68) with the aid of the Bloch theorem and according to Eqs. (H.1) and (H.5), this ACF equation becomes

$$[\mathrm{G}^{\circ}_{II}(t)]^{\circ} = (\mu_0^2)e^{i\omega^{\circ}t}(e^{-i2\alpha^{\circ 2}\Omega t})(e^{i\alpha^{\circ 2}\sin\Omega t})[e^{2\alpha^{\circ 2}(\langle n\rangle+1/2)(\cos\Omega t-1)}] \tag{69}$$

$$\text{with } \langle \mathrm{n}\rangle = \frac{1}{\mathrm{e}^{\tilde{\lambda}} - 1}$$

Besides, Rösch and Ratner [47] treated the quantum direct damping [55], in which the excited of the fast mode relaxes directly toward its ground state via dipole–dipole interaction with the surrounding. They obtained an ACF that is the ACF (69) times an exponential decay $\exp\{-\gamma^{\circ}t\}$ in which γ° is the direct relaxation parameter. This leads us to write

$$[\mathrm{G}^{\circ}{}_{II}(t)] = (\mu_0^2)e^{i\omega^{\circ}t}(e^{-i2\alpha^{\circ 2}\Omega t})(e^{i\alpha^{\circ 2}\sin\Omega t})[e^{2\alpha^{\circ 2}(\langle n\rangle+1/2)(\cos\Omega t-1)}](e^{-\gamma^{\circ}t}) \tag{70}$$

For this parameter, they have obtained an analytical expression that is somewhat formal and that we do not reproduce here, since later we will consider it as an empirical term.

Then, the spectral density may be obtained by Fourier transform of ACF (70), that is,

$$[\mathrm{I}^{\circ}_{II}(\omega)] = (\mu_0^2)2\,\mathrm{Re}\int_0^{\infty} e^{i\omega^{\circ}t}(e^{-i2\alpha^{\circ 2}\Omega t})(e^{i\alpha^{\circ 2}\sin\Omega t})[e^{2\alpha^{\circ 2}(\langle n\rangle+1/2)(\cos\Omega t-1)}](e^{-\gamma^{\circ}t})e^{-i\omega t}\,dt \tag{71}$$

2. *Accuracy [56] of the SD (71) as Compared to the Nonadiabatic One (34)*

It may be of interest to verify if the analytical SD (71) which has been obtained within the adiabatic approximation, may reproduce satisfactorily that (34) which

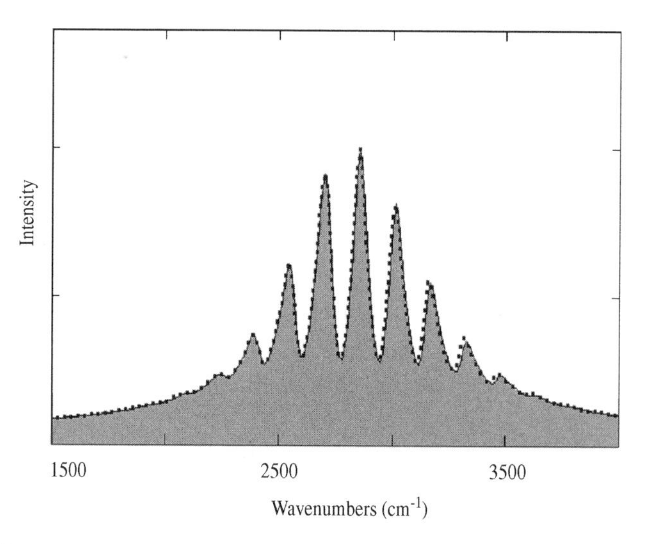

Figure 4. Comparison between the adiabatic and the reference SDs. $\alpha^{\circ} = 1.00$, $T = 300\,\text{K}$, $\gamma^{\circ} = 0.20\Omega$, $\omega^{\circ} = 3000\,\text{cm}^{-1}$, $\Omega = 150\,\text{cm}^{-1}$. Matrix dimension in the adiabatic method: 30.

is working beyond this approximation. Of course, since SD (71) is dealing with the harmonic approximation for the H-bond bridge potential, it is necessary to work in the same way for the reference SD (34). Thus, it is necessary to use for it Eq. (20) for the Hamiltonian in place of Eq. (19). As a consequence, the eigenvalues and the eigenvectors of the full-Hamiltonian given, respectively, by Eqs. (31) and (33) and appearing in the SD (34), are those of the Hamiltonian (19).

Figure 4 gives a comparison between the approached adiabatic SD (71) and the reference one (34) for some basic parameters that may be expected for weak H-bonds. Its examination shows that the adiabatic SD (dots) fits very accurately with the reference nonadiabatic one (grayed).

This good agreement may be considered as an *a posteriori* confirmation of the adiabatic approximation procedure exposed in Appendix B.

3. *Equivalence between the SD (71) and That of Maréchal and Witkowski*

Now, we will show that the SD (71) that is the Fourier transform of the ACF (68) via Eq. (69), is equivalent to that obtained by Maréchal and Witkowski in their pioneering work [18]. For this purpose, look at the ACF (68). It is possible to write the translation operator appearing in it, which has to be averaged on the Boltzmann density operator, as a product of two translation operators. According to Eq. (F.4), it is possible to write

$$(e^{i\alpha^{\circ 2}\sin\Omega t})[e^{\Phi^{\circ}_{II}(t)^{*}\mathrm{a}^{\dagger}-\Phi^{\circ}_{II}(t)\mathrm{a}}] = [e^{-\alpha^{\circ}\mathrm{a}^{\dagger}+\alpha^{\circ}\mathrm{a}}][e^{\Phi^{\circ}_{III}(t)^{*}\mathrm{a}^{\dagger}-\Phi^{\circ}_{III}(t)\mathrm{a}}] \tag{72}$$

Here, $\Phi^\circ_{II}(t)$ and $\Phi^\circ_{III}(t)$ are, respectively, given by Eq. (67) and (F.5), that is,

$$\Phi^\circ_{II}(t) = \alpha^\circ[e^{-i\Omega t} - 1] \qquad \text{and} \qquad \Phi^\circ_{III}(t) = \alpha^\circ e^{-i\Omega t}$$

Thus, the ACF that was given by (68) in representation $\{II\}$, becomes in this new representation, which is $\{III\}$:

$$[\mathrm{G}^\circ_{III}(t)]^\circ = (\mu_0^2)\ e^{i\omega^\circ t}\ (e^{-i2\alpha^{\circ 2}\Omega t})\varepsilon\, \mathrm{tr}_{\mathrm{Slow}}[[e^{-\tilde{\lambda}\mathrm{a}^\dagger\mathrm{a}}][e^{-\alpha^\circ\mathrm{a}^\dagger+\alpha^\circ\mathrm{a}}][e^{\Phi^\circ_{III}(t)^*\mathrm{a}^\dagger-\Phi^\circ_{III}(t)\mathrm{a}}]] \tag{73}$$

Now, observe that, in view of the expression of $\Phi^\circ_{III}(t)$, the last operator appearing on the right-hand side of Eq. (72), may be written as the result of the Heisenberg transformation

$$[e^{\Phi^\circ_{III}(t)^*\mathrm{a}^\dagger-\Phi^\circ_{III}(t)\mathrm{a}}] = (e^{i\mathrm{a}^\dagger\mathrm{a}\Omega t})[e^{\alpha^\circ\mathrm{a}^\dagger-\alpha^\circ\mathrm{a}}](e^{-i\mathrm{a}^\dagger\mathrm{a}\Omega t})$$

As a consequence, the ACF (73) gives

$$[\mathrm{G}^\circ_{III}(t)]^\circ = (\mu_0^2)\ e^{i\omega^\circ t}\ (e^{-i2\alpha^{\circ 2}\Omega t})$$
$$\varepsilon\, \mathrm{tr}_{\mathrm{Slow}}[(e^{-\tilde{\lambda}\mathrm{a}^\dagger\mathrm{a}})[\mathrm{A}(\alpha^\circ)]^{-1}[(e^{i\mathrm{a}^\dagger\mathrm{a}\Omega t})[\mathrm{A}(\alpha^\circ)](e^{-i\mathrm{a}^\dagger\mathrm{a}\Omega t})]] \tag{74}$$

with

$$\mathrm{A}(\alpha^\circ) = e^{\alpha^\circ\mathrm{a}^\dagger-\alpha^\circ\mathrm{a}}$$

Next, the trace involved in Eq. (74) may be performed over the basis built up from the eigenstates of $\mathrm{a}^\dagger\mathrm{a}$, that is,

$$\mathrm{a}^\dagger\mathrm{a}\mid(n)\rangle = n|(n)\rangle \qquad \text{with} \qquad \langle(m)|(n)\rangle = \delta_{m,n} \tag{75}$$
$$\sum|(n)\rangle\langle(n)| = \widehat{1} \tag{76}$$

Then, performing the trace and inserting the closeness relation (76) between the two translation operators, Eq. (74) gives

$$[\mathrm{G}^\circ_{III}(t)]^\circ = (\mu_0^2)e^{i\omega^\circ t}\ (e^{-i2\alpha^{\circ 2}\Omega t})\varepsilon\sum_m\sum_n$$
$$\langle(n)|(e^{-\tilde{\lambda}\mathrm{a}^\dagger\mathrm{a}})[[\mathrm{A}(\alpha^\circ)]^{-1}|(m)\rangle\langle(m)|(e^{i\mathrm{a}^\dagger\mathrm{a}\Omega t})[\mathrm{A}(\alpha^\circ)](e^{-i\mathrm{a}^\dagger\mathrm{a}\Omega t})]]|(n)\rangle \tag{77}$$

Again, owing to the fact that $|(n)\rangle$ are the eigenstates of $a^{\dagger}a$, via Eq. (75), Eq. (77) leads to

$$[G^{\circ}_{III}(t)]^{\circ} = (\mu_0^2)\, e^{i\omega^{\circ}t}\, (e^{-i2\alpha^{\circ 2}\Omega t})\varepsilon \sum_m \sum_n (e^{-\tilde{\lambda} n})$$
$$\langle (n)|[A(\alpha^{\circ})]^{-1}|(m)\rangle\langle (m)|(e^{im\Omega t})[A(\alpha^{\circ})](e^{-in\Omega t})|(n)\rangle \quad (78)$$

Moreover, this last expression of the ACF reads

$$[G^{\circ}_{III}(t)]^{\circ} = (\mu_0^2)e^{i\omega^{\circ}t}\, (e^{-i2\alpha^{\circ 2}\Omega t})\varepsilon \sum_m \sum_n (e^{-\tilde{\lambda} n})[e^{i(m-n)\Omega t}]|A_{mn}(\alpha^{\circ})|^2 \quad (79)$$

with

$$A_{mn}(\alpha^{\circ}) = \langle (m)|[e^{\alpha^{\circ}a^{\dagger} - \alpha^{\circ}a}]|(n)\rangle$$

Here, $A_{mn}(\alpha^{\circ})$ are the Franck–Condon factors, the expression of which are given by Eq. (C.15).

At last, the line shape is the Fourier transform of the ACF (79), that is,

$$[I^{\circ}_{III}(\omega)]^{\circ} = (\mu_0^2)\, \varepsilon \sum_m \sum_n (e^{-n\hbar\Omega/k_B T})|A_{mn}(\alpha^{\circ})|^2 \delta[\omega - [\omega^{\circ} - (n-m) - 2\alpha^{\circ 2}]\Omega] \quad (80)$$

In Eq. (80), one may recognize the Franck–Condon progression appearing in the model of Maréchal and Witkowski [18] dealing with weak H-bonds. Of course, since Eq. (69) and Eq. (79) are two different, but equivalent, expressions of the ACF, the two expressions of the line shape given by Eqs. (80) and (71), which are, respectively, the Fourier transforms of Eq. (69) and Eq. (79) are also equivalent. The advantage of the expression (80) with respect to that (71), is it allows a pictorial representation of the line shape. This is performed in Fig. 5 for absolute zero temperature that implies $e^{-\tilde{\lambda} m} \rightarrow \delta_{0,m}$ and for which the SD (80) reduces to

$$[I^{\circ}_{III}(\omega)]^{\circ} = (\mu_0^2)\, \varepsilon \sum_n |A_{\text{on}}(\alpha^{\circ})|^2 \delta[\omega - [\omega^{\circ} - n - 2\alpha^{\circ 2}]\Omega] \qquad \text{for } T = 0$$

Note that the equivalence between Eqs. (69) and (79) leads us to write the following result, which will be used later:

$$(e^{i\alpha^{\circ 2}\sin\Omega t})[e^{\alpha^{\circ 2}(\langle n\rangle + 1/2)(2\cos\Omega t - 1)}] = \sum_m \sum_n (e^{-\tilde{\lambda} n})(e^{i(m-n)\Omega t})|A_{mn}(\alpha^{\circ})|^2 \quad (81)$$

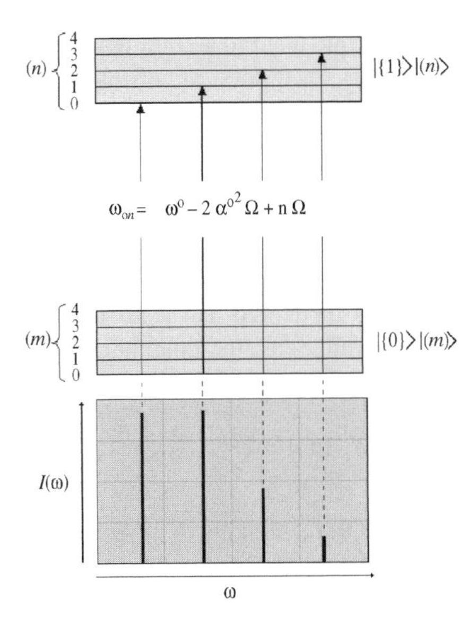

Figure 5. Spectral analysis at $T = 0\,\mathrm{K}$ in the absence of indirect damping $\omega^\circ = 3000\,\mathrm{cm}^{-1}$, $\Omega = 100\,\mathrm{cm}^{-1}$, $\alpha^\circ = 1$, $\gamma^\circ = 0.025\,\Omega$, $\gamma = 0$.

At last, it may be of interest to observe that, according to Eq. (K.2), the expression (80) of the SD may be reproduced using Eqs. (1–3) as starting equations in place of the special ones (4), (5), and (7).

Next, if the direct damping is incorporated, in the spirit of the Rösch and Ratner model, as seen at the end of Section II.E.1, then, owing to the equivalence between Eqs. (69) and (79), the Fourier transform of ACF (70) now involving direct damping, appears to be given by

$$[\mathrm{I}^\circ_{III}(\omega)] = \langle \mu_0^2 \rangle\ \varepsilon \sum_m \sum_n (e^{-n\hbar\Omega/k_\mathrm{B}T}) |\mathrm{A}_{mn}(\alpha^\circ)|^2 \times \frac{\gamma^\circ}{[\omega - (\omega^\circ - (n-m)\Omega - 2\alpha^{\circ 2}\Omega)]^2 + \gamma^{\circ 2}} \tag{82}$$

As seen, the spectral density involving direct damping is the double sum over m and n of Lorentzians centered on $\omega = \omega^\circ - (n-m)\Omega - 2\alpha^{\circ 2}\Omega$ and having the same half-width γ°, but different intensities, given by $e^{-\tilde{\lambda} n}|\mathrm{A}_{mn}(\alpha^\circ)|^2$.

4. *Importance of the Dyson Ordering Operator*

In the undamped situation, we showed that, the adiabatic ACF may be given by Eq. (64). Now, suppose that in this equation the Dyson [57] operator reduces to unity, that is, $\widehat{\mathrm{P}} \to 1$. This is equivalent to assuming that in contrast to Eq. (G.9), the IP time-dependent $\mathrm{Q}(t)$ operators at different times commute; this consequence implies that the Q coordinate commutes with its conjugate momentum P, that is, the Q and P operators become classical quantities, so that the H-bond bridge must be considered as classical.

Besides, the approximation according to which $\widehat{\mathrm{P}} \to 1$, must likewise lead us, to remove the shift $-\alpha^{\circ 2}\Omega/2$, which is a quantum consequence of removing the driven term involved in the effective Hamiltonian (40) corresponding to the ground state $|\{0\}\rangle$ of the fast mode and that leads to Eq. (50).

Then, within these approximations, the quantum ACF (64) transforms into

$$[\mathrm{G}^{\circ}{}_{II}(t)]^{\circ}_{\mathrm{Class}} = (\mu_0^2)\ e^{i\omega^{\circ}t}\left\langle \left[\ e^{ib\int_0^t \mathrm{Q}(t')^{\mathrm{IP}}dt'}\right]\right\rangle^{\mathrm{Class}}_{\mathrm{Slow}} \tag{83}$$

where, as for the ACF (64), $\mathrm{Q}(t)^{\mathrm{IP}}$ remains to be given by

$$\mathrm{Q}(t)^{\mathrm{IP}} = Q^{\circ\circ}[\mathrm{a}^{\dagger}e^{i\Omega t} + \mathrm{a}e^{-i\Omega t}]$$

In view of Eq. (41), one may write

$$\left[e^{ib\int_0^t \mathrm{Q}(t')^{\mathrm{IP}}dt'}\right] = \left[e^{i\alpha^{\circ}\Omega\int_0^t[\mathrm{a}^{\dagger}e^{i\Omega t'}+\mathrm{a}e^{-i\Omega t'}]dt'}\right]$$

Integration of the right-hand side of this last equation leads to

$$\left[e^{i\alpha^{\circ}\Omega\int_0^t[\mathrm{a}^{\dagger}e^{i\Omega t'}+\mathrm{a}e^{-i\Omega t'}]dt'}\right] = \left[e^{\alpha^{\circ}[\mathrm{a}^{\dagger}(e^{i\Omega t}-1)-\mathrm{a}(e^{-i\Omega t}-1)]}\right]$$

Moreover, in view of Eq. (67), the following result holds

$$[e^{\alpha^{\circ}[\mathrm{a}^{\dagger}(e^{i\Omega t}-1)-\mathrm{a}(e^{-i\Omega t}-1)]}] = [e^{\Phi^{\circ}_{II}(t)^{*}\mathrm{a}^{\dagger}-\Phi^{\circ}_{II}(t)\mathrm{a}}]$$

Consequently, the ACF (83) becomes

$$[\mathrm{G}^{\circ}{}_{II}(t)]^{\circ}_{\mathrm{Class}} = (\mu_0^2)e^{i\omega^{\circ}t}\varepsilon\,\mathrm{tr}_{\mathrm{Slow}}[(e^{-\tilde{\lambda}\mathrm{a}^{\dagger}\mathrm{a}})[e^{\Phi^{\circ}_{II}(t)^{*}\mathrm{a}^{\dagger}-\Phi^{\circ}_{II}(t)\mathrm{a}}]] \tag{84}$$

Again, performing the trace with the aid of the Bloch theorem, and passing as from Eqs. (68) to (69), gives

$$[\mathrm{G}^{\circ}{}_{II}(t)]^{\circ}_{\mathrm{Class}} = (\mu_0^2)\ e^{i\omega^{\circ}t}[e^{\alpha^{\circ 2}(\langle n\rangle+1/2)(2\cos\Omega t-1)}] \tag{85}$$

It appears that the passage from the ACF (69) to (85) allows us to write simultaneously:

$$\widehat{\mathrm{P}} \rightarrow 1,\ 2\alpha^{\circ 2}\Omega \rightarrow 0 \qquad \text{and} \qquad (e^{-i\alpha^{\circ 2}\Omega t})[e^{i\alpha^{\circ 2}\sin\Omega t}] \rightarrow 1$$

Note that if the Fourier transform of the phase of the ACF is simply at the origin of the angular shift $-2\alpha^{\circ 2}\Omega$ in the SD (80), on the other hand, that on the phase factor $\exp(i\alpha^{\circ 2}\sin\Omega t)$ generates the asymmetry of this SD (80), which may be verified as follows.

For this purpose, let us look at the Fourier analysis of this phase, that is,

$$\mathrm{I}_{\Phi}(\omega) = \sum \mathrm{I}(n\Omega)\ \delta[(\omega - \omega^{\circ} - n\Omega)]$$

with

$$\mathrm{I}(n\Omega) = \mathrm{Re}\left[\int_{-T/2}^{T/2} [e^{i\alpha^{\circ 2}\sin\Omega t}]e^{-in\Omega t}dt\right]$$

This Fourier analysis gives

$$\mathrm{Re}\left[\int_{-T/2}^{T/2} [e^{i\alpha^{\circ 2}\sin\Omega t}]e^{-in\Omega t}dt\right] \propto \frac{\alpha^{\circ}2n}{n!}$$

Now, observe with the aid of Eq. (C.15) that the right-hand side of this last result corresponds to Franck–Condon factors, that is,

$$\frac{\alpha^{\circ 2n}}{n!} \propto |A_{\mathrm{on}}(\alpha^{\circ})|^2$$

As a consequence, the Fourier analysis gives

$$\mathrm{I}_{\Phi}(\omega) \propto \sum |A_{\mathrm{on}}(\alpha^{\circ})|^2 \delta[(\omega - \omega^{\circ} - n\Omega)]$$

As verified by comparison with Eq. (80) and its zero temperature limit, this Fourier analysis is in agreement with the asymmetric Franck–Condon progression at zero temperature of the Maréchal and Witkowski model [18].

Finally, note that the equivalence of Eqs. (83) and (85), leads us to write the following result that will later appear of interest:

$$\left\langle \left[e^{ib\int_0^t \mathrm{Q}(t')^{\mathrm{IP}}dt'} \right] \right\rangle_{\mathrm{Slow}}^{\mathrm{Class}} = [e^{2\alpha^{\circ 2}(\langle n\rangle + 1/2)(\cos\Omega t - 1)}] \tag{86}$$

5. *The ACF Within Representation* $\{III\}$

Consider the expression (60) of the ACF:

$$[G^{\circ}_{III}(t)]^{\circ} = \left(\frac{\mu_0^2}{Z_{II}^{\{0\}}}\right) \mathrm{tr}_{\mathrm{Slow}}[(e^{-\beta H_{II}^{\{0\}}})[e^{iH_{II}^{\{1\}}t/\hbar}][e^{-iH_{II}^{\{0\}}t/\hbar}]] \tag{87}$$

Now, insert before and after the evolution operators the unity operator given by

$$[A(\alpha^{\circ})]^{-1}[A(\alpha^{\circ})] = \hat{1} \qquad \text{with} \qquad [A(\alpha^{\circ})] = e^{\alpha^{\circ}[a^{\dagger}-a]}$$

Then, we obtain

$$[G^{\circ}_{III}(t)]^{\circ} = \left(\frac{\mu_0^2}{Z_{II}^{\{0\}}}\right)$$
$$\mathrm{tr}_{\mathrm{Slow}}\left[(e^{-\beta H_{II}^{\{0\}}})\ [A(\alpha^{\circ})]^{-1}[A(\alpha^{\circ})][e^{iH_{II}^{\{1\}}t/\hbar}][A(\alpha^{\circ})]^{-1}[A(\alpha^{\circ})]\ [e^{-iH_{II}^{\{0\}}t/\hbar}]\right] \tag{88}$$

Of course, according to Eq. (C.4), the following equation holds

$$[A(\alpha^{\circ})][e^{iH_{II}^{\{1\}}t/\hbar}][A(\alpha^{\circ})]^{-1} = [e^{iH_{III}^{\{1\}}t/\hbar}]$$

Consequently, the ACF (88) becomes

$$[G^{\circ}_{III}(t)]^{\circ} = \left(\frac{\mu_0^2}{Z_{II}^{\{0\}}}\right) \mathrm{tr}_{\mathrm{Slow}}\left[(e^{-\beta H_{II}^{\{0\}}})\ [A(\alpha^{\circ})]^{-1}[e^{iH_{III}^{\{1\}}t/\hbar}][A(\alpha^{\circ})]\ [e^{-iH_{II}^{\{0\}}t/\hbar}]\right] \tag{89}$$

Then, owing to Eqs. (46), (C.3) and (C.7), the ACF takes the form:

$$[G^{\circ}_{III}(t)]^{\circ} = (\mu_0^2)\tilde{\varepsilon}$$
$$\mathrm{tr}_{\mathrm{Slow}}\left[(e^{-\beta(a^{\dagger}a+\frac{1}{2})\hbar\Omega})[A(\alpha^{\circ})]^{-1}\ [e^{i[(a^{\dagger}a+\frac{1}{2})\Omega-2\alpha^{\circ 2}\Omega+\omega^{\circ}]t}]\right.$$
$$\left.\times\ [A(\alpha^{\circ})]\ [e^{-i(a^{\dagger}a+\frac{1}{2})\Omega t}]\right]$$

where $\tilde{\varepsilon}$ is the inverse of the partition function of the H-bond bridge harmonic oscillator when its zero-point energy is taken into account. This last result may be simplified after canceling the zero-point energy of the slow mode involved in the Heisenberg transformation and also ignoring this same zero-point energy in the Boltzmann operator; this gives

$$[G^{\circ}_{III}(t)]^{\circ} = (\mu_0^2)e^{i\omega^{\circ}t}(e^{-i2\alpha^{\circ 2}\Omega t})\varepsilon\, \mathrm{tr}_{\mathrm{Slow}}\left[(e^{-\tilde{\lambda}a^{\dagger}a})\,[A(\alpha^{\circ})]^{-1}\right. \\ \left.\times\,[e^{ia^{\dagger}a\Omega t}]\,[A(\alpha^{\circ})]\,[e^{-ia^{\dagger}a\Omega t}]\right]$$

Again, insert a closeness relation over the eigenstates of $a^{\dagger}a$. Then make explicit the trace within the basis involving these eigenstates. As a consequence, after action of the Boltzmann and of the time evolution operators on these states, one obtains

$$[G^{\circ}_{III}(t)]^{\circ} = (\mu_0^2)e^{i\omega^{\circ}t}\,(e^{-i2\alpha^{\circ 2}\Omega t})\varepsilon \\ \sum_m\sum_n(e^{-\tilde{\lambda}n})\langle(n)|[A(\alpha^{\circ})]^{-1}|(m)\rangle\langle(m)|(e^{im\Omega t})[A(\alpha^{\circ})](e^{-in\Omega t})|(n)\rangle$$

where ε is the inverse of the partition function of the H-bond bridge harmonic oscillator when its zero-point energy is ignored. At last, using the fact that the translation operator is unitary, this ACF reduces to

$$[G^{\circ}_{III}(t)]^{\circ} = (\mu_0^2)\;e^{i\omega^{\circ}t}\;(e^{-i2\alpha^{\circ 2}\Omega t})\varepsilon\;\sum_m\sum_n(e^{-n\hbar\Omega/k_BT})(e^{i(m-n)\Omega t})|A_{mn}(\alpha^{\circ})|$$

which is the same as Eq. (79).

III. INCORPORATION OF THE INDIRECT DAMPING IN THE ADIABATIC REPRESENTATIONS $\{II\}$ AND $\{III\}$ [48–50]

A. Generalities

Now, we study, within the adiabatic approximation, the theoretical question of the action of quantum damping of the H-bond bridge on the IR line shape.

Previously, we found that within the adiabatic quantum representation $\{II\}$, the ACF of the dipole moment operator of the fast mode may be expressed in the absence of damping according to the simple form (64), that is,

$$[G^{\circ}{}_{II}(t)]^{\circ} = (\mu_0^2)e^{i\omega^{\circ}t}(e^{-i\alpha^{\circ}2\Omega t})\left\langle\widehat{P}\left[e^{ib\int_0^{t}Q(t')^{\mathrm{IP}}dt'}\right]\right\rangle_{\mathrm{Slow}}$$

In order to incorporate the indirect relaxation into this ACF, it is necessary to introduce a thermal bath and to interact with it. Then we perform an average on this bath, symbolized by $\langle\cdots\rangle_{\theta}$. This formally leads us to write in place of the precedent expression, the following equation:

$$[\mathrm{G}_{II}(t)]^{\circ} = (\mu_0^2)e^{i\omega^{\circ}t}(e^{-i\alpha^{\circ}2\Omega t})\left\langle\left\langle\widehat{\mathrm{P}}\left[e^{ib\int_0^t \mathrm{Q}(t')^{\mathrm{IP}}dt'}\right]\right\rangle_{\mathrm{Slow}}\right\rangle_{\theta} \tag{90}$$

Here, on the left-hand side of the first subscript "°" has disappeared since it was related to the absence of indirect damping, whereas in the last one, "°" remains since the direct damping is ignored for a time. Recall that $\widehat{\mathrm{P}}$ is the Dyson time-ordering operator, b is the anharmonic coupling between the slow and fast modes, and $\mathrm{Q}(t)^{\mathrm{IP}}$ is the IP time-dependent Q coordinate of the H-bond bridge. The question is to know what is the nature of $\mathrm{Q}(t')^{\mathrm{IP}}$ in the presence of its coupling with the thermal bath and also to precise the nature of the double average. For this purpose, it will be suitable to first glance at the dynamics of the H-bond bridge.

B. Physical Ideas Dealing With the H-Bond Bridge Dynamics

1. *Undamped H-Bond Bridge Dynamics Within the Adiabatic Approximation*

First, consider within the adiabatic approximation the dynamics of the H-bond bridge in the absence of damping.

a. H-Bond Bridge Dynamics When the Fast Mode Is in Its Ground State. When the fast mode is in its ground state. we showed previously that in both quantum representations $X = \{II\}$ and $\{III\}$, the effective Hamiltonian and the initial time-density operator, which are describing the H-bond bridge in this situation, obey (see Table VI)

$$[\mathrm{H}_X^{\{0\}}] = \left[\mathrm{a}^{\dagger}\mathrm{a} + \frac{1}{2}\right]\hbar\Omega \qquad \text{and} \qquad [\rho_X^{\circ\{0\}}(0)] = \varepsilon(e^{-\tilde{\lambda}\mathrm{a}^{\dagger}\mathrm{a}})$$

The dynamics of the density operator obeys the Liouville equation:

$$i\hbar\left(\frac{\partial\rho_X^{\circ\{0\}}(t)}{\partial t}\right) = [[\mathrm{H}_X^{\{0\}}], [\rho_X^{\circ\{0\}}(t)]]$$

Of course, the time derivative involved in this equation is zero, since the density operator and the corresponding Hamiltonian commute:

$$i\hbar\left(\frac{\partial\rho_X^{\circ\{0\}}(t)}{\partial t}\right) = \varepsilon[[\mathrm{a}^{\dagger}\mathrm{a}], (e^{-\tilde{\lambda}\mathrm{a}^{\dagger}\mathrm{a}})] = 0$$

Thus, the density operator , which is Boltzmann-like at initial time, remains so at any time, that is,

$$[\rho_X^{\circ\{0\}}(t)] = \varepsilon(e^{-\tilde{\lambda}\mathrm{a}^\dagger\mathrm{a}})]$$

Of course, the thermal average of the H-bond bridge coordinate obeys

$$\langle \mathrm{Q}_X^{\circ\{0\}} \rangle = \mathrm{tr}_{\mathrm{Slow}}[[\rho_X^{\circ\{0\}}(t)](\mathrm{a}^\dagger + \mathrm{a})]Q^{\circ\circ}$$

Consequently, in view of the above result, this average remains zero at any time in both representation according to

$$\langle \mathrm{Q}_{X\ (t)}^{\circ\{0\}} \rangle = \varepsilon\mathrm{tr}_{\mathrm{Slow}}[(e^{-\tilde{\lambda}\mathrm{a}^\dagger\mathrm{a}})(\mathrm{a}^\dagger + \mathrm{a})]Q^{\circ\circ} = 0$$

As a consequence of these equations, it appears that, when the fast mode is in its ground state, the slow mode does not involve any dynamics that might be the X quantum representation.

b. H-Bond Bridge Dynamics When the Fast Mode Is In Its First Excited State. Now, consider the situation where the fast mode was excited.

QUANTUM REPRESENTATION $\{II\}$. In quantum representation $\{II\}$ and from Table VI, the Hamiltonian and the Boltzmann density operator of the H-bond bridge are given, respectively, by the equations:

$$[\mathrm{H}_{II}^{\{1\}}] = \left[\mathrm{a}^\dagger\mathrm{a} + \frac{1}{2} + \alpha^\circ(\mathrm{a}^\dagger + \mathrm{a}) - \alpha^{\circ 2}\right]\hbar\Omega + \hbar\omega^\circ \quad \text{and} \quad \left[\rho_{II}^{\circ\{1\}}(0)\right] = \varepsilon e^{-\tilde{\lambda}\mathrm{a}^\dagger\mathrm{a}}$$

Then, the Liouville equation governing the density operator of the H-bond bridge, is

$$i\hbar\left(\frac{\partial\rho_{II}^{\circ\{1\}}(t)}{\partial t}\right) = \left[\left[\mathrm{H}_{II}^{\{1\}}\right], \left[\rho_{II}^{\circ\{1\}}(t)\right]\right] = \varepsilon[\mathrm{a}^\dagger\mathrm{a} + \alpha^\circ(\mathrm{a}^\dagger + \mathrm{a})), (e^{-\tilde{\lambda}\mathrm{a}^\dagger\mathrm{a}})]$$

Since the right-hand side commutator is not zero, this finding implies

$$i\hbar\left(\frac{\partial\rho_{II}^{\circ\{1\}}(t)}{\partial t}\right) \neq 0$$

As a consequence, the density operator of the H-bond bridge that was Boltzmann-like at the initial time, cannot remain the same as time evolves. Its time-dependent expression may be obtained with the aid of Eq. 17 that is,

$$[\rho_{II}^{\circ\{1\}}(t)] = \varepsilon\left[e^{-\tilde{\lambda}[a^\dagger - \Phi_{II}^\circ(t)^*][a - \Phi_{II}^\circ(t)]}\right] \quad \text{with} \quad \Phi_{II}^\circ(t) = \alpha^\circ[e^{-i\Omega t} - 1] \tag{91}$$

This operator is that of a coherent state at any temperature T (see Appendix N). Now, for the special situation of zero temperature, this density operator reduces to

$$[\rho_{II}^{\circ\{1\}}(t)] = (e^{-|\Phi_{II}^\circ(t)|^2})\sum \frac{|\Phi_{II}^\circ(t)|^{2n}}{n!}|(n)\rangle\langle(n)| \tag{92}$$

It must be emphasized that in this dynamics, all the states $|\Phi_{II}^\circ(t)\rangle$ evolve as a whole in the same way.

On the other hand, the average of the slow mode coordinate is given by

$$\langle Q^\circ{}_{II}^{\{1\}}(t)\rangle = \varepsilon\ \mathrm{tr}_{\mathrm{Slow}}[(e^{-\tilde{\lambda}[a^\dagger - \Phi^\circ{}_{II}(t)^*][a - \Phi^\circ{}_{II}(t)]})(a^\dagger + a)]Q^{\circ\circ} = 2\alpha^\circ Q^{\circ\circ}\sin\Omega t \tag{93}$$

Note that this dynamics is classical-like, as the coherent state properties density operator.

Quantum Representation $\{III\}$. Now, consider the quantum representation $\{III\}$. Becaused Table VI, the Hamiltonian and the density operator of the H-bond bridge obey the equations:

$$[H_{III}^{\{1\}}] = \left[a^\dagger a + \frac{1}{2} - 2\alpha^{\circ 2}\right]\hbar\Omega + \hbar\omega^\circ \qquad \text{and} \qquad [\rho^\circ{}_{III}^{\{1\}}(0)] = \varepsilon(e^{-\tilde{\lambda}[a^\dagger - \alpha^\circ][a - \alpha^\circ]})$$

Then, the Liouville equation governing the density operator of the H-bond bridge obeys

$$i\hbar\left(\frac{\partial\rho^\circ{}_{III}^{\{1\}}(t)}{\partial t}\right) = [H_{III}^{\{1\}}, \rho^\circ{}_{III}^{\{1\}}(t)] = \varepsilon[[a^\dagger a], (e^{-\tilde{\lambda}[a^\dagger - \alpha^\circ][a - \alpha^\circ]})] \neq 0$$

Consequently, this density operator must evolve with time. It is shown in Appendix I that, with the above boundary condition, the dynamics of this density operator is given by Eq. (I.15), that is,

$$[\rho^\circ{}_{III}^{\{1\}}(t)] = \varepsilon\,[e^{-\tilde{\lambda}[a^\dagger - \Phi^\circ{}_{III}(t)^*][a - \Phi^\circ{}_{III}(t)]}] \qquad \text{with} \qquad \Phi^\circ{}_{III}(t) = \alpha^\circ e^{-i\Omega t} \tag{94}$$

According to Appendix N, this density operator has the structure of a coherent state at any temperature. Again, for the special situation of zero temperature, this density operator reduces to

$$[\rho^{\circ\{1\}}_{III}(t)] = (e^{-|\Phi^{\circ}_{III}(t)|^2}) \sum \frac{|\Phi^{\circ}_{III}(t)|^{2n}}{n!} |(n)\rangle\langle(n)| \tag{95}$$

In this last expression, as for the quantum representation $\{II\}$, the structure of a quasiclassical coherent state minimizing the uncertainty Heisenberg relations may be recognized.

Besides, the average of the slow mode coordinate appears to be given by

$$\langle Q_{III}{}^{\circ\{1\}}(t)\rangle = \varepsilon\ \mathrm{tr}_{\mathrm{Slow}}[e^{-\tilde{\lambda}[a^{\dagger}-\Phi^{\circ}_{III}(t)^{*}][a-\Phi^{\circ}_{III}(t)]}(a^{\dagger}+a)]Q^{\circ\circ} = 2\alpha^{\circ}Q^{\circ\circ}\cos\Omega t \tag{96}$$

Note that, as above, this classical-like dynamics is not without relation to the quasiclassical coherent state properties of the density operator involved in this average.

c. Changes in the H-Bond Bridge Dynamics by Excitation of the Fast Mode. As a consequence of Eqs. (92) and (93), for representation $\{II\}$ and of Eqs. (95) and (96), for representation $\{III\}$, it appears that, when the fast mode has jumped into its first excited state, there is in both quantum representations a dynamics for the wave functions, the density operator, and the average of the position coordinate of the H-bond bridge: the average value of the coordinate oscillates back and forth, whereas the wave function and the density operator evolve in the same way as those of a time-dependent coherent state. This result may be emphasized, keeping in mind that when the fast mode is in its ground state, there is no dynamics for the H-bond bridge.

2. *Damped H-Bond Bridge Dynamics Within the Adiabatic Approximation*

Consider the H-bonded bridge, always within the adiabatic approximation, but now coupled to a thermal bath corresponding to the surroundings. Then, it is possible to infer, from the above conclusions in the situation without damping, some physical ideas on the irreversible action of the surroundings on the H-bond bridge dynamics:

1. When the fast mode is in its ground state, the medium, even present, may be assumed not to affect the H-bond bridge, so that the coupling with the thermal bath may be viewed as in a "off" situation.
2. When the H-bond bridge has jumped into its first excited state, the slow mode must be damped by the medium as a whole because it is described

by a coherent state in which all the eigenstates of the quantum harmonic oscillator are mixed in a coherent fashion.

Both these aspects will be the characteristics of the quantum indirect damping.

Later, it will appear that in the presence of indirect relaxation, Eqs. (91) and (94) Eqs. (92), (95), and Eqs. (93) and (96) respectively, transform into damped forms. Then, at any temperature the density operators in both quantum representations $\{II\}$ and $\{III\}$ appear to be those of coherent states at any temperature (see Appendix N):

$$[\rho_{II}^{\{1\}}(t)] = [e^{-\tilde{\lambda}[a^\dagger - \Phi_{II}(t))^*][a - \Phi_{II}(t)]}] \quad \text{and} \quad [\rho_{III}^{\{1\}}(t)] = [e^{-\tilde{\lambda}[a^\dagger - \Phi_{III}(t))^*][a - \Phi_{III}(t)]}]$$

$$\Phi_{II}(t) = \alpha^\circ e^{-i\Omega t}[e^{-\gamma t/2} - 1] \qquad \text{and} \qquad \Phi_{III}(t) = \alpha^\circ e^{-i\Omega t} e^{-\gamma t/2}$$

These same operators at zero temperature are found to reduce to

$$[\rho_X^{\{1\}}(t)] = (e^{-|\Phi_X(t)|^2}) \sum \frac{|\Phi_X(t)|^{2n}}{n!} |(n)\rangle\langle(n)| \ \text{ at } T \to 0 \ \text{ with } X = \{II\}, \{III\}$$

Finally, the time-dependent thermal averages of the H-bond bridge coordinate, in both quantum representations $\{II\}$ and $\{III\}$, evolve according to

$$\langle \mathrm{Q}_{II}^{\{1\}}(t)\rangle = \varepsilon \, \mathrm{tr}_{\mathrm{Slow}}[(e^{-\tilde{\lambda}[a^\dagger - \Phi_{II}(t)^*][a - \Phi_{II}(t)]})(a^\dagger + a)]Q^{\circ\circ} = 2\alpha^\circ Q^{\circ\circ} e^{-\gamma t/2} \sin \Omega t$$

$$\langle \mathrm{Q}_{III}^{\{1\}}(t)\rangle = \varepsilon \, \mathrm{tr}_{\mathrm{Slow}}[(e^{-\tilde{\lambda}[a^\dagger - \Phi_{III}(t)^*][a - \Phi_{III}(t)]})(a^\dagger + a)]Q^{\circ\circ} = 2\alpha^\circ Q^{\circ\circ} e^{-\gamma t/2} \cos \Omega t$$

Examination of these equations shows that, when the fast mode is in its first excited state, the slow mode eigenstates of the Hamiltonian react as a whole, with respect to the damping, because they form a coherent state.

In conclusion of these physical considerations, it appears that in the present model if, when the fast mode is in its ground state, the slow mode reacts as if it were not influenced by the medium, because it is at equilibrium, at the opposite, the excitation of the fast mode, induces a coherent organization of the H-bond bridge out of equilibrium that must return to equilibrium because of the surrounding.

C. The Adiabatic Effective Hamiltonians Involving Thermal Bath

Now, let us look at the incorporation of the quantum indirect damping in the quantum representation $\{II\}$ of the H-bond bridge. It is necessary to introduce in the model of the weak H-bond working within the strong anharmonic coupling theory, an hypothesis on the nature and on the irreversible action of the medium, that is, on the thermal bath and on its coupling with the H-bond bridge. A suitable assumption is to use the usual quantum theory of damping [54]. In this spirit, the Hamiltonian of the H-bond bridge embedded in the thermal bath playing the role of the surrounding may be written in the following way:

$$\mathbb{H}_{\mathrm{Tot}} = \mathrm{H}_{\mathrm{Tot}} + \mathbb{H}_{\theta} + \mathbb{H}_{\mathrm{Int}} \tag{97}$$

Here, $\mathrm{H}_{\mathrm{Tot}}$ is given by Eq. (35), that is, by the following equation:

$$\mathrm{H}_{\mathrm{Tot}} = \mathrm{H}_{\mathrm{Free}} + \mathrm{H}^{\circ} + \mathrm{H}_{\mathrm{Int}} \tag{98}$$

H° and $\mathrm{H}_{\mathrm{Free}}$ are, respectively, the Hamiltonians of the fast and slow modes viewed as quantum harmonic oscillators, whereas $\mathrm{H}_{\mathrm{Int}}$ is the anharmonic coupling between the two modes, which are given by Eqs. (15), (21), and (22). Besides, $\mathbb{H}_{\theta}$ is the Hamiltonian of the thermal bath, while $\mathbb{H}_{\mathrm{Int}}$ is the Hamiltonian of the interaction of the H-bond bridge with the thermal bath.

As usual in the quantum theory of damping [54], the thermal bath may be figured by an infinite set of harmonic oscillators and its coupling with the H-bond bridge by terms that are linear in the position coordinates of the bridge and of the bath oscillators:

$$\mathbb{H}_{\theta} = \sum_{i} \left(\frac{\tilde{\mathrm{p}}_i^2}{2\tilde{m}_i} + \frac{1}{2} \tilde{m}_i \tilde{\omega}_i^2 \tilde{\mathrm{q}}_i^2 \right) \tag{99}$$

$$\mathbb{H}_{\mathrm{Int}} = \sum \eta_i \tilde{\mathrm{q}}_i \mathrm{Q} \tag{100}$$

Here, $\tilde{\mathrm{q}}_i$ are the position coordinate operators of the oscillators of the bath and $\tilde{\mathrm{p}}_i$ are the conjugate momentum, whereas $\tilde{m}_i$ and $\tilde{\omega}_i$ are the corresponding reduced masses and angular frequencies. At last, the η_i parameters describe the strength of the coupling between the H-bond bridge (described by the coordinate Q) and the oscillators of the bath.

Note that Hamiltonians of the form (99) and (100) allow us to illustrate an irreversible evolution toward equilibrium [58]. Consider a set of linear degenerate oscillators characterized by the Hamiltonian (99), mutually coupled by Hamiltonians, such as those involved in Eq. (100) within the rotating wave approximation and starting from a situation where the first oscillator is a coherent state, whereas the other ones are in their ground state. It is possible to find the dynamics of such a system. Then, performing a coarse-grained analysis, it appears that the system. progressively evolves toward Boltzmann distribution for the oscillator energy levels and maximization of the oscillator chain entropy [58].

Next, owing to Eqs. (15), (21), (22), (99), and (100), the full Hamiltonian (97) takes the form:

$$\mathbb{H}_{\text{Tot}} = \frac{\text{P}^2}{2M} + \frac{1}{2}M\Omega^2\text{Q}^2 + \frac{\text{p}^2}{2m} + \frac{1}{2}m\omega^{\circ 2}\text{q}^2$$
$$+ m\omega^{\circ}b\text{q}^2\text{Q} + \frac{1}{2}mb^2\text{q}^2\text{Q}^2 + \sum_i\left(\frac{\tilde{\text{p}}_i^2}{2\tilde{m}_i} + \frac{1}{2}\tilde{m}_i\tilde{\omega}_i^2\tilde{\text{q}}_i^2\right) + \sum_i \eta_i\tilde{\text{q}}_i\text{Q}$$

Of course, the following commutation rules stand:

$$[\text{q},\text{p}] = i\hbar \qquad [\text{Q},\text{P}] = i\hbar \qquad [\tilde{\text{q}}_i,\tilde{\text{p}}_j] = i\hbar\delta_{i,j}$$

Now, for the same reasons as for the situation without damping, we may perform the adiabatic approximation. Then, proceeding as in the absence of damping, it may be easily shown that the full Hamiltonian $\mathbb{H}_{\text{Tot}}$ may be approximated by an equation similar to that (37), that is,

$$\mathbb{H}_{\text{Tot}} \simeq \mathbb{H}_{\text{Adiab}} = \sum_i [\mathbb{H}_{II}^{\{k\}}]|\{k\}\rangle\langle\{k\}| \tag{101}$$

$\mathbb{H}_{II}^{\{k\}}$ are effective Hamiltonians that play the role of Eqs. (49) and (50) in the absence of damping and which are

$$[\mathbb{H}_{II}^{\{0\}}] = \frac{\text{P}^2}{2M} + \frac{1}{2}M\Omega^2\text{Q}^2 + \sum_i\left(\frac{\tilde{\text{p}}_i^2}{2\tilde{m}_i} + \frac{1}{2}\tilde{m}_i\tilde{\omega}_i^2\tilde{\text{q}}_i^2\right) + \sum_i \eta_i\tilde{\text{q}}_i\text{Q} \tag{102}$$

$$[\mathbb{H}_{II}^{\{1\}}] = \frac{\text{P}^2}{2M} + \frac{1}{2}M\Omega^2\text{Q}^2 + \hbar\omega^{\circ} - \alpha^{\circ 2}\hbar\Omega$$
$$+ \hbar b\text{Q} + \sum_i\left(\frac{\tilde{\text{p}}_i^2}{2\tilde{m}_i} + \frac{1}{2}\tilde{m}_i\tilde{\omega}_i^2\tilde{\text{q}}_i^2\right) + \sum_i \eta_i\tilde{\text{q}}_i\text{Q} \tag{103}$$

Again, refer to Bosons with the help of equations given in Table V. Then, within this description, each term of the coupling between the H-bond bridge and the thermal bath is given by

$$\eta_i\tilde{\text{q}}_i\text{Q} = [\text{a}^{\dagger}\tilde{\text{b}}_i\hbar\tilde{\kappa}_i + \text{hc}] + [\text{a}^{\dagger}\tilde{\text{b}}_i^{\dagger}\hbar\tilde{\kappa}_i + \text{hc}] \tag{104}$$

where, for each bracket, the abbreviation hc stands for the Hermitean conjugate of the term involved in the bracket.

$$\tilde{\kappa}_i = \frac{\eta_i}{\tilde{\omega}_i}\sqrt{\frac{\hbar}{2\tilde{m}_i\tilde{\omega}_i}}$$

Moreover, within the rotating wave approximation, which is usual in the quantum theory of damping, the terms corresponding to double excitations or desexcitations in Eq. (104) are neglected. Thus, within this approximation, Eq. (104) reduces to

$$\eta_i\tilde{q}_i Q \simeq [a^\dagger \tilde{b}_i \hbar\tilde{\kappa}_i + hc]$$

As a consequence, using the Bosons description and performing the rotating wave approximation lead us to write the effective Hamiltonians (102) and (103) describing the H-bond bridge coupled to the thermal bath as follows:

$$[\mathbb{H}_{II}^{\{0\}}] = \left(a^\dagger a + \frac{1}{2}\right)\hbar\Omega + \sum_i \left[\tilde{b}_i^\dagger \tilde{b}_i + \frac{1}{2}\right]\hbar\tilde{\omega}_i + \sum_i [a^\dagger \tilde{b}_i \hbar\tilde{\kappa}_i + hc] \tag{105}$$

$$[\mathbb{H}_{II}^{\{1\}}] = \left(a^\dagger a + \frac{1}{2}\right)\hbar\Omega + [\alpha^\circ(a^\dagger + a) - \alpha^{\circ 2}]\hbar\Omega + \hbar\omega^\circ + \sum_i \left[\tilde{b}_i^\dagger \tilde{b}_i + \frac{1}{2}\right]\hbar\tilde{\omega}_i + \sum_i [a^\dagger \tilde{b}_i \hbar\tilde{\kappa}_i + hc] \tag{106}$$

D. A General Expression for the ACF

Now, consider the ACF of the H-bond bridge. Within the adiabatic approximation, the damped ACF has the same structure as (59) appearing in the absence of damping, that is,

$$[G_{II}(t)]^\circ = \left(\frac{\mu_0^2}{\mathbf{Z}_{\text{Adiab}}}\right) \tilde{\text{tr}}[(e^{-\beta \mathbf{H}_{\text{Adiab}}})|\{0\}\rangle\langle\{1\}|(e^{i\mathbb{H}_{\text{Adiab}}t/\hbar})|\{1\}\rangle\langle\{0\}|(e^{-i\mathbb{H}_{\text{Adiab}}t/\hbar})]$$

Here, $\tilde{\text{tr}}_{\text{Adiab}}$ is the trace to be performed over the base spanned by the eigenstates of the Hamiltonian $\mathbf{H}_{\text{Adiab}}$, whereas $\mathbf{Z}_{\text{Adiab}}$ is the partition function related to this same Hamiltonian, that is,

$$\mathbf{Z}_{\text{Adiab}} = \tilde{\text{tr}}_{\text{Adiab}}[(e^{-\beta \mathbf{H}_{\text{Adiab}}})]$$

With the help of Eq. (101), this equation becomes

$$[\mathbf{G}_{II}(t)]^{\circ} = \mu_0^2 \, \tilde{\mathbf{tr}} \, [\tilde{\boldsymbol{\rho}}_{II}^{\{0\}} |\{0\}\rangle\langle\{1\}| [\mathbf{e}^{i\mathbb{H}_{II}^{\{1\}} t/\hbar}] |\{1\}\rangle\langle\{0\}| [\mathbf{e}^{-i\mathbb{H}_{II}^{\{1\}} t/\hbar}]]$$

Here, $\tilde{\rho}_{II}^{\{0\}}$ is the Boltzmann density operator of the H-bond bridge,

$$[\tilde{\rho}_{II}^{\{0\}}] = \tilde{\varepsilon}(e^{-\tilde{\lambda}(a^{\dagger}a+1/2)}) \quad \text{with} \quad \tilde{\varepsilon} = e^{\tilde{\lambda}/2} - e^{-\tilde{\lambda}/2} \quad \text{and} \quad \tilde{\lambda} = \frac{\hbar\Omega}{k_{\text{B}}T} \tag{107}$$

Now factorize, the trace $\tilde{\text{tr}}$ into a trace tr_{Slow} to be performed over the space of the slow mode Hamiltonian and a trace tr_{θ} to be performed over the thermal bath:

$$\tilde{\text{tr}}_{\text{Slow}} = \text{tr}_{\text{Slow}} \text{tr}_{\theta}$$

As a consequence, the ACF takes the form:

$$[\text{G}_{II}(t)]^{\circ} = (\mu_0^2)\text{tr}_{\text{Slow}}[[\tilde{\rho}_{II}^{\{0\}}]\text{tr}_{\theta}[|\{0\}\rangle\langle\{1\}|[e^{i\mathbb{H}_{II}^{\{1\}} t/\hbar}]|\{1\}\rangle\langle\{0\}|[e^{-i\mathbb{H}_{II}^{\{0\}} t/\hbar}]]$$

After using a circular permutation of the *bra* and *kets* that do not belong to the spaces where the exponential operators work and then, after simplification using the orthogonality property $\langle\{k\}|\{l\}\rangle = \delta_{kl}$, it reads

$$[\text{G}_{II}(t)]^{\circ} = (\mu_0^2)\text{tr}_{\text{Slow}}[[\tilde{\rho}_{II}^{\{0\}}]\text{tr}_{\theta}[[e^{i\mathbb{H}_{II}^{\{1\}} t/\hbar}][e^{-i\mathbb{H}_{II}^{\{0\}} t/\hbar}]] \tag{108}$$

This new expression of the ACF may be also written in view of Eqs. (107):

$$[\text{G}_{II}(t)]^{\circ} = (\mu_0^2)\tilde{\varepsilon} \; \text{tr}_{\text{Slow}}[(e^{-\tilde{\lambda}[a^{\dagger}a+\frac{1}{2}]})\text{tr}_{\theta}[[\mathbb{U}_{II}^{\{1\}}(t)]^{-1}[\mathbb{U}_{II}^{\{0\}}(t)]]] \tag{109}$$

with

$$[\mathbb{U}_{II}^{\{0\}}(t)] = [e^{-i\mathbb{H}_{II}^{\{0\}}t/\hbar}] \quad \text{and} \quad [\mathbb{U}_{II}^{\{1\}}(t)]^{-1} = [e^{i\mathbb{H}_{II}^{\{1\}}t/\hbar}] \tag{110}$$

Note that the expression of the ACF (109), which takes into account the irreversible influence of the surrounding, has the same structure as that (61), which holds in the absence of damping.

Next, owing to Eqs. (105) and (106), the effective time-evolution operators are, respectively, given by

$$[\mathbb{U}_{II}^{\{0\}}(t)] = \left(e^{-i[a^\dagger a+1/2]\Omega t - \sum_j i[\tilde{b}_j^\dagger \tilde{b}_j+1/2]\omega_j t - i\sum_i [a^\dagger \tilde{b}_i \tilde{\kappa}_i t + hc]}\right) \tag{111}$$

$$[\mathbb{U}_{II}^{\{1\}}(t)] = e^{-i\omega^\circ t} e^{i\alpha^{\circ 2}\Omega t}\left[e^{-i[a^\dagger a+1/2]\Omega t - i\alpha^\circ(a^\dagger + a)\Omega t - \sum_j i[a^\dagger \tilde{b}_j \tilde{\kappa}_j t + hc] - \sum_j i[\tilde{b}_j^\dagger \tilde{b}_j + 1/2]\tilde{\omega}_j t}\right] \tag{112}$$

Appendix J shows that the ACF (109) may be transformed into Eq.(J10) of this appendix, that is,

$$[G_{II}(t)]^\circ = (\mu_0^2) e^{i\omega^\circ t} e^{-i\alpha^{\circ 2}\Omega t} \tilde{\varepsilon}\, \mathrm{tr}_{\mathrm{Slow}}[(e^{-\tilde{\lambda}[a^\dagger a+1/2]}) \mathrm{tr}_\theta[[\mathbb{U}_{II}^{\{1\}}(t)^{\mathrm{IP}}]^{-1}]] \tag{113}$$

In Eq. (113), $\mathbb{U}_{II}^{\{1\}}(t)^{\mathrm{IP}}$ is the IP time-evolution operator of the driven quantum harmonic oscillator interacting with the thermal bath,

Next, using Eq. (J.11), the ACF (113) may be in terms of a reduced time-dependent operator with the aid of the following equation:

$$[G_{II}(t)]^\circ = (\mu_0^2) e^{i\omega^\circ t} (e^{-i\alpha^{\circ 2}\Omega t}) \tilde{\varepsilon}\, \mathrm{tr}_{\mathrm{Slow}}[(e^{-\tilde{\lambda}[a^\dagger a+1/2]})[\mathrm{U}_{II}^{\{1\}}(t)^{\mathrm{IP}}]^\dagger] \tag{114}$$

with, respectively,

$$[\mathbf{U}_{II}^{\{1\}}(t)^{\mathrm{IP}}] = [\widehat{\mathrm{P}}[e^{\frac{i}{\hbar}\int_0^t \mathbf{V}^{\mathrm{IP}}(t')dt'}]] \tag{115}$$

$$\mathbf{V}^{\mathrm{IP}}(t) = \alpha^\circ \hbar \Omega \mathrm{tr}_\theta[[e^{i\mathbb{H}_{II}^{\{0\}}t/h}][a^\dagger + a][e^{-i\mathbb{H}_{II}^{\{0\}}t/h}]] \tag{116}$$

A formal possibility to obtain the reduced time-evolution operator involved in Eq. (114) is to solve the following Schrödinger equation within the

Wigner–Weisskopf approximation with the above constraint, after performing the partial trace (117), that is,

$$i\frac{\partial \mathbf{U}_{II}^{\{1\}}(t)^{\mathrm{IP}}}{\partial t} = \alpha^{\circ}\Omega \mathrm{tr}_{\theta}[[e^{i\mathbb{H}_{II}^{\{0\}}t/\hbar}][a^{\dagger} + a][e^{i\mathbb{H}_{II}^{\{0\}}t/\hbar}]\mathbf{U}_{II}^{\{1\}}(t)^{\mathrm{IP}}] \tag{117}$$

with the boundary condition:

$$\mathbf{U}_{II}^{\{1\}}(0)^{\mathrm{IP}} = 1$$

Another possibility is to extract the reduced time evolution operator from the analytical solution obtained by Louisell and Walker for the reduced time-dependent density operator of a driven damped quantum harmonic oscillator.

Now, perform the trace over the eigenstates of the slow mode quantum harmonic oscillator involved in the ACF (114). This leads, after neglecting the zero-point energy of H-bond bridge oscillator, to

$$[G_{II}(t)]^{\circ} = (\mu_0^2)e^{i\omega^{\circ}t}(e^{-i\alpha^{\circ 2}\Omega t})\varepsilon\sum_n(e^{-\tilde{\lambda}n})\langle(n)|[\mathbf{U}_{II}^{\{1\}}(t)^{\mathrm{IP}}]^{-1}|(n)\rangle \tag{118}$$

Of course, at zero temperature, there are

$$e^{-\tilde{\lambda}n} = \delta_{no} \qquad \text{and} \qquad \varepsilon = 1$$

Thus, the ACF reduces to

$$[G_{II}(t)]^{\circ} = (\mu_0^2)e^{i\omega^{\circ}t}(e^{-i\alpha^{\circ 2}\Omega t})\langle(0)|[\mathbf{U}_{II}^{\{1\}}(t)^{\mathrm{IP}}]^{-1}|(0)\rangle \tag{119}$$

1. *Obtainment of the Line Shape*

Appendix D shows that the IP time evolution operator of a driven quantum harmonic oscillator is given by Eq. (D.23), that is,

$$[\mathrm{U}^{\circ}_{II}(t)^{\mathrm{IP}}]^{-1} = (e^{-i\alpha^{\circ 2}\Omega t})(e^{i\alpha^{\circ 2}\sin\Omega t})[e^{\Phi^{\circ}_{II}(t)^{*}a^{\dagger} - \Phi^{\circ}_{II}(t)a}] \tag{120}$$

Besides, Appendix I proves that in the presence of quantum indirect damping, according to Eqs. (I.6) and (I.12), the translation operator involved in Eq. (120) conserves its structure, but in such a way as its argument is modified according to

$$[e^{\Phi^\circ_{II}(t)^* a^\dagger - \Phi^\circ_{II}(t) a}] \rightarrow [e^{\Phi_{II}(t)^* a^\dagger - \Phi_{II}(t) a}]$$

where

$$\Phi^\circ_{II}(t) = \alpha^\circ[e^{-i\Omega t} - 1] \rightarrow \Phi_{II}(t) = \alpha^\circ[e^{-i\Omega t} e^{-\gamma t/2} - 1]$$

Then, it may be assumed that there is, in a similar way, the following change in the simple scalar involved in the argument $i\alpha^\circ 2 \sin \Omega t$ of the phase factor involved in Eq. (120):

$$e^{\pm i\Omega t} \rightarrow e^{\pm i\Omega t} e^{-\gamma t/2} \tag{121}$$

Consequently, the reduced IP time-evolution operator of the driven damped quantum harmonic oscillator takes the form:

$$[U_{II}^{\{1\}}(t)^{\mathrm{IP}}]^{-1} = (e^{-i\alpha^{\circ 2}\Omega t})(e^{i\alpha^{\circ 2} e^{-\gamma t/2} \sin \Omega t})[e^{\Phi_{II}(t)^* a^\dagger - \Phi_{II}(t) a}] \tag{122}$$

As will appear later, this expression may be confirmed *a posteriori*.

We are now able to find the expression of the ACF of the dipole moment operator of the high frequency mode, when the H-bond bridge is damped. Owing to Eq. (122), the ACF (117) becomes

$$[G_{II}(t)]^\circ = (\mu_0^2) e^{i\omega^\circ t}(e^{-i2\alpha^{\circ 2}\Omega t})(e^{i\alpha^{\circ 2} e^{-\gamma t/2} \sin \Omega t})\varepsilon\, \mathrm{tr}_{\mathrm{Slow}}[(e^{-\tilde{\lambda} a^\dagger a})[e^{\Phi_{II}(t)^* a^\dagger - \Phi_{II}(t) a}]] \tag{123}$$

Next, perform the trace over the slow mode with the help of the Bloch theorem. Then, with Eqs. (H.1) and (H.4), the ACF (123) becomes

$$[G_{II}(t)]^\circ = (\mu_0^2) e^{i\omega^\circ t}(e^{-i2\alpha^{\circ 2}\Omega t})(e^{i\alpha^{\circ 2} e^{-\gamma t/2} \sin \Omega t})[e^{\alpha^{\circ 2}[\langle n \rangle + 1/2](2e^{-\gamma t/2}\cos(\Omega t) - e^{-\gamma t} - 1)}] \tag{124}$$

with

$$\langle n \rangle = \frac{1}{e^{\tilde{\lambda}} - 1} \qquad \text{and} \qquad \tilde{\lambda} = \frac{\hbar\Omega}{k_B T}$$

The main lines of the way in which the ACF (124) were obtained are summarized in Table VII.

The ACF (124) takes into account the quantum indirect damping, but not the direct one that may be incorporated, as above [see from Eqs. (69) to (70)], by

TABLE VII
Main Lines for Obtaining the ACF (124) Involving Quantum Indirect Damping

1. The basic equations for the ACF

$$\boxed{[G_{II}(t)]^{\circ} \propto \mathrm{tr}\{p_{B}\ \mu(0)\dagger\mu(t)\}}$$

$$\mu(t) = e^{iH_{\mathrm{Tot}}t/h}\ \mu(0)e^{-iH_{\mathrm{Tot}}t/h} \quad p_n = e^{-\beta H_{\mathrm{Tot}}}$$

$$\boxed{[G_{II}(t)]^{\circ} \propto \mathrm{tr}_{\mathrm{Slow}}[e^{-\beta H_{II}^{\{a\}}}\mathrm{tr}_{\theta}[e^{iH_{II}^{\{1\}}t/h}e^{-iH_{II}^{\{0\}}t/h}]]}$$

$$[\mathbb{H}_{II}^{\{k\}}] = \frac{P^2}{2M} + \frac{1}{2}M\Omega^2 Q^2 + k[bQ - \alpha^{\circ 2}\hbar\Omega] + \hbar\omega^{\circ} + \mathbb{H}_0 + \mathbb{H}_{\mathrm{Int}}$$

2. Equivalent intermediate expression for the ACF

$$\boxed{[G_{II}(t)]^{\circ} \propto e^{i\omega^{\circ}t}\left\langle\left\langle \widehat{P}\left[e^{ib\int_0^t Q^{IP}(t')dt'}\right]\right\rangle_{\theta}\right\rangle_{\mathrm{Slow}}}$$

$$\boxed{[G_{II}(t)]^{\circ} \propto e^{i\omega^{\circ}t}e^{-i\omega^{\circ 2}\Omega t}\mathrm{tr}_{\mathrm{Slow}}[p_{B}\{U_{II}^{\{1\}}(t)^{IP}\}^{-1}]}$$

3. The final expressions for the ACF

$$[G_{II}(t)]^{\circ} \propto e^{i\omega^{\circ}t}(e^{-i2\alpha^{\circ 2}\Omega t})(e^{i\alpha^{\circ 2}e^{-\gamma t/2}\sin\Omega t})\varepsilon\mathrm{tr}_{\mathrm{Slow}}[(e^{-\tilde{\lambda}a^{\dagger}a})[e^{\Phi_{II}(t)^{\circ}a^{\dagger}-\Phi_{II}(t)a}]]$$

$$[G_{II}(t)]^{\circ} \propto e^{i\omega^{\circ}t}(e^{-i2\alpha^{\circ 2}\Omega t})(e^{i\alpha^{\circ 2}e^{-\gamma t/2}\sin\Omega t})(e^{\alpha^{\circ 2}[\langle n\rangle+1/2](2e^{-\gamma t/2}\cos(\Omega t)-e^{-\gamma t}-1)}]$$

multiplying it with a simple exponential decay $e^{-\gamma^{\circ}t}$, where γ° is the direct damping parameter. This leads us to write

$$[G_{II}(t)] = (\mu_0^2)e^{i\omega^{\circ}t}(e^{-i2\alpha^{\circ 2}\Omega t})(e^{i\alpha^{\circ 2}e^{-\gamma t/2}\sin\Omega t})[e^{\alpha^{\circ 2}[\langle n\rangle+1/2](2e^{-\gamma t/2}\cos(\Omega t)-e^{-\gamma t}-1)}](e^{-\gamma^{\circ}t}) \tag{125}$$

Moreover, the spectral density of the system involving both direct and indirect quantum damping, is the Fourier transform of Eq. (125), that is,

$$[I_{II}(\omega)] = (\mu_0^2)2\,\mathrm{Re} \int_0^{\infty} e^{i\omega^{\circ}t}(e^{-i2\alpha^{\circ 2}\Omega t})(e^{i\alpha^{\circ 2}e^{-\gamma t/2}\sin\Omega t})[e^{\alpha^{\circ 2}[\langle n\rangle+1/2](2e^{-\gamma t/2}\cos(\Omega t)-e^{-\gamma t}-1)}](e^{-\gamma^{\circ}t})e^{-i\omega t}dt \tag{126}$$

Observe that, because of Eqs. (90) and (125) there is the following equivalence:

$$\left\langle\left\langle \widehat{\mathrm{P}}e^{ib\int_0^t \mathrm{Q}(t')^{\mathrm{IP}}dt'}\right\rangle_{\mathrm{Slow}}\right\rangle_{\theta} = (e^{-i\alpha^{\circ 2}\Omega t})(e^{i\alpha^{\circ 2}e^{-\gamma t/2}\sin\Omega t})[e^{\alpha^{\circ 2}[\langle n\rangle+1/2](2e^{-\gamma t/2}\cos(\Omega t)-e^{-\gamma t}-1)}] \tag{127}$$

2. *Another Approach of Quantum Indirect Damping Within Representation {III}*

Recall that at the basis of the SD (126), there is the assumption (121). Here, it will be shown that Eq. (126) may be confirmed by another approach that avoids this assumption.

3. *The ACF Expressed in Terms of Damped Translation Operators*

In *the absence of both direct and indirect dampings*, it was shown that the ACF may be given by Eq. (74), that is,

$$[\mathrm{G}^{\circ}{}_{III}(t)]^{\circ} = (\mu_0^2)e^{i\omega^{\circ}t}(e^{-i2\alpha^{\circ 2}\Omega t})$$
$$\varepsilon\ \mathrm{tr}_{\mathrm{Slow}}[(e^{-\tilde{\lambda}\mathrm{a}^{\dagger}\mathrm{a}})[\mathrm{A}^{\circ\{1\}}_{III}(0)]^{-1}(e^{i\mathrm{a}^{\dagger}\mathrm{a}\Omega t})[\mathrm{A}^{\circ\{1\}}_{III}(0)](e^{-i\mathrm{a}^{\dagger}\mathrm{a}\Omega t})] \tag{128}$$

Here, $[\mathrm{A}^{\circ\{1\}}_{III}(0)]$ is the translation operator at initial time given by

$$[\mathrm{A}^{\circ\{1\}}_{III}(0)] = e^{\alpha^{\circ}\mathrm{a}^{\dagger}-\alpha^{\circ}\mathrm{a}} \equiv [\mathrm{A}(\alpha^{\circ})] \tag{129}$$

Next, Eq. (128) may be written

$$[\mathrm{G}^{\circ}{}_{III}(t)]^{\circ} = (\mu_0^2)e^{i\omega^{\circ}t}(e^{-i2\alpha^{\circ 2}\Omega t})\varepsilon\ \mathrm{tr}_{\mathrm{Slow}}[(e^{-\tilde{\lambda}\mathrm{a}^{\dagger}\mathrm{a}})[\mathrm{A}^{\circ\{1\}}_{III}(0)]^{-1}[\mathrm{A}^{\circ\{1\}}_{III}(t)]] \tag{130}$$

with

$$[\mathrm{A}^{\circ\{1\}}_{III}(t)] = (e^{i\mathrm{a}^{\dagger}\mathrm{a}\Omega t})[\mathrm{A}^{\circ\{1\}}_{III}(0)](e^{-i\mathrm{a}^{\dagger}\mathrm{a}\Omega t})$$

Now, consider the *situation involving indirect damping*. Then, it is possible to write a formal expression for the ACF that will be a generalization for damped situations of (130), which holds for undamped cases, that is,

$$[\mathrm{G}_{III}(t)]^{\circ} = (\mu_0^2)e^{i\omega^{\circ}t}(e^{-i2\alpha^{\circ 2}\Omega t})\varepsilon\ \mathrm{tr}_{\mathrm{Slow}}[(e^{-\tilde{\lambda}\mathrm{a}^{\dagger}\mathrm{a}})[\mathrm{A}^{\{1\}}_{III}(0)]^{-1}[\mathrm{A}^{\{1\}}_{III}(t)]] \tag{131}$$

Here, $[\mathrm{A}_{III}^{\{1\}}(t)]$ is the damped time-dependent translation operator to be found. Of course, at the initial time the translation operator $[\mathrm{A}_{III}^{\{1\}}(0)]$ cannot be affected by the damping so that

$$[\mathrm{A}_{III}^{\{1\}}(0)] = [\mathrm{A}(\alpha^\circ)] = [e^{\alpha^\circ \mathrm{a}^\dagger - \alpha^\circ \mathrm{a}}] \tag{132}$$

Besides, Appendix I shows that the damped translation operator is given by Eq. (I.19), that is,

$$[\mathrm{A}_{III}^{\{1\}}(t)] = [e^{\Phi_{III}(t)^* \mathrm{a}^\dagger - \Phi_{III}(t)\mathrm{a}}] \quad \text{with} \quad \Phi_{III}(t) = \alpha^\circ[e^{-i\Omega t}e^{-\gamma t/2}]$$

As a consequence of Eq. (132) and of this last equation, the ACF (131) takes the form:

$$[\mathrm{G}_{III}(t)]^\circ = (\mu_0^2)\, e^{i\omega^\circ t}(e^{-i2\alpha^{\circ 2}\Omega t})\varepsilon\ \mathrm{tr}_{\mathrm{Slow}}[(e^{-\tilde{\lambda}\mathrm{a}^\dagger \mathrm{a}})[e^{-\alpha^\circ(\mathrm{a}^\dagger - \mathrm{a})}][e^{\Phi_{III}(t)^* \mathrm{a}^\dagger - \Phi_{III}(t)\mathrm{a}}]] \tag{133}$$

Besides, Appendix F shows that the product of the translation operators involved in Eq. (133) is given by Eq. (F.4), that is,

$$[e^{-\alpha^\circ(\mathrm{a}^\dagger - \mathrm{a})}][e^{+\Phi_{III}(t)^* \mathrm{a}^\dagger - \Phi_{III}(t)\mathrm{a}}] = (e^{i\alpha^2 e^{-\gamma t/2}\sin \Omega t})[e^{+\Phi_{II}(t)^* \mathrm{a}^\dagger - \Phi_{II}(t)\mathrm{a}}]$$

with

$$\Phi_{II}(t) = \alpha^\circ[e^{-i\Omega t}e^{-\gamma t/2} - 1]$$

Thus, the ACF (133) becomes

$$\begin{aligned} [\mathrm{G}_{III}(t)]^\circ &= [\mathrm{G}_{II}(t)]^\circ \\ &= (\mu_0^2)e^{i\omega^\circ t}(e^{-i2\alpha^{\circ 2}\Omega t})(e^{i\alpha^2 e^{-\gamma t/2}\sin \Omega t})\varepsilon\ \mathrm{tr}_{\mathrm{Slow}}[(e^{-\tilde{\lambda}\mathrm{a}^\dagger \mathrm{a}})[e^{\Phi_{II}(t)^* \mathrm{a}^\dagger - \Phi_{II}(t)\mathrm{a}}]] \end{aligned} \tag{134}$$

As it appears, this result is equivalent to that given by Eq. (123) obtained by another way. Thus, since the new obtainment has not involved any approximation, it may be considered as a confirmation of the precedent approach that lead to Eq. (123).

4. *Physical Conclusions*

We showed that, within the adiabatic approximation, the irreversible influence of the damping on the H-bond bridge may be considered to play a role only after the fast mode has jumped on its first excited state. But the H-bond bridge is coupled to the thermal bath even when the fast mode is in its ground state: The fast mode transition from the ground to the excited states makes the H-bond bridge sensitive to the irreversible influence of the thermal bath.

When the fast mode is in its ground state, the effective Hamiltonian of the H-bond bridge is undriven, whereas the density operator describing it is a Boltzmann one. Thus, when the H-bond bridge is at thermal equilibrium, the bath may be assumed not to play an explicit role on it. On the other hand, when the fast mode has jumped on its first excited state, it induces a change in the physics of the H-bond bridge that may be differently viewed according to the fact that one is working in representations $\{II\}$ or $\{III\}$.

In representation $\{II\}$, the fast mode excitation may be viewed as driving the Hamiltonian of the H-bond bridge without modifying the Boltzmann density operator describing the bridge. Then, the thermal bath is damping the driven quantum harmonic oscillator describing the H-bond bridge. On the other hand, in representation $\{III\}$ the fast mode excitation does not modify the Hamiltonian of the bridge that remains undriven, but transforms the Boltzmann density operator of the bridge into that of a coherent state that is out of equilibrium. Then, the thermal bath damps the coherent state density operator.

IV. CLASSICAL LIMIT OF QUANTUM INDIRECT DAMPING [59, 60]

In Section III, we obtained within the strong anharmonic coupling theory and within the adiabatic approximation, an expression for the ACF of the dipole moment operator that takes into account the irreversible influence of the surrounding on the H-bond bridge. In quantum representation $\{II\}$, this ACF is given by Eq. (124), which is a consequence of Eq. (123) after performing the trace over the slow mode, that is,

$$[\mathrm{G}_{II}(t)]^{\circ} = (\mu_0^2)e^{i\omega^{\circ}t}(e^{-i2\alpha^{\circ 2}\Omega t})(e^{i\alpha^{\circ 2}e^{-\gamma t/2}\sin\Omega t})[e^{\alpha^{\circ 2}[\langle n\rangle+1/2](2e^{-\gamma t/2}\cos\Omega t-e^{-\gamma t}-1)}] \tag{135}$$

Recall that in the presence of quantum direct damping, this ACF must be multiplied by an exponential decay $e^{-\gamma^{\circ}t}$ as it appears by comparison of Eqs. (124) and (125). Also recall that the ACF (123) that is at the origin of Eq. (135), is an explicit expression of the formal one given by Eq. (114). We need to know what is the classical limit of this ACF (135) involving quantum indirect damping.

A. Classical Limit of ACF (135)

1. The Meaning of the Thermal Average

For this purpose, first return to the formal expression (114) of the ACF (135), that is,

$$[\mathrm{G}_{II}(t)]^{\circ} = (\mu_0^2)\varepsilon\ e^{i\omega^{\circ}t}(e^{-i\alpha^{\circ 2}\Omega t})$$

$$\mathrm{tr}_{\mathrm{Slow}}\left[\left(e^{-\tilde{\lambda}[\mathrm{a}^{\dagger}\mathrm{a}+\frac{1}{2}]}\right)\widehat{\mathrm{P}}\left[e^{\frac{i}{\hbar}\int_0^t \mathbb{V}^{\mathrm{IP}}(t')dt'}\right]\right] \tag{136}$$

with according to eq. (116)

$$\mathrm{V}^{\mathrm{IP}}(t) = \alpha^{\circ}\hbar\Omega\mathrm{tr}_{\theta}\left[e^{i\mathbb{H}_{II}^{\{0\}}t/\hbar}(\mathrm{a}^{\dagger}+\mathrm{a})e^{-i\mathbb{H}_{II}^{\{0\}}t/\hbar}\right]$$

In the absence of the thermal bath, $\mathrm{V}^{\mathrm{IP}}(t)$ reduces to

$$\mathrm{V}^{\mathrm{IP}}(t) = \alpha^{\circ}\hbar\Omega\left[e^{i\mathbb{H}_{II}^{\{0\}}t}[\mathrm{a}^{\dagger}+\mathrm{a}]e^{-i\mathbb{H}_{II}^{\{0\}}t/\hbar}\right]$$

or, because of Eq.(46) giving the expression of the effective Hamilonian involved in the canonical transformation:

$$\mathbf{V}^{\mathrm{IP}}(t) = \mathrm{a}^{\dagger}e^{i\Omega t} + \mathrm{a}e^{-i\Omega t}$$

Thus, the ACF (132) simplifies into

$$[\mathrm{G}^{\circ}{}_{II}(t)]^{\circ} = (\mu_0^2)e^{i\omega^{\circ}t}(e^{-i\alpha^{\circ 2}\Omega t})\ \tilde{\varepsilon}\mathrm{tr}_{\mathrm{Slow}}\left[\left(e^{-\tilde{\lambda}[\mathrm{a}^{\dagger}\mathrm{a}+\frac{1}{2}]}\right)\widehat{\mathrm{P}}\left[e^{i\alpha^{\circ}\int_0^t\{\mathrm{a}^{\dagger}e^{i\Omega t'}+\mathrm{a}e^{-i\Omega t'}\}dt'}\right]\right] \tag{137}$$

Again, owing to Eq. (41) and to the definitions of the Bosons given in Table V, the argument of the integral involved in the exponential operator of Eq. (137), is nothing but the time-dependent angular coordinate $\mathrm{Q}(t)$ of the H-bond bridge times the anharmonic coupling parameter b, that is,

$$\alpha^{\circ}[\mathrm{a}^{\dagger}e^{i\Omega t} + \mathrm{a}e^{-i\Omega t}] = b\mathrm{Q}(t)^{\mathrm{IP}}Q^{\circ\circ} \tag{138}$$

Now, in the presence of damping and by analogy with Eq. (138), we may write the argument of the integral involved in Eq. (136) formally according to

$$\mathrm{tr}_\theta\left\{\mathrm{a}^\dagger e^{i\Omega t}\left(1+\sum \tilde{\mathrm{b}}_i \tilde{\kappa}^\circ{}_i e^{-i\omega_i t}\right) + \mathrm{a}e^{-i\Omega t}\left(1+\sum \tilde{\mathrm{b}}_i^\dagger \tilde{\kappa}^{\circ *}_i e^{i\omega_i t}\right)\right\} = b\tilde{\mathrm{Q}}(t)^{\mathrm{IP}}/Q^{\circ\circ}$$

Then, the ACF (136) may be written

$$[\mathrm{G}_{II}(t)]^\circ = (\mu_0^2)e^{i\omega^\circ t}(e^{-i\alpha^{\circ 2}\Omega t})\tilde{\varepsilon}\ \mathrm{tr}_{\mathrm{Slow}}\left[\left(e^{-\tilde{\lambda}\left[\mathrm{a}^\dagger \mathrm{a}+\frac{1}{2}\right]}\right)\mathrm{tr}_\theta\left[\ \widehat{\mathrm{P}}\left[e^{ib\int_0^t \tilde{\mathrm{Q}}(t')^{\mathrm{IP}}dt'}\right]\right]\right] \tag{139}$$

The ACF (139) is analogous in the presence of damping of the ACF (137), which holds in the situation without.

2. *Semiclassical Limit of the Quantum ACF (139)*

Now, it may be suitable to change the notations in Eq. (139) for the traces to be performed over both the thermal bath and the slow mode, according to

$$\tilde{\varepsilon}\mathrm{tr}_{\mathrm{Slow}}\left[\left(e^{-\tilde{\lambda}\left[\mathrm{a}^\dagger \mathrm{a}+\frac{1}{2}\right]}\right)\mathrm{tr}_\theta[\cdots]\right] \rightarrow \langle\langle\cdots\rangle_\theta\rangle_{\mathrm{Slow}}$$

Then, the ACF (139) may be written

$$[\mathrm{G}_{II}(t)]^\circ = (\mu_0^2)e^{i\omega^\circ t}(e^{-i\alpha^{\circ 2}\Omega t})\left\langle\left\langle\widehat{\mathrm{P}}\left[e^{ib\int_0^t \tilde{\mathrm{Q}}(t')^{\mathrm{IP}}dt'}\right]\right\rangle_\theta\right\rangle_{\mathrm{Slow}} \tag{140}$$

The question is to know what is the classical limit of the ACF (140).

For this reason, we must ignore first the noncommutativity at different times of the coordinate $\tilde{\mathrm{Q}}(t)^{\mathrm{IP}}$ involved in Eq. (140) and thus ignore it in the Dyson time-ordering operator $\widehat{\mathrm{P}}$. In the same way, we must ignore the shift $-\alpha^{\circ 2}\Omega$, which is related to the removal of a quantum driven term, resulting from the zero-point energy of the high frequency mode and that is the same for both the effective Hamiltonians of the slow mode corresponding, respectively, to the ground state and to the first excited state of the high frequency mode. This leads us to write in Eq. (140)

$$\widehat{\mathrm{P}} \rightarrow 1, \tilde{\mathrm{Q}}(t)^{\mathrm{IP}} \rightarrow \tilde{\mathrm{Q}}(t) \quad \text{and} \quad \alpha^{\circ 2}\Omega \rightarrow 0 \tag{141}$$

Then, in this spirit, the semiclassical limit of the ACF (141) becomes

$$[\mathrm{G}_{II}(t)]^\circ_{\mathrm{Semi}} = (\mu_0^2)e^{i\omega^\circ t}\left\langle\left\langle\left[e^{ib\int_0^t \tilde{\mathrm{Q}}(t')dt'}\right]\right\rangle_\theta\right\rangle_{\mathrm{Slow}} \tag{142}$$

On the other hand, it may be observed that *in the damped situation*, Eq. (135) and Eq. (140) are equivalent, so that

$$\left\langle\left\langle \widehat{\mathrm{P}}\left[e^{ib\int_0^t \tilde{\mathrm{Q}}(t')^{\mathrm{IP}}dt'}\right]\right\rangle_\theta\right\rangle_{\mathrm{Slow}} = \left(e^{-i\alpha^{\circ 2}\Omega t}\right)\left(e^{i\alpha^{\circ 2}e^{-\gamma t/2}\sin\Omega t}\right)\left[e^{\alpha^{\circ 2}[\langle n\rangle+1/2](2e^{-\gamma t/2}\cos\Omega t-e^{-\gamma t}-1)}\right] \quad (143)$$

Moreover, in the absence of damping, $\tilde{\mathrm{Q}}(t)^{\mathrm{IP}}$ becomes $\mathrm{Q}(t)^{\mathrm{IP}}$ given by Eq. (G.8), whereas the trace over the thermal bath disappears and γ becomes zero. Then, Eq. (143) reduces in this special situation, to

$$\left\langle \widehat{\mathrm{P}}\left[e^{ib\int_0^t \mathrm{Q}(t')^{\mathrm{IP}}dt'}\right]^{\mathrm{Semi}}\right\rangle_{\mathrm{Slow}} = \left(e^{-i\alpha^{\circ 2}\Omega t}\right)\left(e^{i\alpha^{\circ 2}\sin\Omega t}\right)\left[e^{2\alpha^{\circ 2}[\langle n\rangle+1/2](\cos\Omega t-1)}\right]$$

Besides, *in the situation without damping*, it has been shown that the semiclassical limit leads us to write

$$\widehat{\mathrm{P}} \to 1 \qquad \text{and} \qquad \left(e^{-i\alpha^{\circ 2}\Omega t}\right)\left(e^{i\alpha^{\circ 2}\sin\Omega t}\right) \to 1$$

Then, Eq. (86) is reobtained, that is,

$$\left\langle \left[e^{ib\int_0^t \mathrm{Q}(t')dt'}\right]^{\mathrm{Semi}}\right\rangle_{\mathrm{Slow}} = \left[e^{2\alpha^{\circ 2}[\langle n\rangle+1/2](\cos\Omega t-1)}\right]$$

Thus, *in the presence of damping*, it appears that the classical limit of the quantum average (143) may be obtained in writing in a similar way:

$$\widehat{\mathrm{P}} \to 1 \text{ and } \left(e^{-i\alpha^{\circ 2}\Omega t}\right)\left(e^{i\alpha^{\circ 2}e^{-\gamma t/2}\sin\Omega t}\right) \to 1$$

Hence, the average involved in Eq. (143) admits the semiclassical limit:

$$\left\langle\left\langle \left[e^{ib\int_0^t \tilde{\mathrm{Q}}(t')dt'}\right]^{\mathrm{Semi}}\right\rangle_\theta\right\rangle_{\mathrm{Slow}} = \left[e^{\alpha^{\circ 2}[\langle n\rangle+1/2](2e^{-\gamma t/2}\cos\Omega t-e^{-\gamma t}-1)}\right]$$

As a consequence, the semiclassical limit of the ACF (142) involving quantum indirect damping is

$$[\mathrm{G}_{II}(t)]^{\circ}_{\mathrm{Semi}} = (\mu_0^2)e^{i\omega^\circ t}\left[e^{\alpha^{\circ 2}[\langle n\rangle+1/2](2e^{-\gamma t/2}\cos\Omega t-e^{-\gamma t}-1)}\right] \quad (144)$$

Note that if quantum indirect damping has been taken into account in Eq. (144), on the other hand, the direct one has been ignored. Thus, in order to incorporate this last one into the semiclassical ACF, it is necessary, as above, to multiply Eq. (144) by an exponential decay:

$$[G_{II}(t)]_{\text{Semi}} = (\mu_0^2) e^{i\omega^\circ t} [e^{\alpha^{\circ 2}[\langle n \rangle + 1/2](2e^{-\gamma t/2}\cos\Omega t - e^{-\gamma t} - 1)}](e^{-\gamma^\circ t}) \tag{145}$$

Then, the corresponding semiclassical SD involving both direct and indirect quantum dampings is the Fourier transform of this ACF (145):

$$[I_{II}(\omega)]_{\text{Semi}} = 2(\mu_0^2)\,\text{Re} \int_0^\infty e^{i\omega^\circ t} [e^{\alpha^{\circ 2}[\langle n \rangle + 1/2](2e^{-\gamma t/2}\cos\Omega t - e^{-\gamma t} - 1)}] e^{-\gamma^\circ t} e^{-i\omega t} dt \tag{146}$$

3. *Some Further Classical Approximations*

Moreover, since we are interested in the classical limit situation, it is suitable to look at the high temperature limit of the average of the quantum number operator appearing in Eq. (145). The quantum average is

$$\langle n \rangle = \frac{1}{e^{\tilde{\lambda}} - 1} \qquad \text{with} \qquad \tilde{\lambda} = \frac{\hbar\Omega}{k_B T} \tag{147}$$

Again, its high temperature limit is

$$\langle n \rangle \rightarrow \frac{k_B T}{\hbar\Omega} = \bar{n} \tag{148}$$

In the same way, it is suitable to ignore in Eq. (145) the factor 0.5 related to the zero point energy of the H-bond bridge. This last approximation combined together with the high temperature limit, leads us to introduce in the ACF (145), the approximation:

$$\langle n \rangle + 0.5 \rightarrow \bar{n} \tag{149}$$

At last, in order to simplify, it is suitable to perform in Eq. (145), the following approximation that holds for small γt:

$$e^{-\gamma t} \rightarrow 1 \tag{150}$$

Then, owing to approximations (149) and (150), the semiclassical ACF (145) becomes

$$[G_{II}(t)]_{Cl} = (\mu_0^2) e^{i\omega^\circ t} [e^{2\alpha^{\circ 2}\bar{n}(e^{-\gamma t/2}\cos\Omega t - 1)}] e^{-\gamma^\circ t} \tag{151}$$

The corresponding SD is

$$[\mathrm{I}_{II}(\omega)]_{Cl} = 2(\mu_0^2)\,\mathrm{Re}\int_0^{\infty} e^{i\omega^\circ t}[e^{2\alpha^{\circ 2}\bar{n}(e^{-\gamma t/2}\cos\Omega t-1)}]e^{-\gamma^\circ t}e^{-i\omega t}dt \tag{152}$$

Note that when the direct damping is missing in Eq. (151), this equation gives an explicit expression of the classical limit ACF:

$$[\mathrm{G}_{II}(t)]_{Cl}^{\circ} = (\mu_0^2)e^{i\omega^\circ t}\left\langle\left\langle\left[e^{ib\int_0^t \tilde{\mathrm{Q}}(t')dt'}\right]^{\mathrm{Semi}cl}\right\rangle_{\theta}\right\rangle_{\mathrm{Slow}} \quad \text{if} \quad \gamma^\circ = 0 \tag{153}$$

Thus, there is the following expression for the semiclassical limit of quantum indirect damping:

$$\left\langle\left\langle\left[e^{ib\int_0^t \tilde{\mathrm{Q}}(t')dt'}\right]^{\mathrm{Semi}cl}\right\rangle_{\theta}\right\rangle_{\mathrm{Slow}} = [e^{2\alpha^{\circ 2}\bar{n}(e^{-\gamma t/2}\cos\Omega t-1)}] \tag{154}$$

Equation (153) is the semiclassical limit of the quantum approach of indirect damping. Now, the question may arise as to how Eq. (153) may be viewed from the classical theory of relaxation in order to make a connection with the semiclassical approach of Robertson and Yarwood, which used the classical theory of Brownian motion.

B. The ACF (154) Viewed from the Classical Theory of Brownian Motion [60]

Now, for this purpose consider Eq. (153), where the direct damping γ° is missing, from the viewpoint of classical statistical mechanics dealing with the theory of Brownian motion. First, this leads us to remove the trace over the basis related to the H-bond bridge slow mode in Eq. (153).

$$\langle\cdots\rangle_{\mathrm{Slow}} \rightarrow 1 \tag{155}$$

Besides, the notation $\tilde{\mathrm{Q}}(t)$ for the time-dependent H-bond bridge coordinate interacting with the thermal bath may be viewed as the coordinate of a Brownian oscillator $\mathrm{Q}(t)$, which is a time-dependent stochastic variable:

$$\tilde{\mathrm{Q}}(t) \rightarrow \mathrm{Q}(t) \tag{156}$$

Moreover, introduce the time-dependent stochastic variable $S(t)$, such as,

$$\frac{d\mathrm{S}(t)}{dt} = \mathrm{Q}(t) \tag{157}$$

so that

$$S(t) = \int^{t} Q(t')dt'$$

As a consequence, owing to Eqs.(155–157), the calculation of the ACF (153) viewed from the Brownian (Br) motion theory [61, 62] relies on evaluation of the statistical average:

$$[G_{II}(t)]^{\circ}_{Class} = \langle \mu_0^2 \rangle e^{i\omega^{\circ} t} \langle e^{ibS(t)} \rangle_{Br} \tag{158}$$

As it appears, because of Eq. (155), the passage from the average (153) to the last one (158) leads us to write

$$\langle \langle \cdots \rangle_{\theta} \rangle_{Slow} \rightarrow \langle \cdots \rangle_{Br} \tag{159}$$

In order to compute Eq. (158), write the Langevin equations governing the dynamics of the Brownian oscillator. In the present situation that leads us to consider three time-dependent stochastic variables $S(t)$, $Q(t)$, and $v(t)$, described by the following three equations:

$$\frac{dS(t)}{dt} = Q(t) \tag{160}$$

$$\frac{dQ(t)}{dt} = v(t) \tag{161}$$

$$\frac{dv(t)}{dt} = -\gamma v(t) - \Omega^2 Q(t) + F^{\circ}(t) \tag{162}$$

In the last equation, γ is the usual indirect damping parameter, whereas $F^{\circ}(t)$ is the fluctuating force resulting from the action of the surrounding molecules on the oscillator, the average of which and its corresponding ACF, are given, respectively, by

$$\overline{F^{\circ}(t)} = 0$$

$$\overline{F^{\circ}(t)F^{\circ}(t')} = 2\gamma \frac{k_B T}{M} \delta(t - t')$$

Note that Eq. (160) must be incorporated because the ACF (158) is involving the integral over time of the H-bond bridge coordinate. That is the reason for the presence of the new variable $S(t)$ besides the usual ones $Q(t)$ and $v(t)$. Note that following Lax [63], the integral $S(t)$ of the time-dependent random variable $Q(t)$ is itself a time random variable.

The meaning of the notation $\langle e^{ibS(t)} \rangle$ in Eq. (158) is an average over the distribution function $W(S, Q, v, t)$ of the realization of the random variables S, Q, and v at time t.

Now, perform a cumulant expansion of the average involved in the ACF (158). Then, since the random variable is Gaussian, the expansion of Eq. (158) limits us to second order, so that the following relation holds:

$$\langle e^{ibS(t)} \rangle_{Br} = e^{ib\langle S(t) \rangle} \left[e^{-\frac{b^2}{2}\langle S(t)S(t') \rangle} \right] \tag{163}$$

Of course, the average of the stochastic variable $S(t)$ is zero. Thus, the semiclassical ACF (158) reduces to

$$[G_{II}(t)]^{\circ}_{Class} = (\mu_0^2) e^{i\omega^{\circ} t} \left[e^{-\frac{b^2}{2}\langle S(t)S(0) \rangle} \right] \tag{164}$$

C. The Classical Limit Line Shape [60]

Appendix L shows that the ACF of the stochastic variable $S(t)$ is given by Eq. (L.14), that is,

$$\langle S(t)S(0) \rangle = \langle S^2 \rangle + \frac{\langle Q^2 \rangle}{\Omega^4} \left[(\gamma^2 - \Omega^2)[e^{-\gamma t/2} \cos \tilde{\Omega} t - 1] + \frac{\gamma}{2\tilde{\Omega}} (\gamma^2 - 3\Omega^2) e^{-\gamma t/2} \sin \tilde{\Omega} t \right] \tag{165}$$

with

$$\tilde{\Omega} = \sqrt{\Omega^2 - \frac{\gamma^2}{4}}$$

Now, return to the ACF (164). We may observe that in view of Eq. (47), the anharmonic coefficient b appearing in this ACF, is related to the dimensionless coefficient α° by

$$b = \Omega \alpha^{\circ} \sqrt{\frac{2M\Omega}{\hbar}} \tag{166}$$

Besides, the equipartition energy theorem allows us to write for the average of the potential energy of the harmonic oscillator:

$$\frac{1}{2} M \Omega^2 \langle Q^2 \rangle = \frac{1}{2} k_B T \tag{167}$$

Alternatively, according to Eqs. (147) and (148), the classical limit of the mean number occupation of a quantum harmonic oscillator is

$$\bar{n} = \frac{k_B T}{\hbar \Omega} \tag{168}$$

Thus, as a consequence of Eqs. (166–168), the squared anharmonic coupling parameter b^2 involved in the ACF (164), appears to be given by

$$b^2 = 2\bar{n} \frac{\alpha^{\circ 2}}{\Omega^2} \left[\frac{\Omega^4}{\langle Q^2 \rangle} \right] \tag{169}$$

Equation (165) and this last result (169) allows us to write the ACF (164) in the following form:

$$[G_{II}(t)]^{\circ}_{\text{Class}} = (\mu_0^2) e^{i\omega^{\circ} t} [e^{-b^2 \langle S^2 \rangle /2}] (e^{2\tilde{\beta}^2 \bar{n} e^{-\gamma t/2} \sin \tilde{\Omega} t}) [e^{2\tilde{\alpha}^2 \bar{n} (e^{-\gamma t/2} \cos \tilde{\Omega} t - 1)}] \tag{170}$$

with, respectively,

$$\tilde{\alpha} = \alpha^{\circ} \sqrt{\frac{\Omega^2 - \gamma^2}{\Omega^2}} \tag{171}$$

$$\tilde{\beta} = \alpha^{\circ} \frac{1}{\sqrt{2}} \sqrt{\frac{(3\Omega^2 - \gamma^2)}{\Omega^2} \left(\frac{\gamma}{\tilde{\Omega}} \right)} \tag{172}$$

Of course, we may ignore in Eq. (170) the normalizing factor $\exp\{-b^2 \langle S^2 \rangle /2\}$, because the ACF (170) is a proportionality. Besides, in the underdamped situation, where $\Omega \gg \gamma$, Eqs. (171, 172) reduce to

$$\tilde{\alpha} \simeq \alpha^{\circ} \qquad \tilde{\Omega} \simeq \Omega \qquad \text{and} \qquad \tilde{\beta} \simeq \alpha^{\circ}$$

Again, we may neglect the dephasing involving $\sin \Omega t$. Then, the semiclassical ACF (170) simplifies into

$$[G_{II}(t)]^{\circ}_{\text{Class}} = (\mu_0^2) e^{i\omega^{\circ} t} [e^{2\alpha^{\circ 2} \bar{n} [e^{-\gamma t/2} \cos \Omega t - 1]}] \tag{173}$$

Now, if the direct damping is incorporated as above, Eq. (173) becomes

$$[G_{II}(t)]_{\text{Class}} = (\mu_0^2) e^{i\omega^{\circ} t} [e^{2\alpha^{\circ 2} \bar{n} [e^{-\gamma t/2} \cos \Omega t - 1]}] e^{-\gamma^{\circ} t}$$

As a consequence, the classical spectral density taking into account both direct and indirect dampings become

$$[I_{II}(\omega)]_{\text{Class}} = (\mu_0^2) 2\text{Re} \int_{o}^{\infty} e^{i\omega^{\circ} t} [e^{2\alpha^{\circ 2} \bar{n} (e^{-\gamma t/2} \cos \tilde{\Omega} t - 1)}] e^{-\gamma^{\circ} t} e^{-i\omega t} dt \tag{174}$$

As it appears, the classical spectral density (174) is very similar to the semiclassical approached (144) obtained above and it is equivalent to that given by Eq. (151).

D. Recall on the Robertson–Yarwood Model

Now, in order to get the connection between the present classical approach and that of Robertson and Yarwood [46, 64, 65], let us look at the ACF (158), which was the starting semiclassical expression for Eq. (173), that is,

$$[G_{II}(t)]^{\circ}_{\text{Class}} = (\mu_0^2)e^{i\omega^{\circ}t}\left\langle\left[e^{ib\int^{t}Q(t')dt'}\right]\right\rangle_{\text{Br}} \tag{175}$$

Now, perform the following change:

$$\int^{t}Q(t')dt' \rightarrow \int_{o}^{t}Q(t')dt'$$

This leads to the starting equation of Robertson and Yarwood in their semiclassical approach of indirect damping:

$$[G_{II}(t)]^{\circ}_{\text{Rob}} = (\mu_0^2)e^{i\omega^{\circ}t}\left\langle\left[e^{ib\int_{o}^{t}Q(t')dt'}\right]\right\rangle_{\text{Br}} \tag{176}$$

Next, expand the characteristic function in cummulants up to second order. This leads to

$$\left\langle\left[e^{ib\int_{o}^{t}Q(t')dt'}\right]\right\rangle_{\text{Br}} = e^{ib\left\langle\int_{o}^{t}Q(t')dt'\right\rangle}e^{-\frac{b^2}{2}\left\langle\int_{o}^{t}\int_{o}^{t'}Q(t')Q(t'')dt'dt''\right\rangle}$$

Again, since the first right-hand side term of the expansion is unity because the average of the exponential argument is zero, that is, $\langle Q(t')\rangle = 0$, the above expression becomes:

$$\left\langle e^{ib\int_{o}^{t}Q(t')dt'}\right\rangle_{\text{Br}} = e^{-\frac{b^2}{2}\left\langle\int_{o}^{t}\int_{o}^{t'}Q(t')Q(t'')dt'dt''\right\rangle} \tag{177}$$

In their pioneering work, Robertson and Yarwood [46] implicitly assumed that the integration procedure and the thermal average dealing with stochastic variables are commuting, that is,

$$\left\langle\int_{0}^{t}\int_{0}^{t'}Q(t')Q(t'')dt'dt''\right\rangle = \int_{0}^{t}\int_{0}^{t'}\langle Q(t')Q(t'')\rangle dt'dt'' \tag{178}$$

Thus, they have written in place of Eq. (177), the following equation:

$$\left\langle e^{ib\int_0^t Q(t')dt'} \right\rangle_{Br} = \left(e^{-\frac{b^2}{2}\int_0^t \int_0^{t'} \langle Q(t')Q(t'')\rangle dt'dt''} \right) \tag{179}$$

Because of Eq. (179), the authors have transformed the ACF (176) into

$$[G_{II}(t)]^{\circ}_{Rob} = (\mu_0^2)e^{i\omega^{\circ}t}\left(e^{-\frac{b^2}{2}\int_0^t \int_0^{t'} \langle Q(t')Q(t'')\rangle dt'dt''} \right) \tag{180}$$

Next, the authors used the Uhlenbeck and Ornstein results [66] dealing with the ACF $\langle Q(t')Q(t'')\rangle$, which is found in Appendix L, to be given by Eq. (L.4), that is,

$$\langle Q(t')Q(t'')\rangle = \langle Q^2\rangle e^{-\gamma t/2}\left[\cos[\tilde{\Omega}(t'-t'')] + \frac{\gamma}{2\tilde{\Omega}}\sin[\tilde{\Omega}(t'-t'')]\right]$$

because of the stationary character of this ACF:

$$\langle Q(t')Q(t'')\rangle = \langle Q(t)Q(0)\rangle \qquad \text{with} \qquad t = t' - t''$$

Then, the ACF (180) takes the form:

$$[G_{II}(t)]^{\circ}_{Rob} = (\mu_0^2)e^{i\omega^{\circ}t}\left(e^{-\frac{b^2}{2}\langle Q^2\rangle \int_0^t \int_0^{t'} [\cos\tilde{\Omega}(t'-t'') + \frac{\gamma}{2\tilde{\Omega}}\sin\tilde{\Omega}(t'-t'')]dt'dt''} \right) \tag{181}$$

After integration and using Eqs. (167) – (168). (169), the ACF (181) takes the final form:

$$[G_{II}(t)]^{\circ}_{Rob} = (\mu_0^2)e^{i\omega^{\circ}t}(e^{-\Gamma t})[e^{2\tilde{\alpha}^2\bar{n}(e^{-\gamma t/2}\cos\tilde{\Omega}t-1)}][e^{\tilde{\beta}^2 2\bar{n}e^{-\gamma t/2}\sin\tilde{\Omega}t}] \tag{182}$$

where $\tilde{\alpha}$ and $\tilde{\beta}$ are, respectively, given by Eqs.(171) and (172), whereas Γ is an effective damping parameter given by

$$\Gamma = 2[\alpha^{\circ 2}\bar{n}]\gamma \tag{183}$$

Moreover, in the underdamped situation, according to Eqs. (171) and (172), $\tilde{\alpha}^2 > \tilde{\beta}^2$, the ACF (182) simplifies into

$$[G_{II}(t)]^{\circ}_{Rob} = (\mu_0^2)e^{i\omega^{\circ}t}(e^{-\Gamma t})[e^{2\tilde{\alpha}^2\bar{n}(e^{-\gamma t/2}\cos\tilde{\Omega}t-1)}] \tag{184}$$

The ACF (184) holds in the absence of direct damping. Its introduction, as above, requires us to multiply it by an exponential decay $e^{-\gamma^{\circ}t}$ involving the direct damping parameter γ°. This leads to

$$[G_{II}(t)]_{\text{Rob}} = (\mu_0^2)e^{i\omega^\circ t}(e^{-\Gamma t})[e^{2\tilde{\alpha}^2\bar{n}(e^{-\gamma t/2}\cos\tilde{\Omega}t-1)}](e^{-\gamma^\circ t})$$

Then, after Fourier transform, one obtains the following approached expression for the Robertson and Yarwood SD:

$$[I_{II}(\omega)]_{\text{Rob}} = (\mu_0^2)2\text{Re}\int_0^\infty e^{i\omega^\circ t}(e^{-(\Gamma+\gamma^\circ)t})[e^{2\tilde{\alpha}^{\circ 2}\bar{n}(e^{-\gamma t/2}\cos\Omega t-1)}]e^{-i\omega t}dt \qquad (185)$$

As seen, this SD is very near the semiclassical one (174) except for the presence of the decay term $e^{-(\Gamma+\gamma^\circ)t}$, in place of the simple one $e^{-\gamma^\circ t}$. The behavior of the additional decay term $e^{-\Gamma t}$ is the same as that of the direct damping that is $e^{-\gamma^\circ t}$, although this additional term is of indirect nature, Γ being a function of the indirect damping parameter γ via Eq. (183).

E. Comparison of the Approached Semiclassical SDs Involving Indirect Damping to the Quantum One

Now, it may be of interest to compare the different SDs incorporating both direct and indirect dampings we have met, that is, SDs (126), (146), (152), (174) and (185)

This is performed in Fig. 6. For this purpose, two spectral densities are superposed in each case, one of reference involving only some lot of weak direct damping (grayed line shapes) and the other involving both direct and indirect damping (full lines).

The different situations illustrated in Fig. 6 correspond more and more to approximate models when passing up to down. The top of the figure given the case described by Eq. (126) and corresponds to the reference quantum indirect damping. Below, is depicted the situation described by Eq. (146), where the Dyson time ordering operator is ignored in the quantum model. Further below is given the behavior that corresponds to both Eqs. (152) and (174), that is, to weak approximations on the classical limit of the quantum model [Eq. (152)] and to the semiclassical model [Eq. (174)]. At last, at the bottom we visualize the semiclassical model of Robertson and Yarwood [described by Eq. (185)].

For all situations there is superposed the same SD computed by aid of the reference eq. 126 but with $\gamma = 0$

First, one may observe that the asymmetry of the line shape appears only in the quantum approach corresponding to the top of the figure. One may also remark about the similarity in the line shapes that are intermediate between that of the quantum model and that of the semiclassical Robertson and Yarwood model appearing at the bottom.

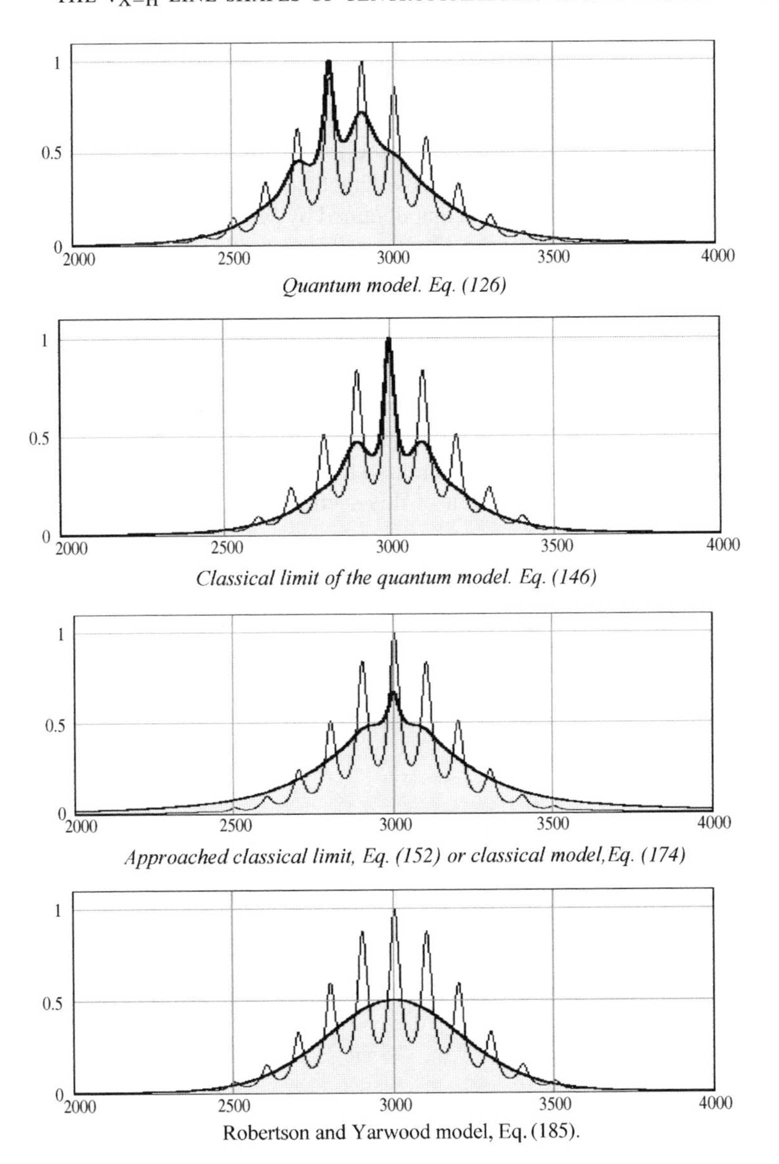

Figure 6. Comparisons of the spectral densities according to the quantum and different semiclassical models. $\omega^\circ = 3000\,\mathrm{cm}^{-1}$, $\Omega = 100\,\mathrm{cm}^{-1}$, $\alpha^\circ = 1$, $T = 300\,\mathrm{K}$, $\gamma^\circ = 0.20\,\Omega$: for all situations there is superposed the same SD computed by aid of eq. 126 with $\gamma = 0$, *grayed*: $\gamma = 0.80\,\Omega$.

V. APPROXIMATIONS FOR QUANTUM INDIRECT DAMPING

Section IV shows that the classical limit of the quantum indirect damping leads to a spectral density that reduces to the semiclassical one obtained by Robertson and Yarwood [46]. Besides, the SD involving quantum indirect damping reduces, when this damping is missing, to that obtained by Maréchal and Witkowski [18]. Both these results show that the quantum approach of indirect damping is well behaved. Unfortunately, however, the quantum theory of indirect damping is somewhat difficult to handle when the H-bonded species are involving, not only the strong anharmonic coupling, but also other mechanisms, such as Fermi resonances.

Thus, owing to the suitability of quantum theory of indirect damping and to its difficulty in being generalized to complex situations, it appears to be of interest to find a suitable approximation to take into account simply the quantum indirect damping. This is the aim of this section [67].

A. A New Expression of Eq. (124)

Previously, it was shown that the ACF of the H-bond with quantum indirect damping, but without direct damping, is given by Eq. (124), that is, by

$$[\mathrm{G}_{II}(t)]^{\circ} = (\mu_0^2)e^{i\omega^{\circ}t}(e^{-i2\alpha^{\circ 2}\Omega t})(e^{i\alpha^{\circ 2}e^{-\gamma t/2}\sin\Omega t})[e^{\alpha^{\circ 2}[\langle n\rangle+1/2](2e^{-\gamma t/2}\cos(\Omega t)-e^{-\gamma t}-1)}] \tag{186}$$

Now, expand the $\sin\Omega t$ and $\cos\Omega t$ functions appearing in Eq. (186) in complex exponentials. This leads us to

$$[\mathrm{G}_{II}(t)]^{\circ} =(\mu_0^2)e^{i\omega^{\circ}t}(e^{-i2\alpha^{\circ 2}\Omega t})$$
$$(e^{\frac{\alpha^{\circ 2}}{2}e^{-\gamma t/2}[e^{i\Omega t}-e^{-i\Omega t}]})[e^{[\alpha^{\circ 2}[\langle n\rangle+1/2]e^{-\gamma t/2}[e^{i\Omega t}+e^{-i\Omega t}]-\alpha^{\circ 2}[\langle n\rangle+1/2]e^{-\gamma t}-\alpha^{\circ 2}[\langle n\rangle+1/2]]}]$$

or

$$[\mathrm{G}_{II}(t)]^{\circ} =(\mu_0^2)e^{i\omega^{\circ}t}(e^{-i2\alpha^{\circ 2}\Omega t})(e^{\frac{\alpha^{\circ 2}}{2}e^{-\gamma t/2}e^{i\Omega t}-\frac{\alpha^{\circ 2}}{2}e^{-\gamma t/2}e^{-i\Omega t}})$$
$$[e^{\alpha^{\circ 2}[\langle n\rangle+1/2]e^{-\gamma t/2}e^{i\Omega t}+\alpha^{\circ 2}[\langle n\rangle+1/2]e^{-\gamma t/2}e^{-i\Omega t}-\alpha^{\circ 2}[\langle n\rangle+1/2]e^{-\gamma t}-\alpha^{\circ 2}[\langle n\rangle+1/2]}]$$

Next, separate each exponential and rearrange. This gives

$$[\mathrm{G}_{II}(t)]^{\circ} =(\mu_0^2)e^{i\omega^{\circ}t}(e^{-i2\alpha^{\circ 2}\Omega t})[e^{-\alpha^{\circ 2}[\langle n\rangle+1/2]e^{-\gamma t}}][e^{-\alpha^{\circ 2}[\langle n\rangle+1/2]}]$$
$$[e^{\alpha^{\circ 2}e^{-\gamma t/2}e^{i\Omega t}[\langle n\rangle+1]}][e^{\alpha^{\circ 2}e^{-\gamma t/2}e^{-i\Omega t}\langle n\rangle}]$$

Moreover, the two last exponentials may be expanded

$$[G_{II}(t)]^{\circ} = (\mu_0^2)e^{i\omega^{\circ}t}(e^{-i2\alpha^{\circ 2}\Omega t})[e^{-\alpha^{\circ 2}[\langle n\rangle+1/2]e^{-\gamma t}}][e^{-\alpha^{\circ 2}[\langle n\rangle+1/2]}]$$
$$\sum_k\sum_l \frac{\alpha^{\circ 2l}e^{-l\gamma t/2}e^{il\Omega t}[\langle n\rangle+1]^l}{l!}\frac{\alpha^{\circ 2k}e^{-k\gamma t/2}e^{-ik\Omega t}\langle n\rangle^k}{k!}$$

Finally, the following is obtained

$$[G_{II}(t)]^{\circ} = (\mu_0^2)e^{i\omega^{\circ}t}(e^{-i2\alpha^{\circ 2}\Omega t})[e^{-\alpha^{\circ 2}[\langle n\rangle+1/2]e^{-\gamma t}}][e^{-\alpha^{\circ 2}[\langle n\rangle+1/2]}]$$
$$\sum_k\sum_l\left[\frac{(\langle n\rangle+1)^l}{l!}\right]\left[\frac{\langle n\rangle^k}{k!}\right]\alpha^{\circ 2(l+k)}(e^{-(k+l)\gamma t/2})(e^{-i(k-l)\Omega t}) \quad (187)$$

Observe that, when $\gamma = 0$ and Eq. (187) reduces to

$$[G^{\circ}{}_{II}(t)]^{\circ} = (\mu_0^2)e^{i\omega^{\circ}t}(e^{-i2\alpha^{\circ 2}\Omega t})$$
$$[e^{-2\alpha^{\circ 2}(\langle n\rangle+1/2)}]\sum_k\sum_l\left[\frac{(\langle n\rangle+1)^l}{l!}\right]\left[\frac{\langle n\rangle^k}{k!}\right]\alpha^{\circ 2(l+k)}(e^{i(l-k)\Omega t}) \quad (188)$$

Of course, this last expression must be equivalent to (186) in which γ is taken to be zero, that is,

$$[G^{\circ}{}_{II}(t)]^{\circ} = (\mu_0^2)e^{i\omega^{\circ}t}(e^{-i2\alpha^{\circ 2}\Omega t})(e^{i\alpha^{\circ 2}\sin\Omega t})[e^{2\alpha^{\circ 2}[\langle n\rangle+1/2](\cos(\Omega t)-1)}] \quad (189)$$

As may be verified, this last expression is the same as Eq. (69). Moreover, this expression of the ACF without damping in representation $\{II\}$ has been shown previously to be equivalent to expression (79) in representation $\{III\}$, which is

$$[G^{\circ}{}_{III}(t)]^{\circ} = (\mu_0^2)e^{i\omega^{\circ}t}(e^{-i2\alpha^{\circ 2}\Omega t})\varepsilon\sum_k\sum_l(e^{-k\tilde{\lambda}})|A_{kl}(\alpha^{\circ})|^2(e^{i(l-k)\Omega t}) \quad (190)$$

Here, $A_{kl}(\alpha)$ are the Franck–Condon factors, that is, the matrix elements of the translation operator in the basis where $a^{\dagger}a$ is diagonal:

$$[A(\alpha^{\circ})] = [e^{\alpha^{\circ}a^{\dagger}-\alpha^{\circ}a}] \quad (191)$$

Since in the absence of damping the three expressions (188–190) are equivalent, we get by identification of Eqs. (188) and (190):

$$[e^{-2\alpha^{\circ 2}(\langle n\rangle+1/2)}]\sum_k\sum_l\left[\frac{(\langle n\rangle+1)^l}{l!}\right]\left[\frac{\langle n\rangle^k}{k!}\right]\alpha^{\circ 2(l+k)}(e^{i(l-k)\Omega t})$$
$$=\varepsilon\sum_k\sum_l(e^{-k\tilde{\lambda}})|A_{kl}(\alpha^\circ)|^2(e^{i(l-k)\Omega t}) \quad (192)$$

This result shows that the coefficients are the same for each component of angular frequency $(l-k)\Omega$, that is,

$$[e^{-2\alpha^{\circ 2}(\langle n\rangle+1/2)}]\left[\frac{(\langle n\rangle+1)^l}{l!}\right]\left[\frac{\langle n\rangle^k}{k!}\right]\alpha^{\circ 2(l+k)}=\varepsilon(e^{-k\tilde{\lambda}})|A_{kl}(\alpha^\circ)|^2 \quad (193)$$

Note that at zero temperature, $\tilde{\lambda}\to\infty, \varepsilon\to 1$, $e^{-k\tilde{\lambda}}\to\delta_{k,0}$, and $\langle n\rangle\to 0$, so that, as required by Eq. (C.15) giving the Franck–Condon factors, Eq. (193) reduces to

$$|A_{0l}(\alpha^\circ)|^2=e^{-\alpha^{\circ 2}}\frac{\alpha^{\circ 2l}}{l!} \quad (194)$$

B. An Approximation for Eq. (187)

Now, return to the situation where the quantum indirect damping is taken into account.

1. *Basic Assumption*

The comparison of Eq. (182) corresponding to the situation with damping and of Eq. (188) corresponding to the situation without damping, leads to the following conclusion: The incorporation of the quantum indirect damping in the model introduces in to the energy levels of the H-bond bridge viewed as a quantum harmonic oscillator, an imaginary part reflecting this damping. This imaginary part is linearly increasing with the quantum number characterizing the levels. Thus, owing to this fact and to the equivalence between Eqs. (188) and (190), it appears to be suitable to suppose that the quantum indirect damping may induce in the same way the same consequence in the ACF (187). Thus, in order to *incorporate the quantum indirect damping*, write.

$$[\mathrm{G}_{III}^{\mathrm{eff}}(t)]^\circ\simeq(\mu_0^2)e^{i\omega^\circ t}(e^{-i2\alpha^{\circ 2}\Omega t})\varepsilon\sum_k\sum_l(e^{-k\tilde{\lambda}})|A_{kl}(\alpha^\circ)|^2(e^{-(k+l)\gamma t/2})(e^{-i(k-l)\Omega t}) \quad (195)$$

Of course, the validity of this assumption has to be verified *a posteriori*. Besides, such an assumption implies a near equivalence between the exact Eq. (187) and its approximation given by Eq. (195), that is,

$$[e^{-\alpha^{\circ 2}[\langle n\rangle+1/2]e^{-\gamma t}}][e^{-\alpha^2(\langle n\rangle+1/2)}]$$

$$\sum_k\sum_l\left[\frac{(\langle n\rangle+1)^l}{l!}\right]\left[\frac{\langle n\rangle^k}{k!}\right]\alpha^{\circ 2(l+k)}(e^{-(k+l)\gamma t/2})(e^{-i(k-l)\Omega t})$$

$$=\varepsilon\sum_k\sum_l(e^{-k\tilde{\lambda}})|A_{kl}(\alpha^\circ)|^2(e^{-(k+l)\gamma t/2})(e^{-i(k-l)\Omega t})$$

This last expression may be viewed as the basic approximation in order to treat more easily the quantum indirect damping.

2. *Implication of the Approximation*

Next, if the above approximation is verified, each term of the double sums must be equivalent. Thus, after simplification, each term of the double summation has to verify:

$$[e^{-\alpha^{\circ 2}(\langle n\rangle+1/2)}]\left[\frac{(\langle n\rangle+1)^l}{l!}\right]\left[\frac{\langle n\rangle^k}{k!}\right](\alpha^{\circ 2(l+k)})\{e^{-\alpha^{\circ 2}[\langle n\rangle+1/2]e^{-\gamma t}}\}\simeq\varepsilon(e^{-k\tilde{\lambda}})|A_{kl}(\alpha^\circ)|^2 \tag{196}$$

The right-hand sides of Eqs. (196) and (193) are the same so that, by identification, the following result is obtained

$$[e^{-\alpha^{\circ 2}(\langle n\rangle+1/2)}]\left[\frac{(\langle n\rangle+1)^l}{l!}\right]\left[\frac{\langle n\rangle^k}{k!}\right](\alpha^{\circ 2(l+k)})\{e^{-\alpha^{\circ 2}[\langle n\rangle+1/2]e^{-\gamma t}}\}$$

$$\simeq\{e^{-2\alpha^{\circ 2}(\langle n\rangle+1/2)}\}\left[\frac{(\langle n\rangle+1)^l}{l!}\right]\left[\frac{\langle n\rangle^k}{k!}\right](\alpha^{\circ 2(l+k)})$$

After simplification this leads to

$$\{e^{-\alpha^{\circ 2}[\langle n\rangle+1/2]e^{-\gamma t}}\}\simeq\{e^{-\alpha^{\circ 2}(\langle n\rangle+1/2)}\}$$

As it appears, this result is as well verified as the product γt is small.

3. *The ACF in Terms of Complex Energy Levels*

Note that the approached ACF (195) may be rewritten in terms of complex energy levels, with the aid of the following equations:

$$[\mathrm{G}_{III}^{\mathrm{eff}}(t)]^{\circ} = (\mu_0^2)\varepsilon \sum_m \sum_n (e^{-n\tilde{\lambda}}) \left[e^{\frac{it}{\hbar}E_m^{\{1\}*}} \right] |A_{mn}(\alpha^{\circ})|^2 \left[e^{-\frac{it}{\hbar}E_n^{\{0\}}} \right] \tag{197}$$

$$E_n^{\{0\}} = n\hbar\Omega\left(1 - i\frac{\gamma}{2\Omega}\right) \tag{198}$$

$$E_m^{\{1\}} = \hbar\omega^{\circ} + m\hbar\Omega\left(1 - i\frac{\gamma}{2\Omega}\right) - 2\alpha^{\circ 2}\hbar\Omega \tag{199}$$

As it appears by inspection of Eqs. (198) and (199), the imaginary part of the slow mode sublevels, and therefore their thickness, are linearly increasing with their excitation degree.

Note that, as required, the ACF (197) reduces *in the absence of indirect damping*, that is, when $\gamma \to 0$, to that obtained above (79), which is at the basis of the Maréchal and Witkowski Franck–Condon progression, that is,

$$[\mathrm{G}^{\circ}{}_{III}(t)]^{\circ} = (\mu_0^2)e^{i\omega^{\circ}t}(e^{-i2\alpha^{\circ 2}\Omega t})\varepsilon \sum_m \sum_n (e^{-\tilde{\lambda}n})|A_{mn}(\alpha^{\circ})|^2(e^{-i(m-n)\Omega t})$$

4. *The SD Involving Indirect Damping in Terms of Lorentzian of Different Half-Widthes*

The approached ACF (197) takes into account only the indirect damping. In order to introduce direct damping, it is necessary to multiply it by the usual exponential decay $e^{-\gamma^{\circ}t}$.

Thus, after using Eqs. (198) and (199), the approached ACF (197) *involving both direct and indirect dampings* takes the form:

$$[\mathrm{G}_{III}^{\mathrm{eff}}(t)] = (\mu_0^2)\varepsilon \sum_m \sum_n |A_{mn}(\alpha^{\circ})|^2 (e^{-n\tilde{\lambda}})(e^{i(\omega^{\circ}+m\Omega-2\alpha^{\circ 2}\Omega/2)t})(e^{-in\Omega t})\left(e^{-(m+n)\frac{\gamma}{2}t}\right)e^{-\gamma^{\circ}t} \tag{200}$$

The corresponding approached SD involving indirect damping is the Fourier transform of (200), that is,

$$[\mathrm{I}_{III}^{\mathrm{eff}}(\omega)] = (\mu_0^2)\varepsilon \sum_m \sum_n |A_{mn}(\alpha^{\circ})|^2 (e^{-n\hbar\Omega/k_B T}) \frac{\left((m+n)\frac{\gamma}{2} + \gamma^{\circ}\right)}{[\omega - (\omega^{\circ} + (m-n)\Omega - 2\alpha^{\circ 2}\Omega)]^2 + \left((m+n)\frac{\gamma}{2} + \gamma^{\circ}\right)^2} \tag{201}$$

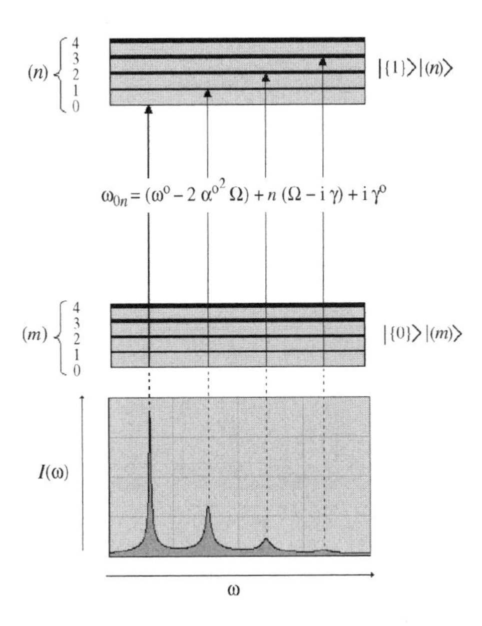

Figure 7. Spectral analysis at $T = 0\,\mathrm{K}$ in the presence of indirect damping. $\omega^{\circ} = 3000\,\mathrm{cm}^{-1}$ $\Omega = 100\,\mathrm{cm}^{-1}$, $\alpha^{\circ} = 0.025\,\Omega$, $\gamma = 0.10\,\Omega$.

Of course, when the indirect damping is missing, this SD reduces to that given by Eq. (82), which we obtained above for the situation involving only direct damping.

Besides, examination of Eq. (201) shows that the spectral density is the sum of different components, each of them being a superposition of Lorentzians involving different half-widthes and intensities. Note that this result differs deeply from the situation without indirect damping given by Eq. (82), for which all the Lorentzians forming the line shape have the same half-width. This finding may be verified by taking $\gamma = 0$ in Eq. (201).

This is visualized in Fig. 7 for zero temperature and in Fig. 8 for room temperature.

This figure may be compared to Fig. 5, which illustrates the corresponding situation where the indirect damping is missing.

C. Component Analysis of the Whole Line Shape

According to Eqs. (198) and (199), the indirect damping introduces in the energy sublevels $|\{0\}\rangle|(m)\rangle$ and $|\{1\}\rangle|(n)\rangle$ of the H-bond bridge imaginary parts that are increasing with the excitation degree (m) or (n) of the H-bond bridge. This leads to a broadening of the $|\{0\}\rangle|(m)\rangle \rightarrow |\{1\}\rangle|(n)\rangle$ transitions that linearly increases with (m) and (n). Therefore, the half-widths of the $|\{0\}\rangle|(m)\rangle \rightarrow |\{1\}\rangle|(n)\rangle$

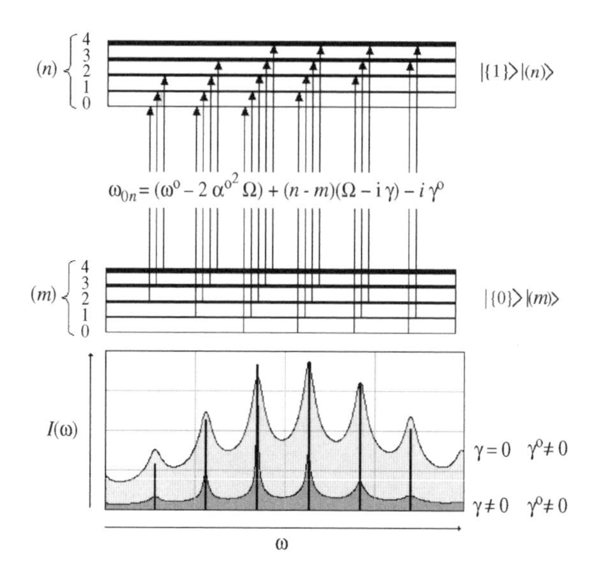

Figure 8. Spectral analysis involving direct and indirect dampings at $T = 300\,\text{K}$. The direct damping parameter has been chosen greater ($\gamma^\circ = 0.25\,\Omega$) when the indirect damping is missing, than ($\gamma^\circ = 0.025\,\Omega$) when it is present ($\gamma = 0.1\,\Omega$) in order to distinguish clearly the spectral densities. Dirac delta peaks are corresponding to the situation without any damping. $\omega^\circ = 3000\,\text{cm}^{-1}$, $\Omega = 100\,\text{cm}^{-1}$, $\omega^\circ = 1$.

transitions are increasing with (m) and (n), in a $(m+n)\gamma/2$ form. Of course, the $|\{0\}\rangle|(0)\rangle \rightarrow |\{1\}\rangle|(0)\rangle$ transition cannot be affected by indirect damping, so that in the absence of direct damping , it is given by a Dirac delta peak. Thus, caused by this specific broadening of the lines corresponding to these transitions, there is a concomitant specific lowering of each component intensity that is deeply correlated to this broadening. This component is more and more sensitive when passing from the central $|\{0\}\rangle|(0)\rangle \rightarrow |\{1\}\rangle|(0)\rangle$ transition to its low and high frequency tails.

As a consequence, when both direct and quantum indirect dampings are occurring together, the combination of the two relaxation mechanisms induces, when the indirect damping is increased at constant direct damping, a progressive collapse of the low and high frequency tails of the spectrum evoking some narrowing [15, 68].

The component analysis of the whole line shape is illustrated in Figs. 7 and 8.

These figures need some comments: If they have the merit to allow a pictorial representation of the whole line shape as a sum of transition between damped energy levels, unfortunately they may lead us to think erroneously that the H-bond bridge is damped through an irreversible process, even when the fast mode is in its ground state.

It remains more suitable to look simply at the SD pictured on Figs. 7 and 8 as the Fourier transform of the ACF (118), that is,

$$[\mathrm{I}_{II}(\omega)]^{\circ} = (\mu_0^2)\varepsilon \sum_n e^{-\tilde{\lambda}n} 2\mathrm{Re} \int_0^{\infty} e^{i\omega^{\circ}t}(e^{-i\alpha^{\circ 2}\Omega t})(U_{II}^{\{1\}}(t)_{nn}^{\mathrm{IP}})^* e^{-i\omega t} dt$$

In this last equation, the right-hand side matrix elements are those of the IP time evolution operator of the driven damped quantum harmonic oscillator describing the H-bond bridge when the fast mode is in its first excited state:

$$(U_{II}^{\{1\}}(t)_{nn}^{\mathrm{IP}})^* = \langle(n)|[\mathrm{U}_{II}^{\{1\}}(t)^{\mathrm{IP}}]^{-1}|(n)\rangle$$

From this viewpoint, which is the most fundamental, the line shape as a whole is the sum of the diagonal matrix elements of the time evolution operator of the driven damped quantum harmonic oscillator in the IP representation with respect to the diagonal part of the Hamiltonian of this oscillator. According to Eq. (120), each diagonal element is a sum of time-dependent terms

$$(U_{II}^{\{1\}}(t)_{nn}^{\mathrm{IP}})^* = (e^{-i\alpha^{\circ 2}\Omega t})(e^{i\alpha^{\circ 2}\sin\Omega t})\langle(n)|[e^{\Phi^{\circ}_{II}(t)^*\mathrm{a}^{\dagger} - \Phi^{\circ}_{II}(t)\mathrm{a}}]|(n)\rangle$$

D. Effective Non-Hermitean Hamiltonians

Now, we show that the approximation that has been performed to treat the quantum indirect damping allows us to find other approximations for handling the quantum indirect relaxation in which the damping of the H-bond bridge is taken into account by aid of non-Hermitean effective Hamiltonians.

1. *Effective Non-Hermitean Hamiltonians in Representation* $\{III\}$

The complex energy levels (198) and (199) characterizing the initial and final states may be considered as the eigenvalues of the effective non-Hermitean Hamiltonians characterizing the H-bond bridge:

$$[\tilde{\mathrm{H}}_{III}^{\{k\}}]|(n)\rangle = E_n^{\{k\}}|(n)\rangle \tag{202}$$

Thus, in view of Eqs. (198) and (199), these eigenvalue equations are

$$[\tilde{\mathrm{H}}_{III}^{\{0\}}]|(n)\rangle = n\left(1 - i\frac{\gamma}{2\Omega}\right)\hbar\Omega|(n)\rangle$$

$$[\tilde{\mathrm{H}}_{III}^{\{1\}}]|(n)\rangle = \left[\hbar\omega^{\circ} + n\left(1 - i\frac{\gamma}{2\Omega}\right)\hbar\Omega - 2\alpha^{\circ 2}\hbar\Omega\right]|(n)\rangle$$

Consequently, because of these eigenvalue equations, it is clear that the effective non-Hermitean Hamiltonians are, respectively, given by

$$[\tilde{\mathrm{H}}_{III}^{\{0\}}] = \mathrm{a}^{\dagger}\mathrm{a}\left(1 - i\frac{\gamma}{2\Omega}\right)\hbar\Omega \tag{203}$$

$$[\tilde{\mathrm{H}}_{III}^{\{1\}}] = \hbar\omega^{\circ} + \mathrm{a}^{\dagger}\mathrm{a}\left(1 - i\frac{\gamma}{2\Omega}\right)\hbar\Omega - 2\alpha^{\circ 2}\hbar\Omega \tag{204}$$

The reason for the subscript *III* is that when the indirect damping is missing, the non-Hermitean Hamiltonians (203) and (204) reduce to the Hermitean Hamiltonians (C.8) and (C.7).

Note that in the eigenvalue equation (204) the zero-point energy was not taken into account, and so the corresponding factor 0.5 in the expression of the Hamiltonians in terms of Bosons.

2. *The ACF Within Representation* $\{III\}$ *for Non-Hermitean Hamiltonians*

Now, return to the ACF (197), that is,

$$[\tilde{\mathrm{G}}_{III}^{\mathrm{eff}}(t)]^{\circ} = (\mu_0^2)\varepsilon \sum_m \sum_n (e^{-n\tilde{\lambda}}) \left[e^{\frac{it}{\hbar}E_m^{\{1\}*}}\right] |A_{mn}(\alpha^{\circ})|^2 \left[e^{-\frac{it}{\hbar}E_n^{\{0\}}}\right]$$

Next, explicit the square of the absolute value of the translation operator matrix elements according to

$$|A_{mn}(\alpha^{\circ})|^2 = \langle (n)|\mathrm{A}(\alpha^{\circ})^{-1}|(m)\rangle\langle (m)|\mathrm{A}(\alpha^{\circ})|(n)\rangle$$

Then, the above ACF takes the form:

$$[\tilde{\mathrm{G}}_{III}^{\mathrm{eff}}(t)]^{\circ} = (\mu_0^2)\,\varepsilon \sum_m \sum_n (e^{-n\tilde{\lambda}}) \left[e^{\frac{it}{\hbar}E_m^{\{1\}*}}\right] \langle (n)|[\mathrm{A}(\alpha^{\circ})]^{-1}|(m)\rangle\langle (m)|[\mathrm{A}(\alpha^{\circ})]|(n)\rangle \left[e^{-\frac{it}{\hbar}E_n^{\{0\}}}\right]$$

Again, displace the phase factors with the translation operator and also the *bras*$\langle (n)|$ and the *kets*$|(n)\rangle$ in the following way:

$$[\tilde{\mathrm{G}}_{III}^{\mathrm{eff}}(t)]^{\circ} = (\mu_0^2)\varepsilon \sum_m \sum_n \langle (n)|(e^{-n\tilde{\lambda}})[\mathrm{A}(\alpha^{\circ})]^{-1} \left[e^{\frac{it}{\hbar}E_m^{\{1\}*}}\right] |(m)\rangle\langle (m)|[\mathrm{A}(\alpha^{\circ})] \left[e^{-\frac{it}{\hbar}E_n^{\{0\}}}\right] |(n)\rangle$$

Then, using the eigenvalue equations (202) it becomes

$$[\tilde{G}^{\text{eff}}_{III}(t)]^{\circ} = (\mu_0^2)\varepsilon \sum_m \sum_n \langle(n)|(e^{-\tilde{\lambda} a^{\dagger} a})[A(\alpha^{\circ})]^{-1}\left[e^{\frac{it}{\hbar}\tilde{H}^{\{1\}\dagger}_{III}}\right]|(m)\rangle\langle(m)|[A(\alpha^{\circ})]\left[e^{-\frac{it}{\hbar}\tilde{H}^{\{0\}}_{III}}\right]|(n)\rangle$$

At last, the closeness relation may be equated to unity, and then the trace operation over the basis $|(n)\rangle$ may be made implicit to give

$$[\tilde{G}^{\text{eff}}_{III}(t)]^{\circ} = (\mu_0^2)\varepsilon \, \text{tr}_{\text{Slow}}\left\{(e^{-\tilde{\lambda} a^{\dagger} a})[A(\alpha^{\circ})]^{-1}\left[e^{\frac{it}{\hbar}\tilde{H}^{\{1\}\dagger}_{III}}\right][A(\alpha^{\circ})]\left[e^{-\frac{it}{\hbar}\tilde{H}^{\{0\}}_{III}}\right]\right\} \quad (205)$$

3. *The ACF in Terms of Non-Hermitean Hamiltonians*

Now, the ACF (205) may be transformed, because of (191), into

$$[\tilde{G}^{\text{eff}}_{III}(t)]^{\circ} = (\mu_0^2) \varepsilon \, \text{tr}_{\text{Slow}}\left\{(e^{-\tilde{\lambda} a^{\dagger} a})(e^{-\alpha^{\circ} a^{\dagger} + \alpha^{\circ} a})\left[e^{\frac{it}{\hbar}\tilde{H}^{\{1\}\dagger}_{III}}\right](e^{\alpha^{\circ} a^{\dagger} - \alpha^{\circ} a})\left[e^{-\frac{it}{\hbar}\tilde{H}^{\{0\}}_{III}}\right]\right\} \quad (206)$$

Next, it may be seen that the effective Hamiltonians in representations $\{III\}$ are related to those in representation $\{II\}$ by equations similar to those given by Eqs. (C.3) and (C.4),

$$[\tilde{H}^{\{0\}}_{II}] = [\tilde{H}^{\{0\}}_{III}]$$
$$[\tilde{H}^{\{1\}}_{II}] = (e^{-\alpha^{\circ} a^{\dagger} + \alpha^{\circ} a})[\tilde{H}^{\{1\}}_{III}](e^{\alpha^{\circ} a^{\dagger} - \alpha^{\circ} a}) \quad (207)$$

As a consequence, the Hamiltonians (203) and (204) become

$$[\tilde{H}^{\{0\}}_{II}] = a^{\dagger} a\left(1 - i\frac{\gamma}{2\Omega}\right)\hbar\Omega \quad (208)$$

$$[\tilde{H}^{\{1\}}_{II}] = \hbar\omega^{\circ} + [a^{\dagger} a + (a^{\dagger} + a)\alpha^{\circ}]\hbar\Omega\left(1 - i\frac{\gamma}{2\Omega}\right) - \alpha^{\circ 2}\hbar\Omega - i\alpha^{\circ 2}\frac{\hbar\gamma}{2} \quad (209)$$

Next, in order to make the transition between the ground and the first excited state of the fast mode to be real in the absence of direct damping, the last imaginary part appearing in Eq. (209) has to be ignored so that Eq. (209) simplifies into

$$[\tilde{\mathrm{H}}_{II}^{\{1\}}] = \hbar\omega^{\circ} + \hbar\Omega[\mathrm{a}^{\dagger}\mathrm{a} + (\mathrm{a}^{\dagger} + \mathrm{a})\alpha^{\circ}]\left(1 - i\frac{\gamma}{2\Omega}\right) - \alpha^{\circ 2}\hbar\Omega \tag{210}$$

Moreover, in view of Eq. (207), it appears that

$$\left[e^{\frac{it}{\hbar}\tilde{\mathrm{H}}_{II}^{\{1\}\dagger}}\right] = (e^{-\alpha^{\circ}\mathrm{a}^{\dagger}+\alpha^{\circ}\mathrm{a}})\left[e^{\frac{it}{\hbar}\tilde{\mathrm{H}}_{III}^{\{1\}\dagger}}\right](e^{\alpha^{\circ}\mathrm{a}^{\dagger}-\alpha^{\circ}\mathrm{a}}) \tag{211}$$

Now, it is possible to observe that since the passage from the representation $\{III\}$ to representation $\{II\}$ does not affect the effective Hamiltonian of the slow mode corresponding to the ground-state fast mode, when the Hamiltonian is Hermitean, it must be so if it is non-Hermitean, so that one may write, according to Eq. (203):

$$[\tilde{\mathrm{H}}_{II}^{\{0\}}] = \mathrm{a}^{\dagger}\mathrm{a}\left(1 - i\frac{\gamma}{2\Omega}\right)\hbar\Omega = [\tilde{\mathrm{H}}_{III}^{\{0\}}] \tag{212}$$

Thus, according to Eqs. (211) and (212), the damped ACF (206) in representation $\{III\}$ takes in representation $\{II\}$ the following form:

$$[\tilde{\mathrm{G}}_{II}^{\mathrm{eff}}(t)]^{\circ} = (\mu_0^2)\ \varepsilon\ \mathrm{tr}_{\mathrm{Slow}}\left\{(e^{-\tilde{\lambda}\mathrm{a}^{\dagger}\mathrm{a}})\left[e^{\frac{it}{\hbar}\tilde{\mathrm{H}}_{II}^{\{1\}\dagger}}\right]\left[e^{-\frac{it}{\hbar}\tilde{\mathrm{H}}_{II}^{\{0\}}}\right]\right\} \tag{213}$$

Again, it is necessary to use the following eigenvalue equations:

$$[\tilde{\mathrm{H}}_{II}^{\{0\}}]|(n)\rangle = n\hbar\Omega\left(1 - i\frac{\gamma}{2\Omega}\right)|(n)\rangle$$

$$[\tilde{\mathrm{H}}_{II}^{\{1\}}]|\tilde{\Psi}_{\nu}\rangle = \hbar\tilde{\omega}_{\nu}|\tilde{\Psi}_{\nu}\rangle \quad \text{with} \quad \tilde{\omega}_{\nu} = [\omega^{\circ} + \omega^{\circ}{}_{\nu} - i\gamma_{\nu}]$$

The eigenvectors that cannot be orthogonal because of the non-Hermiticity of the Hamiltonian, are given by the expansion:

$$|\tilde{\Psi}_{\nu}\rangle = \sum a_{n\nu}|(n)\rangle \qquad \text{with} \qquad \langle\tilde{\Psi}_{\mu}|\tilde{\Psi}_{\nu}\rangle \neq \delta_{\mu\nu} \tag{214}$$

Now, we may *introduce* in the ACF (213) the *quantum direct damping* by multiplying it by the usual direct damping factor $e^{-\gamma^{\circ}t}$. Then, inserting a pseudocloseness relation on the "base" $\{|\tilde{\Psi}_{\nu}\rangle\}$ and making explicit the trace over this same basis, eq. (213) becomes

$$[\tilde{\mathrm{G}}_{II}^{\mathrm{eff}}(t)] = (\mu_0^2)\varepsilon\sum_m\sum_n\langle(n)|(e^{-\tilde{\lambda}\mathrm{a}^{\dagger}\mathrm{a}})\left[e^{\frac{it}{\hbar}\tilde{\mathrm{H}}_{II}^{\{1\}\dagger}}\right]|\tilde{\Psi}_{\nu}\rangle\langle\tilde{\Psi}_{\nu}|\left[e^{-\frac{it}{\hbar}\tilde{\mathrm{H}}_{II}^{\{0\}}}\right]|(n)\rangle(e^{-\gamma^{\circ}t})$$

Again, performing the trace on the eigenvectors of the number operator $a^{\dagger}a$ and inserting the closeness relation on the eigenvectors (214), the above expression of the ACF takes the final form:

$$[\tilde{G}^{\text{eff}}_{II}(t)] = (\mu_0^2)(e^{i\omega^\circ t})\varepsilon \sum_n \sum_\nu (e^{-\tilde{\lambda}n})(e^{-i(n\Omega-\omega^\circ{}_\nu)t})(e^{-(n\gamma/2+\gamma_\nu)t})|a_{n\nu}|^2(e^{-\gamma^\circ t}) \tag{215}$$

At last, the SD is the Fourier transform of this ACF, that is,

$$[\tilde{I}^{\text{eff}}_{II}(\omega)] = 2\text{Re}\int_0^\infty [\tilde{G}^{\text{eff}}_{III}(t)]e^{-i\omega t}\text{d}t \tag{216}$$

E. Validity of the Approximations: Numerical Tests [67]

The validity of the approximation proposed in Section V. D has to be verified, by comparison of the approached SD (216) to the exact one (126), that is,

$$[I_{II}(\omega)] = (\mu_0^2)\ 2\text{Re} \int_0^\infty e^{i\omega^\circ t}(e^{-i2\alpha^{\circ 2}\Omega t})(e^{i\alpha^{\circ 2}e^{-\gamma t/2}\sin\Omega t})[e^{\alpha^{\circ 2}[\langle n\rangle+1/2](2e^{-\gamma t/2}\cos(\Omega t)-e^{-\gamma t}-1)}](e^{-\gamma^\circ t})e^{-i\omega t}dt \tag{217}$$

This is performed in Fig. 9 for two different temperatures. In this figure, the grayed SDs are the reference temperatures calculated with the aid of the "exact"

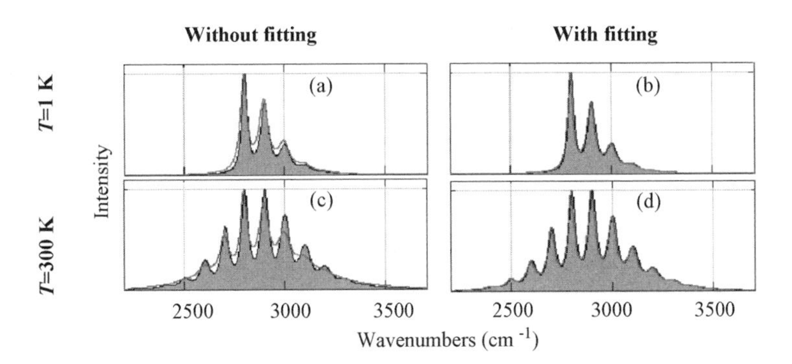

Figure 9. Validity of the approximation. $\omega^\circ = 3000\,\text{cm}^{-1}$ $\Omega = 100\,\text{cm}^{-1}$ $\alpha^\circ = 1$ (a) $\gamma^\circ = 0.2\,\Omega$ $\gamma = 0.15\,\Omega$, (b)$\gamma^\circ_{\text{eff}} = 0.16\,\Omega$, $\gamma_{\text{eff}} = 0.15\,\Omega$, (c) $\gamma^\circ = 0.2\,\Omega$, $\gamma = 0.15\,\Omega$, (d) $\gamma^\circ_{\text{eff}} = 0.18\,\Omega$ $\gamma_{\text{eff}} = 0.05\,\Omega$ *grayed*: Reference SDs; *Full line*: approached SDs.

Eq. (217), whereas the full lines are corresponding to the approximate SDs obtained by use of Eq. (216).

The situations in Fig. 9a and c deal with approximate SDs computed without fitting the damping parameters, γ° and γ, whereas those in Fig. 9b and d are concerning approximate SDs computed with the aid of effective damping parameters, $\gamma^\circ{}_{\text{eff}}$ and γ_{eff} chosen in such a way as to get the best fit with the *exact* ones.

Inspection of Fig. 9 shows that if the approximated SDs are slightly different from the reference ones, when γ° and γ are the same, it is, however, possible to get a good fit after scaling the damping parameters in the approached procedure. This result is general since it has been verified that, for any values of ω°, Ω, α°, and T, by scaling γ° and γ, one obtains SDs that are well fit with the *exact* ones.

F. Fermi Resonance Within Adiabatic and Exchange Approximations

Now, let us apply the approached model of quantum indirect damping to a situation where a Fermi resonance is possible to occur when the first excited state of the fast mode and the first overtones of some bending modes are close. The basic physical terms are defined in Table VIII.

In the absence of damping, the Hamiltonian involving Fermi resonances is

$$[\mathrm{H}_{\text{Fermi}}] = [\mathrm{H}_{\text{Free}} + \mathrm{H}^\circ + \mathrm{H}_{\text{Int}}] + \mathrm{H}_{\text{Bend}} + \mathrm{V}_{\text{Bend}} \tag{218}$$

Here, the three first right-hand side Hamiltonians are the components of the bare H-bond Hamiltonians without Fermi resonance given, respectively, by Eqs. (15), (21), and (22). Besides, the Hamiltonians corresponding to the bending mode and to the interaction between the fast and bending modes are, respectively, according to the definitions of table VIII

$$\mathrm{H}_{\text{Bend}} = \frac{\mathrm{p}_\delta^2}{2m_1^\delta} + \frac{1}{2} m_1^\delta (\omega_1^\delta)^2 \mathrm{q}_\delta^2 \tag{219}$$

$$\mathrm{V}_{\text{Bend}} = L_1^\delta \mathrm{q} \mathrm{q}_\delta^2 \tag{220}$$

TABLE VIII
Physical Terms Involved in the Fermi Resonance Model

High Frequency Mode	Low Frequency Mode	ith Bending Mode
Coordinate position = q	Coordinate position = Q	Coordinate position = q_i^δ
Conjugate momentum = p	Conjugate momentum = P	Conjugate momentum = p_i^δ
$[\mathrm{q}, \mathrm{p}] = i\hbar$	$[\mathrm{Q}, \mathrm{P}] = i\hbar$	$[\mathrm{q}_i^\delta, \mathrm{p}_i^\delta] = i\hbar$
Reference angular frequency = ω°	Angular frequency = Ω	Angular frequency = ω_i^δ
Reduced mass = m	Reduced mass = M	Reduced mass = m_i^δ

where L_1^δ is a coupling parameter. The eigenvalue equations of the harmonic Hamiltonians corresponding, to the fast and slow modes are given by Eqs. (23) and (24), whereas those dealing with the bending modes is

$$\mathrm{H}_{\mathrm{Bend}}|[l_1^\delta]\rangle = \left(l+\frac{1}{2}\right)\hbar\omega_1^\delta|l_1^\delta]\rangle \tag{221}$$

Then, from the eigenstates appearing in Eqs. (23), (24), and (221), it is possible to build up the tensorial states:

$$\begin{pmatrix} |\Psi_a(\{0\}(m)[0])\rangle \\ |\Psi_b(\{1\}(m)[0])\rangle \\ |\Psi_c(\{0\}(m)[2])\rangle \end{pmatrix} = \begin{pmatrix} |\{0\}\rangle\otimes|(m)\rangle\otimes|[0]\rangle \\ |\{1\}\rangle\otimes|(m)\rangle\otimes|[0]\rangle \\ |\{0\}\rangle\otimes|(m)\rangle\otimes|[2]\rangle \end{pmatrix} \equiv \begin{pmatrix} |\{0\}(m)[0]\rangle \\ |\{1\}(m)[0]\rangle \\ |\{0\}(m)[2]\rangle \end{pmatrix} \tag{222}$$

where $m = 0, 1, 2\cdots$.

Now, within the adiabatic approximation, the Hamiltonian (218) may be written

$$[\mathrm{H}_{\mathrm{Fermi}}^{\mathrm{Adiab}}] = [\mathrm{H}_{\mathrm{Adiab}}] + [\mathrm{V}_{\mathrm{Bend}}] + [\mathrm{H}_{\mathrm{Bend}}] \tag{223}$$

The adiabatic term $\mathrm{H}_{\mathrm{Adiab}}$ is the same as (48) in the situation without Fermi resonance, that is,

$$[\mathrm{H}_{\mathrm{Adiab}}] = [\mathrm{H}_l^{\{1\}}]|\{1\}\ \rangle\langle\{1\}| + [\mathrm{H}_l^{\{0\}}]|\{0\}\rangle\langle\{0\}|$$

Of course, the effective Hamiltonians involved in this last equation are (49) and (50) met in the situation without damping and Fermi resonance:

$$[\mathrm{H}_l^{\{k\}}] = \frac{\mathrm{P}^2}{2M} + \frac{1}{2}M\Omega^2\mathrm{Q}^2 + kb\mathrm{Q} + k\hbar\omega^\circ$$

Besides, the Hamiltonian of the bending mode may be written in terms of the projectors on the ground state and on the second excited state of the bending mode:

$$[\mathrm{H}_{\mathrm{Bend}}] = \frac{5}{2}\hbar\omega_1^\delta|[2]\rangle\ \langle[2]| + \frac{1}{2}\hbar\omega_1^\delta|[0]\rangle\ \langle[0]|$$

Here, ω_1^δ is the angular frequency of the bending mode involved in the Fermi resonance. Next, within the exchange approximation [69], the coupling Hamiltonian (220) involved in the Fermi resonance is

$$[\mathrm{V}_{\mathrm{Bend}}] = [|\{0\}\rangle|[2]\rangle\ \langle[0]|\ \langle\{1\}|]\ \hbar f_1^\delta + [|\{1\}\rangle|[0]\rangle\ \langle[2]|\ \langle\{0\}|]\ \hbar f_1^\delta$$

Here, f_1^{δ} is the anharmonic coupling parameter involved in the Fermi resonance, which is related to the above coupling L_1^{δ} through:

$$f_1^{\delta} = \frac{L_1^{\delta}}{2m_1^{\delta}\omega_1^{\delta}}\sqrt{\frac{\hbar}{2m\omega^{\circ}}}$$

As a consequence of the above equations, the full Hamiltonian describing the fast mode coupled to the H-bond bridge (via the strong anharmonic coupling theory) and to the bending mode (via the Fermi resonance process) may be written within the tensorial basis (222) according to [24]:

$$[\mathrm{H}_{\mathrm{Fermi}}^{\mathrm{Adiab}}] = \begin{pmatrix} \mathrm{H}_{II}^{\{0\}} & 0 & 0 \\ 0 & \mathrm{H}_{II}^{\{1\}} & \hbar f_1^{\delta} \\ 0 & \hbar f_1^{\delta} & \mathrm{H}_{II}^{\{0\}} + 2\hbar\omega_1^{\delta} \end{pmatrix} \tag{224}$$

This physical model involving Fermi resonance is illustrated by Fig. 10.

Now, in order to incorporate the indirect damping in the model, we may introduce [70] in the Hamiltonian (224) the indirect relaxations, according to

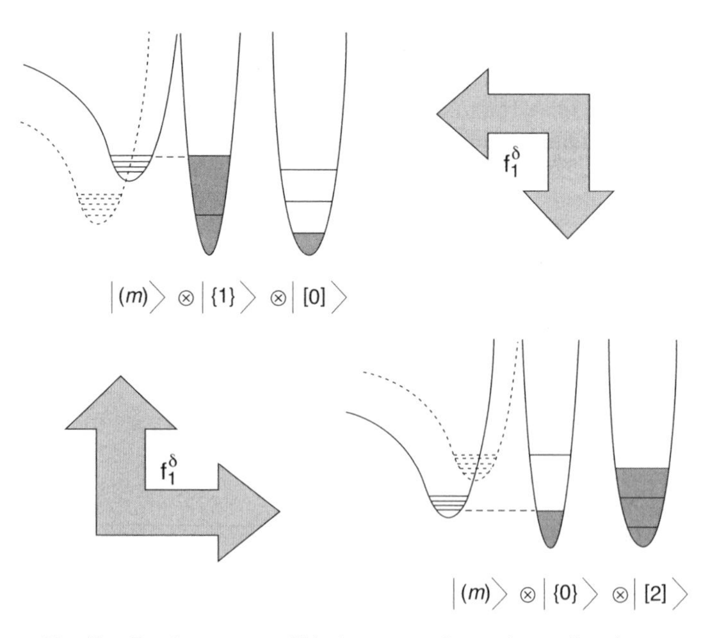

Fig. 10. Fermi resonance within the strong anharmonic coupling theory.

the results of Section V. D, via non-Hermitean effective Hamiltonians. This leads us to write in place of Eq. (224) the following one:

$$[\tilde{H}_{Fermi}^{Adiab}] = \begin{pmatrix} \tilde{H}_{II}^{\{0\}} & 0 & 0 \\ 0 & \tilde{H}_{II}^{\{1\}} & \hbar f_1^{\delta} \\ 0 & \hbar f_1^{\delta} & \tilde{H}_{II}^{\{0\}} + 2\hbar\omega_1^{\delta} - i\hbar\gamma_1^{\delta} \end{pmatrix} \tag{225}$$

Here, γ_1^{δ} is the damping parameter of the first harmonics of the bending mode. Besides, the effective non-Hermitean Hamiltonians appearing in this last equation are those given by Eqs. (208) and (209), that is,

$$\begin{aligned} [\tilde{H}_{II}^{\{0\}}] &= \hbar\Omega\left(1 - i\frac{\gamma}{2\Omega}\right)a^{\dagger}a \\ [\tilde{H}_{II}^{\{1\}}] &= \hbar\omega^{\circ} + \hbar\Omega[a^{\dagger}a + (a^{\dagger} + a)\alpha^{\circ}]\left(1 - i\frac{\gamma}{2\Omega}\right) - \alpha^{\circ 2}\hbar\Omega\left(1 - i\frac{\gamma}{2\Omega}\right) \end{aligned} \tag{226}$$

Here, γ° is the direct damping parameter, whereas γ is the indirect one. Note that the imaginary term $i\hbar\gamma_1^{\delta}$ has been incorporated in Eq. (225), to take into account the finite lifetime of the second excited state of the bending mode via the damping parameter γ_1^{δ}.

In the same spirit, the Hamiltonian of the damped bending mode must be written

$$[H_{Bend}] = \left[\frac{5}{2}\hbar\omega_1^{\delta} - i\hbar\gamma_1^{\delta}\right]|[2_1^{\delta}]\rangle\langle[2_1^{\delta}]| + \frac{1}{2}\hbar\omega_1^{\delta}|[0_1^{\delta}]\rangle\ \langle[0_1^{\delta}]|$$

The eigenvalue equation of the non-Hermitean Fermi Hamiltonian and the corresponding eigenvector, which are not orthonormal because of the non-Hermiticity are, respectively,

$$[\tilde{H}_{Fermi}^{Adiab}]|\tilde{\Phi}_{\nu}\rangle = \hbar(\omega^{\circ} + \omega_{\nu} - i\gamma_{\nu}t)|\tilde{\Phi}_{\nu}\rangle \qquad \text{with} \qquad \langle\tilde{\Phi}_{\mu}|\tilde{\Phi}_{\nu}\rangle \neq \delta_{\mu\nu}$$

$$|\tilde{\Phi}_{\nu}\rangle = \sum_k\sum_m\sum_l C_{kml}^{\nu}|\{k\}\rangle|(m)\rangle|[l_1^{\delta}]\rangle$$

Here, ω_{ν} and γ_{ν} are, respectively, the real and imaginary parts of the νth eigenvalue, whereas C_{kml}^{ν} are the corresponding expansion coefficients. In the last equation, k, l, and m run, according to the definition of the basis (222).

The ACF of the dipole moment operator of the fast mode may be written in the presence of Fermi resonances by aid of Eq. (10). Besides, the dipole moment operator at time t appearing in this equation is given by a Heisenberg equation involving the full Hamiltonian (225). The thermal average involved in the ACF must be performed on the Boltzmann operator of the system involving the real part of this same Hamiltonian (225). Again since the full Hamiltonian (225) has a block structure and since, even at room temperature, only the real part of the energy spectrum plays a physical role in the Boltzmann operator, one may write for the density operator:

$$\rho_B \propto (e^{-\beta\,[\tilde{\mathrm{H}}^{\mathrm{Adiab}}_{\mathrm{Fermi}}]}) \rightarrow (e^{-\beta\,[\mathrm{H}^{\{0\}}_{II}]})$$

As a consequence, making explicit the trace over the tensorial eigenstates $|[r_1^{\delta}](m)\rangle$ of the two combined harmonic oscillators corresponding to the H-bond bridge and to the bending mode, the ACF takes the form:

$$[\tilde{\mathrm{G}}^{\mathrm{eff}}_{\mathrm{Fermi}}(t)] = (\mu_0^2)\tilde{\varepsilon}\sum_m\sum_r \langle [r_1^{\delta}](m)\{0\}|(e^{-\beta\,[\mathrm{H}^{\{0,0\}}_{II}]})$$
$$|\{0\}\rangle\ \langle\{1\}|[e^{i[\tilde{\mathrm{H}}^{\mathrm{Adiab}}_{\mathrm{Fermi}}]^{\dagger}t/\hbar}]|\{1\}\rangle\langle\{0\}|$$
$$[e^{-i[\tilde{\mathrm{H}}^{\mathrm{Adiab}}_{\mathrm{Fermi}}]t/\hbar}]|\{0\}(m)[r_1^{\delta}]\rangle$$

Again, taking the real part for the Boltzmann operator and then performing standard calculations as in the above sections, one obtains for the spectral density obtained by Fourier transform of the final expression of the ACF:

$$[\tilde{\mathrm{I}}^{\mathrm{eff}}_{\mathrm{Fermi}}(\omega)] = (\mu_0^2)\varepsilon\sum_m\sum_{\nu}(e^{-\beta m\hbar\Omega})|C^{\nu}_{1mo}|^2$$
$$\int_o^{\infty}(e^{i\omega^{\circ}t})(e^{-im\Omega t}e^{i\omega^{\circ}{}_{\nu}})[e^{-m\gamma t}e^{-\gamma_{\nu}t}e^{-\gamma^{\circ}t}]e^{-i\omega t}dt$$

G. Other Approximations for Quantum Indirect Damping [71]

1. *General Considerations*

In the last sections, the approached model for quantum indirect damping involving non-Hermitean Hamiltonians has been applied to a situation involving Fermi resonance, but within the adiabatic, the exchange and the harmonic approximations. Unfortunately, this approximate method for handling the quantum indirect damping is not susceptible to generalizing the situation working beyond the adiabatic approximation and involving anharmonic H-bond bridge Morse-like potentials.

Thus, in order to treat the indirect damping of H-bond bridge beyond the adiabatic and harmonic approximations, it is necessary to propose new simple approaches. This is the aim of this section, which is dealing with two new approximate methods [71]. One of them, the *non-Hermitean Diagonal Hamiltonians (NHDH)* method uses non-Hermitean diagonal Hamiltonians.

The other one, the *Complex Energy Levels (CEL)* method, directly postulates complex energy levels. As will be seen, these two methods lead to equivalent line shapes when both the adiabatic and harmonic approximations are simultaneously removed.

In both methods, the full adiabatic Hamiltonian will be partitioned into a diagonal part (Hamiltonians of the quantum harmonic oscillators corresponding to the fast, slow, and bending modes) and a nondiagonal part (perturbations incorporating all the anharmonic terms, i.e., the anharmonic correction to the potential of the H-bond bridge, a possible anharmonicity of the fast mode, the anharmonic coupling between the fast mode and the H-bond bridge and at last, the anharmonic couplings related to Fermi resonances $\cdots$)

2. *The Non-Hermitean Diagonal Hamiltonian Method*

Observed that within the adiabatic approximation, the total Hamiltonian of the NHDH method may be written

$$[\mathrm{H_{Tot}}] \simeq [\mathrm{H_{Adiab}}] = \sum [\tilde{\mathrm{H}}_{II}^{\{k\}}]|\{k\}\rangle\langle\{k\}|$$

Here, the effective Hamiltonians involved in the adiabatic Hamiltonian that are considered as non-Hermitean in order to take into account the damping, are partitioned according to

$$[\tilde{\mathrm{H}}_{II}^{\{k\}}] = [\tilde{\mathrm{H}}^{\circ}] + [\mathrm{V}^{\{k\}}]$$

Here, $\tilde{\mathrm{H}}^{\circ}$ is the diagonal Hamiltonian of the harmonic oscillator describing the bond bridge, which will be considered as non-Hermitean, whereas $\mathrm{V}^{\{k\}}$ is an Hermitean perturbation taking into account the driving term for which the damping is neglected, that is,

$$[\mathrm{V}^{\{k\}}] = k\ [\hbar\Omega\alpha(\mathrm{a}^{\dagger} + \mathrm{a}) + \hbar\omega^{\circ}]$$

In the spirit of the previous sections, the non-Hermitean diagonal Hamiltonian are, respectively,

$$[\tilde{\mathrm{H}}_{II}^{\{0\}}] = \mathrm{a}^{\dagger}\mathrm{a}\left(1 - i\frac{\gamma}{2\Omega}\right)\hbar\Omega \tag{227}$$

$$[\tilde{\mathrm{H}}_{II}^{\{1\}}] = \hbar\Omega\left(1 - i\frac{\gamma}{2\Omega}\right)\mathrm{a}^{\dagger}\mathrm{a} + \hbar\Omega\alpha^{\circ}(\mathrm{a}^{\dagger} + \mathrm{a}) + \hbar(\omega^{\circ} - i\gamma^{\circ}) - \alpha^{\circ 2}\hbar\Omega \tag{228}$$

Besides, the eigenvalue equations of these non-Hermitean Hamiltonians are:

$$[\tilde{\mathrm{H}}_{II}^{\{0\}}]|(n)\rangle = n\left(1 - i\frac{\gamma}{2\Omega}\right)\hbar\Omega|(n)\rangle$$

$$[\tilde{\mathrm{H}}_{II}^{\{1\}}]|\tilde{\Phi}_{\mathrm{v}}^{\{1\}}\rangle = (\hbar\omega^{\circ}{}_{\mathrm{v}} + i\hbar\gamma_{\mathrm{v}})|\tilde{\Phi}_{\mathrm{v}}^{\{1\}}\rangle \tag{229}$$

The expansions of the eigenvectors appearing in Eq. (229), in terms of the kets $|(n)\rangle$ are

$$|\tilde{\Phi}_{\nu}^{\{k\}}\rangle = \sum_{m} b_{m\nu}^{\{k\}}|(m)\rangle \qquad \text{with} \qquad \langle\tilde{\Phi}_{\mu}^{\{k\}}|\tilde{\Phi}_{\nu}^{\{k\}}\rangle \neq \delta_{\mu\nu} \tag{230}$$

Now, use the effective non-Hermitean Hamiltonians (228) to get the time dependence of the transition moment operator through Heisenberg transformation, but ignore non-Hermitean part for the Boltzmann operator. Then, the ACF taking into account both direct and indirect dampings, appears to be given by

$$[\tilde{G}_{II}(t)_{\mathrm{NHDH}}] = (\mu_0^2)\tilde{\varepsilon}\ \mathrm{tr}_{\mathrm{Slow}}[(e^{-\beta H_{II}^{\{0\}}})|\ \{0\}\rangle\langle\{1\}|[e^{i\tilde{H}_{II}^{\{1\}\dagger}t/\hbar}]|\{1\}\rangle\ \langle\{0\}|[e^{-i\tilde{H}_{II}^{\{0\}}t/\hbar}]]$$

Again, after a circular permutation inside the trace and simplifications using orthonormal properties of the kets $|\{k\}\rangle$, one obtains

$$[\tilde{G}_{II}(t)_{\mathrm{NHDH}}] = (\mu_0^2)\ \tilde{\varepsilon}\ \mathrm{tr}_{\mathrm{Slow}}[(e^{-\beta H_{II}^{\{0\}}})[e^{i\tilde{H}_{II}^{\{1\}\dagger}t/\hbar}][e^{-i\tilde{H}_{II}^{\{0\}}t/\hbar}]]$$

On the other hand, although the eigenvectors (230) do not form a base because they are eigenvectors of the Hamiltonians (227) and (228), which are non-Hermitean, it is reasonable to admit that they may lead to a base as far as the indirect damping γ is small with respect to Ω. Thus, keeping that in mind, let us introduce the *pseudocloseness relation* involving the eigenstates (230):

$$\sum |\tilde{\Phi}_{\nu}^{\{1\}}\rangle\langle\tilde{\Phi}_{\nu}^{\{1\}}| \simeq \hat{1}$$

Then, after performing the trace over the base $\{|(n)\rangle\}$ of the eigenstates of $a^{\dagger}a$, the ACF becomes approximately

$$[\tilde{G}_{II}(t)_{\mathrm{NHDH}}] \simeq (\mu_0^2)\tilde{\varepsilon}\sum_{\nu}\sum_{n}\langle(n)|(e^{-\beta H_{II}^{\{0\}}})[e^{i\tilde{H}_{II}^{\{1\}}t/\hbar}]|\tilde{\Phi}_{\nu}^{\{1\}}\rangle\langle\tilde{\Phi}_{\nu}^{\{1\}}|[e^{-i\tilde{H}_{II}^{\{0\}}t/\hbar}]|(n)\rangle$$

By using the eigenvalue equations (229) and then neglecting the zero-point energy of the Boltzmann density operator, this ACF transforms into

$$[\tilde{G}_{II}(t)_{\mathrm{NHDH}}] = (\mu_0^2)\varepsilon\sum_{\nu}\sum_{n}\langle(n)|(e^{-\beta n\hbar\Omega})[e^{i(\omega^{\circ}{}_{\nu}+i\gamma_{\nu})t}]|\tilde{\Phi}_{\nu}^{\{1\}}\rangle\langle\tilde{\Phi}_{\nu}^{\{1\}}|[e^{-in(\Omega-i\gamma)t}]|(n)\rangle \tag{231}$$

Recall that the passage from $\tilde{\varepsilon}$ to ε is linked to the neglect of the zero-point energy in the density operator.

Then, owing to Eqs. (230), Eq. (231) leads to

$$[\tilde{G}_{II}(t)_{\text{NHDH}}] = (\mu_0^2)\varepsilon \sum_{\nu} \sum_{n} (e^{-\lambda n}) |b_{\nu n}^{\{1\}}|^2 [e^{-i(n\Omega - \omega^{\circ}{}_{\nu})t}][e^{-(n\gamma + \gamma_{\nu})t}] \tag{232}$$

3. *The Complex Energy Levels Method (CEL)*

Now, consider the other simple method for the quantum indirect damping, where the Hamiltonians are Hermitean, but the eigenvalues become complex by addition of imaginary parts.

a. Situation Without Damping. Start from the adiabatic approximation for the full Hamiltonian *in the situation without damping*:

$$[H_{\text{Tot}}] \simeq [H_{\text{Adiab}}] = \sum [H_{II}^{\{k\}}] |\{k\}\rangle\langle\{k\}|$$

Here, the effective Hamiltonians are those of the situation in which all the dampings are ignored, that is,

$$[H_{II}^{\{k\}}] = \left(a^{\dagger}a + \frac{1}{2}\right)\hbar\Omega + \hbar\Omega k[\alpha^{\circ}(a^{\dagger} + a) - \alpha^{\circ 2}] + k\hbar\omega^{\circ}$$

Besides, *for situations involving indirect damping*, the eigenvalue equation for the non-Hermitean effective Hamiltonian corresponding to the ground state of the high frequency mode is

$$[\tilde{H}_{II}^{\{0\}}]|(n)\rangle = n\hbar\left[\Omega - i\frac{\gamma}{2}\right]|(n)\rangle \tag{233}$$

On the other hand, for situations *without indirect damping*, the eigenvalue equation for the Hermitean effective Hamiltonian corresponding to the first excited state of the high frequency mode is

$$[H_{II}^{\{1\}}]|\Phi^{\circ}{}_{\nu}\rangle = \hbar\omega^{\circ}{}_{\nu}|\Phi^{\circ}{}_{\nu}\rangle \qquad \text{with} \qquad \langle\Phi^{\circ}{}_{\mu}|\Phi^{\circ}{}_{\nu}\rangle = \delta_{\mu\nu} \tag{234}$$

Again, the expansions of the corresponding eigenvectors on the basis in which the quantum harmonic Hamiltonian of the H-bond bridge is diagonal, are

$$|\Phi^{\circ}{}_{\nu}\rangle = \sum_{m} [b^{\circ}{}_{m\nu}]|(m)\rangle \tag{235}$$

b. Incorporation of the Direct and Indirect Dampings in the Energy Levels $E^{\circ}{}_{\nu}^{\{1\}}$. Now, assume that an imaginary part, the meaning of which is a time decay taking into account both the direct and indirect relaxations, is associated to each real eigenvalue of the Hermitean effective Hamiltonians involved in Eq. (234). More precisely, it is supposed that each complex energy level of the system *in the presence of the two kinds of damping* contains two imaginary contributions, that is,

$$\hbar\omega_{\nu} = \hbar\omega^{\circ}{}_{\nu} - i\hbar[\gamma^{\circ} + \gamma_{\nu}] \tag{236}$$

Here, γ° is the direct damping parameter, whereas the indirect damping contribution γ_{ν} is assumed to be the sum over all the eigenstates $|(n)\rangle$ of the H-bond bridge oscillator, which is proportional to

1. The probability for the system initially in the energy levels $\hbar\omega^{\circ}_{\nu}$ to jump on the $|(n)\rangle$ state.
2. The quantum number n.
3. The indirect damping parameter $\gamma/2$.

All this leads us to assume that for the indirect damping contribution the following expression is valid

$$\gamma_{\nu} = \sum [b^{\circ}{}_{n\nu}]^2 n \frac{\gamma}{2} \tag{237}$$

In Eq. (237), the expansion components given by Eq. (235) appear, which determine the eigenvectors of the effective Hamiltonians of the H-bond bridge in the absence of damping.

c. Obtainment of the ACF. Now, consider in a first place the ACF *without direct and indirect dampings*. It is given by Eq. (60), that is,

$$[\mathrm{G}^{\circ}{}_{II}(t)]^{\circ} = (\mu_0^2)\tilde{\varepsilon}\ \mathrm{tr}_{\mathrm{Slow}}[(e^{-\beta \mathrm{H}_{II}^{\{0\}}})[e^{i\mathrm{H}_{II}^{\{1\}\dagger} t/\hbar}][e^{-i\mathrm{H}_{II}^{\{0\}} t/\hbar}]]$$

Next, insert the closeness relation involving the eigenstates appearing in Eq. (233) and perform the trace over the states $|(n)\rangle$. Then, the ACF becomes

$$[\mathrm{G}^{\circ}{}_{II}(t)]^{\circ} = (\mu_0^2)\tilde{\varepsilon}\sum_n\sum_\nu \langle (n)|(e^{-\beta \mathrm{H}_{II}^{\{0\}}})[e^{i\mathrm{H}_{II}^{\{1\}} t/\hbar}]|\Phi_{\nu}{}^{\circ}\rangle\langle\Phi_{\nu}{}^{\circ}|[e^{-i\mathrm{H}_{II}^{\{0\}} t/\hbar}]|(n)\rangle$$

Moreover, using the eigenvalue equation (234) leads, after neglecting the zero-point energy in the Boltzmann operator, to

$$[\mathrm{G}^{\circ}{}_{II}(t)]^{\circ} = (\mu_0^2)\tilde{\varepsilon}\sum_n\sum_\nu \langle (n)|(e^{-\beta \mathrm{H}_{II}^{\{0\}}})[e^{i\mathrm{H}_{II}^{\{1\}} t/\hbar}]|\Phi^{\circ}{}_{\nu}\rangle\langle\Phi^{\circ}{}_{\nu}|[e^{-i\mathrm{H}_{II}^{\{0\}} t/\hbar}]|(n)\rangle$$

Again, *introduce the direct and indirect dampings*. This may be performed by transforming the exponential argument $i\omega^\circ{}_\nu$ into $i\omega^*_\nu$ where ω^*_ν, is the complex conjugate of ω_ν given by Eq. (236), whereas the Hermitean Hamiltonian $H_{II}^{\{0\}}$ is transformed into the non-Hermitean $\tilde{H}_{II}^{\{0\}}$. This leads to

$$[\tilde{G}_{II}(t)_{CEL}] = (\mu_0^2)\varepsilon \sum_n \sum_\nu \langle (n)|(e^{-\beta n\hbar\Omega})(e^{i\omega^*_\nu t})|\Phi^\circ{}_\nu\rangle\langle\Phi^\circ{}_\nu|[e^{-i\tilde{H}_{II}^{\{0\}}t/\hbar}]|(n)\rangle$$

Then, eigenvalue equation (233) the ACF transforms into:

$$[\tilde{G}_{II}(t)_{CEL}] = (\mu_0^2)\varepsilon \sum_n \sum_\nu \langle (n)|(e^{-\beta n\hbar\Omega})(e^{i(\omega_\nu{}^\circ - \gamma_\nu)t})|\Phi^\circ{}_\nu\rangle\langle\Phi^\circ{}_\nu|(e^{-in\Omega t})(e^{-n\gamma t/2})|(n)\rangle$$

As a consequence, in view of Eq. (235), the CEL ACF appears to be given by

$$[\tilde{G}_{II}(t)_{CEL}] = (\mu_0^2)\varepsilon \sum_n \sum_\nu (e^{-\tilde{\lambda} n})[b^\circ{}_{\nu m}]^2(e^{i\omega_\nu{}^\circ t})(e^{-in\Omega t})(e^{-\gamma_\nu t})(e^{-n\gamma t/2}) \tag{238}$$

4. *Comparison of the NHDH and CEL Spectral Densities With the Reference Ones [71]*

The suitability of the two above approaches has to be tested by comparison with the reference analytical ones. Keep in mind that, up to now, the direct and indirect damping parameters are only adjustable, so what is required is only the ability for these approaches to fit the most exact theoretical SDs.

In Figs. 11 and 12, the NHDH and CEL SDs (dots) obtained, respectively, by Fourier transform of Eqs. (232) and (238) are compared with the corresponding reference ones (grayed) obtained by Eq. (126):

Inspection of this figure shows that the two approached methods are able to fit the theoretical SDs obtained in the straightforward way, the only cost is being able to use effective damping parameters that are weakly different from those used for the reference approach.

5. *SD Beyond Adiabatic and Harmonic Approximations [71]*

Now, owing to the well-behaved character of the approached NHDH and CEL methods, it appears reasonable to apply them to more complex situations where, in the absence of Fermi resonance, the quantum indirect damping, and not only the direct damping [72], has to be treated beyond the adiabatic and harmonic approximations [71].

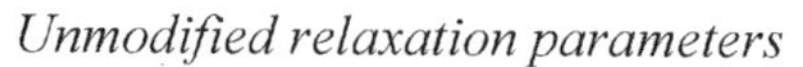

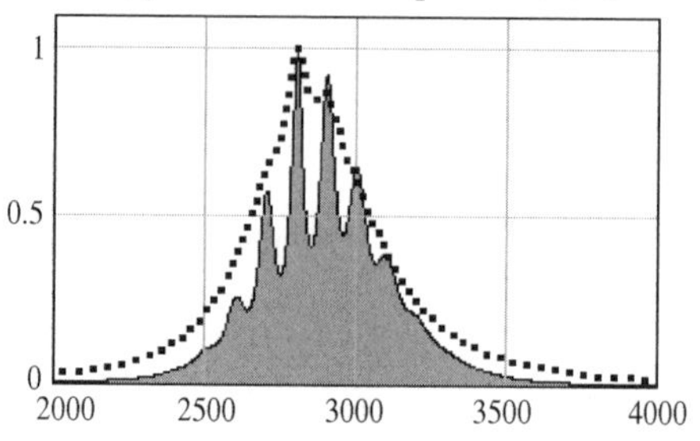

Adjusted relaxation parameters

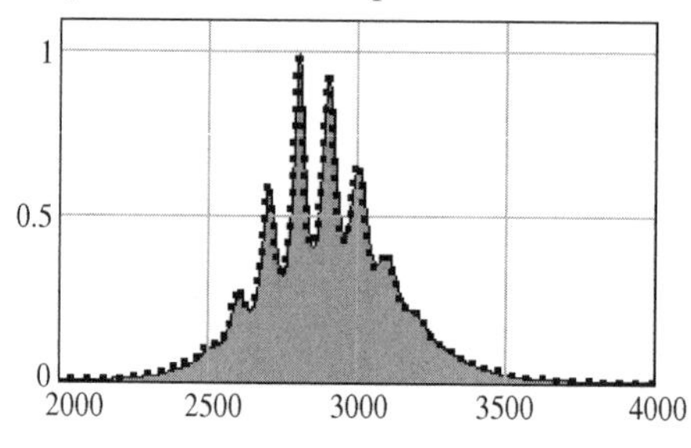

Figure 11. SDs *within adiabatic* and *harmonic approximations*. Comparison of the approached NHDH method (dots: $\gamma_{\text{dot}}^{\circ} = 0.585\ \gamma^{\circ}; \Omega_{\text{dot}} = 0.30\ \Omega$) with the reference (grayed: $\gamma^{\circ} = 0.2\ \Omega$, $\gamma = 0.3\ \Omega$). Common parameters: $\alpha^{\circ} = 1$, $\omega^{\circ} = 3000\,\text{cm}^{-1}$, $\Omega = 100\,\text{cm}^{-1}$, $T = 300$ K.

a. CEL Method. First, consider the CEL approach. Beyond the harmonic and adiabatic approximations, the full Hamiltonian of the bare H-bond bridge in the absence of damping, is given by Eq. (19):

$$\mathrm{H}_{\text{Tot}} = \frac{\mathrm{P}^2}{2M} + D_e\left[1 - e^{-\Omega\sqrt{\frac{M}{2D_e}}\mathrm{Q}}\right]^2 + \frac{\mathrm{p}^2}{2m} + \frac{1}{2}m\omega^{\circ 2}\mathrm{q}^2 + m\omega^{\circ}b\mathrm{q}^2\mathrm{Q} + \frac{1}{2}mb^2\mathrm{q}^2\mathrm{Q}^2 \tag{239}$$

where D_e is the dissociation energy of the H-bond bridge. Now, write the Hamiltonian of the H-bond bridge involving the Morse potential, as the sum of the diagonal part H° plus an anharmonic correction, according to

$$\frac{\mathrm{P}^2}{2M} + D_e\left[1 - e^{-\Omega\sqrt{\frac{M}{2D_e}}\mathrm{Q}}\right]^2 = \mathrm{H}^{\circ} - \left(\frac{1}{2}M\Omega^2\mathrm{Q}^2\right) + D_e\left[1 - e^{-\Omega\sqrt{\frac{M}{2D_e}}\mathrm{Q}}\right]^2$$

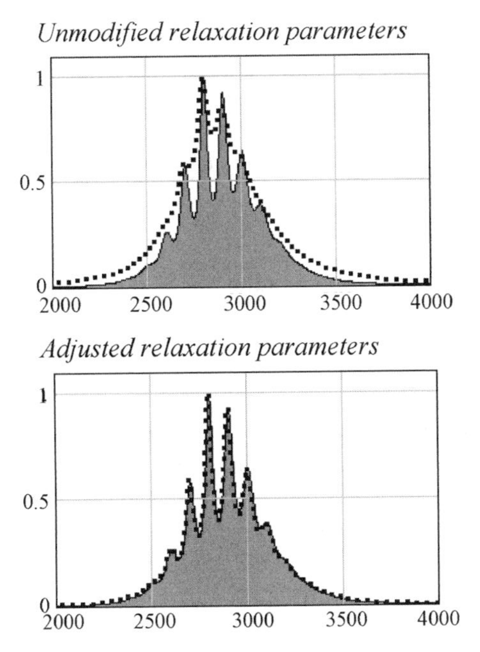

Figure 12. SDs within adiabatic and harmonic approximations. Comparison of the approached CEL method (dots: $\gamma_{dot}{}^{\circ} = 0.75\gamma^{\circ}; \Omega^{\circ\circ}_{dot} = 0.33\,\Omega$) with the reference (grayed: $\gamma^{\circ} = 0.2\,\Omega$, $\gamma = 0.3\,\Omega$). Common parameters: $\alpha^{\circ} = 1$, $\omega^{\circ} = 3000\,\text{cm}^{-1}$, $\Omega = 100\,\text{cm}^{-1}$, $T = 300$ K.

Next, pass to Bosons according to Table V.

$$\mathrm{H_{Tot}} = \left(\mathrm{b^{\dagger}b} + \frac{1}{2}\right)\hbar\omega^{\circ} + \left(\mathrm{a^{\dagger}a} + \frac{1}{2}\right)\hbar\Omega + D_e(e^{-2z[\mathrm{a^{\dagger}+a}]} - 2e^{-z[\mathrm{a^{\dagger}+a}]} + 1)$$
$$- \hbar\Omega\left\{\frac{1}{4}[\mathrm{a^{\dagger}+a}]^2 - \frac{1}{2}[\alpha^{\circ}[\mathrm{b^{\dagger}+b}]^2[\mathrm{a^{\dagger}+a}]] - \frac{1}{4}\left[\alpha^{\circ 2}\frac{\Omega}{\omega^{\circ}}[\mathrm{b^{\dagger}+b}]^2[\mathrm{a^{\dagger}+a}]^2\right]\right\} \tag{240}$$

with

$$\alpha^{\circ} = \frac{b}{\Omega}\sqrt{\frac{\hbar}{2M\Omega}} \qquad \text{and} \qquad z = \frac{1}{2}\sqrt{\frac{\hbar\Omega}{D_e}}$$

The eigenvalue equation of this full Hamiltonian (240) is

$$\mathrm{H_{Tot}}|\Theta_{\mu}{}^{\circ}\rangle = \hbar\omega^{\circ}{}_{\mu}|\Theta_{\mu}{}^{\circ}\rangle$$

Within the tensorial product basis $\{|\{k\},(m)\rangle\}$ built up from the eigenvectors of the Hermitean harmonic Hamiltonians of the slow and fast modes, the eigenvectors are given by the expansion:

$$|\Theta^{\circ}{}_{\mu}\rangle = \sum_{k}\sum_{m}[D^{\circ\{k\}}_{\mu m}]|\{k\}(m)\rangle$$

As for the special situations involved in our above adiabatic approach and dealing with the CEL method, we may write the complex energy levels as

$$\hbar\omega_{\mu} = \hbar\omega^{\circ}_{\mu} - i\hbar[\gamma^{\circ}_{\mu} + \gamma_{\mu}]$$

$\hbar\omega^{\circ}_{\mu}$ is the real part of the complex energy levels $\hbar\omega_{\mu}$, whereas the corresponding imaginary part is the reflect of two damping parameters $\gamma^{\circ}{}_{\mu}$ and γ_{μ}.

The first contribution, $\gamma^{\circ}{}_{\mu}$, which is a reflection of the quantum direct damping of the high frequency mode, is supposed to be proportional to the direct damping parameter γ° times the excitation degree of the fast mode k through:

$$\gamma^{\circ}{}_{\mu} = \sum_{k}\sum_{m}[[D^{\circ\{k\}}_{\mu m}]^2 - [D^{\circ\{k\}}_{om}]^2]k\gamma^{\circ}$$

Here, the right-hand side substracted terms have been introduced in order to remove the imaginary part $\gamma^{\circ}{}_{o}$ for the ground state ($\mu = 0$).

Besides, in a similar way, the second contribution, γ_{μ}, which is a reflection of the quantum indirect damping, is assumed to be proportional to the indirect damping parameter γ, times the excitation degree of the slow mode via:

$$\gamma_{\mu} = \sum_{k}\sum_{m}[[D^{\circ\{k\}}_{\mu m}]^2 - [D^{\circ\{k\}}_{om}]^2]m\gamma$$

where the right-hand side substracted terms have been introduced for the same reason as for γ°.

In the *absence of damping*, the ACF of the bare weak H-bond may be written as the following trace $\mathrm{tr}_{\mathrm{Tot}}$ over the base spanned by the total Hamiltonian $\mathrm{H}_{\mathrm{Tot}}$:

$$[\mathrm{G}^{\circ}_{\mathrm{Mono}}(t)]^{\circ} = \left(\frac{\mu_0^2}{Z_{\mathrm{Tot}}}\right)\mathrm{tr}_{\mathrm{Tot}}[(e^{-\beta\mathrm{H}_{\mathrm{Tot}}})|\{0\}\rangle\langle\{1\}|[e^{i\mathrm{H}_{\mathrm{Tot}}t/\hbar}]|\{1\}\rangle\langle\{0\}|[e^{-i\mathrm{H}_{\mathrm{Tot}}t/\hbar}]]$$

By using the expression of the Hamiltonian H_{Tot} and with the aid of closeness relations involving its eigenstates, one obtains

$$[G^{\circ}_{Mono}(t)]^{\circ} = \left(\frac{\mu_0^2}{Z_{Tot}}\right)$$

$$\sum_{\nu}\sum_{m}\sum_{\mu}\sum_{n}(e^{-\beta\hbar\omega^{\circ}{}_{\upsilon}})[e^{-i(\omega^{\circ}{}_{\upsilon}-\omega^{\circ}{}_{\mu})t}][D^{\circ}{}^{\{0\}}_{\nu m}][D^{\circ}{}^{\{1\}}_{m\mu}][D^{\circ}{}^{\{1\}}_{\mu n}][D^{\circ}{}^{\{0\}}_{n\upsilon}]$$

Now, the direct and indirect dampings may be incorporated by transforming in the real energy levels in the phase factor involved in the ACF without damping,

$$\omega^{\circ}_{\nu} \rightarrow \omega_{\nu} \qquad \omega_{\mu} \rightarrow \omega^{*}_{\mu}$$

Consequently, the damped SD, which is the Fourier transform of the CEL ACF in which the above transformations have been performed, takes the form:

$$[\tilde{I}_{Mono}(\omega)_{CEL}] = \left(\frac{\mu_0^2}{Z_{Tot}}\right)\sum_{\nu}\sum_{m}\sum_{\mu}\sum_{n}(e^{-\beta\hbar\omega^{\circ}_{\nu}})$$

$$\mathrm{Re}\int_{o}^{\infty} dt\; e^{-i\omega t}[e^{-i(\omega^{\circ}_{\nu}-\omega^{\circ}_{\mu})t}][e^{-(\gamma_{\nu}+\gamma^{\circ}_{\nu}+\gamma_{\mu}+\gamma^{\circ}_{\mu})t}][D^{\circ\{0\}}_{\nu m}][D^{\circ\{1\}}_{m\mu}][D^{\circ\{1\}}_{\mu n}][D^{\circ\{0\}}_{n\upsilon}] \tag{241}$$

b. NHDH Method. Now, pass to the NHDH approach. Start from the Hermitean diagonal parts involved in the full Hamiltonian (240). Then, replace them by their corresponding non-Hermitean forms. This leads us to write, for the full non-Hermitean Hamiltonian *involving both direct and indirect dampings,*

$$\tilde{H}_{Tot} = \left(b^{\dagger}b + \frac{1}{2}\right)\left[1 - i\frac{\gamma^{\circ}}{2\omega^{\circ}}\right]\hbar\omega^{\circ}$$
$$+ \left(a^{\dagger}a + \frac{1}{2}\right)\left[1 - i\frac{\gamma}{2\Omega}\right]\hbar\Omega + D_e(e^{-2z[a^{\dagger}+a]} - 2e^{-z[a^{\dagger}+a]} + 1)$$
$$- \hbar\Omega\left\{\frac{1}{4}[a^{\dagger}+a]^2 - \frac{1}{2}[\alpha^{\circ}[b^{\dagger}+b]^2[a^{\dagger}+a]] - \frac{1}{4}\left[\alpha^{\circ 2}\frac{\Omega}{\omega^{\circ}}[b^{\dagger}+b]^2[a^{\dagger}+a]^2\right]\right\}$$

As seen in Section V.D.I, the zero-point energy of the non-Hermitean diagonal Hamiltonians has been ignored.

The eigenvalue equation of the non-Hermitean full Hamiltonian is

$$\tilde{\mathrm{H}}_{\mathrm{Tot}}|\tilde{\Theta}_\mu\rangle = \hbar\omega_\mu|\tilde{\Theta}_\mu\rangle \qquad \text{with} \qquad \hbar\omega_\mu = \hbar\omega^\circ{}_\mu - i\hbar\gamma_\mu$$

Within the tensorial product basis $\{|\{k\},(m)\rangle\}$ built up from the eigenkets of $\mathrm{a}^\dagger\mathrm{a}$ and $\mathrm{b}^\dagger\mathrm{b}$, the corresponding eigenvectors are given by the expansion:

$$|\tilde{\Theta}_\mu\rangle = \sum_k \sum_m [C^{\{k\}}_{m\,\mu}]|\{k\}(m)\rangle \qquad \text{with} \qquad \langle\tilde{\Theta}_\mu|\tilde{\Theta}_\nu\rangle \neq \delta_{\mu\nu}$$

Then, the damped ACF takes the form:

$$[\tilde{\mathrm{G}}_{\mathrm{Mono}}(t)_{\mathrm{NHDH}}] = \left(\frac{\mu_0^2}{Z_{\mathrm{Tot}}}\right)\mathrm{tr}_{\mathrm{Tot}}(e^{-\beta\mathrm{H}_{\mathrm{Tot}}})|\{0\}\rangle\langle\{1\}|[e^{i\tilde{\mathrm{H}}^\dagger_{\mathrm{Tot}}t/\hbar}]|\{1\}\rangle\langle\{0\}|[e^{-i\tilde{\mathrm{H}}_{\mathrm{Tot}}t/\hbar}]$$

Next, it use the pseudo-"closeness relation", which may be approximately verified when γ and γ° are weak with respect to Ω.

$$\sum |\tilde{\Theta}_\mu\rangle\langle\tilde{\Theta}_\mu| \simeq \widehat{1}$$

Then, following inference of the same kind as those used for the NHDH method, the ACF becomes

$$[\tilde{\mathrm{G}}_{\mathrm{Mono}}(t)_{\mathrm{NHDH}}] = \left(\frac{\mu_0^2}{Z_{\mathrm{Tot}}}\right)\sum_\upsilon\sum_m\sum_n\sum_\mu (e^{-\beta\hbar\omega^\circ{}_\nu})[C^{\{0\}}_{\nu\,m}][C^{\{1\}}_{m\,\mu}][C^{\{1\}}_{\mu\,n}][C^{\{0\}}_{n\,\upsilon}](e^{-i(\omega_\nu-\omega^*_\mu)t})$$

At last, perform a shift in the imaginary part of the complex eigenvalues in the same spirit as above, in order to make real the ground-state energy $\hbar\omega_0$ corresponding to $\nu = 0$. This leads us to write

$$\Gamma_\mu = \gamma_\mu + i\gamma_o$$

γ_o is the imaginary part corresponding to $\nu = 0$. Then, after Fourier transform the SD taking into account both direct and indirect dampings is

$$[\tilde{\mathrm{I}}_{\mathrm{Mono}}(\omega)_{\mathrm{NHDH}}] = \left(\frac{\mu_0^2}{Z_{\mathrm{Tot}}}\right)\sum_\nu\sum_m\sum_n\sum_\mu (e^{-\beta\hbar\omega^\circ{}_\nu})\mathrm{Re}\int_o^\infty dt\; e^{-i\omega t}[C^{\{0\}}_{\nu\,m}][C^{\{1\}}_{m\,\mu}][C^{\{1\}}_{\mu\,n}][C^{\{0\}}_{n\,\nu}](e^{-i(\omega^\circ{}_\nu-\omega^\circ{}_\mu)t})(e^{-(\Gamma_\nu+\Gamma_\mu)t}) \qquad (242)$$

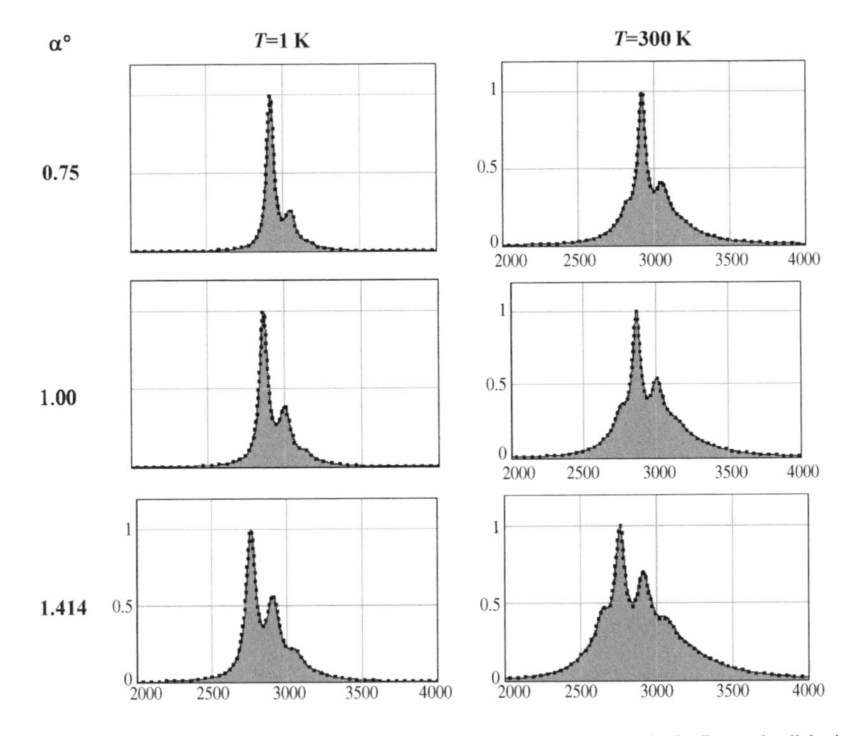

Figure 13. Comparison of SDs calculated by CEL and NHDH methods. Beyond adiabatic approximation. H-bond bridge Morse potential. Grayed: NHDH. Dots: CEL $\omega^{\circ} = 3000\,\text{cm}^{-1}$, $\Omega = 100\,\text{cm}^{-1}$, $\gamma^{\circ} = 0.2\,\Omega$, $\gamma = 0.3\,\Omega$, De $= 0.3\,$eV.

c. Comparisons between NHDH and CEL Methods. Now, because there is no SD reference in this situation working beyond adiabatic and harmonic approximations, a possibility is to look at the reliability of these approaches and to mutually compare the approached CEL and NHDH SDs (241) and (242). This is performed in Fig. 13 for two temperatures and different values of the anharmonic coupling parameter α°.

Inspection of this figures shows that the two approaches lead to identical line shapes, although the approximation hypothesis for indirect damping are different. This is in favor of the suitability of these approached methods.

6. *SD With Fermi Resonance, Beyond Adiabatic, Harmonic, and Exchange Approximations [59]*

Because of the agreement between the two approached NHDH and CEL methods for the situation dealing without Fermi resonance, but beyond adiabatic and harmonic approximations, it appears reasonable to apply these methods to many more complex physical situations involving Fermi resonances, beyond the exchange, adiabatic, and harmonic approximations. This is the aim of this section. Here, the relaxation will be treated with the aid of the NHDH model, that is a generalization of the study performed by Chamma [73] for a situation with direct, but without indirect, damping.

In the absence of damping, the full Hamiltonian may be written

$$\mathrm{H_{Fermi}} = \mathrm{H_{Tot}} + \mathrm{H_{Bend}} + \mathrm{V^{\circ}}_{\mathrm{Fermi}}$$

Here, $\mathrm{H_{Tot}}$ is given by Eq. (239), whereas $\mathrm{H_{Bend}}$ and $\mathrm{V^{\circ}}_{\mathrm{Fermi}}$ are, respectively, given by Eqs. (219) and (220).

Now, consider the situation where the dampings of the fast, slow, and bending modes have to be taken into account. These dampings may be introduced in the high frequency mode, the H-bond bridge, and the bending mode by assuming the corresponding harmonic Hamiltonians to be non-Hermitean via complex angular frequencies.

Then, passing to Bosons with the aid of the expressions of Table V, the non-Hermitean full Hamiltonian takes the form:

$$[\tilde{\mathrm{H}}_{\mathrm{Fermi}}] = [\tilde{\mathrm{H}}_{\mathrm{Free}} + \tilde{\mathrm{H}}_{\mathrm{Bend}} + \tilde{\mathrm{H}}^{\circ}] + [\mathrm{H_{Int}} + \mathrm{V_{Anh}} + \mathrm{V_{Fermi}}] \tag{243}$$

The different diagonal Hamiltonians $\tilde{\mathrm{H}}_{\mathrm{Free}}, \tilde{\mathrm{H}}_{\mathrm{Bend}}$, and $\tilde{\mathrm{H}}^{\circ}$, dealing, respectively, with the high frequency mode, the bending mode, and the H-bond bridge may be assumed to be the non-Hermitean ones given in Table IX, in which the zero-point

TABLE IX
Hamiltonians Involved in Eq. (243)

$\tilde{\mathrm{H}}_{\mathrm{Free}} = \left(\mathrm{b}^{\dagger}\mathrm{b} + \frac{1}{2}\right)\left[1 - i\frac{\gamma^{\circ}}{2\omega^{\circ}}\right]\hbar\omega^{\circ} - i\frac{\hbar\gamma^{\circ}}{2}$	$[\mathrm{b}, \mathrm{b}^{\dagger}] = 1$
$\tilde{\mathrm{H}}^{\circ} = \left(\mathrm{a}^{\dagger}\mathrm{a} + \frac{1}{2}\right)\left[1 - i\frac{\gamma}{2\Omega}\right]\hbar\Omega - i\frac{\hbar\gamma}{2}$	$[\mathrm{a}, \mathrm{a}^{\dagger}] = 1$
$\tilde{\mathrm{H}}_{\mathrm{Bend}} = \left(\mathrm{b}_1^{\delta\dagger}\mathrm{b}_1^{\delta} + \frac{1}{2}\right)\left[1 - i\hbar\frac{\gamma_1^{\delta}}{2\omega_1^{\delta}}\right]\hbar\omega_1^{\delta} - i\frac{\hbar\gamma_1^{\delta}}{2}$	$[\mathrm{b}_1^{\delta}, \mathrm{b}_1^{\delta\dagger}] = 1$
$\mathrm{H_{Int}} = \frac{1}{2}\hbar\Omega[\alpha^{\circ}[\mathrm{b}^{\dagger} + \mathrm{b}]^2[\mathrm{a}^{\dagger} + \mathrm{a}]] + \frac{1}{4}\hbar\Omega\left[\alpha^{\circ}2\frac{\Omega}{\omega^{\circ}}[\mathrm{b}^{\dagger} + \mathrm{b}]^2[\mathrm{a}^{\dagger} + \mathrm{a}]^2\right]$	
$\mathrm{V}^{\circ}_{\mathrm{Anh}} = D_e(e^{-2z[\mathrm{a}^{\dagger}+\mathrm{a}]} - 2e^{-z}[\mathrm{a}^{\dagger} + \mathrm{a}] + 1) - \frac{1}{4}\hbar\Omega[\mathrm{a}^{\dagger} + \mathrm{a}]^2$	$z = \frac{1}{2}\sqrt{\frac{\hbar\Omega}{D_z}}$
$\mathrm{V}^{\circ}_{\mathrm{Fermi}} = [\mathrm{b}_1^{\delta\dagger} + \mathrm{b}_1^{\delta}]^2[\mathrm{b}^{\dagger} + \mathrm{b}]\hbar f_1^{\delta}$	

energies have been ignored. Besides, the Hermitean terms H_{Int} and V_{Anh} are, respectively, the anharmonic coupling between the H-bond bridge and the high frequency mode, and the anharmonic correction to the intrinsic potential of the H-bond bridge introduced via Morse-like potentials. At last, V_{Fermi} is the anharmonic coupling between the high frequency mode and the bending modes involved in the Fermi resonances and considered beyond the exchange approximation.

Next, the eigenvalue equation of the Hamiltonian (243) is

$$[\tilde{H}_{Fermi}]|\tilde{\xi}_\nu\rangle = \hbar(\omega^\circ{}_\nu - i\gamma_\nu)|\tilde{\xi}_\nu\rangle \tag{244}$$

Besides, the expansion of the corresponding eigenvectors on the basis built up from the eigenvectors of $a^\dagger a$, $b^\dagger b$, and $b_\delta^\dagger b_\delta$ are

$$|\tilde{\xi}_\nu\rangle = \sum_k \sum_m \sum_l [C_{m\,l\,\nu}^{\{k\}}]|\{k\}\rangle|(m)\rangle|[l_1^\delta]\rangle \qquad \text{with} \qquad \langle\tilde{\xi}_\mu|\tilde{\xi}_\nu\rangle \neq \delta_{\mu\nu} \tag{245}$$

Again, keeping in mind that the Hamiltonian involved in the Boltzmann density operator has no reason to be non-Hermitean, whereas the Hamiltonians involved in the evolution operators must be non-Hermitean in order to take into account the damping, the ACF appears to be given by the following trace tr_{Fermi} over the base spanned by the Hamiltonian H_{Fermi}:

$$[\tilde{G}_{Fermi}^{Beyond}(t)_{NHDH}] = \left(\frac{\mu_0^2}{Z_{Fermi}}\right)$$
$$tr_{Fermi}[(e^{-\beta H_{Fermi}})|\{0\}\rangle\langle\{1\}|[e^{i\tilde{H}_{Fermi}^\dagger t/\hbar}]|\{1\}\rangle\langle\{0\}|[e^{-i\tilde{H}_{Fermi}t/\hbar}]]$$

with the ad hoc definition of the partition function

Next, perform a circular permutation within the trace:

$$[\tilde{G}_{Fermi}^{Beyond}(t)_{NHDH}] = \left(\frac{\mu_0^2}{Z_{Fermi}}\right)$$
$$tr_{Fermi}[\langle\{0\}|[e^{-i\tilde{H}_{Fermi}t/\hbar}](e^{-\beta H_{Fermi}})|\{0\}\rangle \ \langle\{1\}|[e^{i\tilde{H}_{Fermi}^\dagger t/\hbar}]|\{1\}\rangle]$$

Moreover, it is suitable to use the following pseudocloseness relation, which is as well verified as the damping factors are small with respect to the corresponding angular frequencies:

$$\sum |\tilde{\xi}_\nu\rangle\langle\tilde{\xi}_\nu| \simeq \hat{1}_\nu$$

Again, insert twice this *pseudocloseness relation* in the following way:

$$[\tilde{G}^{\text{Beyond}}_{\text{Fermi}}(t)_{\text{NHDH}}] \simeq \left(\frac{\mu_0^2}{Z_{\text{Fermi}}}\right)\sum_{\nu}\sum_{\mu}\text{tr}_{\text{Fermi}}[\langle\{0\}|(e^{-i\tilde{H}_{\text{Fermi}}t/\hbar})(e^{-\beta H_{\text{Fermi}}})$$
$$|\tilde{\xi}_{\nu}\rangle\langle\tilde{\xi}_{\nu}|\{0\}\rangle\langle\{1\}|(e^{i\tilde{H}^{\dagger}_{\text{Fermi}}t/\hbar})|\tilde{\xi}_{\mu}\rangle\langle\tilde{\xi}_{\mu}|\{1\}\rangle]$$

Then, owing to the eigenvalue equation (244), the ACF takes the form:

$$[\tilde{G}^{\text{Beyond}}_{\text{Fermi}}(t)_{\text{NHDH}}] = \left(\frac{\mu_0^2}{Z_{\text{Fermi}}}\right)\sum_{\nu}\sum_{\mu}$$
$$\text{tr}_{\text{Fermi}}[\langle\{0\}|[(e^{-i(\omega^{\circ}{}_{\nu}-i\gamma_{\nu})t})(e^{-\beta\hbar\omega^{\circ}{}_{\nu}})|\tilde{\xi}_{\nu}\rangle\langle\tilde{\xi}_{\nu}|]|\{0\}\rangle$$
$$\langle\{1\}|[(e^{i(\omega^{\circ}{}_{\mu}+i\gamma_{\mu})t})|\tilde{\xi}_{\mu}\rangle\langle\tilde{\xi}_{\mu}|]|\{1\}\rangle]$$
$$[\tilde{G}^{\text{Beyond}}_{\text{Fermi}}(t)_{\text{NHDH}}] = \left(\frac{\mu_0^2}{Z_{\text{Fermi}}}\right)\sum_{\nu}\sum_{\mu}$$
$$\text{tr}_{\text{Fermi}}[(e^{-i(\omega^{\circ}{}_{\nu}-i\gamma_{\nu})t})(e^{-\beta\hbar\omega^{\circ}{}_{\nu}})|\langle\{0\}\,|\tilde{\xi}_{\nu}\rangle|^2|\langle\{1\}\,|\tilde{\xi}_{\nu}\rangle|^2(e^{i(\omega^{\circ}{}_{\mu}+i\gamma_{\mu})t})] \tag{246}$$

Observe that because of Eq. (245), the scalar products involved in this ACF are

$$\langle\{r\}|\tilde{\xi}_{\nu}\rangle = \sum_{k}\sum_{m}\sum_{l}$$
$$[C^{\{k\}}_{m\,l\,\nu}]\langle\{r\}|\{k\}\rangle|(m)\rangle|[l_1^{\delta}]\rangle = \sum_{m}\sum_{l}[C^{\{r\}}_{m\,l\,\nu}]|(m)\rangle|[l_1^{\delta}]\rangle$$

Thus, owing to Eq. (245) the ACF takes a final form, which after Fourier transform gives for the SD:

$$[\tilde{I}^{\text{Beyond}}_{\text{Fermi}}(\omega)_{\text{NHDH}}] = \left(\frac{\mu_0^2}{Z_{\text{Fermi}}}\right)\sum_{\nu}\sum_{\mu}(e^{-\beta\hbar\omega^{\circ}{}_{\nu}})$$
$$\left|\sum_{m}\sum_{l}[C^{\{0\}}_{m\,l\,\nu}]\right|^2\left|\sum_{n}\sum_{r}[C^{\{1\}}_{n\,l\,\mu}]\right|^2\text{Re}\int_{o}^{\infty}(e^{i(\omega^{\circ}{}_{\mu}-\omega^{\circ}{}_{\nu})t})(e^{-(\gamma_{\mu}+\gamma_{\nu})t})e^{-i\omega t}dt$$

This spectral density holds for a H-bonded species involving a Fermi resonance between the high frequency and bending modes, where the potential of the H-bond bridge is Morse-like, and when working beyond the adiabatic approximation for the strong anharmonic coupling between the high frequency mode and the H-bond bridge, beyond the exchange approximation for the anharmonic coupling between the high frequency and bending modes, and taking into account the damping of the high frequency mode of the H-bond bridge and of the bending mode.

VI. CYCLIC HYDROGEN-BONDED DIMERS

In Section III, a theoretical approach of the quantum indirect damping of the H-bond bridge was exposed within the strong anharmonic coupling theory, with the aid of the adiabatic approximation. In Section III, this theory was shown to reduce to the Maréchal and Witkowski and Rösch and Ratner quantum approaches. In Section IV, this quantum theory of indirect damping was shown to admit as an approximate semiclassical limit the approach of Robertson and Yarwood.

This section now deals with H-bonded species, where together with the strong anharmonic coupling and the quantum indirect damping, Davydov coupling and Fermi resonances may occur, that is, centrosymmetric H-bonded cyclic dimers the theory of which, for situations without damping,was first performed by Maréchal and Witkowski [18].

A. Dimers Involving Davydov Coupling and Direct and Indirect Dampings

1. *The Hamiltonian in the Presence of Damping, But in the Absence of Davydov Coupling*

Consider a cyclic dimer of carboxylic acid involving two H-bond bridges, as depicted on Fig.14. The two parts of the dimer are labeled $r = \text{a, b}$. According to Sections VI.B and VI.C, for such dimers, there are two degenerate high frequency modes and also two degenerate low frequency H-bond vibrations.

The adiabatic approximation leads to a description of each moiety by effective Hamiltonians of the H-bond bridge: for a single H-bond bridge, this effective Hamiltonian is, either that of an harmonic oscillator, if the fast mode is in its ground state, or that of a driven harmonic oscillator, if the fast mode is

$C_2Q_1 = Q_2 \qquad C_2q_1 = q_2$

$C_2Q_2 = Q_1 \qquad C_2q_2 = q_1$

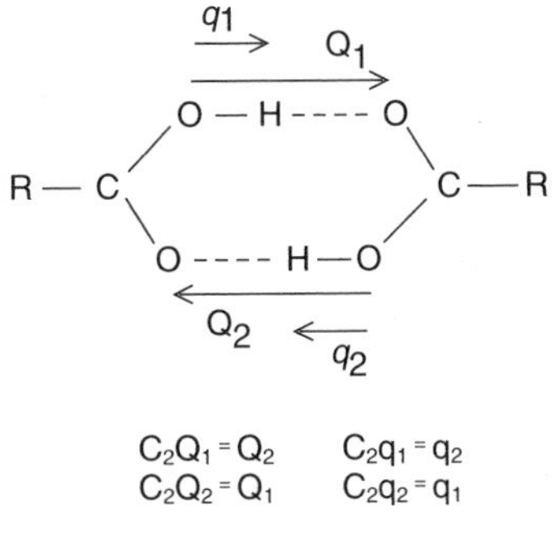

Figure 14. Cyclic H-bonded dimers. The action of the parity operator C_2 on the high and low frequency coordinates of the centrosymmetric cyclic dimer exchange the coordinates. (The subscripts 1 and 2 refer to moiety a and b of the centrosymmetric cyclic dimer, respectively).

excited. When one of the two identical fast modes is excited, then, because of the symmetry of the cyclic dimer and of a possible coupling $\hbar V^{\circ}$ between the two degenerate excited states of the fast mode, an interaction may occur (Davydov coupling) leading to an exchange between the two identical excited parts of the dimer. Of course, this interaction between degenerate excited states is of a nonadiabatic nature, although the adiabatic approximation was performed to separate the high and low frequency motions of each moiety.

Note that because of the symmetry of the dimer, there is a $\widehat{C}_2$ operator (with $\widehat{C}_2^2 = \widehat{1}$), which exchanges the coordinates Q_i of the two H-bond bridges of the cyclic dimer according to

$$\widehat{C}_2 Q_a = Q_b \qquad \widehat{C}_2 Q_b = Q_a \tag{247}$$

Of course, the same symmetry properties hold for the conjugate momenta, that is,

$$\widehat{C}_2 P_a = \widehat{P}_b \qquad \widehat{C}_2 P_b = P_a \tag{248}$$

For now, ignoring the interaction between the two moieties and assuming that, within each moiety, the adiabatic approximation may be performed as for a single H-bond, the Hamiltonian of the symmetric dimer embedded in the thermal bath, is

$$[\mathbb{H}^{\{\mathrm{Adiab}\}}] = [\mathbb{H}^{\{\mathrm{Adiab}\}}]_a + [\mathbb{H}^{\{\mathrm{Adiab}\}}]_b \tag{249}$$

In Eq.(249), the two first right-hand side terms are the adiabatic Hamiltonians of each moiety. Owing to Eq. (101), they are given by an expression of the same form that is, by the following equations

$$[\mathbb{H}^{\{\mathrm{Adiab}\}}]_i = [\mathbb{H}_{II}^{\{0\}}]_i |\{0\}_i\rangle\langle\{0\}_i| + [\mathbb{H}_{II}^{\{1\}}]_i |\{1\}_i\rangle\langle\{1\}_i|$$

$$\begin{aligned}
&[\mathbb{H}_{II}^{\circ\,\{0\}}]_i \equiv [\mathrm{H}^{\{\mathrm{slow}\}}]_i + [\mathbb{H}_{II}^{\{\mathrm{int}\}}]_i + [\mathbb{H}^{\{\theta\}}] \\
&[\mathbb{H}_{II}^{\circ\,\{1\}}]_i \equiv [\mathrm{H}^{\{\mathrm{slow}\}}]_i + b Q_i + (\hbar\omega^{\circ} - \alpha^{\circ^2}\hbar\Omega) + [\mathbb{H}_{II}^{\{\mathrm{int}\}}]_i + [\mathbb{H}^{\{\theta\}}]
\end{aligned} \tag{250}$$

$[\mathrm{H}^{\{\mathrm{Slow}\}}]_i$ are the quantum harmonic oscillators of the two H-bond bridges that are of the same kind as those given by Eqs. (102) and (103) and are reproduced in Table X with their corresponding eigenvalue equations. Besides, $|\{k\}_i\rangle$ are the eigenkets of the Hamiltonians of the fast modes harmonic oscillators given in the same table. Moreover, $[\ \mathbb{H}^{\{\mathrm{int}\}}]_i$ are the interaction Hamiltonians coupling the H-bond bridges with the thermal bath, which are

$$[\mathbb{H}^{\{\mathrm{Int}\}}]_i = \sum \tilde{q}_r Q_i \eta_r \tag{251}$$

TABLE X
Harmonic Hamiltonians Describing the Slow and High Frequency Modes and the Thermal Bath

$[H^\circ]_i = \frac{P_i^2}{2M} + \frac{M\Omega^2 Q_i^2}{2}$	$[Q_i, P_j] = i\delta_{i,j}$	$[H^\circ]_i\|(n)_i\rangle = \left(n_i + \frac{1}{2}\right)\hbar\Omega\|(n)_i\rangle$
$[H_{Free}]_i = \frac{p_i^2}{2m} + \frac{m\omega^{\circ 2} q_i^2}{2}$	$[q_i, p_j] = i\delta_{i,j}$	$[H_{Free}]_i\|\{k\}_i\rangle = \left(k_i + \frac{1}{2}\right)\hbar\omega^\circ\|\{k\}_i\rangle$
$H_r^{\{\theta\}} = \frac{\widetilde{P}_r^2}{2m_r} + \frac{m_r\widetilde{\omega}_r^2\widetilde{q}_r^2}{2}$	$[\widetilde{q}_r, \widetilde{P}_s] = i\delta_{r,s}$	$H_r^{\{\theta\}}\|(\widetilde{n})_r\rangle = ((\widetilde{n})_r + \frac{1}{2})\hbar\widetilde{\omega}_r\|(\widetilde{n})_r\rangle$

$\mathbb{H}^{\{\theta\}}$ is the Hamiltonian of the thermal bath given by

$$[\mathbb{H}^{\{\theta\}}] = \sum [H_r^{\{\theta\}}] \tag{252}$$

where $H_r^{\{\theta\}}$ are given in Table X with their corresponding eigenvalue equations.

In these equations, P_i are the conjugate momenta of the coordinates Q_i of the H-bond bridges of the two moieties, whereas q_i and p_i are the coordinates and the conjugate momenta of the two degenerate high frequency modes of the two moieties. Besides, Ω is the angular frequency of the two degenerate slow modes and ω° is that of the fast modes when the H-bond bridges are at equilibrium. At last, b is the scalar, measuring in each moiety the strength of the coupling between the high and low frequency modes.

2. *Introduction of the Davydov Coupling in the Hamiltonian*

Consider an excitation of the fast mode of one moiety of the dimer. The corresponding excited state is resonant with the state corresponding to the situation where it is the fast mode of the other moiety that is excited. A resonant exchange mechanism must occur when one of the fast modes has been excited. This mechanism, which is of a nonadiabatic nature, is at the origin of the Davydov coupling [74], which has been introduced by Maréchal and Witkowski [18] in their pioneering works.

The Hamiltonian of the cyclic dimer involving Davydov coupling between the first excited state of the high frequency oscillator a of one moiety and the excited state of the oscillator b of the other moiety, and vice versa, is

$$[\mathbb{H}_{Dav}] = [\mathbb{H}^{\{Adiab\}}]_a + [\mathbb{H}^{\{Adiab\}}]_b + [V_{Dav}] \tag{253}$$

The Davydov coupling Hamiltonian V_{Dav} appearing in this equation is

$$[V_{Dav}] = V^\circ[|\{1\}_a\rangle\langle\{0\}_b| + |\{0\}_a\rangle\langle\{1\}_b|]$$

Again, we may build the following basis:

$$\begin{pmatrix} |\Phi^{\{0,0\}}_{\{a,b\}}\rangle \\ |\Phi^{\{1,0\}}_{\{a,b\}}\rangle \\ |\Phi^{\{0,1\}}_{\{a,b\}}\rangle \end{pmatrix} = \begin{pmatrix} |\{0\}_a\rangle \otimes |\{0\}_b\rangle \\ |\{1\}_a\rangle \otimes |\{0\}_b\rangle \\ |\{0\}_a\rangle \otimes |\{1\}_b\rangle \end{pmatrix} \tag{254}$$

Then, within this basis, the representation of the Hamiltonian (253) is

$$[\mathbb{H}_{\text{Dav}}] = \begin{pmatrix} [\mathbb{H}_{II}^{\{0,0\}}] & 0 & 0 \\ 0 & [\mathbb{H}_{II}^{\{1,0\}}] & \text{V}^{\circ} \\ 0 & \text{V}^{\circ} & [\mathbb{H}_{II}^{\{0,1\}}] \end{pmatrix} \tag{255}$$

with, respectively,

$$[\mathbb{H}_{II}^{\{0,0\}}] = \sum_i \left(\frac{\text{P}_i^2}{2M} + \frac{M\Omega^2\text{Q}_i^2}{2}\right) + \sum_i [\mathbb{H}_{II}^{\{Int\}}]_i + \mathbb{H}^{\{\theta\}}$$

$$[\mathbb{H}_{II}^{\{1,0\}}] = \left(\frac{\text{P}_a^2}{2M} + \frac{M\Omega^2\text{Q}_a^2}{2}\right)$$

$$+ \left(\frac{\text{P}_b^2}{2M} + \frac{M\Omega^2\text{Q}_b^2}{2}\right) + b\text{Q}_a + [\mathbb{H}_{II}^{\{\text{Int}\}}]_a + [\mathbb{H}_{II}^{\{\text{Int}\}}]_b + \mathbb{H}^{\{\theta\}} + (\hbar\omega^{\circ} - \alpha^{\circ^2}\hbar r) \tag{256}$$

Later, the following notation will be used

$$\left[\mathbb{H}_{II}^{\{k,l\}}\right] = \left[\mathbb{H}_{II}^{\{k\}}\right]_a + \left[\mathbb{H}_{II}^{\{l\}}\right]_b \tag{257}$$

Then, because of the symmetry properties given by Eqs. (247) and (248), it appears that the parity operator exchanges the two Hamiltonians:

$$\widehat{\text{C}}_2[\mathbb{H}_{II}^{\{1,0\}}] = [\mathbb{H}_{II}^{\{0,1\}}] \tag{258}$$

The physics corresponding to the situation described by the Hamiltonian (255) is depicted in Fig. 15.

3. *The ACF and SD Without Indirect Damping [75]*

Now, look at the special situation with direct damping, but without indirect damping. Then, the Hamiltonian (255) reduces to

$$[\text{H}_{\text{Dav}}] = \begin{pmatrix} [\text{H}_{II}^{\{0,0\}}] & 0 & 0 \\ 0 & [\text{H}_{II}^{\{1,0\}}] & \text{V}^{\circ} \\ 0 & \text{V}^{\circ} & [\text{H}_{II}^{\{0,1\}}] \end{pmatrix}$$

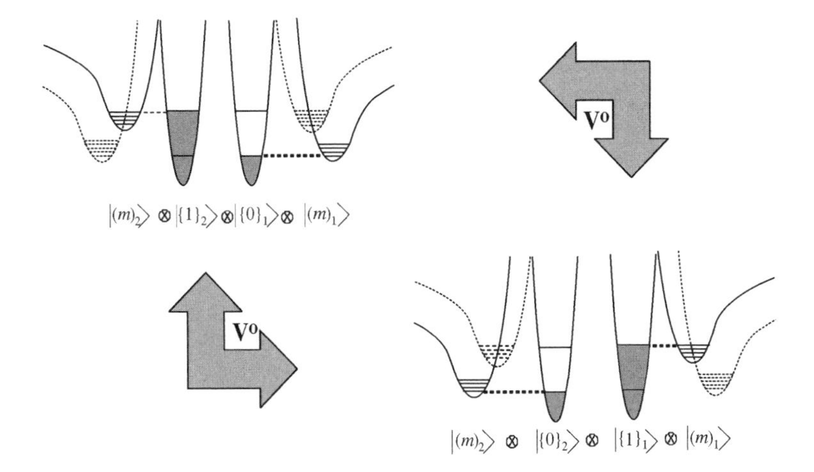

Figure 15. Davydov effect within strong anharmonic coupling theory. (The subscripts 1 and 2 refer, respectively, to the a and b moieties of the centrosymmetric cyclic dimer)

with, respectively,

$$[\mathrm{H}_{II}^{\{0,0\}}] = \sum_i \left(\frac{\mathrm{P}_i^2}{2M} + \frac{M\Omega^2 \mathrm{Q}_i^2}{2} \right)$$

$$[\mathrm{H}_{II}^{\{1,0\}}] = \left(\frac{\mathrm{P}_a^2}{2M} + \frac{M\Omega^2 \mathrm{Q}_a^2}{2} \right) + \left(\frac{\mathrm{P}_b^2}{2M} + \frac{M\Omega^2 \mathrm{Q}_b^2}{2} \right) + b\mathrm{Q}_a + (\hbar\omega^\circ - \alpha^\circ 2\hbar\Omega)$$

Then, it is possible to get the SD in a straightforward way without using the symmetry of the system.

The ACF of the dipole moment operator may be written formally as the following trace over the base $\mathrm{tr}_{\mathrm{Dav}}$ spanned by the Hamiltonian $\mathrm{H}_{\mathrm{Dav}}$:

$$[\mathrm{G}^\circ_{\mathrm{Dav}}(t)]^\circ = \mathrm{tr}_{\mathrm{Dav}}[\rho_{\mathrm{Dav}}[\widehat{\mu}_{\mathrm{Dim}}(0)][\widehat{\mu}_{\mathrm{Dim}}(t)]]$$

Here, $\widehat{\mu}_{\mathrm{Dim}}(0)$ is the dipole moment operator of the dimer high frequency mode at initial time, whereas $\widehat{\mu}_{\mathrm{Dim}}(t)$ is the corresponding operator at time t, which is obtained by a Heisenberg transformation over the first one with the aid of the Hamiltonian $\mathrm{H}_{\mathrm{Dav}}$.

This formal ACF may take two different forms according to the fact that one considers the allowed (u) or forbidden (g) IR transitions. These two forms are, respectively,

$$[\mathrm{G}^\circ_{\mathrm{Dav}}(t)^{(u)}] = \mathrm{tr}_{\mathrm{Dav}}[\rho_{\mathrm{Dav}}[\widehat{\mu}_u(\mathrm{q}_a, \mathrm{q}_b, 0)][\widehat{\mu}_u(\mathrm{q}_a, \mathrm{q}_b, t)]]$$

$$[\mathrm{G}^\circ_{\mathrm{Dav}}(t)^{(g)}] = \mathrm{tr}_{\mathrm{Dav}}[\rho_{\mathrm{Dav}}[\widehat{\mu}_g(\mathrm{q}_a, \mathrm{q}_b, 0)][\widehat{\mu}_g(\mathrm{q}_a, \mathrm{q}_b, t)]]$$

In these equations, $\mu_u(q_a, q_b, 0)$ and $\widehat{\mu}_g(q_a, q_b, 0)$ are the dipole moment operators at initial time belonging, respectively, to the irreducible representations B and A of the C_2 symmetry group that transforms themselves, the first one, according to x and y, and the last one, according to x^2 and y^2(allowed Raman transition), where x and y and the Cartesian coordinates that are perpendicular to the C_2 symmetry axis. Here, we prefer the notations "g" and "u" in place of "A" and "B" of group theory.

At last, ρ_{Dav} is the Boltzmann density operator given by

$$\rho_{\mathrm{Dav}} = \frac{1}{Z_{\mathrm{Dav}}}\left(e^{-\mathrm{H}_{\mathrm{Dav}}/k_{\mathrm{B}}T}\right)$$

with

$$Z_{\mathrm{Dav}} = \mathrm{tr}_{\mathrm{Dav}}\left[e^{-\mathrm{H}_{\mathrm{Dav}}/k_{\mathrm{B}}T}\right]$$

By expanding the above dipole moment operators around the origins of q_a and q_b, in the Taylor series gives

$$\widehat{\mu}_g(q_a, q_b, 0) = \mu(0,0,0) + \left[\left(\frac{\partial\widehat{\mu}}{\partial q_b}\right)_{q_a=q_b=0}\right] q_b + \left[\left(\frac{\partial\widehat{\mu}}{\partial q_a}\right)_{q_a=q_b=0}\right] q_a$$

$$\widehat{\mu}_u(q_a, q_b, 0) = \mu(0,0,0) + \left[\left(\frac{\partial\widehat{\mu}}{\partial q_b}\right)_{q_a=q_b=0}\right] q_b - \left[\left(\frac{\partial\widehat{\mu}}{\partial q_a}\right)_{q_a=q_b=0}\right] q_a$$

Observe that for absorption it is possible to write, in the spirit of Section I.B

$$\left[\left(\frac{\partial\widehat{\mu}}{\partial q_b}\right)_{q_a=q_b=0}\right] q_b = \mu_b^{\circ}|\{0\}_a\{1\}_b\rangle\langle\{0\}_b\{0\}_a|$$

$$\left[\left(\frac{\partial\widehat{\mu}}{\partial q_a}\right)_{q_a=q_b=0}\right] q_a = \mu_a^{\circ}|\{1\}_a\{0\}_b\rangle\langle\{0\}_b\{0\}_a|$$

Here, the terms $\mu^{\circ}a$ and $\mu^{\circ}b$ are, respectively, given by

$$\mu_a^{\circ} = q^{\circ\circ}\left(\frac{\partial\hat{\mu}}{\partial q_a}\right)_{q_a=0} \quad \text{and} \quad \mu_b^{\circ} = q^{\circ\circ}\left(\frac{\partial\hat{\mu}}{\partial q_b}\right)_{q_b=0} \quad \text{with} \quad q^{\circ\circ} = \sqrt{\frac{\hbar}{2m\omega}}$$

Besides, in this same equation the kets are

$$|\{k\}_a\{l\}_b\rangle = |\{k\}_a\rangle \otimes |\{l\}_b\rangle \qquad \text{with} \qquad k, l = 0, 1$$

As a consequence, it is possible to write the symmetrized Hermitean dipole moment operators in terms of symmetrized non-Hermitean absorption transition operators, according to

$$\widehat{\mu}_g(q_a, q_b, 0) = \mu(0,0,0) + [\mu(0)_g]$$
$$\widehat{\mu}_u(q_a, q_b, 0) = \mu(0,0,0) + [\mu(0)_u]$$

These symmetrized non-Hermitean operators are, respectively, [75]

$$[\mu(0)_g] = \frac{1}{\sqrt{2}}[(\mu_b^\circ)|\{0\}_a\{1\}_b\rangle + (\mu_a^\circ)|\{1\}_a\{0\}_b\rangle]\langle\{0\}_b\{0\}_a|$$
$$[\mu(0)_u] = \frac{1}{\sqrt{2}}[(\mu_b^\circ)|\{0\}_a\{1\}_b\rangle - (\mu_a^\circ)|\{1\}_a\{0\}_b\rangle]\langle\{0\}_b\{0\}_a|$$

The IR absorption ACF is

$$[G_{\mathrm{Dav}}^\circ(t)^{(u)}]^\circ = \mathrm{tr}_{\mathrm{Dav}}[\rho_{\mathrm{Dav}}[\mu(0)_u]^\dagger[\mu(t)_u]] \tag{259}$$

At time t, the u transition operator is obtained by the Heisenberg transformation:

$$[\mu(t)_u] = \frac{1}{\sqrt{2}}[e^{i\mathrm{H}_{\mathrm{Dav}}t/\hbar}][(\mu_b^\circ)|\{0\}_a\{1\}_b\rangle - (\mu_a^\circ)|\{1\}_a\{0\}_b\rangle][\langle\{0\}_b\{0\}_a|][e^{-i\mathrm{H}_{\mathrm{Dav}}t/\hbar}]$$

Of course, for the Raman absorption the ACF is

$$[G_{\mathrm{Dav}}^\circ(t)^{(g)}]^\circ = \mathrm{tr}_{\mathrm{Dav}}[\rho_{\mathrm{Dav}}[\mu(0)_g]^\dagger[\mu(t)_g]]$$

Appendix O shows that the ACF (259) takes the form given by Eq. (O.15), that is,

$$[G_{\mathrm{Dav}}^\circ(t)^{(u)}]^\circ = \frac{1}{2}\varepsilon^2 \sum_{n_a}\sum_{m_b}\sum_{\mu}\left(e^{-(n_a+m_b)\hbar\Omega t/k_BT}\right)\left[e^{-i(n_a+m_b)\Omega t/\hbar}\right]\left[e^{i\omega_\mu t/\hbar}\right]$$
$$\left\{[\mu_a^\circ]^2\left[C_{n_a m_b \mu}^{\{1\}\{0\}}\right]^2 + [\mu_b^\circ]^2\left[C_{n_a m_b \mu}^{\{0\}\{1\}}\right]^2 - 2[\mu_a^\circ\mu_b^\circ]\left[C_{n_a m_b \mu}^{\{1\}\{0\}}\right]\left[C_{n_a m_b \mu}^{\{0\}\{1\}}\right]\right\}$$

Here, ω_μ are the eigenvalues appearing in the following eigenvalue equation:

$$\left[\mathrm{H}_{\{1,0\}}^{\{0,1\}}\right]|\Psi_\mu\rangle = \hbar\omega_\mu|\Psi_\mu\rangle \qquad \text{with} \qquad \langle\Psi_\mu|\Psi_\nu\rangle = \delta_{\mu\nu}$$

whereas the coefficients are those involved in the expansion of the corresponding eigenvectors:

$$|\Psi_\mu\rangle = \sum_{n_a}\sum_{m_b}\left[\left[C_{n_a m_b \mu}^{\{1\}\{0\}}\right]|\{1\}_b(m)_b\{0\}_a(n)_a\rangle + \left[C_{n_a m_b \mu}^{\{0\}\{1\}}\right]|\{0\}_b(m)_b\{1\}_a(n)_a\rangle\right]$$

with

$$|\{1\}_b(m)_b\{1\}_a(n)_a\rangle \equiv |\{1\}_b\rangle|(m)_b\rangle\; |\{1\}_a\rangle|(n)_a\rangle$$

Besides, the Hamiltonian involved in the eigenvalue equation is given by Eq. (O.2), that is,

$$\left[\mathrm{H}_{\{1,0\}}^{\{0,1\}}\right] = \begin{pmatrix} \left[\mathrm{H}_{II}^{\{1,0\}}\right] & \mathrm{V}^\circ \\ \mathrm{V}^\circ & \left[\mathrm{H}_{II}^{\{0,1\}}\right] \end{pmatrix} \tag{260}$$

Note that, up to now, the damping has been ignored. Incorporation of the direct one may be performed, as usually, by multiplying the ACF by the exponential decay $e^{-\gamma^\circ t}$.

As a consequence, for allowed IR absorption, the SD involving both Davydov coupling and *direct damping* is [75]

$$\left[\mathrm{I}_{\mathrm{Dav}}^{\circ}(\omega)^{(u)}\right] = \int_{-\infty}^{\infty} \left[\mathrm{G}_{\mathrm{Dav}}^{\circ}(t)^{(u)}\right]^{\circ} e^{-i\omega t}(e^{-\gamma^\circ t})dt \tag{261}$$

4. *Taking into Account Quantum Indirect Damping by Aid of $\widehat{C}_2$ Symmetry [76]*

Note that, if it was possible to introduce the quantum direct damping easily, it is not so for the indirect one, without using the symmetry of the problem involving the $\widehat{\mathrm{C}}_2$ parity operator.

a. Diagonalization of the Davydov Hamiltonian (255). For this purpose, it is suitable to diagonalize the Davydov Hamiltonian. Previously, we met the basis (254) that is

$$\begin{pmatrix} |\Phi_{\{a,b\}}^{\{0,0\}}\rangle \\ |\Phi_{\{a,b\}}^{\{1,0\}}\rangle \\ |\Phi_{\{a,b\}}^{\{0,1\}}\rangle \end{pmatrix} = \begin{pmatrix} |\{0\}_a\rangle \otimes \{0\}_b\rangle \\ |\{1\}_a\rangle \otimes \{0\}_b\rangle \\ |\{0\}_a\rangle \otimes \{1\}_b\rangle \end{pmatrix} \tag{262}$$

Besides, it was shown that within this basis, the Davydov Hamiltonian matrix, is given by Eq. (255), that is,

$$[\mathbb{H}_{\text{Dav}}] = \begin{pmatrix} \left[\mathbb{H}_{II}^{\{0,0\}}\right] & 0 & 0 \\ 0 & \left[\mathbb{H}_{II}^{\{1,0\}}\right] & V^\circ \\ 0 & V^\circ & \left[\mathbb{H}_{II}^{\{0,1\}}\right] \end{pmatrix} \tag{263}$$

Now, in order to diagonalize Hamiltonian (263), it is possible to perform the following linear transformation [18]:

$$\begin{pmatrix} |\Phi_{\{a,b\}}^{\{0,0\}}\rangle \\ |\beta_{\{1,0\}\leftrightarrow\{0,1\}}^{(+)}\rangle \\ |\beta_{\{1,0\}\leftrightarrow\{0,1\}}^{(-)}\rangle \end{pmatrix} = \begin{pmatrix} |\Phi_{\{a,b\}}^{\{0,0\}}\rangle \\ |\Phi_{\{a,b\}}^{\{1,0\}}\rangle + \widehat{C}_2|\Phi_{\{a,b\}}^{\{0,1\}}\rangle \\ |\Phi_{\{a,b\}}^{\{1,0\}}\rangle - \widehat{C}_2|\Phi_{\{a,b\}}^{\{0,1\}}\rangle \end{pmatrix} \tag{264}$$

Then, Appendix P show that within the basis defined by Eq. (264) the Davydov Hamiltonian (263) takes the diagonal form given by Eq. (P 21), that is,

$$[\mathbb{H}_{\text{Dav}}] = \begin{pmatrix} [\mathbb{H}_{II}^{\{0,0\}}] & 0 & 0 \\ 0 & [\mathbb{H}_{II}^{\{1,0\}} + V^\circ\widehat{C}_2] & 0 \\ 0 & 0 & [\mathbb{H}_{II}^{\{1,0\}} - V^\circ\widehat{C}_2] \end{pmatrix} \tag{265}$$

b. Separation of the g and u Hamiltonians in the Davydov Hamiltonian (265). Now, in order to make tractable the action of the $\widehat{C}_2$ parity operator, it is suitable to pass to the symmetrized coordinates and their conjugate momenta according to Table XI:

The action of the parity operator is given in Fig. 16.

Next, we may consider the *kets* (264) as tensorial products of states, according to

$$\begin{pmatrix} |\Phi_{\{g,u\}}^{\{0,0\}}\rangle \\ |\beta_{\{1,0\}\leftrightarrow\{0,1\}}^{(+)}\rangle \\ |\beta_{\{1,0\}\leftrightarrow\{0,1\}}^{(-)}\rangle \end{pmatrix} = \begin{pmatrix} |\{0\}_g\rangle \otimes |\{0\}_u\rangle \\ |\{1\}_g\rangle \otimes |\{\beta^{(+)}\}_{u^+}^{\{1\}}\rangle \\ |\{1\}_g\rangle \otimes |\{\beta^{(-)}\}_{u-}^{\{1\}}\rangle \end{pmatrix} \tag{266}$$

TABLE XI
Symmetrized Coordinates

$Q_g = \frac{1}{\sqrt{2}}[Q_a + Q_b]$	$Q_u = \frac{1}{\sqrt{2}}[Q_a - Q_b]$	$P_g = \frac{1}{\sqrt{2}}[P_a + P_b]$	$P_u = \frac{1}{\sqrt{2}}[P_a - P_b]$
$C_2Q_g = Q_g$	$C_2Q_u = -Q_u$	$C_2P_g = P_g$	$C_2P_u = -P_u$

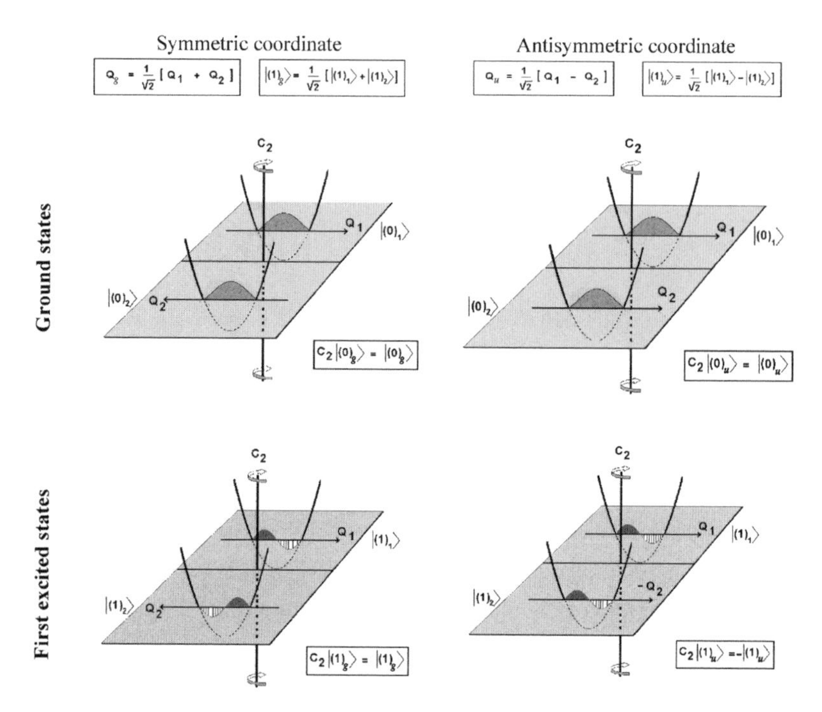

Figure 16. The effects of the parity operator C_2 on the ground and the first excited states of the symmetrized g and u eigenfunctions of the g and u quantum harmonic oscillators involved in the centrosymmetric cyclic dimer. (The subscripts 1 and 2 refer, respectively, to the a and b moieties of the centrosymmetric cyclic dimer).

with

$$\begin{pmatrix} |\{0\}_g\rangle \\ |\{0\}_u\rangle \end{pmatrix} = \begin{pmatrix} \frac{1}{\sqrt{2}}[|\{0\}_a\rangle + |\{0\}_b\rangle] \\ \frac{1}{\sqrt{2}}[|\{0\}_a\rangle - |\{0\}_b\rangle] \end{pmatrix} \tag{267}$$

Appendix P shows that, within the framework of the symmetrized coordinates, the Hamiltonian (265) takes the block form (P.35), that is,

$$[\mathbb{H}_{\mathrm{Dav}}] = \begin{pmatrix} [\mathbb{H}_{II}^{\{0\}}]_g & 0 & 0 & 0 & 0 \\ 0 & [\mathbb{H}_{II}^{\{1\}}]_g & 0 & 0 & 0 \\ 0 & 0 & [\mathrm{H}_{II}^{\{0\}}]_u & 0 & 0 \\ 0 & 0 & 0 & [\mathrm{H}_{(+)}^{\{1\}}]_{u^+} & 0 \\ 0 & 0 & 0 & 0 & [\mathrm{H}_{(-)}^{\{1\}}]_{u^-} \end{pmatrix} \text{ in } \begin{pmatrix} |\{0\}_g\rangle \\ |\{1\}_g\rangle \\ |\{0\}_u\rangle \\ |\{\beta^{(+)}\}_{u^+}^{\{1\}}\rangle \\ |\{\beta^{(-)}\}_{u^-}^{\{1\}}\rangle \end{pmatrix} \tag{268}$$

with, respectively,

$$\left[\mathbb{H}_{II}^{\{0\}}\right]_g = \frac{P_g^2}{2M} + \frac{M\Omega^2 Q_g^2}{2} + \sqrt{2}\sum_r \tilde{q}_r Q_g \hbar\tilde{\kappa}_r + [\mathbb{H}^{\{\theta\}}]$$

$$[H_{II}^{\{0\}}]_u = \frac{P_u^2}{2M} + \frac{M\Omega^2 Q_u^2}{2} \tag{269}$$

$$[\mathbb{H}_{II}^{\{1\}}]_g = \hbar\omega^\circ - \alpha^{\circ 2}\hbar\Omega + \frac{P_g^2}{2M} + \frac{M\Omega^2 Q_g^2}{2} + \frac{1}{\sqrt{2}} bQ_g + \sqrt{2}\sum_r \tilde{q}_r Q_g \hbar\tilde{\kappa}_r + [\mathbb{H}^{\{\theta\}}] \tag{270}$$

$$[H_{(\pm)}^{\{1\}}]_u = \frac{P_u^2}{2M} + \frac{M\Omega^2 Q_u^2}{2} + \frac{1}{\sqrt{2}} bQ_u \pm V^\circ(\widehat{C}_2)_u \tag{271}$$

It may be observed that the two Hamiltonians (269) are those of quantum harmonic oscillators, whereas Hamiltonian (270) is that of a driven damped quantum harmonic oscillator, and Hamiltonians (271) are those of driven undamped quantum harmonic oscillators perturbed by the Davydov coupling term $\pm V^\circ(\widehat{C}_2)_u$.

5. *The ACF and SD With Quantum Indirect Damping*

Appendix R shows that the ACF (259) of the transition moment operator of the fast mode, involving indirect damping, but without direct damping ignored, takes the form given by Eq. (R.3). Of course, the direct damping may be taken into account with the aid of a damped exponential factor, as performed in the simpler situation without Davydov coupling. Consequently, in the present situation, the ACF is

$$[G_{Dav}(t)] = [G(t)]^\circ{}_g[[G^{(+)}(t)]_u + [G^{(-)}(t)]_u](e^{-\gamma^\circ t}) \tag{272}$$

Here, $[G(t)]_g^\circ$ is the ACF of the g part of the system related to Hamiltonian (270) whereas $[G^\circ(\pm)(t)]_u^\circ$ are the ACFs corresponding to the u parts dealing with Hamiltonians (271). At last, γ° is the natural width of the excited state of the high frequency mode.

The g ACF is shown in Appendix R to be given by Eq. (R.8), that is, by the following trace $\mathrm{tr}_{\mathrm{Slow}_g}$ over the base spanned by the symmetrized g Hamiltonian $[\mathbb{H}_{II}^{\{0\}}]_g$ of the corresponding H–bond bridge:

$$[G(t)]_g^\circ = \tilde{\varepsilon}(\mu^\circ{}_u)^2 \mathrm{tr}_{\mathrm{Slow}_g}[(e^{-[\mathbb{H}_{II}^{\{0\}}]_g/k_B T})\mathrm{tr}_\theta[[e^{i[\mathbb{H}_{II}^{\{1\}}]_g t/\hbar}][e^{-i[\mathbb{H}_{II}^{\{0\}}]_g t/\hbar}]]]$$

Here, the Hamiltonians are given by Eqs. (269) and (270). This last expression may be written also as:

$$[\mathrm{G}(t)]_g^{\circ} = (\mu^{\circ}{}_u)^2 \mathrm{tr}_{\mathrm{Slow}_g}[[\tilde{\rho}_{II}^{\{1\}}]_g \mathrm{tr}_{\theta}[[e^{i[\mathbb{H}_{II}^{\{1\}}]_g t/\hbar}][e^{-i[\mathbb{H}_{II}^{\{0\}}]_g t/\hbar}]]]$$

with

$$[\tilde{\rho}_{II}^{\{1\}}]_g = \tilde{\varepsilon}(e^{-[\mathrm{H}_{II}^{\{0\}}]_g/k_{\mathrm{B}}T})$$

As it appears, this ACF has the same structure as (108). This equation was found for a single H-bond bridge involving quantum indirect damping, except for the fact that $\tilde{b} = b/\sqrt{2}$ in the Hamiltonian $[\mathbb{H}_{II}^{\{1\}}]_g$ given by Eq. (270) in place of b, which appears in the effective Hamiltonian $\mathbb{H}_{II}^{\{1\}}$ given by Eq. (106).

Thus, owing to the result of Section III.D.1, which leads us to write

$$[\mathrm{G}(t)]_g^{\circ} = (\mu^{\circ}{}_u)^2 e^{i\omega^{\circ} t}(e^{-i\tilde{\alpha}^{\circ 2}\Omega t})\left\langle\left\langle \widehat{\mathrm{P}}\left[e^{i\tilde{b}\int_0^t \mathrm{Q}_g(t')^{\mathrm{IP}}dt'}\right]\right\rangle_{\theta}\right\rangle_{\mathrm{Slow}_g} \tag{273}$$

with

$$\left\langle\left\langle \widehat{\mathrm{P}}\left[e^{i\tilde{b}\int_0^t \mathrm{Q}_g(t')^{\mathrm{IP}}dt'}\right]\right\rangle_{\theta}\right\rangle_{\mathrm{Slow}_g} = (e^{-i\tilde{\alpha}^{\circ 2}\Omega t})(e^{i\tilde{\alpha}^{\circ 2}e^{-\tilde{\gamma}t/2}\sin\Omega t})[e^{\tilde{\alpha}^{\circ 2}[\langle n\rangle + 1/2](2e^{-\tilde{\gamma}t/2}\cos(\Omega t) - e^{-\tilde{\gamma}t} - 1)}]$$

with, respectively,

$$\tilde{\alpha}^{\circ} = \frac{\alpha^{\circ}}{\sqrt{2}} \quad \text{and} \quad \tilde{\gamma} = \gamma\sqrt{2} \tag{274}$$

Here, $\widehat{\mathrm{P}}$ is the Dyson time-ordering operator [57], $\mathrm{Q}_g(t)^{\mathrm{IP}}$ is the Q_g coordinate in the interaction picture with respect to the thermal bath and to the diagonal part of the Hamiltonian of the H-bond bridge, and the notation $\langle\langle\ldots\rangle_{\theta}\rangle_{\mathrm{Slow}}$ has the meaning of a partial trace on the thermal bath and on the H-bond bridge coordinates.

Consequently, since Eq. (108), which is analogous to Eq. (273), leads to Eq. (125), it follows that Eq.(273) must lead to the following equation, which is analogous to Eq.(125),

$$[\mathrm{G}(t)]_g^{\circ} = (\mu^{\circ}{}_u)^2 e^{i\omega^{\circ} t}(e^{-i2\tilde{\alpha}^{\circ 2}\Omega t})(e^{i\tilde{\alpha}^{\circ 2}e^{-\tilde{\gamma}t/2}\sin\Omega t})[e^{\tilde{\alpha}^{\circ 2}[\langle n\rangle + 1/2](2e^{-\tilde{\gamma}t/2}\cos(\Omega t) - e^{-\tilde{\gamma}t} - 1)}] \tag{275}$$

Recall that $\langle n \rangle$ is the thermal average of the occupation number of the H-bond bridge harmonic oscillator.

On the other hand, Appendix R shows that the $\pm u$ ACFs appearing in Eq. (272) are given by Eqs. (R.19), that is,

$$[\mathrm{G}^{\circ}(\pm)(t)]_u^{\circ} = \varepsilon \sum_{\mu} \sum_{n_u}$$

$$|B^{\pm}_{\mu\, n_u}|^2 e^{-n_u \hbar\Omega\, / k_B T} e^{i\omega^{\pm}_{\mu} t} e^{-i n_u \Omega t} [[1 \pm (-1)^{n_u+1}] + \eta^{\circ}[1 \mp (-1)^{n_u+1}]]^2 \tag{276}$$

In Equation (276), $\hbar\omega^{\pm}_{\mu}$ are the eigenvalues of the Schrödinger equation:

$$[\mathrm{H}^{\{1\}}_{\pm}]_u |\beta^{(\pm)}_{\mu}\rangle = \hbar\omega^{\pm}_{\mu} |\beta^{(\pm)}_{\mu}\rangle \tag{277}$$

whereas $B^{\pm}_{n_u,\mu}$ are the expansion coefficients of the corresponding eigenvectors:

$$|\beta^{(\pm)}_{\mu}\rangle = \sum_{n_u} B^{\pm}_{n_u \mu} |(n)_u\rangle$$

Besides, η° is a dimensionless parameter running between 0 and 1, susceptible to reflect some break of the IR selection rule probably following from some lack of symmetry of the dimer or perhaps from dynamic vibronic structure of the H-bond system [77].

Finally, within the linear response theory, the spectral density is

$$[\mathrm{I}_{\mathrm{Dav}}(\omega)] = 2\mathrm{Re} \int_0^{\infty} [\mathrm{G}_{\mathrm{Dav}}(t)] e^{-i\omega t} dt \tag{278}$$

Thus, according to Eq. (272) it becomes

$$[\mathrm{I}_{\mathrm{Dav}}(\omega)] = 2\mathrm{Re} \int_0^{\infty} [\mathrm{G}(t)]_g^{\circ} [[\mathrm{G}^{(+)}(t)]^{\circ}{}_u + [\mathrm{G}^{(-)}(t)]_u^{\circ}] (e^{-\gamma^{\circ} t}) e^{-i\omega t} dt$$

Again, by using Eqs. (275) and (276), yields [76]:

$$[\mathrm{I}_{\mathrm{Dav}}(\omega)] = (\mu^{\circ}{}_u)^2 [[\mathrm{I}^{+}_{\mathrm{Dav}}(\omega)] + [\mathrm{I}^{-}_{\mathrm{Dav}}(\omega)]] \tag{279}$$

TABLE XII
SD of Damped Centrosymmetric Cyclic Dimers

$$[\mathrm{I}^{\pm}_{Dav}(\omega)] \propto \sum_{m_g}\sum_{n_g}[\mathrm{P}_{m_g n_g}]\sum_{n_u}\sum_{\mu}$$

$$e^{-\widetilde{\lambda} n_u}[[1 \pm (-1)^{n_u+1}] + \eta^{\circ}[1 \mp (-1)^{n_u+1}]]^2 |B^{\pm}_{n_u\mu}|[^2\mathrm{I}^{\pm}_{m_g n_g n_u \mu}(\omega)]$$

$$[\mathrm{I}^{\pm}_{m_g n_g n_u \mu}(\omega)] \propto \frac{\gamma_{m_g n_g}}{(\omega - \Omega^{\pm}_{m_g n_g n_u \mu})^2 + (\gamma_{m_g n_g})^2}$$

$$\Omega^{\pm}_{m_g n_g n_u \mu} = \omega^{\circ} - [(m_g - n_g + n_u)\Omega - \omega^{\pm}_{\mu}] - 2\widetilde{\alpha}^{\circ 2}\Omega$$

$$\gamma_{m_g n_g} = (m_g + n_g)\widetilde{\gamma} + \gamma^{\circ} \quad \widetilde{\gamma} = \gamma\sqrt{2}$$

$$[\mathrm{P}_{m_g n_g}] = \frac{[1+\langle n\rangle]^{m_g}\langle n\rangle^{n_g}\widetilde{\alpha}^{\circ 2(m_g+n_g)}}{m_g! n_g!} \quad \widetilde{\alpha}^{\circ} = \alpha^{\circ}/\sqrt{2}$$

$$\langle n\rangle = \frac{1}{e^{\widetilde{\lambda}}-1} \qquad \widetilde{\lambda} = \frac{\hbar\Omega}{k_B T}$$

$$\left[\frac{\mathrm{P}^2_{\mu}}{2M} + \frac{M\Omega^2\mathrm{Q}^2_u}{2} + \frac{1}{\sqrt{2}}b\mathrm{Q}_u \pm \mathrm{V}^{\circ}\mathrm{C}_2\right]|\beta^{\pm}_{\mu}\rangle = \hbar\omega^{\pm}_{\mu}|\beta^{\pm}_{\mu}\rangle$$

$$|\beta^{\pm}_{\mu}\rangle = \sum_{n_u}[B^{\pm}_{n_u\mu}]|(n)_u\rangle$$

$$\left[\frac{\mathrm{P}^2_x}{2M} + \frac{M\Omega^2\mathrm{Q}^2_x}{2}\right]|(n)_x\rangle = \left(n_x + \tfrac{1}{2}\right)\hbar\Omega|(n)_x\rangle, x = g, u$$

The components appearing in this equation are given by the equations given in Table XII:

6. *Special Situations for the SD (279)*

Now, we will show that the spectral density (279) satisfactorily reduces to many special situations of the literature, especially those obtained by

1. Maréchal and Witkowski [18] in the absence of damping.
2. Rösch and Ratner [47] in the absence of Davydov coupling and indirect damping.
3. Boulil et al. [48] in the absence of Davydov coupling and direct damping.
4. Robertson and Yarwood [46], in the semiclassical limit, without Davydov coupling and direct damping.
5. Chamma et al. [75], that is, Eq. (261) when the indirect damping is missing.

a. Quantum Limit With $\eta^{\circ} = 0$, *for Situations Without Damping.* First, consider a special situation in which the IR selection rule is strictly verified, that is, where $\eta^{\circ} = 0$ and where *both the direct and indirect damping are missing*. Then, the ACF given by Eq. (272) reduces to

$$[\mathrm{G}^{\circ}_{\mathrm{Dav}}(t)]^{\circ} = [\mathrm{G}^{\circ}(t)]^{\circ}_g[[\mathrm{G}^{\circ(+)}(t)]^{\circ}_u + [\mathrm{G}^{\circ(-)}(t)]^{\circ}_u]$$

Here, the two u ACFs are always given by Eq.(276), because they are not influenced by damping, whereas the single g ACF is now given by Eq. (275) in which the damping γ is taken to be zero, the only difference being that $\tilde{\alpha}^\circ = \alpha^\circ/\sqrt{2}$ in place of α°, that is,

$$[G^\circ(t)]_g^\circ = (\mu_u^\circ)^2 e^{i\omega^\circ t}(e^{-i2\tilde{\alpha}^{\circ 2}\Omega t})(e^{i\tilde{\alpha}^{\circ 2}\sin\Omega t})[e^{2\tilde{\alpha}^{\circ 2}[\langle n\rangle+1/2](\cos(\Omega t)-1)}] \tag{280}$$

Again, use Eq. (81) which, in the present situation, takes the form

$$(e^{i\tilde{\alpha}^{\circ 2}\sin\Omega t})[e^{2\tilde{\alpha}^{\circ 2}[\langle n\rangle+1/2](\cos(\Omega t)-1)}] = \varepsilon\sum_{m_g}\sum_{n_g}(e^{-m_g\tilde{\lambda}})|A_{m_g n_g}(\tilde{\alpha}^\circ)|^2(e^{i(n_g-m_g)\Omega t}) \tag{281}$$

Moreover, note that when η° is zero, μ_u° reduces to μ_o. Thus, as a consequence of Eqs. (276) and (281), Eq. (280) becomes

$$\begin{aligned}[G^\circ_{\text{Dav}}(t)]^\circ &= \mu_o^2 e^{i\omega^\circ t}(e^{-i2\tilde{\alpha}^{\circ 2}\Omega t})\\ &\quad\varepsilon\sum_{m_g}\sum_{n_g}(e^{-m_g\hbar\Omega/k_BT})|A_{m_g n_g}(\tilde{\alpha}^\circ)|^2\ (e^{i(n_g-m_g)\Omega t})\varepsilon\sum_{\mu}\sum_{n_u}(e^{-n_u\hbar\Omega/k_BT})\\ &\quad\{|B^+_{\mu\, n_u}|^2(e^{i\omega^+_\mu t})[1+(-1)^{n_u+1}]^2+|B^-_{\mu\, n_u}|^2(e^{i\omega^-_\mu t})[1-(-1)^{n_u+1}]^2\}\end{aligned}$$

Then, after Fourier transform we obtain the following expression for the SD:

$$\begin{aligned}[I^\circ{}_{\text{Dav}}(\omega)]^\circ &= \mu_o^2\varepsilon^2\sum_{m_g}\sum_{n_g}\sum_{\mu}\sum_{n_u}(e^{-(m_g+n_u)\hbar\Omega/k_BT})|A_{m_g\, n_g}(\tilde{\alpha}^\circ)|^2\\ &\left\{|B^+_{\mu\, n_u}|^2[1+(-1)^{n_u+1}]^2\delta(\omega-[\omega^\circ+(n_g-m_g-2\tilde{\alpha}^\circ 2)\Omega+\omega^+_\mu])\right.\\ &\left.+|B^-_{\mu\, n_u}|^2[1-(-1)^{n_u+1}]^2\delta(\omega-[\omega^\circ+(n_g-m_g-2\tilde{\alpha}^\circ 2)\Omega+\omega^-_\mu])\right\}\end{aligned} \tag{282}$$

Of course, when the Davydov coupling is in turn missing, the SD (282) reduces to that of a monomer involving the anharmonic coupling α°:

$$\begin{aligned}&[I^\circ{}_{\text{Dav}}(\omega)]^\circ \to [I^\circ{}_{III}(\omega)]^\circ\\ &= \mu_o^2\ \varepsilon\sum_m\sum_n(e^{-\tilde{\lambda}n})|A_{m\,n}(\alpha^\circ)|^2\delta[\omega-(\omega^\circ-(m-n)\Omega-2\alpha^{\circ 2}\Omega t)]\end{aligned}$$

This is the result of Maréchal and Witkowski [18] dealing with monomers.

b. Quantum Limit With $\eta^\circ = 0$, for Situations With Direct and Indirect Damping But Without Davydov Coupling. Now, consider the limit situation in the absence of Davydov coupling, but with direct and indirect damping. Then, the two $\pm u$ ACFs have the same structure, since the two $\pm u$ Hamiltonians (271) become the same when $V^\circ = 0$. Moreover, these two Hamiltonians are then equivalent to Hamiltonian (270), except for the fact that the last one is coupled to the thermal bath and takes into account the scalar $\hbar(\omega^\circ - \alpha^\circ 2\Omega)$. Consequently, the two ACFs $[G^\circ(\pm)(t)]_u^\circ$ become equivalent to $[G^\circ(t)]_u^\circ$, so that the full ACF of the centrosymmetric cyclic dimer takes the form:

$$[G_{\mathrm{Dav}}(t)] \to [G_{\mathrm{Dim}}(t)] = 2[G(t)]_g^\circ [G^\circ(t)]_u^\circ e^{-\gamma^\circ t} \tag{283}$$

Now, we must observe that because of the absence of the Davydov coupling, it is indifferent to consider the symmetry, so that it is possible to come back to the nonsymmetrized ACFs, and thus to write for the nonsymmetrized coordinate:

$$Q_a = \frac{Q_g + Q_u}{\sqrt{2}} \qquad \text{and} \qquad Q_b = \frac{Q_g - Q_u}{\sqrt{2}}$$

Within these nonsymmetrized coordinates, the indirect damping parameter is γ and not $\tilde{\gamma}$, whereas the dimensionless anharmonic coupling parameter $\tilde{\alpha}^\circ$ remains to be given by

$$\tilde{\alpha}^\circ = \frac{\alpha^\circ}{\sqrt{2}}$$

Thus, the full ACF (283) becomes the product of the two moieties ACFs, which are the same and both involve indirect damping, times the damped exponential term corresponding to the direct damping,

$$[G_{\mathrm{Dim}}(t)] = 2e^{i\omega^\circ t}(e^{-i2\alpha^{\circ 2}\Omega t})[G_{II}(t)]_a^\circ \; [G_{II}(t)]_b^\circ e^{-\gamma^\circ t} \tag{284}$$

Now, owing to Eq. (124), each moiety ACF $[G_{II}(t)_a]$ or $[G_{II}(t)_b]$ is given, with $x = a, b$, by

$$[G_{II}(t)]_x^\circ = (\mu_0^2)(e^{i\tilde{\alpha}^{\circ 2}\sin\Omega t e^{-\gamma t/2}})[e^{\tilde{\alpha}^{\circ 2}[\langle n\rangle + 1/2](2e^{-\gamma t/2}\cos\Omega t - e^{-\gamma t} - 1)}] \tag{285}$$

Equation (287) shows that multiplying the a ACF by the b ACF leads to a new ACF of the same structure in which $\tilde{\alpha}^\circ$ transforms into α°. Besides, when η° is zero, μ_u° reduces to μ_o. As a consequence, the full ACF (284) of the dimer in the absence of Davydov coupling becomes

$$\begin{aligned}[G_{\mathrm{Dim}}(t)] &= 2(\mu_0^2)(e^{i\omega^\circ t})(e^{-i2\alpha^{\circ 2}\Omega t}) \\ &(e^{i\alpha^{\circ 2}\sin\Omega t e^{-\gamma t/2}})[e^{\alpha^{\circ 2}[\langle n\rangle + 1/2][2e^{-\gamma t/2}\cos\Omega t - e^{-\gamma t} - 1]}]e^{-\gamma^\circ t}\end{aligned} \tag{286}$$

It appears that, as required, this ACF is twice that of the single H-bond bridge involving both direct and indirect damping which, after Fourier transform, leads to the spectral density (285). Note that this last SD reduces in turn to that of Boulil et al. [49] when the direct damping is missing, that is, when $\gamma^\circ = 0$:

$$[G_{Dim}(t)]^\circ \rightarrow 2[G_{II}(t)]^\circ$$

c. Quantum Limit With $\eta^\circ = 0$, for Situations Without Davydov Coupling and Indirect Damping. In the *absence* of Davydov coupling and *of indirect damping*, the ACF of the dimer is given by Eq. (286) in which γ is taken to be zero, that is,

$$[G^\circ_{Dim}(t)] = 2(\mu_0^2)e^{i\omega^\circ t}(e^{-i2\alpha^{\circ 2}\Omega t})[(e^{i\alpha^{\circ 2}\sin\Omega t})[e^{2\alpha^{\circ 2}[\langle n\rangle+1/2][\cos\Omega t-1]}]]e^{-\gamma^\circ t}$$

After Fourier transform of this ACF, the SD obtained is that (71) of Rösch and Ratner [47] in their quantum theory of direct damping.

$$[G^\circ_{Dim}(t)] \rightarrow 2[G^\circ_{II}(t)]$$

d. Classical Limit for Situations Without Davydov Coupling and Direct Damping. Next, recall that Eq. (143) is,

$$(e^{-i\alpha^{\circ 2}\Omega t})(e^{i\alpha^{\circ 2}e^{-\gamma t/2}\sin\Omega t})[e^{\alpha^{\circ 2}[\langle n\rangle+1/2](2e^{-\gamma t/2}\cos\Omega t-e^{-\gamma t}-1)}] = \left\langle\left\langle \widehat{P}\left[e^{ib\int_0^t \tilde{q}(t')dt'}\right]\right\rangle_\theta\right\rangle_{Slow} \tag{287}$$

where $\widehat{P}$ is the Dyson time-ordering operator. Again, in the classical limit,

$$(e^{-i\alpha^{\circ 2}\Omega t}) \rightarrow 1 \qquad \text{and} \qquad \widehat{P} \rightarrow 1$$

Besides, in the classical limit, the IP coordinate $Q(t)^{IP}$ takes into account the influence of the surroundings and becomes a stochastic variable to be averaged over a Fokker–Planck equation. Moreover, the integral over this stochastic variable must become indefinite. Then, owing to Eq. (287), in the classical limit, the dimer ACF (286) becomes

$$[G_{Dim}(t)]^\circ = 2(\mu_0^2)e^{i\omega^\circ t}\left\langle\left[e^{ib\int_0^t Q(t')dt'}\right]\right\rangle_\theta \tag{288}$$

At last, after (1) cummulant expansion up to second order of the characteristic function, (2) taking finite integral instead of indefinite, and (3) assuming that the thermal average on stochastic variable commutes with integration over time, the

ACF (288) reduces to (180) appearing in the Robertson and Yarwood model [46], that is,

$$[G_{\mathrm{Dim}}(t)]^{\circ} \rightarrow 2[G_{II}(t)]^{\circ}_{\mathrm{Rob}}$$

or

$$[G_{\mathrm{Dim}}(t)]^{\circ} \rightarrow 2(\mu_0^2)\, e^{i\omega^{\circ} t} \left(e^{-\frac{b^2}{2}\int_0^t \int_0^{t'} \langle Q_g(t')Q_g(t'')\rangle \, dt' dt''} \right)$$

Recall that in the slow modulation limit, the Robertson and Yarwood ACF reduces in turn to that used by Bratos, in his pioneering work dealing with the SD of H-bonded species within the linear response theory. Also recall that there are two generalizations of the semiclassical model of Robertson and Yarwood, one by Sakun [78] and the other by Abramczyk [79]. The first is incorporating memory functions and the second rotational structure.

e. Recapitulation. Figure 17 recapitulates how the SD (279) reduces to all of the special situations given previously in the literature.

Note therefore that most theories dealing with the SD of H-bonded species are special situations of the general SD (279). This good behavior is encouraging for the future.

7. *An Approximation for Quantum Indirect Damping [80] When Davydov Coupling is Occurring*

In the expression of the Davydov Hamiltonian (268), the Hamiltonians appearing in the diagonal blocks are given by Eqs. (269–271). Express them in terms of Bosons with the aid of Table V:

Then, they become

$$[\mathbb{H}_{II}^{\{0\}}]_g = \left[a_g^{\dagger}a_g + \frac{1}{2}\right]\hbar\Omega + \left[\sum_r \tilde{b}_r^{\dagger}\tilde{b}_r + \frac{1}{2}\right]\hbar\omega_r + \sqrt{2}\sum_r [a_g^{\dagger}\tilde{b}_r\hbar\tilde{\kappa}_r + \mathrm{hc}] \qquad (289)$$

$$[H_{II}^{\{0\}}]_u = \left[a_u^{\dagger}a_u + \frac{1}{2}\right]\hbar\Omega \qquad (290)$$

$$[H_{II}^{\{1\}}]_u = \left[a_u^{\dagger}a_u + \frac{1}{2}\right]\hbar\Omega + \tilde{\alpha}^{\circ}[a_u^{\dagger} + a_u] \pm V^{\circ}C_2$$

$$[\mathbb{H}_{II}^{\{1\}}]_g = \hbar\omega^{\circ} - \alpha^{\circ}\hbar\Omega + \left[a_g^{\dagger}a_g + \frac{1}{2} + \tilde{\alpha}^{\circ}[a_g^{\dagger} + a_g]\right]\hbar\,\Omega$$
$$+ \sqrt{2}\sum_r [a_g^{\dagger}\tilde{b}_r\hbar\tilde{\kappa}_r + \mathrm{hc}] + \left[\sum_r \tilde{b}_r^{\dagger}\tilde{b}_r + \frac{1}{2}\right]\hbar\omega_r \qquad (291)$$

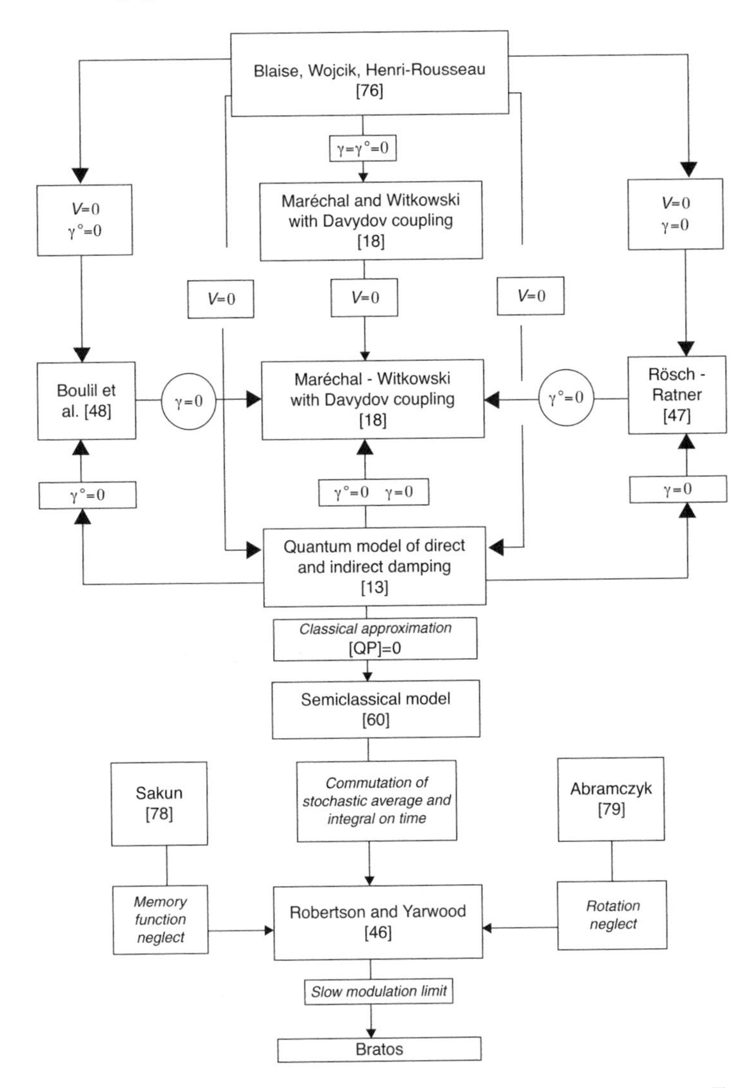

Figure 17. Connections between the present theory and different older models. The arrows denote the connection between a general theory and a more special one to which it reduces when some physical terms are vanishing or some approximations are performed. Approximations (italics) or vanishing parameters are indicated in non grayed boxes.

Now, in order to get an approach for Davydov coupling susceptible to generalization to more complex situations, it may be suitable, according to Eqs. (208) and (210), to replace the effective Hamiltonians (289) and (291) by the corresponding non-Hermitean ones, where, as above, the zero-point energy may be ignored, that is,

$$[\tilde{H}_{II}^{\{0\}}]_g \simeq a_g^\dagger a_g \left[1 - i\frac{\gamma}{2\Omega}\right]\hbar\Omega$$
$$[\tilde{H}_{II}^{\{1\}}]_g \simeq \hbar\omega^\circ - \alpha^\circ\hbar\Omega + \left[a_g^\dagger a_g + \frac{1}{\sqrt{2}}\alpha^\circ[a_g^\dagger + a_g]\right]\left[1 - i\frac{\gamma}{2\Omega}\right]\hbar\Omega \tag{292}$$

The eigenvalue equation of this last non-Hermitean Hamiltonian is

$$[\tilde{H}_{II}^{\{1\}}]_g|\tilde{\Psi}_\nu\rangle = \hbar[\omega_\nu^\circ - i\gamma_\nu]|\Psi_\nu\rangle \quad \text{and} \quad |\tilde{\Psi}_\nu\rangle = \sum [a_{n_g\nu}]|(n)_g\rangle$$

Of course, the full ACF remains given by an equation that is similar to Eq. (272). In the present situation, it is

$$[\tilde{G}_{\mathrm{Dav}}(t)] = [\tilde{G}(t)]_g^\circ[[G^{\circ(+)}(t)]_u^\circ + [G^{\circ(-)}(t)]_g^\circ]e^{-\gamma^\circ t} \tag{293}$$

Here, $[\tilde{G}(t)]_g^\circ$ plays the role of $[G(t)]_g^\circ$, whereas the two u ACF appearing in Eq. (272) are not modified since they do not involve the influence of the damping. Thus, for the u ACF, Eq. (276) still holds. On the other hand, the g ACF has the same structure as that obtained when we approached the quantum indirect relaxation with the aid of non-Hermitean Hamiltonians. Therefore it is obtained by Eq. (232). Consequently, the full ACF (293) becomes

$$[\tilde{G}_{\mathrm{Dav}}(t)] = (\mu_0^2)e^{i\omega^\circ t}e^{-\gamma^\circ t}$$
$$\varepsilon\sum_{n_g}\sum_{\nu}(e^{-\tilde{\lambda}n_g})(e^{-i(n_g\Omega-\omega_\nu^\circ)t})(e^{-(n_g\gamma/2+\gamma_\nu)t})|a_{n_g\nu}|^2[[G^{\circ(+)}(t)]_u^\circ + [G^{\circ(-)}(t)]_u^\circ] \tag{294}$$

Then, it is possible to compare the approximate SD obtained by Fourier transform of the ACF (294) with the reference SD (279). This is performed in Fig. 18.

Its examination shows that it is possible to reproduce rather satisfactorily the reference line shape at the condition of the introduction of effective parameters.

B. Dimer Involving Damping, Davydov Coupling, and Fermi Resonances

Now, it is possible to introduce Fermi resonances in the precedent model taking into account direct and indirect dampings, together with the relaxation of the bending modes [81].

1. *The Hamiltonian*

If one considers only the first harmonic terms of the bending modes that belong to the g symmetry, then Fermi resonances will only affect the g excited states of the

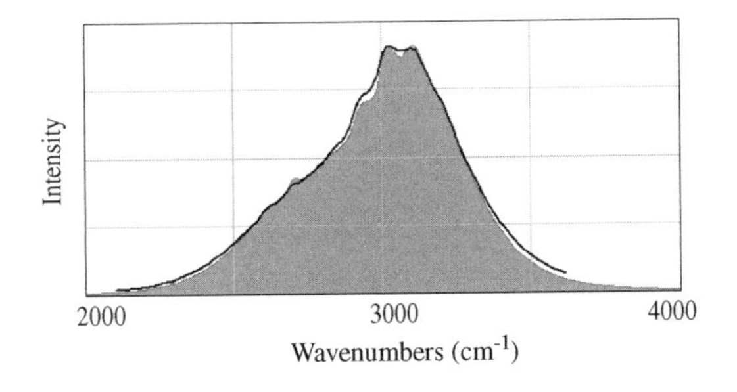

Parameters / SD	ω° (cm^{-1})	Ω (cm^{-1})	α°	γ° (Ω)	γ (Ω)	V° (Ω)	η
Reference	3100	88	1.19	0.30	0.24	-1.55	0.25
Approached	3035	88	1.210	0.60	0.15	-1.55	0.20

Figure 18. Comparison between the reference (grayed) SD (279) and the approached (thick line) SD obtained by Fourier transform of ACF (294) at room temperature.

fast mode. To introduce the Fermi resonances, begin with the Hamiltonian (268) of the system involving Davydov coupling, but without Fermi resonance, that is,

$$[\mathbb{H}_{\mathrm{Dav}}] = \begin{pmatrix} \left[\mathbb{H}_{II}^{\{0\}}\right]_g & 0 & 0 & 0 & 0 \\ 0 & [\mathbb{H}_{II}^{\{1\}}]_g & 0 & 0 & 0 \\ 0 & 0 & [\mathrm{H}_{II}^{\{0\}}]_u & 0 & 0 \\ 0 & 0 & 0 & [\mathrm{H}_{(+)}^{\{1\}}]_u & 0 \\ 0 & 0 & 0 & 0 & [\mathrm{H}_{(-)}^{\{1\}}]_u \end{pmatrix}$$

Then, perform the following transformation in this Hamiltonian:

$$[\mathbb{H}_{II}^{\{1\}}]_g \rightarrow [\mathbb{H}_{\mathrm{Fermi}}^{\{1\}}]_g$$

Here, the right-hand side Hamiltonian is that dealing with the Fermi resonances occurring between the g excited state of the fast mode and the g first harmonics of the bending mode. For three Fermi resonances, this Hamiltonian is

$$[\mathbb{H}_{\mathrm{Fermi}}^{\{1\}}]_g = \begin{pmatrix} [\mathbb{H}_{II}^{\{1\}}]_g & \hbar f_1^{\delta} & \hbar f_2^{\delta} & \hbar f_3^{\delta} \\ \hbar f_1^{\delta} & \breve{\mathrm{H}}^{\circ} + \hbar\Delta_1^{\delta} & 0 & 0 \\ \hbar f_2^{\delta} & 0 & \breve{\mathrm{H}}^{\circ} + \hbar\Delta_2^{\delta} & 0 \\ \hbar f_3^{\delta} & 0 & 0 & \breve{\mathrm{H}}^{\circ} + \hbar\Delta_3^{\delta} \end{pmatrix}$$

Here, $\breve{H}^{\circ}$ is given by

$$\breve{H}^{\circ} = [H_{II}^{\{0\}}]_g + \hbar\omega^{\circ} - \alpha^{\circ}2\hbar\Omega$$

Besides, the f_i^{δ} are the coupling parameters involved in the Fermi resonances expressed as angular frequencies, whereas the Δ_i^{δ} are the angular frequency gap:

$$\Delta_i^{\delta} = \Delta_i^{\circ} - i\gamma_i^{\delta} \tag{295}$$

$$\Delta_i^{\circ}\delta = -(\omega^{\circ} - 2\omega_i^{\delta}) \tag{296}$$

The imaginary parts of these gaps are related to the lifetime of the corresponding excited states. Note that for a single Fermi resonance and neither Davydov coupling nor damping, when $\gamma_i^{\delta} = \gamma = 0$, this Hamiltonian is the one used by Witkowski and Wojcik [24] in their pioneering work on Fermi resonances within the strong anharmonic coupling theory.

Consequently, the Hamiltonian of the dimer that involves Davydov coupling, Fermi resonances between the g excited state of the fast mode and the g first harmonics of the bending mode, together with the damping of is

$$[\mathbb{H}_{\mathrm{Dav}}^{\mathrm{Fermi}}] = \begin{pmatrix} \left[\mathbb{H}_{II}^{\{0\}}\right]_g & 0 & 0 & 0 & 0 \\ 0 & [\mathbb{H}_{\mathrm{Fermi}}^{\{1\}}]_g & 0 & 0 & 0 \\ 0 & 0 & [H_{II}^{\{0\}}]_u & 0 & 0 \\ 0 & 0 & 0 & [H_{(+)}^{\{1\}}]_u & 0 \\ 0 & 0 & 0 & 0 & [H_{(-)}^{\{1\}}]_u \end{pmatrix}$$

Unfortunately, the Hamiltonian is not susceptible to diagonalization because of the infinite set of oscillators involved in the thermal bath. Thus, for the Hamiltonians appearing in the above Davydov Hamiltonian, it is suitable to take their corresponding non-Hermitean form, given by Eq. (292). This leads us

$$[\tilde{H}_{\mathrm{Dav}}^{\mathrm{Fermi}}] = \begin{pmatrix} [\tilde{H}_{II}^{\{0\}}]_g & 0 & 0 & 0 & 0 \\ 0 & [\tilde{H}_{\mathrm{Fermi}}^{\{1\}}]_g & 0 & 0 & 0 \\ 0 & 0 & [H_{II}^{\{0\}}]_u & 0 & 0 \\ 0 & 0 & 0 & [H_{(+)}^{\{1\}}]_u & 0 \\ 0 & 0 & 0 & 0 & [H_{(-)}^{\{1\}}]_u \end{pmatrix}$$

with

$$[\widetilde{H}^{\{1\}}_{\text{Fermi}}]_g = \begin{pmatrix} [\widetilde{H}^{\{1\}}_{II}]_g & \hbar f_1^\delta & \hbar f_2^\delta & \hbar f_3^\delta \\ \hbar f_1^\delta & \widetilde{H}^\circ + \hbar\Delta_1^\delta & 0 & 0 \\ \hbar f_2^\delta & 0 & \widetilde{H}^\circ + \hbar\Delta_2^\delta & 0 \\ \hbar f_3^\delta & 0 & 0 & \widetilde{H}^\circ + \hbar\Delta_3^\delta \end{pmatrix} \tag{297}$$

Of course, in view of Eq. (292) the effective non-Hermitean Hamiltonians are given by the following equations, where, as usual, for the non-Hermitean Hamiltonians, the zero-point energies have been ignored

$$\tilde{H}^\circ = [\tilde{H}^{\{0\}}_{II}]_g + \hbar\omega^\circ - \alpha^\circ 2\hbar\Omega$$

$$[\tilde{H}^{\{0\}}_{II}]_g \simeq a_g^\dagger a_g \left[1 - i\frac{\gamma}{2\Omega}\right]\hbar\Omega$$

$$[\tilde{H}^{\{1\}}_{II}]_g \simeq \hbar\omega^\circ - \alpha^\circ 2\hbar\Omega + [a_g^\dagger a_g + \frac{1}{\sqrt{2}}\alpha^\circ [a_g^\dagger + a_g]\,]\left[1 - i\frac{\gamma}{2\Omega}\right]\hbar\Omega - i\hbar\gamma^\circ$$

For the special case of a single Fermi resonance, the physics related to this Hamiltonian is depicted in Fig. 19:

The eigenvalue equation of the non-Hermitean Hamiltonian (297) may be written in order to make explicit the angular frequency ω° of the fast mode and the corresponding direct damping parameter γ°:

$$[\tilde{H}^{\text{Fermi}}]_g|\tilde{\Phi}^{\text{Fermi}}_\mu\rangle_g = \hbar(\omega^\circ + \omega^\circ_\mu - i\gamma^\circ - i\gamma_\mu)|\tilde{\Phi}^{\text{Fermi}}_\mu\rangle_g \tag{298}$$

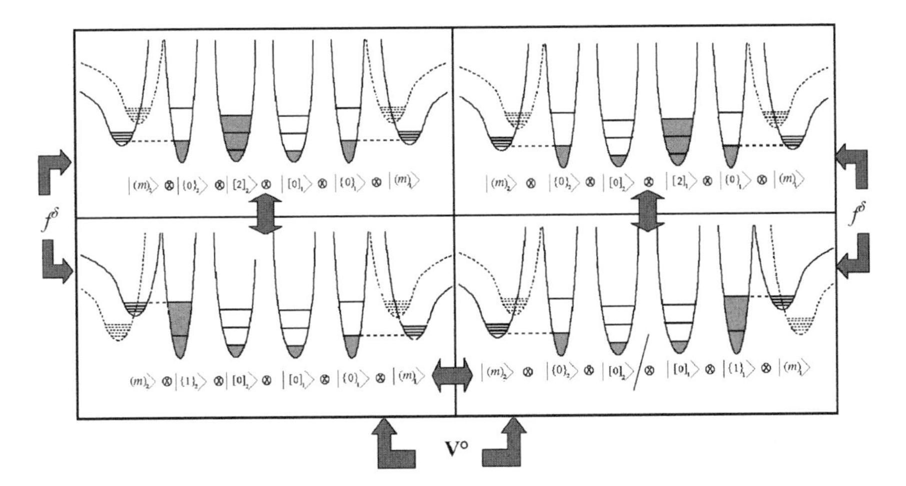

Figure 19. Davydov coupling and 1 Fermi resonance, for centrosymmetric cyclic dimer within the strong anharmonic coupling theory. (The subscripts 1 and 2 refer, respectively, to the a and b moieties of the centrosymmetric cyclic dimer).

where ω_μ° and γ_μ are, respectively, the real and imaginary part of the eigenvalues. The corresponding eigenvectors are given by

$$|\tilde{\Phi}_\mu^{\text{Fermi}}\rangle_g = \sum_X \sum_{n_g} [a^\mu{}_{X\, m_g}]|\Psi_{X\, m}\rangle_g \quad \text{with} \quad \langle\tilde{\Phi}_\mu^{\text{Fermi}} \mid \tilde{\Phi}_\nu^{\text{Fermi}}\rangle \neq \delta_{\mu\nu} \qquad (299)$$

The g basis used for this expansion is defined by

$$\begin{pmatrix} |\Psi_{I\, m}\rangle_g \\ |\Psi_{II\, m}\rangle_g \\ |\Psi_{III\, m}\rangle_g \\ |\Psi_{IV\, m}\rangle_g \end{pmatrix} = \begin{pmatrix} |\{1\}_g\rangle|(m)_g\rangle|[0_1^\delta]_g\rangle|[0_2^\delta]_g\rangle|[0_3^\delta]_g\rangle \\ |\{0\}_g\rangle|(m)_g\rangle|[2_1^\delta]_g\rangle|[0_2^\delta]_g\rangle|[0_3^\delta]_g\rangle \\ |\{0\}_g\rangle|(m)_g\rangle|[0_1^\delta]_g\rangle|[2_2^\delta]_g\rangle|[0_3^\delta]_g\rangle \\ |\{0\}_g\rangle|(m)_g\rangle|[0_1^\delta]_g\rangle|[0_2^\delta]_g\rangle|[2_3^\delta]_g\rangle \end{pmatrix}$$

$|[0_i^\delta]_g\rangle$ and $|[2_i^\delta]_g\rangle$ are the ground state and the first harmonics of the ith bending mode (δ) of g symmetry, involved in the Fermi resonances.

2. *The ACF and the Corresponding SD*

In the presence of Davydov coupling and Fermi resonances, the ACF $[\mathrm{G}_{\text{Dav}}^{\text{Fermi}}(t)]$ of the dipole moment operator of the fast mode *involving both direct and indirect dampings* and also *damping of the bending modes*, has the same structure as Eq. (272), that is,

$$[\mathrm{G}_{\text{Dav}}^{\text{Fermi}}(t)] = [\mathrm{G}_{II}^{\text{Fermi}}(t)]_g[[\mathrm{G}^{\circ(+)}(t)]_u^\circ + [\mathrm{G}^{\circ(-)}(t)]_u^\circ] \qquad (300)$$

Here, $[\mathrm{G}_{II}^{\text{Fermi}}(t)]_g$ is the ACF of the g part of the system involving Fermi resonances and quantum indirect damping, whereas $[\mathrm{G}^{(\pm)}(t)]_u$ are the ACFs corresponding to the u parts involving neither Fermi resonance nor indirect damping that are given by Eq. (276).

If we take only the Hermitean part of the Hamiltonian in the Boltzmann operator, the g ACF is the following trace, $\text{tr}_{\text{Fermi}g}$, to be performed over the space spanned by the Hamiltonian $[\mathrm{H}^{\text{Fermi}}]_g$:

$$[\mathrm{G}_{II}^{\text{Fermi}}(t)]_g = \left(\frac{\mu_u^{o\,2}}{[Z]_g}\right)\text{tr}_{\text{Fermi}\,g}[(e^{-\beta[\mathrm{H}^{\text{Fermi}}]_g})[\mu(0)]_g^\dagger[\mu(t)]_g] \qquad (301)$$

$[\mu(0)]_g$ is the transition moment operator of the g fast mode at the initial time,

$$[\mu(0)]_g^\dagger = \mu_u^o|\{0\}_g\rangle\langle\{1\}_g|$$

Besides, $[\mu(t)]_g$ is the transition moment operator of the g fast mode at time t, which is given by

$$[\mu(t)]_g = \mu_u^o [e^{i[\tilde{H}^{\text{Fermi}}]_g t/\hbar}] |\{1\}_g\rangle \langle \{0\}_g | [e^{-i[\tilde{H}^{\text{Fermi}}]_g t/\hbar}] \tag{302}$$

At last, the partition function is

$$[Z]_g = \text{tr}_{\text{Fermi}\, g}\left[\left(e^{-\beta[H_{II}^{\{0\}}]_g}\right)\right]$$

Owing to these equations, the ACF (301) becomes

$$[G_{II}^{\text{Fermi}}(t)]_g = \left(\frac{\mu_u^{o\,2}}{[Z]_g}\right)$$
$$\text{tr}_{\text{Fermi}\, g}\left[\left(e^{-\beta[H_{II}^{\{0\}}]_g}\right)\left|\{0\}_g\right\rangle \left\langle \{1\}_g\right|\left[e^{i[\tilde{H}_{\text{Fermi}}^{\{1\}}]_g t/\hbar}\right]\left|\{1\}_g\right\rangle \left\langle \{0\}_g\right| \left[e^{-i[\tilde{H}_{II}^{\{0\}}]_g t/\hbar}\right]\right]$$

Again, owing to the expression of the Hermitean Hamiltonian involved in the Boltzmann operator, the ACF may be obtained by performing the trace $\text{tr}_{\text{Slow}_g}$ over the base spanned by this Hamiltonian:

$$[G_{II}^{\text{Fermi}}(t)]_g = (\mu_u^{o2})$$
$$\tilde{\varepsilon}\text{tr}_{\text{Slow}\, g}\left[\left(e^{-\beta[H_{II}^{\{0\}}]_g}\right)\left|\{0\}_g\right\rangle \left\langle \{1\}_g\right| \left[e^{i[\tilde{H}_{\text{Fermi}}^{\{1\}}]_g t/\hbar}\right]\left|\{1\}_g\right\rangle \left\langle\{0\}_g\right|\left[e^{-i[H_{II}^{\{0\}}]_g t/\hbar}\right]\right]$$

Of course, using the fact that the operators involved in the ACF do not work on the space of the fast mode, because of the orthonormality, the above ACF reduces after a circular permutation within the trace and making the trace explicit, to

$$[G_{II}^{\text{Fermi}}(t)]_g = (\mu_u^{o2})\tilde{\varepsilon}\sum_{n_g}\left\langle (n)_g\right|\left[e^{i[\tilde{H}_{\text{Fermi}}^{\{1\}}]_g t/\hbar}\right]\left[e^{-i[\tilde{H}_{II}^{\{0\}}]_g t/\hbar}\right]\left(e^{-\beta[H_{II}^{\{0\}}]_g}\right)\left|(n)_g\right\rangle$$

At last, insert between the two evolution operators the pseudocloseness relation:

$$\sum_{\mu}\left|\tilde{\Phi}_{\mu}^{\text{Fermi}}\right\rangle_g \left\langle \tilde{\Phi}_{\mu}^{\text{Fermi}}\right|_g \simeq \hat{1} \tag{303}$$

Then, using the eigenvalue equations dealing with the two Hamiltonians of interest and neglecting the zero-point energy involved in the Boltzmann operator, the ACF takes the final form:

$$[\mathrm{G}_{II}^{\mathrm{Fermi}}(t)]_g = (\mu_u^o)^2 e^{i\omega^\circ t} e^{-\gamma^\circ t} \varepsilon \sum_{n_g} \sum_{\mu} \left(e^{-\tilde{\lambda} n_g}\right)\left(e^{i\omega_\mu^\circ t}\right)\left(e^{-in_g\Omega t}\right)\left(e^{-n_g\gamma t/2}\right)\left(e^{-\gamma_\mu t}\right)\left|a_{I\,n_g}^{\mu}\right|^2 \tag{304}$$

where $\omega^\circ{}_\mu$ and γ_μ are the real and imaginary parts of the eigenvalues appearing in Eq. (298), whereas $a_{I\,n_g}^{\mu}$ are the expansion coefficients defined by Eq. (299). As it appears, the ACF of the system involving both Davydov coupling and Fermi resonances has the same structure as that (294) in the absence of Fermi resonance, except for the fact that the g part must be given in the present situation by Eq. (304).

a. Remarks on the Approximations. This approach deals with Davydov coupling and Fermi resonance and involves some approximations, such as the harmonic approximation for the H-bond bridge, the non-Hermitean Hamiltonian approach for the quantum indirect damping, and the exchange approximation for Fermi resonances. Among these approximations, the *exchange approximation* is probably that which mostly affects the line shape: It is questionable since it has been shown [104] that removing it leads to weakly modify both frequencies and intensities. Unfortunately, its removal, as done in the section taking into account Fermi resonances, but ignoring Davydov coupling, would require us to work beyond the adiabatic approximation. Then, for a single Fermi resonance in each moiety, it is easy to show that, in the spirit of the work of Chamma et al. [82], the treatment would lead to diagonalize very large Hamiltonian matrices of dimension around at least 10^4–10^5.

3. *Special Situations*

Of course, of the ACF (300), although slightly less rigorous than that, the Fourier transform of which is the SD (279), reduces to the same special situations. Besides, it is deeply connected to the ACF used by Maréchal in his "peeling-off" approach of Fermi resonances that in turn reduce to many approaches of Fermi resonances appearing in situations without a H-bond.

a. Connection of the ACF (300) With the "Peeling-Off" Approach of Maréchal [83].

Consider a special situation *where the indirect damping is missing* and *where the direct dampings* affecting the first excited state of the fast mode and the first harmonics of the bending mode are the same. Then, Appendix S shows that the g

ACF of this approach involving Fermi resonances may transform according to Eq. (S.22) into

$$\left[G_{II}^{\circ\,\mathrm{Fermi}}(t)\right]_g \simeq \left[\breve{G}^{\circ}(t)\right] = (\mu_0^2)\left(e^{i\omega^{\circ}t}\right)\left(e^{-\gamma^{\circ}t}\right)$$
$$\left[O_0(t)^{\mathrm{IP}}\right]^{\dagger}\left(e^{-i2\alpha^{\circ 2}\Omega t}\right)\left(e^{i\alpha^{\circ 2}\sin\Omega t}\right)\left[e^{2\alpha^{\circ 2}[\langle n\rangle+1/2](\cos(\Omega t)-1)}\right] \tag{305}$$

In Eq. (305), $O_0(t)^{\mathrm{IP}}$ may be obtained by resolution of the following set of coupled differential equations,

$$i\hbar\frac{\partial}{\partial t}\begin{pmatrix} O_0(t)^{\mathrm{IP}} \\ O_1(t)^{\mathrm{IP}} \\ O_2(t)^{\mathrm{IP}} \end{pmatrix} = \begin{pmatrix} 0 & \hbar f_1^{\delta}(t) & \hbar f_2^{\delta}(t) \\ \hbar f_1^{\delta}(t)^{\dagger} & \hbar\Delta_1^{\delta} & 0 \\ \hbar f_2^{\delta}(t)^{\dagger} & 0 & \hbar\Delta_2^{\delta} \end{pmatrix}\begin{pmatrix} O_0(t)^{\mathrm{IP}} \\ O_1(t)^{\mathrm{IP}} \\ O_2(t)^{\mathrm{IP}} \end{pmatrix} \tag{306}$$

subject to the boundary conditions:

$$O_k(0)^{\mathrm{IP}} = \delta_{0\,k} \tag{307}$$

In Eq. (306), $\hbar f\,_1^{\delta}(t)$ is given by

$$\hbar f_i^{\delta}(t) = \hbar f_1^{\delta}\left[\breve{G}^{\circ}(t)\right]$$

Next, as a consequence of Eq. (305), the full ACF (300) in this situation becomes

$$\left[G_{\mathrm{Dav}}^{\circ\,\mathrm{Fermi}}(t)\right] = \left[O_0(t)^{\mathrm{IP}}\right]^{\dagger}\left[\left[G^{\circ}(+)(t)\right]_u^{\circ} + \left[G^{\circ}(-)(t)\right]_u^{\circ}\right]$$
$$(\mu_0^2)(e^{i\omega^{\circ}t})(e^{-\gamma^{\circ}t})(e^{-i2\alpha^{\circ 2}\Omega t})(e^{i\alpha^{\circ 2}\sin\Omega t})[e^{2\alpha^{\circ 2}[\langle n\rangle+1/2](\cos(\Omega t)-1)}] \tag{308}$$

where the ACFs $[G^{\circ}(\pm)(t)]_u^{\circ}$ remain to be given by Eq. (276). Besides, the ACF (309) may be written simply as:

$$\left[G_{\mathrm{Dav}}^{\mathrm{Fermi}}(t)\right] = \left[O_k(0)^{\mathrm{IP}}\right]^{\dagger}\left[G_{\mathrm{Dav}}^{\circ}(t)\right] \tag{309}$$

where $[G_{\mathrm{Dav}}^{\circ}(t)]$ is the Davydov ACF $[G_{\mathrm{Dav}}(t)]$ involved in the SD (279) in which the *indirect damping is missing*. After Fourier transform, the ACF (308) leads to the SD:

$$\left[I_{\mathrm{Dav}}^{\circ\,\mathrm{Fermi}}(\omega)\right] = 2\mathrm{Re}\int_0^{\infty}\left[G_{\mathrm{Dav}}^{\circ\,\mathrm{Fermi}}(t)\right]e^{-i\omega t}dt$$

Now, note that the ACF (309) has the same formal structure as that used by Maréchal in his "peeling-off" approach [83] of centrosymmetric cyclic dimers involving Fermi resonances: this "peeling-off" ACF is, as for Eq. (309), the product of an ACF times a function that is the solution of a linear set of time-dependent differential equations having the structure of Eq. (306), but in which the matrix elements are constant. But, the Maréchal procedure works in a way different from the present one:

1. In his model, the ACF $G_{exp}(t)$ playing the role of $[G^{\circ\,Fermi}_{Dav}(t)]$ given by Eq. (309), is obtained by the Fourier transform of the experimental spectral density $[I_{exp}(\omega)]$, whereas in this approach, the theoretical SD $[I^{\circ Fermi}_{Dav}(\omega)]$ is obtained by the above Fourier transform of the theoretical ACF$[G^{\circ\,Fermi}_{u_{Dav}}(t)]$.
2. In his model, the time-dependent term playing the role of $[O_0(t)^{IP}]^{\dagger}$ in the present situation, is obtained by solving a set of linear equations, which is Eq. (306), except the fact that the matrix elements are time independent.
3. At last, in his model, the ACF playing the role of $[G^{\circ}_{Dav}(t)]$ in Eq. (309), is extracted from the two other models playing the role of $[G^{\circ Fermi}_{Dav}(t)]$ and $[O_0(t)^{IP}]^{\dagger}$. They give a numerical expression, which after Fourier transform, is assumed to give the "peeled-off" spectral density, that is, the SD of the H-bonded centrosymmetric cyclic dimer that would be observed if the Fermi resonances were missing.

Recall that the "peeling-off" ACF of Maréchal admits to special situations of many important works dealing with Fermi resonances and taking into account the surrounding [83–89].

b. Other Special Situations. The coupled set of differential equations (306) taking into account multiple Fermi resonances, is well behaved with respect to important limit situations. Neglect the time dependence of the matrix elements in Eq. (306) and consider only a single Fermi resonance. Then, Eq. (306) reduces to, omitting the specific IP notation

$$i\frac{\partial}{\partial t}\begin{pmatrix} O_0(t) \\ O_1(t) \end{pmatrix} = \begin{pmatrix} \omega^{\circ} - i\gamma^{\circ} & f_1^{\delta} \\ f_1^{\delta} & 2\omega_i^{\delta} - i\gamma_1^{\delta} \end{pmatrix}\begin{pmatrix} O_0(t) \\ O_1(t) \end{pmatrix} \tag{310}$$

Equation (310) admits to an analytical solution that was obtained by Giry et al. [90, 91]. Again, in the situation of degenerate diagonal matrix elements, this equation reduces to that solved by Davydov and Serikov [92], that is,

$$i\frac{\partial}{\partial t}\begin{pmatrix} O_0(t) \\ O_1(t) \end{pmatrix} = \begin{pmatrix} -i\gamma^{\circ} & f_1^{\delta} \\ f_1^{\delta} & -i\gamma_1^{\delta} \end{pmatrix}\begin{pmatrix} O_0(t) \\ O_1(t) \end{pmatrix}$$

Besides, when $\gamma^{\circ} = 0$, the Davydov–Serikov equation reduces the Bethe equation [93], that is,

$$i\frac{\partial}{\partial t}\begin{pmatrix} O_0(t) \\ O_1(t) \end{pmatrix} = \begin{pmatrix} 0 & f_1^{\delta} \\ f_1^{\delta} & -i\gamma_1^{\delta} \end{pmatrix}\begin{pmatrix} O_0(t) \\ O_1(t) \end{pmatrix}$$

At last, when the damping parameter γ_1^{δ} is vanishing, Eq. (310) reduces to the Rabi equation [93]:

$$i\frac{\partial}{\partial t}\begin{pmatrix} O_0(t) \\ O_1(t) \end{pmatrix} = \begin{pmatrix} \omega^{\circ} & f_1^{\delta} \\ f_1^{\delta} & 2\omega_i^{\delta} \end{pmatrix}\begin{pmatrix} O_0(t) \\ O_1(t) \end{pmatrix}$$

VII. LINE SHAPES OF CARBOXYLIC ACIDS

Now we apply the theoretical methods presented above to reproduce the experimental spectra of some carboxylic acids. This section is devoted to such tests dealing with carboxylic acids. These acids are acetic, acrylic, propynoic, adipic, glutaric, and naphtoic acid. The different phases will be the gaseous, liquid, and crystalline states.

A. Liquid or Gaseous Carboxylic Acids

1. *Gaseous Acetic Acid Dimers CD_3CO_2H and CH_3CO_2H*

First, consider the line shapes of $(CD_3CO_2H)_2$ and $(CD_3CO_2D)_2$ in the gas phase. The experimental spectra were published by Haurie and Novak [94]. Maréchal and Witkowski [18] proposed a theoretical model to understand the spectra that is equivalent to this one, except it ignores the relaxations of the fast mode and H-bond bridge. They calculated stick spectra that were roughly in agreement with the experimental profiles and satisfactorily reproduce the change in the line shape caused by the D isotopic substitution. Leviel and Maréchal [95] introduced in the initial Maréchal and Witkowski model Morse potentials in place of harmonic models to describe the H-bond bridge. The results were similar, except for the fact that the angular frequency of the H-bond bridge vibration used to fit the experimental line shape, $160\,\text{cm}^{-1}$, was closer to the experimental frequency (in the range 170–$180\,\text{cm}^{-1}$) [96]. Maréchal studied the experimental line shape of acetic acid assuming the presence of several Fermi resonances. Using his "peeling-off" approach [83], where allows us to eliminate the effects due to Fermi resonances, he was able [97] to simulate model line shapes that were very near the experimental models. The "peeled-off" spectra were assumed to concentrate the basic physics of the H-bond.

Recently [76], we proposed an approach in which we introduced the quantum indirect damping in the Maréchal and Witkowski model involving Davydov coupling. More precisely, this model equally takes into account the strong

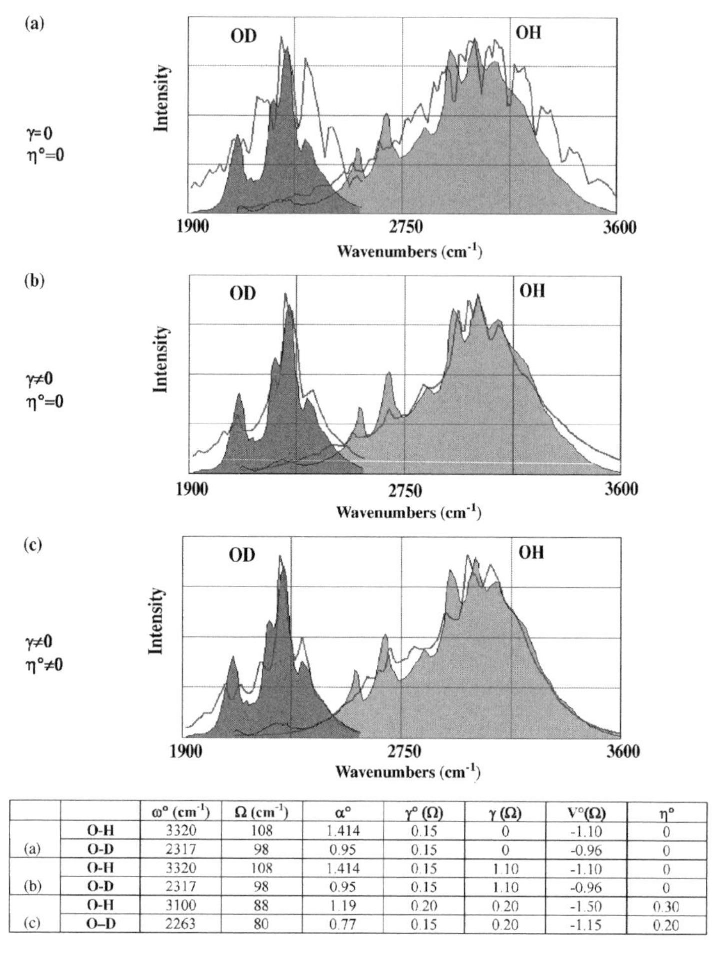

		ω° (cm^{-1})	Ω (cm^{-1})	α°	$\gamma^\circ(\Omega)$	$\gamma(\Omega)$	$V^\circ(\Omega)$	η°
	O-H	3320	108	1.414	0.15	0	-1.10	0
(a)	O-D	2317	98	0.95	0.15	0	-0.96	0
	O-H	3320	108	1.414	0.15	1.10	-1.10	0
(b)	O-D	2317	98	0.95	0.15	1.10	-0.96	0
	O-H	3100	88	1.19	0.20	0.20	-1.50	0.30
(c)	O–D	2263	80	0.77	0.15	0.20	-1.15	0.20

Figure 20. The IR $\nu_{X\text{–}H}$ line shapes of gaseous cyclic acetic acid dimers and isotope effect at room temperature.

anharmonic coupling between the high and low frequency modes, the quantum direct damping of the high frequency mode, the quantum indirect damping of the H-bond bridges, and the resonance exchange between the excited states of the two high frequency modes.

Figure 20 gives spectra for the CD_3CO_2H dimer in the gas phase at room temperature. The grayed lines represent the experimental line shapes taken from Novak et al. [94], for the O–H (right part) and the O–D (left part) substitutions. On the top, solid lines give the theoretical line shape computed with Eq. (279):.

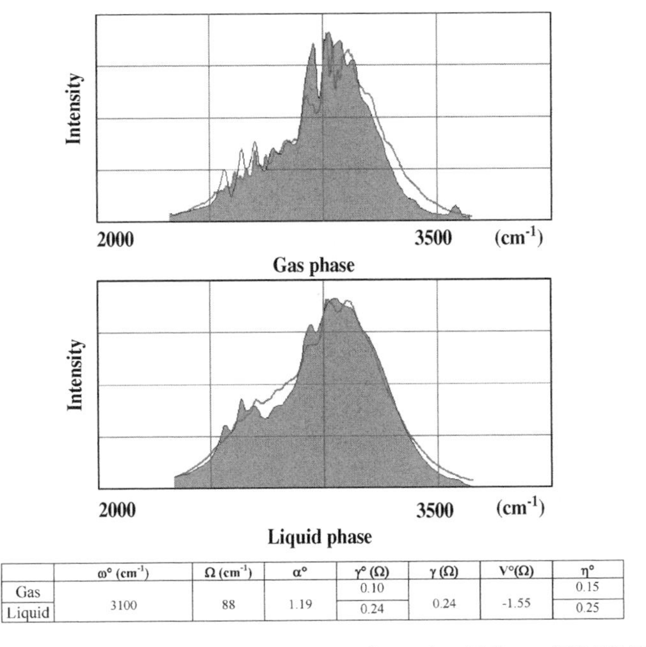

	ω° (cm^{-1})	Ω (cm^{-1})	α°	$\gamma^\circ(\Omega)$	$\gamma(\Omega)$	$V^\circ(\Omega)$	η°
Gas	3100	88	1.19	0.10	0.24	-1.55	0.15
Liquid				0.24			0.25

Figure 21. The IR ν_{X-H} line shapes of gaseous cyclic acetic acid dimers $[CH_3CO_2H]_2$ at room temperature Grayed: experimental spectrum according to Ref. [98].

(a) gives the fitting when the indirect relaxation is omitted, while (b) gives the result when this damping is taken into account [72]. At last, (c) gives the fitting when the IR selection rule is transgressed [98], that is, when $\eta^\circ \neq 0$. The parameters used for the calculations are given in the caption of Fig. 20.

Note that the parameters used in the calculations are the same for the O—H and O—D species, except the angular frequency of the fast mode and the anharmonic coupling that have to be reduced by a factor $\sqrt{2}$.

As seen, the introduction of the η° parameter improves the fitting of the theoretical line shape on the low frequency side.

Now, consider the CH_3CO_2H isotopomer in the gas and liquid phase measured by Flakus and theoretically studied [98] by aid of Eq. (279). Comparisons between the theoretical (thick line) and experimental line shapes (grayed) are given in Fig. 21:

2. *Gaseous and Liquid Propynoic and Acrylic Acids*

Now, look at the line shapes of O—H and O—D gaseous propynoic and acrylic acids, as measured by Bournay and Maréchal [99], and at the corresponding theoretical approach [100] using Eq. (279). Figures 22 and 23 compare the experimental and theoretical line shapes.

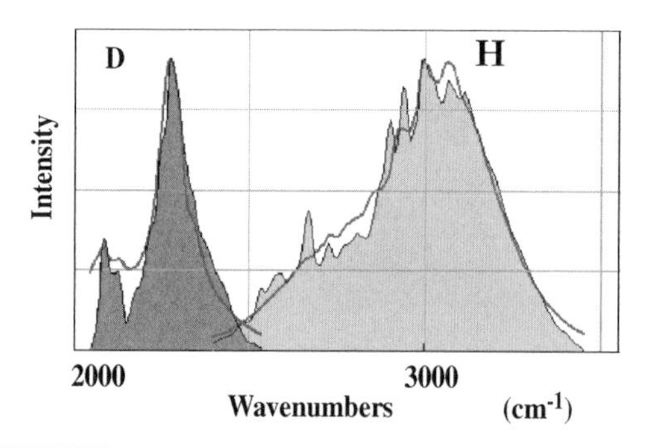

	ω° (cm^{-1})	Ω (cm^{-1})	α°	$\gamma^\circ(\Omega)$	$\gamma(\Omega)$	$V^\circ(\Omega)$	η°
H	3020	71	1.09	0.22	0.27	-1.66	0.19
D	2237	71	0.77	0.24	1.80	-1.08	0.18

Figure 22. Spectra in the gas phase at 323 K of hydrogenated (right part) and deuterated (left part) of acrylic acid dimer. Grayed: Experimental line shape according to ref. [99]. Thick line: Theoretical line shape according to Eq. (279).

B. Crystalline Carboxylic Acids

1. Adipic Acid

In this section, we consider the line shapes of O—H and O—D crystalline adipic acid as measured by Auvert [101] at different temperatures. These spectra have

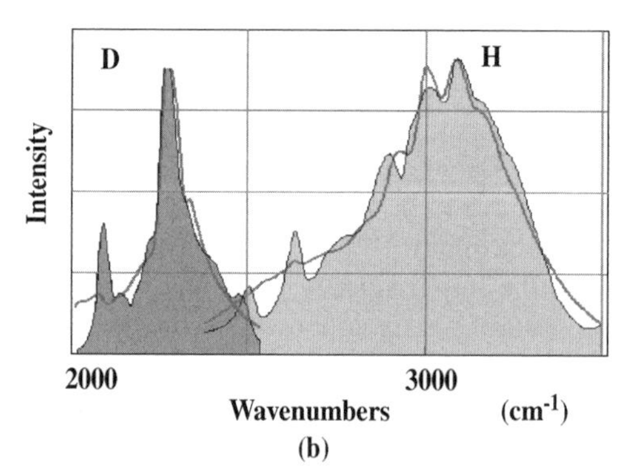

	ω° (cm^{-1})	Ω (cm^{-1})	α°	$\gamma^\circ(\Omega)$	$\gamma(\Omega)$	$V^\circ(\Omega)$	η°
H	3032	86	1.25	0.27	0.36	-1.94	0.16
D	2260	86	0.88	0.18	1.40	-1.25	0.05

Figure 23. Spectra in the gas phase at 300 K of hydrogenated (right part) and deuterated (left part) of propynoic acid dimer. Grayed: Experimental line shape according to Ref. [99]. Thick line: Theoretical line shape according to Eq. (279).

been theoretically studied by Maréchal [102] within his "peeling-off" approach [83]. Besides, since the space symmetry group to which the crystalline adipic acid belongs is $P2_1/c$, and since there are two centrosymmetric cyclic dimers in each cell, the straightforward approach without indirect damping would be that of Wojcik [103].

Nevertheless, the possible interactions between the two centrosymmetric cyclic dimers belonging to a given cell of the crystal can be ignored, since Flakus [104] showed in his theoretical work on these spectra that these interactions are too weak to influence the band profile.

Since it was admitted that Fermi resonances may play a role in crystalline adipic acid [102], they were incorporated in the model by using the Fourier transform of Eq. (304) in place of Eq. (279). Note that Yaremko et al. [105] also studied these spectra in a rather formal way with the aid of a general formalism [106–108], which is not without connection with that used by the Witkowski school. The left column of Fig. 24 reports the comparisons between the experimental line shapes measured by Auvert and Maréchal [101] and the corresponding theoretical ones calculated by Eq. (279) in this way [98]. In this figure, the line shapes are reproduced up to down, for, respectively, 10, 100, 200, and 300 K. For each temperature, the right part is devoted to the O—H species and the left one to the O—D isotopomer. The values of the parameters used for the calculations are reported in the caption of the figure.

In these computations, it was assumed, following Flakus [104, 109], that there is some breaking of the IR selection rule (introduced via the η° parameter), which is increasing when the crystal order is increased, that is, when the temperature is lowered.

The left column of Fig. 24 illustrates the role played by this possible breaking of the selection rule, via introduction of the η° parameter.

In both sets of calculations, the parameters used are those used in the captions except for the fact that the η° parameter is zero for the right set of theoretical line shapes. Comparison of the two sets clearly shows that, if the high frequency wing of each experimental line shape continues to be well reproduced when η° is zero, at the opposite, the corresponding low frequency wing cannot be reproduced at all (the theoretical low frequency wing collapses). The values of the parameters used in the computations are of the same magnitude as those used above in the study dealing with acetic acid in the gas phase, except for the indirect damping γ at 300 K, which is 0.30 for adipic acid, whereas it is 1 for acetic acid.

As a complement, Fig. 25 gives the comparison of the theoretical and experimental dependence on the temperature of the second moments of the left set of spectra of Fig. 24, for both isotopic species [81].

The evolution of the half-width of the spectra with temperature appears to follow that observed by Sandorfi et al. [110], [111] in the area of the H-bond.

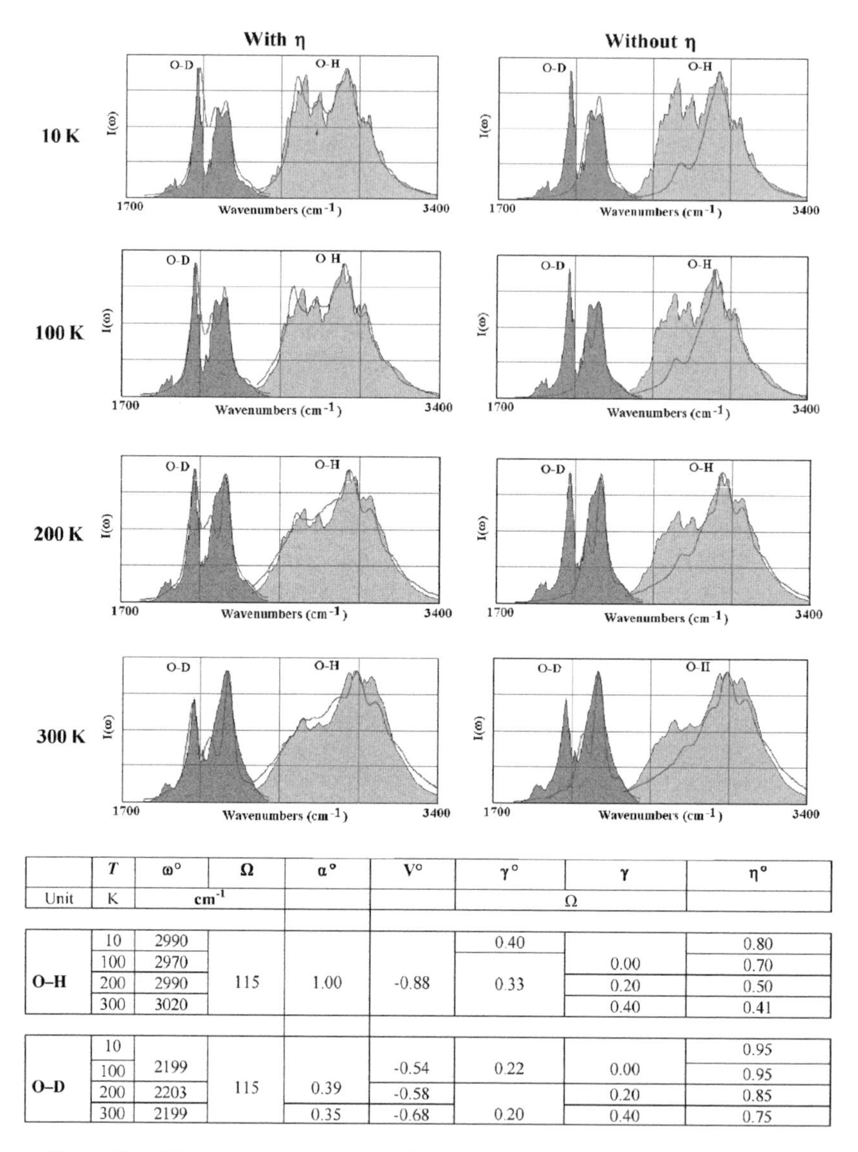

	T	ω°	Ω	α°	V°	γ°	γ	η°
Unit	K	cm^{-1}				Ω		
O–H	10	2990	115	1.00	-0.88	0.40	0.00	0.80
	100	2970				0.33		0.70
	200	2990					0.20	0.50
	300	3020					0.40	0.41
O–D	10	2199	115	0.39	-0.54	0.22	0.00	0.95
	100							0.95
	200	2203			-0.58	0.20	0.20	0.85
	300	2199		0.35	-0.68		0.40	0.75

Figure 24. Effects of temperature and isotopic substitution on the spectral densities of crystalline adipic acid in the absence of Fermi resonance. Comparison between theory (Eq. (279)) (thick line) and experiment [101] (grayed spectra). Left column: calculations using the breaking of the IR selection rule ($\eta^\circ = 0$). Right column: same calculations but without the breaking of the IR selection rule ($\eta^\circ = 0$).

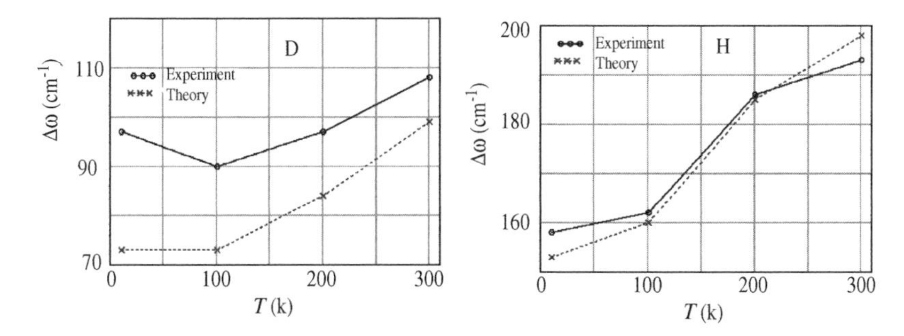

Figure 25. Second moments of the spectra of adipic acid in the absence of Fermi resonance corresponding to the left set of spectra of Fig. 24.

Figure 26 is devoted to the incorporation of Fermi resonances in the model of adipic acid. The values of the parameters involved in the calculations are reported in Table XIII. Comparison of Figs. 25 and 26 shows that the improvement in accuracy resulting from the incorporation of Fermi resonances is rather weak. Nevertheless, note that this incorporation allows us to "stabilize" the fitted parameters that ought to be the same, when passing from one temperature to another. From our viewpoint, this stabilization behavior is more revealing for the presence of Fermi resonances than the weak improvement of the fit.

In conclusion, note that the quality of the theoretical SD with respect to the experimental line shape is not as good for crystalline adipic acid as for the gaseous and liquid carboxylic acids studied above. The reason is that if Fermi resonances seem to be unavoidable in order to reproduce all features of the experimental line shapes and to conserve a good stability of the basic physical parameters when changing the temperature, however, the way in which the Fermi resonances are taken into account is very sentitive to the used adiabatic and exchange approximations [82].

2. *Glutaric Acid*

This section, considers the line shapes of O—H and O—D crystalline glutaric acid, as measured by Flakus and Miros [112], at room (298 K) and liquid nitrogen (77 K) temperatures, as well as with two different polarizations: By using the IR beam of normal incidence with respect to the crystalline *ac* plane, the polarized spectra were measured by these authors for two orientations of the electric field vector E: parallel (Pol $= 0°$) and perpendicular (Pol $= 90°$) to the *c* axis. Comparison is made between the experiment (grayed spectra) and theory (thick line).

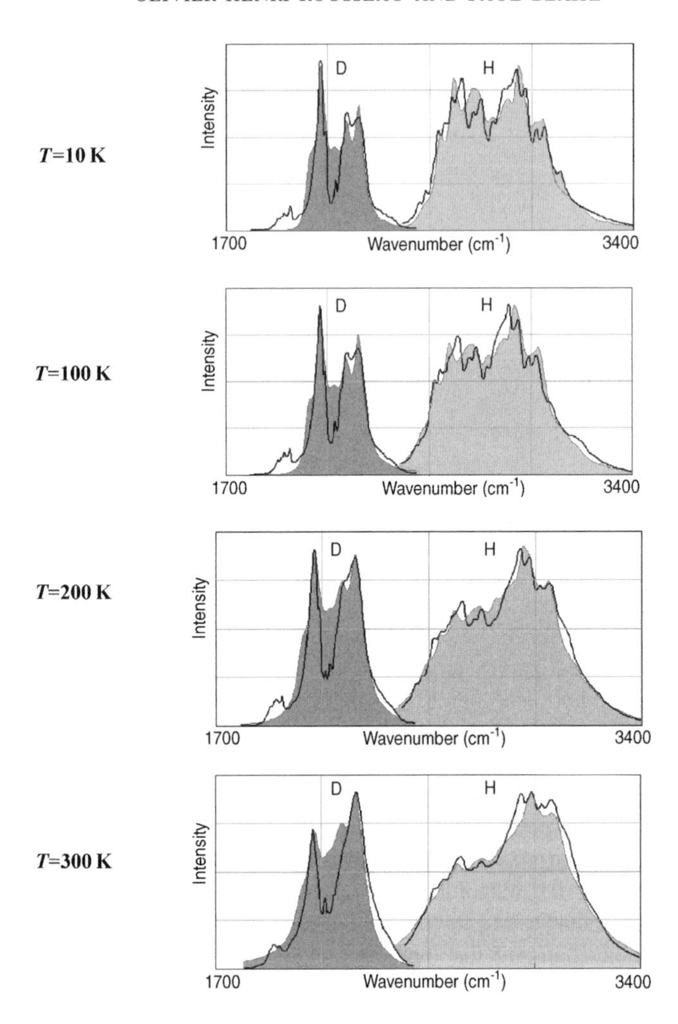

Figure 26. Adipic acid with four Fermi resonances and with η°. Comparison between theory (thick line) and experiment (grayed spectra). The theoretical parameters are given in Table XIII.

Figure 27 gives the results obtained by Eq. (279) in the case where Fermi resonances are ignored, whereas Fig. 28 gives the corresponding results when five Fermi resonances are taken into account via Eq. (304).

As it appears, and as seen in Section VII.B.1 dealing with adipic acid, the introduction of Fermi resonances weakly improves the accuracy of the fitting and allows a stabilization of the physical parameters when passing from one temperature to another.

TABLE XIII
Parameters Involved in the Calculations of Fig. 26

	T (K)	ω° (cm^{-1})	Ω (cm^{-1})	α°	V° (Ω)	γ° (Ω)	γ (Ω)	f^{δ}_{1} (cm^{-1})	f^{δ}_{2} (cm^{-1})	f^{δ}_{3} (cm^{-1})	f^{δ}_{4} (cm^{-1})	Δ^{δ}_{1} (cm^{-1})	Δ^{δ}_{2} (cm^{-1})	Δ^{δ}_{3} (cm^{-1})	Δ^{δ}_{4} (cm^{-1})	γ^{δ}_{1} (Ω)	γ^{δ}_{2} (Ω)	γ^{δ}_{3} (Ω)	γ^{δ}_{4} (Ω)	η°
O-H	10	2865	108	0.88	-0.88	0.11	0.16	30	37	36	36	-185	-80	20	220	0.10	0.10	0.10	0.05	0.80
	100	2850			-0.89		0.18													0.60
	200	2870				0.10	0.34													0.40
	300	2890			-0.93	0.12	0.30													0.35
O-D	10	2170	108	0.29	-0.62	0.10	0.00	17	15	13	13	-33	-21	-0.84	126	0.07	0.04	0.17	0.10	0.90
	100						0.10													0.80
	200	2172				0.12	0.20													0.66
	300						0.35													0.45

3. *Naphtoic Acid*

Consider the crystalline spectra of the naphtoic acid dimer measured by Flakus et al. [113]. Comparison between the theoretical line shapes (thick line) calculated [114] with the aid of Eq. (304) and the experimental ones (grayed) is given in Fig. 29, where the experimental line shapes are the grayed lines while the theoretical are figured by thick lines. The right and left parts of the figure

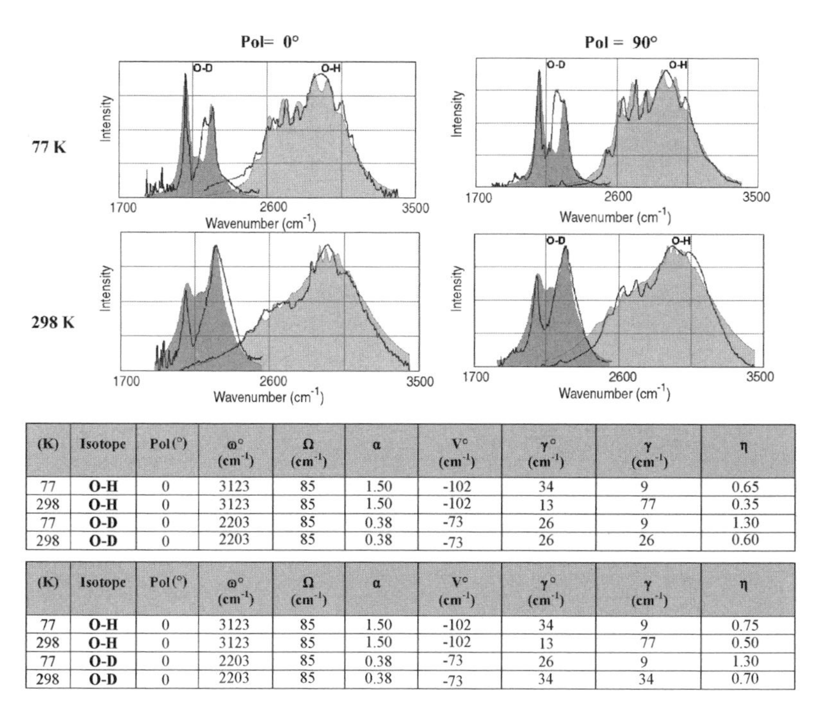

(K)	Isotope	Pol (°)	ω° (cm^{-1})	Ω (cm^{-1})	α	V° (cm^{-1})	γ° (cm^{-1})	γ (cm^{-1})	η
77	**O-H**	0	3123	85	1.50	-102	34	9	0.65
298	**O-H**	0	3123	85	1.50	-102	13	77	0.35
77	**O-D**	0	2203	85	0.38	-73	26	9	1.30
298	**O-D**	0	2203	85	0.38	-73	26	26	0.60

(K)	Isotope	Pol (°)	ω° (cm^{-1})	Ω (cm^{-1})	α	V° (cm^{-1})	γ° (cm^{-1})	γ (cm^{-1})	η
77	**O-H**	0	3123	85	1.50	-102	34	9	0.75
298	**O-H**	0	3123	85	1.50	-102	13	77	0.50
77	**O-D**	0	2203	85	0.38	-73	26	9	1.30
298	**O-D**	0	2203	85	0.38	-73	34	34	0.70

Figure 27. Spectral densities of crystalline glutaric acid theory without Fermi resonances. *Experiment:* Thick line to [112] *Theoretical*: grayed spectra.

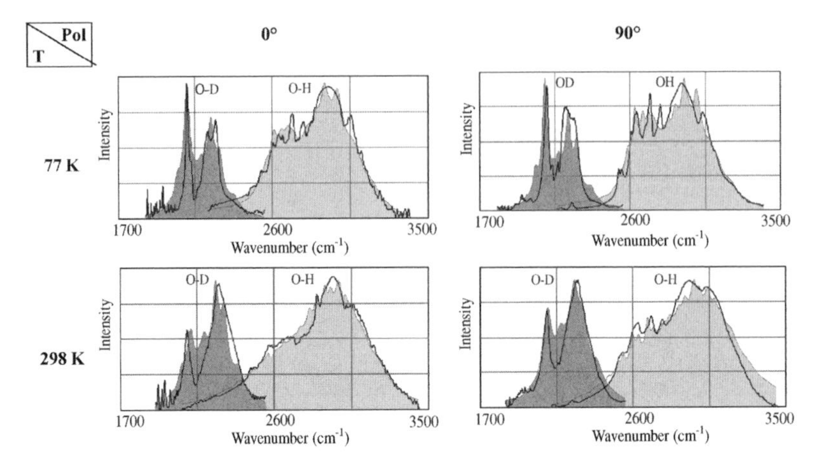

Figure 28. Spectral densities of crystalline glutaric acid theory with 5 Fermi resonances. *Experiment:* Thick line [112] *Theoretical*: grayed spectra Eq. (304).

respectively, show, the spectra of the hydrogenated (O–H) and deuterated (O–D) compounds.

The line shapes of acetic acid, [116] studied by Wojcik [117] that of β oxalic acid, [118] studied theoretically by Witkowski and Wojcik [117], that of formic acid, [119] studied by Flakus [120], and that of benzoic acid, [121] studied by Flakus [122] would be studied in this spirit.

TABLE XIV
Parameters Involved in the Calculations of Fig. 28

(K)	Isotope	Pol.(°)	ω° (cm^{-1})	Ω (cm^{-1})	α	V° (cm^{-1})	γ° (cm^{-1})	γ (cm^{-1})	η
77	**O-H**	0	3063	90	1.30	-103	18	9	0.55
298	**O-H**	0	3063	90	1.30	-103	18	18	0.12
77	**O-D**	0	2202	90	0.33	-74	14	18	1.00
298	**O-D**	0	2217	90	0.33	-74	18	18	0.40

(K)	Isotope	Pol.(°)	ω° (cm^{-1})	Ω (cm^{-1})	α	V° (cm^{-1})	γ° (cm^{-1})	γ (cm^{-1})	η
77	**O-H**	90	3083	90	1.30	-103	18	14	0.70
298	**O-H**	90	3090	90	1.30	-103	18	18	0.40
77	**O-D**	90	2188	90	0.33	-74	14	18	1.00
298	**O-D**	90	2217	90	0.33	-74	9	23	0.50

Fermi parameters in cm^{-1} (common to both isotopic species)

f_1	f_2	f_3	f_4	f_5	Δ_1	Δ_2	Δ_3	Δ_4	Δ_5	$\gamma^\delta{}_1$	$\gamma^\delta{}_2$	$\gamma^\delta{}_3$	$\gamma^\delta{}_4$	$\gamma^\delta{}_5$
40	25	20	30	30	-430	-120	30	100	200	9	9	9	9	9

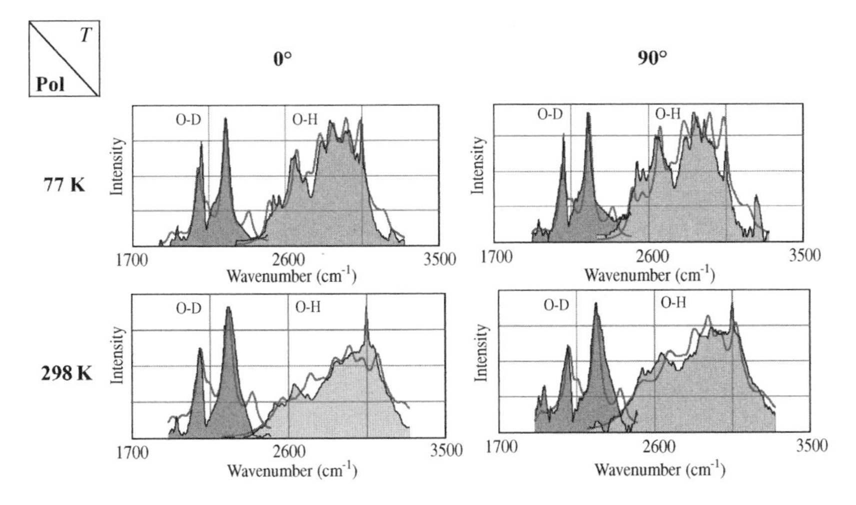

Figure 29. Spectral densities of crystalline naphtoic acid with four Fermi resonances are acting. *Experiment:* grayed spectra [113].

TABLE XV
Parameters Involved in the Calculations of Fig. 29

(K)	Isotope	Pol.(°)	ω° (cm^{-1})	Ω (cm^{-1})	α	V° (cm^{-1})	γ° (cm^{-1})	γ (cm^{-1})	η
77	**O-H**	0	2788	90	0.84	-113	18	9	0.40
298	**O-H**	0	2853	90	0.84	-113	18	45	0.40
77	**O-D**	0	2071	90	0.21	-23	18	72	1.00
298	**O-D**	0	2075	90	0.21	-23	18	45	0.10

(K)	Isotope	Pol.(°)	ω° (cm^{-1})	Ω (cm^{-1})	α	V° (cm^{-1})	γ° (cm^{-1})	γ (cm^{-1})	η
77	**O-H**	90	2788	90	0.84	-113	18	9	0.50
77	**O-H**	90	2753	90	0.84	-113	18	45	0.40
298	**O-D**	90	2082	90	0.21	-23	18	9	1.00
298	**O-D**	90	1981	90	0.21	-23	18	45	0.20

Fermi parameters in cm^{-1} (common to both isotopic species)

f_1	f_2	f_3	f_4	Δ_1	Δ_2	Δ_3	Δ_4	$\gamma^\delta{}_1$	$\gamma^\delta{}_2$	$\gamma^\delta{}_3$	$\gamma^\delta{}_4$
50	50	70	50	-72	140	550	-115	9	14	14	9

C. Conclusion

This section was devoted to the applications of the theoretical model previously presented to experimental line shapes of carboxylic acids.

From these comparisons between the theoretical SDs and the experimental ones in gas phase, it seems difficult, particularly in the case of the acetic acid, that the agreement would occur only by chance (see Figs. 20 b, c). Besides, it has been shown that the model used for the theoretical SDs is very general since it reduces in special situations to all the theoretical approaches existing in the literature, particularly those of Maréchal and Witkowski, Rösch and Ratner, Robertson and Yarwood, and Boulil et al. Then, the generality of the model and its good agreement with experiment, lead us to infer that the strong anharmonic coupling theory working within the adiabatic approximation and in which the Davydov coupling and the quantum indirect damping were incorporated, appears to be the basic tool for the physical understanding of the IR ν_{X-H} line shape of centrosymmetric cyclic dimers of gaseous carboxylic acids.

On the other hand, in the crystalline state, Fermi resonances probably play a subsidiary role. The arguments in favor of the Fermi resonances are that they are deeply assisted by the strong anharmonic coupling and that they have to be introduced in the model in order to stabilize the basic parameters when passing from a temperature to another one. Unfortunately, the introduction of the Fermi resonances do not improve sensitively the fit of the line shape. The explanation lies probably in the fact that the methods used above for handling the Fermi resonances, together with the Davydov coupling and the quantum indirect damping, avoided neither the adiabatic nor the exchange approximations. The removal of these approximations appears to be necessary. This would be a good direction for further studies in this area. But the cost will be to multiply the dimension of the Hamiltonian matrices to be diagonalized, by several orders of magnitude.

Note that the theoretical results used above are working within the linear response theory, according to which the spectral density is the Fourier transform of the autocorrelation function of the dipole moment operator. Within this approach, the combination of the strong anharmonic coupling, of the Davydov coupling, of the Fermi resonances, and of the quantum indirect damping, equally lead to a subtle complex time-dependent autocorrelation function. The corresponding spectral density cannot be anymore analyzed in terms of transitions of defined intensities between well-defined energy levels. The reason is that the intensities and the center of gravity of the components of the whole line shape, depend on the relative magnitude of the different damping parameters, so that this line shape must be viewed as it really is, that is, the whole result of the response of the dipole moment operator to the electric field. As a matter of fact, the dynamics of the dipole moment operator is a whole that depends, through Heisenberg transformations, on the total Hamiltonian involving all the oscillators of the H-bonded species that are anharmonically coupled, the surrounding thermal bath and, at finally, the coupling of the system with this bath. This is an example of the holistic character of quantum mechanics.

APPENDIX A: THE IR SPECTRAL DENSITY AS THE FOURIER TRANSFORM OF THE ACF [28]

Consider a group of molecules of dipole moment operator $\widehat{\mu}$ that are shined by a monochromatic isotropic electromagnetic radiation.

Let H_{Tot} be the total Hamiltonian of each molecule, the eigenvalue equation of which is

$$H_{Tot}| k\rangle = E_k| k\rangle \qquad \text{with} \qquad \langle f \mid k\rangle = \delta_{fk} \tag{A.1}$$

The interaction Hamiltonian $V(t)$ of each molecule with the monochromatic electric field of angular frequency ω and of strength E° is

$$V(t) = \widehat{\mu}.E^{\circ}(e^{i\omega t} + e^{-i\omega t}) \tag{A.2}$$

In the following we will assume that the strength E° of the electric field is weak. Next, the full Hamiltonian, $H_{Full}(t)$, of the system formed by a single molecule interacting with the electromagnetic field is

$$H_{Full}(t) = H_{Tot} + V(t)$$

In the interaction picture with respect to the unperturbed part, the full-time evolution operator $U_{Tot}(t)$ is

$$[U_{Tot}(t)] = (e^{-iH_{Tot}t/\hbar})[U_{Tot}(t)^{IP}] \tag{A.3}$$

where $U_{Tot}(t)^{IP}$ is the interaction picture time-evolution operator obeying the Schrödinger equation:

$$i\hbar\frac{\partial}{\partial t}[U_{Tot}(t)^{IP}] = [V(t)^{IP}][U_{Tot}(t)^{IP}] \tag{A.4}$$

with

$$[V(t)^{IP}] = (e^{iH_{Tot}t/\hbar})V(t)(e^{-iH_{Tot}t/\hbar}) \tag{A.5}$$

If the coupling Hamiltonian is small, that is, when the electromagnetic field is weak, the Schrödinger equation (A.4) may be solved in an iterative way. Then limit up to first order, that is,

$$[\mathrm{U_{Tot}}(t)^{\mathrm{IP}}] = 1 + \frac{1}{i\hbar}\int_0^t [\mathrm{V}(t')^{\mathrm{IP}}]dt'$$

The full time-evolution operator (A.3) is therefore:

$$[\mathrm{U_{Tot}}(t)] = (e^{-i\mathrm{H_{Tot}}t/\hbar})\left[1 + \frac{1}{i\hbar}\left[\int_0^t \mathrm{V}(t')^{\mathrm{IP}}\right]dt'\right] \tag{A.6}$$

Now, consider the probability P_{kf} for passing from the eigenstate $|\,k\rangle$ of the Hamiltonian H° at time $-(\tau/2)$ to the corresponding eigenstate $|\,f\rangle$ at time $+(\tau/2)$, that is,

$$P_{kf} = |\langle k\,|\mathrm{U_{Tot}}(\tau)|\,f\rangle|^2$$

Then, according to Eqs. (A.5) and (A.6), this probability becomes

$$P_{kf} = \left|\langle k\,|e^{-iE_k t/\hbar}\left[1 + \frac{1}{i\hbar}\int_{-\tau/2}^{\tau/2} e^{i\mathrm{H_{Tot}}t'/\hbar}\mathrm{V}(t')e^{-i\mathrm{H_{Tot}}t'/\hbar}dt'\right]|\,f\rangle\right|^2$$

Again, in view of the orthogonality of the kets, this probability reduces to

$$P_{kf} = \frac{1}{\hbar^2}\left|\langle f\,|\int_{-\tau/2}^{\tau/2} e^{i\Omega_{kf}t'}\mathrm{V}(t')dt'|\,k\,\rangle\right|^2$$

where

$$\Omega_{kf} = (E_k - E_f)/\hbar \tag{A.7}$$

Next, using Eq. (A.2) and neglecting the antiresonant terms of the form $\exp(i(\Omega_{kf} + \Omega)t)$, gives

$$P_{kf} = \left(\frac{\mathrm{E}^\circ}{\hbar}\right)^2 |\langle k\,|\widehat{\mu}\,|\,f\rangle|^2\left|\int_{-\tau/2}^{\tau/2} e^{i(\Omega_{kf}-\Omega)t'}dt'\right|^2$$

This last result may be expressed in terms of the transverse Dirac function, that is,

$$P_{kf} = \sqrt{2\pi}\left(\frac{\mathrm{E}^\circ}{\hbar}\right)^2 |\langle k\,|\widehat{\mu}\,|\,f\rangle|^2[\delta^\tau(\Omega_{kf} - \Omega)]^2$$

or

$$P_{kf} = \sqrt{2\pi}\left(\frac{E^{\circ}}{\hbar}\right)^2 |\langle k\,|\widehat{\mu}\,|\,f\rangle|^2 \frac{[\delta^{\tau}(\Omega_{kf}-\Omega)]}{\sqrt{2\pi\hbar}}\tau \tag{A.8}$$

Besides, the rate of energy transfer by unit time τ is

$$W = \left[\frac{1}{\tau}\right]\sum_k \sum_f \rho_{kk}\, P_{kf} \quad \hbar\Omega_{kf}$$

where the ρ_{kk} are the diagonal elements of the Boltzmann density operator, that is, the occupation probabilities of the energy levels E_k. Then, in view of Eq. (A.8), this rate becomes

$$W = \frac{1}{\sqrt{\hbar}}\left(\frac{E^{\circ}}{\hbar}\right)^2 \sum_k \sum_f \rho_{kk}\, |\langle k|\widehat{\mu}\,|\,f\rangle|^2 [\delta^{\tau}(\Omega_{kf}-\Omega)]\hbar\Omega_{kf}$$

Besides, the spectral density $I(\omega)$ is simply related to W by

$$W = I(\omega)\hbar\Omega$$

As a consequence, the SD takes the form:

$$I(\omega) = \frac{1}{\sqrt{\hbar}}\left(\frac{E^{\circ}}{\hbar}\right)^2 \sum_k \sum_f \rho_{kk}\, |\langle k\,|\widehat{\mu}\,|\,f\rangle|^2 [\delta^{\tau}(\Omega_{kf}-\Omega)]$$

If:

$$\tau \gg \frac{1}{\Omega_{kf}-\Omega}$$

the transverse Dirac function, that is, the diffraction function $\delta^{\tau}(\Omega_{kf}-\Omega)$, reduces to the Dirac function $\delta(\Omega_{kf}-\Omega)$; then, after making explicit the absolute value of the matrix element of $\widehat{\mu}$, the SD reads

$$I(\omega) = \frac{1}{\sqrt{\hbar}}\left(\frac{E^{\circ}}{\hbar}\right)^2 \sum_k \sum_f \rho_{kk}\langle\, k|\widehat{\mu}\,|\,f\rangle\ \langle f\,|\widehat{\mu}\,|\,k\rangle[\delta(\Omega_{kf}-\Omega)]$$

Again, passing from the Dirac function to its corresponding integral expression, it becomes

$$I(\omega) = \frac{1}{\sqrt{2\pi\hbar}}\left(\frac{E^{\circ}}{\hbar}\right)^2 \sum_k \sum_f \rho_{kk}\langle\, k|\widehat{\mu}\,|\,f\rangle\ \langle f\,|\widehat{\mu}\,|\,k\rangle\left[\int_{-\infty}^{\infty} e^{i(\boldsymbol{\Omega}_{kf}-\boldsymbol{\Omega})t}dt\right]$$

This last result may be expressed by aid of Eq. (A.7) as:

$$\mathrm{I}(\omega)=\frac{1}{\sqrt{2\pi\hbar}}\left(\frac{\mathrm{E}^{\circ}}{\hbar}\right)^{2}\int_{-\infty}^{\infty}\sum_{k}\sum_{f}\rho_{kk}\,\langle f\,|\widehat{\mu}\mid k\rangle\langle k\,|(e^{iE_{k}t/\hbar})\widehat{\mu}\;(e^{-iE_{f}t/\hbar})|\,f\rangle\;\;e^{-i\Omega t}dt$$

Moreover, using the eigenvalue equation (A.1) of the molecular Hamiltonian, and then suppressing the closeness relation between the dipole moment operator, the SD is transformed into

$$\mathrm{I}(\omega)=\frac{1}{\sqrt{2\pi\hbar}}\left(\frac{\mathrm{E}^{\circ}}{\hbar}\right)^{2}\int_{-\infty}^{\infty}\sum_{f}\rho_{kk}\,\langle k\,|\widehat{\mu}(e^{i\mathrm{H}_{\mathrm{Tot}}t/\hbar})\widehat{\mu}\;(e^{-i\mathrm{H}_{\mathrm{Tot}}t/\hbar})\mid k\rangle\;e^{-i\Omega t}dt$$

Besides, observe that the probabilities ρ_k are at thermal equilibrium with the matrix elements of the Boltzmann density operator that is diagonal with respect to $|\,k\rangle$,

$$\rho_{kk}=\langle k\mid\rho_{\mathrm{Tot}}|\,k\rangle\quad\text{with}\quad\rho_{\mathrm{Tot}}=\frac{1}{Z}(e^{-\beta\mathrm{H}_{\mathrm{Tot}}}),\;Z_{\mathrm{Tot}}=\mathrm{tr}_{\mathrm{Tot}}[(\rho_{\mathrm{Tot}})]\quad\text{and}\quad\beta=\frac{1}{k_{\mathrm{B}}T}$$

As a consequence, the SD becomes

$$\mathrm{I}(\omega)=\frac{1}{\sqrt{2\pi\hbar}}\left(\frac{\mathrm{E}^{\circ}}{\hbar}\right)^{2}\int_{-\infty}^{\infty}\sum_{k}\langle k\mid\rho_{\mathrm{Tot}}\;\widehat{\mu}\;\;\widehat{\mu}(t)\,|k\rangle\,e^{-i\Omega t}dt$$

with

$$\widehat{\mu}\;(t)=(e^{i\mathrm{H}_{\mathrm{Tot}}t/\hbar})\widehat{\mu}\;(e^{-i\mathrm{H}_{\mathrm{Tot}}t/\hbar})$$

As it appears, this SD is the Fourier transform of the ACF of the dipole moment operator, that is,

$$\mathrm{I}(\omega)=\frac{1}{\sqrt{2\pi\hbar}}\int_{-\infty}^{\infty}\mathrm{G}(t)e^{-i\Omega t}dt\tag{A.9}$$

Here, the ACF is given by

$$\mathrm{G}(t)=\left(\frac{\mathrm{E}}{\hbar}\right)^{2}\mathrm{tr}_{\mathrm{Tot}}[\rho_{\mathrm{Tot}}\;\widehat{\mu}\;(0)\;\widehat{\mu}\;(t)]\tag{A.10}$$

with $\widehat{\mu}(0)\equiv\widehat{\mu}$

APPENDIX B: THE ADIABATIC APPROXIMATION [51] DEALING WITH THE STRONG ANHARMONIC COUPLING THEORY OF WEAK H-BONDS

B.1 PARTITION OF THE FULL HAMILTONIAN INTO DIABATIC AND ADIABATIC PARTS

Consider the time-independent Schrödinger equation (28), governing the dynamics of the high frequency mode:

$$\mathrm{H_{Fast}}|\Phi_k(\mathrm{Q})\rangle = \hbar\omega(\mathrm{Q})\left(k+\tfrac{1}{2}\right)|\Phi_k(\mathrm{Q})\rangle \tag{B.1}$$

Here, Q is the H-bond bridge coordinate, whereas $\mathrm{H_{Fast}}$ is the Hamiltonian of the high frequency mode:

$$\mathrm{H_{Fast}} = \left[\frac{\mathrm{p}^2}{2m} + \tfrac{1}{2}m[\omega(\mathrm{Q})]^2\,\mathrm{q}^2\right] \tag{B.2}$$

Besides, $\omega(\mathrm{Q})$ is the angular frequency of the fast mode, which is assumed to depend linearly on the coordinate Q.

$$\omega(\mathrm{Q}) = \omega^\circ + b\mathrm{Q} \tag{B.3}$$

At last, $|\Phi_k(\mathrm{Q})\rangle$ are the eigenstates of the fast mode Hamiltonian that depend parametrically on Q. Moreover, owing to Eqs. (B.1) and (B.2), the explicit expression of the eigenvalue equation of the Hamiltonian (B.2) is

$$\left[\frac{\mathrm{p}^2}{2m} + \tfrac{1}{2}m[\omega(\mathrm{Q})]^2\,\mathrm{q}^2\right]|\Phi_k(\mathrm{Q})\rangle = \left(k+\tfrac{1}{2}\right)\,\hbar\,\omega(\mathrm{Q})|\Phi_k(\mathrm{Q})\rangle \tag{B.4}$$

Note that the wave functions $\Phi_k(\mathrm{q},\mathrm{Q})$ corresponding to the kets $|\Phi_k(\mathrm{Q})\rangle$ are given by the scalar products

$$\Phi_k(q,\mathrm{Q}) = \langle \mathrm{q} \mid \Phi_k(\mathrm{Q})\rangle$$

where $|\,\mathrm{q}\rangle$ is an eigenket of the coordinate operator q of the fast mode, corresponding to the eigenvalue q. Of course, the wave function $\Phi_k(q, \mathrm{Q})$ of the fast mode is a function of q that depends parametrically on the coordinate Q of the H-bond bridge.

Note also that Eq. (B.4) holds whatever the value of Q may be. It is therefore true for $\mathrm{Q} = 0$. In this last case, one may write

$$|\Phi_k(\mathrm{Q}=0)\rangle \equiv |\,\{k\}\rangle$$

Now, let us look at the Hamiltonian of the H-bond bridge that is assumed to be driven

$$\mathrm{H}_k = \frac{\mathrm{P}^2}{2M} + \tfrac{1}{2}M\Omega^2\mathrm{Q}^2 + kg\mathrm{Q}$$

Here g is a constant and k is a quantum number taking the values $0, 1, \ldots$. The eigenvalue equations of these Hamiltonians are

$$\mathrm{H}_k|\chi_n^{\{k\}}\rangle = E_n^{\{k\}}|\chi_n^{\{k\}}\rangle \tag{B.5}$$

Of course, when $k = 0$, these Hamiltonians reduce to that of a free harmonic oscillator, that is,

$$\mathrm{H}^\circ = \frac{\mathrm{P}^2}{2M} + \frac{1}{2}M\Omega^2\mathrm{Q}^2$$

with the usual eigenvalue equation for such a quantum harmonic oscillator:

$$\mathrm{H}^\circ|(n)\rangle = \hbar\Omega\left(n+\tfrac{1}{2}\right)|(n)\rangle \quad \text{with} \quad \langle(m)\mid(n)\rangle = \delta_{mn}$$

Thus, the following equivalence holds:

$$|\chi_n^{\{0\}}\rangle = |(n)\rangle$$

The kets $|\Phi_k(\mathrm{Q})\rangle$ and $|\chi_n^{\{k\}}\rangle$ form orthonormal bases:

$$\langle\Phi_k(\mathrm{Q})|\Phi_l(\mathrm{Q})\rangle = \boldsymbol{\delta}_{k,l} \qquad \text{and} \qquad \sum_k |\Phi_k(\mathrm{Q})\rangle\langle\Phi_k(\mathrm{Q})| = 1 \tag{B.6}$$

$$\langle\chi_m^{\{k\}}|\chi_n^{\{k\}}\rangle = \boldsymbol{\delta}_{m,n} \qquad \text{and} \qquad \sum_n |\chi_n^{\{k\}}\rangle\,\langle\chi_n^{\{k\}}| = 1$$

Now, consider the following tensorial product of *kets* and the corresponding tensorial product of wave functions:

$$|\Phi_k(Q)\rangle|\chi_n^{\{k\}}\rangle \qquad \text{and} \qquad \Phi_k(q, Q)\,\chi_n^{\{k\}}(Q)$$

with

$$\langle\chi_m^{\{l\}}|\langle\Phi_l(Q)\mid\Phi_k(Q)\rangle|\chi_n^{\{k\}}\rangle = \delta_{kl}\delta_{mn} \tag{B.7}$$

For the basis obtained from this tensorial product kets, it is possible to define the closeness relation:

$$\sum_k\sum_n |\Phi_k(Q)\rangle|\chi_n^{\{k\}}\rangle\ \langle\chi_n^{\{k\}}|\langle\Phi_k(Q)| = 1 \tag{B.8}$$

Now, let us look at the total Hamiltonian given by Eq. (36), which is

$$H_{Tot} = \left[\frac{P^2}{2M} + \frac{1}{2}M\Omega^2Q^2 + \frac{p^2}{2m} + \frac{1}{2}m[\omega(Q)]^2\,q^2\right] \tag{B.9}$$

Now, in the above relation, use the closeness relation (B.8) in the following way:

$$H_{Tot} = \left\{\sum_k\sum_n |\Phi_k(Q)\rangle|\chi_n^{\{k\}}\rangle\langle\chi_n^{\{k\}}|\langle\Phi_k(Q)|\right\}H_{Tot} \times \left\{\sum_l\sum_m |\Phi_l(Q)\rangle|\chi_m^{\{l\}}\rangle\langle\chi_m^{\{l\}}|\langle\Phi_l(Q)|\right\}$$

Next, perform the following partition:

$$H_{Tot} = H_{Adiab} + H_{Diab} \tag{B.10}$$

with, respectively, the following adiabatic and diabatic parts:

$$H_{Adiab} = \sum_k\sum_n |\Phi_k(Q)\rangle|\chi_n^{\{k\}}\rangle\langle\chi_n^{\{k\}}|\langle\Phi_k(Q)|H_{Tot} \times \sum_j\sum_m\ |\Phi_j(Q)\rangle|\chi_m^{\{j\}}\rangle\langle\chi_m^{\{j\}}|\langle\Phi_j(Q)|\delta_{jk} \tag{B.11}$$

$$H_{Diab} = \sum_k\sum_n\sum_{l\neq k}\sum_{m\neq n} |\Phi_k(Q)\rangle|\chi_n^{\{k\}}\rangle\langle\chi_n^{\{k\}}|\langle\Phi_k(Q)|H_{Tot}\ |\Phi_l(Q)\rangle \times |\chi_m^{\{l\}}\rangle\langle\chi_m^{\{l\}}|\langle\Phi_l(Q)| \tag{B.12}$$

B.2 WEAKNESS OF THE DIABATIC HAMILTONIAN [123]

Because eq. (B.9) the diabatic elements involved in Eq. (B.12) are

$$\langle\chi_n^{\{k\}}|\langle\Phi_k(\mathrm{Q})|H_{\mathrm{Tot}}\;|\Phi_l(\mathrm{Q})\rangle|\chi_m^{\{l\}}\rangle$$
$$=\left\langle\chi_n^{\{k\}}\middle|\left\langle\Phi_k(\mathrm{Q})\middle|\left[\frac{\mathrm{P}^2}{2M}+\frac{1}{2}M\Omega^2\mathrm{Q}^2+\frac{\mathrm{p}^2}{2m}+\frac{1}{2}m[\omega(\mathrm{Q})]^2\mathrm{q}^2\right]\middle|\Phi_l(\mathrm{Q})\right\rangle\middle|\chi_m^{\{l\}}\right\rangle \tag{B.13}$$

Thus, owing to Eq. (B.4), (B.13) becomes

$$\langle\chi_n^{\{k\}}|\langle\Phi_k(\mathrm{Q})|H_{\mathrm{Tot}}\;|\Phi_l(\mathrm{Q})\rangle|\chi_m^{\{l\}}\rangle$$
$$=\left\langle\chi_n^{\{k\}}\middle|\left\langle\Phi_k(\mathrm{Q})\middle|\left[\frac{\mathrm{P}^2}{2M}+\frac{1}{2}\,M\Omega^2\mathrm{Q}^2+\left(l+\frac{1}{2}\right)\hbar\omega(\mathrm{Q})\right]\;\middle|\Phi_l(\mathrm{Q})\right\rangle\middle|\chi_m^{\{l\}}\right\rangle \tag{B.14}$$

Next, because of Eq. (B.6), and since $k \neq l$, some of the matrix elements involved in Eq. (B.14) are zero.

$$\langle\chi_n^{\{k\}}|\langle\Phi_k(\mathrm{Q})|\left[\left(l+\tfrac{1}{2}\right)\hbar\,\omega(\mathrm{Q})\right]\;|\Phi_l(\mathrm{Q})\rangle|\chi_m^{\{l\}}\rangle$$
$$=\langle\chi_n^{\{k\}}|\langle\Phi_k(\mathrm{Q})\;|\;\Phi_l(\mathrm{Q})\rangle\left[\left(l+\tfrac{1}{2}\right)\hbar\,\omega(\mathrm{Q})\right]\;|\chi_m^{\{l\}}\rangle=0 \tag{B.15}$$

Besides, since the dependence of the ket $|\Phi_l(\mathrm{Q})\rangle$ on the Q coordinate is parametric, Q does not act on this ket as an operator, but as a scalar. As a consequence:

$$\langle\chi_n^{\{k\}}|\langle\Phi_k(\mathrm{Q})|\mathrm{Q}^2\;|\Phi_l(\mathrm{Q})\rangle|\chi_m^{\{l\}}\rangle$$
$$=\langle\chi_n^{\{k\}}|\mathrm{Q}^2|\chi_m^{\{l\}}\rangle\;\langle\Phi_k(\mathrm{Q})\;|\;\Phi_l(\mathrm{Q})\rangle=0\quad\text{since}\quad k\neq l \tag{B.16}$$

Moreover, the transition matrix elements of the kinetic energy operator involved in Eq. (B.14) read

$$\left\langle\chi_n^{\{k\}}\middle|\left\langle\Phi_k(\mathrm{Q})\middle|\frac{\mathrm{P}^2}{2M}\middle|\Phi_l(\mathrm{Q})\right\rangle\middle|\chi_m^{\{l\}}\right\rangle$$
$$=\frac{1}{2M}\left\langle\chi_n^{\{k\}}\middle|\left\langle\Phi_k(\mathrm{Q})\middle|\mathrm{P}\middle|\Phi_l(\mathrm{Q})\right\rangle\mathrm{P}\middle|\chi_m^{\{l\}}\right\rangle+\left\langle\Phi_k(\mathrm{Q})\middle|\frac{\mathrm{P}^2}{2M}\middle|\Phi_l(\mathrm{Q})\right\rangle\left\langle\chi_n^{\{k\}}\middle|\chi_m^{\{l\}}\right\rangle \tag{B.17}$$

At last, the P operator may be expressed in the Q representation according to

$$\mathrm{P} = -i\hbar \frac{\partial}{\partial \mathrm{Q}} \tag{B.18}$$

Then, with the aid of Eqs. (B.15) (B.18), a matrix element of the diabatic Hamiltonian part (B.12) of the full Hamiltonian becomes

$$\begin{aligned} &\langle \chi_n^{\{k\}} | \langle \Phi_k(\mathrm{Q}) | \mathrm{H}_{\mathrm{Tot}} \, | \Phi_l(\mathrm{Q}) \rangle | \chi_m^{\{l\}} \rangle \\ &\quad = -\frac{\hbar^2}{2M} \left\langle \Phi_k(\mathrm{Q}) \right| \left(\frac{\partial}{\partial \mathrm{Q}} \right) \left| \Phi_l(\mathrm{Q}) \right\rangle \left\langle \chi_n^{\{k\}} \right| \left(\frac{\partial}{\partial \mathrm{Q}} \right) \left| \chi_m^{\{l\}} \right\rangle \\ &\qquad - \frac{\hbar^2}{2M} \left\langle \Phi_k(\mathrm{Q}) \right| \left(\frac{\partial^2}{\partial \mathrm{Q}^2} \right) \left| \Phi_l(\mathrm{Q}) \right\rangle \left\langle \chi_n^{\{k\}} \middle| \chi_m^{\{l\}} \right\rangle \end{aligned} \tag{B.19}$$

In order to evaluate the transition matrix elements appearing in Eq. (B.19), it is of interest to look at the commutator of $\partial/\partial \mathrm{Q}$ with the Hamiltonian (B.2), that is,

$$\begin{aligned} &\left\langle \Phi_k(\mathrm{Q}) \right| \left[\mathrm{H}_{\mathrm{Fast}}, \left(\frac{\partial}{\partial \mathrm{Q}} \right) \right] \left| \Phi_l(\mathrm{Q}) \right\rangle \\ &\quad = \left\langle \Phi_k(\mathrm{Q}) \right| \mathrm{H}_{\mathrm{Fast}} \left(\frac{\partial}{\partial \mathrm{Q}} \right) \left| \Phi_l(\mathrm{Q}) \right\rangle - \left\langle \Phi_k(\mathrm{Q}) \right| \left(\frac{\partial}{\partial \mathrm{Q}} \right) \mathrm{H}_{\mathrm{Fast}} \left| \Phi_l(\mathrm{Q}) \right\rangle \end{aligned} \tag{B.20}$$

Then, according to Eq. (B.1), this commutator appears to be

$$\begin{aligned} &\left\langle \Phi_k(\mathrm{Q}) \right| \mathrm{H}_{\mathrm{Fast}} \left(\frac{\partial}{\partial \mathrm{Q}} \right) \left| \Phi_l(\mathrm{Q}) \right\rangle - \left\langle \Phi_k(\mathrm{Q}) \right| \left(\frac{\partial}{\partial \mathrm{Q}} \right) \mathrm{H}_{\mathrm{Fast}} \left| \Phi_l(\mathrm{Q}) \right\rangle \\ &\quad = \hbar\omega(\mathrm{Q})(k - l) \left\langle \Phi_k(\mathrm{Q}) \right| \left(\frac{\partial}{\partial \mathrm{Q}} \right) \left| \Phi_l(\mathrm{Q}) \right\rangle \end{aligned} \tag{B.21}$$

On the other hand, the Hamiltonian (B.2) may be split into its kinetic and potential parts according to

$$\mathrm{H}_{\mathrm{Fast}} = \mathrm{T} + \mathrm{V}^{\circ}(\mathrm{q}, \mathrm{Q})$$

with

$$\mathrm{T} = \frac{\mathrm{p}^2}{2m} \quad \text{and} \quad \mathrm{V}^{\circ}(\mathrm{q}, \mathrm{Q}) = \tfrac{1}{2} m [\omega(\mathrm{Q})]^2 \, \mathrm{q}^2 \tag{B.22}$$

Of course, the kinetic operator of the fast mode commutes with the partial derivative of the H-bond bridge coordinate, that is,

$$\left[\mathrm{T}, \left(\frac{\partial}{\partial \mathrm{Q}}\right)\right] = 0$$

As a consequence, the commutator (B.20) may be written as:

$$\left\langle \Phi_k(\mathrm{Q}) \right| \left[\mathrm{H}_{\mathrm{Fast}}, \left(\frac{\partial}{\partial \mathrm{Q}}\right)\right] \left| \Phi_l(\mathrm{Q}) \right\rangle = \left\langle \Phi_k(\mathrm{Q}) \right| \mathrm{V}^{\circ}(\mathrm{q}, \mathrm{Q}) \left(\frac{\partial}{\partial \mathrm{Q}}\right) \left| \Phi_l(\mathrm{Q}) \right\rangle - \left\langle \Phi_k(\mathrm{Q}) \right| \left(\frac{\partial}{\partial \mathrm{Q}}\right) \mathrm{V}^{\circ}(\mathrm{q}, \mathrm{Q}) \left| \Phi_l(\mathrm{Q}) \right\rangle \quad \text{(B.23)}$$

After simplification this gives

$$\left\langle \Phi_k(\mathrm{Q}) \right| \left[\mathrm{H}_{\mathrm{Fast}}, \frac{\partial}{\partial \mathrm{Q}}\right] \left| \Phi_l(\mathrm{Q}) \right\rangle = -\left\langle \Phi_k(\mathrm{Q}) \right| \left(\frac{\partial \mathrm{V}^{\circ}(\mathrm{q}, \mathrm{Q})}{\partial \mathrm{Q}}\right) \left| \Phi_l(\mathrm{Q}) \right\rangle \quad \text{(B.24)}$$

Then, combination of Eqs. (B.3), (B.20), (B.21), and (B.24), gives:

$$\left\langle \Phi_k(\mathrm{Q}) \right| \left(\frac{\partial}{\partial \mathrm{Q}}\right) \left| \Phi_l(\mathrm{Q}) \right\rangle = -\frac{\left\langle \Phi_k(\mathrm{Q}) \right| \left(\frac{\partial \mathrm{V}^{\circ}(\mathrm{q}, \mathrm{Q})}{\partial \mathrm{Q}}\right) \left| \Phi_l(\mathrm{Q}) \right\rangle}{(k - l)\hbar[\omega^{\circ} + b\mathrm{Q}]} \quad \text{(B.25)}$$

Again, according to Eqs. (B.3) and (B.22), the operator appearing on the numerator of the right-hand side of Eq. (B.25), appears to be

$$\left(\frac{\partial \mathrm{V}^{\circ}(\mathrm{q}, \mathrm{Q})}{\partial \mathrm{Q}}\right) = mb[\omega^{\circ} + b\mathrm{Q}]\mathrm{q}^2 \quad \text{(B.26)}$$

Consequently, Eq. (B.25) reduces to

$$\left\langle \Phi_k(\mathrm{Q}) \right| \left(\frac{\partial}{\partial \mathrm{Q}}\right) \left| \Phi_l(\mathrm{Q}) \right\rangle = -\frac{mb}{\hbar} \frac{\langle \Phi_k(\mathrm{Q}) |\, \mathrm{q}^2 | \Phi_l(\mathrm{Q}) \rangle}{(k - l)} \quad \text{(B.27)}$$

Besides, according to Eq. (41), the b anharmonic parameter is related to the corresponding dimensionless one α° by

$$b = \alpha^{\circ}\Omega \frac{1}{Q^{\circ\circ}} \qquad \text{with} \qquad Q^{\circ\circ} = \sqrt{\frac{\hbar}{2M\Omega}}$$

Then, Eq. (B.27) becomes

$$\left\langle \Phi_k(Q) \right| \left(\frac{\partial}{\partial Q}\right) \left| \Phi_l(Q) \right\rangle = -\left(\frac{\alpha^\circ}{2}\right)\left(\frac{\Omega}{\omega^\circ}\right)\left(\frac{1}{Q^{\circ\circ}}\right) \frac{\langle \Phi_k(Q)|\, \tilde{q}^2 |\Phi_l(Q)\rangle}{(k-l)} \quad \text{(B.28)}$$

where $\tilde{q}$ is the dimensionless coordinate operator of the fast mode related to the operator q by

$$q = \sqrt{\frac{\hbar}{2m\omega^\circ}}\,\tilde{q}$$

After evaluation of the right-hand side matrix elements, Eq. (B.28) gives

$$\left\langle \Phi_k(Q) \right| \left(\frac{\partial}{\partial Q}\right) \left| \Phi_l(Q) \right\rangle = -\left(\frac{\alpha^\circ}{2}\right)\left(\frac{\Omega}{\omega^\circ}\right)\left(\frac{1}{Q^{\circ\circ}}\right) C_{kl} \quad \text{(B.29)}$$

with

$$C_{kl} = \left[\frac{\sqrt{l+1}\sqrt{l+2}\,\delta_{k\,l+2} + \sqrt{l}\sqrt{l-1}\,\delta_{k\,l-2}}{(k-l)}\right] \quad \text{with} \quad k \neq l \quad \text{(B.30)}$$

As a consequence, the first right-hand side term of Eq. (B.19), after using the expression of $Q^{\circ\circ}$, becomes

$$\frac{\hbar^2}{2M}\left\langle \Phi_k(Q) \right| \left(\frac{\partial}{\partial Q}\right) \left| \Phi_l(Q) \right\rangle \left\langle \chi_n^{\{k\}} \right| \left(\frac{\partial}{\partial Q}\right) \left| \chi_m^{\{l\}} \right\rangle$$
$$= \frac{\alpha^\circ}{4}\hbar\Omega\left(\frac{\Omega}{\omega^\circ}\right) C_{kl}\left[\sqrt{m+1}\,\tilde{S}^{kl}_{n,m+1} - \sqrt{m}\,\tilde{S}^{kl}_{n,m-1}\right]$$

Here, there appears on the right-hand side the overlap matrix elements given by

$$\tilde{S}^{kl}_{m,m} = \langle \chi_m^{\{k\}}(Q)\, | \chi_m^{\{l\}}(Q) \rangle$$

Now, observe that the quantum numbers k or l must be small, because the theory dealing with weak H-bonds involvs only the ground and first excited state of the fast mode and because $\alpha^\circ \simeq 1$. Thus, the first nonzero terms defined by Eq. (B.30) and playing a role for weak H-bonds are

$$C_{02} = -C_{20} = 0.707 \quad C_{13} = -C_{31} = 1.225 \quad C_{24} = -C_{42} = 1.732$$

Besides, for weak or intermediate H-bonds, the ratio of the slow and fast modes angular frequencies roughly obeys

$$\frac{\Omega}{\omega^{\circ}} \sim 1/20 \tag{B.31}$$

At last, the overlap involved in the above equations must obey for normalized wave functions:

$$|\widetilde{S}^{kl}_{m,n}| \leq 1$$

Consequently, it appears that

$$\frac{\hbar^2}{2M}\left\langle \Phi_k(\mathrm{Q}) \right| \left(\frac{\partial}{\partial \mathrm{Q}}\right) \left| \Phi_l(\mathrm{Q}) \right\rangle \left\langle \chi_n^{\{k\}}(\mathrm{Q}) \right| \left(\frac{\partial}{\partial \mathrm{Q}}\right) \left| \chi_m^{\{l\}}(\mathrm{Q}) \right\rangle \simeq 0 \tag{B.32}$$

Now, consider the transition matrix elements of $\partial^2/\partial \mathrm{Q}^2$ appearing in Eq. (B.19). Inserting a closeness relation leads to

$$\left\langle \Phi_k(\mathrm{Q}) \right| \left(\frac{\partial^2}{\partial \mathrm{Q}^2}\right) \left| \Phi_l(\mathrm{Q}) \right\rangle = \sum_j \left\langle \Phi_k(\mathrm{Q}) \right| \left(\frac{\partial}{\partial \mathrm{Q}}\right) \left| \Phi_j(\mathrm{Q}) \right\rangle \left\langle \Phi_j(\mathrm{Q}) \right| \left(\frac{\partial}{\partial \mathrm{Q}}\right) \left| \Phi_l(\mathrm{Q}) \right\rangle$$

In view of Eq. (B.29), the second terms involved on the right-hand side of Eq. (B.19) is

$$\left\langle \Phi_k(\mathrm{Q}) \right| \left(\frac{\partial^2}{\partial \mathrm{Q}^2}\right) \left| \Phi_l(\mathrm{Q}) \right\rangle = \left(\frac{\alpha^{\circ}}{2}\right)^2 \left(\frac{\Omega}{\omega^{\circ}}\right)^2 \left(\frac{1}{Q^{\circ\circ}}\right)^2 \sum_j C_{kj} C_{jl}$$

Again, premultiplying both terms by $\hbar^2/2M$, and then using the above expression of $Q^{\circ\circ}$, it appears that:

$$\frac{\hbar^2}{2M}\left\langle \Phi_k(\mathrm{Q}) \right| \left(\frac{\partial^2}{\partial \mathrm{Q}^2}\right) \left| \Phi_l(\mathrm{Q}) \right\rangle = \hbar\Omega \left(\frac{\alpha^{\circ}}{2}\right)^2 \left(\frac{\Omega}{\omega^{\circ}}\right)^2 \sum_j C_{kj} C_{jl} \tag{B.33}$$

Next, observe that, because of the Kronecker symbols involved in Eq. (B.30), the summation over j appearing in Eq. (B.33) is dealing only with two terms not far from unity. Therefore, according to Eq. (B.31), it is possible to perform the following approximation:

$$\frac{\hbar^2}{2M}\left\langle \Phi_k(\mathrm{Q}) \right| \left(\frac{\partial^2}{\partial \mathrm{Q}^2}\right) \left| \Phi_l(\mathrm{Q}) \right\rangle \simeq 0 \tag{B.34}$$

Thus, as a consequence of Eqs. (B.32) and (B.34), it appears that for weak-to-medium H-bonds, Eq. (B.12), which is the explicit expression of the diabatic Hamiltonian (B.19), may be neglected. The full Hamiltonian (B.10) therefore reduces to its adiabatic part:

$$\mathrm{H_{Tot}} = \mathrm{H_{Adiab}} \tag{B.35}$$

B.3 THE ADIABATIC HAMILTONIAN AND THE EFFECTIVE HAMILTONIANS

Now, let us look at the adiabatic Hamiltonian (B.11). Owing to Eq. (B.9), it is:

$$\begin{aligned} H_{\mathrm{Adiab}} = &\sum_k \sum_n |\Phi_k(\mathrm{Q})\rangle |\chi_n^{\{k\}}\rangle \langle \chi_n^{\{k\}}| \langle \Phi_k(\mathrm{Q})| \\ &\left[\frac{\mathrm{P}^2}{2M} + \frac{1}{2} M\Omega^2 \mathrm{Q}^2 + \frac{\mathrm{p}^2}{2m} + \tfrac{1}{2} m[\omega(\mathrm{Q})]^2 \, \mathrm{q}^2\right] \\ &\sum_j \sum_m |\Phi_j(\mathrm{Q})\rangle |\chi_m^{\{j\}}\rangle \langle \chi_m^{\{j\}}| \langle \Phi_j(\mathrm{Q})| \delta_{mn} \delta_{jk} \end{aligned} \tag{B.36}$$

Besides, in view of Eqs. (B.4) and (B.3), the matrix elements involved on the right-hand side of Eq. (B.36) become:

$$\begin{aligned} &\left\langle \chi_n^{\{k\}} \middle| \left\langle \Phi_k(\mathrm{Q}) \middle| \left[\frac{\mathrm{P}^2}{2M} + \frac{1}{2} M\Omega^2 \mathrm{Q}^2 + \frac{\mathrm{p}^2}{2m} + \frac{1}{2} m[\omega(\mathrm{Q})]^2 \, \mathrm{q}^2\right] \middle| \Phi_k(\mathrm{Q}) \right\rangle \middle| \chi_m^{\{k\}} \right\rangle \\ &\quad = \left\langle \chi_n^{\{k\}} \middle| \left\langle \Phi_k(\mathrm{Q}) \middle| \left[\frac{\mathrm{P}^2}{2M} + \frac{1}{2} M\Omega^2 \mathrm{Q}^2 + \left(k + \frac{1}{2}\right) \hbar \, (\omega^\circ + b\mathrm{Q})\right] \middle| \Phi_k(\mathrm{Q}) \right\rangle \middle| \chi_m^{\{k\}} \right\rangle \end{aligned}$$

Moreover, since $(k + 1/2)\hbar\,\omega^\circ$ is a scalar, one obtains, owing to Eq. (B.5):

$$\begin{aligned} &\left\langle \chi_n^{\{k\}} \middle| \left\langle \Phi_k(\mathrm{Q}) \middle| \left[\frac{\mathrm{P}^2}{2M} + \frac{1}{2} M\Omega^2 \mathrm{Q}^2 + \left(k + \frac{1}{2}\right) \hbar \, (\omega^\circ + b\mathrm{Q})\right] \middle| \Phi_k(\mathrm{Q}) \right\rangle \middle| \chi_n^{\{k\}} \right\rangle \\ &\quad = (k + \tfrac{1}{2})\hbar\,\omega^\circ + \left\langle \chi_n^{\{k\}} \middle| \left[\frac{\mathrm{P}^2}{2M} + \frac{1}{2} M\Omega^2 \mathrm{Q}^2 + (k + \tfrac{1}{2}) \, \hbar \, b\mathrm{Q}\right] \middle| \chi_n^{\{k\}} \right\rangle \end{aligned} \tag{B.37}$$

As a consequence of Eq. (B.37), the adiabatic Hamiltonian (B.36) becomes:

$$H_{\text{Adiab}} = \sum_k |\Phi_k(\mathrm{Q})\rangle \sum_n |\chi_n^{\{k\}}\rangle\langle\chi_n^{\{k\}}|$$

$$\left[\left(k+\frac{1}{2}\right)\hbar\,\omega^\circ + \frac{\mathrm{P}^2}{2M} + \frac{1}{2}\,M\Omega^2\mathrm{Q}^2 + \left(k+\tfrac{1}{2}\right)\hbar\,b\mathrm{Q}\right]\sum_m |\chi_m^{\{k\}}\rangle\langle\chi_m^{\{k\}}|\langle\Phi_k(\mathrm{Q})|\delta_{nm}$$

Moreover, use the closeness relation dealing with the H-bond bridge kets, i.e:

$$H_{\text{Adiab}} = \sum_m |\chi_m^{\{k\}}\rangle\langle\chi_m^{\{k\}}| = 1$$

Then, the adiabatic Hamiltonian reduces to:

$$\mathrm{H}_{\text{Adiab}} = \sum_k |\Phi_k(\mathrm{Q})\rangle$$
$$\times\left[\left(k+\frac{1}{2}\right)\hbar\,\omega^\circ + \sum_m\left[\frac{\mathrm{P}^2}{2M} + \frac{1}{2}\,M\Omega^2\mathrm{Q}^2 + \left(k+\tfrac{1}{2}\right)\hbar\,b\mathrm{Q}\right]\langle\Phi_k(\mathbf{Q})|\right.$$

Now, observe that since, for each value of $\mathbf{Q}$, $|\,\Phi_k(\mathbf{Q})\rangle$ commutes with the terms involved on the right handside of this last equation, it is possible to write:

$$\mathrm{H}_{\text{Adiab}} = \sum_k\left[\left(k+\tfrac{1}{2}\right)\hbar\omega^\circ + \frac{\mathrm{P}^2}{2M} + \frac{1}{2}\,M\Omega^2\mathrm{Q}^2 + \left(k+\tfrac{1}{2}\right)\hbar\,b\mathrm{Q}\right]|\Phi_k(\mathrm{Q})\rangle\langle\Phi_k(\mathrm{Q})| \tag{B.38}$$

Now, observe that this adiabatic Hamiltonian is the sum of the effective Hamiltonians of the H-bond bridge corresponding to the different degrees of excitation of the fast mode via the projectors $|\,\Phi_k(\mathrm{Q})\rangle\,\langle\,\Phi_k(\mathrm{Q})|$. Since the parametric dependence on Q does not modify the structure of Eq. (B.38), we may simply write the adiabatic Hamiltonian (B.38), with the aid of Eq. (B.5), according to the equations:

which may be also written, owing to the equivalence of notations between the notations $|\Phi_k(\mathrm{Q})\rangle$ and $|\{k\}\rangle$:

$$\mathrm{H}_{\text{Adiab}} = \sum_k \mathrm{H}_I^{\{k\}}\,|\,\{k\}\rangle\,\langle\,\{k\}| \tag{B.39}$$

where the effective Hamiltonians are given by:

$$\mathrm{H}_I^{\{k\}} = \frac{\mathrm{P}^2}{2M} + \frac{1}{2}M\Omega^2\mathrm{Q}^2 + \left(k+\tfrac{1}{2}\right)\hbar b\mathrm{Q} + \left(k+\tfrac{1}{2}\right)\hbar\omega^\circ \tag{B.40}$$

APPENDIX C: QUANTUM REPRESENTATIONS $\{II\}$ AND $\{III\}$

Consider the effective Hamiltonian (47) of a driven quantum harmonic oscillator. Since it is not diagonal, it may be suitable to diagonalize it with the aid of a canonical transformation that will affect it or its equivalent form (50), but not that of (46) or its equivalent expression (49), which is yet to be diagonal [13].

C.1 FROM REPRESENTATION $\{II\}$ TO $\{III\}$

For this purpose, consider selective canonical transformations leading to a new quantum representation that we name $\{III\}$ in order to diagonalize the effective Hamiltonian corresponding to $\{k\} = 1$, without affecting that corresponding to $\{k\} = 0$. This may be performed on the different effective operators $\mathrm{B}_{II}^{\{k\}}$ dealing with $\{k\} = 0, 1$ with the aid of

$$\{\mathrm{B}_{III}^{\{k\}}\} = [\mathrm{A}(k\alpha^\circ)]\{\mathrm{B}_{II}^{\{k\}}\}[\mathrm{A}(k\alpha^\circ)]^{-1} \tag{C.1}$$

with

$$[\mathrm{A}(k\alpha^\circ)] = [e^{k\alpha^\circ[\mathrm{a}^\dagger - \mathrm{a}]}] \tag{C.2}$$

The operator $\mathrm{A}(k\alpha^\circ)$ is the translation operator that is unitary according to Eq. (E.8). In the following, the $\{\mathrm{B}_{II}^{\{k\}}\}$ operators will be either the effective Hamiltonians $\mathrm{H}_{II}^{\{k\}}$ or the Boltzmann density operators $\{\rho_{II}^{\{k\}}\}$ built up from these Hamiltonians.

Before we go further, we must keep in mind that the matrix elements of an operator remains unmodified by changing the representation. This may be verified by inserting a unity operator at the right and at the left of the operator B_{II}, to give

$$\langle \xi_{II}|\{\mathrm{B}_{II}^{\{1\}}\}|\zeta_{II}\rangle = \langle \xi_{II}|[\mathrm{A}(\alpha^\circ)]^{-1}[\mathrm{A}(\alpha^\circ)]\{\mathrm{B}_{II}^{\{1\}}\}[\mathrm{A}(\alpha^\circ)]^{-1}[\mathrm{A}(\alpha^\circ)]|\zeta_{II}\rangle$$

The result is

$$\langle \xi_{II} | \{ \mathrm{B}_{II}^{\{1\}} \} | \zeta_{II} \rangle = \langle \xi_{III} | \{ \mathrm{B}_{III}^{\{1\}} \} | \zeta_{III} \rangle$$

with, respectively

$$\{ \mathrm{B}_{III}^{\{1\}} \} = [\mathrm{A}(\alpha^\circ)] \{ \mathrm{B}_{II}^{\{1\}} \} [\mathrm{A}(\alpha^\circ)]^{-1}$$
$$| \zeta_{III} \rangle = [\mathrm{A}(\alpha^\circ)] | \zeta_{II} \rangle \qquad \text{and} \qquad \langle \xi_{III} | = \langle \xi_{II} | [\mathrm{A}(\alpha^\circ)]^{-1}$$

C.2 EFFECTIVE HAMILTONIANS WITHIN REPRESENTATION {III}

In the new representation $\{III\}$ generated by the unitary transformation (C.1), the effective Hamiltonians of the slow mode (46) corresponding to the ground state $|\{0\}\rangle$ of the fast mode is unmodified

$$[\mathrm{H}_{III}^{\{0\}}] = [\mathrm{H}_{II}^{\{0\}}] \tag{C.3}$$

On the other hand, the slow mode effective Hamiltonian (47), related to the situation where the fast mode has jumped into its first excited state $|\{1\}\rangle$, becomes

$$[\mathrm{H}_{III}^{\{1\}}] = [e^{\alpha^\circ [\mathrm{a}^\dagger - \mathrm{a}]}][\mathrm{H}_{II}^{\{1\}}][e^{-\alpha^\circ [\mathrm{a}^\dagger - \mathrm{a}]}] \tag{C.4}$$

In view of Eq. (47), the canonical transformation takes the form:

$$[\mathrm{H}_{III}^{\{1\}}] = [e^{\alpha^\circ [\mathrm{a}^\dagger - \mathrm{a}]}] \left[\left(\mathrm{a}^\dagger \mathrm{a} + \frac{1}{2} \right) \hbar\Omega + [\alpha^\circ (\mathrm{a}^\dagger + \mathrm{a}) - \alpha^{\circ 2}] \hbar\Omega + \hbar\omega^\circ \right] [e^{-\alpha^\circ [\mathrm{a}^\dagger - \mathrm{a}]}] \tag{C.5}$$

Now, again use the theorem [54]:

$$[e^{+\alpha \mathrm{a}^\dagger - \alpha \mathrm{a}}]\{ \mathrm{f}(\mathrm{a}^\dagger, \mathrm{a}) \}[e^{-\alpha \mathrm{a}^\dagger + \alpha \mathrm{a}}] = \{ \mathrm{f}(\mathrm{a}^\dagger - \alpha, \mathrm{a} - \alpha) \} \tag{C.6}$$

Then, the canonical transformation makes the driven Hamiltonian diagonal, leading to

$$[\mathrm{H}_{III}^{\{1\}}] = \left(\mathrm{a}^\dagger \mathrm{a} + \tfrac{1}{2} \right) \hbar\Omega - 2\alpha^{\circ 2} \hbar\Omega + \hbar\omega^\circ \tag{C.7}$$

Note that, owing to Eq. (C.3), the ground-state effective Hamiltonian appears to be given in the new representation by the same expression as in the initial, so that it remains given by Eq. (46), that is,

$$[\mathrm{H}_{III}^{\{0\}}] = \left(\mathrm{a}^{\dagger}\mathrm{a} + \tfrac{1}{2}\right)\hbar\Omega \equiv \mathrm{H}^{\circ} \tag{C.8}$$

C.3 SLOW MODE STATES WITHIN REPRESENTATION $\{III\}$

The passage from representation $\{II\}$ to $\{III\}$ does not affect the eigenstates of the slow mode harmonic Hamiltonian when the fast mode is in its ground state $|\{0\}\rangle$, but it affects them when this mode has jumped on its first excited state $|\{1\}\rangle$. Accordingly, there are the relations:

$$|(n)_{III}^{\{0\}}\rangle = |(n)\rangle \tag{C.9}$$

$$|(n)_{III}^{\{1\}}\rangle = [\mathrm{A}(\alpha^{\circ})]|(n)\rangle \tag{C.10}$$

Of course, since the operator involved in the canonical transformation is unitary, the *bra* corresponding to the last ket is

$$\langle(n)_{III}^{\{1\}}| = \langle(n)|[\mathrm{A}(\alpha^{\circ})]^{-1} \tag{C.11}$$

Owing to Eqs. (C.9) and (25), the wave functions corresponding to the kets $|(n)_{III}^{\{0\}}\rangle$ and $|(n)_{III}^{\{1\}}\rangle$ are, respectively,

$$\langle\mathrm{Q}|(n)_{III}^{\{0\}}\rangle = \langle\mathrm{Q}|\widehat{1}|(n)\rangle = \chi_n(\mathrm{Q}) \tag{C.12}$$

$$\langle\mathrm{Q}|(n)_{III}^{\{1\}}\rangle = \langle\mathrm{Q}|[\mathrm{A}(\alpha^{\circ})]|(n)\rangle = \chi_n(\mathrm{Q} - \alpha^{\circ}Q^{\circ\circ}) \tag{C.13}$$

with

$$Q^{\circ\circ} = \sqrt{\frac{\hbar}{2M\Omega}}$$

The examination of Eq. (C.12) shows that when $\{k\} = 1$, the operator given by Eq. (C.2) displaces the origin used for the kets, thus the name "translation operator". In order to visualize this displacement of the position of the H-bond bridge potential in representation $\{III\}$ when the fast mode has jumped from its ground state $|\{0\}\rangle$ to its first excited state $|\{1\}\rangle$, it is more suitable to represent it by a Morse-like one. This is performed in Fig. 2.

C.4 FRANCK–CONDON FACTORS [124]

Besides, because of Eq. (C.10), the matrix elements of the operator given by Eq. (C.2) are

$$\langle (m)|[e^{\alpha^\circ [a^\dagger - a]}]|(n)\rangle = \langle (m) \mid (n)^{\{1\}}_{III}\rangle = \{A_{m\,n}(\alpha^\circ)\} \tag{C.14}$$

Again, insert the closeness relation on the eigenkets of the coordinate operator given by

$$\int_{-\infty}^{\infty} \mid \mathrm{Q}\rangle\langle \mathrm{Q}|d\mathrm{Q} = \widehat{1}$$

Then, the above matrix elements become

$$\langle (m)|[e^{\alpha^\circ [a^\dagger - a]}]|\ (n)\rangle = \int_{-\infty}^{\infty} \langle (m) \ \mid \mathrm{Q}\rangle \, \langle \mathrm{Q} \mid (n)^{\{1\}}_{III}\rangle d\mathrm{Q}$$

Again, in view of Eqs. (C.12) and (C.13), these elements give

$$\langle (m)|[e^{\alpha^\circ [a^\dagger - a]}]|(n)\rangle = \int_{-\infty}^{\infty} \chi_m(\mathrm{Q})\chi_n(\mathrm{Q} - \alpha^\circ Q^{\circ\circ})d\mathrm{Q} = \{A_{m\,n}(\alpha^\circ)\}$$

It appears that the matrix elements of the translation operator are the Franck–Condon factors [124]:

$$\{A_{m\,n}(\alpha^\circ)\} = e^{-\frac{\alpha^{\circ 2}}{2}} \sum_{\lambda=0}^{n} \frac{\sqrt{m!n!}}{(n-\lambda)!\,\lambda!(m-n+\lambda)!} \frac{(-1)^{m-n+\lambda}}{} \alpha^{\circ 2\lambda+m-n}$$
$$\text{with} \quad m - n > 0 \tag{C.15}$$
$$\{A_{m\,n}(\alpha^\circ)\} = e^{-\frac{\alpha^{\circ 2}}{2}} \sum_{\lambda=0}^{m} \frac{\sqrt{m!n!}}{(m-\lambda)!\,\lambda!(n-m+\lambda)!} \frac{(-1)^{n-m+\lambda}}{} \alpha^{\circ 2\lambda+n-m}$$
$$\text{with} \quad m - n \leq 0$$

C.5 GENERATION OF A COHERENT STATE [125]

Note that, for the ground state, the linear transformation (C.10) reduces to

$$|(0)_{III}^{\{1\}}\rangle = [A(\alpha^{\circ})]|(0)\rangle = [e^{\alpha^{\circ}[a^{\dagger}-a]}]|(0)\rangle$$

The right-hand side is a coherent state, that is,

$$[e^{\alpha^{\circ}[a^{\dagger}-a]}]|(0)\rangle = |\alpha^{\circ}\rangle$$

Recall that the coherent states are by definition the eigenkets of the lowering operator:

$$a|\alpha\rangle = \alpha|\alpha\rangle$$

The expansion of a coherent state on the eigenstates of the number occupation operator are well known to be given by

$$|\alpha\rangle = \left(e^{-\frac{\alpha^2}{2}}\right)\sum \frac{\alpha^n}{\sqrt{n!}}|(n)\rangle$$

As a consequence, if, when the fast mode is in its ground state $|\{0\}\rangle$, the ground state of the slow mode remains a ground state when passing from representation $\{II\}$ to representation $\{III\}$. The opposite occurs when the fast mode has jumped into its first excited state $|\{1\}\rangle$, where the ground state of the slow mode in representation $\{II\}$ becomes in representation $\{III\}$ a coherent state. This leads to

$$|(0)_{III}^{\{0\}}\rangle = |(0)\rangle$$
$$|(0)_{III}^{\{1\}}\rangle = \left(e^{-\frac{\alpha^{\circ 2}}{2}}\right)\sum \frac{\alpha^{\circ n}}{\sqrt{n!}}|(n)\rangle \tag{C.16}$$

C.6 BOLTZMANN DENSITY OPERATORS IN BOTH REPRESENTATIONS

In quantum representation $\{II\}$, the Boltzmann density operator corresponding to the H-bond bridge viewed as a quantum harmonic oscillator may be written, neglecting the zero-point energy, according to

$$[\rho_{II}^{\{k\}}] = \varepsilon(e^{-\tilde{\lambda}a^{\dagger}a}) \quad \text{with} \quad \tilde{\lambda} = \frac{\hbar\Omega}{k_B T} \quad \text{and} \quad \varepsilon = 1 - e^{-\tilde{\lambda}} \quad \text{and} \quad k = 0, 1 \tag{C.17}$$

In quantum representation $\{III\}$, the density operator of the slow mode corresponding to the situation where the fast mode is in its ground state $|\{0\}\rangle$ is unchanged with respect to representation $\{II\}$:

$$[\rho_{III}^{\{0\}}] = [\rho_{II}^{\{0\}}] = [\rho_{II}^{\{k\}}] = \varepsilon\ (e^{-\tilde{\lambda} a^\dagger a})$$

On the other hand, when passing from representation $\{II\}$ to representation $\{III\}$, the density operator of the slow mode corresponding to the situation where the fast mode is in its excited state $|\{1\}\rangle$ is changed according to the canonical transformation (C.1):

$$[\rho_{III}^{\{1\}}] = [A(0)][\rho_{II}^{\{1\}}][A(0)]^{-1} \tag{C.18}$$

Then, in view of Eqs. (C.2) and (C.17), we have

$$[\rho_{III}^{\{1\}}] = [e^{\alpha^\circ [a^\dagger - a]}](e^{-\tilde{\lambda} a^\dagger a})[e^{-\alpha^\circ [a^\dagger - a]}]$$

Moreover, use theorem (C.6), that is,

$$[e^{\alpha a^\dagger - \alpha a}]\{f(a^\dagger, a)\}[e^{-\alpha a^\dagger + \alpha a}] = \{f(a^\dagger - \alpha, a - \alpha)\}$$

Because of this theorem, the density operator takes the form:

$$[\rho_{III}^{\{1\}}] = \varepsilon(e^{-\tilde{\lambda}(a^\dagger - \alpha^\circ)(a - \alpha^\circ)})$$

We see that this density operator is that of a coherent state at temperature T which, according to a theorem given in the Louisell book [54], reduces at zero temperature to

$$[\rho_{III}^{\{1\}}] \to |\alpha\rangle\langle\alpha| \quad \text{for} \quad T \to 0 \quad \text{with} \quad a|\alpha\rangle = \alpha|\ \alpha\rangle$$

APPENDIX D: THE EVOLUTION OPERATOR OF A DRIVEN QUANTUM HARMONIC OSCILLATOR [54]

Consider the effective Hamiltonian (50) of the H-bond bridge when the fast mode is in its first excited state. It is a driven quantum harmonic oscillator, that is,

$$[\mathrm{H}_{II}^{\{1\}}]=\frac{\mathrm{P}^2}{2M}+\frac{1}{2}M\Omega^2\mathrm{Q}^2+b\mathrm{Q}+\hbar\omega^\circ-\alpha^{\circ 2}\hbar\Omega$$

where the different terms have the usual meaning of the main text with $[\mathrm{Q},\mathrm{P}]=i\hbar$. The aim of this appendix is to find the expression of its time-evolution operator.

This time-evolution operator is the solution of the Schrödinger equation:

$$i\hbar\frac{\partial[\mathrm{U}^\circ{}_{II}^{\{1\}}(t)]}{\partial t}=[\mathrm{H}_{II}^{\{1\}}]\,[\mathrm{U}^\circ{}_{II}^{\{1\}}(t)]\quad\text{with}\quad[\mathrm{U}^\circ{}_{II}^{\{1\}}(0)]=1$$

Next, perform the partition:

$$[\mathrm{H}_{II}^{\{1\}}]=\mathrm{H}_c+b\mathrm{Q}$$

with

$$\mathrm{H}_c=\frac{\mathrm{P}^2}{2M}+\frac{1}{2}M\Omega^2\mathrm{Q}^2+\hbar\omega^\circ-\alpha^{\circ 2}\hbar\Omega \tag{D.1}$$

Within the interaction picture, the time-evolution operator is

$$[\mathrm{U}^\circ{}_{II}^{\{1\}}(t)]=[\mathrm{U}^\circ{}_{II}^{\{0\}}(t)]\,[\mathrm{U}^\circ{}_{II}^{\{1\}}(t)^{\mathrm{IP}}] \tag{D.2}$$

Here,

$$[\mathrm{U}^\circ{}_{II}^{\{0\}}(t)]=(e^{-i\mathrm{H}_c t/\hbar}) \tag{D.3}$$

Besides, the IP time-evolution operator is a solution of the Schrödinger equation:

$$i\frac{\partial[\mathrm{U}^{\circ\{1\}}_{II}(t)^{\mathrm{IP}}]}{\partial t} = b\mathrm{Q}(t)^{\mathrm{IP}}\;[\mathrm{U}^{\circ\{1\}}_{II}(t)^{\mathrm{IP}}] \tag{D.4}$$

with the boundary condition:

$$[\mathrm{U}^{\circ\{1\}}_{II}(0)^{\mathrm{IP}}] = 1 \tag{D.5}$$

The coordinate in the interaction picture is

$$\mathrm{Q}(t)^{\mathrm{IP}} = (e^{i\mathrm{H}_c t/\hbar})\mathrm{Q}(0)(e^{-i\mathrm{H}_c t/\hbar})$$

The solution of Eq. (D.4) is

$$[\mathrm{U}^{\circ\{1\}}_{II}(t)^{\mathrm{IP}}] \equiv 1 + \frac{b}{i\hbar}\int_0^t \mathrm{Q}(t')^{\mathrm{IP}}dt + \left(\frac{b}{i\hbar}\right)^2\int_0^t \mathrm{Q}(t')^{\mathrm{IP}}dt'\int_0^{t'}\mathrm{Q}(t'')^{\mathrm{IP}}dt'' + \ldots$$

This solution may be written formally as:

$$[\mathrm{U}^{\circ\{1\}}_{II}(t)^{\mathrm{IP}}] \equiv \widehat{\mathrm{P}}\left[e^{-ib\int_0^t \mathrm{Q}(t')^{\mathrm{IP}}dt'}\right] \tag{D.6}$$

Here, $\widehat{\mathrm{P}}$ is the Dyson time-ordering operator. At this step, note that the operator $\mathrm{Q}(t)^{\mathrm{IP}}$ does not commute with itself at different times [see Eq. (G.9)]

Now, pass to Bosons according to

$$\mathrm{Q} = \sqrt{\frac{\hbar}{2M\Omega}}[\mathrm{a}+\mathrm{a}^\dagger] \quad \mathrm{P} = i\sqrt{\frac{\hbar M\Omega}{2}}[\mathrm{a}^\dagger - \mathrm{a}] \quad \text{with} \quad [\mathrm{a},\mathrm{a}^\dagger] = 1$$

Then, owing to Eq. (D.1), within the Bosons representation, the time-evolution operator (D.3) becomes

$$[\mathrm{U}^{\circ\{0\}}_{II}(t)] = (e^{-i(\mathrm{a}^\dagger\mathrm{a}+1/2)\Omega t})(e^{i\alpha^{\circ 2}\Omega t})(e^{-i\omega^\circ t}) \tag{D.7}$$

Besides, within this Bosons representation, one may denote

$$[\mathrm{U}_{II}(\mathrm{a}^\dagger,\mathrm{a},t)^{\mathrm{IP}}] \equiv [\mathrm{U}^{\circ\{1\}}_{II}(t)^{\mathrm{IP}}]$$

Then, within this representation, the dynamic equation (D.4) of the IP time-evolution operator takes the form:

$$i\frac{\partial[\mathrm{U}_{II}(\mathrm{a}^\dagger,\mathrm{a},t)^{\mathrm{IP}}]}{\partial t} = \alpha^\circ\Omega[\mathrm{a}^\dagger(t)^{\mathrm{IP}} + \mathrm{a}(t)^{\mathrm{IP}}]\ [\mathrm{U}_{II}(\mathrm{a}^\dagger,\mathrm{a},t)^{\mathrm{IP}}] \tag{D.8}$$

with

$$\mathrm{a}(t)^{\mathrm{IP}} = [\mathrm{U}^\circ{}_{II}^{\{0\}}(t)]^{-1}\mathrm{a}\ [\mathrm{U}^\circ{}_{II}^{\{0\}}(t)]$$

The Boson in the Heisenberg picture becomes, with the aid of Eq. (D.7):

$$\begin{aligned}\mathrm{a}(t)^{\mathrm{IP}} &= (e^{i\mathrm{a}^\dagger\mathrm{a}\Omega t})\ \mathrm{a}\ (e^{-i\mathrm{a}^\dagger\mathrm{a}\Omega t})\\ \alpha^\circ &= \frac{b}{\Omega}\sqrt{\frac{\hbar}{2M\Omega}}\end{aligned} \tag{D.9}$$

Performing the canonical transformation on the lowering operator a gives

$$\mathrm{a}(t)^{\mathrm{IP}} = \mathrm{a}\ e^{-i\Omega t} \tag{D.10}$$

As a consequence, Eq. (D.8) becomes

$$i\frac{\partial[\mathrm{U}_{II}(\mathrm{a}^\dagger,\mathrm{a},t)^{\mathrm{IP}}]}{\partial t} = \alpha^\circ\Omega\ [\mathrm{a}^\dagger e^{i\Omega t} + \mathrm{a}e^{-i\Omega t}]\ [\mathrm{U}_{II}(\mathrm{a}^\dagger,\mathrm{a},t)^{\mathrm{IP}}] \tag{D.11}$$

Solve the differential equation (D.11), with the aid of the normal ordering procedure according to which it is possible to pass from operators equations that are functions of the noncommutative Bosons to scalar equations. This is possible with the help of the $\widehat{\mathrm{N}}^{-1}$ operator, which us allow to write the following transformations [54]:

$$\begin{aligned}\widehat{\mathrm{N}}^{-1}\{\mathrm{a}^\dagger[\mathrm{U}_{II}(\mathrm{a}^\dagger,\mathrm{a},t)^{\mathrm{IP}}]\} &= \alpha^*\{U(\alpha,\alpha^*,t)\}^{(n)}\\ \widehat{\mathrm{N}}^{-1}\{\mathrm{a}[\mathrm{U}_{II}(\mathrm{a}^\dagger,\mathrm{a},t)^{\mathrm{IP}}]\} &= \left(\alpha+\frac{\partial}{\partial\alpha^*}\right)\{U(\alpha,\alpha^*,t)\}^{(n)}\\ \widehat{\mathrm{N}}^{-1}\left\{\frac{\partial[\mathrm{U}_{II}(\mathrm{a}^\dagger,\mathrm{a},t)^{\mathrm{IP}}]}{\partial t}\right\} &= i\left(\frac{\partial U(\alpha,\alpha^*,t)}{\partial t}\right)^{(n)}\end{aligned}$$

As a consequence, it is possible to pass from Eq. (D.11) to the following:

$$i\left(\frac{\partial U(\alpha,\alpha^*,t)}{\partial t}\right)^{(n)} = \alpha^\circ\,\Omega\left[\alpha^* e^{i\Omega t} + \left(\alpha + \frac{\partial}{\partial\alpha^*}\right)e^{-i\Omega t}\right]\{U(\alpha,\alpha^*,t)\}^{(n)} \quad \text{(D.12)}$$

with, for the boundary condition (D.5), the following expression:

$$\{U(\alpha,\alpha^*,0)\}^{(n)} = 1$$

Now, in order to solve the partial derivative equation (D.12), let us write

$$\{U(\alpha,\alpha^*,t)\}^{(n)} = e^{G^{(n)}(t)} \quad \text{(D.13)}$$

Next, in terms of the new scalar function $G^{(n)}(t)$, the partial derivatives of $U^{(n)}(t)$ with respect to the scalars t and α^* are, respectively,

$$\left(\frac{\partial\{U(\alpha,\alpha^*,t)\}}{\partial t}\right)^{(n)} = \left(\frac{\partial G^{(n)}(t)}{\partial t}\right)\{U(\alpha,\alpha^*,t)\}^{(n)}$$

$$\left(\frac{\partial\{U(\alpha,\alpha^*,t)\}}{\partial\alpha^*}\right)^{(n)} = \left(\frac{\partial G^{(n)}(t)}{\partial\alpha^*}\right)\{U(\alpha,\alpha^*,t)\}^{(n)}$$

Thus, owing to these equations and after simplification by $\{U(\alpha,\alpha^*,t)\}^{(n)}$, Eq. (D.12) becomes

$$i\left(\frac{\partial G^{(n)}}{\partial t}\right) = \alpha^\circ\Omega\, e^{i\Omega t}\alpha^* + \alpha^\circ\Omega e^{-i\Omega t}\alpha + \alpha^\circ\Omega e^{-i\Omega t}\left(\frac{\partial G^{(n)}}{\partial\alpha^*}\right) \quad \text{(D.14)}$$

Again, assume that $G(t)$ is of the form:

$$G^{(n)}(t) = A(t) + B(t)\alpha + C(t)\alpha^* \quad \text{(D.15)}$$

Here, $A(t)$, $B(t)$, and $C(t)$ are unknown functions to be found. Then, in terms of these new functions, the partial derivatives involved in Eq. (D.14) are, respectively,

$$\frac{\partial G^{(n)}}{\partial\alpha^*} = C(t)$$

$$i\frac{\partial G^{(n)}}{\partial t} = i\frac{\partial A(t)}{\partial t} + i\frac{\partial B(t)}{\partial t}\alpha + i\frac{\partial C(t)}{\partial t}\alpha^*$$

Thus, Eq. (D.14) becomes

$$i\left(\frac{\partial A(t)}{\partial t}+\frac{\partial B(t)}{\partial t}\alpha+\frac{\partial C(t)}{\partial t}\alpha^*\right)=\alpha^\circ\Omega e^{i\Omega t}\alpha^*+\alpha^\circ\Omega e^{-i\Omega t}\alpha+\alpha^\circ\Omega e^{-i\Omega t}C(t)$$

By identification that gives, respectively,

$$\begin{aligned}i\frac{\partial A(t)}{\partial t}&=\alpha^\circ\Omega e^{-i\Omega t}C(t)\\ i\frac{\partial B(t)}{\partial t}&=\alpha^\circ\Omega e^{-i\Omega t}\\ i\frac{\partial C(t)}{\partial t}&=\alpha^\circ\Omega e^{i\Omega t}\end{aligned}$$

Solving these equations, leads, respectively, to

$$C(t)=-\alpha^\circ[e^{i\Omega t}-1] \tag{D.16}$$

$$B(t)=\alpha^\circ[e^{-i\Omega t}-1] \tag{D.17}$$

$$A(t)=i\alpha^{\circ 2}\Omega t+\alpha^\circ B(t) \tag{D.18}$$

Then, in view of Eqs. (D.16–D.18) and of Eqs. (D.13) and (D.15), the solution of the differential equation (D.12) is

$$\begin{aligned}&\{U(\alpha,\alpha^*,t)\}^{(n)}=(e^{-\Phi^\circ{}_{II}(t)^*\alpha^*})(e^{+\Phi^\circ{}_{II}(t)\alpha})(e^{-\Phi^\circ{}_{II}(t)^*\alpha^*})(e^{+\Phi^\circ{}_{II}(t)\alpha})\\ &\Phi^\circ{}_{II}(t)\equiv\alpha^\circ[e^{-i\Omega t}-1]=B(t)=-C(t)^*\end{aligned} \tag{D.19}$$

Now, return to the time-evolution operator with the aid of the $\widehat{\mathrm{N}}$ operator:

$$\widehat{\mathrm{N}}\ [\{U(\alpha,\alpha^*,t)\}^{(n)}]=(e^{-\Phi^\circ{}_{II}(t)^*\alpha^*})(e^{+\Phi^\circ{}_{II}(t)\alpha})\widehat{\mathrm{N}}\ [(e^{-\Phi^\circ{}_{II}(t)^*\alpha^*})(e^{+\Phi^\circ{}_{II}(t)\alpha})] \tag{D.20}$$

This leads, respectively, to

$$\widehat{\mathrm{N}}\ [\{U(\alpha,\alpha^*,t)\}^{(n)}]=[\mathrm{U}^\circ{}_{II}(\mathrm{a}^\dagger,\mathrm{a},t)^{\mathrm{IP}}]\equiv[\mathrm{U}^{\circ\{1\}}_{II}(t)^{\mathrm{IP}}] \tag{D.21}$$

$$\widehat{\mathrm{N}}\ [(e^{-\Phi^\circ{}_{II}(t)^*\alpha^*})(e^{+\Phi^\circ{}_{II}(t)\alpha})]=[e^{-\Phi^\circ{}_{II}(t)^*\mathrm{a}^\dagger}e^{\Phi^\circ{}_{II}(t)\mathrm{a}}] \tag{D.22}$$

As a consequence, because of Eqs. (D.20–D.22), Eq. (D.19) allows us to obtain the IP time-evolution operator in the form:

$$[\mathrm{U}^{\circ}{}_{II}(\mathrm{a}^{\dagger},\mathrm{a},t)^{\mathrm{IP}}] = (e^{i\alpha^{\circ 2}\Omega t})(e^{\alpha^{\circ}\Phi^{\circ}{}_{II}(t)})[e^{-\Phi^{\circ}{}_{II}(t)^{*}\mathrm{a}^{\dagger}}e^{\Phi^{\circ}{}_{II}(t)\mathrm{a}}]$$

Moreover, the Glauber–Weyl theorem leads us to write

$$e^{\alpha^{\circ}\Phi^{\circ}{}_{II}(t)}[e^{-\Phi^{\circ}{}_{II}(t)^{*}\mathrm{a}^{\dagger}}e^{\Phi^{\circ}{}_{II}(t)\mathrm{a}}] = (e^{-i\alpha^{\circ 2}\sin\Omega t})[e^{-\Phi^{\circ}{}_{II}(t)^{*}\mathrm{a}^{\dagger}+\Phi^{\circ}{}_{II}(t)\mathrm{a}}]$$

Thus, the IP time-evolution operator becomes

$$[\mathrm{U}^{\circ}{}_{II}(\mathrm{a}^{\dagger},\mathrm{a},t)^{\mathrm{IP}}] = (e^{i\alpha^{\circ 2}\Omega t})(e^{-i\alpha^{\circ 2}\sin\Omega t})[e^{-\Phi^{\circ}{}_{II}(t)^{*}\mathrm{a}^{\dagger}+\Phi^{\circ}{}_{II}(t)\mathrm{a}}] = [\mathrm{U}^{\circ\{1\}}_{II}(t)^{\mathrm{IP}}] \tag{D.23}$$

As a consequence, because of Eqs. (D.7) and (D.23), the full-time evolution operator (D.2) of the driven quantum harmonic oscillator takes the form:

$$[\mathrm{U}^{\circ\{1\}}_{II}(t)] = (e^{i\alpha^{\circ 2}\Omega t})(e^{-i\alpha^{\circ 2}\sin\Omega t})[e^{-i(\mathrm{a}^{\dagger}\mathrm{a}+1/2)\Omega t}][e^{-\Phi^{\circ}{}_{II}(t)^{*}\mathrm{a}^{\dagger}+\Phi^{\circ}{}_{II}(t)\mathrm{a}}](e^{-i\omega^{\circ}t})$$

Note that because of Eqs. (D.6) and (D.23), the following expression is verified

$$[\mathrm{U}^{\circ\{1\}}_{II}(t)^{\mathrm{IP}}]^{-1} = \widehat{\mathrm{P}}\left\{e^{ib\int_0^t \mathrm{Q}(t')^{\mathrm{IP}}dt'}\right\} = (e^{-i\alpha^{\circ 2}\Omega t})(e^{i\alpha^{\circ 2}\sin\Omega t})[e^{\Phi^{\circ}{}_{II}(t)^{*}\mathrm{a}^{\dagger}-\Phi^{\circ}{}_{II}(t)\mathrm{a}}] \tag{D.24}$$

As a matter of fact, the Boson normal-ordering procedure allows us to get the formal solution of the IP time-evolution operator involving the Dyson time-ordering operator $\widehat{\mathrm{P}}$. Also, observe that within the Bosons representation, the IP coordinate is

$$\mathrm{Q}(t)^{\mathrm{IP}} = \sqrt{\frac{\hbar}{2M\Omega}}(\mathrm{a}^{\dagger}(t)^{\mathrm{IP}} + \mathrm{a}(t)^{\mathrm{IP}})$$

Thus, in view of Eqs. (D.9) and (D.10), there is the relation:

$$b\mathrm{Q}(t)^{\mathrm{IP}} = \alpha^{\circ}(\mathrm{a}^{\dagger}e^{i\Omega t} + \mathrm{a}e^{-i\Omega t})\Omega$$

Therefore, Eq. (D.24) takes the form:

$$\widehat{\mathrm{P}}\left\{e^{i\alpha^\circ \int_0^t (\mathrm{a}^\dagger \mathrm{e}^{i\Omega t'} + \mathrm{a}\mathrm{e}^{-i\Omega t'})dt'}\right\}$$
$$= e^{-i\alpha^{\circ 2}\Omega t}\left(e^{i\alpha^{\circ 2}\sin \Omega t}\right)\left[e^{\Phi^\circ{}_{II}(t)^*\mathrm{a}^\dagger - \Phi^\circ{}_{II}(t)\mathrm{a}}\right] = \left[\mathrm{U}^\circ{}_{II}^{\{1\}}(t)^{\mathrm{IP}}\right]^{-1} \qquad \text{(D.25)}$$

Consequently, the full-time evolution operator is given by Eq. (D.2) with the help of Eqs. (D.7) and (D.25).

APPENDIX E: PROPERTIES OF THE TRANSLATION OPERATOR [54], [125]

Consider the canonical transformation allowing us to write [54]

$$[e^{+\alpha^* a}]\{f(a^\dagger, a)\}[e^{-\alpha^* a}] = \{f(a^\dagger + \alpha, a)\}$$

where a and $a^\dagger$ are the usual Bosons obeying $[a, a^\dagger] = 1$, and where $\{f(a^\dagger, a)$ is any operator function of these Bosons.

Premultiplying it by $e^{-\alpha a^\dagger}$ and postmutiplying it by $e^{+\alpha a^\dagger}$ gives

$$[e^{-\alpha a^\dagger}][e^{+\alpha^* a}]\{f(a^\dagger, a)\}[e^{-\alpha^* a}][e^{+\alpha a^\dagger}] = [e^{-\alpha a^\dagger}]\{f(a^\dagger + \alpha^*, a)\}[e^{+\alpha a^\dagger}] \tag{E.1}$$

an addition, one may apply the following theorem [54]

$$[e^{-\alpha a^\dagger}]\{f(a^\dagger + \alpha^*, a)\}[e^{+\alpha a^\dagger}] = \{f(a^\dagger + \alpha^*, a + \alpha)\}$$

Thus, Eq. (E.1) transforms into

$$[e^{-\alpha a^\dagger}][e^{+\alpha^* a}]\{f(a^\dagger, a)\}[e^{-\alpha^* a}][e^{+\alpha a^\dagger}] = \{f(a^\dagger + \alpha^*, a + \alpha)\}$$

Now, the Glauber–Weyl theorem leads us to write, respectively,

$$\begin{aligned} [e^{-\alpha a^\dagger}][e^{+\alpha^* a}] &= [e^{-\alpha a^\dagger + \alpha^* a}](e^{\frac{1}{2}|\alpha|^2}) \\ [e^{-\alpha^* a}][e^{+\alpha a^\dagger}] &= [e^{+\alpha a^\dagger - \alpha^* a}](e^{-\frac{1}{2}|\alpha|^2}) \end{aligned} \tag{E.2}$$

As a consequence, Eq. (E.1) gives

$$[e^{-\alpha a^\dagger + \alpha^* a}]\{f(a^\dagger, a)\}[e^{+\alpha a^\dagger - \alpha^* a}] = \{f(a^\dagger + \alpha^*, a + \alpha)\} \tag{E.3}$$

This equation may be written

$$[A(\alpha^*, \alpha)]^{-1}\{f(a^\dagger, a)\}[A(\alpha^*, \alpha)] = \{f(a^\dagger + \alpha^*, a + \alpha)\} \tag{E.4}$$

with

$$[A(\alpha^*, \alpha)] \equiv [e^{+\alpha a^\dagger - \alpha^* a}] \tag{E.5}$$

Now, suppose that the operator $\{f(a^\dagger, a)\}$ is the position operator $Q(a^\dagger, a)$, as a function of the Boson representation:

$$Q(a^\dagger, a) = Q^\circ(a^\dagger + a) \quad \text{with} \quad Q^{\circ\circ} \equiv \sqrt{\frac{\hbar}{2M\Omega}}$$

Then, as a consequence of Eq. (E.4), it reads

$$[A(\alpha^*, \alpha)]^{-1}\{Q(a^\dagger, a)\}[A(\alpha^*, \alpha)]$$

$$= Q(a^\dagger, a) + Q^{\circ\circ}\,(\alpha + \alpha^*) \tag{E.6}$$

Again, if $\alpha = \alpha^* = \alpha^\circ$ is real, the above result reduces to

$$[A(\alpha)]^{-1}\{Q(a^\dagger, a)\}[A(\alpha)] = Q(a^\dagger, a) + \alpha^\circ Q^{\circ\circ} \tag{E.7}$$

As it appears, the transformation (E,6) is corresponding to a translation of the origin of the position operator Q. This is the reason why the operator involved in this transformation is called the "translation operator" [125].

Note that the translation operator $A(\alpha^*, \alpha)$ is unitary, that is, its inverse is equal to its Hermitean conjugate:

$$[A(\alpha^*, \alpha)]^{-1} = [A(\alpha^*, \alpha)]^\dagger \tag{E.8}$$

Besides, because of Eq. (E.2), note that the translation operator (E.5) may be also written

$$[A(\alpha^*, \alpha)] = (e^{-\frac{1}{2}|\alpha|^2})[e^{+\alpha a^\dagger}][e^{-\alpha^* a}] \tag{E.9}$$

Note also that when $\alpha = \alpha^* = \alpha^\circ$, the translation operator may be expressed in terms of the momentum operator by aid of

$$\mathrm{P} = i\sqrt{\frac{M\hbar\Omega}{2}}(\mathrm{a}^\dagger - \mathrm{a})$$

Then, the operator (E.5) becomes

$$\mathrm{A}(Q^{\circ\circ}) = e^{-i\alpha^\circ Q^{\circ\circ}\mathrm{P}/\hbar} \tag{E.10}$$

This operator translates the eigenkets of the position operator according to

$$\mathrm{A}(Q^{\circ\circ})|\mathrm{Q}\rangle = e^{-i\alpha^\circ Q^{\circ\circ}\mathrm{P}/\hbar}|\mathrm{Q}\rangle = |\mathrm{Q} + Q^{\circ\circ}\rangle$$

It is shown in Appendix N that the action of the translation operator on the ground state $|\ (0)\ \rangle$ of the Hamiltonian of the quantum harmonic oscillator gives a coherent state $|\ \alpha\ \rangle$:

$$\mathrm{a}\ |\ \alpha\ \rangle = \alpha\ |\ \alpha\ \rangle \tag{E.11}$$

with the following expansion of the coherent state on the eigenstates of the Hamiltonian of the quantum harmonic oscillator:

$$|\ \alpha\ \rangle = e^{-\frac{1}{2}|\alpha|^2} \sum \frac{\alpha^n}{\sqrt{n!}}|\ (n)\ \rangle \tag{E.12}$$

APPENDIX F: PRODUCT OF TWO TRANSLATION OPERATORS

Consider the following two translation operators where $[a, a^\dagger] = 1$ and ξ and ζ are complex scalars

$$[A(\xi)] = [e^{\xi a^\dagger - \xi^* a}] \quad [A(\zeta)] = [e^{\zeta a^\dagger - \zeta^* a}]$$

Now, let us look at the product of these two translation operators that are unitary:

$$[A(\zeta)]\,[A(\xi)] = [e^{\zeta a^\dagger - \zeta^* a}][e^{\xi a^\dagger - \xi^* a}]$$

Owing to the Glauber theorem, the right-hand side term of this last equation becomes

$$[e^{\zeta a^\dagger - \zeta^* a}][e^{\xi a^\dagger - \xi^* a}] = [e^{(\zeta a^\dagger - \zeta^* a)+(\xi a^\dagger - \xi^* a)}](e^{-\frac{1}{2}[(\xi a^\dagger - \xi^* a),(\zeta a^\dagger - \zeta^* a)]}) \tag{F.1}$$

Now, observe that by using $[a, a^\dagger] = 1$, the commutator is

$$[(\xi a^\dagger - \xi^* a), (\zeta a^\dagger - \zeta^* a)] = \xi\zeta^* - \xi^*\zeta$$

Thus, because of this result, Eq. (F.1) becomes

$$[e^{\zeta a^\dagger - \zeta^* a}][e^{\xi a^\dagger - \xi^* a}] = (e^{-\frac{1}{2}[\xi\zeta^* - \xi^*\zeta]})[e^{(\xi+\zeta)a^\dagger - (\xi^*+\zeta^*)a}] \tag{F.2}$$

Now, let us look at the translation operator:

$$[e^{\Phi^\circ_{II}(t)^* a^\dagger - \Phi^\circ_{II}(t) a}]$$

with

$$\Phi^\circ_{II}(t) \equiv \alpha^\circ[e^{-i\Omega t} - 1] \quad \text{and} \quad \alpha^\circ = \alpha^{\circ *} \tag{F.3}$$

This translation operator has the same form as that involved on the right-hand side of Eq. (F.2)

$$[e^{\Phi_{II}^{\circ}(t)^{*}\mathrm{a}^{\dagger}-\Phi_{II}^{\circ}(t)\mathrm{a}}] = [e^{(\xi+\zeta)\mathrm{a}^{\dagger}-(\xi^{*}+\zeta^{*})\mathrm{a}}]$$

This leads us to write

$$\xi+\zeta = \Phi_{II}(t)^{*}$$
$$\xi = \alpha^{\circ}e^{i\Omega t} \quad \text{and} \quad \zeta = \zeta^{*} = -\alpha^{\circ}$$

Therefore, in view of Eq. (F.3), there is

$$\frac{\xi\zeta^{*}-\xi^{*}\zeta}{2} = -i\alpha^{\circ}\sin\Omega t$$

Consequenctly, Eq. (F.2), allows us to write

$$e^{-\alpha^{o}\mathrm{a}^{\dagger}+\alpha^{o}\mathrm{a}}[e^{\Phi_{III}^{\circ}(t)^{*}\mathrm{a}^{\dagger}-\Phi_{III}^{\circ}(t)\mathrm{a}}] = (e^{i\alpha^{\circ}\sin\Omega t})[e^{\Phi_{III}^{\circ}(t)^{*}\mathrm{a}^{\dagger}-\Phi_{III}^{\circ}(t)\mathrm{a}}] \tag{F.4}$$

with

$$\Phi_{III}^{\circ}(t) \equiv \alpha^{\circ}e^{-i\Omega t} \tag{F.5}$$

APPENDIX G: DISENTANGLING THE DRIVEN TERM[59]

Consider the product of the two evolution operators appearing in Eq. (61) that is,

$$[\mathrm{U}^{\circ\{1\}}_{II}(t)]^{-1}[\mathrm{U}^{\circ\{0\}}_{II}(t)] = \left[e^{-i\left(\frac{\mathrm{P}^2}{2M}+\frac{1}{2}M\Omega^2\mathrm{Q}^2+\hbar b\mathrm{Q}\right)t/\hbar}\right]\left[e^{-i\left(\frac{\mathrm{P}^2}{2M}+\frac{1}{2}M\Omega^2\mathrm{Q}^2\right)t/\hbar}\right] \quad (\text{G.1})$$

with, respectively

$$[\mathrm{U}^{\circ\{1\}}{}_{II}(t)] = \left[e^{-i\left(\frac{\mathrm{P}^2}{2M}+\frac{1}{2}M\Omega^2\mathrm{Q}^2+\hbar b\mathrm{Q}\right)t/\hbar}\right] \quad (\text{G.2})$$

$$[\mathrm{U}^{\circ\{0\}}_{II}(t)] = \left[e^{-i\left(\frac{\mathrm{P}^2}{2M}+\frac{1}{2}M\Omega^2\mathrm{Q}^2\right)t/\hbar}\right] \quad (\text{G.3})$$

where the different parameters have their usual meaning and with, of course, $[\mathrm{Q},\mathrm{P}] = i\hbar$.

The aim of this appendix is to disentangle the driven term $\hbar b\mathrm{Q}$ appearing in (G.1). That may be performed with the aid of the IP [54], which allows us to write the time-evolution operator (G.2) according to

$$[\mathrm{U}^{\circ\{1\}}_{II}(t)] = [\mathrm{U}^{\circ\{0\}}_{II}(t)][\mathrm{U}^{\circ\{1\}}_{II}(t)^{\mathrm{IP}}] \quad (\text{G.4})$$

with

$$[\mathrm{U}^{\circ\{1\}}_{II}(t)^{\mathrm{IP}}] = \hat{\mathrm{P}}[e^{-ib\int_0^{t'}\mathrm{Q}(t')^{\mathrm{IP}}dt'}] \quad (\text{G.5})$$

Here, $\mathrm{Q}(t)^{\mathrm{IP}}$ is the H-bond bridge coordinate in the interaction picture that is given by

$$\mathrm{Q}(t)^{\mathrm{IP}} = \left[e^{i\left(\frac{\mathrm{P}^2}{2M}+\frac{1}{2}M\Omega^2\mathrm{Q}^2\right)t/\hbar}\right]\mathrm{Q}(0)\left[e^{-i\left(\frac{\mathrm{P}^2}{2M}+\frac{1}{2}M\Omega^2\mathrm{Q}^2\right)t/\hbar}\right] \quad (\text{G.6})$$

Besides $\widehat{\mathrm{P}}$, which showed not be confused with the momentum operator P of the **H**-bond bridge, is the Dyson time-ordering operator [57] acting on the Taylor expansion terms of the exponential operator in such a way so that the time arguments involved in the different integrals will be $t > t' > t''$.

$$\hat{\mathrm{P}}[e^{-ib\int_0^t \mathrm{Q}(t')^{\mathrm{IP}}dt'}] \equiv 1 + \frac{b}{i}\int_0^t \mathrm{Q}(t')^{\mathrm{IP}}dt' + \left(\frac{b}{i}\right)^2 \int_0^t \mathrm{Q}(t')^{\mathrm{IP}}dt' \int_0^{t'} \mathrm{Q}(t'')^{\mathrm{IP}}dt'' + \cdots \tag{G.7}$$

The use of the Dyson operator is required because the time dependant IP Q coordinates at different times does not commute as it will appear below.

By performing the canonical transformation involved in Eq. (G.6), we obtain for the coordinate in the interaction picture:

$$\mathrm{Q}(t)^{\mathrm{IP}} = \mathrm{Q}\cos\Omega t + i\mathrm{P}\sin\Omega t \tag{G.8}$$

Note that this IP coordinate does not commute at different times since:

$$[\mathrm{Q}(t)^{\mathrm{IP}}, \mathrm{Q}(t')^{\mathrm{IP}}] = i\frac{\hbar}{M\Omega}\sin(\Omega(t - t')) \tag{G.9}$$

This is the reason for the use of the Dyson time-ordering operator.

Within the IP procedure and in view of Eqs. (G.3–G.5), the time-evolution operator (G.2) leads to

$$[\mathrm{U}^{\circ\{1\}}_{II}(t)] = \left[e^{-i\left(\frac{\mathrm{P}^2}{2M}+\frac{1}{2}M\Omega^2\mathrm{Q}^2\right)t/\hbar}\right]\hat{\mathrm{P}}\,[e^{-ib\int_0^t \mathrm{Q}(t')^{\mathrm{IP}}dt'}] \tag{G.10}$$

The disentangling procedure therefore allows us to write for the inverse of Eq. (G.10),

$$[\mathrm{U}^{\circ\{1\}}_{II}(t)]^{-1} = \hat{\mathrm{P}}[e^{ib\int_0^t \mathrm{Q}(t')^{\mathrm{IP}}dt'}]\left[e^{i\left(\frac{\mathrm{P}^2}{2M}+\frac{1}{2}M\Omega^2\mathrm{Q}^2\right)t/\hbar}\right]$$

Thus, Eq. (G.1) takes the form:

$$[\mathrm{U}^{\circ\{1\}}_{II}(t)]^{-1}[\mathrm{U}^{\circ\{0\}}_{II}(t)]$$
$$= \hat{\mathrm{P}}[e^{ib\int_0^t \mathrm{Q}(t')^{\mathrm{IP}}dt'}]\left[e^{i\left(\frac{\mathrm{P}^2}{2M}+\frac{1}{2}M\Omega^2\mathrm{Q}^2\right)t/\hbar}\right]\left[e^{-i\left(\frac{\mathrm{P}^2}{2M}+\frac{1}{2}M\Omega^2\mathrm{Q}^2\right)t/\hbar}\right]$$

After simplification, this product reduces to

$$[\mathrm{U}^{\circ\{1\}}_{II}(t)]^{-1}[\mathrm{U}^{\circ\{0\}}_{II}(t)] = \hat{\mathrm{P}}[e^{ib\int_0^t \mathrm{Q}(t')^{\mathrm{IP}}dt'}] \quad \text{(G.11)}$$

Of course, the left-hand side term of this equation is nothing but the inverse of the IP time-evolution operator, that is,

$$[\mathrm{U}^{\circ\{1\}}_{II}(t)]^{-1}[\mathrm{U}^{\circ\{0\}}_{II}(t)] = [\mathrm{U}^{\circ\{1\}}_{II}(t)^{\mathrm{IP}}]^{-1}$$

Hence,

$$[\mathrm{U}^{\circ\{1\}}_{II}(t)^{\mathrm{IP}}]^{-1} = \hat{\mathrm{P}}[e^{ib\int_0^t \mathrm{Q}(t')^{\mathrm{IP}}dt'}]$$

APPENDIX H: THERMAL AVERAGE OF THE TRANSLATION OPERATOR

Consider the thermal average involved in Eq. (123):

$$\langle A(\Phi_{II}(t))\rangle = \varepsilon \, \mathrm{tr}_{\mathrm{Slow}}[(e^{-\tilde{\lambda} a^\dagger a})[e^{\Phi_{II}(t)^* a^\dagger - \Phi_{II}(t) a}]] \tag{H.1}$$

Here, $a^\dagger$ and a are the usual Bosons obeying

$$[a, a^\dagger] = 1$$

Besides, $\tilde{\lambda}$ is the Boltzmann factor related to the angular frequency of the H-bond bridge, that is,

$$\tilde{\lambda} = \frac{\hbar\Omega}{k_B T}$$

ε is the inverse of the partition function,

$$\varepsilon = 1 - e^{-\tilde{\lambda}}$$

Finally, $\Phi_{II}(t)$ is the scalar involved in the time-dependent displacement operator:

$$\Phi_{II}(t) = \alpha^\circ [e^{-i\Omega t} e^{-\gamma t/2} - 1]$$

To calculate this thermal average, it is suitable to use the following theorem, the demonstration of which is given in the book by Louisell [54] which, deals with the quantum theory of light. Another possibility is to use the Bloch theorem [123] according to which

$$\varepsilon \, \mathrm{tr}_{\mathrm{Slow}}[e^{-\tilde{\lambda} a^\dagger a}[e^{[\Phi_{II}(t)^* a^\dagger - \Phi_{II}(t) a]}]] = e^{\frac{1}{2}\varepsilon \, \mathrm{tr}_{\mathrm{Slow}}[e^{-\tilde{\lambda} a^\dagger a}[\Phi_{II}(t)^* a^\dagger - \Phi_{II}(t) a]^2]} \tag{H.2}$$

By using the commutation rule of the Bosons, it appears that

$$[\Phi_{II}(t)^* a^\dagger - \Phi_{II}(t) a]^2 = [\Phi_{III}(t)^* a^\dagger]^2 + [\Phi_{III}(t) a]^2 - |\Phi_{III}(t)|^2 [2a^\dagger a + 1]$$

Of course, there is the following result:

$$\varepsilon \ \mathrm{tr}_{\mathrm{Slow}}[e^{-\tilde{\lambda}a^\dagger a}[a^\dagger]^2] = \varepsilon \ \mathrm{tr}_{\mathrm{Slow}}[e^{-\tilde{\lambda}a^\dagger a}[a]^2] = 0$$

As a consequence of the above equations, the thermal average involved on the right-hand side of Eq. (H.2) is

$$\varepsilon \ \mathrm{tr}_{\mathrm{Slow}}[e^{-\tilde{\lambda}a^\dagger a}[\Phi_{II}(t)^* a^\dagger - \Phi_{II}(t)a]^2] = -|\Phi_{III}(t)|^2[2\langle n\rangle + 1]$$

where $\langle n\rangle$ is the thermal average of the quantum number of the Hamiltonian of the H-bond bridge, that is,

$$\langle n\rangle = \varepsilon \ \mathrm{tr}_{\mathrm{Slow}}[(e^{-\tilde{\lambda}a^\dagger a})a^\dagger a] = \frac{1}{e^{\tilde{\lambda}} - 1} \tag{H.3}$$

Then, by using these last results, the thermal average (H.1) of the translation operator appears to be

$$\langle A(\Phi_{II}(t))\rangle = (e^{-(\langle n\rangle + 1/2)\ |\Phi_{II}(t)|^2})$$

As a consequence, because of the above expression of the time-dependent scalar,

$$\langle A(\Phi_{II}(t))\rangle = [e^{\alpha^{\circ 2}(\langle n\rangle + 1/2)(2e^{-\gamma t/2}\cos\Omega t - e^{-\gamma t} - 1)}] \tag{H.4}$$

On the other hand, when the damping is missing, that is, when $\gamma = 0$, this equation reduces to

$$\langle A(\Phi^{\circ}{}_{II}(t))\rangle = [e^{2\alpha^{\circ 2}(\langle n\rangle + 1/2)(\cos\Omega t - 1)}] \tag{H.5}$$

APPENDIX I: THE DRIVEN DAMPED QUANTUM HARMONIC OSCILLATOR

I.1 QUANTUM REPRESENTATION $\{II\}$

Consider, within representation $\{II\}$, the effective Hamiltonian $\mathbb{H}_{II}^{\{1\}}$ of the H-bond bridge embedded in the thermal bath, after excitation of the fast mode. It is given by Eq. (106),

$$\begin{aligned}[\mathbb{H}_{II}^{\{1\}}]=&(\mathrm{a}^{\dagger}\mathrm{a}+\frac{1}{2})\hbar\Omega\\&+\alpha^{\circ}[\mathrm{a}^{\dagger}+\mathrm{a}]\hbar\Omega+\sum_{j}\left(\mathrm{b}_j^{\dagger}\mathrm{b}_j+\frac{1}{2}\right)\hbar\omega_i+\sum_{j}[\mathrm{a}^{\dagger}\mathrm{b}_j\hbar\widetilde{\kappa}_j+\mathrm{a}\mathrm{b}_j^{\dagger}\hbar\widetilde{\kappa}_j^{*}]+\hbar\omega^{\circ}-\alpha^{\circ 2}\hbar\Omega\end{aligned} \tag{I.1}$$

Also consider, within the same representation, the full density operator $\mathbb{R}_{II}(t)$ of the H-bond bridge and of the thermal bath, that is,

$$[\mathbb{R}_{II}(t)] = \rho_{\theta}[\rho_{II}^{\{1\}}(t)]$$

Here, $\rho_{II}^{\{1\}}(t)$ is the density operator of the driven harmonic oscillator that has to be found. Besides, ρ_{θ} is the density operator of the thermal bath that is viewed as a product of Boltzmann operators characterizing the different oscillators describing this bath and conserving their structure along time. Neglecting the zero-point energies, this density operator is

$$\rho_{\theta} = \prod_{j} \varepsilon_j(e^{-\tilde{\lambda}_j \mathrm{b}_j^{\dagger}\mathrm{b}_j}) \tag{I.2}$$

with

$$\varepsilon_j = 1 - e^{-\tilde{\lambda}_j} \qquad \text{and} \qquad \widetilde{\lambda}_j = \frac{\hbar\omega_j}{k_{\mathrm{B}}T}$$

Now, let us look at the Liouville equation for the total density operator $\mathbb{R}_{II}^{\{1\}}(t)$:

$$i\hbar \frac{\partial[\mathbb{R}_{II}^{\{1\}}(t)]}{\partial t} = [[\mathbb{H}_{II}^{\{1\}}],[\mathbb{R}_{II}^{\{1\}}(t)]] \tag{I.3}$$

subject to the boundary conditions for initial time:

$$[\mathbb{R}_{II}^{\{1\}}(0)] = [\rho_{II}^{\{1\}}(0)] \prod_j \varepsilon_j(e^{-\tilde{\lambda}_j b_j^\dagger b_j})$$

Here, $\rho_{II}^{\{1\}}(0)$ is the density operator of the driven oscillator at initial time (see Table VI), which is assumed to be at this step that of equilibrium, that is, a Boltzmann operator. Neglecting the zero-point energy, it is

$$[\rho_{II}^{\{1\}}(0)] = \varepsilon\ (e^{-\tilde{\lambda} a^\dagger a}) \qquad \text{with} \qquad \varepsilon = 1 - e^{-\tilde{\lambda}} \qquad \text{and} \qquad \widetilde{\lambda} = \frac{\hbar\Omega}{k_B T} \tag{I.4}$$

Of course, the reduced-density operator of the driven damped quantum harmonic oscillator at time t is the partial trace over the thermal bath of the full density operator:

$$[\rho_{II}^{\{1\}}(t)] = tr_\theta[\mathbb{R}_{II}^{\{1\}}(t)]$$

Louisell and Walker [22] showed that in the underdamped situation, the reduced-density operator of the driven damped quantum harmonic oscillator is given by

$$[\rho_{II}^{\{1\}}(t)] = \varepsilon[e^{-\tilde{\lambda}[a^\dagger - \Phi_{II}(t)^*][a - \Phi_{II}(t)]}] \tag{I.5}$$

with, respectively,

$$\Phi_{II}(t) = \widetilde{\beta}[e^{-i\Omega t} e^{-\gamma t/2} - 1]$$
$$\gamma = 2\pi \sum |\widetilde{\kappa}_r|^2 \delta(\omega - \omega_r)$$
$$\widetilde{\beta} = \alpha^\circ \frac{[4\Omega^4 + \gamma^2\Omega^2]^{\frac{1}{2}}}{2\left(\Omega^2 + \frac{\gamma^2}{4}\right)}$$

In the following, to simplify, it will be assumed

$$\widetilde{\beta} \simeq \alpha^\circ$$

so that

$$\Phi_{II}(t) = \alpha^{\circ}[e^{-i\Omega t}e^{-\gamma t/2} - 1] \tag{I.6}$$

Note that when the damping is missing, the density operator reduces to

$$[\rho^{\circ}{}^{\{1\}}_{II}(t)] = \varepsilon[e^{-\tilde{\lambda}[a^{\dagger} - \Phi^{\circ}{}_{II}(t)^{*}][a - \Phi^{\circ}{}_{II}(t)]}] \qquad \text{with} \qquad \Phi^{\circ}{}_{II}(t) \equiv \alpha^{\circ}[e^{-i\Omega t} - 1] \tag{I.7}$$

The damped density operator (I.5) at time t is necessarily related to this same operator at initial time through the canonical transformation:

$$[\rho^{\{1\}}_{II}(t)] = [A^{\{1\}}_{II}(t)][\rho^{\{1\}}_{II}(0)][A^{\{1\}}_{II}(t)]^{-1} \tag{I.8}$$

By using Eqs. (I.4) and (I.5), this equation becomes

$$(e^{-\tilde{\lambda}[a^{\dagger} - \Phi_{II}(t)^{*}][a - \Phi_{II}(t)]}) = [A^{\{1\}}_{II}(t)](e^{-\tilde{\lambda}a^{\dagger}a})[A^{\{1\}}_{II}(t)]^{-1} \tag{I.9}$$

In this equation, the unknown is the time-dependent translation operator $\{A_{II}(t)\}$. Then, the following theorem may be used

$$f(a^{\dagger} - \alpha^{*}, a - \alpha) = [e^{\alpha a^{\dagger} - \alpha^{*}a}]\{f(a^{\dagger}, a)\}[e^{-\alpha a^{\dagger} + \alpha^{*}a}] \tag{I.10}$$

Then, taking $\Phi_{II}(t)^{*}$ in place of α^{*}, and $\Phi_{II}(t)$ in place of α, this theorem leads us to write the left-right-hand side of Eq. (I.9) in the following way:

$$(e^{-\tilde{\lambda}[a^{\dagger} - \Phi_{II}(t)^{*}][a - \Phi_{II}(t)]}) = [e^{\Phi_{II}(t)a^{\dagger} - \Phi_{II}(t)^{*}a}](e^{-\tilde{\lambda}a^{\dagger}a})[e^{-\Phi_{II}(t)a^{\dagger} + \Phi_{II}(t)^{*}a}] \tag{I.11}$$

As a consequence, the unitary operator involved on the right-hand side of Eq. (I.9) is found to be given by

$$[A^{\{1\}}_{II}(t)] = [e^{\Phi_{II}(t)a^{\dagger} - \Phi_{II}(t)^{*}a}] \tag{I.12}$$

I.2 QUANTUM REPRESENTATION $\{III\}$

On the other hand, Louisell and Walker [22] studied the dynamics of a damped coherent-state density operator. At the initial time they considered the full density operator $\mathbb{R}^{\{1\}}_{III}(0)$, as formed by the product of the density operator

$[\rho_{III}^{\{1\}}(0)]$ given in Section III.B.2, times that of the thermal bath supposed at thermal equilibrium:

$$[\mathbb{R}_{III}^{\{1\}}(0)] = \varepsilon(e^{-\tilde{\lambda}[a^\dagger - \alpha^\circ][a - \alpha^\circ]}) \prod_k \varepsilon_k(e^{-\tilde{\lambda}_k b_k^\dagger b_k}) \tag{I.13}$$

The authors solved the Liouville equation:

$$i\hbar\left(\frac{\partial[\mathbb{R}_{III}^{\{1\}}(t)]}{\partial t}\right) = [[\mathbb{H}_{III}^{\{1\}}], [\mathbb{R}_{III}^{\{1\}}(t)]]$$

Here, $\mathbb{H}_{III}^{\{1\}}$ is the Hamiltonian of the quantum harmonic oscillator coupled to the thermal bath, which is that given by Eq. (I.1).

$$[\mathbb{H}_{III}^{\{1\}}] = \left(a^\dagger a + \frac{1}{2}\right)\hbar\Omega + \sum\left(b_i^\dagger b_i + \frac{1}{2}\right)\hbar\omega_i + \sum[\, a^\dagger b_i \hbar\widetilde{\kappa}_i + a\, b_i^\dagger\, \hbar\widetilde{\kappa}_i] \\ - \alpha^\circ(a^\dagger + a)\hbar\Omega$$

Next, the reduced density operator at time t may be obtained by performing the partial trace tr_θ over the thermal bath:

$$[\rho_{III}^{\{1\}}(t)] = \mathrm{tr}_\theta[\mathbb{R}_{III}^{\{1\}}(t)]$$

Louisell and Walker showed that this density operator starting at initial time from (I.13) evolves during time, because of the presence of the surrounding, into the following one:

$$[\rho_{III}^{\{1\}}(t)] = \varepsilon(e^{-\tilde{\lambda}[a^\dagger - \Phi_{III}(t)^*][a - \Phi_{III}(t)]}) \tag{I.14}$$

with

$$\Phi_{III}(t) \equiv \alpha^\circ e^{-i\Omega t} e^{-\gamma t/2}$$

Of course, when the damping is missing, that is, when $\gamma \to 0$, the density operator at time t reduces to

$$[\rho^{\circ}{}_{III}^{\{1\}}(t)] = \varepsilon(e^{-\tilde{\lambda}[a^\dagger - \Phi^\circ{}_{III}(t)^*][a - \Phi^\circ{}_{III}(t)]}) \qquad \text{with} \qquad \Phi^\circ{}_{III}(t) \equiv \alpha^\circ e^{-i\Omega t} \tag{I.15}$$

Now, the damped time-dependent density operator (I.14) may be viewed as resulting from the following canonical transformation on the Boltzmann density operator, involving the damped translation operator appearing in Eq. (131), that is,

$$[\rho_{III}^{\{1\}}(t)] = \varepsilon[\mathrm{A}_{III}^{\{1\}}(t)](e^{-\tilde{\lambda}\mathrm{a}^{\dagger}\mathrm{a}})[\mathrm{A}_{III}^{\{1\}}(t)]^{-1}$$

In view of Eq. (I.14), this canonical transformation is

$$e^{-\tilde{\lambda}[\mathrm{a}^{\dagger}-\Phi_{III}(t)^{*}][\mathrm{a}-\Phi_{III}(t)]} = [\mathrm{A}_{III}^{\{1\}}(t)](e^{-\tilde{\lambda}\mathrm{a}^{\dagger}\mathrm{a}})[\mathrm{A}_{III}^{\{1\}}(t)]^{-1} \tag{I.16}$$

The left-hand side of this equation is of the form:

$$e^{-\tilde{\lambda}[\mathrm{a}^{\dagger}-\Phi_{III}(t)^{*}][\mathrm{a}-\Phi_{III}(t)]} = \mathrm{f}(\mathrm{a}^{\dagger} - \Phi_{III}(t)^{*}, \mathrm{a} - \Phi_{III}(t)) \tag{I.17}$$

Moreover, because of theorem (I.10), the right-hand side of this last equation is

$$\mathrm{f}(\mathrm{a}^{\dagger} - \Phi_{III}(t)^{*}, \mathrm{a} - \Phi_{III}(t)) = [e^{\Phi_{III}(t)\mathrm{a}^{\dagger}-\Phi_{III}(t)^{*}\mathrm{a}}]\{\mathrm{f}(\mathrm{a}^{\dagger}, \mathrm{a})\}[e^{-\Phi_{III}(t)\mathrm{a}^{\dagger}+\Phi_{III}(t)^{*}\mathrm{a}}] \tag{I.18}$$

Therefore, in view of Eqs. (I.17) and (I.18), Eq. (I.16) becomes

$$e^{-\tilde{\lambda}[\mathrm{a}^{\dagger}-\Phi_{III}(t)^{*}][\mathrm{a}-\Phi_{III}(t)]} = [e^{\Phi_{III}(t)\mathrm{a}^{\dagger}-\Phi_{III}(t)^{*}\mathrm{a}}](e^{-\tilde{\lambda}\mathrm{a}^{\dagger}\mathrm{a}})[e^{-\Phi_{III}(t)\mathrm{a}^{\dagger}+\Phi_{III}(t)^{*}\mathrm{a}}]$$

Thus, it appears that the damped translation operator involved in Eq. (I.16), is

$$[\mathrm{A}_{III}^{\{1\}}(t)] = [e^{\Phi_{III}(t)\mathrm{a}^{\dagger}-\Phi_{III}(t)^{*}\mathrm{a}}] \tag{I.19}$$

APPENDIX J: A FORMAL EXPRESSION FOR THE ACF (109)

In the main text, we see that the ACF $[G_{II}(t)]^{\circ}$ of a bare H-bond embedded in a thermal bath, is given by Eq. (109):

$$[G_{II}(t)]^{\circ} = (\mu_0^2)e^{i\omega^{\circ}t}(e^{-\alpha^{\circ 2}\Omega t})\mathrm{tr}_{\mathrm{Slow}}[\widetilde{\rho}_{II}^{\{0\}}\mathrm{tr}_{\theta}[\mathbb{U}_{II}^{\{1\}}(t)]^{-1}\ [\mathbb{U}_{II}^{\{0\}}(t)]] \qquad (\mathrm{J}.1)$$

Here, $\widetilde{\rho}_{II}^{\{0\}}$ is the Boltzmann density operator of the H-bond bridge viewed as a quantum harmonic oscillator, $\widetilde{\rho}_{\theta}$ is the Boltzmann density operator of the thermal bath, and $\mathbb{U}_{II}^{\{k\}}(t)$ are effective time-evolution operators governing the dynamics of the H-bond bridge depending on the excited-state degree $\{k\}$ of the fast mode. They are given by Eq. (110), that is,

$$\mathbb{U}_{II}^{\{k\}}(t) = e^{-i\mathbb{H}_{II}^{\{k\}}t/\hbar} \qquad (\mathrm{J}.2)$$

In the last equations (J.2) appear the effective Hamiltonians (105), (106), that is,

$$[\mathbb{H}_{II}^{\{0\}}] = [a^{\dagger}a + \tfrac{1}{2}]\hbar\Omega + \sum_i [\widetilde{b}_i^{\dagger}\widetilde{b}_i + \tfrac{1}{2}]\hbar\omega_i + \sum_i [a^{\dagger}\widetilde{b}_i\widetilde{\kappa}_i\hbar + \mathrm{hc}] \qquad (\mathrm{J}.3)$$

$$[\mathbb{H}_{II}^{\{1\}}] = \hbar\omega^{\circ} + [a^{\dagger}a + \tfrac{1}{2}]\ \hbar\Omega$$
$$+ k\alpha^{\circ}(a^{\dagger} + a)\hbar\Omega - \alpha^{\circ 2}\hbar\Omega + \sum_i [\widetilde{b}_i^{\dagger}\widetilde{b}_i + \tfrac{1}{2}]\hbar\omega_i \sum_i [a^{\dagger}\widetilde{b}_i\widetilde{\kappa}_i\hbar + \mathrm{hc}] \qquad (\mathrm{J}.4)$$

$$[a, a^{\dagger}] = 1 \quad [\widetilde{b}_i, \widetilde{b}_j^{\dagger}] = \delta_{i,j}$$

Where hc stands for "Hermitean conjugate". The expression of the density operator $\widetilde{\rho}_{\theta}$ of the bath is given by Eq. (I.2) and that of the density operator $\widetilde{\rho}_{II}^{\{0\}}$ of the H-bond bridge is

$$[\widetilde{\rho}_{II}^{\{0\}}] = \varepsilon(e^{-\tilde{\lambda}[a^{\dagger}a+\frac{1}{2}]})$$

The aim of this appendix is to find a disentangled expression for the ACF (J.1), that is, the ACF (109), in which the diagonal parts of the effective Hamiltonians (J.3) and (J.4) are involved in the effective time-evolution operators (J.2) are removed.

Because of Eqs. (J.3) and (J.4), the time-evolution operators (J.2) involved in Eq. (J.1) are, respectively, given by

$$[\mathbb{U}_{II}^{\{0\}}(t)] = \left(e^{-i[a^\dagger a+1/2]\Omega t - \sum i[\tilde{b}_i^\dagger \tilde{b}_i+1/2]\omega_i t - \sum i[a^\dagger \tilde{b}_i \tilde{\kappa}_i t + hc]}\right) \tag{J.5}$$

$$[\mathbb{U}_{II}^{\{1\}}(t)] = \left(e^{-i[a^\dagger a+1/2]\Omega t - i\alpha^\circ(a^\dagger + a)\Omega t - \sum i[b_i^\dagger b_i+1/2]\omega_i t - \sum i[a^\dagger b_i \tilde{\kappa}_i t + hc]}\right) \tag{J.6}$$

Now, let us disentangle the diagonal part in Eq. (J.6) as follows. Perform an interaction picture within the following partition of the Hamiltonian (J.4):

$$\mathbb{H}_{II}^{\{1\}} = \mathbb{H}_{II}^{\{0\}} + \mathbb{V}$$
$$\mathbb{V} = \alpha^\circ(a^\dagger + a)\hbar\Omega$$

Then, within the IP formalism, the time-evolution operator (J.6) becomes

$$[\mathbb{U}_{II}^{\{1\}}(t)] = [\mathbb{U}_{II}^{\{0\}}(t)][\mathbb{U}_{II}^{\{1\}}(t)^{\text{IP}}] \tag{J.7}$$

Here, the IP time-evolution operator is

$$[\mathbb{U}_{II}^{\{1\}}(t)^{\text{IP}}] = \widehat{\text{P}}[e^{-\frac{i}{\hbar}\int_0^t \{\mathbb{V}(t')\}^{\text{IP}} dt'}] \tag{J.8}$$

where $\widehat{\text{P}}$ is the Dyson time-ordering operator, whereas $\mathbb{V}(t)^{\text{IP}}$ is the IP potential given by

$$\mathbb{V}(t)^{\text{IP}} = [\mathbb{U}_{II}^{\{0\}}(t)]^{-1}\mathbb{V}(0)[\mathbb{U}_{II}^{\{0\}}(t)]$$

or

$$[\mathbb{V}(t)^{\text{IP}}] = \left(e^{i[a^\dagger a+1/2]\Omega t + \sum i[\tilde{b}_i^\dagger \tilde{b}_i+1/2]\omega_i t + \sum i[a^\dagger \tilde{b}_i \tilde{\kappa}_i t + hc]}\right)$$
$$\mathbb{V}(0)\left(e^{-i[a^\dagger a+1/2]\Omega t - \sum i[\tilde{b}_i^\dagger \tilde{b}_i+1/2]\omega_i t - i[a^\dagger \tilde{b}_i \tilde{\kappa}_i t + hc]}\right)$$

Now, according to Eq. (J.7), and since the time-evolution operators are unitary, the product of the two time-evolution operators involved in the ACF (J.1) is

$$[\mathbb{U}_{II}^{\{1\}}(t)]^{-1}[\mathbb{U}_{II}^{\{0\}}(t)] = [\mathbb{U}_{II}^{\{1\}}(t)^{\mathrm{IP}}]^{-1}[\mathbb{U}_{II}^{\{0\}}(t)]^{-1}[\mathbb{U}_{II}^{\{0\}}(t)] = [\mathbb{U}_{II}^{\{1\}}(t)^{\mathrm{IP}}]^{-1}$$

As a consequence, in view of Eq. (J.8), the ACF (J.1) becomes

$$[\mathrm{G}_{II}(t)]^{\circ} = (\mu_0^2)e^{i\omega^{\circ}t}(e^{-i\alpha_{\circ}^2\Omega t})\mathrm{tr}_{\mathrm{Slow}}[\tilde{\rho}_{II}^{\{0\}}\mathrm{tr}_{\theta}[\mathbb{U}_{\mathrm{II}}^{\{1\}}(t)^{\mathrm{IP}}]^{-1}] \tag{J.9}$$

Now, consider the following reduced form of the time evolution operator defined by:

$$[\mathbf{U}_{II}^{\{1\}}(t)^{IP}] = \widehat{P}\left[\mathbf{e}^{-\frac{i}{\hbar}\int_0^t\{\mathbf{V}(t')^{IP}\}dt'}\right]$$

with:

$$\mathbf{V}(t)^{IP} = \mathbf{tr}_{\theta}[\mathbb{V}(t)^{IP}]$$

Now, the precedent expression of the ACF may be considered to be given by:

$$[\mathbf{G}_{II}(t)]^{\circ} = (\mu_0^2)\mathbf{e}^{i\omega^{\circ}t}(\mathbf{e}^{-\alpha^{\circ 2}\Omega t})\mathbf{tr}_{Slow}[\widetilde{\boldsymbol{\rho}}_{II}^{\{0\}}[\mathbf{U}_{II}^{\{1\}}(t)^{IP}]^{-1}] \tag{J.10}$$

Such an approach may be confirmed [127]. Therefore the ACF becomes:

$$[\mathbf{G}_{II}(t)]^{\circ} = (\mu_0^2)\mathbf{e}^{i\omega^{\circ}t}(\mathbf{e}^{-\alpha^{\circ 2}\Omega t})\widetilde{\varepsilon}\,\mathbf{tr}_{Slow}\left[(\mathbf{e}^{-\tilde{\lambda}[\mathbf{a}^{\dagger}\mathbf{a}+1/2]})\left[\widehat{P}\left[\mathbf{e}^{\frac{i}{\hbar}\int_0^t\{\mathbb{V}(t')^{IP}\}dt'}\right]\right]\right] \tag{J.11}$$

To get the evolution operator involved on the right handside of this last equation, it is necessary to solve the Schrödinger equation within the Wigner-Weiskopf approximation:

$$i\hbar\frac{\partial}{\partial t}[\mathbf{U}_{II}^{\{1\}}(t)^{IP}] = [\mathbf{V}(t)^{IP}][\mathbf{U}_{II}^{\{1\}}(t)^{IP}] \tag{J.12}$$

That is performed in Ref. [127].

APPENDIX K: ANOTHER APPROACH OF THE SPECTRAL DENSITY (80) USING EQS. (1–3) IN PLACE OF EQS. (4–6)

The spectral density (80), dealing with the bare H-bond without indirect damping was found with the aid of the set of simplified basic Eqs. (4–6). The aim of this appendix is to show that in this situation without direct and indirect dampings, the result (80) remains the same if the set of basic Eqs. (1–3) is used.

For this purpose, start from the expression (2) of the ACF, that is, the following trace, tr over the base spanned by the operators describing the system:

$$[G^{\circ II}(t)]^{\circ} = \frac{1}{\beta}\mathrm{tr}\left[\rho_B \int_0^{\beta} \{\widehat{\mu}(0)\}\{\widehat{\mu}(t+i\lambda\hbar)\}d\lambda\right]$$

Here, β has its usual statistical meaning, whereas λ is a scalar, and ρ_B is the Boltzmann density operator describing the system. It was been written $[G^{\circ II}(t)]^{\circ}$ in place of $[G^{\circ II}(t)]^{\circ}$ in order to make clear the differences in the fashion so we could calculate the ACF.

In terms of the non-Hermitean IR absorption transition moments, this ACF becomes

$$[G^{\circ II}(t)]^{\circ} = \frac{1}{\beta}\mathrm{tr}\left[\rho_B \int_0^{\beta} \{\mu(0)^{\dagger}\}\{\mu(t+i\lambda\hbar)\}d\lambda\right]$$

Within the adiabatic approximation, and within the quantum representation $\{II\}$, the Boltzmann density operator playing the physical role is

$$\rho_B = \tilde{\varepsilon}(e^{-\beta H_{II}^{\{0\}}}) = \rho_{II}^{\{0\}}$$

In the same spirit, because of Eq. (3), the IR non-Hermitean absorption moment operators at initial time and at a later time t, are given, respectively, by

$$\{\mu\ (0)^{\dagger}\} = \mu_0|\{0\}\rangle\ \langle\ \{1\}|\{\mu(t+i\lambda\hbar)\}$$
$$= \mu_0[e^{iH_{II}^{\{1\}}(t+i\lambda\hbar)/\hbar}]|\{1\}\ \rangle\ \langle\ \{0\}|\ [e^{-iH_{II}^{\{0\}}(t+i\lambda\hbar)/\hbar}]$$

In these equations, the parameters $H_{II}^{\{1\}}$ are the effective Hamiltonians of the slow mode appearing in Table VI. Then, in this representation, the ACF without indirect damping becomes

$$[G^{\circ II(t)}]^{\circ} = (\mu_0^2)\tilde{\varepsilon}\mathrm{tr}_{\mathrm{Slow}}\,[(e^{-\beta H_{II}^{\{0\}}})$$
$$\int_0^{\beta} d\lambda|\{0\}\,\rangle\,\langle\,\{1\}|\,[e^{iH_{II}^{\{1\}}(t+i\lambda\hbar)/\hbar}]|\{1\}\,\rangle\,\langle\,\{0\}|\,[e^{-iH_{II}^{\{0\}}(t+i\lambda\hbar)/\hbar}]] \tag{K.1}$$

Now, pass to quantum representation $\{III\}$. Within it, the effective Hamiltonians corresponding to the ground-state fast mode is unchanged, whereas that corresponding to the first excited state is modified. Owing to Table VI, the following equations hold:

$$[H_{II}^{\{0\}}] = [H_{III}^{\{0\}}] \quad \text{and} \quad [H_{II}^{\{1\}}] = [A^{\{1\}}(\alpha^{\circ})]^{-1}[H_{III}^{\{1\}}][A^{\{1\}}(\alpha^{\circ})]$$

Here, $A^{\{1\}}(\alpha^{\circ})$ is the translation operator given by Eq. (C.2) with $k = 1$.

Besides, it is possible to use the fact that the kets and the bras of the fast mode do not act on the operators belonging to the space of the slow mode. Then, because of the orthonormality properties leading to simplifications, the ACF in representation $\{III\}$ takes the following form:

$$[G^{\circ III}(t)]^{\circ} = (\mu_0^2)$$
$$\tilde{\varepsilon}\mathrm{tr}_{\mathrm{Slow}}(e^{-\beta H_{III}^{\{0\}}})\int_0^{\beta} d\lambda[A^{\{1\}}(\alpha^{\circ})]^{-1}\,[e^{iH_{III}^{\{1\}}(t+i\lambda\hbar)/\hbar}][A^{\{1\}}(\alpha^{\circ})]\,[e^{-iH_{III}^{\{0\}}(t+i\lambda\hbar)/\hbar}]$$

Again, introduce the closeness relation (76) on the eigenstates of the slow mode harmonic oscillator and neglect the zero-point energy of the Boltzmann operator. This gives

$$[G^{\circ III}(t)]^{\circ} = (\mu_0^2)\varepsilon\ \mathrm{tr}_{\mathrm{Slow}}\ \{(e^{-\beta a^{\dagger}a\hbar\Omega})\sum_n$$
$$\int_o^{\beta} d\lambda[A^{\{1\}}(\alpha^{\circ})]^{-1}|(n)\rangle\langle(n)|\,[e^{iH_{III}^{\{1\}}(t+i\lambda\hbar)/\hbar}][A^{\{1\}}(\alpha^{\circ})][\,e^{-iH_{III}^{\{0\}}(t+i\lambda\hbar)/\hbar}]\}$$

Moreover, performing the trace over the eigenstates of the harmonic slow mode and by using the definition (C.14) of the Franck–Condon factors, the ACF takes the form:

$$[G^{\circ III}(t)]^{\circ} = (\mu_0^2)\varepsilon\sum_m\sum_n$$
$$(e^{-m\beta\hbar\Omega})|A_{mn}(\alpha^{\circ})|^2\int_o^{\beta}(e^{i(n-m)\Omega(t+i\lambda)})(e^{-i2\alpha^{\circ 2}\Omega(t+i\lambda)})(e^{i\omega^{\circ}(t+i\lambda)})d\lambda$$

Again, by integration over λ, we obtain

$$[G^{\circ}_{III}(t)]^{\circ} = (\mu_0^2)\, e^{i\omega^{\circ}t}\, (e^{-i2\alpha^{\circ 2}\Omega t})\varepsilon \sum_m \sum_n |A_{mn}(\alpha^{\circ})|^2 (e^{i(n-m)\Omega t})$$
$$\frac{1}{[(m-n)+2\alpha^{\circ 2}]\hbar\Omega - \hbar\omega^{\circ}} [(e^{-\beta\hbar\omega^{\circ}+\beta(2\alpha^{\circ 2}-n)\hbar\Omega}) - (e^{-m\beta\hbar\Omega})]$$

Now, observe that, even at room temperature, the following inequality holds:

$$\hbar\omega^{\circ} \gg k_B T \qquad \text{or} \qquad \beta\hbar\omega^{\circ} \gg 1$$

Besides, for weak H-bonds, the following inequality holds $\omega^{\circ} \gg \Omega$. Moreover, only small values of n are playing a physical role in view of the Franck–Condon factors $A_{mn}(\alpha^{\circ})$, which are quickly narrowing to zero, for α° around unity, when the quantum number n is increasing. As a consequence, it is possible to perform the following approximation for weak H-bonds.

$$(e^{-\beta\hbar\omega^{\circ}+\beta(2\alpha^{\circ 2}-n)\hbar\Omega}) \simeq 0$$

Thus, the above ACF may be approached very satisfactorily by the following equation:

$$[G^{\circ}_{III}(t)]^{\circ} = (\mu_0^2)\, e^{i\omega^{\circ}t}\, (e^{-i2\alpha^{\circ 2}\Omega t})$$
$$\varepsilon \sum_m \sum_n (e^{-m\beta\hbar\Omega}) |A_{mn}(\alpha^{\circ})|^2 (e^{i(n-m)\Omega t}) \frac{1}{[(m-n) + 2\alpha^{\circ 2}]\hbar\Omega - \hbar\omega^{\circ}}$$

After Fourier transform and multiplication by ω according to Eq. (1.1), we obtain an SD of the form:

$$[I^{\circ}_{III}(\omega)]^{\circ} = (\mu_0^2)\varepsilon \sum_m \sum_n (e^{-m\beta\hbar\Omega})\, |A_{mn}(\alpha^{\circ})|^2$$
$$\frac{1}{[(n-m) - 2\alpha^{\circ 2}]\hbar\Omega + \hbar\omega^{\circ}}\, \omega \int_{-\infty}^{+\infty} (e^{i(n-m)\Omega t})\, e^{i\omega^{\circ}t}\, (e^{-i2\alpha^{\circ 2}\Omega t}) e^{-i\omega t} dt$$

Then, performing the Fourier transform, leads to

$$[I^{\circ}_{III}(\omega)]^{\circ} = (\mu_0^2)\varepsilon \sum_m \sum_n (e^{-m\beta\hbar\Omega})\, |A_{mn}(\alpha^{\circ})|^2$$
$$\frac{\omega}{[(n-m) - 2\alpha^{\circ 2}]\hbar\Omega + \hbar\omega^{\circ}} \boldsymbol{\delta}[\omega - ([(n-m) - 2\alpha^{\circ 2}]\Omega + \omega^{\circ})]$$

Moreover, the argument of the Dirac δ function allows us to write

$$\omega = [(n - m) - 2\alpha^{\circ 2}]\Omega + \omega^{\circ}$$

Consequently, the SD takes the final form:

$$[\mathrm{I}^{\circ III}(\omega)]^{\circ} = (\mu_0^2)\varepsilon \sum_m \sum_n (e^{-m\hbar\Omega/k_B T})|A_{mn}(\alpha^{\circ})|^2 \delta[\omega - ([(n-m) - 2\alpha^{\circ 2}]\Omega + \omega^{\circ})] \tag{K.2}$$

In Eq. (K.2), one may recognize Eq. (80), that is, the Franck–Condon progression appearing in the model of Maréchal and Witkowski,

$$[\mathrm{I}^{\circ III}(\omega)]^{\circ} = [\mathrm{I}^{\circ}{}_{III}(\omega)]^{\circ}$$

As a consequence, it appears that in the absence of damping and within the linear response theory, the SD corresponding to the IR $\nu_{X\text{–}H}$ stretching mode of a weak H-bond may be obtained in an equivalent way, either by Eqs. (1.1–1.3) or by the simple equations (1.4–1.6). This is the reason for the choice of the simplest method involving the later set of three basic equations, even in situations where direct or indirect damping are occurring.

APPENDIX L: CLASSICAL CALCULATION FOR $\langle S(t)S(0)\rangle$ [60]

The calculation of the ACF (164) requires knowledge of the ACF $\langle S(t)S(0)\rangle$ of the random variable $S(t)$ appearing on the right-hand side. This is the object of Appendix L.

Following Coffey et al. [62], we may use the stationary properties of this ACF as follows:

$$\langle S(t')S(t'')\rangle = \langle S(t'-\tau)\ S(t''-\tau)\rangle = \langle S(t)S(0)\rangle \tag{L.1}$$

From stationary properties (L.1) and those of the correlation function of the stochastic variable $S(t)$, as well as of its time derivative $Q(t)$, it is possible to write

$$\frac{d}{dt}\langle S(t)S(0)\rangle = \langle Q(t)S(0)\rangle \tag{L.2}$$

$$\frac{d^2}{dt^2}\langle S(t)S(0)\rangle = -\langle Q(t)Q(0)\rangle \tag{L.3}$$

The explicit expression of the ACF appearing on the right-hand side of Eq. (L.3), was obtained by Uhlenbeck–Ornstein [66]. Its proof is given in Appendix M. It is found to be given by Eq. (M.17), that is,

$$\langle Q(t)Q(0)\rangle = \langle Q^2\rangle e^{-\gamma t/2}\left[\cos\tilde{\Omega}t + \frac{\gamma}{2\tilde{\Omega}}\sin\tilde{\Omega}t\right] \tag{L.4}$$

$$\tilde{\Omega}^2 \equiv \Omega^2 - \frac{\gamma^2}{4} \tag{L.5}$$

Here, $\langle Q^2\rangle$ is the fluctuation of the Brownian oscillator coordinate at equilibrium, the classical and quantum statistical expressions of which are given in Table L.1:

Now, take the Laplace transform of both sides of Eq. (L.3),

$$\int_0^\infty \frac{d^2}{dt^2}\langle S(t)S(0)\rangle\ e^{-st}dt = -\int_0^\infty \langle Q(t)Q(0)\rangle\ e^{-st}dt \tag{L.6}$$

TABLE L.1
Classical and Quantum Statistical $\langle Q^2 \rangle$

$\langle Q^2 \rangle_{Cl}$	$\langle Q^2 \rangle_{Qu}$
$\langle Q_{Cl}(0)Q_{Cl}(0) \rangle$	$\langle Q_{Qu}(0)Q_{Qu}(0) \rangle$
$\frac{1}{Z_{Cl}} \int_{-\infty}^{+\infty} Q^2 e^{-M\Omega^2 Q^2/2k_B T} dQ$	$\frac{\hbar}{2M\Omega} \varepsilon \mathrm{tr}[[a^\dagger + a]^2 e^{-a^\dagger a \hbar\Omega/k_B T}]$
$\frac{k_B T}{M\Omega^2}$	$\frac{\hbar}{2M\Omega} \coth \frac{\hbar\Omega}{2k_B T}$

Owing to Eq. (L.4), the Laplace transform appearing in the right-hand side of Eq. (L.6) is

$$\int_0^\infty \langle Q(t)Q(0) \rangle \; e^{-st} dt = \langle Q^2 \rangle \int_0^\infty e^{-\gamma t/2} \left[\cos \tilde{\Omega} t + \frac{\gamma}{2\tilde{\Omega}} \sin \tilde{\Omega} t \right] \; e^{-st} dt \tag{L.7}$$

with

$$\tilde{\Omega} = \sqrt{\Omega^2 - \frac{\gamma^2}{4}}$$

The result is

$$\int_0^\infty \langle Q(t)Q(0) \rangle e^{-st} dt = \langle Q^2 \rangle \frac{\left[2\left(s + \frac{1}{2}\gamma\right) + \gamma \right]}{2\left[\left(s + \frac{1}{2}\gamma\right)^2 + \tilde{\Omega}^2 \right]} \tag{L.8}$$

On the other hand, the Laplace transform appearing on the left-hand side of Eq. (L.6) is

$$\begin{aligned} &\int_0^\infty \frac{d^2}{dt^2} \langle S(t)S(0) \rangle \; e^{-st} dt \\ &= s^2 \int_0^\infty \langle S(t)S(0) \rangle \; e^{-st} dt - s\langle S(0)S(0) \rangle - \left[\frac{d}{dt} \langle S(t)S(0) \rangle \right]_{t=0} \end{aligned} \tag{L.9}$$

Of course, because of Eq. (L.2), the last right-hand side term of Eq. (L.9) is

$$\left[\frac{d}{dt} \langle S(t)S(0) \rangle \right]_{t=0} = \langle S(0)Q(0) \rangle$$

Then, as a consequence of Eqs. (L.8) and (L.9), Eq. (L.6) takes the form:

$$\int_0^\infty \langle S(t)S(0)\rangle\, e^{-st}dt - \frac{\langle S(0)S(0)\rangle}{s} - \frac{\langle S(0)Q(0)\rangle}{s^2} = -\frac{\langle Q^2\rangle}{2s^2}\frac{\left[2\left(s+\frac{1}{2}\gamma\right)+\gamma\right]}{\left[\left(s+\frac{1}{2}\gamma\right)^2+\tilde{\Omega}^2\right]} \tag{L.10}$$

Now, in order to get the ACF of the stochastic variable $S(t)$, take the inverse Laplace transform of Eq. (L.10). This gives

$$\langle S(t)S(0)\rangle = \langle S^2\rangle + \langle S(0)Q(0)\rangle t - W(t) \tag{L.11}$$

$$\langle S(0)S(0)\rangle \equiv \langle S^2\rangle$$

In Eq. (L.11), $W(t)$ is the inverse Laplace transform of the right-hand side of Eq. (L.10). Performing this transform leads to

$$W(t) = \langle Q^2\rangle \frac{\gamma t}{\Omega^2} \tag{L.12}$$
$$+ \langle Q^2\rangle \frac{1}{\Omega^4}\left[(\gamma^2-\Omega^2)[e^{-\gamma t/2}\cos\tilde{\Omega}t - 1] + \frac{\gamma}{2\tilde{\Omega}}(\gamma^2-3\Omega^2)e^{-\gamma t/2}\sin\tilde{\Omega}t\right]$$

Besides, it is shown in Appendix M that the middle right-hand side term involved in Eq. (L.11) is given by Eq. (M.21), that is,

$$\langle\, S(0)Q(0)\rangle = \frac{\gamma}{\Omega^2}\langle Q^2\rangle \tag{L.13}$$

This last result shows that S and Q are correlated at initial time $t = 0$.

Consequently, in view of Eqs. (L.12) and (L.13), it appears that the ACF (L.11) takes the form:

$$\langle S(t)S(0)\rangle = \langle S^2\rangle + \frac{\langle Q^2\rangle}{\Omega^4}$$
$$\left[(\gamma^2-\Omega^2)[e^{-\gamma t/2}\cos\tilde{\Omega}t - 1] + \frac{\gamma}{2\tilde{\Omega}}(\gamma^2-3\Omega^2)e^{-\gamma t/2}\sin\tilde{\Omega}t\right] \tag{L.14}$$

APPENDIX M: DEMONSTRATION OF $\langle Q(0)S(0)\rangle$

This appendix is devoted [60] to the calculation of the correlation function, at a given time, of the two stochastic coordinates that are, the position Q, and its indefinite integral S over time, that is, $\langle Q(0)S(0)\rangle$.

M.1 LINEAR TRANSFORMATION ON LANGEVIN EQUATIONS

Let us consider the two usual Langevin equations (161) and (162), dealing with the Brownian oscillator together with the definition of the stochastic coordinate S given by Eq. (160), that is,

$$\frac{d}{dt}\mathrm{Q}(t) = \mathrm{v}(t) \tag{M.1}$$

$$M\frac{d}{dt}\mathrm{v}(t) = -M\Omega^2\mathrm{Q}(t) - \gamma M\mathrm{v}(t) + \mathrm{F}^\circ(t) \tag{M.2}$$

$$\mathrm{S}(t) = \int^t \mathrm{Q}(t')dt' \tag{M.3}$$

The Langevin force obeys the following equations specifying its average, which is zero, and its ACF, which is a delta function:

$$\langle \mathrm{F}^\circ(t)\rangle = 0 \qquad \text{and} \qquad \langle \mathrm{F}^\circ(t)\mathrm{F}^\circ(t')\rangle = 2\gamma k_\mathrm{B}T\delta(t-t')$$

The two first stochastic Langevin equations (M.1) and (M.2) may be written in the matrix form:

$$\frac{d}{dt}\begin{pmatrix}\mathrm{Q}(t)\\ \mathrm{v}(t)\end{pmatrix} = \begin{pmatrix}0 & 1\\ -\Omega^2 & -\gamma\end{pmatrix}\begin{pmatrix}\mathrm{Q}(t)\\ \mathrm{v}(t)\end{pmatrix} + \begin{pmatrix}0\\ \tilde{\mathrm{F}}^\circ(\mathrm{t})\end{pmatrix} \tag{M.4}$$

Here $\tilde{\mathrm{F}}^\circ(t)$ is the Langevin stochastic force, the average and the autocorrelation of which are given by the equations:

$$\tilde{\mathrm{F}}^\circ(t) = \mathrm{F}^\circ(t)/M \quad \text{and} \quad \langle\tilde{\mathrm{F}}^\circ(t)\rangle = 0 \quad \text{and} \quad \langle\tilde{\mathrm{F}}^\circ(t)\tilde{\mathrm{F}}^\circ(t')\rangle = 2\gamma\frac{k_\mathrm{B}T}{M^2}\delta(t-t') \tag{M.5}$$

Now, in order to decouple the time dependence of the stochastic variables governed by the Langevin equations (M.4), introduce new stochastic variables $\mathrm{R}(t)$ and $\mathrm{R}(t)^*$, which are linear combinations of $\mathrm{Q}(t)$ and $\mathrm{v}(t)$, in a way that is not far from that used in quantum mechanics, when passing from Q and P to a and $\mathrm{a}^\dagger$. This may be performed according to

$$\begin{pmatrix} \mathrm{Q}(t) \\ \mathrm{v}(t) \end{pmatrix} = \begin{pmatrix} 1 & 1 \\ \lambda_\gamma^* & \lambda_\gamma \end{pmatrix} \begin{pmatrix} \mathrm{R}(t) \\ \mathrm{R}(t)^* \end{pmatrix} \tag{M.6}$$

with

$$\lambda_\gamma = -\frac{\gamma}{2} + i\tilde{\Omega} \qquad \text{and} \qquad \tilde{\Omega} = \sqrt{\Omega^2 - \frac{\gamma^2}{4}} \tag{M.7}$$

The inverse transformation of Eq. (M.6), is

$$\begin{pmatrix} \mathrm{R}(t) \\ \mathrm{R}(t)^* \end{pmatrix} = \frac{1}{\lambda_\gamma - \lambda_\gamma^*} \begin{pmatrix} \lambda_\gamma & -1 \\ -\lambda_\gamma^* & 1 \end{pmatrix} \begin{pmatrix} \mathrm{Q}(t) \\ \mathrm{v}(t) \end{pmatrix} \tag{M.8}$$

Expressed in terms of these new stochastic coordinates given by Eq. (M.8), the set of Eqs. (M.4) becomes

$$\frac{d}{dt} \begin{pmatrix} \mathrm{R}(t) \\ \mathrm{R}(t)^* \end{pmatrix} = \begin{pmatrix} \lambda_\gamma^* & 0 \\ 0 & \lambda_\gamma \end{pmatrix} \begin{pmatrix} \mathrm{R}(t) \\ \mathrm{R}(t)^* \end{pmatrix} + \begin{pmatrix} \tilde{\mathrm{F}}(t) \\ \tilde{\mathrm{F}}(t)^* \end{pmatrix} \tag{M.9}$$

with

$$\begin{pmatrix} \tilde{\mathrm{F}}(t) \\ \tilde{\mathrm{F}}(t)^* \end{pmatrix} = \frac{1}{\lambda_\gamma - \lambda_\gamma^*} \begin{pmatrix} \lambda_\gamma & -1 \\ -\lambda_\gamma^* & 1 \end{pmatrix} \begin{pmatrix} 0 \\ \tilde{\mathrm{F}}(t)^* \end{pmatrix}$$

and

$$\tilde{\mathrm{F}}(t) = \frac{\tilde{\mathrm{F}}^\circ(t)}{2i\tilde{\Omega}} \qquad \text{with} \qquad \langle \tilde{\mathrm{F}}(t) \rangle = 0 \tag{M.10}$$

Note that the set of Eqs. (M.9) is equivalent to that governing the Bosons of a damped quantum harmonic oscillator.

Now, let us look at the complex stochastic variable $\mathrm{R}(0)$ at initial time. According to Eqs. (M.7) and (M.8), it is

$$\mathrm{R}(0) = \frac{1}{2i\tilde{\Omega}} [\lambda_\gamma \mathrm{Q}(0) - \mathrm{v}(0)] \tag{M.11}$$

Then, owing to Eq. (M.9), the stochastic variable appears to be given at time t by

$$R(t) = R(0)e^{\lambda_\gamma^* t} + \int_0^t \tilde{F}(t')e^{\lambda_\gamma^*(t-t')}dt' \tag{M.12}$$

M.2 CALCULATION OF $\langle Q(t)Q(0)\rangle$

Because of Eq. (M.6), the stochastic coordinate $Q(t)$, is

$$Q(t) = R(t) + R(t)^* \tag{M.13}$$

Then, from Eq. (M.12) and its complex conjugate coordinate, Eq. (M.13) becomes

$$\begin{aligned} Q(t) = {} & \frac{Q(0)}{2i\tilde{\Omega}}[\lambda_\gamma e^{\lambda_\gamma^* t} - \lambda_\gamma^* e^{\lambda_\gamma t}] + \frac{v(0)}{2i\tilde{\Omega}}[e^{\lambda_\gamma t} - e^{\lambda_\gamma^* t}] \\ & + e^{\lambda_\gamma^* t}\int \tilde{F}(t')e^{-\lambda_\gamma^* t'}dt' + e^{\lambda_\gamma t}\int \tilde{F}(t')^* e^{-\lambda_\gamma t'}dt' \end{aligned} \tag{M.14}$$

Again, multiplying this result by $Q(0)$ and averaging the stochastic variables, the ACF of the Q coordinate is obtained

$$\begin{aligned} \langle Q(t)Q(0)\rangle = {} & \frac{\langle Q(0)Q(0)\rangle}{2i\tilde{\Omega}}[\lambda_\gamma e^{\lambda_\gamma^* t} - \lambda_\gamma^* e^{\lambda_\gamma t}] + \frac{\langle v(0)Q(0)\rangle}{2i\tilde{\Omega}}[e^{\lambda_\gamma t} - e^{\lambda_\gamma^* t}] \\ & + \langle Q(0)\rangle\, e^{\lambda_\gamma^* t}\int \langle\tilde{F}(t')\rangle e^{-\lambda_\gamma^* t'}dt' + \langle Q(0)\rangle e^{\lambda_\gamma t}\int \langle\tilde{F}(t')^*\rangle e^{-\lambda_\gamma t'}dt' \end{aligned} \tag{M.15}$$

Next, according to statistical mechanics of equilibrium, the correlation function of Q and v is zero since the distribution of the velocities is assumed Maxwellian, that is,

$$\langle v(0)Q(0)\rangle = 0 \tag{M.16}$$

As a consequence, by using Eq. (M.10), that is, the fact that the average of the Langevin force is zero, the ACF (M.15) simplifies into

$$\langle Q(t)Q(0)\rangle = \frac{\langle Q^2\rangle}{2i\tilde{\Omega}}[\lambda_\gamma e^{\lambda_\gamma^* t} - \lambda_\gamma^* e^{\lambda_\gamma t}] \qquad \text{with} \qquad \langle Q^2\rangle \equiv \langle Q(0)Q(0)\rangle$$

At last, owing to Eq. (M.7), this ACF takes the final form:

$$\langle\, \mathrm{Q}(t)\mathrm{Q}(0)\rangle = \langle \mathrm{Q}^2\rangle e^{-\gamma t/2}\left[\cos(\tilde{\Omega}t) + \frac{\gamma}{2\tilde{\Omega}}\sin(\tilde{\Omega}t)\right] \tag{M.17}$$

This is the result of Ornstein and Uhlenbeck, which is used in Eq. (L.4).

M.3 CALCULATION OF $\langle\, \mathbf{Q(0)S(0)}\rangle$

On the other hand, let us look at the stochastic variable $\mathrm{S}(t)$ given by Eq. (M.3). It must be emphasized that this stochastic variable, the time derivative of which is the stochastic coordinate $\mathrm{Q}(t)$, is given by an indefinite integral which is

$$\mathrm{S}(t) \equiv \int \mathrm{Q}(t)dt \tag{M.18}$$

By using Eq. (M.14), the indefinite integral (M.18) takes the form:

$$\begin{aligned}\mathrm{S}(t) = {} & \frac{\mathrm{Q}(0)}{2i\tilde{\Omega}}\int [\lambda_\gamma e^{\lambda_\gamma^* t} - \lambda_\gamma^* e^{\lambda_\gamma t}]dt + \frac{\mathrm{v}(0)}{2i\tilde{\Omega}}\int [e^{\lambda_\gamma t} - e^{\lambda_\gamma^* t}\,]dt \\ & + \int\left[e^{\lambda_\gamma^* t}\int \tilde{\mathrm{F}}(t')e^{-\lambda_\gamma^* t'}dt' + \mathrm{hc}\right]dt\end{aligned} \tag{M.19}$$

The double integration involved in the last two right-hand side terms may be performed by parts. For this purpose, let us write, respectively,

$$\int u dv = uv - \int v du$$

$$u = \int \tilde{\mathrm{F}}(t')e^{-\lambda_\gamma^* t'}dt' \quad dv = e^{\lambda_\gamma^* t}dt$$

$$du = \tilde{\mathrm{F}}(t')e^{-\lambda_\gamma^* t'}dt' \quad dv = \int e^{\lambda_\gamma^* t}dt \;= \frac{e^{\lambda_\gamma^* t}}{\lambda_\gamma^*}$$

As a consequence, Eq. (M.19) gives

$$\begin{aligned}S(t) = {} & \frac{\mathrm{Q}(0)}{2i\tilde{\Omega}}\left[\frac{\lambda_\gamma}{\lambda_\gamma^*}e^{\lambda_\gamma^* t} - \frac{\lambda_\gamma^*}{\lambda_\gamma}e^{\lambda_\gamma t}\right] + \frac{\mathrm{v}(0)}{2i\tilde{\Omega}}\left[\frac{1}{\lambda_\gamma}e^{\lambda_\gamma t} - \frac{1}{\lambda_\gamma^*}e^{\lambda_\gamma^* t}\right] \\ & + \left[\frac{1}{\lambda_\gamma^*}e^{\lambda_\gamma^* t}\int \tilde{\mathrm{F}}(t)e^{-\lambda_\gamma^* t}dt + \mathrm{hc}\right] - \left[\frac{1}{\lambda_\gamma^*}\int \tilde{\mathrm{F}}(t)dt - \mathrm{hc}\right]\end{aligned} \tag{M.20}$$

Next, multiply this result by Q(0) and perform the average over the stochastic variables. Then, the following result is obtained

$$\langle Q(0)S(t)\rangle = \frac{\langle Q(0)Q(0)\rangle}{2i\tilde{\Omega}}\left[\frac{\lambda_\gamma}{\lambda_\gamma^*}e^{\lambda_\gamma^* t} - \frac{\lambda_\gamma^*}{\lambda_\gamma}e^{\lambda_\gamma t}\right] + \frac{\langle Q(0)v(0)\rangle}{2i\tilde{\Omega}}\left[\frac{1}{\lambda_\gamma}e^{\lambda_\gamma t} - \frac{1}{\lambda_\gamma^*}e^{\lambda_\gamma^* t}\right]$$
$$+\left[\frac{1}{\lambda_\gamma^*}e^{\lambda_\gamma^* t}\int \langle\tilde{F}(t)\rangle e^{-\lambda_\gamma^* t}dt + \mathrm{hc}\right] - \left[\frac{1}{\lambda_\gamma^*}\int \langle\tilde{F}(t)\rangle dt - \mathrm{hc}\right]$$

Again, owing to Eq. (M.10), that is to the fact that the average of the Langevin force is zero, this correlation function reduces to

$$\langle Q(0)S(t)\rangle = \frac{\langle Q(0)\rangle^2}{2i\tilde{\Omega}}\left[\frac{\lambda_\gamma}{\lambda_\gamma^*}e^{\lambda_\gamma^* t} - \frac{\lambda_\gamma^*}{\lambda_\gamma}e^{\lambda_\gamma t}\right] + \frac{\langle Q(0)v(0)\rangle}{2i\tilde{\Omega}}\left[\frac{1}{\lambda_\gamma}e^{\lambda_\gamma t} - \frac{1}{\lambda_\gamma^*}e^{\lambda_\gamma^* t}\right]$$

Of course, at $t = 0$, the correlation function becomes

$$\langle\, Q(0)S(0)\rangle = \frac{\langle Q^2\rangle}{2i\tilde{\Omega}}\left[\frac{\lambda_\gamma}{\lambda_\gamma^*} - \frac{\lambda_\gamma^*}{\lambda_\gamma}\right] + \frac{\langle Q(0)v(0)\rangle}{2i\tilde{\Omega}}\left[\frac{1}{\lambda_\gamma} - \frac{1}{\lambda_\gamma^*}\right]$$

Because of Eq. (M.16), this equation reads

$$\langle\, Q(0)S(0)\rangle = \frac{\langle Q^2\rangle}{2i\tilde{\Omega}}\left[\frac{\lambda_\gamma}{\lambda_\gamma^*} - \frac{\lambda_\gamma^*}{\lambda_\gamma}\right]$$

At last, in view of Eq. (M.7) giving the expression of λ_γ simplifies into

$$\langle\, Q(0)S(0)\rangle = \frac{\gamma}{\Omega^2}\langle Q^2\rangle \tag{M.21}$$

APPENDIX N: SOME PROPERTIES OF COHERENT STATES

This appendix deals with some properties of coherent states that are playing an important physical role in the physics of weak H-bond species.

N.1 EXPANSION OF THE COHERENT STATE ON THE EIGENVECTORS OF THE QUANTUM HARMONIC OSCILLATOR HAMILTONIAN

By definition, a coherent state $|\alpha\rangle$ is the eigenvector of the non-Hermitean lowering operator a of the quantum harmonic oscillator. Thus, the basic equation and its conjugate are, respectively,

$$\mathrm{a}|\alpha\rangle = \alpha|\alpha\rangle \qquad \text{and} \qquad \langle\alpha|\mathrm{a}^{\dagger} = \alpha^{*}\langle\alpha| \tag{N.1}$$

Now, we may insert in front of a coherent state, the closeness relation on the eigenstates of the number operator $\mathrm{a}^{\dagger}\mathrm{a}$ of the quantum harmonic oscillator (with $[\mathrm{a}, \mathrm{a}^{\dagger}] = 1$), in the following way:

$$|\alpha\rangle = \left[\sum_{n=0}^{\infty} |n\rangle\langle n|\right]|\alpha\rangle$$

with

$$\mathrm{a}^{\dagger}\mathrm{a}|n\rangle = n|n\rangle \qquad \text{and} \qquad \langle n|m\rangle = \delta_{nm}$$

Then, one obtains

$$|\alpha\rangle = \sum_{n=0}^{\infty} C_n(\alpha)|n\rangle \qquad \text{with} \qquad C_n(\alpha) = \langle n|\alpha\rangle \tag{N.2}$$

Again, with the aid of this expansion, the definition relation of the coherent state becomes

$$\mathrm{a}|\alpha\rangle = \sum_{n=0}^{\infty} \mathrm{a}C_n(\alpha)|n\rangle$$

Then, one may use the eigenvalue equation of the lowering operator, that is,

$$a|n\rangle = \sqrt{n}|n-1\rangle$$

Consequently, the definition relation of the coherent state takes the form:

$$a|\alpha\rangle = \sum_{n=0}^{\infty} C_n(\alpha)\sqrt{n}|n-1\rangle \tag{N.3}$$

Now, with the help of Eqs. (N.2) and (N.3), the eigenvalue equation (N.1) of the lowering operator appears to be given by

$$\sum_{n=0}^{\infty} C_n(\alpha)\sqrt{n}|n-1\rangle = \sum_{n=0}^{\infty} C_n(\alpha)\alpha|n\rangle$$

In this last equation, the first element corresponding to $n = 0$ involved on the left-hand side term disappears, so that the summation must run from the situation corresponding to $n = 1$. Then, by reorganization of this sum, the above result becomes

$$\sum_{n=0}^{\infty} C_{n+1}(\alpha)\sqrt{n+1}|n\rangle = \sum_{n=0}^{\infty} C_n(\alpha)\alpha|n\rangle$$

Of course, in this last equation the equation must be verified for each term of the sum, so that it is possible to write

$$C_{n+1}(\alpha)\sqrt{n+1} = C_n(\alpha)\alpha$$

For $n = 0$ and $n = 1$, one obtains, respectively,

$$\sqrt{1}C_1(\alpha) = \alpha C_0(\alpha) \qquad \text{and} \qquad \sqrt{2}C_2(\alpha) = \alpha C_1(\alpha)$$

Then, replacing the first equation in the second gives

$$C_2(\alpha) = \frac{\alpha^2}{\sqrt{2}} C_0(\alpha)$$

Next, for $n = 2$ it is found that

$$\sqrt{3}C_3(\alpha) = \alpha C_2(\alpha) \tag{N.4}$$

Now, replacing $C_2(\alpha)$ by its expression in terms of $C_0(\alpha)$, one obtains

$$C_3(\alpha) = \frac{\alpha^3}{\sqrt{3}\sqrt{2}} C_0(\alpha)$$

Moreover, one gets by recurrence:

$$C_n(\alpha) = \frac{\alpha^n}{\sqrt{n!}} C_0(\alpha)$$

As a consequence, the expansion (N.2) of the coherent state on the eigenkets of the quantum harmonic oscillator Hamiltonian, gives

$$|\alpha\rangle = \sum_{n=0}^{\infty} \frac{\alpha^n}{\sqrt{n!}} C_0(\alpha)|n\rangle \tag{N.5}$$

In order to get the coefficient $C_0(\alpha)$, it is necessary to normalize the coherent state. This leads us to write

$$\langle\alpha|\alpha\rangle = \sum_{n}^{\infty}\sum_{m}^{\infty} \frac{(\alpha^*)^n(\alpha)^n}{\sqrt{n!}\sqrt{m!}} |C_0(\alpha)|^2 \langle n|m\rangle = 1$$

Then, owing to the orthonormality $\langle n|m\rangle = \delta_{nm}$ of the eigenstates of the quantum harmonic oscillator Hamiltonian, the above expression reduces to

$$\sum_{m}^{\infty} \frac{|\alpha|^{2n}}{n!} |C_0(\alpha)|^2 = 1$$

Then, passing to the exponential and after reorganization, the coefficient to be found appears to be

$$|C_0(\alpha)|^2 = e^{-|\alpha|^2}$$

Thus, the expansion of the coherent state of the quantum harmonic oscillator Hamiltonian, becomes

$$|\alpha\rangle = e^{-|\alpha|^2/2} \sum_{n} \frac{\alpha^n}{\sqrt{n!}} |n\rangle \tag{N.6}$$

It is easy to find that the scalar product of two coherent states is not zero:

$$\langle\beta|\alpha\rangle = e^{-\frac{1}{2}|\alpha-\beta|^2}$$

N.2 THE COHERENT STATE AS THE RESULT OF THE ACTION OF THE TRANSLATION OPERATOR ON THE GROUND STATE OF THE NUMBER OPERATOR

Let us consider the unitary translation operator, which with the aid of the Glauber theorem, may be written according to the two forms:

$$A(\alpha) = e^{\alpha a^\dagger - \alpha^* a} = e^{-\frac{1}{2}|\alpha|^2} e^{\alpha a^\dagger} e^{-\alpha^* a} \tag{N.7}$$

By the action of this operator on the ground state of the number occupation operator $a^\dagger a$ of the quantum harmonic oscillator, one obtains

$$A(\alpha)|0\rangle = e^{-\frac{1}{2}|\alpha|^2} e^{\alpha a^\dagger} e^{-\alpha^* a}|0\rangle$$

Of course, since the action of the lowering operator a on the ground–state $|0\rangle$ is zero, this equation reduces to

$$A(\alpha)|0\rangle = e^{-\frac{1}{2}|\alpha|^2} e^{\alpha a^\dagger}|0\rangle \tag{N.8}$$

Besides, by expansion of the exponential operator, one finds

$$e^{\alpha a^\dagger}|0\rangle = \sum_n \frac{\alpha^n}{n!}(a^\dagger)^n|0\rangle \tag{N.9}$$

Now, owing to the action of $a^\dagger$ on the eigenstates $|n\rangle$ of $a^\dagger a$, that is, $a^\dagger|\ n\rangle = \sqrt{(n+1)}|n+1\rangle$, there is the following property:

$$(a^\dagger)^n|0\rangle = \sqrt{n!}|n\rangle \tag{N.10}$$

As a consequence, in view of Eqs. (N.11), (N.9), and (N.10), Eq. (N.8) becomes

$$A(\alpha)|0\rangle = e^{-\frac{1}{2}|\alpha|^2} \sum_n \frac{\alpha^n}{\sqrt{n!}}|n\rangle \tag{N.11}$$

Owing to Eq. (N.6), this last result shows that the action of the translation operator on the ground state of the quantum harmonic oscillator Hamiltonian, generates a coherent state:

$$A(\alpha)|0\rangle = |\alpha\rangle \tag{N.12}$$

N.3 COHERENT STATES AS MINIMIZING THE HEISENBERG UNCERTAINTY RELATION

Now, we look at the average values of the coordinate Q and its conjugate momentum P, which are given in terms of the Bosons by

$$Q = \sqrt{\frac{\hbar}{2m\omega}}(a^{\dagger} + a) \qquad P = i\sqrt{\frac{\hbar m\omega}{2}}(a^{\dagger} - a) \tag{N.13}$$

Next, first consider the average value of the coordinate of a quantum harmonic oscillator performed on a coherent state,

$$\langle\alpha|Q|\alpha\rangle = \sqrt{\frac{\hbar}{2M\Omega}}\langle\alpha|(a^{\dagger} + a)|\alpha\rangle$$

Using the eigenvalue equation of the coherent state gives

$$\langle\alpha|Q|\alpha\rangle = \sqrt{\frac{\hbar}{2M\Omega}}\langle\alpha|(\alpha^{*} + \alpha)|\alpha\rangle$$

Then, after simplification, one finds

$$\langle\alpha|Q|\alpha\rangle = \sqrt{\frac{\hbar}{2M\Omega}}\{\alpha^{*} + \alpha\}$$

In a similar way, for the corresponding average of the conjugate momentum of the coordinate, one obtains

$$\langle\alpha|P|\alpha\rangle = i\sqrt{\frac{\hbar M\Omega}{2}}\{\alpha^{*} - \alpha\}$$

Now, consider the average of Q^2 on the coherent state, that is,

$$\langle\alpha|Q^2|\alpha\rangle = \frac{\hbar}{2M\Omega}\alpha|(a^{\dagger} + a)^2|\alpha\rangle$$

After expansion of the right-hand side square, with the aid of the commutator $[a, a^{\dagger}] = 1$, this average value takes the form:

$$\langle\alpha|Q^2|\alpha\rangle = \frac{\hbar}{2M\Omega}\langle\alpha|[a^{\dagger}a^{\dagger} + aa + 2a^{\dagger}a + 1]|\alpha\rangle \tag{N.14}$$

Then, using the eigenvalue equation of the coherent state and its conjugate equation, one obtains after using the fact that the coherent state is normalized

$$\langle\alpha|Q^2|\alpha\rangle = \frac{\hbar}{2M\Omega}[(\alpha^\dagger)^2 + (\alpha)^2 + (2|\alpha|^2 + 1)] \tag{N.15}$$

In a similar way, it may be found that the corresponding average value of the square of the conjugate momentum of Q is

$$\langle\alpha|P^2|\alpha\rangle = -\frac{\hbar M\Omega}{2}[(\alpha^*)^2 + (\alpha)^2 - (2|\alpha|^2 + 1)] \tag{N.16}$$

The uncertainties on the coordinate and on its conjugate momentum are

$$\Delta Q_\alpha = \sqrt{\langle\alpha|Q^2|\alpha\rangle - \langle\alpha|Q|\alpha\rangle^2} \quad \text{and} \quad \Delta P_\alpha = \sqrt{\langle\alpha|P^2|\alpha\rangle - \langle\alpha|P|\alpha\rangle^2} \tag{N.17}$$

Because of the above equations, one obtains, respectively,

$$\Delta Q_\alpha = \sqrt{\frac{\hbar}{2M\Omega}} \qquad \text{and} \qquad \Delta P_\alpha = \sqrt{\frac{\hbar M\Omega}{2}}$$

Consequently, the product of the two uncertainties is a minimum of what the coherent state may be, since

$$\Delta Q_\alpha \Delta P_\alpha = \frac{\hbar}{2}$$

N.4 STATISTICAL EQUILIBRIUM DENSITY OPERATOR OF A COHERENT STATE

Now, consider the normalized density operator ρ_α of a system of equivalent quantum harmonic oscillators embedded in a thermal bath at temperature **T** owing to the fact that the average values of the Hamiltonian H, of the coordinate Q and of the conjugate momentum P, of these oscillators (with $[Q, P] = i\hbar$) are known. The equations governing the statistical entropy S,

$$\mathrm{tr}\{\rho_\alpha \ln \rho_\alpha\} = S$$

with the four constraints dealing with the normalization of the density operator and the three mean values $\langle H\rangle$, $\langle Q\rangle$, and $\langle P\rangle$ averaged on the density operator:

$$\mathrm{tr}\{\rho_\alpha\} = 1 \qquad \mathrm{tr}\{\rho_\alpha H\} = \langle H\rangle \qquad \mathrm{tr}\{\rho_\alpha Q\} = \langle Q\rangle \qquad \text{and} \qquad \mathrm{tr}\{\rho_\alpha P\} = \langle P\rangle$$

Now, by using the Lagrange multiplicators method, the equation governing the maximalization of the entropy required in order to obtain the statistical equilibrium, in view of the four constraints:

$$\mathrm{tr}\{1 + \ln\{\rho_\alpha\} + \lambda_0 + \beta H - \lambda_1 P - \lambda_2 Q\}\delta\rho_\alpha = 0$$

where β, λ_0, λ_1, and λ_2 are, respectively, the Lagrange parameters related to the four constraints (recall that β is the usual Lagrange parameter involved in canonical distributions). Thus, since $\delta\rho_\alpha \neq 0$, the condition of maximalization of the entropy is therefore:

$$1 + \ln\{\rho_\alpha\} + \lambda_0 + \beta H - \lambda_1 P - \lambda_2 Q = 0$$

Rearranging, after using the fact that the Lagrange parameter is a scalar

$$\rho_\alpha = e^{-(1+\lambda_0)} e^{-\beta H + \lambda_1 P + \lambda_2 Q}$$

Now, one may introduce two complex conjugate dimensionless constants α and α^*, with the aid of

$$\alpha = \frac{1}{\lambda}\sqrt{\frac{\hbar}{2M\Omega}}(\lambda_1 + i\Omega\lambda_2) \quad \text{with} \quad i^2 = -1$$

and then, use the expression of Q and P in terms of $a^\dagger$ and a. Then, the density operator takes the form:

$$\rho_\alpha = \frac{1}{Z} e^{-\lambda(a^\dagger - \alpha^*)(a-\alpha)} \tag{N.18}$$

where the partition function of the density operator given by

$$Z = e^{-(1+\lambda_0)} = \mathrm{tr} e^{-\lambda(a^\dagger - \alpha^*)(a-\alpha)} \quad \text{with} \quad \lambda = \beta\hbar\Omega = \frac{\hbar\Omega}{k_B T}$$

Now, note that the density operator (N.18) may be viewed as the result of the following canonical transformation involving the translation operator (N.7):

$$\rho_\alpha = [A(\alpha)][e^{-\lambda a^\dagger a}][A(\alpha)^{-1}] \tag{N.19}$$

This may be verified by explicitly making the translation operator:

$$\rho_\alpha = \frac{1}{Z}\left[e^{-\frac{1}{2}|\alpha|^2} e^{\alpha a^\dagger} e^{-\alpha^* a}\right][e^{-\lambda a^\dagger a}]\left[e^{\alpha^* a} e^{-\alpha a^\dagger} e^{-\frac{1}{2}|\alpha|^2}\right]$$

and then, using the following equation appearing in Appendix E:

$$[e^{\alpha a^\dagger}][e^{-\alpha^* a}]\{f(a^\dagger, a)\}[e^{\alpha^* a}][e^{-\alpha a^\dagger}] = \{f(a^\dagger - \alpha^*, a - \alpha)\}$$

This last result shows that the canonical transformation over the Boltzmann density operator and involving the translation operator, gives the density operator ρ_α. Now, one may observe that owing to Eq. (N.12):

$$A(\alpha)|0\rangle\langle 0|A(\alpha)^{-1} = |\alpha\rangle\langle\alpha| \tag{N.20}$$

Again, note that when the absolute temperature is vanishing, the Boltzmann density operator reduces to that of the ground state of the Hamiltonian of the quantum harmonic oscillator, that is,

$$\frac{1}{Z}[e^{-\lambda a^\dagger a}] \rightarrow |0\rangle\langle 0| \qquad \text{when} \qquad T \rightarrow 0$$

Thus, it appears by comparison of the canonical transformations (N.19) and (N.20), that the density operator of the coherent state $|\alpha\rangle\langle\alpha|$ may be viewed as the limit for zero temperature of the density operator ρ_α,

$$\rho_\alpha \rightarrow |\alpha\rangle\langle\alpha| \qquad \text{when} \qquad T \rightarrow 0$$

As a consequence, one may inversely consider the equilibrium density operator ρ_α as that of a coherent state at any temperature.

APPENDIX O: CYCLIC DIMER ACF APPROACH AVOIDING SYMMETRY [75]

When the *indirect damping is missing*, the Hamiltonian (255) of a cyclic dimer involving Davydov coupling reduces to

$$[\mathrm{H_{Dav}}] = \begin{pmatrix} [\mathrm{H}_{II}^{\{0,0\}}] & 0 \\ 0 & [\mathrm{H}_{\{1,0\}}^{\{0,1\}}] \end{pmatrix} \tag{O.1}$$

with

$$[\mathrm{H}_{\{1,0\}}^{\{0,1\}}] = \begin{pmatrix} [\mathrm{H}_{II}^{\{1,0\}}] & \mathrm{V}^{\circ} \\ \mathrm{V}^{\circ} & [\mathrm{H}_{II}^{\{0,1\}}] \end{pmatrix} \tag{O.2}$$

and

$$[\mathrm{H}_{II}^{\{0,0\}}] = \left(\frac{\mathrm{P}_a^2}{2M} + \frac{M\Omega^2\mathrm{Q}_a^2}{2}\right) + \left(\frac{\mathrm{P}_b^2}{2M} + \frac{M\Omega^2\mathrm{Q}_b^2}{2}\right)$$
$$[\mathrm{H}_{II}^{\{1,0\}}] = [\mathrm{H}_{II}^{\{0,0\}}] + b\mathrm{Q}_a + (\hbar\omega^{\circ} - \alpha^{\circ 2}\hbar\Omega)$$

Here, the parameters have their usual meaning, and the sum has to be performed over the two moieties of the dimer. Next, in order to diagonalize the Hamiltonian (O.1) it is suitable to define the following column vector of kets:

$$\begin{pmatrix} |\{0\}_a\ \{0\}_b\rangle \otimes |(n)_a\ (m)_b\rangle \\ |\{0\}_a\ \{1\}_b\rangle \otimes |(n)_a\ (m)_b\rangle \\ |\{1\}_a\ \{0\}_b\rangle \otimes |(n)_a\ (m)_b\rangle \end{pmatrix} \equiv \begin{pmatrix} |\{0\}_a(n)_a\{0\}_b(m)_b\rangle \\ |\{0\}_a(n)_a\{1\}_b(m)_b\rangle \\ |\{1\}_a(n)_a\{0\}_b(m)_b\rangle \end{pmatrix} \tag{O.3}$$

Then, it is possible to get the SD in a straightforward way, without using the symmetry of the system. The ACF governing, respectively, the allowed (u) and forbidden (g) IR absorption involving the fast modes, are given by the following traces over the base spanned by the Hamiltonian $\mathrm{H_{Dav}}$:

$$[\mathrm{G_{Dav}}^{\circ}(t)^{(u)}]^{\circ} = \mathrm{tr_{Dav}}[\rho_{\mathrm{Dav}}[\mu_u(0)]^{\dagger}[\mu_u(t)]] \tag{O.4}$$
$$[\mathrm{G_{Dav}}^{\circ}(t)^{(g)}]^{\circ} = \mathrm{tr_{Dav}}\ [\rho_{\mathrm{Dav}}[\mu_g(0)]^{\dagger}[\mu_g(t)]]$$

In these equations, $\mu_u(0)$ and $\mu_g(0)$ are, respectively, the IR and Raman transition moment operators sat initial time and at time t, whereas $\mu_u(t)$ and $\mu_g(t)$ are the corresponding moments at time t. They are respectively

$$[\mu_u(0)] = [\mu_b{}^\circ|\{0\}_a\{1\}_b\rangle - \mu_a{}^\circ|\{1\}_a\{0\}_b\rangle]\langle\{0\}\{0\}_a|$$

$$[\mu_g(0)] = [\mu_b{}^\circ|\{0\}_a\{1\}_b\rangle + \mu_a{}^\circ|\{1\}_a\{0\}_b\rangle]\langle\{0\}_b\{0\}_a|$$

In these last equations, the kets are given by

$$|\{k\}_a\{l\}_b\rangle = |\{k\}_a\rangle \otimes |\{l\}_b\rangle \quad \text{with} \quad k, l = 0, 1$$

The IR transition moment operator at time t is obtained by the Heisenberg transformation involving the Hamiltonian (255):

$$[\mu_u(t)] = [e^{i\text{H}_{\text{Dav}}t/\hbar}][\mu_b{}^\circ|\{0\}_a\{1\}_b\rangle - \mu_a{}^\circ|\{1\}_a\{0\}_b\rangle][\langle\ \{0\}_b\{0\}_a|][e^{-i\text{H}_{\text{Dav}}t/\hbar}]$$

At last, ρ_{Dav} is the Boltzmann density operator given by

$$\rho_{\text{Dav}} = \frac{1}{Z_{\text{Dav}}}(e^{-\text{H}_{\text{Dav}}/k_BT}) \quad \text{with} \quad Z_{\text{Dav}} = \text{tr}[(e^{-\text{H}_{\text{Dav}}/k_BT})]$$

Then, by using the above equations, it is possible to get for the (μ) ACF (O.4) an expression that, after a circular permutation within the trace, gives

$$[\text{G}^{\circ(u)}_{\text{Dav}}(t)]^\circ = \frac{1}{Z_{\text{Dav}}}\text{tr}_{\text{Dav}}$$
$$\{\langle\ \{0\}_a\{0\}_b|[e^{-i\text{H}_{\text{Dav}}t/\hbar}](e^{-\text{H}_{\text{Dav}}/k_BT})|\{0\}_b\{0\}_a\rangle$$
$$[\langle\{1\}_a\{0\}_b|\mu_a{}^\circ - \langle\ \{0\}_a\{1\}_b|\mu_b{}^\circ][e^{i\text{H}_{\text{Dav}}t/\hbar}][\mu_b{}^\circ|\{1\}_b\{0\}_a\rangle - \mu_a{}^\circ|\{0\}_b\{1\}_a\rangle]\}$$

Next, observe that the trace tr_{Dav} over the base spanned by the Hamiltonian reduces in the present situation to the product of traces tr_a and tr_b over the bases spanned by the H-bond bridge harmonic Hamiltonians of the two moieties.

$$\text{tr}_{\text{Dav}} \rightarrow \text{tr}_a\ \text{tr}_b$$

Then, owing to Eq. (O.1), this expression simplifies into

$$[\mathrm{G}^{\circ(u)}_{\mathrm{Dav}}(t)]^{\circ} = \frac{1}{Z_{II}^{\{0,0\}}} \mathrm{tr}_a \ \mathrm{tr}_b \\ \{\langle \{0\}_a\{0\}_b| e^{-i[\mathrm{H}_{II}^{\{0,0\}}]t/\hbar} e^{[-\mathrm{H}_{II}^{\{0,0\}}]/k_{\mathrm{B}}T} |\{0\}_b\{0\}_a\rangle \\ [\langle \{1\}_a\{0\}_b|\mu_a{}^{\circ} - \langle \{0\}_a\{1\}_b|\mu_b{}^{\circ}] \\ e^{i[\mathrm{H}_{\{1,0\}}^{\{0,1\}}]t/\hbar}[\mu_b{}^{\circ}|\{1\}_b\{0\}_a\rangle - \mu_a{}^{\circ}|\{0\}_b\{1\}_a\rangle]\} \tag{O.5}$$

with

$$Z_{II}^{\{0,0\}} \equiv \mathrm{tr}_a\mathrm{tr}_b[e^{[-\mathrm{H}_{II}^{\{0,0\}}]/k_{\mathrm{B}}T}]$$

Moreover, because of the expression of the Hamiltonian involved in it, the partition function $Z_{II}^{\{0,0\}}$ takes the form:

$$Z_{II}^{\{0,0\}} = \mathrm{tr}_a\mathrm{tr}_b[(e^{[-\mathrm{H}_{II}^{\{0\}}]_a/k_{\mathrm{B}}T})(e^{[-\mathrm{H}_{II}^{\{0\}}]_b/k_{\mathrm{B}}T})] = \frac{1}{\tilde{\varepsilon}^2}$$

Again, perform in Eq. (O.5) the traces over the bases and insert a closeness relation on this base. That leads to

$$[\mathrm{G}^{\circ(u)}_{\mathrm{Dav}}(t)]^{\circ} = \tilde{\varepsilon}^2 \sum_{p_a}\sum_{p_b} \langle (p)_b(q)_a| \langle \{0\}_a\{0\}_b| \\ [e^{-i[\mathrm{H}_{II}^{\{0,0\}}]t/\hbar}](e^{[-\mathrm{H}_{II}^{\{0,0\}}]/k_{\mathrm{B}}T})|\{0\}_b\{0\}_a\rangle \\ \left(\sum_{n_a}\sum_{m_b} |(n)_a(m)_b\rangle\langle (m)_b(n)_a|\right)[\langle \{1\}_a\{0\}_b|\mu_a{}^{\circ} - \langle \{0\}_a\{1\}_b|\mu_b{}^{\circ}] \\ [e^{i[\mathrm{H}_{\{1,0\}}^{\{0,1\}}]t/\hbar}][\mu_b{}^{\circ}|\{1\}_b\{0\}_a\rangle - \mu_a{}^{\circ}|\{0\}_b\{1\}_a\rangle]|(q)_a(p)_b\rangle$$

Then, using the notation involved in Eq. (O.3), this ACF takes the form:

$$[\mathrm{G}^{\circ(u)}_{\mathrm{Dav}}(t)]^{\circ} = \tilde{\varepsilon}^2 \sum_{q_a}\sum_{p_b} \\ \langle (p)_b\{0\}_b(q)_a\{0\}_a|[e^{-i[\mathrm{H}_{II}^{\{0,0\}}]t/\hbar}](e^{[-\mathrm{H}_{II}^{\{0,0\}}]/k_{\mathrm{B}}T})|\{0\}_a(n)_a\{0\}_b(m)_b\rangle \\ \sum_{n_a}\sum_{m_b}[\langle (m)_a\{1\}_a(n)_b\{0\}_b|\mu_a{}^{\circ} - \langle (m)_b\{0\}_b(n)_a\{1\}_a|\mu_b{}^{\circ}] \\ [e^{i[\mathrm{H}_{\{1,0\}}^{\{0,1\}}]t/\hbar}][\mu_a{}^{\circ}|\{1\}_a(q)_a\{0\}_b(p)_b\rangle - \mu_b{}^{\circ}|\{0\}_a(q)_a\{1\}_b(p)_b\rangle]$$

Now observe that

$$[\mathrm{H}_{II}^{\{0,0\}}]|\{0\}_a(n)_a\{0\}_b(m)_b\rangle = \left[n_a + \tfrac{1}{2} + m_b + \tfrac{1}{2}\right]\hbar\Omega|\{0\}_a(n)_a\{0\}_b(m)_b\rangle \quad (O.6)$$

As a consequence, the following eigenvalue equation holds

$$\begin{aligned}&(e^{[-\mathrm{H}_{II}^{\{0,0\}}]/k_B T})[e^{-i[\mathrm{H}_{II}^{\{0,0\}}]t/\hbar}]|\{0\}_a(n)_a\{0\}_b(m)_b\rangle \\ &\quad = (e^{-\hbar(n_a+m_b)\Omega/k_B T})[e^{-i(n_a+m_b)\Omega t}]|\{0\}_a(n)_a\{0\}_b(m)_b\rangle \end{aligned} \quad (O.7)$$

Then, after using the orthonormality properties and neglecting the zero-point energy of the Hamiltonians involved in the Boltzmann operator, the ACF may be written

$$\begin{aligned}[\mathrm{G}^{\circ(u)}_{\mathrm{Dav}}(t)]^{\circ} &= \varepsilon^2 \sum_{n_a}\sum_{m_b}(e^{-(n_a+m_b)\hbar\Omega/k_B T})[e^{-i(n_a+m_b)\Omega t}] \\ &\quad [[\mu_a^{\circ}]^2\{^{\{n_a\{1\}_a\}}_{\{m_b\{0\}_b\}}U^{\{n_a\{1\}_a\}}_{\{m_b\{0\}_b\}}(t)\} + [\mu_b^{\circ}]^2\{^{\{n_a\{0\}_a\}}_{\{m_b\{1\}_b\}}U^{\{n_a\{0\}_a\}}_{\{m_b\{1\}_b\}}(t)\} \\ &\quad - \mu_a^{\circ}\,\mu_b^{\circ}[\{^{\{n_a\{1\}_a\}}_{\{m_b\{0\}_b\}}U^{\{n_a\{0\}_a\}}_{\{m_b\{1\}_b\}}(t)\} + \{^{\{n_a\{0\}_a\}}_{\{m_b\{1\}_b\}}U^{\{n_a\{1\}_a\}}_{\{m_b\{0\}_b\}}(t)\}]]\end{aligned} \quad (O.8)$$

$$\{^{\{n_a\{k\}_a\}}_{\{m_b\{l\}_b\}}U^{\{n_a\{r\}_a\}}_{\{m_b\{s\}_b\}}(t)\} \equiv \langle\ \{k\}_a(n)_a\{l\}_b(m)_b|[e^{i[\mathrm{H}^{\{1,0\}}_{\{0,1\}}]t/\hbar}]|\{r\}_a(n)_a\{s\}_b(m)_b\rangle \quad (O.9)$$

Next, write the eigenvalue equation of the Hamiltonian (O.2):

$$[\mathrm{H}^{\{1,0\}}_{\{0,1\}}]|\Psi_\mu\rangle = \hbar\omega_\mu{}^{\circ}|\Psi_\mu\rangle \quad (O.10)$$

Expressed in terms of the base used for the description of the matrix elements (O.9), the corresponding eigenvectors are

$$|\Psi_\mu\rangle = \sum_{n_a}\sum_{m_b}[[C^{\{0\}\{1\}}_{n_a\,m_b\,\mu}]|\{0\}_a(n)_a\{1\}_b(m)_b\rangle + [C^{\{1\}\{0\}}_{n_a\,m_b\,\mu}]|\{1\}_a(n)_a\{0\}_b(m)_b\rangle] \quad (O.11)$$

$$\langle\Psi_\mu|\Psi_\nu\rangle = 0 \quad \text{and} \quad \sum|\Psi_\mu\rangle\langle\Psi_\mu| = \hat{1} \quad (O.12)$$

Now, consider one of the matrix element (O.9) involved in the ACF (O.8). Then, postmultiply the involved exponential operator by the closeness relation (O.12),

$$\{{}^{\{n_a\{k\}_a\}}_{\{m_b\{l\}_b\}}U^{\{n_a\{r\}_a\}}_{\{m_b\{s\}_b\}}(t)\}$$
$$= \sum_{\mu} \langle \{k\}_a(n)_a\{l\}_b(m)_b|[e^{i[\mathrm{H}^{\{1,0\}}_{\{0,1\}}]t/\hbar}]|\Psi_\mu\rangle\langle\Psi_\mu||\{r\}_a(n)_a\{s\}_b(m)_b\rangle \quad (O.13)$$

Next, according to Eq. (O.10), the following eigenvalue equation holds

$$[e^{i[\mathrm{H}^{\{1,0\}}_{\{0,1\}}]t/\hbar}]|\Psi_\mu\rangle = (e^{i\omega_\mu t})|\Psi_\mu\rangle$$

As a consequence, Eq. (O.13) becomes

$$\{{}^{\{n_a\{k\}_a\}}_{\{m_b\{l\}_b\}}U^{\{n_a\{r\}_a\}}_{\{m_b\{s\}_b\}}(t)\} = \sum_{\mu}[C^{\{k\}\{l\}}_{n_a\ m_b\ \mu}]e^{i\omega_\mu t}[C^{\{r\}\{s\}}_{n_a\ m_b\ \mu}] \quad (O.14)$$

$$\langle\Psi_\mu|\{k\}_a(n)_a\{l\}_b(m)_b\rangle \equiv [C^{\{k\}\{l\}}_{n_a\ m_b\ \mu}]$$

Hence, in view of Eqs. (O.7) and (O.14), the ACF (O.8) corresponding to the IR transition becomes

$$[\mathrm{G}^{\circ(u)}_{\mathrm{Dav}}(t)]^{\circ} = \frac{1}{2}\,\varepsilon^2 \sum_{n_a}\sum_{m_b}\sum_{\mu}(e^{-(n_a+mb)\hbar\Omega t/k_\mathrm{B}T})[e^{-i(n_a+mb)\Omega t}](e^{i\omega_\mu t})$$
$$\{[\mu_a{}^{\circ}]^2[C^{\{0\}\{1\}}_{n_a\ m_b\ \mu}]^2 + [\mu_b{}^{\circ}]^2[C^{\{1\}\{0\}}_{n_a\ m_b\ \mu}]^2 - 2[\mu_a{}^{\circ}\ \mu_b{}^{\circ}][C^{\{0\}\{1\}}_{n_a\ m_b\ \mu}][C^{\{1\}\{0\}}_{n_a\ m_b\ \mu}]\} \quad (O.15)$$

Of course, owing to the above equations, the ACF corresponding to the Raman transition, may be obtained from Eq. (O.15) by replacing the "−" term by the corresponding "+" one.

At last, the IR SD, in the situation *with direct damping*, but *without indirect damping*, is the Fourier transform of (O.15), that is,

$$[\mathrm{I_{Dav}}{}^{\circ(u)}(t)] = 2\mathrm{Re}\int_0^{\infty}[\mathrm{G_{Dav}}{}^{\circ(u)}(t)]^{\circ}e^{i\omega t}e^{-\gamma^{\circ}t}dt$$

APPENDIX P: DIAGONALIZATION OF HAMILTONIAN (255) INTO (265) USING $\widehat{C}_2$ SYMMETRY [51]

P.1 STARTING EQUATIONS

Consider a cyclic dimer involving Davydov coupling. Build up the following kets:

$$\begin{pmatrix} |\Phi^{\{0,0\}}_{\{a,b\}}\rangle \\ |\Phi^{\{1,0\}}_{\{a,b\}}\rangle \\ |\Phi^{\{0,1\}}_{\{a,b\}}\rangle \end{pmatrix} \equiv \begin{pmatrix} |\{0\}_a\rangle \otimes |\{0\}_b\rangle \\ |\{1\}_a\rangle \otimes |\{0\}_b\rangle \\ |\{0\}_a\rangle \otimes |\{1\}_b\rangle \end{pmatrix} \tag{P.1}$$

Next, within the basis (P.1), the Hamiltonian (255) is

$$\mathbb{H}_{\text{Dav}} = \begin{pmatrix} \mathbb{H}^{\{0,0\}}_{II} & 0 & 0 \\ 0 & \mathbb{H}^{\{1,0\}}_{II} & \mathrm{V}^{\circ} \\ 0 & \mathrm{V}^{\circ} & \mathbb{H}^{\{0,1\}}_{II} \end{pmatrix} \tag{P.2}$$

Owing to Eq.(258), the action of the parity operator on the Hamiltonian $\mathbb{H}^{\{1,0\}}_{II}$ transforms it into $\mathbb{H}^{\{0,1\}}_{II}$ and vice-versa, whereas this same operator does not affect the Hamiltonian V°.

$$\widehat{\mathrm{C}}_2[\mathbb{H}^{\{i,j\}}_{II}] = [\mathbb{H}^{\{j,i\}}_{II}] \tag{P.3}$$

$$\widehat{\mathrm{C}}_2\mathrm{V}^{\circ} = \mathrm{V}^{\circ} \tag{P.4}$$

Note that the following explicit expressions hold for any quantity $\{A\}$, such as operators or kets:

$$\widehat{\mathrm{C}}_2[\mathbb{H}^{\{i,j\}}_{II}]\{A\} = [\mathbb{H}^{\{j,i\}}_{II}]\widehat{\mathrm{C}}_2\{A\} \qquad \widehat{\mathrm{C}}_2\mathrm{V}^{\circ}\{A\} = \mathrm{V}^{\circ}\widehat{\mathrm{C}}_2\{A\} \tag{P.5}$$

Besides, in the following equation the fact that the square of the parity operator is unity will be used, that is,

$$(\widehat{C}_2)^2 = \widehat{1} \tag{P.6}$$

P.2 COMMUTATIVITY OF TWO MATRICES

The Hamiltonian (P.2) also may be written

$$\mathbb{H}_{\text{Dav}} = \begin{pmatrix} \mathbb{H}_{II}^{\{0,0\}} & 0 \\ 0 & \mathbb{H}_{\{0,1\}}^{\{1,0\}} \end{pmatrix} \tag{P.7}$$

with

$$\left[\mathbb{H}_{\{0,1\}}^{\{1,0\}}\right] = \begin{pmatrix} \mathbb{H}_{II}^{\{1,0\}} & V^\circ \\ V^\circ & \mathbb{H}_{II}^{\{0,1\}} \end{pmatrix} \tag{P.8}$$

Now, let us look at the following 2×2 matrix

$$[\widehat{M}^{(\pm C_2)}] = \begin{pmatrix} \widehat{1} & \pm\widehat{C}_2 \\ \pm\widehat{C}_2 & \widehat{1} \end{pmatrix} \tag{P.9}$$

First, we show that the two matrices (P.8) and (P.9) commute. For this purpose, first premultiply the matrix (P.8) by (P.9). That leads to

$$\begin{pmatrix} \widehat{1} & \pm\widehat{C}_2 \\ \pm\widehat{C}_2 & \widehat{1} \end{pmatrix} \begin{pmatrix} \mathbb{H}_{II}^{\{1,0\}} & V^\circ \\ V^\circ & \mathbb{H}_{II}^{\{0,1\}} \end{pmatrix} = \begin{pmatrix} \widehat{1}\mathbb{H}_{II}^{\{1,0\}} \pm \widehat{C}_2 V^\circ & \widehat{1}V^\circ \pm \widehat{C}_2\mathbb{H}_{II}^{\{0,1\}} \\ \pm\widehat{C}_2\mathbb{H}_{II}^{\{1,0\}} + \widehat{1}V^\circ & \pm\widehat{C}_2 V^\circ + \widehat{1}\mathbb{H}_{II}^{\{0,1\}} \end{pmatrix}$$

Then, using Eq. (P.5), gives

$$\begin{pmatrix} \widehat{1} & \pm\widehat{C}_2 \\ \pm\widehat{C}_2 & \widehat{1} \end{pmatrix} \begin{pmatrix} \mathbb{H}_{II}^{\{1,0\}} & V^\circ \\ V^\circ & \mathbb{H}_{II}^{\{0,1\}} \end{pmatrix} = \begin{pmatrix} \widehat{1}\mathbb{H}_{II}^{\{1,0\}} \pm V^\circ\widehat{C}_2 & \widehat{1}V^\circ \pm \mathbb{H}_{II}^{\{1,0\}}\widehat{C}_2 \\ \pm\mathbb{H}^{\{0,1\}}\widehat{C}_2 + \widehat{1}V^\circ & \pm V^\circ\widehat{C}_2 + \widehat{1}\mathbb{H}_{II}^{\{0,1\}} \end{pmatrix} \tag{P.10}$$

On the other hand, postmultiply the matrix (P.8) by (P.9). This leads to

$$\begin{pmatrix} \mathbb{H}_{II}^{\{1,0\}} & V^\circ \\ V^\circ & \mathbb{H}_{II}^{\{0,1\}} \end{pmatrix} \begin{pmatrix} \widehat{1} & \pm\widehat{C}_2 \\ \pm\widehat{C}_2 & \widehat{1} \end{pmatrix} = \begin{pmatrix} \widehat{1}\mathbb{H}_{II}^{\{1,0\}} \pm V^\circ\widehat{C}_2 & \pm\mathbb{H}_{II}^{\{1,0\}}\widehat{C}_2 + \widehat{1}V^\circ \\ \widehat{1}V^\circ \pm \mathbb{H}_{II}^{\{0,1\}}\widehat{C}_2 & \pm V^\circ\widehat{C}_2 + \widehat{1}\mathbb{H}_{II}^{\{0,1\}} \end{pmatrix} \tag{P.11}$$

As a consequence of the equivalence between the right-hand sides of Eqs.(P.10) and (P.11), it appears that the two matrices (P.9) and (P.9) commute, that is,

$$\left[\begin{pmatrix} \widehat{1} & \pm\widehat{C}_2 \\ \pm\widehat{C}_2 & \widehat{1} \end{pmatrix}, \begin{pmatrix} \mathbb{H}_{II}^{\{1,0\}} & V^\circ \\ V^\circ & \mathbb{H}_{II}^{\{0,1\}} \end{pmatrix}\right] = 0 \tag{P.12}$$

P.3 EIGENVECTORS OF THE MATRIX (P.9)

Now, consider the action of the matrix (P.9) on the column vector built up from the two last components of Eq. (P.1). This leads to

$$\begin{pmatrix} \widehat{1} & \pm\widehat{C}_2 \\ \pm\widehat{C}_2 & \widehat{1} \end{pmatrix}\begin{pmatrix} |\Phi_{\{a,b\}}^{\{1,0\}}\rangle \\ |\Phi_{\{a,b\}}^{\{0,1\}}\rangle \end{pmatrix} = \begin{pmatrix} \widehat{1}|\Phi_{\{a,b\}}^{\{1,0\}}\rangle \pm \widehat{C}_2|\Phi_{\{a,b\}}^{\{0,1\}}\rangle \\ \pm\widehat{C}_2|\Phi_{\{a,b\}}^{\{1,0\}}\rangle + \widehat{1}|\Phi_{\{a,b\}}^{\{0,1\}}\rangle \end{pmatrix}$$

Next, in view of Eq. (P.6) replace the unity operator in the bottom component of the resulting right-hand side vector by $\widehat{C}_2^2$; this gives

$$\begin{pmatrix} \widehat{1} & \pm\widehat{C}_2 \\ \pm\widehat{C}_2 & \widehat{1} \end{pmatrix}\begin{pmatrix} |\Phi_{\{a,b\}}^{\{1,0\}}\rangle \\ |\Phi_{\{a,b\}}^{\{0,1\}}\rangle \end{pmatrix} = \begin{pmatrix} \widehat{1}|\Phi_{\{a,b\}}^{\{1,0\}}\rangle \pm \widehat{C}_2|\Phi_{\{a,b\}}^{\{0,1\}}\rangle \\ \pm\widehat{C}_2|\Phi_{\{a,b\}}^{\{1,0\}}\rangle + (\widehat{C}_2)^2|\Phi_{\{a,b\}}^{\{0,1\}}\rangle \end{pmatrix}$$

Again, use

$$\binom{+}{-} \times \binom{+}{-} = \binom{+}{+} \tag{P.13}$$

Then, the linear transformation leads to

$$\begin{pmatrix} \widehat{1} & \pm\widehat{C}_2 \\ \pm\widehat{C}_2 & \widehat{1} \end{pmatrix}\begin{pmatrix} |\Phi_{\{a,b\}}^{\{1,0\}}\rangle \\ |\Phi_{\{a,b\}}^{\{0,1\}}\rangle \end{pmatrix} = \begin{pmatrix} \{|\Phi_{\{a,b\}}^{\{1,0\}}\rangle \pm \widehat{C}_2|\Phi_{\{a,b\}}^{\{0,1\}}\rangle\} \\ \pm\widehat{C}_2\{|\Phi_{\{a,b\}}^{\{1,0\}}\rangle \pm \widehat{C}_2|\Phi_{\{a,b\}}^{\{0,1\}}\rangle\} \end{pmatrix} \tag{P.14}$$

or

$$\begin{pmatrix} \widehat{1} & \pm\widehat{C}_2 \\ \pm\widehat{C}_2 & \widehat{1} \end{pmatrix}\begin{pmatrix} |\Phi_{\{a,b\}}^{\{1,0\}}\rangle \\ |\Phi_{\{a,b\}}^{\{0,1\}}\rangle \end{pmatrix} = \begin{pmatrix} |\beta_{\{1,0\}\leftrightarrow\{0,1\}}^{(\pm)}\rangle \\ \pm\widehat{C}_2|\beta_{\{1,0\}\leftrightarrow\{0,1\}}^{(\pm)}\rangle \end{pmatrix}$$

with

$$|\beta_{\{1,0\}\leftrightarrow\{0,1\}}^{(\pm)}\rangle \equiv \{|\Phi_{\{a,b\}}^{\{1,0\}}\rangle \pm \widehat{C}_2|\Phi_{\{a,b\}}^{\{0,1\}}\rangle\} \tag{P.15}$$

Again, let us define

$$|\{\beta_{\{1,0\}}^{\{1,0\}}(\pm)\}\rangle \equiv \begin{pmatrix} |\beta_{\{1,0\}\leftrightarrow\{0,1\}}^{(\pm)}\rangle \\ \pm\widehat{C}_2|\beta_{\{1,0\}\leftrightarrow\{0,1\}}^{(\pm)}\rangle \end{pmatrix} \tag{P.16}$$

Then, the action of the matrix (P.9) on the ket defined in Eq. (P.16) is

$$[\widehat{M}^{(\pm C_2)}]|\{\beta_{\{1,0\}}^{\{1,0\}}(\pm)\}\rangle = \begin{pmatrix} \widehat{1}|\beta_{\{1,0\}\leftrightarrow\{0,1\}}^{(\pm)}\rangle \pm \widehat{C}_2(\pm\widehat{C}_2)|\beta_{\{1,0\}\leftrightarrow\{0,1\}}^{(\pm)}\rangle \\ \pm\widehat{C}_2|\beta_{\{1,0\}\leftrightarrow\{0,1\}}^{(\pm)}\rangle \pm \widehat{1}\widehat{C}_2|\beta_{\{1,0\}\leftrightarrow\{0,1\}}^{(\pm)}\rangle \end{pmatrix}$$

Again, removing the unity operators and by using Eqs. (P.6) and (P.13), this last equation gives

$$[\widehat{M}^{(\pm C_2)}]|\{\beta_{\{1,0\}}^{\{1,0\}}\ (\pm)\}\rangle = \begin{pmatrix} |\beta_{\{1,0\}\leftrightarrow\{0,1\}}^{(\pm)}\rangle + |\beta_{\{1,0\}\leftrightarrow\{0,1\}}^{(\pm)}\rangle \\ \pm\widehat{C}_2[|\beta_{\{1,0\}\leftrightarrow\{0,1\}}^{(\pm)}\rangle + |\beta_{\{1,0\}\leftrightarrow\{0,1\}}^{(\pm)}\rangle] \end{pmatrix}$$

As a consequence, after simplification and owing to Eq. (P.16), it appears that

$$[\widehat{M}^{(\pm C_2)}]|\{\beta_{\{1,0\}}^{\{1,0\}}(\pm)\}\rangle = 2\widehat{1}|\{\beta_{\{1,0\}}^{\{1,0\}}(\pm)\}\rangle$$

This shows that the spinors defined in Eq. (P.16) are eigenvectors of matrix (P.9). As a consequence, and in view of the commutation property (P.12), these spinors, which are eigenvectors of matrix (P.9) also must be eigenvectors of matrix (P.8).

P.4 EIGENVALUES OF MATRIX (P.8)

In order to get the matrix (P.8) in a diagonal form, we may proceed in the following way. First, postmultiply the matrix (P.8) by the *ket* (P.16):

$$[\mathbb{H}_{\{0,1\}}^{\{1,0\}}]|\{\beta_{\{1,0\}}^{\{1,0\}}(\pm)\}\rangle = \begin{pmatrix} \mathbb{H}_{II}^{\{1,0\}} & V^\circ \\ V^\circ & \mathbb{H}_{II}^{\{0,1\}} \end{pmatrix} \begin{pmatrix} |\beta_{\{1,0\}\leftrightarrow\{0,1\}}^{(\pm)}\rangle \\ \pm\widehat{C}_2|\beta_{\{1,0\}\leftrightarrow\{0,1\}}^{(\pm)}\rangle \end{pmatrix} \tag{P.17}$$

This gives

$$\begin{pmatrix} \mathbb{H}_{II}^{\{1,0\}} & V^\circ \\ V^\circ & \mathbb{H}_{II}^{\{0,1\}} \end{pmatrix} \begin{pmatrix} |\beta_{\{1,0\}\leftrightarrow\{0,1\}}^{(\pm)}\rangle \\ \pm\widehat{C}_2|\beta_{\{1,0\}\leftrightarrow\{0,1\}}^{(\pm)}\rangle \end{pmatrix} = \begin{pmatrix} [\mathbb{H}_{II}\{1,0\} \pm V^\circ\widehat{C}_2]|\beta_{\{1,0\}\leftrightarrow\{0,1\}}^{(\pm)}\rangle \\ [V^\circ \pm \mathbb{H}_{II}^{\{0,1\}}\widehat{C}_2]|\beta_{\{1,0\}\leftrightarrow\{0,1\}}^{(\pm)}\rangle \end{pmatrix}$$

Now, insert $\widehat{1} = \widehat{C}_2^2 = \widehat{C}_2\widehat{C}_2$ in the following way:

$$\begin{pmatrix} \mathbb{H}_{II}^{\{1,0\}} & V^\circ \\ V^\circ & \mathbb{H}_{II}^{\{0,1\}} \end{pmatrix} \begin{pmatrix} |\beta_{\{1,0\}\leftrightarrow\{0,1\}}^{(\pm)}\rangle \\ \pm\widehat{C}_2|\beta_{\{1,0\}\leftrightarrow\{0,1\}}^{(\pm)}\rangle \end{pmatrix} = \begin{pmatrix} [\mathbb{H}_{II}^{\{1,0\}} \pm V^\circ\widehat{C}_2]|\beta_{\{1,0\}\leftrightarrow\{0,1\}}^{(\pm)}\rangle \\ [V^\circ \pm \mathbb{H}_{II}^{\{0,1\}}\widehat{C}_2]\widehat{C}_2\widehat{C}_2|\beta_{\{1,0\}\leftrightarrow\{0,1\}}^{(\pm)}\rangle \end{pmatrix}$$

This reads

$$\begin{pmatrix} \mathbb{H}_{II}^{\{1,0\}} & V^\circ \\ V^\circ & \mathbb{H}_{II}^{\{0,1\}} \end{pmatrix} \begin{pmatrix} |\beta_{\{1,0\}\leftrightarrow\{0,1\}}^{(\pm)}\rangle \\ (\pm\widehat{C}_2)|\beta_{\{1,0\}\leftrightarrow\{0,1\}}^{(\pm)}\rangle \end{pmatrix}$$
$$= \begin{pmatrix} [\mathbb{H}_{II}^{\{1,0\}} \pm V^\circ\widehat{C}_2]|\beta_{\{1,0\}\leftrightarrow\{0,1\}}^{(\pm)}\rangle \\ [V^\circ\widehat{C}_2 \pm \mathbb{H}_{II}^{\{0,1\}}\widehat{1}]\widehat{C}_2|\beta_{\{1,0\}\leftrightarrow\{0,1\}}^{(\pm)}\rangle \end{pmatrix} \tag{P.18}$$

Again, using the rule $\binom{+}{-} \times \binom{+}{-} = \binom{+}{+}$, this result transforms into

$$\begin{pmatrix} \mathbb{H}_{II}^{\{1,0\}} & V^\circ \\ V^\circ & \mathbb{H}_{II}^{\{0,1\}} \end{pmatrix} \begin{pmatrix} |\beta_{\{1,0\}\leftrightarrow\{0,1\}}^{(\pm)}\rangle \\ (\pm\widehat{C}_2)|\beta_{\{1,0\}\leftrightarrow\{0,1\}}^{(\pm)}\rangle \end{pmatrix}$$
$$= \begin{pmatrix} [\mathbb{H}_{II}^{\{1,0\}} \pm V^\circ\widehat{C}_2] & 0 \\ 0 & [\mathbb{H}_{II}^{\{0,1\}} \pm V^\circ\widehat{C}_2] \end{pmatrix} \begin{pmatrix} |\beta_{\{1,0\}\leftrightarrow\{0,1\}}^{(\pm)}\rangle \\ (\pm\widehat{C}_2)|\beta_{\{1,0\}\leftrightarrow\{0,1\}}^{(\pm)}\rangle \end{pmatrix} \tag{P.19}$$

P.5 DIAGONAL FORM FOR THE MATRIX (P.2)

Keeping in mind Eqs. (P.1) and (P.15), now consider the following base:

$$\begin{pmatrix} |\Phi_{\{g,u\}}^{\{0,0\}}\rangle \\ |\beta_{\{1,0\}\leftrightarrow\{0,1\}}^{(+)}\rangle \\ |\beta_{\{1,0\}\leftrightarrow\{0,1\}}^{(-)}\rangle \end{pmatrix} = \begin{pmatrix} |\Phi_{\{g,u\}}^{\{0,0\}}\rangle \\ |\Phi_{\{a,b\}}^{\{1,0\}}\rangle + \widehat{C}_2|\Phi_{\{a,b\}}^{\{0,1\}}\rangle \\ |\Phi_{\{a,b\}}^{\{1,0\}}\rangle - \widehat{C}_2|\Phi_{\{a,b\}}^{\{0,1\}}\rangle \end{pmatrix} = \begin{pmatrix} |\{0\}_g\{0\}_u\rangle \\ |\{1\}_a\{0\}_b\rangle + \widehat{C}_2|\{0\}_a\{1\}_b\rangle \\ |\{1\}_a\{0\}_b\rangle - \widehat{C}_2|\{0\}_a\{1\}_b\rangle \end{pmatrix} \tag{P.20}$$

$$|\{k\}_r\{l\}_s\rangle \equiv |\{k\}_r\rangle \otimes |\{l\}_s\rangle$$

where $|\{0\}_g\rangle$ and $|\{0\}_u\rangle$ are the symmetrized g and u ground states of the high frequency Hamiltonians, that is,

$$|\{0\}_g\rangle = \frac{1}{\sqrt{2}}[|\{0\}_a\rangle + |\{0\}_b\rangle] \quad \text{and} \quad |\{0\}_u\rangle = \frac{1}{\sqrt{2}}[|\{0\}_a\rangle - |\{0\}_b\rangle]$$

whereas $|\{k\}_a\rangle$ and $|\{k\}_b\rangle$ are the kth states of the high frequency Hamiltonians of the two a and b moieties.

Then, it appears that the Davydov Hamiltonian may be put in a diagonal form as follows:

$$\mathbb{H}_{\text{Dav}} = \begin{pmatrix} \mathbb{H}_{II}^{\{0,0\}} & 0 & 0 \\ 0 & \mathbb{H}_{II}^{\{1,0\}} + \text{V}^{\circ}\widehat{\text{C}}_2 & 0 \\ 0 & 0 & \mathbb{H}_{II}^{\{0,1\}} - \text{V}^{\circ}\widehat{\text{C}}_2 \end{pmatrix} \tag{P.21}$$

P.6 PASSING TO SYMMETRY COORDINATES

Note that Eq. (P.21) may be also written by the equations

$$\mathbb{H}_{\text{Dav}} = [\mathbb{H}_{II}^{\{0,0\}}]|\Phi_{\{a,b\}}^{\{0,0\}}\rangle\langle\Phi_{\{a,b\}}^{\{0,0\}}|$$
$$+ [\mathbb{H}_{\{+\}}^{\{1,1\}}]|\beta_{\{1,0\}\leftrightarrow\{0,1\}}^{(+)}\rangle\langle\beta_{\{1,0\}\leftrightarrow\{0,1\}}^{(+)}| + [\mathbb{H}_{\{-\}}^{\{1,1\}}]|\beta_{\{1,0\}\leftrightarrow\{0,1\}}^{(-)}\rangle\langle\beta_{\{1,0\}\leftrightarrow\{0,1\}}^{(-)}| \tag{P.22}$$

$$[\mathbb{H}_{\{\pm\}}^{\{1,1\}}]\equiv[\mathbb{H}_{II}^{\{1,0\}} \pm \text{V}^{\circ}\widehat{\text{C}}_2] \tag{P.23}$$

Now, according to Eq. (257) the effective Hamiltonian describing the two H-bond bridge moieties, when both the fast modes are in their ground states, is

$$[\mathbb{H}_{II}^{\{0,0\}}] = [\mathbb{H}_{II}^{\{0\}}]_a + [\mathbb{H}_{II}^{\{0\}}]_b \tag{P.24}$$

Besides, owing to Eqs. (249–252), Eq. (257) becomes

$$[\mathbb{H}_{II}^{\{1,0\}}] = [\mathbb{H}^{\circ}{}_{II}^{\{1\}}]_a + [\mathbb{H}^{\circ}{}_{II}^{\{0\}}]_b \tag{P.25}$$

with, respectively,

$$[\mathbb{H}^{\circ}{}_{II}^{\{0\}}]_x = \frac{\text{P}_x^2}{2M} + \frac{M\Omega^2\text{Q}_x^2}{2} + [\mathbb{H}_{II}^{\{Int\}}]_x + [\mathbb{H}^{\{\theta\}}] \qquad x = a, b \tag{P.26}$$

$$[\mathbb{H}^{\circ}{}_{II}^{\{1\}}]_a = \left[\frac{P_a^2}{2M} + \frac{M\Omega^2 Q_a^2}{2} + bQ_a + \hbar\omega^{\circ} - \alpha^{\circ 2}\hbar\Omega\right] + [\mathbb{H}_{II}^{\{Int\}}]_a + [\mathbb{H}^{\{\theta\}}] \quad (P.27)$$

$$[\mathbb{H}_{II}^{\{Int\}}]_x = \sum_r \tilde{q}_r Q_x \eta_r$$

$$[\mathbb{H}^{\{\theta\}}] = \sum_r \left[\frac{\tilde{p}_r^2}{2m_r} + \frac{m_r \omega_r^2 \tilde{q}_r^2}{2}\right] \quad (P.28)$$

Observe that

$$[\mathbb{H}_{II}^{\{Int\}}]_a + [\mathbb{H}_{II}^{\{Int\}}]_b = \sum_r \tilde{q}_r [Q_a + Q_b] \eta_r \quad (P.29)$$

Hamiltonians (P.26) are those of the undriven quantum harmonic oscillator describing the H-bond bridge moieties a and b. Hamiltonian (P.27) is that of the driven quantum harmonic oscillators describing the a H-bond bridge moiety. Finally, Hamiltonians (P.29) are dealing with the coupling of the H-bond bridge with the thermal bath, whereas Hamiltonian (P.28) is that of the thermal bath.

Next, owing to Eqs. (P.26) and (P.28), Hamiltonian (P.24) becomes

$$[\mathbb{H}_{II}^{\{0,0\}}] = \left[\frac{P_a^2}{2M} + \frac{M\Omega^2 Q_a^2}{2}\right] + \left[\frac{P_b^2}{2M} + \frac{M\Omega^2 Q_b^2}{2}\right] + \sum_r \left[\frac{\tilde{p}_r^2}{2m_r} + \frac{m_r \omega_r^2 \tilde{q}_r^2}{2}\right] + \sum_r \tilde{q}_r [Q_a + Q_b] \eta_r$$

On the other hand, owing to Eqs. (P.25), (P.27), and (P.28), Eq. (P.23) becomes

$$[\mathbb{H}_{\{\pm\}}^{\{1,1\}}] = \left[\frac{P_a^2}{2M} + \frac{M\Omega^2 Q_a^2}{2} + bQ_a + \hbar\omega^{\circ} - \alpha^{\circ 2}\hbar\Omega\right] + \left[\frac{P_b^2}{2M} + \frac{M\Omega^2 Q_b^2}{2}\right] + \sum_r \left[\frac{\tilde{p}_r^2}{2m_r} + \frac{m_r \omega_r^2 \tilde{q}_r^2}{2}\right] + \sum_r \tilde{q}_r [Q_a + Q_b] \eta_r \pm V^{\circ} \widehat{C}_2 \quad (P.30)$$

Now, recall that the action of the parity operator transforms one coordinate of the H-bond bridge into the other one:

$$\widehat{C}_2 Q_a = Q_b \qquad \widehat{C}_2 Q_b = Q_a$$

Then, in order to use the symmetry properties of the system, let us pass to symmetry coordinates according to Table P.1.

TABLE P.1
Symmetry Coordinates

$Q_g = \dfrac{Q_a + Q_b}{\sqrt{2}}$	$Q_u = \dfrac{Q_a - Q_b}{\sqrt{2}}$
$P_g = \dfrac{P_a + P_b}{\sqrt{2}}$	$P_u = \dfrac{P_a - P_b}{\sqrt{2}}$
$C_2 Q_g = Q_g$	$C_2 Q_u = -Q_u$

Within the symmetry coordinates, Eq. (P.29) gives

$$[\mathbb{H}_{II}^{\{\mathrm{Int}\}}]_a + [\mathbb{H}_{II}^{\{\mathrm{Int}\}}]_b = \sqrt{2}\sum_r \tilde{q}_r Q_g \hbar \tilde{\kappa}_r \equiv [\mathbb{H}_{II}^{\{\mathrm{Int}\}}]_g$$

Besides, within these symmetry coordinates, there are:

$$P_a^2 + P_b^2 = P_g^2 + P_u^2$$
$$Q_a^2 + Q_b^2 = Q_g^2 + Q_u^2$$

As a consequence of these results and in view of Eq. (P.26), Hamiltonian (P.24) becomes

$$[\mathbb{H}_{II}^{\{0,0\}}] = [\mathbb{H}_{II}^{\{0\}}]_g + [\mathrm{H}_{II}^{\{0\}}]_u$$

with

$$[\mathbb{H}_{II}^{\{0\}}]_g = \left[\frac{P_g^2}{2M} + \frac{M\Omega^2 Q_g^2}{2}\right] + \sum_r \left[\frac{\tilde{p}_r^2}{2m_r} + \frac{m_r \omega_r^2 \tilde{q}_r^2}{2}\right] + \sqrt{2}\sum_r \tilde{q}_r [Q_g]\eta_r$$

and

$$[\mathrm{H}_{II}^{\{0\}}]_u = \left[\frac{P_u^2}{2M} + \frac{M\Omega^2 Q_u^2}{2}\right]$$

On the other hand, in view of Eqs. (P.29) and (P.28), Hamiltonian (P.25) transforms into

$$[\mathbb{H}_{II}^{\{1,0\}}] = [\mathbb{H}_{II}^{\{1\}}]_g + [\mathrm{H}_{II}^{\{1\}}]_u + \hbar\omega^\circ - \alpha^{\circ 2}\hbar\Omega \qquad \text{(P.31)}$$

with, respectively,

$$[\mathbb{H}_{II}^{\{1\}}]_g = \left[\frac{P_g^2}{2M} + \frac{M\Omega^2 Q_g^2}{2} + b\frac{Q_g}{\sqrt{2}}\right] + \sum_r \left[\frac{\tilde{p}_r^2}{2m_r} + \frac{m_r\omega_r^2\tilde{q}_r^2}{2}\right] + \sqrt{2}\sum_r \tilde{q}_r Q_g \eta_r \quad \text{(P.32)}$$

$$[H_{II}^{\{1\}}]_u = \left[\frac{P_u^2}{2M} + \frac{M\Omega^2 Q_u^2}{2} + b\frac{Q_u}{\sqrt{2}}\right] \quad \text{(P.33)}$$

Now, let us look more deeply at the action of the parity operator $\widehat{C}_2$. It cannot modify anything in the situation corresponding to the g symmetry since the g operators and their corresponding eigenkets cannot be modified by this operator. Thus, this operator must be explicitly written in the following form:

$$\widehat{C}_2 = [\widehat{1}]_g \otimes [\widehat{C}_2]_u$$

Here, $[\widehat{1}]_g$ is the unity operator acting in the g subspace.

As a consequence, the last right-hand side term appearing in Eq. (P.30) may be written

$$V^\circ\widehat{C}_2 = V^\circ[[\widehat{1}]_g \otimes [\widehat{C}_2]_u]$$

Then, making the notations dealing with the tensorial product of space explicit, the Hamiltonian (P.30) becomes

$$[\mathbb{H}_{\{\pm\}}^{\{1,1\}}] = \hbar\omega^\circ - \alpha^{\circ 2}\hbar\Omega + \left(\frac{P_g^2}{2M} + \frac{M\Omega^2 Q_g^2}{2} + b\frac{Q_g}{\sqrt{2}}\right) \otimes [\widehat{1}]_\theta \otimes [\widehat{1}]_u + \sqrt{2}\sum_r \tilde{q}_r Q_g \eta_r \otimes [\widehat{1}]_u + \sum_r \left[\frac{\tilde{p}_r^2}{2m_r} + \frac{m_r\omega_r^2\tilde{q}_r^2}{2}\right][\widehat{1}]_g \otimes [\widehat{1}]_u + \left[\frac{P_u^2}{2M} + \frac{M\Omega^2 Q_u^2}{2} + b\frac{Q_u}{\sqrt{2}}\right] \otimes [\widehat{1}]_g \otimes [\widehat{1}]_\theta \pm [\widehat{1}]_\theta \otimes [\widehat{1}]_g \otimes V^\circ[\widehat{C}_2]_u \quad \text{(P.34)}$$

Here, $[\widehat{1}]_u$ and $[\widehat{1}]_\theta$ are the unity operator acting, respectively, in the u and θ spaces.

P.7 OBTAINMENT OF HAMILTONIAN (268)

After separation of the different actions within the different subspaces, Hamiltonian (P.34) simplifies into

$$[\mathbb{H}_{\{\pm\}}^{\{1,1\}}] = [\mathbb{H}_{II}^{\{1\}}]_g + [\mathrm{H}_{(+)}^{\{1\}}]_{u^+} + [\mathrm{H}_{(-)}^{\{1\}}]_{u^-}$$

with, respectively,

$$\begin{aligned}[\mathbb{H}_{II}^{\{1\}}]_g &= \hbar\omega^\circ + \frac{\mathrm{P}_g^2}{2M} + \frac{M\Omega^2\mathrm{Q}_g^2}{2} + b\frac{\mathrm{Q}_g}{\sqrt{2}} - \alpha^{\circ 2}\hbar\Omega \\ &\quad + \sum_r \left[\frac{\tilde{\mathrm{p}}_r^2}{2m_r} + \frac{m_r\omega_r^2\tilde{\mathrm{q}}_r^2}{2}\right] + \sqrt{2}\sum_r \tilde{\mathrm{q}}_r\mathrm{Q}_g\eta_r \\ [\mathrm{H}_{(\pm)}^{\{1\}}]_u &= \frac{\mathrm{P}_u^2}{2M} + \frac{M\Omega^2\mathrm{Q}_u^2}{2} + b\frac{\mathrm{Q}_u}{\sqrt{2}} \pm \mathrm{V}^\circ\widehat{\mathrm{C}}_2\end{aligned}$$

Because of the above results, it appears that the Davydov Hamiltonian (P.21) takes a block form, according to

$$[\mathrm{H}_{\mathrm{Dav}}] = \begin{pmatrix} [\mathbb{H}_{II}^{\{0\}}]_g & 0 & 0 & 0 & 0 \\ 0 & [\mathbb{H}_{II}^{\{1\}}]_g & 0 & 0 & 0 \\ 0 & 0 & [\mathrm{H}_{II}^{\{0\}}]_u & 0 & 0 \\ 0 & 0 & 0 & [\mathrm{H}_{(+)}^{\{1\}}]_{u^+} & 0 \\ 0 & 0 & 0 & 0 & [\mathrm{H}_{(-)}^{\{1\}}]_{u^-} \end{pmatrix} \quad \text{in} \quad \begin{pmatrix} |\{0\}_g\rangle \\ |\{1\}_g\rangle \\ |\{0\}_u\rangle \\ |\{\beta^{(+)}\}_{u^+}\rangle \\ |\{\beta^{(-)}\}_{u^-}\rangle \end{pmatrix} \tag{P.35}$$

Here,

$$\begin{pmatrix} |\{0\}_g\rangle \otimes |\{0\}_u\rangle \\ |\{1\}_g\rangle \otimes |\{\beta^{(+)}\}_{u^+}\rangle \\ |\{1\}_g\rangle \otimes |\{\beta^{(-)}\}_{u^-}\rangle \end{pmatrix} = \begin{pmatrix} |\Phi_{\{g,u\}}^{\{0,0\}}\rangle \\ |\beta_{\{1,0\}\leftrightarrow\{0,1\}}^{(+)}\rangle \\ |\beta_{\{1,0\}\leftrightarrow\{0,1\}}^{(-)}\rangle \end{pmatrix} \tag{P.36}$$

APPENDIX Q: SYMMETRIZATION OF THE TRANSITION MOMENT FOR CYCLIC DIMERS [51, 126]

Consider a centrosymmetric cyclic dimer. When the two high frequency moieties are in their ground states, the system is described by the following ket:

$$|\Phi_{\{a,b\}}^{\{0,0\}}\rangle = |\{0\}_a\rangle \otimes |\{0\}_b\rangle \tag{Q.1}$$

On the other hand, when the excitation is, either on one of the high frequency modes, or on the other one, there are the two degenerate excited states:

$$\begin{pmatrix} |\Phi_{\{a,b\}}^{\{1,0\}}\rangle \\ |\Phi_{\{a,b\}}^{\{0,1\}}\rangle \end{pmatrix} = \begin{pmatrix} |\{1\}_a\rangle \otimes |\{0\}_b\rangle \\ |\{0\}_a\rangle \otimes |\{1\}_b\rangle \end{pmatrix} \tag{Q.2}$$

Next, consider the Hermitean dipole moment operator $[\widehat{\mu}_{\text{Dim}}(0)]$ of the dimer at initial time $t = 0$, we met in Section VI.A.3. It may be split into two Hermitean conjugate parts, the first one corresponding to absorption and the last one to emission, that is,

$$[\widehat{\mu}_{\text{Dim}}(0)] = [\tilde{\mu}_{\text{Dim}}(0)] + [\tilde{\mu}_{\text{Dim}}(0)]^{\dagger}$$

with

$$[\tilde{\mu}_{\text{Dim}}(0)] = \mu^{\circ}{}_a|\Phi_{\{a,b\}}^{\{1,0\}}\rangle\langle\Phi_{\{a,b\}}^{\{0,0\}}| + \mu^{\circ}{}_b|\Phi_{\{a,b\}}^{\{0,1\}}\rangle\langle\Phi_{\{a,b\}}^{\{0,0\}}| \tag{Q.3}$$

Here, $\mu^{\circ}{}_a$ and $\mu^{\circ}{}_b$ are of the form

$$\mu^{\circ}{}_a = q^{\circ\circ}\left[\frac{\partial\mu}{\partial \mathrm{q}_a}\right]_{\mathrm{q}_a=0} \quad \text{and} \quad \mu^{\circ}{}_b = q^{\circ\circ}\left[\frac{\partial\mu}{\partial \mathrm{q}_b}\right]_{\mathrm{q}_b=0}$$

with

$$q^{\circ\circ} = \sqrt{\frac{\hbar}{2m\omega^{\circ}}}$$

Now, pass to the symmetrized coordinates of the two moieties high frequency modes according to

$$q_g = \frac{1}{\sqrt{2}}[q_a + q_b] \quad \text{and} \quad q_u = \frac{1}{\sqrt{2}}[q_a - q_b]$$

Within the symmetrized coordinates, there are

$$\mu^\circ{}_a = q^{\circ\circ}\sqrt{2}\left[\frac{\partial\mu}{\partial q_g} + \frac{\partial\mu}{\partial q_u}\right]_{q_g=q_u=0} \quad \text{and} \quad \mu^\circ{}_b = q^{\circ\circ}\sqrt{2}\left[\frac{\partial\mu}{\partial q_g} - \frac{\partial\mu}{\partial q_u}\right]_{q_g=q_u=0}$$

Besides, recall Eq. (P.15), i.e.

$$|\beta^{(\pm)}_{\{1,0\}\leftrightarrow\{0,1\}}\rangle = \frac{1}{\sqrt{2}}[|\Phi^{\{1,0\}}_{\{a,b\}}\rangle \pm \widehat{C}_2|\Phi^{\{0,1\}}_{\{a,b\}}\rangle] \tag{Q.4}$$

Again, it may be suitable to write:

$$|\beta^{(+)}_{\{1,0\}\leftrightarrow\{0,1\}}\rangle = \tfrac{1}{2}[|\beta^{(+)}_{\{1,0\}\leftrightarrow\{0,1\}}\rangle + |\beta^{(-)}_{\{1,0\}\leftrightarrow\{0,1\}}\rangle + |\beta^{(+)}_{\{1,0\}\leftrightarrow\{0,1\}}\rangle - |\beta^{(-)}_{\{1,0\}\leftrightarrow\{0,1\}}\rangle]$$

Then, Eq. (Q.4) reads

$$|\Phi^{\{1,0\}}_{\{a,b\}}\rangle \pm \widehat{C}_2|\Phi^{\{0,1\}}_{\{a,b\}}\rangle$$
$$= \frac{1}{\sqrt{2}}[|\beta^{(+)}_{\{1,0\}\leftrightarrow\{0,1\}}\rangle + |\beta^{(-)}_{\{1,0\}\leftrightarrow\{0,1\}}\rangle + \widehat{1}[|\beta^{(+)}_{\{1,0\}\leftrightarrow\{0,1\}}\rangle - |\beta^{(-)}_{\{1,0\}\leftrightarrow\{0,1\}}\rangle]]$$

Now, use $(\widehat{C}_2)^2 = \widehat{1}$ in order to get

$$|\Phi^{\{1,0\}}_{\{a,b\}}\rangle + \widehat{C}_2|\Phi^{\{0,1\}}_{\{a,b\}}\rangle$$
$$= \frac{1}{\sqrt{2}}[|\beta^{(+)}_{\{1,0\}\leftrightarrow\{0,1\}}\rangle + |\beta^{(-)}_{\{1,0\}\leftrightarrow\{0,1\}}\rangle + (\widehat{C}_2)^2[|\beta^{(+)}_{\{1,0\}\leftrightarrow\{0,1\}}\rangle - |\beta^{(-)}_{\{1,0\}\leftrightarrow\{0,1\}}\rangle]]$$

This last result may be written in the following way:

$$\begin{pmatrix} |\Phi^{\{1,0\}}_{\{a,b\}}\rangle \\ \widehat{C}_2|\Phi^{\{0,1\}}_{\{a,b\}}\rangle \end{pmatrix} = \frac{1}{\sqrt{2}}\begin{pmatrix} |\beta^{(+)}_{\{1,0\}\leftrightarrow\{0,1\}}\rangle + |\beta^{(-)}_{\{1,0\}\leftrightarrow\{0,1\}}\rangle \\ (\widehat{C}_2)^2[|\beta^{(+)}_{\{1,0\}\leftrightarrow\{0,1\}}\rangle - |\beta^{(-)}_{\{1,0\}\leftrightarrow\{0,1\}}\rangle] \end{pmatrix}$$

Then, dividing the second component by $\widehat{C}_2$ leads to

$$\begin{pmatrix} |\Phi^{\{1,0\}}_{\{a,b\}}\rangle \\ |\Phi^{\{0,1\}}_{\{a,b\}}\rangle \end{pmatrix} = \frac{1}{\sqrt{2}} \begin{pmatrix} |\beta^{(+)}_{\{1,0\}\leftrightarrow\{0,1\}}\rangle + |\beta^{(-)}_{\{1,0\}\leftrightarrow\{0,1\}}\rangle \\ \widehat{C}_2[|\beta^{(+)}_{\{1,0\}\leftrightarrow\{0,1\}}\rangle - |\beta^{(-)}_{\{1,0\}\leftrightarrow\{0,1\}}\rangle] \end{pmatrix} \tag{Q.5}$$

Moreover, in view of Eq. (Q.5), the transition operator (Q.3) becomes

$$[\tilde{\mu}_{\text{Dim}}(0)] = \frac{1}{\sqrt{2}}[\mu^\circ{}_a|\beta^{(+)}_{\{1,0\}\leftrightarrow\{0,1\}}\rangle\langle\Phi^{\{0,0\}}_{\{a,b\}}| + \mu^\circ{}_b\widehat{C}_2|\beta^{(+)}_{\{1,0\}\leftrightarrow\{0,1\}}\rangle\,\langle\Phi^{\{0,0\}}_{\{a,b\}}|]$$
$$+\frac{1}{\sqrt{2}}[\mu^\circ{}_a|\beta^{(-)}_{\{1,0\}\leftrightarrow\{0,1\}}\rangle\,\langle\Phi^{\{0,0\}}_{\{a,b\}}| - \mu^\circ{}_b\widehat{C}_2|\beta^{(-)}_{\{1,0\}\leftrightarrow\{0,1\}}\rangle\,\langle\Phi^{\{0,0\}}_{\{a,b\}}|] \tag{Q.6}$$

Now, observe that Eq. (Q.6) is of the form:

$$[\tilde{\mu}_{\text{Dim}}(0)] = \frac{1}{\sqrt{2}}[\mu^\circ{}_a C^{(+)} + \mu^\circ{}_b D^{(+)}] + \frac{1}{\sqrt{2}}[\mu^\circ{}_a C^{(-)} - \mu^\circ{}_b D^{(-)}] \tag{Q.7}$$

with, respectively,

$$B^{(+)} \equiv |\beta^{(+)}_{\{1,0\}\rightarrow\{0,1\}}\rangle\,\langle\Phi^{\{0,0\}}_{\{a,b\}}| D^{(+)} = \widehat{C}_2|\beta^{(+)}_{\{1,0\}\leftrightarrow\{0,1\}}\rangle\,\langle\Phi^{\{0,0\}}_{\{a,b\}}\,|$$
$$B^{(-)} \equiv |\beta^{(-)}_{\{1,0\}\leftrightarrow\{0,1\}}\rangle\,\langle\Phi^{\{0,0\}}_{\{a,b\}}| D^{(-)} = \widehat{C}_2|\beta^{(-)}_{\{1,0\}\leftrightarrow\{0,1\}}\rangle\,\langle\Phi^{\{0,0\}}_{\{a,b\}}| \tag{Q.8}$$

Next, it is possible to write the identity:

$$\begin{aligned} &\mu^\circ{}_a B^{(+)} + \mu^\circ{}_b D^{(+)} + \mu^\circ{}_a B^{(-)} - \mu^\circ{}_b D^{(-)} \\ &= \tfrac{1}{2}[(\mu^\circ{}_a + \mu^\circ{}_b)(B^{(+)} + D^{(+)}) + (\mu^\circ{}_a - \mu^\circ{}_b)(B^{(+)} - D^{(+)})] \\ &+ \tfrac{1}{2}[(\mu^\circ{}_a + \mu^\circ{}_b)(B^{(-)} - D^{(-)}) + (\mu^\circ{}_a - \mu^\circ{}_b)(B^{(-)} + D^{(-)})] \end{aligned}$$

As a consequence, the transition operator (Q.7) takes the form:

$$[\tilde{\mu}_{\text{Dim}}(0)] = \frac{1}{2\sqrt{2}}[(\mu^\circ{}_a + \mu^\circ{}_b)(B^{(+)} + D^{(+)}) + (\mu^\circ{}_a - \mu^\circ{}_b)(B^{(+)} - D^{(+)})]$$
$$+\frac{1}{2\sqrt{2}}[(\mu^\circ{}_a + \mu^\circ{}_b)(B^{(-)} - D^{(-)}) + (\mu^\circ{}_a - \mu^\circ{}_b)(B^{(-)} + D^{(-)})] \tag{Q.9}$$

On the other hand, it is necessary to use for the ground state a symmetrized one in place of that dealing with the two moieties. This leads us to write for the starting *bra*:

$$\langle\Phi^{\{0,0\}}_{\{g,u\}}| = \langle\Phi^{\{0,0\}}_{\{a,b\}}| \tag{Q.10}$$

The right-hand side of Eq. (Q.10) is given by Eq. (Q.1). Then, in view of Eqs. (Q.8) and (Q.10), (Q.9) of becomes

$$\begin{aligned}
[\tilde{\mu}_{\text{Dim}}(0)] = &\\
\times &\left[\frac{\mu^\circ{}_a + \mu^\circ{}_b}{2}\right]\left[\frac{|\beta^{(+)}_{\{1,0\}\leftrightarrow\{0,1\}}\rangle + \widehat{C}_2|\beta^{(+)}_{\{1,0\}\leftrightarrow\{0,1\}}\rangle}{\sqrt{2}}\right]\langle\Phi^{\{0,0\}}_{\{g,u\}}| \\
+ &\left[\frac{\mu^\circ{}_a - \mu^\circ{}_b}{2}\right]\left[\frac{|\beta^{(+)}_{\{1,0\}\leftrightarrow\{0,1\}}\rangle - \widehat{C}_2|\beta^{(+)}_{\{1,0\}\leftrightarrow\{0,1\}}\rangle}{\sqrt{2}}\right]\langle\Phi^{\{0,0\}}_{\{g,u\}}| \\
+ &\left[\frac{\mu^\circ{}_a + \mu^\circ{}_b}{2}\right]\left[\frac{|\beta^{(-)}_{\{1,0\}\leftrightarrow\{0,1\}}\rangle - \widehat{C}_2|\beta^{(-)}_{\{1,0\}\leftrightarrow\{0,1\}}\rangle}{\sqrt{2}}\right]\langle\Phi^{\{0,0\}}_{\{g,u\}}| \\
+ &\left[\frac{\mu^\circ{}_a - \mu^\circ{}_b}{2}\right]\left[\frac{|\beta^{(-)}_{\{1,0\}\leftrightarrow\{0,1\}}\rangle + \widehat{C}_2|\beta^{(-)}_{\{1,0\}\leftrightarrow\{0,1\}}\rangle}{\sqrt{2}}\right]\langle\Phi^{\{0,0\}}_{\{g,u\}}|
\end{aligned} \tag{Q.11}$$

Now, we may define the following symmetrized parameters characterizing, respectively, the strength of the IR and Raman transitions:

$$\mu^\circ{}_u = \left[\frac{\mu^\circ{}_a - \mu^\circ{}_b}{2}\right] \quad \text{and} \quad \mu^\circ{}_g = \left[\frac{\mu^\circ{}_a + \mu^\circ{}_b}{2}\right]$$

Owing to the expressions of $\mu^\circ{}_a$ and $\mu^\circ{}_b$, these symmetrized parameters are

$$\mu^\circ{}_u = q^{\circ\circ}\sqrt{2}\left[\frac{\partial\mu}{\partial \mathrm{q}_u}\right]_{\mathrm{q}_u=0} \quad \text{and} \quad \mu^\circ{}_g = q^{\circ\circ}\sqrt{2}\left[\frac{\partial\mu}{\partial \mathrm{q}_g}\right]_{\mathrm{q}_g=0}$$

According to these last equations, the transition operator (Q.11) becomes

$$[\tilde{\mu}_{\text{Dim}}(0)] = [\tilde{\mu}^{(+)}_{\text{Dim}}(0)] + [\tilde{\mu}^{(-)}_{\text{Dim}}(0)] \tag{Q.12}$$

with

$$[\tilde{\mu}_{\text{Dim}}^{(\pm)}(0)] = \mu°_g[1 \pm \widehat{C}_2]|\beta^{(\pm)}_{\{1,0\}\leftrightarrow\{0,1\}}\rangle\langle\Phi^{\{0,0\}}_{\{g,u\}}| + \mu°_u[1 \mp \widehat{C}_2]|\beta^{(\pm)}_{\{1,0\}\leftrightarrow\{0,1\}}\rangle\langle\Phi^{\{0,0\}}_{\{g,u\}}| \tag{Q.13}$$

As it appears, the moment operator (Q.13) is the sum of two terms, the first one is forbidden in IR and the last one is active.

Recall that the spinors involved in Eq.(Q.13) are given by Eq.(P.36), that is,

$$\begin{pmatrix} |\Phi^{\{0,0\}}_{\{g,u\}}\rangle \\ |\beta^{(+)}_{\{1,0\}\leftrightarrow\{0,1\}}\rangle \\ |\beta^{(-)}_{\{1,0\}\leftrightarrow\{0,1\}}\rangle \end{pmatrix} = \begin{pmatrix} |\{0\}_g\rangle \otimes |\{0\}_u\rangle \\ |\{1\}_g\rangle \otimes |\{\beta^{(+)}\}^{\{1\}}_{u^+}\rangle \\ |\{1\}_g\rangle \otimes |\{\beta^{(-)}\}^{\{1\}}_{u^-}\rangle \end{pmatrix} \tag{Q.14}$$

At last, we may observe that the operators $[\widehat{1} \pm \widehat{C}_2]$ and $[\widehat{1} \mp \widehat{C}_2]$ appearing in Eq. (Q.13) and acting on the spinors (Q.14), cannot affect the g part of the system, so that

$$[\widehat{1} \pm \widehat{C}_2] = (\widehat{1})_g \otimes [\widehat{1} \pm \widehat{C}_2]_u$$

Thus, owing to the above equation, the two $(\pm)$ symmetrized transition moments (Q.13) become

$$[\tilde{\mu}_{\text{Dim}}^{(\pm)}(0)] = [\tilde{\mu}(0)]_g \otimes [[\widehat{1} \mp (\widehat{C}_2)_u][\tilde{\mu}^{(\pm)}(0)]_{u^\pm} + \eta°[\widehat{1} \pm (\widehat{C}_2)_u][\tilde{\mu}^{(\pm)}(0)]_{u^\pm}]$$

where $\eta°$ is a dimensionless parameter, less than unity, measuring the amount of the IR forbidden transition, and given by

$$\eta° = \frac{\mu°_g}{\mu°_u}$$

and where the transition moments are given by

$$[\tilde{\mu}(0)]_g = \mu°_u|\{1\}_g\rangle\langle\{0\}_g| \tag{Q.15}$$

$$[\tilde{\mu}^{(\pm)}(0)]_{u^\pm} = |\{\beta^{(\pm)}\}^{\{1\}}_{u^\pm}\rangle\langle\{0\}_u| \tag{Q.16}$$

Note that Eq. (Q.15) also may be written in terms of the nonsymmetrized states $|\{k\}_a\rangle$ and $|\{k\}_b\rangle$ of the two moieties.

$$[\tilde{\mu}(0)]_g = \mu°_u \tfrac{1}{\sqrt{2}}[|\{1\}_a\rangle + |\{1\}_b\rangle]\tfrac{1}{\sqrt{2}}[\langle\{0\}_a| + \langle\{0\}_b|]$$

This leads to

$$[\tilde{\mu}(0)]_g = \tfrac{1}{2}\mu^\circ{}_u[|\{1\}_a\rangle\langle\{0\}_a| + |\{1\}_b\rangle\langle\{0\}_b| + |\{1\}_a\rangle\langle\{0\}_b| + |\{1\}_b\rangle\langle\{0\}_a|]$$

Now, the relations between the symmetrized antisymmetric term $\mu^\circ{}_u$ and the unsymmetrized $\mu^\circ{}_a$ and $\mu^\circ{}_b$ are

$$\mu^\circ{}_u = \frac{1}{2}[\mu^\circ{}_a - \mu^\circ{}_b]$$

Thus, the above equation reads

$$[\tilde{\mu}(0)]_g = \tfrac{1}{4}[\mu^\circ{}_a - \mu^\circ{}_b] \times [|\{1\}_a\rangle\langle\{0\}_a| + |\{1\}_b\rangle\langle\{0\}_b| + |\{1\}_a\rangle\langle\{0\}_b| + |\{1\}_b\rangle\langle\{0\}_a|]$$

This last expression for $[\tilde{\mu}(0)]_g$ may be compared to that of $[\mu(0)]_g$, which we met in Section VI.A.3, p. 345, that is

$$[\mu(0)]_g = \frac{1}{\sqrt{2}}[\mu^\circ{}_b|\{0\}_a\{1\}_b\rangle\langle\{0\}_b\{0\}_a| + \mu^\circ{}_a|\{1\}_a\{0\}_b\rangle\langle\{0\}_b\{0\}_a|]$$

For some special physical situations involving centrosymmetric cyclic H-bonded systems, it seems that there is some lack in the IR selection rule, particularly in the crystalline state.

Thus, for such situations, it is suitable to use for the transition operator (Q.12) an expression given by the equations:

$$[\tilde{\mu}_{\mathrm{Dim}}(0)] = [\tilde{\mu}(0)_g] \otimes [\{F^{(+)}(\eta^\circ)\}[\tilde{\mu}^{(+)}(0)]_{u^+} + \{F^{(-)}(\eta^\circ)\}[\tilde{\mu}^{(-)}(0)]_{u^-}] \tag{Q.17}$$

$$\{\widehat{F}^{(\pm)}(\eta^\circ)\} = [[\widehat{1} \mp (\widehat{C}_2)_u] + \eta^\circ[\widehat{1} \pm (\widehat{C}_2)_u]] \tag{Q.18}$$

$$\eta^\circ = \frac{\left[\frac{\partial\mu}{\partial q_g}\right]_{q_g=0}}{\left[\frac{\partial\mu}{\partial q_u}\right]_{q_u=0}}$$

APPENDIX R: SYMMETRIZED ACF FOR CENTROSYMMETRIC CYCLIC DIMERS [76]

This appendix is devoted to the calculation of the symmetrized ACFs of centrosymmetric cyclic dimers *involving indirect damping*.

R.1 THE ACF OF THE CENTROSYMMETRIC CYCLIC DIMER AS A PRODUCT OF g AND u ACFs

Owing to Eq. (10), the ACF of a dimer is

$$[\mathrm{G}_{\mathrm{Dav}}(t)]^{\circ} = \tilde{\mathrm{tr}}_{\mathrm{Dav}}[\mathbf{R}_{\mathrm{Dav}}[\tilde{\mu}_{\mathrm{Dim}}(0)]^{\dagger}\ [\tilde{\mu}_{\mathrm{Dim}}(t)]] \tag{R.1}$$

Besides, the density operator $\mathbf{R}_{\mathrm{Dav}}$ involved in this last equation is

$$\mathbf{R}_{\mathrm{Dav}} = \frac{1}{\mathrm{Z}_{\mathrm{Dav}}}(e^{-\beta[\mathbf{H}_{\mathrm{Dav}}]}) \quad \text{with} \quad \mathrm{Z}_{\mathrm{Dav}} = \tilde{\mathrm{tr}}_{\mathrm{Dav}}[(e^{-\beta[\mathbf{H}_{\mathrm{Dav}}]})]$$

Owing to the "diagonal" form of the Hamiltonian $[\mathbb{H}_{\mathrm{Dav}}]$, the density operator factorizes into

$$\mathbf{R}_{\mathrm{Dav}} = \frac{1}{\mathrm{Z}_{\mathrm{Dav}}}(e^{-\beta[\mathbf{H}_{II}^{\{0\}}]_g})(e^{-\beta[\mathbf{H}_{II}^{\{1\}}]_g})(e^{-\beta[\mathrm{H}_{II}^{\{0\}}]_u})(e^{-\beta[\mathrm{H}_{(+)}^{\{1\}}]_{u^+}})(e^{-\beta[\mathrm{H}_{(-)}^{\{1\}}]_{u^-}})$$

with

$$\mathrm{Z}_{\mathrm{Dav}} = [\mathrm{Z}_{II}^{\{0\}}]_g[\mathrm{Z}_{II}^{\{1\}}]_g[\mathrm{Z}_{II}^{\{0\}}]_u[\mathrm{Z}_{II}^{\{1\}}]_{u^+}[\mathrm{Z}_{II}^{\{1\}}]_{u^-}$$

In this last expression of the partition function, the different terms are given by

$$[\mathrm{Z}_{II}^{\{k\}}]_g = \tilde{\mathrm{tr}}_g[(e^{-\beta[\mathbf{H}_{II}^{\{k\}}]_g})] \quad \text{and} \quad [\mathrm{Z}_{II}^{\{0\}}]_{u^\pm} = \mathrm{tr}_{u^\pm}\ [(e^{-\beta[\mathrm{H}_{II}^{\{0\}}]_{u^\pm}})]$$

and by other expressions of the same form. The trace operations $\tilde{\mathrm{tr}}_g$ and tr_u have to be performed, respectively, over the g and u symmetrized bases spanned by the Hamiltonians $[\mathrm{H}_{II}^{\{k\}}]_g$ and $[\mathrm{H}_{II}^{\{0\}}]_u$. Of course, only the lower spectrum of this density operator is susceptible to play a significant role, since the higher part is negligible at any temperature because of the energy gaps involved in the excitation of the fast mode. Thus, it is possible to write

$$\mathbf{R}_{\text{Dav}} \simeq [\tilde{\rho}_{II}^{\{0\}}]_g \ [\tilde{\rho}_{II}^{\{0\}}]_u \tag{R.2}$$

with, respectively,

$$[\tilde{\rho}_{II}^{\{0\}}]_g = \frac{1}{[Z_{II}^{\{0\}}]_g}(e^{-\beta[\mathbf{H}_{II}^{\{0\}}]_g}) \quad \text{and} \quad [\tilde{\rho}_{II}^{\{0\}}]_u = \frac{1}{[Z_{II}^{\{0\}}]_u}\ (e^{-\beta[\mathbf{H}_{II}^{\{0\}}]_u})$$

Next, owing to the above equations, the ACF (R.1) takes the form:

$$\begin{aligned}&[\mathrm{G}_{\text{Dav}}(t)]^{\circ} = \ \tilde{\mathrm{tr}}_g \ \ \mathrm{tr}_{u^+}\mathrm{tr}_{u^-}[\tilde{\rho}_{II}^{\{0\}}]_g \ [\tilde{\rho}_{II}^{\{0\}}]_u \\ &[[\tilde{\mu}(0)]_g^{\dagger}\{\mathrm{F}^{(+)}(\eta^{\circ})\}[\tilde{\mu}^{(+)}(0)]_{u^+}^{\dagger} + [\tilde{\mu}(0)]_g^{\dagger}\{\mathrm{F}^{(-)}(\eta^{\circ})\}[\tilde{\mu}^{(-)}(0)]_{u^-}^{\dagger} \\ &[\tilde{\mu}(t)]_g\{\mathrm{F}^{(+)}(\eta^{\circ})\}[\tilde{\mu}^{(+)}(t)]_{u^+} + [\tilde{\mu}(t)]_g\{\mathrm{F}^{(-)}(\eta^{\circ})\}[\tilde{\mu}^{(-)}(t)]_{u^-}]\end{aligned}$$

This may be factorized within the g and the u^+ and u^- subspaces according to

$$\begin{aligned}&[\mathrm{G}_{\text{Dav}}(t)]^{\circ} \\ &\quad = \tilde{\mathrm{tr}}_g[[\tilde{\rho}_{II}^{\{0\}}]_g[\tilde{\mu}(0)]_g^{\dagger}[\tilde{\mu}(0)]_g]\mathrm{tr}_{u^+}\ [[\tilde{\rho}_{II}^{\{0\}}]_u\{\mathrm{F}^{(+)}(\eta^{\circ})\}^2[\tilde{\mu}^{(+)}(0)]_{u^+}^{\dagger}[\tilde{\mu}^{(+)}(t)]_{u^+}] \\ &+ \tilde{\mathrm{tr}}_g[[\mathbb{R}_g][\tilde{\mu}(0)]_g^{\dagger}[\tilde{\mu}(0)]_g]\mathrm{tr}_{u^-}\ [[\tilde{\rho}_{II}^{\{0\}}]_u\{\mathrm{F}^{(-)}(\eta^{\circ})\}^2[\tilde{\mu}^{(-)}(0)]_{u^-}^{\dagger}[\tilde{\mu}^{(-)}(t)]_{u^-}]\end{aligned}$$

At this step of factorization of the u^+ and u^- subspaces, it appears that the traces $\mathrm{tr}_{\text{Slow}\ u^{\pm}}$ must be the same because of the presence of the same Boltzmann density operator $[\tilde{\rho}_{II}^{\{0\}}]_u$ on the right-hand side addition, so that they may be written more simply $\mathrm{tr}_{\text{Slow}\ u}$.

Then, using Eqs. (Q.17) and (R.2), the ACF takes the form:

$$[\mathrm{G}_{\text{Dav}}(t)]^{\circ} = [\mathrm{G}(t)]_g^{\circ}[[\mathrm{G}^{\circ(+)}(t)]_u^{\circ} + [\mathrm{G}^{\circ(-)}(t)]_u^{\circ}] \tag{R.3}$$

In Eq. (R.3), $[\mathrm{G}(t)]_g^{\circ}$ and $[\mathrm{G}^{\circ(\pm)}(t)]_u^{\circ}$ are, respectively, the ACFs of the dipole moment in the g and u subspaces. Which are given by:

$$[\mathrm{G}(t)]_g^{\circ} = \tilde{\mathrm{tr}}_g[\tilde{\rho}_{II}^{\{0\}}]_g[\tilde{\mu}(0)]_g^{\dagger}\ [\tilde{\mu}(t)]_g] \tag{R.4}$$

and:

$$[\mathrm{G}^{\circ(\pm)}(t)]_u^{\circ} = \mathrm{tr}_{u^{\pm}}\{[[\tilde{\rho}_{II}^{\{0\}}]_u[\tilde{\mu}^{(\pm)}(0)]_{u^{\pm}}^{\dagger}\ \{\widehat{\mathrm{F}}^{(\pm)}(\eta^{\circ})\}^2\ [\tilde{\mu}^{(\pm)}(t)]_{u^{\pm}}\} \tag{R.5}$$

with:

$$\{\widehat{\mathrm{F}}^{(\pm)}(\eta^{\circ})\} = [[\widehat{1} \mp\ (\widehat{\mathrm{C}}_2)_u] + \eta^{\circ}[\widehat{1} \pm\ (\widehat{\mathrm{C}}_2)_u]] \tag{R.6}$$

The notation $[G^{\circ(\pm)}(t)]_u^\circ$ has been used for these ACFs because it will appear later that they are not affected by the quantum indirect damping, because of symmetry.

R.2 THE *g* ACF

Now, consider the g ACF (R.4). According to Eq. (Q.15), the Hermitean conjugate of the g transition moment involved in ACF (R.4) is

$$[\tilde{\mu}(0)]_g^\dagger = \mu^\circ{}_u|\{0\}_g\rangle\langle\{1\}_g|$$

In the Heisenberg picture, the transition moment at time t is governed by the following Heisenberg transformation involving the Davydov Hamiltonian:

$$[\tilde{\mu}(0)]_g = \mu^\circ{}_u[e^{i\mathbb{H}_{\mathrm{Dav}}\, t/\hbar}]|\{1\}_g\rangle\langle\{0\}_g|\,[e^{-i\mathbb{H}_{\mathrm{Dav}}\, t/\hbar}]$$

Owing to these expressions, the ACF (R.4) takes the form:

$$[G(t)]_g^\circ = (\mu_u^\circ)^2\ \tilde{\mathrm{tr}}_g\ \{[\tilde{\rho}_{II}^{\{0\}}]_g|\{0\}_g\rangle\ \langle\{1\}_g|[e^{i\mathbb{H}_{Dav}t/\hbar}]|\{1\}_g\rangle\ \langle\{0\}_g|[e^{-i\mathbb{H}_{Dav}t/\hbar}]\}$$

Then, using the expression of the density operator, it becomes:

$$[\mathrm{G}(t)]_g^\circ = \frac{(\mu^\circ{}_u)^2}{[\mathrm{Z}_{II}^{\{0\}}]_g}$$
$$\tilde{\mathrm{tr}}_g\ \{(e^{-\beta[\mathbf{H}_{II}^{\{0\}}]_g})|\{0\}_g\rangle\langle\{1\}_g|[e^{i\mathbb{H}_{\mathrm{Dav}}\, t/\hbar}]|\{1\}_g\rangle\langle\{0\}_g|\,[e^{-i\mathbb{H}_{\mathrm{Dav}}\, t/\hbar}]\} \qquad \text{(R.7)}$$

Next, according to Eq. (268), the following equations are verified

$$[e^{i\mathbb{H}_{\mathrm{Dav}}\, t/\hbar}]|\{1\}_g\rangle = [e^{i[\mathbb{H}_{II}^{\{1\}}]_g\, t/\hbar}]|\{1\}_g\rangle$$
$$\langle\{0\}_g|[\, e^{-i\mathbb{H}_{\mathrm{Dav}}\, t/\hbar}] = \langle\{0\}_g|\,[e^{-i[\mathbb{H}_{II}^{\{0\}}]_g\, t/\hbar}]$$

As a consequence, the g ACF (R.7) becomes

$$[\mathrm{G}(t)]_g^\circ = \frac{(\mu^\circ{}_u)^2}{[\mathrm{Z}_{II}^{\{0\}}]_g}$$
$$\tilde{\mathrm{tr}}_g\ \{(e^{-[\mathbf{H}_{II}^{\{0\}}]_g/k_BT})|\{0\}_g\rangle\langle\{1\}_g|[\, e^{i[\mathbb{H}_{II}^{\{1\}}]_g\, t/\hbar}]|\{1\}_g\rangle\langle\{0\}_g|\,[e^{-i[\mathbb{H}_{II}^{\{0\}}{}_g\, t/\hbar}]\}$$

Then, using the fact that the operators do not work within the space to which belong the different *kets* and *bras* belong, and after simplifications with the aid of the orthonormality properties, this ACF reduces to

$$[G(t)]_g^\circ = \frac{(\mu^\circ{}_u)^2}{[Z_{II}^{\{0\}}]_g} \tilde{tr}_g \; \{(e^{-[\mathbf{H}_{II}^{\{0\}}]_g/k_B T}) \; [\; e^{i[\mathbb{H}_{II}^{\{1\}}]_g \, t/\hbar}] \; [e^{-i[\mathbb{H}_{II}^{\{0\}}]_g \, t/\hbar}]\}$$

Now, it appears that, in the same spirit, it is possible to write

$$\tilde{tr}_g = tr_{\text{Slow } g} tr_\theta$$

where $tr_{\text{Slow } g}$ is the trace operation to be performed over the base spanned by the eigenvectors of the Hamiltonian of the g quantum harmonic oscillator describing the corresponding H-bond bridge, whereas tr_θ is dealing with the base of the thermal bath.

As a consequence, the above ACF transforms into

$$[G(t)]_g^\circ = \tilde{\varepsilon} \; (\mu^\circ{}_u)^2 \; tr_{\text{Slow } g} \; [(e^{-[\mathbb{H}_{II}^{\{0\}}]_g/k_B T}) tr_\theta[[e^{i[\mathbb{H}_{II}^{\{1\}}]_g \, t/\hbar}] \; [e^{-i[\mathbb{H}_{II}^{\{0\}}]_g \, t/\hbar}]]] \tag{R.8}$$

R.3 THE u ACF

Next, consider the u ACFs (R.5), that is,

$$[G^{\circ(\pm)}(t)]_u^\circ = tr_u \; [\; [\tilde{\rho}_{II}^{\{0\}}]_u \; [\tilde{\mu}^{(\pm)}(0)]_{u^\pm}^\dagger \; \{\widehat{F}^{(\pm)}(\eta^\circ)\}^2 [\tilde{\mu}^{(\pm)}(t)]_{u^\pm}] \tag{R.9}$$

Then, according to Eq. (Q.16), the Hermitean conjugate of the u transition moment is

$$\left[\tilde{\mu}^{(\pm)}(0)\right]_{u^\pm}^\dagger = |\{0\}_u\rangle\langle\{\beta^{(\pm)}\}_u^{\{1\}}|$$

In the Heisenberg picture, the transition moment at time t is governed by the Davydov Hamiltonian:

$$[\tilde{\mu}^{(\pm)}(t)]_{u^\pm} = [e^{i\mathbb{H}_{\text{Dav}} \, t/\hbar}] |\{\beta^{(\pm)}\}_u^{\{1\}}\rangle\langle\{0\}_u| \; [e^{-i\mathbb{H}_{\text{Dav}} \, t/\hbar}]$$

Besides, the Boltzmann operator of the *u* subsystem is given by the equations

$$[\tilde{\rho}_{II}^{\{0\}}]_u = \tilde{\varepsilon}\ (e^{-[\mathrm{H}_{II}^{\{0\}}]_u/k_\mathrm{B}T})$$

$$\frac{1}{\tilde{\varepsilon}} = \mathrm{tr}_u\ [(e^{-[\mathrm{H}_{II}^{\{0\}}]_u/k_\mathrm{B}T})] = \sum_{n_u}(e^{-\left[n_u+\frac{1}{2}\right]\hbar\Omega/k_\mathrm{B}T})$$

Then, the *u* ACFs (R.9) become

$$[\mathrm{G}^{\circ(\pm)}(t)]_u^{\circ} = \tilde{\varepsilon}\ \mathrm{tr}_u\ \{(e^{-[\mathrm{H}_{II}^{\{0\}}]_u/k_\mathrm{B}T})|\{0\}_u\rangle\langle\{\beta^{(\pm)}\}_u^{\{1\}}| \\ \{\widehat{\mathrm{F}}^{(\pm)}(\eta^{\circ})\}^2[e^{i\mathbb{H}_{\mathrm{Dav}}\,t/\hbar}]|\{\beta^{(\pm)}\}_u^{\{1\}}\rangle\langle\{0\}_u|\ [e^{-i\mathbb{H}_{\mathrm{Dav}}\,t/\hbar}]\} \tag{R.10}$$

Again, owing to Eq. (268) giving the expression of the Hamiltonian $\mathrm{H}_{\mathrm{Dav}}$, the following equations hold for the different terms involved in the ACF (R.10):

$$[e^{i[\mathbb{H}_{\mathrm{Dav}}]\,t/\hbar}]|\{\beta^{(\pm)}\}_u^{\{1\}}\rangle = [e^{i[\mathrm{H}_{(\pm)}^{\{1\}}]_u\,t/\hbar}]|\{\beta^{(\pm)}\}_{u^\pm}^{\{1\}}\rangle$$

$$\langle\{0\}_u|\ [e^{-[i\mathbb{H}_{\mathrm{Dav}}]\,t/\hbar}] = \langle\{0\}_u|\ e^{-i[\mathrm{H}_{II}^{\{0\}}]_u\,t/\hbar}$$

$$[\tilde{\mu}^{(\pm)}(t)]_{u^\pm} = [e^{i[\mathrm{H}_{(\pm)}^{\{1\}}]_u\,t/\hbar}]|\{\beta^{(\pm)}\}_{u^\pm}^{\{1\}}\rangle\langle\{0\}_u|\ [e^{-i[\mathrm{H}_{II}^{\{0\}}]_u\,t/\hbar}]$$

Thus, owing to these equations, the *u* ACFs (R.10) become

$$\left[\mathrm{G}^{\circ(\pm)}(t)\right]_u^{\circ} = \tilde{\varepsilon}\ \mathrm{tr}_u\ \{(e^{-[\mathrm{H}_{II}^{\{0\}}]_u/k_\mathrm{B}T}) \\ |\{0\}_u\rangle\left\langle\left\{\beta^{(\pm)}\right\}_{u^\pm}^{\{1\}}\right|\left\{\widehat{\mathrm{F}}^{(\pm)}(\eta^{\circ})\right\}^2\left[e^{i[\mathrm{H}_{(\pm)}^{\{1\}}]_u\,t/\hbar}\right]\left|\left\{\beta^{(\pm)}\right\}_{u^\pm}^{\{1\}}\right\rangle\langle\{0\}_u|[e^{-i[\mathrm{H}_{II}^{\{0\}}]_u\,t/\hbar}]\}$$

Next, performing a circular permutation within the trace, this equation gives

$$\left[\mathrm{G}^{\circ(\pm)}(t)\right]_u^{\circ} = \tilde{\varepsilon}\ \mathrm{tr}\ \{\{\widehat{\mathrm{F}}^{(\pm)}(\eta^{\circ})\}^2 \\ \left[e^{i[\mathrm{H}_{(\pm)}^{\{1\}}]_u\,t/\hbar}\right]\left|\left\{\beta^{(\pm)}\right\}_{u^\pm}^{\{1\}}\right\rangle\langle\{0\}_u|\ [e^{-i[\mathrm{H}_{II}^{\{0\}}]_u\,t/\hbar}](e^{-[\mathrm{H}_{II}^{\{0\}}]_u/k_\mathrm{B}T})|\{0\}_u\rangle\langle\{\beta^{(\pm)}\}_{u^\pm}^{\{1\}}|\}$$

Again, use the fact that the exponential operators commute with the *kets* and the *bras* appearing explicitly in this equation. Then use the orthonormality properties dealing with these *kets* and these *bras*. At last make explicit the trace involving the eigenstates of the *u* quantum harmonic oscillator.

Then the u ACFs reduce to

$$[\mathrm{G}^{\circ(\pm)}(t)]_u^{\circ} = \tilde{\varepsilon}\sum_n \langle(n)_u|\{\widehat{\mathrm{F}}^{(\pm)}(\eta^{\circ})\}^2[e^{i[\mathrm{H}^{\{1\}}_{(\pm)}]_u\, t/\hbar}][e^{-i[\mathrm{H}^{\{0\}}_{II}]_u t/\hbar}]\,(e^{-[\mathrm{H}^{\{0\}}_{II}]_u\,/k_\mathrm{B}T})|\,(n)_u\rangle \tag{R.11}$$

Now, observe that Eq. (R.6) allows us to write

$$\{\widehat{\mathrm{F}}^{(\pm)}(\eta^{\circ})\}|(n)_u\rangle = \{[\widehat{1} \mp \widehat{\mathrm{C}}_2] + \eta^{\circ}[\widehat{1} \pm \widehat{\mathrm{C}}_2]\}|(n)_u\rangle \tag{R.12}$$

Again, it appears by inspection of Fig. 16 that

$$\widehat{\mathrm{C}}_2|(n)_u\rangle = (-1)^{n_u}|(n)_u\rangle$$

Thus, Eq. (R.12) leads to

$$\{\widehat{\mathrm{F}}^{(\pm)}(\eta^{\circ})\}|(n)_u\rangle = [F^{(\pm)}_{n_u}]|(n)_u\rangle \tag{R.13}$$

with:

$$[F^{(\pm)}_{n_u}] = [1 \pm (-1)^{n_u+1}] + \eta^{\circ}[1 \mp (-1)^{n_u+1}] \tag{R.14}$$

Moreover, use the eigenvalue equation

$$[\mathrm{H}^{\{0\}}_{II}]_u\,|\,(n)_u\rangle = [n_u + \frac{1}{2}]\hbar\Omega|\,(n)_u\rangle \tag{R.15}$$

As a consequence of Eqs. (R.13) and (R.15) and after neglecting the zeropoint energy, the ACFs (R.11) take the form:

$$[\mathrm{G}^{\circ(\pm)}(t)]_u^{\circ} = \tilde{\varepsilon}\sum_{n_u}(e^{-n_u\hbar\Omega/k_\mathrm{B}T})(e^{-in_u\Omega t})[F^{(\pm)}_{n_u}]^2\langle(n)_u|(e^{i[\mathrm{H}^{\{1\}}_{(\pm)}]_u\,t/\hbar})|\,(n)_u\rangle \tag{R.16}$$

The eigenvalue equation of the effective Hamiltonian involved in the exponential operator is

$$[\mathrm{H}^{\{1\}}_{(\pm)}]|\,\beta^{(\pm)}_{\mu}\rangle = \hbar\omega^{\pm}_{\mu}\,|\,\beta^{(\pm)}_{\mu}\rangle \tag{R.17}$$

with

$$|\,\beta^{(\pm)}_{\mu}\rangle = \sum[B^{\pm}_{n_u\,\mu}]|\,(n)_u\rangle \tag{R.18}$$

Next, introduce into Eq. (R.16) just after the evolution operator, a closeness relation on the eigenstates involved in Eq. (R.17). Then, the *u* ACFs (R.16) become

$$[G^{\circ(\pm)}(t)]_u^{\circ} = \varepsilon \sum_{\mu} \sum_{n_u}$$

$$(e^{-n_u \hbar\Omega/k_B T})(e^{-in_u\Omega t})[F_{n_u}^{(\pm)}]^2 \langle (n)_u | (e^{i[H_{(\pm)}^{\{1\}}]_u \, t/\hbar}) | \, \beta_{\mu}^{(\pm)} \rangle \langle \beta_{\mu}^{(\pm)} || \, (n)_u \rangle$$

At last, using in turn Eq. (R.18), the *u* ACFs take the final form:

$$[G^{\circ(\pm)}(t)]_u^{\circ} = \varepsilon \sum_{\mu} \sum_{n_u}$$

$$|B_{\mu\, n_u}^{\pm}|^2 (e^{-n_u \hbar\Omega \,/k_B T}) \; (e^{i\omega_{\mu}^{\pm} t})(e^{-in_u\Omega t})\{[1 \pm (-1)^{n_u+1}] + \eta^{\circ}[1 \mp (-1)^{n_u+1}]\}^2 \qquad \text{(R.19)}$$

APPENDIX S: FROM EQ. (304) TO EQ. (309) DEALING WITH THE "PEELING-OFF" APPROACH OF MARÉCHAL [83]

The aim of this appendix is to show that the g ACF (304) involving Davydov coupling, Fermi resonances, and damping , may be viewed, after some simplifications, as formally equivalent to that used by Maréchal [83] in his "peeling-off" approach of Fermi resonances.

For this aim, purpose let us look at the Hamiltonian $[\tilde{\mathrm{H}}^{\mathrm{Fermi}}]_g$ given by Eq.(297). For simplicity, we are here limit to the special case of two Fermi resonances and the situation where the *indirect damping is missing* and where the *direct damping* of the first excited state of the high frequency mode *is the same* as those of the first harmonics of the two bending modes. Then, the Hamiltonian (297) reduces to

$$[\mathrm{H}^{\mathrm{Fermi}}]_g = \begin{pmatrix} [\mathrm{H}_{II}^{\{1\}}]_g & \hbar f_1^{\delta} & \hbar f_1^{\delta} \\ \hbar f_1^{\delta} & \mathrm{H}^{\circ} + \hbar\Delta_1^{\delta} & 0 \\ \hbar f_2^{\delta} & 0 & \mathrm{H}^{\circ} + \hbar\Delta_2^{\delta} \end{pmatrix} - i\hbar\gamma^{\circ} \begin{pmatrix} 1 & 0 & 0 \\ 0 & 1 & 0 \\ 0 & 0 & 1 \end{pmatrix} \tag{S.1}$$

with

$$[\mathrm{H}_{II}^{\{1\}}]_g = \breve{\mathrm{H}}^{\circ} + \mathrm{V}_{\mathrm{Dr}} \tag{S.2}$$

$\breve{\mathrm{H}}^{\circ}$ and V_{Dr}, are, respectively, given by

$$\breve{\mathrm{H}}^{\circ} = [\mathrm{H}_{II}^{\{0\}}]_g + \hbar\omega^{\circ} - \tilde{\alpha}^{\circ 2}\hbar\Omega \tag{S.3}$$

$$[\mathrm{H}_{II}^{\{0\}}]_g = \left[\mathrm{a}_g^{\dagger}\mathrm{a}_g + \frac{1}{2}\right]\hbar\Omega \tag{S.4}$$

$$\mathrm{V}_{\mathrm{Dr}} = \tilde{\alpha}^{\circ}[\mathrm{a}_g^{\dagger} + \mathrm{a}_g]\hbar\Omega \tag{S.4}$$

In these equation, the basic parameters have the same meaning as in the section of reference. Besides, f_i^{δ} and Δ_i^{δ} are, respectively, the couplings and the complex gaps expressed in angular frequencies and involved in the Fermi resonances.

Now, look at the nonunitary time-evolution operator governed by the non-Hermitean Hamiltonian (S.1), which is the solution of the Shrödinger equation:

$$i\hbar \frac{\partial}{\partial t}[\tilde{\text{U}}^{\text{Fermi}}(t)]_g = [\text{H}^{\text{Fermi}}]_g[\tilde{\text{U}}^{\text{Fermi}}(t)]_g$$

The solution is

$$[\tilde{\text{U}}^{\text{Fermi}}(t)]_g = [e^{-i[\text{H}^{\text{Fermi}}]_g t/\hbar}](e^{-\gamma^\circ t}) \tag{S.6}$$

Now, in order to make this result in connection with the Maréchal model [83], it is suitable to work within the IP, and thus perform the following partition:

$$[\text{H}^{\text{Fermi}}]_g = [\text{H}_I^{\text{Fermi}}]_g + \breve{\text{H}}^\circ \tag{S.7}$$

with

$$[\text{H}_I^{\text{Fermi}}]_g = \begin{pmatrix} \text{V}_{\text{Dr}} & \hbar f_1^\delta & \hbar f_2^\delta \\ \hbar f_1^\delta & \hbar\Delta_1^\delta & 0 \\ \hbar f_2^\delta & 0 & \hbar\Delta_2^\delta \end{pmatrix} \tag{S.8}$$

In the IP with respect to H°, the first right Hamiltonian involved in this equation is given by

$$[\text{H}_I^{\text{Fermi}}(t)]_g^{\text{IP}} = (e^{i\breve{\text{H}}^\circ t/\hbar})[\text{H}_I^{\text{Fermi}}]_g(e^{-i\breve{\text{H}}^\circ t/\hbar}) \tag{S.9}$$

Thus, in view of Eq. (S.1), the IP Hamiltonian (S.9) takes the form:

$$[\text{H}_I^{\text{Fermi}}(t)]_g^{\text{IP}} = (e^{i\breve{\text{H}}^\circ t/\hbar})\begin{pmatrix} 1 & 0 & 0 \\ 0 & 1 & 0 \\ 0 & 0 & 1 \end{pmatrix}\begin{pmatrix} \text{V}_{\text{Dr}} & \hbar f_1^\delta & \hbar f_2^\delta \\ \hbar f_1^\delta & \hbar\Delta_1^\delta & 0 \\ \hbar f_2^\delta & 0 & \hbar\Delta_2^\delta \end{pmatrix}\begin{pmatrix} 1 & 0 & 0 \\ 0 & 1 & 0 \\ 0 & 0 & 1 \end{pmatrix}(e^{-i\breve{\text{H}}^\circ t/\hbar})$$

Again, performing the matricial product, it becomes

$$\text{H}_I^{\text{Fermi}}(t)]_g^{\text{IP}} \begin{pmatrix} [\text{V}_{\text{Dr}}(t)]^{\text{IP}} & \hbar f_1^\delta & \hbar f_2^\delta \\ \hbar f_1^\delta & \hbar\Delta_1^\delta & 0 \\ \hbar f_2^\delta & 0 & \hbar\Delta_2^\delta \end{pmatrix} \tag{S.10}$$

with

$$[\text{V}_{\text{Dr}}(t)]^{\text{IP}} = (e^{i\breve{\text{H}}^\circ t/\hbar})[\text{V}_{\text{Dr}}](e^{-i\breve{\text{H}}^\circ t/\hbar})$$

As a consequence, in this IP, the time-evolution operator (S.6) of the system is

$$[\mathrm{U}^{\text{Fermi}}(t)]_g = (e^{-i\breve{\mathrm{H}}^\circ t/\hbar})\ [\mathrm{S}(t)]^{\mathrm{IP}}(e^{-\gamma^\circ t}) \tag{S.11}$$

where $[\mathrm{S}(t)]^{\mathrm{IP}}$ is the IP time-evolution operator that is the solution of the Schrödinger equation:

$$i\hbar\frac{\partial[\mathrm{S}(t)]^{\mathrm{IP}}}{\partial t} = [\mathrm{H}_I^{\text{Fermi}}(t)]_g^{\mathrm{IP}}[\mathrm{S}(t)]^{\mathrm{IP}} \tag{S.12}$$

Now, perform a new partition, dealing in this case with the Hamiltonian (S.10), according to

$$[\mathrm{H}_I^{\text{Fermi}}(t)]_g^{\mathrm{IP}} = [\mathrm{H}_{II}^{\text{Fermi}}]_g + [\mathrm{V}_{\mathrm{Dr}}(t)]^{\mathrm{IP}}$$

with

$$[\mathrm{H}_{II}^{\text{Fermi}}]_g = \begin{pmatrix} 0 & \hbar f_1^{\delta} & \hbar f_2^{\delta} \\ \hbar f_1^{\delta} & \hbar\Delta_1^{\delta} & 0 \\ \hbar f_2^{\delta} & 0 & \hbar\Delta_2^{\delta} \end{pmatrix} \tag{S.13}$$

In this IP, which will be working with respect to $[\mathrm{V}_{\mathrm{Dr}}(t)]^{\mathrm{IP}}$, the Hamiltonian (S.13) is given by

$$[\mathrm{H}_{II}^{\text{Fermi}}(t)]_g^{\mathrm{IP}} = [\mathrm{K}(t)^{\mathrm{IP}}]^{-1}[\mathrm{H}_{II}^{\text{Fermi}}]_g^{\mathrm{IP}}[\mathrm{K}(t)^{\mathrm{IP}}]^{-1} \tag{S.14}$$

Here, $k(t)^{Ie}$ is the time-evolution operator that is the solution of the Schrödinger equation:

$$i\hbar\frac{\partial[\mathrm{K}(t)^{\mathrm{IP}}]}{\partial t} = [\mathrm{V}_{\mathrm{Dr}}(t)^{\mathrm{IP}}][\mathrm{K}(t)^{\mathrm{IP}}]$$

The formal solution of this Schrödinger equation is

$$[\mathrm{K}(t)^{\mathrm{IP}}] = \widehat{\mathrm{P}}\left[e^{-\frac{i}{\hbar}\int_0^t[\mathrm{V}_{\mathrm{Dr}}(t')]^{\mathrm{IP}}dt'}\right] \tag{S.15}$$

where $\widehat{\mathrm{P}}$ is the Dyson time-ordering operator. As a consequence of Eq. (S.15), Eq. (S.14) may be written

$$[\mathrm{H}_{II}^{\text{Fermi}}(t)]_g^{\mathrm{IP}} = \widehat{\mathrm{P}}\left[e^{\frac{i}{\hbar}\int_0^t[\mathrm{V}_{\mathrm{Dr}}(t')^{\mathrm{IP}}]^{\dagger}dt'}\right]\ [\mathrm{H}_{II}^{\text{Fermi}}]_g\widehat{\mathrm{P}}\ [e^{-\frac{i}{\hbar}\int_0^t[\mathrm{V}_{\mathrm{Dr}}(t')^{\mathrm{IP}}]dt'}]$$

Then, owing to Eq. (S.13), it becomes

$$[\mathrm{H}_{II}^{\mathrm{Fermi}}(t)]_g^{\mathrm{IP}} = \widehat{\mathrm{P}}\left[e^{\frac{i}{\hbar}\int_0^t [\mathrm{V}_{\mathrm{Dr}}(t')^{\mathrm{IP}}]^\dagger dt'}\right]\begin{pmatrix} 1 & 0 & 0 \\ 0 & 0 & 0 \\ 0 & 0 & 0 \end{pmatrix}$$

$$\begin{pmatrix} 0 & \hbar f_1^\delta & \hbar f_2^\delta \\ \hbar f_1^\delta & \hbar\Delta_1^\delta & 0 \\ \hbar f_2^\delta & 0 & \hbar\Delta_2^\delta \end{pmatrix}\begin{pmatrix} 1 & 0 & 0 \\ 0 & 0 & 0 \\ 0 & 0 & 0 \end{pmatrix}\widehat{\mathrm{P}}[e^{-\frac{i}{\hbar}\int_0^t [\mathrm{V}_{\mathrm{Dr}}(t')^{\mathrm{IP}}]dt'}]$$

Again, performing the matrix product, gives

$$[\mathrm{H}_{II}^{\mathrm{Fermi}}(t)]_g^{\mathrm{IP}} = \begin{pmatrix} 0 & \hbar f_1^\delta(t) & \hbar f_2^\delta(t) \\ \hbar f_1^\delta(t)^\dagger & \hbar\Delta_1^\delta & 0 \\ \hbar f_2^\delta(t)^\dagger & 0 & \hbar\Delta_2^\delta \end{pmatrix}$$

with

$$f_i^\delta(t) = f_i^\delta\ \widehat{\mathrm{P}}\left[e^{-\frac{i}{\hbar}\int_0^t [\mathrm{V}_{\mathrm{Dr}}(t')^{\mathrm{IP}}]dt'}\right] \tag{S.16}$$

In this new interaction picture, the time-evolution operator $\mathrm{S}^{\{1\}}(t)^{\mathrm{IP}}$, solution of Eq. (S.12), becomes

$$[\mathrm{S}(t)]^{\mathrm{IP}} = \widehat{\mathrm{P}}\left[e^{-\frac{i}{\hbar}\int_0^t [\mathrm{V}_{\mathrm{Dr}}(t')^{\mathrm{IP}}]dt'}\right][\mathrm{O}_0(t)^{\mathrm{IP}}] \tag{S.17}$$

Here, $\mathrm{O}_0(t)^{\mathrm{IP}}$ is the solution of the following linear set of differential equations:

$$i\hbar\frac{\partial}{\partial t}\begin{pmatrix} \mathrm{O}_0(t)^{\mathrm{IP}} \\ \mathrm{O}_1(t)^{\mathrm{IP}} \\ \mathrm{O}_2(t)^{\mathrm{IP}} \end{pmatrix} = \begin{pmatrix} 0 & \hbar f_1^\delta(t) & \hbar f_2^\delta(t) \\ \hbar f_1^\delta(t)^\dagger & \hbar\Delta_1^\delta & 0 \\ \hbar f_2^\delta(t)^\dagger & 0 & \hbar\Delta_2^\delta \end{pmatrix}\begin{pmatrix} \mathrm{O}_0(t)^{\mathrm{IP}} \\ \mathrm{O}_1(t)^{\mathrm{IP}} \\ \mathrm{O}_2(t)^{\mathrm{IP}} \end{pmatrix} \tag{S.18}$$

involving the boundary conditions:

$$\mathrm{O}_k(0)^{\mathrm{IP}} = \delta_{0,k} \tag{S.19}$$

As a consequence of Eq. (S.17), the full-time evolution operator (S.11) takes the form:

$$[\tilde{\mathrm{U}}^{\mathrm{Fermi}}(t)]_g = (e^{-\gamma^\circ t})[e^{-i\mathrm{H}^\circ t/\hbar}]\widehat{\mathrm{P}}\left[e^{-\frac{i}{\hbar}\int_0^t [\mathrm{V}_{\mathrm{Dr}}(t')^{\mathrm{IP}}]dt'}\right][\mathrm{O}_k(t)^{\mathrm{IP}}] \tag{S.20}$$

Now, let us look at the g ACF (301), that is,

$$[\breve{G}(t)]_g = \mu_0^2 \mathrm{tr}_{\mathrm{Fermi}\ g}[[\tilde{\rho}_{II}^{\{0\}}]_g\ [\mu(0)]_g^\dagger\ [\mu(t)]_g]$$

In this last equation, $[\tilde{\rho}_{II}^{\{0\}}]_g$ is the Boltzmann density operator of the g H-bond bridge. At last $[\mu(t)]_g$ is the transition moment operator of the g fast mode at time t that are given by

$$[\mu(t)]_g = [e^{i[\tilde{H}^{\mathrm{Fermi}}]_g t/\hbar}]|\{1\}_g\ \rangle\ \langle\ \{0\}_g|\ [e^{-i[H_{II}^{\{0\}}]_g t/\hbar}]$$
$$[\mu(0)]_g^\dagger = |\{0\}_g\ \rangle\ \langle\ \{1\}_g|$$

Proceeding in the usual way, and using Eqs. (S.3) and (S.6), the g ACF takes the form:

$$[\breve{G}(t)]_g = (\mu_0^2)\mathrm{tr}_{\mathrm{Fermi}\ g}\{[\tilde{\rho}_{II}^{\{0\}}]_g\ [\tilde{U}^{\mathrm{Fermi}}(t)]_g^\dagger[e^{-i[H_{II}^{\{0\}}]_g t/\hbar}]\}$$

with

$$[\tilde{\rho}_{II}^{\{0\}}]_g = \tilde{\varepsilon}(e^{-\beta[H_{II}^{\{0\}}]_g})$$

Again, with the aid of Eq. (S.20), it becomes

$$[\breve{G}^\circ(t)]_g = (\mu_0^2)(e^{-\gamma^\circ t})[O_0(t)^{\mathrm{IP}}]^\dagger$$
$$\tilde{\varepsilon}\mathrm{tr}_{\mathrm{Slow}\ g}\left\{(e^{-\beta[H_{II}^{\{0\}}]_g})\ \widehat{P}\ \left[e^{\frac{i}{\hbar}\int_0^t [V_{\mathrm{Dr}}(t')^{\mathrm{IP}}]dt'}\right](e^{iH^\circ t/\hbar})[e^{-i[H_{II}^{\{0\}}]_g t/\hbar}]\right\}$$

Thus, owing to Eqs. (S.3) and (S.4), it appears that

$$[e^{i\tilde{H}^\circ t/\hbar}][e^{-i[H_{II}^{\{0\}}]_g t/\hbar}] = (e^{i\omega^\circ t})(e^{-i^\circ\alpha^{\circ 2}\Omega t})$$

As a consequence, using this result, the above ACF reads

$$[\breve{G}^\circ(t)]_g = (\mu_0^2)e^{(i\omega^\circ t)}(e^{-\gamma^\circ t}\ )(e^{-i\alpha^{\circ 2}t})[O_0(t)^{\mathrm{IP}}]^\dagger$$
$$\tilde{\varepsilon}\mathrm{tr}\ _{Slow\ g}\left\{(e^{-\beta[H_{II}^{\{0\}}]_g})\ P\left[e^{\frac{i}{\hbar}\int_0^t [V_{\mathrm{Dr}}(t')^{\mathrm{IP}}]dt'}\right]\right\} \quad \text{(S.21)}$$

After calculating $[V_{Dr}(t')^{IP}]$ and passing to Bosons, the ACF (S.21) becomes (see passage from Eq. (64) to Eq. (65):

$$[\breve{G}^{\circ}(t)]_g = (\mu_0^2)(e^{i\omega^{\circ}t})(e^{-\gamma^{\circ}t})(e^{-i\alpha^{\circ 2}\Omega t})[O_0(t)^{IP}]^{\dagger}$$
$$\tilde{\varepsilon}\ \mathrm{tr}_{Slow}\left[(e^{-\tilde{\lambda}[a^{\dagger}a+1/2]})\widehat{P}\left[e^{i\alpha^{\circ}\int_0^t [a^{\dagger}e^{it'}+ae^{-i\Omega t'}]dt'}\right]\right]$$

Then, working in the same way as for passing from Eqs. (65) to (66), and after neglecting the zero-point energy, the ACF becomes

$$[\breve{G}^{\circ}(t)]_g = (\mu_0^2)(e^{i\omega^{\circ}t})(e^{-\gamma^{\circ}t})(e^{-i\alpha^{\circ 2}t})[O_0(t)^{IP}]^{\dagger}$$
$$\mathrm{tr}_{Slow}[(e^{-\tilde{\lambda}a^{\dagger}a})(e^{-i\alpha^{\circ 2}\Omega t})(e^{i\alpha^{\circ 2}\sin\Omega t})[e^{\Phi^{\circ}{}_{II}(t)^{*}a^{\dagger}-\Phi^{\circ}{}_{II}(t)a}]]$$

At last, performing the trace as in Section II.E.1, the above ACF takes the final form:

$$[\breve{G}^{\circ}(t)]_g = (\mu_0^2)(e^{i\omega^{\circ}t})(e^{-\gamma^{\circ}t})[O_0(t)^{IP}]^{\dagger}(e^{-i2\alpha^{\circ 2}\Omega t})(e^{\,i\alpha^{\circ 2}\sin\Omega t})$$
$$e^{2\alpha^{\circ 2}[\langle n\rangle+1/2](\cos(\Omega t)-1)} \tag{S.22}$$

This last result is Eq. (309) used in order to make a connection with the "peeling-off" model of Maréchal [102].

APPENDIX T: GLOSSARY

T.1. BASIC TERMS

T.1.A. Notations and Abbreviations

$\hbar$ and k_B:	Respectively, Planck constant divided by 2π and Boltzmann constant.
$[A, B]$:	Commutator of operators A and B.
$A^{\dagger}$ and A^{-1}:	Respectively, Hermitean conjugate (*hc*) and inverse of the operator A.
tr[A]:	Trace operation on operator A over a base belonging to the Hilbert space in which *A* is working.
$\langle A \rangle$:	Thermal average of the operator A.
$[\delta(\omega - \omega')]$ and δ_{ij}:	Respectively, Dirac distribution and Kronecker symbol.
$[\delta^{\tau}(\omega - \omega')]$:	Transverse Dirac function.
g and *u*:	Symmetric and antisymmetric with respect to the C_2 symmetry operation.

T.1.B. Basic Physical Observables and Corresponding Scalars

T.1.B.1. Monomer Observables

Q and P:	Coordinate operator of the *H-bond bridge* and its conjugate momentum with $[Q, P] = i\hbar$
$Q(t)$:	*Brownian* H-bond bridge coordinate.
$Q(t)^{IP}$:	Coordinate operator describing the H-bond bridge in the interaction picture with respect to H°, in the absence of thermal bath.
$\tilde{Q}(t)^{IP}$:	Operator Q in the interaction picture with respect to H°, taking into account the interaction with the thermal bath.
q and p:	Coordinate operator describing the *high frequency* mode and its conjugate momentum with $[q, p] = i\hbar$.

$\tilde{q}_r$ and $\tilde{p}_r$:	Coordinate operator describing the *r*-th oscillator of the *thermal bath* and its conjugate momentum, with $[\tilde{q}_r \tilde{p}_r] = i\hbar$.

T.1.B.2. Centrosymmetric Dimer Observables

Q_a, P_a, and Q_b, P_b:	Coordinates and conjugate momentum of the *a* and *b* *H-bond bridge* moieties.
Q_g, P_g, and Q_u, P_u:	Coordinates and conjugate momentum of the symmetrized *g* and *u* *H-bond bridge* moieties.
q_a, p_a and q_b, p_b:	Coordinates and conjugate momentum of the *a* and *b* *high frequency* oscillators moieties.
q_g, p_g, and q_u, p_u:	Coordinates and conjugate momentum of the *g* and *u* *high frequency* oscillators moieties.

T.1.B.3. Basic Scalars Common to Monomers and Centrosymmetric Dimers

D_e:	Dissociation energy of the H-bond bridge.
$Q^{\circ\circ}$ and $q^{\circ\circ}$:	Fundamental unit length for the H-bond bridge and of the high frequency mode.
M:	Reduced mass of the $H - bond\ bridge$.
m:	Reduced mass of the *high frequency* mode.
m_i^{δ} :	Reduced mass of the *i*-th *bending* mode.
$\tilde{m}_i$:	Reduced mass of the *i*-th harmonic oscillator of the *thermal bath*.

T.1.B.3.a. Angular Frequencies and Related Gaps Ω: Harmonic H-bond bridge.

$\tilde{\Omega}$:	Brownian harmonic *H-bond bridge*.
ω°:	*High frequency* mode when the H-bond bridge length is at equilibrium.
$\omega(Q)$:	*High frequency* mode that reduces to ω° when $Q = 0$.
ω_i^{δ}:	*i*-th *Bending* mode.
$\Delta_i^{\circ\delta}$:	*Gap* between the first excited state of the fast mode and the first harmonic of the *i*th bending mode.

$\tilde{\omega}_i$:	*i*-th mode of the *thermal bath.*
Δ_i^{δ}:	*Gap* between the first excited state of the fast mode and the damped first harmonic of the *i*th bending mode.
ω_{μ}°:	Real angular frequency involved in the μth eigenvalue $\hbar\omega_{\mu}^{\circ}$ of a non-diagonal Hamiltonian.
ω_{μ}:	Complex angular frequency involved in the μth eigenvalue $\hbar\omega_{\mu}$ of a nondiagonal Hamiltonian.
$\omega_{\mu}^{\pm}$:	Real angular frequency involved in the μth eigenvalue $\hbar\omega_{\mu}^{\pm}$ of the Hamiltonian $[H_{(\pm)}^{\{1\}}]_u$.

T.1.B.3.b. Damping Parameters (in Angular Frequencies)}

γ°:	Hydrogen-bond bridge (indirect damping).
$\tilde{\gamma}$:	*Indirect* one equal to $\gamma/\sqrt{2}$.
γ°:	High frequency mode(*direct* damping).
γ_i^{δ}:	*i*th bending mode.
Γ:	Artificial *Direct* damping-like of *indirect nature* appearing in the Robertson and Yarwood model.

T.1.B.4. Anharmonic Couplings Parameters

b:	Basic one between the *H-bond bridge* and the *high frequency mode.*
α°:	Dimensionless one corresponding to b.
$\tilde{\alpha}^{\circ}$:	Dimensionless one equal to $\alpha^{\circ}/\sqrt{2}$.
$\tilde{\alpha}$:	Effective dimensionless one related to α° within the semiclassical limit.
$\tilde{\beta}$:	Other dimensionless one related to α° within the semiclassical limit.
η_i:	Between the *H-bond bridge* and the *i*th oscillator of the *thermal bath.*
$\tilde{\kappa}_i$:	Dimensionless between the *H-bond bridge* coordinate and the *i*th oscillator of the *thermal bath.*
$\tilde{\kappa}_i^{\circ}$:	Dimensionless one equal to $\tilde{\kappa}_i/\alpha^{\circ}$.
f_i^{δ}:	Between the high frequency and the ith bending mode (Fermi resonance).

T.1.B.5. Parameters Dealing with Statistical Thermal Physics

T: Absolute temperature.

β: Statistical temperature parameter $1/k_B T$, where k_B is the Boltzmann constant.

λ: Variable having the dimension of β.

$\tilde{\lambda}$: Dimensionless ratio $\hbar\Omega/k_B T$, where $\hbar$ is the Planck constant and k_B is the Boltzmann constant.

$\tilde{\varepsilon}$: Partition function of the H-bond bridge quantum harmonic oscillator when taking into account the zero-point energy.

ε: Partition function of the H-bond bridge quantum harmonic oscillator when neglecting the zero-point energy..

$\langle n \rangle$: Thermal average of the excitation degree of the H-bond bridge oscillator.

$\bar{n}$: Classical limit of $\langle n \rangle$.

T.1.B.6. Stochastic Functions

$\mathrm{S}(t)$: Classical stochastic variable corresponding to the *indefinite integral* of stochastic coordinate $\mathrm{Q}(t)$.

$\mathrm{Q}(t)$: *Brownian coordinate* of the H-bond bridge.

$\mathrm{v}(t)$: Classical stochastic vibrational velocity of the H-bond bridge.

$\mathrm{R}(t)$: Complex linear combination of the stochastic coordinates $\mathrm{Q}(t)$ and $\mathrm{v}(t)$.

$\mathrm{F}(t), \tilde{\mathrm{F}}(t)$ and $\tilde{\mathrm{F}}^{\circ}(t)$: Different *Langevin forces* dealing with the Brownian H-bond bridge and having different dimensions.

T.1.C. Physical Parameters Dealing With Spectroscopy

T.1.C.1. Dipole Moment Operators $\widehat{\mu}$ of the High Frequency Mode and Related Terms

T.1.C.1.a. Monomers

$\widehat{\mu}(q, 0)$: *Hermitean* dipole moment *operator* at position q and at time $t = 0$.

μ: Dimensionless *non-Hermitean* absorption transition moment *operator.*

μ_0: *Scalar* relating $\widehat{\mu}$ to μ.

$\mu(t)$:	Dimensionless non-Hermitean absorption transition operator at time t.
$\mu(t + i\lambda\hbar)$:	Dimensionless non-Hermitean absorption transition operator at temporal situation $t + i\lambda\hbar$ with $i^2 = -1$. and $\lambda\hbar$ having the dimension of a time.

T.1.C.1.b Centrosymmetric Cyclic Dimers

$[\widehat{\mu}(q_a, q_b, 0)]$:	*Hermitean* Dipole Moment *Operator* at Position q_a and q_b and at Time $t = 0$.
$[\widehat{\mu}_g(q_a, q_b, 0)]$ and $[\widehat{\mu}_u(q_a, q_b, 0)]$:	*Symmetrized g* and *u Hermitean* dipole moment *operators* at $t = 0$.
$[\mu(t)_g]$ and $[\mu(t)_u]$:	Dimensionless *non-Hermitean* Raman and IR absorption transition operators at time t.
$[\widehat{\mu}_{Dim}(t)]$:	*Hermitean* dipole moment operator at time t as a whole.
$[\mu^{(\pm)}(t)]$:	Transition moments operator at time t corresponding to the $\pm$ components related to $H_{\{\pm\}}^{\{1,1\}}$.
$\{\mu(t)\}_g$:	Transition moment operator at time t corresponding to the symmetrized g subspace.
$\{\mu^{(\pm)}(t)\}_{u^\pm}$:	Dimensionless transition moments operators at time t corresponding to the $\pm$ components of the u subspace.
μ_a° and μ_b°,:	*Scalar* transition moments characterizing, respectively the a and b moieties.
μ_g° and μ_u°:	*Scalar* transition moments transforming, respectively, according to the A and B irreducible representations.
η°:	Dimensionless positive parameter ≤ 1 giving the *amount of IR forbidden transition.*
$F_n^{(\pm)}(\eta^\circ)$:	IR scalar selection rule characterizing the $(\pm)$ n^{th} excited state of the$u$ H–bond bridge, as a function of η°.

T.1.C.2. Complex Time-dependent Scalar Parameters Dealing With The Translation and The Time Evolution Operators

$\Phi_{II}^\circ(t)$ and $\Phi_{III}^\circ(t)$:	In quantum representation $\{II\}$ and $\{III\}$ characterizing the dynamics *without indirect damping* of the H-bond bridge when the fast mode is in its first excited state.

$\Phi_{II}(t)$ and $\Phi_{III}(t)$: in quantum representation $\{II\}$ and $\{III\}$ characterizing the dynamics *with indirect damping* of the H-bond bridge when the fast mode is in its first excited state.

T.1.D. Scalar Functions Dealing With the Spectral Density

T.1.D.1. Complex Autocorrelation Function (ACF) G(t) of the Dipole Moment Operator of the High Frequency Mode

T.1.D.1.a. Monomers

$[G^{\circ}_{II}(t)]^{\circ}$ and $[G^{\circ}_{III}(t)]^{\circ}$: The ACFs of a monomer *without direct and indirect damping,* respectively, in representation $\{II\}$ and $\{III\}$.

$[G^{\circ}_{II}(t)]$ and $[G^{\circ}_{III}(t)]$: The ACF *with direct damping* and *without indirect damping,* respectively, in representation $\{II\}$ or $\{III\}$.

$[G_{II}(t)]^{\circ}$ and $[G_{III}(t)]^{\circ}$: The ACF *with indirect damping* and *without direct damping* in representation $\{II\}$ or $\{III\}$.

$[G_{II}(t)]$ and $[G_{III}(t)]$: The ACFs *with direct* and *indirect dampings*, respectively, in representation $\{II\}$ and $\{II\}$.

$[G_{II}(t)]_{\text{Cl}}$: Classical limit of the ACF *with direct* and *indirect dampings.*

$[G_{II}(t)]_{\text{Semi}}$: Semiclassical limit of the ACF *with direct* and *indirect dampings.*

$[\tilde{G}(t)]^{\circ}$: Different kinds of ACFs *involving indirect damping* via non-Hermitean Hamiltonian formalisms.

T.1.D.1.b. Centrosymmetric Cyclic Dimers

$[G^{\circ}_{\text{Dav}}(t)]^{\circ}$: The ACF involving Davydov coupling, without Fermi resonance and *without direct* and *indirect dampings.*

$[G_{\text{Dav}}(t)]$: The ACF involving Davydov coupling without Fermi resonance, *involving direct* and *indirect dampings.*

$[G^{\circ}_{\text{Dav}}(t)]$: The ACF involving Davydov coupling without Fermi resonance, *involving direct damping* but *without indirect damping.*

$[G(t)]_g^\circ$: The ACF of the g part *involving indirect damping.*

$[G^\circ(t)]_g^\circ$: The ACF of the g part *without indirect damping.*

$[G^\circ(\pm)(t)]_u^\circ$: The ACFs of the u part of a centrosymmetric cyclic dimer corresponding to the $(\pm)$ components.

T.1.D.2. Real IR Spectral Densities (SDs) $I(\omega)$ of the High Frequency Mode

T.1.D.2.a. Monomers

$[I_{II}^\circ(\omega)]^\circ$ and $[I_{III}^\circ(\omega)]^\circ$: The SDs *without direct* and *indirect dampings*, respectively, in representation $\{II\}$ and $\{III\}$.

$[I_{II}(\omega)]^\circ$ and $[I_{III}(\omega)]^\circ$ The SDs *without direct* and *with indirect dampings*, respectively, in representation $\{II\}$ and $\{III\}$.

$[I_{II}^\circ(\omega)]$ and $[I_{III}^\circ(\omega)]$: The SDs *with direct* and *without indirect dampings*, respectively, in representation $\{II\}$ and $\{III\}$.

$[I_{II}(\omega)]$ and $[I_{III}(\omega)]$: The SDs *with indirect* and *direct dampings*, respectively in representation $\{II\}$ and $\{III\}$.

$[\tilde{I}(\omega)]^\circ$: Different kinds of SDs *involving indirect damping* via non Hermitean Hamiltonian formalisms.

T.1.D.2.b. Centrosymmetric Cyclic Dimers

$[\mathbf{I}_{\mathrm{Dav}}^\circ(\omega)]$: The SD involving Davydov coupling, but without Fermi resonance and indirect damping.

$[I_{\mathrm{Dav}}(\omega)]$: The SD involving Davydov coupling and direct and indirect dampings, but without Fermi resonance.

T.2. OPERATORS, KETS, AND DENSITY OPERATORS

T.2.A. Operators That Are Not Observable

T.2.A.1. Special Operators

$\widehat{C}_2$: *Parity operator* exchanging the two moieties of a cyclic dimer.

$\widehat{N}$: *Normal ordering operator* allowing to pass from function of Bosons to function of complex scalars.

$\widehat{P}$: *Dyson time ordering operator.*

$[\widehat{F}^{(\pm)}(\eta)]$: *Infrared selection rule operators* acting in the u $(\pm)$ H-bond bridge subspaces of centrosymmetric cyclic dimers taking into account some degree η of forbidden transition.

$[\widehat{M}^{(\pm C_2)}]$: The 2×2 matrices the diagonal elements of which are unity and the off ones are $\pm\widehat{C}_2$.

$\widehat{1}_\theta$: *Unity operator* built up from the closeness relation on the eigenstates of the infinite set of harmonic oscillators of the thermal bath.

$\widehat{1}_g$ and $\widehat{1}_u$: *Unity operators* involved in the theory of centrosymmetric cyclic dimers and built up, respectively, from the closeness relations on the eigenstates of the g and u symmetrized quantum harmonic Hamiltonians of the H-bond bridge.

T.2.A.2. Non-Hermitean Lowering and Raising Operators (Bosons)

a and $a^\dagger$: Bosons of the quantum harmonic oscillator describing the *H-bond bridge* with $[a, a^\dagger] = 1$.

a_g and a_u: Bosons corresponding to the g and u *symmetrized H-bond bridge coordinates with* $[a_g, a_g^\dagger] = 1$ *and* $[a_u, a_u^\dagger] = 1$

b and $b^\dagger$: Bosons of the quantum harmonic oscillator describing the *high frequency mode* with $[b, b^\dagger] = 1$.

$\tilde{b}_i$ and $\tilde{b}_i^\dagger$: Bosons of the ith quantum harmonic oscillator of the *thermal bath,* with $[\tilde{b}_i, \tilde{b}_i^\dagger] = 1$.

b_i^δ and $b_i^{\delta\dagger}$: Bosons of the quantum harmonic oscillator describing the ith *bending mode,* with $[b_i^\delta, b_i^{\delta\dagger}] = 1$.

T.2.A.3. H-Bond Bridge Unitary Time Evolution Operators $U(t)$

$[U_{II}^{\circ\{0\}}(t)]$: *Undriven undamped* situation corresponding to $|\{0\}\rangle$.

$[U^{\circ\{1\}}_{II}(t)]$: *Driven undamped* situation corresponding to $|\{1\}\rangle$.

$[U^{\{1\}}_{II}(t)]$: *Driven damped* situation corresponding to $|\{1\}\rangle$

$[U^{\circ\{1\}}_{II}(t)^{IP}]$: IP *driven undamped* situation corresponding to $|\{1\}\rangle$

$[U^{\{1\}}_{II}(t)^{IP}]$: *Reduced* IP *driven damped* situation corresponding to $|\{1\}\rangle$.

$[U^{\{1\}}_{II}(t)]$: *Full driven* situation *embedded in the thermal bath* and corresponding to $|\{1\}\rangle$.

$[U^{\{1\}}_{II}(t)^{IP}]$: Corresponding IP situation

T.2.A.4. H-Bond Bridge Unitary Translation Operators A and Related Matrix Elements

$[A(l\alpha^\circ)]$: Time-independent translation operators of dimensionless argument α° with $l = 1/2, 1$.

$[A^{\circ\{1\}}_{II}(t)]$ and $[A^{\circ\{1\}}_{III}(t)]$: *Undamped* time-dependent situations corresponding to representations$\{II\}$and $\{III\}$.

$[A^{\{1\}}_{II}(t)]$ and $[A^{\{1\}}_{III}(t)]$: *Damped* time-dependent situations corresponding to representations $\{II\}$ and $\{III\}$.

$A_{mn}(\alpha^\circ)$: Matrix elements of the translation operator (Franck–Condon factors).

T.2.B. Hermitean Hamiltonians

T.2.B.1. Monomer Hamiltonians Without Quantum Indirect Damping

$[H_{Tot}]$: Total Hamiltonian of the *H-bond Bridge* Within the Strong Anharmonic Coupling Theory for Situation.

$[H_{Diab}]$ and $[H_{Adiab}]$: *Diabatic* and *adiabatic* part of H_{Tot}.

H° and $[H_{Slow}]$: Hydrogen-bond bridge *harmonic* and *Morse-like* Hamiltonians.

$[H_{Fast}]$: Hamiltonian of the high frequency mode, involving an angular frequency depending on the H-bond bridge coordinate.

$[H_{Free}]$ and $[V_{Anh}]$: *Harmonic* and *anharmonic* parts of the high frequency mode Hamiltonian.

$[\breve{H}^{\{k\}}_{II}]$: Hydrogen-bond bridge effective Hamiltonian in representation $\{II\}$ and corresponding to $|\{k\}\rangle$.

$[H_I^{\{k\}}]$, $[H_{II}^{\{k\}}]$ and $[H_{III}^{\{k\}}]$: Hydrogen-bond bridge effective Hamiltonians corresponding to representations $\{I\}$, $\{II\}$, and $\{III\}$ and situation $|\{k\}\rangle$.

T.2.B.2. Monomer Hamiltonians Taking into Account the Indirect Damping

$\tilde{V}_{Dr}$: Non, Hermitean Operator Driving the H-bond Bridge Harmonic Hamiltonian.

$[\mathbb{H}^{\circ}_{Tot}]$: Total Hamiltonian of the H-bond bridge within the strong anharmonic coupling theory for situation *with indirect damping.*

$[\mathbb{H}_{II}^{\{k\}}]$ and $[\mathbb{H}_{III}^{\{k\}}]$: Effective Hamiltonians of the thermal bath and of the H-bond bridge quantum harmonic oscillator, respectively, in representation $\{II\}$ and $\{III\}$, when the fast mode is in its eigenstate $|\{k\}\rangle$.

T.2.B.3. Centrosymmetric Dimer Hamiltonians Without Indirect Damping

V°: Davydov coupling Hamiltonian.

$[H_{Dav}]$: Hamiltonian involving Davydov coupling and quantum indirect damping within the adiabatic approximation for the two moieties.

$[H_{II}^{\{k,l\}}]$: Effective Hamiltonian of the two H-bond bridge quantum harmonic oscillator in representation $\{II\}$ when the two moieties fast modes are in their eigenstates $|\{k\}_a\rangle$ and $|\{l\}_b\rangle$.

$[H_{\{1,0\}}^{\{0,1\}}]$: Effective Hamiltonian corresponding to the situation where there is an exchange between the two effective ones $H_{II}^{\{1,0\}}$ and $H_{II}^{\{0,1\}}$ via the V° Davydov coupling.

T.2.B.4. Centrosymmetric Dimer Hamiltonians Taking into Account the Indirect Damping

$[\mathbb{H}_{Dav}]$: Hamiltonian involving Davydov Coupling and Quantum Indirect Damping Within the Adiabatic Approximation for the Two Moieties.

$[\mathbb{H}_{\{0,1\}}^{\{1,0\}}]$: Exchange degenerate part of $[\mathbb{H}_{Dav}]$ involving the Davydov coupling between $|\{1\}_a\rangle \otimes |\{0\}_b\rangle$ and $|\{0\}_a\rangle \otimes |\{1\}_b\rangle$.

$[\mathbb{H}_{II}^{\{k,l\}}]$: Effective Hamiltonian of the two H-bond bridge quantum harmonic oscillator in representation $\{II\}$ when the two moieties fast modes are in their eigenstates $|\{k\}_a\rangle$ and $|\{l\}_b\rangle$.

$[\mathbb{H}_{\{\pm\}}^{\{1,1\}}]$: Effective Hamiltonians involving the $\pm$ $V°\widehat{C}_2$ Davydov coupling correction.

$[\mathbb{H}_{II}^{\{1\}}]_g$: Effective Hamiltonian of the g symmetrized part, coupled to the thermal bath, when the g fast mode is in $|\{1\}_g\rangle$.

$[H_{(\pm)}^{\{1\}}]_u$: Effective Hamiltonians of the u symmetrized part involving the $\pm V°\widehat{C}_2$ perturbation, but which is decoupled from the thermal bath, when the u fast mode is in $|\{1\}_u\rangle$.

T.2.C. Non-Hermitean Hamiltonians Taking into Account the Indirect Damping

$[\tilde{H}_{II}^{\{1\}}]_g$: Effective Hamiltonian of the g symmetrized part, when the g fast mode is in $|\{1\}_g\rangle$.

$[\tilde{H}^{\mathrm{Fermi}}]_g$: Effective Hamiltonian involving Fermi resonance exchange and corresponding to the g part.

$[\tilde{H}_{\mathrm{Dav}}^{\mathrm{Fermi}}]_g$: Hamiltonian involving Fermi resonance and Davydov coupling.

$[\tilde{H}_{\mathrm{Fermi}}^{\{1\}}]_g$: Effective Hamiltonian involving Fermi resonance exchange of the g symmetrized part, when the g fast mode is in $|\{1\}_g\rangle$

T.2.D. Kets, Spinors, Density Operators, and Wave Functions

T.2.D.1. Orthonormal Kets and Wave functions

T.2.D.1.a. Monomers

$|\{k\}\rangle$: the kth state of the harmonic Hamiltonian of the fast mode for H-bonded monomer with $\{k\} = 0, 1$.

$|\Phi_k(Q)\rangle$ and $\Phi_k(q, Q)$ k^{th} eigenket and corresponding kth eigenfunction of the anharmonic high frequency mode parametrically depending on the Q coordinate of the H-bond bridge.

$\|(n)\rangle$ and $\chi_n(Q)$:	The nth eigenket and corresponding n^{th} eigenfunction of the harmonic Hamiltonian of the H-bond bridge.
$\|(n)_{II}\rangle$ and $\|(n)_{III}\rangle$:	The nth eigenstates of the harmonic Hamiltonian of the H-bond bridge in quantum representations $\{II\}$ and $\{III\}$,
$\langle (n)_{II} \|(n)_{III}\rangle$:	*Franck–Condon* factors.
$\| \alpha^\circ \rangle = \|(0)_{II}\rangle$:	*Coherent state* (of eigenvalue α°).
$\|[l]_j\rangle$:	lth eigenstate of the jth harmonic Hamiltonian of the jth *bending mode.*

T.2.D.1.b. Centrosymmetric Cyclic Dimer before Symmetrization

$\|\{k\}_a\rangle$ and $\|\{k\}_b\rangle$:	Fast mode kth eigenstates of the two moieties a and b *harmonic Hamiltonians.*
$\|(n)_a\rangle$ and $\|(n)_b\rangle$:	Hydrogen-bond bridge nth eigenstates of the two moieties a and b *harmonic Hamiltonians.*
$\|\{k\}_a\{l\}_b\rangle$:	Tensorial products $\|\{k\}_a\rangle \otimes \|\{l\}_b\rangle$
$\|(n)_a(m)_b\rangle$:	Tensorial products $\|(n)_a\rangle \otimes \|(m)_b\rangle$
$\|\{k\}_a(n)_a\{l\}_b(m)_b\rangle$:	Tensorial products of the states $\|\{k\}_a\rangle \otimes \| (n)_a\rangle \otimes \|\{l\}_b\rangle \otimes \| (m)_b\rangle$
$\|\Phi^{\{0,k\}}_{\{a,b\}}\rangle \equiv \|\{0\}_a\{k\}_b\rangle$:	Tensorial product $\|\{0\}_a\rangle \otimes \|\{k\}_b\rangle$

T.2.D.1.C. Centrosymmetric Cyclic Dimer After Symmetrization

$\|\{k\}_g\rangle$ and $\|\{k\}_u\rangle$:	Symmetric g and Antisymmetric u kth States of the Symmetrized *High Frequency Mode* Harmonic Hamiltonians.
$\|(n)_g\rangle$ and $\|(n)_u\rangle$	Symmetric g and antisymmetric u nth states of the symmetrized *H-bond bridge* harmonic Hamiltonians.
$\|\Phi^{\{0,0\}}_{\{g,u\}}\rangle \equiv \|\{0\}_g\{0\}_u\rangle$:	Tensorial product $\|\{0\}_g\rangle \otimes \|\{0\}_u\rangle$.
$\|\{\beta\}^{\{1\}}_g\rangle$:	Spinor dealing with the g H-bond bridge, when the g high frequency mode is in $\|\{1\}_g\rangle$.
$\|\{\beta^{(\pm)}\}^{\{1\}}_u\rangle$:	The $(\pm)$ spinors dealing with the u H-bond bridge, when the u high frequency mode is in $\|\{1\}_u\rangle$.
$\|\{\beta^{\{1,0\}}_{\{1,0\}}(\pm)\}\rangle$:	Eigenvector of Hamiltonian$[\mathbb{H}^{\{1,0\}}_{\{0,1\}}]$.
$\| \beta^{(\pm)}_{\{1,0\}\leftrightarrow\{0,1\}}\rangle$:	Symmetrized spinors dealing with the tensorial product of the H-bond bridge

	states, in the situation where the excitation of the fast modes is exchanged between $\|\{1\}_a\rangle$ and $\|\{1\}_b\rangle$.
$\|\beta_\mu^{(\pm)}\rangle$:	The μth eigenket of the effective Hamiltonian of the u symmetrized H-bond bridge.

T.2.D.2. Density Operators (Hermitean)

T.2.D.2.a. Boltzmann Density Operators

ρ_B:	General expression.
ρ_{kk}:	Diagonal matrix elements of the general expression, that is, Boltzmann probabilities.
$\boldsymbol{\rho}_\theta$:	Density operator of the thermal bath.
$[\boldsymbol{\rho}_{II}^{\{0\}}]$, $[\boldsymbol{\rho}_{II}^{\{0\}}]_g$ and $[\boldsymbol{\rho}_{II}^{\{0\}}]_u$:	Monomer and g and u density operators of the H-bond bridge quantum harmonic oscillator for situation $\|\{0\}\rangle$.
$[\mathbb{R}_{II}^{\{0\}}] = [\boldsymbol{\rho}_{II}^{\{0\}}]\boldsymbol{\rho}_\theta$:	Full expression.

T.2.D.2.b. Nonequilibrium Density Operators

$[\boldsymbol{\rho}_{II}^{\circ\{k\}}(t)]$ and $[\boldsymbol{\rho}_{III}^{\circ\{k\}}(t)]$:	Undamped H-bond bridge operators, respectively, in representations $\{II\}$ and $\{III\}$, for situations $\|\{k\}\rangle$.
$[\boldsymbol{\rho}_{II}^{\{k\}}(t)]$and $[\boldsymbol{\rho}_{III}^{\{k\}}(t)]$:	Damped H-bond bridge reduced forms, respectively, in representations $\{II\}$ and $\{III\}$, for situations $\|\{k\}\rangle$.
$[\mathbb{R}_{II}^{\{1\}}(t)]$ and $[\mathbb{R}_{III}^{\{1\}}(t)]$:	Full-time dependent damped operators for H-bond bridge and thermal bath, respectively, in representations $\{II\}$ and $\{III\}$, for situation $\|\{1\}\rangle$.

T.1.D.4. Nonorthonormal Kets

$\|\tilde{\Phi}_\nu^{\{k\}}\rangle$, $\|\tilde{\Phi}_\nu\rangle$, $\|\tilde{\Theta}_\nu\rangle$, $\|\tilde{\xi}_\nu\rangle$:	Kets involved in the non-Hermitean approaches of quantum indirect damping.

Acknowledgments

Here, we express all our gratitude to Prof. A. Witkowski (Jagellion University, Cracow) who is at the origin of our works on the quantum theory of indirect damping by suggesting us the interest of damping in the area of cyclic carboxylic acids IR spectra and to Prof. Y. Maréchal (CEA, Grenoble) for many helpful discussions. Besides, we thank Prof. D. Hadzi (Ljublijana University, Slovenia) for encouraging us in our theoretical investigations and Prof. S. Bratos (Université Pierre et Marie Curie,

Paris) and Prof. C. Sandorfy (University of Montreal) for their remarks on some of our quoted works. At last, we would not forget Prof. M. Wojcik, Prof. H. Flakus and Prof. F. Fillaux, for our fructuous exchanges at the occasion of their visits. Also, we thank Dr. P-M. Déjardin, Dr. D. Chamma, Dr. A. Ceausu-Velcescu, and Dr. M. E-A. Benmalti for their kind remarks on the manuscript.

References

1. S. Bratos, J.C. Leicknam, G. Gallot, H. Ratajczak, in T. Elsaesser, H.J. Bakker (eds.), *Ultrafast Hydrogen Bonding Dynamics and Proton Transfer Processes in the Condensed Phase*, Kluwer Academic Publishers, Dordrecht, The Netherlands, 2002.
2. E. T. J. Nibbering and T. Elsaesser, *Chem. Rev.* **104**, 1887 (2004).
3. K. Heyne, N. Huse, J. Dreyer, E. T.J. Nibbering, T. Elsaesser, and S. Mukamel, *J. Chem. Phys.* **121**, 902 (2004).
4. G. Pimentel and A. Mc Clellan, *The Hydrogen Bond*, Freeman, San Francisco, 1960.
5. A. Novak, *Structure Bonding*, (Berlin) **18**, 177 (1974).
6. G. Hofacker, Y. Maréchal, and M. Ratner, in P. Schuster,G. Zundel and C. Sandorfy (eds.), The hydrogen Bond Theory, North-Holland Publishing. Amsterdam, the Netherlands 1976 p. 297.
7. D. Hadi and S. Bratos, in P.Schuster, G. Zundel, and C. Sandorfy (eds.), The Hydrogen Bond Theory, North-Holland Publishing, Amsterdam The Netherlands, 1976 p. 567.
8. C. Sandorfy, in P. Schuster, G. Zundel, and C. Sandorfy (eds.), The Hydrogen Bond Theory, North-Holland Publishing Amsterdam, The Netherlands 1976, p. 616.
9. C. Sandorfy, *Topics in Current Chemistry*, Vol. 120 Springer-Verlag, New York 1984, p. 41.
10. Y. Maréchal, in H. Ratajczak and W.Orville-Thomas (eds.) John Wiley & Sons, Inc., New York, 1980, p. 230.
11. Y. Maréchal, in J. Durig (ed.),*Vibrational Spectra and Structure*, Vol.16, Elsevier, Amsterdam, the Netherlands, 1987 p. 312.
12. O. Henri-Rousseau and P. Blaise, *Theoretical Treatment of Hydrogen Bonding*, in D. Hadzi, (ed.), John Wiley & Sons, Inc., New York, 1997, p. 165.
13. O. Henri-Rousseau and P. Blaise, in: I. Prigogine and S.A. Rice, (eds.) Advances in Chemical Physis Vol. 103, John Wiley & Sons Inc., New York, 1998 p. 1.
14. O. Henri-Rousseau and P. Blaise, Recent Res and Development in Chemical Physics, Vol. 2, Trivandrum, 1998, p.181.
15. O. Henri-Rousseau, P. Blaise, and D. Chamma, in I. Prigogine and S.A. Rice (eds.), *Advances in Chemical Physics* John Wiley & Sons, Inc., New York, 2002) Vol. 121, p. 241.
16. Y. Maréchal, *J. Chem. Phys.* **83**, 247 (1985).
17. H. Flakus, *Polish J. Chem.* **77**, 489 (2003).
18. Y. Maréchal and A. Witkowski, *J. Chem. Phys.* **48**, 3637 (1968).
19. S. Fischer, G. Hofacker, and M. Ratner, *J. Chem. Phys.* **52**, 1934 (1970).
20. J. Witkowski, *J. Chem. Phys.* **79**, 852 (1983) ; *Phys. Rev.* **A41**, 3511 (1990) ; A. Witkowski, O. Henri-Rousseau, and P. Blaise, *Acta. Phys. Pol.* **A91**, 495 (1997).
21. R. Feynmann and F. Vernon, *Ann. Phys.* **24**, 1181 (1963).
22. W. Louisell and L. Walker, *Phys. Rev.* **137**, 204 (1965).
23. S. Bratos and D. Hadzi, *J. Chem. Phys.* **27**, 991 (1957).
24. A. Witkowski and M. Wojcik, *Chem. Phys.* **1**, 9 (1973).
25. M. Wojcik, *Mol. Phys.* **36**, 1757 (1978).

26. O. Henri-Rousseau and D. Chamma, *Chem. Phys.* **229**, 37 (1998).
27. D. Chamma and O. Henri-Rousseau,*Chem. Phys.* **229**, 51 (1998).
28. R. Gordon, *Adv.Magn. Res.* **3**, 1 (1968).
29. S. Bratos, J. Rios, and Y. Guissani, *J. Chem. Phys.* **52** 439 (1970).
30. C. Coulson and G. Robertson, *Proc. R. Soc. London* **A337**, 167 (1974).
31. C. Coulson and G. Robertson, *Proc. R. Soc. London* **A342**, 380 (1975).
32. G. Robertson, *J. Chem. Soc. Faraday Trans.* **72**, 1153 (1976).
33. S. Mülhe, K. Süsse, and D., Welsch, *Phys. Lett.* **66-A**, 25 (1978).
34. S., Mülhe, K., Süsse, and D. Welsch, *Ann. Phys.* **7**, 213 (1980).
35. G. Ewing, *Chem. Phys.* **29**, 253 (1978), *ibid.* **72**, 2096 (1980).
36. G. Robertson and M. Lawrence, *Chem. Phys.* **62**, 131 (1981).
37. G. Robertson and M. Lawrence, *Mol. Phys.* **43**, 193 (1981).
38. G. Robertson and M. Lawrence, *J. Chem. Phys.* **89**, 5352 (1988).
39. N. Rösch, *Chem. Phys.* **1**, 220, (1973).
40. H. Abramczyk, *Chem. Phys.* **144**, 305; 319 (1990).
41. N. Rekik, A. Velcescu, P. Blaise, and O. Henri-Rousseau, *Chem. Phys.* **273**, 11 (2001).
42. G. Robertson, *Phil. Trans. R. Soc. London* **A286**, 25 (1977).
43. H. Abramczyk, *Chem. Phys.* **116**, 249 (1987).
44. Kh. Belhayara, D. Chamma, A. Velcescu, and O. Henri-Rousseau, *J. Mol. Struct.* in press
45. S. Bratos, *J. Chem. Phys.* **63**, (1975) 3499, see also S. Bratos and H. Ratajczak, *J. Chem. Phys.* **76**, 77 (1982).
46. G. Robertson and J. Yarwood, *Chem. Phys.* **32**, 267 (1978).
47. N. Rösch and M. Ratner, *J. Chem. Phys.* **61**, 3344 (1974).
48. B. Boulil, O. Henri-Rousseau, and P. Blaise, *Chem. Phys.* **126**, 263 (1988).
49. B. Boulil, J-L. Déjardin, N. El-Ghandour, and O. Henri-Rousseau, *J. Mol. Struct. Theochem*, **314**, 83 (1994).
50. B. Boulil, P. Blaise and O. Henri-Rousseau, *J. Mol. Struc. Theochem*, **314**, 101 (1994).
51. Y. Maréchal, Ph.D. Dissertation, Grenoble, 1968.
52. R. Badger and S.Bauer, *J. Chem. Phys.* **5**, 839 (1937).
53. S. Barton, and W. Thorson, *J. Chem. Phys.* **71**, 4263 (1979), see also J. Beswick and J. Jortner, *J. Chem. Phys.* **68**, 2277 (1978), *ibid.* **69**, 512 (1978).
54. H. Louisell, *Quantum statistical Properties of Radiations*, John Wiley & Sons, Inc., New York, 1973, p. 64.
55. R. Janoschek, G. Weidemann, and G. Zundel, *J. C. S. Faraday II* **69**, 505 (1973).
56. P. Blaise and O. Henri-Rousseau, *Chem. Phys.* **243**, 229 (1999).
57. F. Dyson, *Phys. Rev.*, **75**, 486 (1949) and 1736.
58. P. Blaise, Ph. Durand and O. Henri-Rousseau, *Phy.* **A. 209**, 51 (1994).
59. A. Velcescu, P. Blaise, and O. Henri-Rousseau, Trends in Chemical Physics, Vol. 11, Trivandrum, 2004.
60. P. Blaise, P-M. Déjardin, and O. Henri-Rousseau, *Chem. Phys.* **313**, 177 (2005).
61. H. Risken, The Fokker–Planck Equation, 2nd ed. Springer, Berlin, 1989.
62. W. T. Coffey, Yu.P. Kalmykov, and J. T. Waldron, *The Langevin Equation*, 2nd ed., World Scientific, London, 2004.

63. M. Lax, *Rev. Mod. Phys.* **38**, 359 (1966).
64. J. Yarwood and G. Robertson, Nature (London) **257**, 41 (1975).
65. J. Yarwood, R. Ackroyd, and G. Robertson, *Chem. Phys.* **32**, 28 (1978).
66. G. Uhlenbeck and L. Ornstein, *Phys. Rev.* **36**, 823 (1930).
67. Kh. Belharaya, P. Blaise, and O. Henri-Rousseau, *Chem. Phys.* **293**, 9 (2003).
68. P. Blaise, O. Henri-Rousseau, and A. Grandjean, *Chem. Phys.* **244**, 405 (1999).
69. D. Chamma and O. Henri-Rousseau, *Chem. Phys.* **248**, 91 (1998).
70. Kh. Belhayara, D.Chamma, and O. Henri-Rousseau, *Chem. Phys.* **293**, 31 (2003).
71. N. Rekik, B. Ouari, P. Blaise, and O. Henri-Rousseau, *J. Mol. Struc.* **687**, 125 (2004).
72. P. Blaise and O.Henri-Rousseau, *Chem. Phys.* **256**, 85 (2000).
73. D. Chamma, *J. Mol. Struct* **552**, 187 (2000).
74. A. Davydov, *Theory of Molecular Excitons*, Mc-Graw Hill, New York 1962.
75. D. Chamma and O. Henri-Rousseau, *Chem. Phys.* **248**, 53 (1999).
76. P. Blaise, M. J. Wojcik, and Olivier Henri-Rousseau *J. Chem. Phys.* **122**, 064306 (2005).
77. H.T. Flakus, *Chem. Phys.* **62**, 103 (1981).
78. V. Sakun, *Chem. Phys.* **99**, 457 (1985).
79. H. Abramczyk, *Chem. Phys.* **94**, 91 (1985).
80. A. Velcescu, P. Blaise, and O. Henri-Rousseau, *Chem. Phys.* **293**, 23 (2003).
81. P. Blaise, M. E-A. Benmalti, and O. Henri-Rousseau, *J. Chem. Phys.* **124**, 024514 (2006).
82. D. Chamma and O. Henri-Rousseau, *Chem. Phys.* **248**, 91 (1999).
83. Y. Maréchal, *Chem. Phys.* **79**, 69 (1983).
84. M. Schwartz and C. H. Wang, *J. Chem. Phys.* **59**, 5258 (1973).
85. E. Weidemann and A. Hayd, *J. Chem. Phys.* , **67**, 3713 (1977).
86. J. L. Mc Hale and C. H. Wang, *J. Chem. Phys.* **73**, 3601 (1980).
87. K. Fujita and M. Kimura, *Mol. Phys.* **41**, 1203 (1980).
88. P. Piaggio, G. Dellepiane, R. Turbino, and L. Piseri, *Chem. Phys.* **77**, 185 (1983).
89. A. Brodka and B. Stryczek, *Mol. Phys.* **54**, 677 (1985).
90. B. Boulil, M. Giry, and O. Henri-Rousseau, *Phys. Stat. Sol.*(b) **158**, 629 (1990).
91. M. Giry, B. Boulil, and O. Henri-Rousseau, *C. R. Acad. Sc. Paris* **316**, 455 (1993).
92. A. Davydov and A. Serikov, *Phys. Stat. Sol.* (b) **51**, 57 (1972).
93. C. Cohen-Tannoudji, J. Dupont-Roc, and G. Grinberg, *Atom Photon interactions, Basic processes and Applications*, John Wiley & Sons, Inc., 1992.
94. M. Haurie and A. Novak, *J. Chim. Phys.* **62**, 146 (1965).
95. J. L. Leviel and Y. Maréchal, *J. Chem. Phys.* **54**, 1104 (1971).
96. R. J. Jakobson, Y. Mirawa, and J.W. Brasch, *Spectrochim. Acta*, **23A**, 2199.L (1967).
97. Y. Maréchal, *J. Chem. Phys.* **87**, 6344 (1987).
98. M. E-A. Benmalti, P. Blaise , H.T. Flakus, and O. Henri-Rousseau, *Chem. Phys.* **320**, 267 (2006).
99. J. Bournay and Y. Maréchal, *J. Chem. Phys.* **55**, 1230 (1971).
100. M. E-A. Benmalti, D. Chamma, P. Blaise, and O. Henri-Rousseau, *J. Mol. Struct.* **785**, 27 (2006).
101. G. Auvert and Y. Maréchal, *Chem. Phys.* **40**, 51 (1979); *ibid*, p.61.
102. Y. Maréchal, *Chem. Phys.* **79**, 85 (1983).

103. M. J. Wojcik, *Int. J. Quant. Chem.* **X**, 747 (1976).
104. H. T. Flakus, *J. Mol. Struct. (Theochem)* **285**, 281 (1993).
105. A. M. Yaremko, B. Silvi, and H. R. Zelsmann, *J. Mol. Struct.* **520**, 125 (2000).
106. H. Ratajczak and A.M. Yaremko, *Chem.Phys. Lett.*, **243**, 348 (1995).
107. A. M. Yaremko, D. I. Ostrovskii, H. Ratajczak, and B. Silvi, *J. Mol. Struct.* **482**, 665 (1999).
108. A.M. Yaremko, H. Ratajczak, J. Baran, A.J. Barnes, E.V. Mozdor, and B. Silvi, *Chem. Phys.* **306**, 57 (2004).
109. H. T. Flakus, *J. Mol. Struct. THEOCHEM* **187**, 35 (1989).
110. M. Asselin and C. Sandorfy, *Chem. Phys. Lett.* **8**, 601 (1971).
111. M. Asselin and C. Sandorfy, *Can. J. Spectrose.* **17**, 24 (1972).
112. H. T. Flakus and A. Miros, *J. Mol. Struct.* **484**, 103 (1999).
113. H.T. Flakus and M.Chełmecki, *J. Mol. Struct.* **659**, 103 (2003).
114. P. Blaise, H.T. Flakus, M. E-A. Benmalti and O. Henri-Rousseau (to be published).
115. M. J. Wojcik, *J. Mol. Struct.* **62**, 71 (1980).
116. A. Novak, *J. Chim. Phys.* **72**, 981 (1975).
117. A. Witkowski and M.J. Wojcik, *Chem. Phys.* **21**, 385 (1977).
118. L. J. Bellamy and R. J. Pace, *Spectr. Chim. Acta* **19**, 435 (1963).
119. H. R. Zelsmann and Y. Maréchal, *Chem. Phys.* **5**, 367 (1974), *ibid.* **20**, 445 (1977), *ibid.* **20**, 459 (1977).
120. H. T. Flakus, *Chem. Phys.* **50**, 79 (1980).
121. F. Fillaux, M. H. Limage, and F. Romain, *Chem. Phys.* **276**, 181 (2002).
122. H. T. Flakus and M. Chełmecki, *Spectrochimica Acta*, *Part* **A 58**, 179 (2002).
123. F. Fong, Theory of Molecular Relaxation, John Wiley & Sons, Inc., New York, 1975.
124. S. Von Koide, *Z. Naturforschg.* **15a**, 123 (1960).
125. P. Carruthers and M. Nieto, *Am. J. Phys.* **33**, 537 (1965).
126. R. L. Fulton and M. Gouterman, *J. Chem. Phys.* **35**, 1059 (1961).
127. P. Blaise and O. Henri-Rousseau (to be published).

TIME-RESOLVED PHOTOELECTRON SPECTROSCOPY OF NONADIABATIC DYNAMICS IN POLYATOMIC MOLECULES

ALBERT STOLOW

Steacie Institute for Molecular Sciences, National Research Council Canada, Ottawa, Ontario, K1A 0R6, Canada

JONATHAN G. UNDERWOOD

Department of Physics & Astronomy, University College London, Gower Street, London WC1E 6BT UK

CONTENTS

Advances in Chemical Physics, Volume 139, edited by Stuart A. Rice

I. INTRODUCTION

The photodynamics of polyatomic molecules generally involves complex intramolecular processes that rapidly redistribute both charge and vibrational energy within the molecule. The coupling of vibrational and electronic degrees of freedom leads to the processes known as radiationless transitions, internal conversion, isomerization, proton and electron transfer, and so on [1–8]. These nonadiabatic dynamics underlie the photochemistry of almost all polyatomic molecules [9] and are important in photobiological processes, such as vision and photosynthesis [10], and underlie many concepts in active molecular electronics [11]. The coupling of charge and energy flow is often understood in terms of the breakdown of the Born–Oppenheimer approximation (BOA), an adiabatic separation of electronic from nuclear motions. The BOA allows the definition of the nuclear potential energy surfaces that describe both molecular structures and nuclear trajectories, thereby permitting a mechanistic picture of molecular dynamics. The breakdown of the BOA is uniquely due to nuclear dynamics and occurs at the intersections or near intersections of potential energy surfaces belonging to different electronic states. Nonadiabatic coupling often leads to complex, broadened absorption spectra due to the high density of nuclear states and strong variations of transition dipole with nuclear coordinate. In this situation, the very notion of distinct and observable vibrational and electronic states is obscured. The general treatment of these problems remains one of the most challenging problems in molecular physics, particularly when the state density becomes high and multimode vibronic couplings are involved. Our interest is in developing time-resolved methods for the experimental study of nonadiabatic molecular dynamics. The development of femtosecond methods for the study of gas-phase chemical dynamics is founded upon the seminal studies of A.H. Zewail and co-workers, as recognized in 1999 by the Nobel Prize in Chemistry [12]. This methodology has been applied to chemical reactions ranging in complexity from bond breaking in diatomic molecules to dynamics in larger organic and biological molecules.

Femtosecond time-resolved methods involve a pump-probe configuration in which an ultrafast pump pulse initiates a reaction or, more generally, creates a nonstationary state or wave packet, the evolution of which is monitored as a function of time by means of a suitable probe pulse. Time-resolved or wave packet methods offer a view complementary to the usual spectroscopic approach and often yield a physically intuitive picture. Wave packets can behave as zeroth-order or even classical-like states and are therefore very helpful in discerning underlying dynamics. The information obtained from these experiments is very much dependent on the nature of the final state chosen in a given probe scheme. Transient absorption and nonlinear wave mixing are often the methods of choice in condensed-phase experiments because of their generality. In studies of molecules and clusters in the gas phase, the most popular methods, laser-induced fluorescence and resonant multiphoton ionization, usually require the probe laser to be resonant with an electronic transition in the species being monitored. However, as a chemical reaction initiated by the pump pulse evolves toward products, one expects that both the electronic and vibrational structures of the species under observation will change significantly and some of these probe methods may be restricted to observation of the dynamics within a small region of the reaction coordinate.

Here, we focus upon gas-phase time-resolved photoelectron spectroscopy (TRPES) of neutral polyatomic molecules. This spectroscopy is particularly well suited to the study of ultrafast nonadiabatic processes because photoelectron spectroscopy is sensitive to both electronic configurations and vibrational dynamics [13]. Due to the universal nature of ionization detection, TRPES has been demonstrated to be able to follow dynamics along the entire reaction coordinate. In TRPES experiments, a time-delayed probe laser generates free electrons via photoionization of the evolving excited state, and the electron kinetic energy and/or angular distribution is measured as a function of time. As a probe, TRPES has several practical and conceptual advantages [14]: (1) Ionization is always an allowed process, with relaxed selection rules due to the range of symmetries of the outgoing electron. Any molecular state can be ionized. There are no "dark" states in photoionization; (2) Highly detailed, multiplexed information can be obtained by differentially analyzing the outgoing photoelectron as to its kinetic energy and angular distribution; (3) Charged-particle detection is extremely sensitive; (4) Detection of the ion provides mass information on the carrier of the spectrum; (5) Higher order (multiphoton) processes, which can be difficult to discern in femtosecond experiments, are readily revealed; (6) Photoelectron–photoion coincidence measurements can allow for studies of cluster solvation effects as a function of cluster size and for time-resolved studies of scalar and vector correlations in photodissociation dynamics. Beginning in 1996, TRPES has been the subject of a number of reviews [15–29] and these cover various aspects of the field. An exhaustive review of the TRPES literature, including dynamics in both neutrals and anions, was published recently [30]. Therefore, rather than a survey, our emphasis here will be on the conceptual foundations of TRPES and the advantages of this approach in solving problems of nonadiabatic molecular dynamics, amplified by examples of applications of TRPES chosen mainly from our own work.

In the following sections, we begin with a review of wave packet dynamics. We emphasize the aspects of creating and detecting wave packets and the special role of the final state that acts as a "template" onto which the dynamics are projected. We then discuss aspects of the dynamical problem of interest here, namely, the nonadiabatic excited-state dynamics of isolated polyatomic molecules. We believe that the molecular ionization continuum is a particularly interesting final state for studying time-resolved nonadiabatic dynamics. Therefore, in some detail, we consider the general process of photoionization and discuss features of single-photon photoionization dynamics of an excited molecular state and its energy and angle-resolved detection. We briefly review the experimental techniques that are required for laboratory studies of TRPES. As TRPES is more involved than ion detection, we felt it important to motivate the use of photoelectron spectroscopy as a probe by comparing mass-resolved ion yield measurements with TRPES, using the example of internal conversion dynamics in a linear hydrocarbon molecule. Finally, we consider various applications of TRPES, with examples selected to illustrate the general issues that have been addressed.

II. WAVE PACKET DYNAMICS

A. Frequency and Time Domain Perspectives

Time-resolved experiments on isolated systems involve the creation and detection of wave packets that we define to be coherent superpositions of exact molecular eigenstates $|N\rangle$. By definition, the exact (non-Born–Oppenheimer) eigenstates are the solutions to the time-independent Schrödinger equation and are stationary. Time dependence, therefore, can only come from superposition and originates in the differing quantum mechanical energy phase factors $e^{-iE_N t/\hbar}$ associated with each eigenstate. Conceptually, there are three steps to a pump–probe wave packet experiment: (1) the preparation or pump step; (2) the dynamical evolution; and (3) the probing of the nonstationary superposition state.

From a frequency domain point of view, a femtosecond pump–probe experiment, shown schematically in Fig. 1, is a sum of coherent two-photon transition amplitudes constrained by the pump and probe laser bandwidths. The measured signal is proportional to the population in the final state $|\Psi_f\rangle$ at the end of the two-pulse sequence. As these two-photon transitions are coherent, we must therefore add the transition amplitudes and then square in order to obtain the probability. As discussed below, the signal contains interferences between all degenerate two-photon transitions. When the time delay between the two laser fields is varied, the

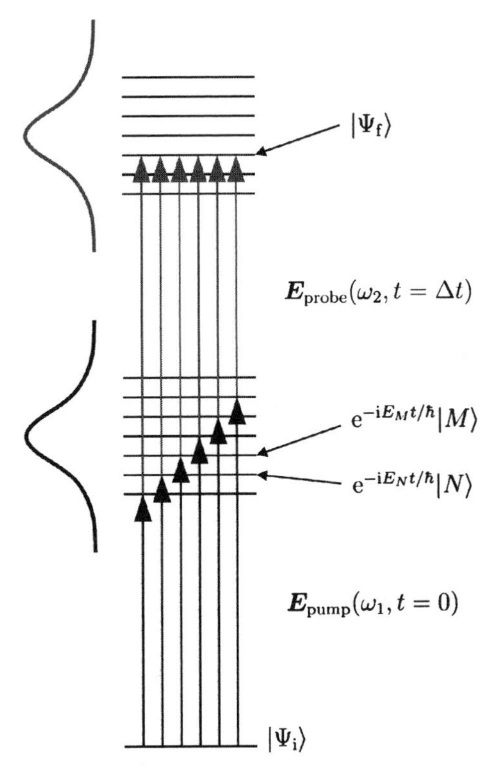

Figure 1. The creation, evolution, and detection of wave packets. The pump laser pulse $\boldsymbol{E}_{\text{pump}}$ (black) creates a coherent superposition of molecular eigenstates at $t = 0$ from the ground state $|\Psi_i\rangle$. The set of excited-state eigenstates $|N\rangle$ in the superposition (wave packet) have different energy-phase factors, leading to nonstationary behavior (wave packet evolution). At time $t = \Delta t$ the wave packet is projected by a probe pulse $\boldsymbol{E}_{\text{probe}}$ (gray) onto a set of final states $|\Psi_f\rangle$ that act as a "template" for the dynamics. The time-dependent probability of being in a given final state $|\Psi_f\rangle$ is modulated by the interferences between all degenerate coherent two-photon transition amplitudes leading to that final state.

phase relationships between the two-photon transition amplitudes changes, modifying the interference in the final state. The amplitudes and initial phases of the set of the initially prepared excited eigenstates are determined by the amplitudes and phases of the pump laser field frequencies, and the transition dipole amplitudes between the initial and the excited state of interest. Once the pump laser pulse is over, the wave packet $\Psi(t)$ evolves freely according to relative energy-phase factors in the superposition as given by

$$|\Psi(t)\rangle = \sum_N A_N e^{-iE_N t/\hbar}|N\rangle \tag{1}$$

The complex coefficients A_N contain both the amplitudes and initial phases of the exact molecular eigenstates $|N\rangle$ that are prepared by the pump laser, and the E_N are the excited-state eigenenergies. The probe laser field interacts with the wave packet after the pump pulse is over, projecting it onto a specific final state $|\Psi_f\rangle$ at some time delay Δt. This final state is the "template" onto which the wave packet dynamics are projected. The time dependence of the differential signal, $S_f(\Delta t)$, for projection onto a single final state can be written as

$$S_f(t) = |\langle\Psi_f|\boldsymbol{E}_{\text{probe}}(\omega)\cdot\boldsymbol{d}|\Psi(t)\rangle|^2 = \left|\sum_N B_N e^{-iE_N t/\hbar}\right|^2 \tag{2}$$

where the complex coefficients B_N contain both the wave packet amplitudes A_N and the (complex) probe transition dipole matrix elements connecting each eigenstate in the superposition $|N\rangle$ to the final state,

$$B_N = A_N\langle\Psi_f|\boldsymbol{E}_{\text{probe}}(\omega)\cdot\boldsymbol{d}|N\rangle \tag{3}$$

Equation (2) may be rewritten as:

$$S_f(t) = 2\sum_N\sum_{M\leq N}|B_N||B_M|\cos[(E_N - E_M)t/\hbar + \Phi_{NM}] \tag{4}$$

where the phase factor Φ_{NM} contains the initial phase differences of the molecular eigenstates, and the phase difference of the probe transition dipole matrix elements connecting the states $|N\rangle$ and $|M\rangle$ to the final state. The most detailed information is in this final-state resolved differential signal $S_f(t)$. It arises from the coherent sum over all two-photon transition amplitudes consistent with the pump and probe laser bandwidths and contains interferences between all degenerate two-photon transitions. It can be seen that the signal as a function of Δt contains modulations at frequencies $(E_N - E_M)/\hbar$, corresponding to the set of all level spacings in the superposition. This is the relationship between the wave packet dynamics and observed pump–probe signal. It is the interference between individual two-photon transitions arising from the initial state, through different excited eigenstates and terminating in the same single final state, which leads to these modulations. The Fourier transform power spectrum of this time domain signal therefore contains frequencies that give information about the set of level spacings in the excited state. The transform, however, also yields the Fourier amplitudes at these frequencies, each corresponding to a modulation depth seen in the time-domain data at that frequency. These Fourier amplitudes relate to the overlaps of each excited-state eigenfunction within the wave packet with a specific, chosen final state. Different final states will generally have differing transition dipole moment matrix elements with the eigenstates $|N\rangle$ comprising the wave packet, and so in general each final state will produce a signal S_f that has different Fourier amplitudes in its power spectrum. For example, if two interfering transitions have very similar overlaps with the final state, they will interfere constructively or destructively with nearly 100% modulation and, hence, have a very large Fourier amplitude at that frequency. Conversely, if one transition has much smaller overlap with the final state (due to, e.g., a "forbidden" transition or negligible Franck–Condon overlap) than the other, then the interference term will be small and the modulation amplitude at that frequency will be negligible. Clearly, the form of the pump–probe signal will depend on how the final state "views" the various eigenstates comprising the wave packet. An important point is that by carefully choosing different final states, it is possible for the experimentalist to emphasize and probe particular aspects of the wave packet dynamics. In general, there will be a set of final states that fall within the probe laser bandwidth. We must differentiate, therefore, between integral and differential detection techniques. With integral detection techniques (total fluorescence, ion yield, etc.), the experimentally measured total signal, $S(\Delta t)$, is proportional to the total population in the set of all energetically allowed final states, $\sum_f S_f(\Delta t)$, created at the end of the two-pulse sequence. Information is clearly lost in carrying out this sum since the individual final states may each have different overlaps with the wave packet. Therefore, differential techniques such as dispersed fluorescence, translational energy spectroscopy or photoelectron spectroscopy, which can disperse the observed signal with respect to final state, will be important. The choice of the final state is of great importance as it determines the experimental technique and significantly determines the information content of an experiment.

We now consider a pump–probe experiment from a time-domain perspective. The coherent superposition of exact molecular eigenstates constructs, for a short time, a zeroth-order state. Zeroth-order states are often physically intuitive solutions to a simpler Hamiltonian H_0, and can give a picture of the basic dynamics of the problem. The full Hamiltonian is then given by $H = H_0 + V$. Suppose we choose to expand the molecular eigenstates in a complete zeroth-order basis of H_0 that we denote by $|n\rangle$

$$|N\rangle = \sum_n a_n^N |n\rangle \tag{5}$$

then the wave packet described in Eq. (1) may be written in terms of these basis states as

$$|\Psi(t)\rangle = \sum_n C_n \mathrm{e}^{-\mathrm{i}(E_n + E_n^{\mathrm{int}})t/\hbar} |n\rangle \tag{6}$$

where the coefficients in the expansion are given by $C_n = \sum_N a_n^N A_N$ [with A_N the eigenstate coefficients in the wave packet in Eq. (1)]. To zeroth order, the eigenstate $|N\rangle$ is approximated by $|n\rangle$. The time dependence of the wave packet expressed in the zeroth-order basis reflects the couplings between the basis states $|n\rangle$, which are caused by terms in the full molecular Hamiltonian that are not included in the model Hamiltonian, H_0. In writing Eq. (6), the eigenenergies of the true molecular eigenstates have been expressed in terms of the eigenenergies of the zeroth-order basis as $E_N = E_n + E_n^{\text{int}}$, where E_n^{int} is the interaction energy of zeroth-order state $|n\rangle$ with all other zeroth-order states. The wave packet evolution, when considered in terms of the zeroth-order basis, contains frequency components corresponding to the couplings between states, as well as frequency components corresponding to the energies of the zeroth-order states. To second order in perturbation theory, the interaction energy (coupling strength) E_n^{int} between zeroth-order states is given in terms of the matrix elements of V by

$$E_n^{\text{int}} = \langle n|V|n\rangle + \sum_{m\neq n} \frac{\langle m|V|n\rangle^2}{E_m - E_n} \tag{7}$$

Just as the expansion in the zeroth-order states can describe the exact molecular eigenstates, likewise an expansion in the exact states can be used to prepare, for a short time, a zeroth-order state. If the perturbation V is small, and the model Hamiltonian H_0 is a good approximation to H, then the initially prepared superposition of eigenstates will resemble a zeroth-order state. The dephasing of the exact molecular eigenstates in the wave packet superposition subsequently leads to an evolution of the initial zeroth-order electronic character, transforming into a different zeroth-order electronic state as a function of time.

A well-known example is found in the problem of intramolecular vibrational energy redistribution (IVR). The exact vibrational states are eigenstates of the full rovibrational Hamiltonian that includes all orders of couplings and are, of course, stationary. An example of a zeroth-order state would be a normal mode, the solution to a parabolic potential. A short pulse could create a superposition of exact vibrational eigenstates which, for a short time, would behave as a normal mode (e.g., stretching). However, due to the dephasing of the exact vibrational eigenstates in the wave packet, this zeroth-order stretching state would evolve into a superposition of other zeroth-order states (e.g., other normal modes like bending). Examples of using TRPES to study such vibrational dynamics will be given in Section VI.B.

B. Nonadiabatic Molecular Dynamics

As discussed in Section III.A, wave packets allow for the development of a picture of the time evolution of the zeroth-order states, and with a suitably chosen basis this provides a view of both charge and energy flow in the molecule. For the case of interest here, excited-state nonadiabatic dynamics, the appropriate zeroth-order states are the Born–Oppenheimer (BO) states [1–8]. These are obtained by invoking an adiabatic approximation that the electrons, being much lighter than the nuclei, can rapidly adjust to the slower time-dependent fields due to the vibrational motion of the atoms. The molecular Hamiltonian can be separated into kinetic energy operators of the nuclei $T_n(\boldsymbol{R})$ and electrons $T_e(\boldsymbol{r})$, and the potential energy of the electrons and nuclei, $V(\boldsymbol{R}, \boldsymbol{r})$,

$$H(\boldsymbol{r}, \boldsymbol{R}) = T_n(\boldsymbol{R}) + T_e(\boldsymbol{r}) + V(\boldsymbol{R}, \boldsymbol{r}) \tag{8}$$

where $\boldsymbol{R}$ denotes the nuclear coordinates, and $\boldsymbol{r}$ denotes the electronic coordinates. The BO basis is obtained by setting $T_n(\boldsymbol{R}) = 0$, such that H describes the electronic motion in a molecule with fixed nuclei, and solving the time-independent Schrödinger equation treating the nuclear coordinates $\boldsymbol{R}$ as a parameter [6]. In this approximation, the adiabatic BO electronic states $\Phi_\alpha(\boldsymbol{r}; \boldsymbol{R})$ and potential energy surfaces $V_\alpha(\boldsymbol{R})$ are defined by

$$[H_e(\boldsymbol{r}; \boldsymbol{R}) - V_\alpha(\boldsymbol{R})]\Phi_\alpha(\boldsymbol{r}; \boldsymbol{R}) = 0 \tag{9}$$

where the "clamped nuclei" electronic Hamiltonian is defined by $H_e(\boldsymbol{r}; \boldsymbol{R}) = T_e(\boldsymbol{r}) + V(\boldsymbol{r}, \boldsymbol{R})$. The eigenstates of the full molecular Hamiltonian [Eq. (8)] may be expanded in the complete eigenbasis of BO electronic states defined by Eq. (9),

$$\langle \boldsymbol{r}; \boldsymbol{R} | N \rangle = \sum_\alpha \chi_\alpha(\boldsymbol{R}) \Phi_\alpha(\boldsymbol{r}; \boldsymbol{R}) \tag{10}$$

where the expansion coefficients $\chi_\alpha(\boldsymbol{R})$ are functions of the nuclear coordinates. The zeroth-order BO electronic states $\Phi_\alpha(\boldsymbol{r}; \boldsymbol{R})$ have been obtained by neglecting the nuclear kinetic energy operator $T_n(\boldsymbol{R})$, and so will be coupled by this term in the Hamiltonian. Substitution of the expansion Eq. (10) into the Schrödinger equation $[H(\boldsymbol{r}, \boldsymbol{R}) - E_N]|N\rangle = 0$ gives a system of coupled differential equations for the nuclear wave functions [5, 6, 8, 31]

$$[T_n(\boldsymbol{R}) + V_\alpha(\boldsymbol{R}) - E_N]\chi_\alpha(\boldsymbol{R}) = \sum_\beta \Lambda_{\alpha\beta}(\boldsymbol{R})\chi_\beta(\boldsymbol{R}) \tag{11}$$

where E_N is the eigenenergy of the exact moleculer eigenstate $|N\rangle$. The nonadiabatic coupling parameters $\Lambda_{\alpha\beta}(\boldsymbol{R})$ are defined as:

$$\Lambda_{\alpha\beta}(\boldsymbol{R}) = T_n(\boldsymbol{R})\delta_{\alpha\beta} - \int d\boldsymbol{r} \Phi_\alpha^*(\boldsymbol{r}) T_n(\boldsymbol{r}) \Phi_\beta(\boldsymbol{r}) \tag{12}$$

The diagonal terms $\alpha = \beta$ are corrections to the frozen nuclei potentials $V_\alpha(\boldsymbol{R})$ and together form the nuclear zeroth-order states of interest here. The off-diagonal terms $\alpha \neq \beta$ are the operators that lead to transitions (evolution) between zeroth-order states. The kinetic energy is a derivative operator of the nuclear coordinates and, hence, it is the motion of the nuclei that leads to electronic transitions. One could picture that it is the time-dependent electric field of the oscillating (vibrating) charged nuclei that can lead to electronic transitions. When the Fourier components of this time-dependent field match electronic level spacings, transitions can occur. As nuclei move slowly, usually these frequencies are too small to induce any electronic transitions. When the adiabatic electronic states become close in energy, the coupling between them can be extremely large, the adiabatic approximation breaks down, and the nuclear and electronic motions become strongly coupled [1–8]. A striking example of the result of the nonadiabatic coupling of nuclear and electronic motions is a conical intersection between electronic states, which provide pathways for interstate crossing on the femtosecond time scale and have been termed "photochemical funnels" [5]. Conical intersections occur when adiabatic electronic states become degenerate in one or more nuclear coordinates, and the nonadiabatic coupling becomes infinite. This divergence of the coupling and the pronounced anharmonicity of the adiabatic potential energy surfaces in the region of a conical intersection causes very strong electronic couplings, as well as strong coupling between vibrational modes. Such nonadiabatic couplings can have pronounced effects. For example, analysis of the the absorption band corresponding to the S_2 electronic state of pyrazine demonstrated that the vibronic bands in this region of the spectrum have a very short lifetime due to coupling of the S_2 electronic state with the S_1 electronic state [32, 33], and an early demonstration of the effect of a conical intersection was made in the study of an unexpected band in the photoelectron spectrum of butatriene [34, 35]. Detailed examples are given in Section VI.

The nuclear function $\chi_\alpha(\boldsymbol{R})$ is usually expanded in terms of a wave function describing the vibrational motion of the nuclei, and a rotational wave function [36, 37]. Analysis of the vibrational part of the wave function usually assumes that the vibrational motion is harmonic, such that a normal mode analysis can be applied [36, 38]. The breakdown of this approximation leads to vibrational coupling, commonly termed intramolecular vibrational energy redistribution, IVR. The rotational basis is usually taken as the rigid rotor basis [36, 38–40]. This separation between vibrational and rotational motions neglects centrifugal and Coriolis coupling of rotation and vibration [36, 38–40]. Next, we will write the wave packet prepared by the pump laser in terms of the zeroth-order BO basis as

$$|\Psi(t)\rangle = \sum_{J_\alpha M_\alpha \tau_\alpha v_\alpha \alpha} C_{J_\alpha M_\alpha \tau_\alpha v_\alpha \alpha}(t) |J_\alpha M_\alpha \tau_\alpha\rangle |v_\alpha\rangle |\alpha\rangle \tag{13}$$

The three kets in this expansion describe the rotational, vibrational, and electronic states of the molecule, respectively,

$$\langle \phi, \theta, \chi | J_\alpha M_\alpha \tau_\alpha \rangle = \psi_{J_\alpha M_\alpha \tau_\alpha}(\phi, \theta, \chi) \quad (14)$$

$$\langle \boldsymbol{R} | v_\alpha \rangle = \psi_{v_\alpha}(\boldsymbol{R}) \quad (15)$$

$$\langle \boldsymbol{r}; \boldsymbol{R} | \alpha \rangle = \Phi_\alpha(\boldsymbol{r}; \boldsymbol{R}) \quad (16)$$

where (ϕ, θ, χ) are the Euler angles [40] connecting the lab fixed frame (LF) to the molecular frame (MF). The quantum numbers J_α and M_α denote the total angular momentum and its projection on the lab-frame z axis, and τ_α labels the $(2J_\alpha + 1)$ eigenstates corresponding to each (J_α, M_α) [38–40]. The vibrational state label v_α is a shorthand label that denotes the vibrational quanta in each of the vibrational modes of the molecule. The time-dependent coefficients $C_{J_\alpha M_\alpha \tau_\alpha v_\alpha}(t)$ will in general include exponential phase factors that reflect all of the couplings described above, as well as the details of the pump step.

For a vibrational mode of the molecule to induce coupling between adiabatic electronic states $\Phi_\alpha(\boldsymbol{r}; \boldsymbol{R})$ and $\Phi_\beta(\boldsymbol{r}; \boldsymbol{R})$, the direct product of the irreducible representations of $\Phi_\alpha(\boldsymbol{r}; \boldsymbol{R})$, $\Phi_\beta(\boldsymbol{r}; \boldsymbol{R})$, and the vibrational mode must contain the totally symmetric representation of the molecular point group,

$$\Gamma_\alpha \otimes \Gamma_v \otimes \Gamma_\beta \supset A_1 \quad (17)$$

where Γ_v is the irreducible representation of the vibrational mode causing the nonadiabatic coupling. As discussed in Section II.A, an initially prepared superposition of the molecular eigenstates will tend to resemble a zeroth-order BO state. This BO state will then evolve due to the coupling provided by the nuclear kinetic energy operator that leads to this evolution: a process that is often called a radiationless transition. For example, a short pulse may prepare the S_2 (zeroth-order) BO state which, via nonadiabatic coupling, evolves into the S_1 (zeroth-order) BO state, a process that is referred to as "internal conversion". For the remainder of this chapter we will adopt the language of zeroth-order states and their evolution due to intramolecular couplings.

III. PROBING NONADIABATIC DYNAMICS WITH PHOTOELECTRON SPECTROSCOPY

As discussed in Section II, the excited-state dynamics of polyatomic molecules is dictated by the coupled flow of both charge and energy within the molecule. As such, a probe technique that is sensitive to both nuclear (vibrational) and electronic configuration is required in order to elucidate the mechanisms of such processes. Photoelectron spectroscopy provides such a technique, allowing for the disentangling of electronic and nuclear motions, and in principle leaving no configuration of the molecule unobserved, since ionization may occur for all molecular configurations. This is in contrast to other techniques, such as absorption or fluorescence spectroscopy, which sample only certain areas of the potential energy surfaces involved, as dictated by oscillator strengths, selection rules, and Franck–Condon factors.

The molecular ionization continuum provides a template for observing both excited-state vibrational dynamics, via Franck–Condon distributions, and evolving excited-state electronic configurations. The latter are understood to be projected out via electronic structures in the continuum, of which there are two kinds: that of the cation and that of the free electron. The electronic states of the cation can provide a map of evolving electronic structures in the neutral state prior to ionization—in the independent electron approximation emission of an independent outer electron occurs without simultaneous electronic reorganization of the "core" (be it cation or neutral)—this is called the "molecular orbital" or Koopmans picture [13, 41, 42]. These simple correlation rules indicate the cation electronic state expected to be formed upon single-photon single active electron ionization of a given neutral state. The probabilities of partial ionization into specific cation electronic states can differ drastically with respect to the molecular orbital nature of the probed electronic state. If a given probed electronic configuration correlates, upon removal of a single active outer electron, to the ground electronic configuration of the continuum, then the photoionization probability is generally higher than if it does not. The electronic states of the free electron, commonly described as scattering states, form the other electronic structure in the continuum. The free electron states populated upon photoionization reflect angular momentum correlations, and are therefore sensitive to neutral electronic configurations and symmetries. This sensitivity is expressed in the form of the photoelectron angular distribution (PAD). Furthermore, since the active molecular frame ionization dipole moment components are geometrically determined by the orientation of the molecular frame within the laboratory frame, and since the free electron scattering states are dependent on the direction of the molecular frame ionization dipole, the form of the laboratory frame PAD is sensitive to the molecular orientation, and so will reflect the rotational dynamics of the neutral molecules.

First, we consider a schematic example to illustrate how the cation electronic structures can be used in (angle integrated) TRPES to disentangle electronic from vibrational dynamics in ultrafast nonadiabatic processes, depicted in Fig. 2. A zeroth- order bright state, α, is coherently prepared with a femtosecond pump pulse. According to the Koopmans picture [13, 41, 42], it should ionize into the α^+ continuum, the electronic state of the cation obtained upon removal of the outermost valence electron (here chosen to be

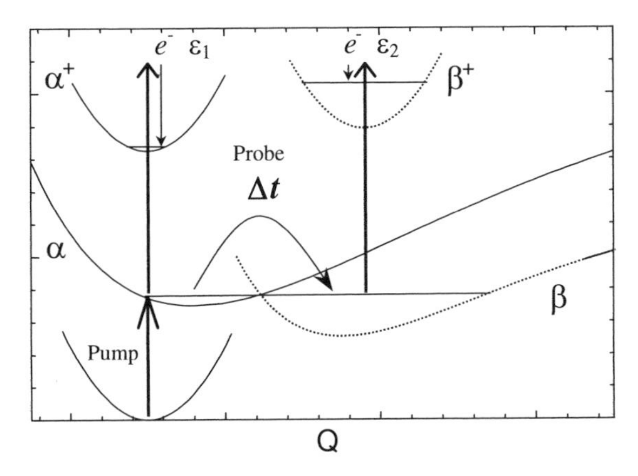

Figure 2. A TRPES scheme for disentangling electronic from vibrational dynamics in excited polyatomic molecules. A zeroth-order electronic state α is prepared by a femtosecond pump pulse. Via a nonadiabatic process it converts to a vibrationally hot lower lying electronic state, β. The Koopmans-type ionization correlations suggest that these two states will ionize into different electronic continua: $\alpha \rightarrow \alpha^+ + e^-(\varepsilon_1)$ and $\beta \rightarrow \beta^+ + e^-(\varepsilon_2)$. When the wave packet has zeroth-order α character, any vibrational dynamics in the α state will be reflected in the structure of the ε_1 photoelectron band. After the nonadiabatic process, the wave packet has zeroth-order β electronic character; any vibrational dynamics in the state will be reflected in the ε_2 band. This allows for the simultaneous monitoring of both electronic and vibrational excited-state dynamics.

the ground electronic state of the ion). This process produces a photoelectron band ε_1. We now consider any nonadiabatic coupling process that transforms the zeroth-order bright state α into a lower lying zeroth order dark state β, as induced by promoting vibrational modes of appropriate symmetry. Again, according to the Koopmans picture, the state should ionize into the β^+ ionization continuum (here assumed to be an electronically excited state of the ion), producing a photoelectron band ε_2. Therefore, for a sufficiently energetic probe photon (i.e., with both ionization channels open), we expect a switching of the electronic photoionization channel from ε_1 to ε_2 during the nonadiabatic process. This simple picture suggests that one can directly monitor the evolving excited-state electronic configurations (i.e., the electronic population dynamics) during nonadiabatic processes while simultaneously following the coupled nuclear dynamics via the vibrational structure within each photoelectron band. The cation electronic structures can act as a "template" for the disentangling of electronic from vibrational dynamics in the excited state [43–47].

More specifically, the BO electronic state $\Phi_\alpha(\boldsymbol{r};\boldsymbol{R})$ (which is an eigenfunction of the electronic Hamiltonian H_e) is a complex multielectron wave function. It can be expressed in terms of self-consistent field (SCF) wave functions $|\Phi_n\rangle$, which are comprised of a Slater determinant of single electron molecular spin–orbitals [42],

$$|\alpha\rangle = \sum_n A_n^\alpha |\Phi_n\rangle \tag{18}$$

Each $|\Phi_n\rangle$ corresponds to a single configuration, and the fractional parentage coefficients A_n^α reflect the configuration interaction (caused by electron correlation) for each BO electronic state. The configuration interaction "mixes in" SCF wave functions of the same overall symmetry, but different configurations. The correlations between the neutral electronic state and the ion electronic state formed upon ionization are readily understood in this independent electron picture [42, 44, 48, 49]. In the Koopmans' picture of photoionization, a single active electron approximation is adopted, ionization occurs out of a single molecular orbital, and the remaining core electron configuration is assumed to remain unchanged upon ionization. As such, a multielectron matrix element reduces to a single electron matrix element for each configuration that contributes to the electronic state, weighted by the fractional parentage coefficients.

The two limiting cases for Koopmans-type correlations in TRPES experiments, as initially proposed by Domcke [48, 49], were demonstrated experimentally [47, 50] and will be further discussed in Section V.B. The first case, Type (I), is when the neutral excited states α and β clearly correlate to different cation electronic states, as in Fig. 2. Even if there are large geometry changes upon internal conversion and/or ionization, producing vibrational progressions, the electronic correlations should favor a disentangling of the vibrational dynamics from the electronic population dynamics. An example of this situation is discussed in Section V.B. The other limiting case, Type (II), is when the neutral excited states α and β correlate equally strongly to the same cation electronic states, and so produce overlapping photoelectron bands. An example of a Type (II) situation in which vastly different Franck–Condon factors allow the states α and β to be distinguished in the PES is given in Section V.B, but more generally Type (II) ionization correlations are expected to hinder the disentangling of electronic from vibrational dynamics purely from the PES. It is under these Type (II) situations when measuring the time-resolved PAD is expected to be of most utility—as discussed below, the PAD will reflect the evolution of the molecular electronic symmetry under situations where electronic states are not readily resolved in the PES. The continuum state accessed by the probe transition may be written as a direct product of the cation and free electron states. As with any optical transition, there are symmetry based "selection rules" for the photoionization step. In the case of photoionization, there is the requirement that the direct product of the irreducible representations of the state of the ion, the free electron wave function, the molecular frame transition dipole moment, and the neutral state contains the totally symmetric irreducible representation of the molecular point group [51, 52]. Since the symmetry of the free electron wave function determines the form of the PAD, the shape of the PAD will reflect the electronic symmetry of the neutral molecule and the symmetries of the contributing molecular frame transition dipole moment components. Since the relative contributions of the molecular frame transition dipole moments are geometrically determined by the orientation of the molecule relative to the ionizing laser-field polarization, the form of the laboratory frame PAD will reflect the distribution of molecular axis in the laboratory frame, and so will reflect the rotational dynamics of the molecule [53–59].

We turn now to a more detailed description of the photoionization probe step in order to clarify the ideas presented above. Time-resolved photoelectron spectroscopy probes the excited-state dynamics using a time-delayed probe laser pulse that brings about ionization of the excited-state wave packet, usually with a single photon

$$\mathrm{AB}(|\Psi(t)\rangle) + h\nu \rightarrow \mathrm{AB}^{+}(|\Psi_{+}\rangle) + e^{-}(\epsilon\hat{\boldsymbol{k}}_{\mathrm{L}}) \tag{19}$$

Here and in what follows we use a subscript + to denote the quantum numbers of the ion core, ϵ to denote the kinetic energy of the electron, and $\hat{\boldsymbol{k}}_{\mathrm{L}}$ to denote the LF direction of the emitted photoelectron, with magnitude $k = \sqrt{2m_e\epsilon}$. In the following treatment, we assume that the probe laser intensity remains low enough that a perturbative description of the probe process is valid, and that the pump and probe laser pulses are separated temporally. Full nonperturbative treatments were given in the literature for situations in which these approximations are not appropriate [22, 58, 60–62].

The single-particle wave function for the free photoelectron may be expressed as an expansion in angular momentum partial waves characterized by an orbital angular momentum quantum number l and and associated quantum number λ for the projection of l on the molecular frame (MF) z axis [22, 23, 63–66],

$$\langle \boldsymbol{r}'; \boldsymbol{R}|\hat{\boldsymbol{k}}_{\mathrm{M}}\epsilon\rangle = \sum_{l\lambda} \mathrm{i}^{l} e^{-\mathrm{i}\sigma_l(\epsilon)} Y_{l\lambda}(\hat{\boldsymbol{k}}_{\mathrm{M}})\psi_{l\lambda}(\boldsymbol{r}'; \epsilon, \boldsymbol{R}) \tag{20}$$

where the asymptotic recoil momentum vector of the photoelectron in the MF is denoted by $\hat{\boldsymbol{k}}_{\mathrm{M}}$, and $Y_{l\lambda}(\hat{\boldsymbol{k}}_{\mathrm{M}})$ is a spherical harmonic [40]. The radial wave function in this expansion, $\psi_{l\lambda}(\boldsymbol{r}'; \epsilon, \boldsymbol{R})$ depends on the MF position vector of the free photoelectron $\boldsymbol{r}'$, and parametrically upon the nuclear coordinates $\boldsymbol{R}$, and also the photoelectron energy ϵ. The energy-dependent scattering phase shift $\sigma_l(\epsilon)$ depends on the potential of the ion core, and contains the Coulomb phase shift. This radial wave function contains all details of the scattering of the photoelectron from the nonspherical potential of the molecule [66]. In this discussion, we neglect the spin of the free electron, assuming it to be uncoupled from the other (orbital and rotational) angular momenta – the results derived here are unaffected by other angular momenta coupling schemes.

When considering molecular photoionization, it is useful to keep in mind the conceptually simpler case of atomic ionization [63, 65]. For atomic ionization, the ionic potential experienced by the photoelectron is a central field within the independent electron approximation: close to the ion core, the electron experiences a potential that is partially shielded due to the presence of the other electrons [67]. Far from the ion core, in the asymptotic region, the Coulombic potential dominates. The spherically symmetric nature of this situation means that the angular momentum partial waves of orbital angular momentum l form a complete set of independent ionization channels (i.e., l remains a good quantum number throughout the scattering process). Single photon ionization from a single electronic state of an atom produces a free electron wave function comprising only two partial waves with angular momenta $l_0 \pm 1$, where l_0 is the angular momentum quantum number of the electron prior to ionization. In the molecular case, however, the potential experienced by the photoelectron in the region of the ion core is noncentral. As a result, l is no longer a good quantum number and scattering from the ion core potential can cause changes in l. For linear and symmetric top molecules, λ remains a good quantum number, but for asymmetric top molecules λ also ceases to be conserved during scattering. The multipolar potential felt by the electron in the ion core region falls off rapidly such that in the asymptotic region, the Coulombic potential dominates. As such, a partial wave description of the free electron remains useful in the molecular case [64, 68], but the partial waves are no longer eigenstates of the scattering potential, resulting in multichannel scattering among the partial wave states and a much richer partial wave composition when compared to the atomic case [66]. To add to this richness of partial waves, the molecular electronic state is no longer described by a single value of l_0. Nonetheless, a partial wave description of the free electron wave function remains a useful description since, despite the complex scattering processes, the expansion is truncated at relatively low values of l.

For polyatomic molecules, Chandra showed that it is useful to reexpress the photoelectron wave function in terms of symmetry adapted spherical harmonics [51, 52, 55, 69–72],

$$\langle \boldsymbol{r}'; \boldsymbol{R} | \hat{\boldsymbol{k}}_{\mathrm{M}} \epsilon \rangle = \sum_{\Gamma\mu h l} \mathrm{i}^l e^{-\mathrm{i}\sigma_l(\epsilon)} X_{hl}^{\Gamma\mu *}(\hat{\boldsymbol{k}}_{\mathrm{M}}) \psi_{\Gamma\mu h l}(\boldsymbol{r}'; \epsilon, \boldsymbol{R}) \tag{21}$$

The symmetry adapted spherical harmonics (also referred to as generalized harmonics), $X_{hl}^{\Gamma\mu}(\hat{\boldsymbol{k}}_{\mathrm{M}})$, satisfy the symmetries of the molecular point group [51] and are defined as:

$$X_{hl}^{\Gamma\mu}(\hat{\boldsymbol{k}}) = \sum_{\lambda} b_{hl\lambda}^{\Gamma\mu} Y_{l\lambda}(\hat{\boldsymbol{k}}) \tag{22}$$

where Γ defines an irreducible representation (IR) of the molecular point group of the molecule plus electron system, μ is a degeneracy index, and h distinguishes harmonics with the same values of $\Gamma\mu l$ induces. The $b_{hl\lambda}^{\Gamma\mu}$ symmetry coefficients are found by constructing generalized harmonics using the projection theorem [73–76] employing the spherical harmonics $Y_{lm}(\theta, \phi)$ as the generating function. In using the molecular point group, rather than the symmetry group of the full molecular Hamiltonian, we are assuming rigid behavior. To go beyond this assumption, it is necessary to consider the full molecular symmetry group [38]. Such a treatment has been given by Signorell and Merkt [77].

Combining Eqs. (21) and (22), the free electron wave function Eq. (21) may be reexpressed in the LF using the properties of the spherical harmonics under rotation as:

$$\langle \boldsymbol{r}'; \boldsymbol{R} | \hat{\boldsymbol{k}}_{\mathrm{L}} \epsilon \rangle = \sum_{l\lambda m} \sum_{\Gamma\mu h} \mathrm{i}^l e^{-\mathrm{i}\sigma_l(\epsilon)} b_{hl\lambda}^{\Gamma\mu} D_{m\lambda}^{l*}(\phi, \theta, \chi) Y_{lm}^*(\hat{\boldsymbol{k}}_L) \psi_{\Gamma\mu hl}(\boldsymbol{r}'; \epsilon, \boldsymbol{R}) \tag{23}$$

where $D_{m\lambda}^{l}(\phi, \theta, \chi)$ is a Wigner rotation matrix element [40].

The partial differential photoionization cross-section for producing photoelectrons with a kinetic energy ϵ at time t ejected in the LF direction $\hat{\boldsymbol{k}}_{\mathrm{L}}$ is then

$$\sigma(\epsilon, \hat{\boldsymbol{k}}_{\mathrm{L}}; t) \propto \sum_{n_{\alpha_+} M_{\alpha_+}} \left| \sum_{n_\alpha M_\alpha} C_{n_\alpha M_\alpha}(t) \langle \hat{\boldsymbol{k}}_{\mathrm{L}} \epsilon n_{\alpha_+} M_{\alpha_+} | \boldsymbol{d} \cdot \hat{\boldsymbol{e}} | n_\alpha M_\alpha \rangle \mathcal{E}(n_{\alpha_+}, n_\alpha, \epsilon) \right|^2 \tag{24}$$

where we have introduced the shorthand notation for quantum numbers $n_\alpha = J_\alpha \tau_\alpha v_\alpha \alpha$. We have implicitly assumed that the coefficients $C_{n_\alpha M_\alpha}(t)$ do not vary over the duration of the probe pulse. We have taken the laser field of the probe pulse to be of the form

$$E(t) = \hat{\boldsymbol{e}} f(t) \cos(\omega_0 t + \phi(t)) \tag{25}$$

where $f(t)$ is the pulse envelope, $\hat{\boldsymbol{e}}$ is the probe pulse polarization vector, ω_0 is the carrier frequency, and $\phi(t)$ is a time-dependent phase. In Eq. (24), $\mathcal{E}(n_{\alpha_+}, n_\alpha, \epsilon)$ is the Fourier transform of the probe pulse at the frequency $2\pi(E_{n_{\alpha_+}} - E_{n_\alpha} + \epsilon)/h$, as defined by

$$E(\omega) = \int e^{i\omega t} E(t)\, dt \tag{26}$$

In order to evaluate the matrix elements of the dipole moment operator in Eq. (24), it is convenient to separate out the geometrical aspects of the problem from the dynamical parameters. To that end, it is convenient to decompose the LF scalar product of the transition dipole moment $\boldsymbol{d}$ with the polarization vector of the probe laser field $\hat{\boldsymbol{e}}$ in terms of the spherical tensor components as [40]

$$\boldsymbol{d} \cdot \hat{\boldsymbol{e}} = \sum_{p=-1}^{1} (-1)^p d_p e_{-p}. \tag{27}$$

The LF spherical tensor components of the electric-field polarization are defined as:

$$e_0 = e_z \qquad e_{\pm 1} = \frac{\mp 1}{\sqrt{2}}(e_x \pm i e_y) \tag{28}$$

For linearly polarized light, it is convenient to define the lab frame z axis along the polarization vector, such that the only nonzero component is e_0. For circularly polarized light, the propagation direction of the light is usually chosen to define the LF z axis such that the nonzero components are e_1 for right circularly polarized light and e_{-1} for left circularly polarized light. Other polarizations states of the probe pulse are described by more than a single nonzero component e_p, and, for generality, in what follows we will not make any assumptions about the polarization state of the ionizing pulse. The LF components of the dipole moment d_p are related to the MF components through a rotation,

$$d_p = \sum_{q=-1}^{1} D_{pq}^{1*}(\phi, \theta, \chi) d_q \tag{29}$$

The rotational wave functions appearing in Eq. (24) may be expressed in terms of the symmetric top basis as [40]

$$|JM\tau\rangle = \sum_K a_K^{J\tau} |JKM\rangle \tag{30}$$

where the symmetric top rotational basis functions are defined in terms of the Wigner rotation matrices as:

$$\langle \mathbf{\Omega}|JKM\rangle = \left(\frac{2J+1}{8\pi^2}\right)^{1/2} D^{J*}_{MK}(\phi,\theta,\chi) \tag{31}$$

By using Eqs. (27)–(31), the matrix elements of the dipole moment operator in Eq. (24) may be written as

$$\begin{aligned}
&\langle \hat{\boldsymbol{k}}_{\mathrm{L}}\epsilon n_{\alpha_+}M_{\alpha_+}|\boldsymbol{d}\cdot\hat{\boldsymbol{e}}|n_\alpha M_\alpha\rangle \\
&= \frac{1}{8\pi^2}[J_\alpha, J_{\alpha_+}]^{1/2}\sum_{l\lambda m}(-\mathrm{i})^l e^{\mathrm{i}\sigma_l(\epsilon)}Y_{lm}(\hat{\boldsymbol{k}}_L)\sum_{K_\alpha K_{\alpha_+}} a^{J_\alpha\tau_\alpha}_{K_\alpha} a^{J_{\alpha_+}\tau_{\alpha_+}}_{K_{\alpha_+}}\sum_{pq}(-1)^p e_{-p} \\
&\times \int D^l_{m\lambda}(\phi,\theta,\chi)D^{J_{\alpha_+}}_{M_{\alpha_+}K_{\alpha_+}}(\phi,\theta,\chi)D^{1*}_{pq}(\phi,\theta,\chi)D^{J_\alpha *}_{M_\alpha K_\alpha}(\phi,\theta,\chi)\,\mathrm{d}\mathbf{\Omega} \\
&\times \sum_{\Gamma\mu h} b^{\Gamma\mu}_{hl\lambda} D^{\alpha v_\alpha \alpha_+ v_{\alpha_+}}_{\Gamma\mu hl}(q)
\end{aligned} \tag{32}$$

where we have introduced the shorthand $[X, Y, \ldots] = (2X+1)(2Y+1)\ldots$. The dynamical functions in Eq. (32) are defined as:

$$D^{\alpha v_\alpha \alpha_+ v_{\alpha_+}}_{\Gamma\mu hl}(q) = \int \psi^*_{v_{\alpha_+}}(\boldsymbol{R})\psi_{v_\alpha}(\boldsymbol{R}) d^{\alpha\alpha_+}_{\Gamma\mu hl}(q;\boldsymbol{R})\,\mathrm{d}\boldsymbol{R} \tag{33}$$

These dynamical parameters are integrals over the internuclear separations $\boldsymbol{R}$, as well as the electronic coordinates $\boldsymbol{r}$ through the electronic transition dipole matrix elements, $d^{\alpha\alpha_+}_{\Gamma\mu hl}(q;\boldsymbol{R})$. These electronic transition dipole matrix elements are evaluated at fixed internuclear configurations [68] and are defined as:

$$d^{\alpha\alpha_+}_{\Gamma\mu hl}(q;\boldsymbol{R}) = \int \Phi^*_{\alpha_+\Gamma\mu hl}(\boldsymbol{r};\epsilon,\boldsymbol{R}) d_q \Phi_\alpha(\boldsymbol{r};\boldsymbol{R})\,\mathrm{d}\boldsymbol{r} \tag{34}$$

Here, $\Phi_{\alpha_+\Gamma\mu hl}(\boldsymbol{r};\epsilon,\boldsymbol{R})$ is the antisymmetrized electronic wave function that includes the free electron radial wave function $\psi_{\Gamma\mu hl}(\boldsymbol{r}';\epsilon,\boldsymbol{R})$ and the electronic wave function of the ion $\Phi_{\alpha_+}(\boldsymbol{r}'';\boldsymbol{R})$ [51, 66, 68, 78] (where $\boldsymbol{r}''$ are the position vectors of the ion electrons). For the integral in Eq. (34) to be nonzero, the following condition must be met

$$\Gamma \otimes \Gamma_{\alpha_+} \otimes \Gamma_q \otimes \Gamma_\alpha \subset A_1 \tag{35}$$

That is, the direct product of the IRs of the free electron, the ion, the transition dipole moment, and the neutral electronic state, respectively, must contain the totally symmetric IR of the molecular point group, A_1. Clearly, the symmetries of the contributing photoelectron partial waves will be determined by the electronic symmetry of the BO electronic state undergoing ionization, as well as the molecular frame direction of the ionization transition dipole moment, which determines the possible Γ_q, and the electronic symmetry of the cation. As such, the evolution of the photoelectron angular distribution, which directly reflects the allowed symmetries of the partial waves, will reflect the evolution of the molecular electronic symmetry.

It is frequently the case that the electronic transition dipole matrix element $d^{\alpha\alpha_+}_{\Gamma\mu hl}(q;\boldsymbol{R})$ is only weakly dependent on the nuclear coordinates $\boldsymbol{R}$ such that the Franck–Condon approximation [37] may be employed. Within this approximation,

$$D^{\alpha v_\alpha \alpha_+ v_{\alpha_+}}_{\Gamma\mu hl}(q) = \bar{d}^{\alpha\alpha_+}_{\Gamma\mu hl}(q) \int \psi^*_{v_{\alpha_+}}(\boldsymbol{R})\psi_{v_\alpha}(\boldsymbol{R})\, \mathrm{d}\boldsymbol{R} \tag{36}$$

where $\bar{d}^{\alpha\alpha_+}_{\Gamma\mu hl}(q)$ is the value of $d^{\alpha\alpha_+}_{\Gamma\mu hl}(q;\boldsymbol{R})$ averaged over $\boldsymbol{R}$. Within this approximation, the overlap integral between the molecular vibrational state and the cation vibrational state determines the ionization efficiency to each cation vibrational state [13, 42, 79–85]. The Franck–Condon factors are determined by the relative equilibrium geometries of the electronic states of the neutral $(\Phi_\alpha(\boldsymbol{r};\boldsymbol{R}))$ and cation $(\Phi_{\alpha_+}(\boldsymbol{r}'';\epsilon,\boldsymbol{R}))$ [13, 42]. If the neutral and cation electronic states have similar equilibrium geometries, each neutral vibronic state will produce a single photoelectron peak for each vibrational mode corresponding to $\Delta v = 0$ transitions upon ionization. However, if there is a substantial difference in the equilibrium geometries, a vibrational progression in the PES results from ionization of each neutral vibronic state, corresponding to $\Delta v = 0, 1, 2\ldots$ transitions upon ionization for each populated vibrational mode. In either case, the photoelectron spectrum will reflect the vibronic composition of the molecular wave packet, and the time dependence of the vibrational structure in the photoelectron spectrum directly reflects the nuclear motion of the molecule. Of course, this Franck–Condon mapping of the vibrational dynamics onto the PES will break down if the variation of the electronic ionization dipole matrix elements varies significantly with $\boldsymbol{R}$, for example, in a region in which vibrational autoionization is active [13, 42, 84, 86].

In the Koopmans' picture of photoionization [13, 41, 42], a single active electron approximation is adopted, ionization occurs out of a single molecular orbital, and the remaining core electron configuration is assumed to remain unchanged upon ionization. As such, the multielectron matrix element in Eq. (34) reduces to a single electron matrix element for each configuration that contributes to the electronic state, weighted by the fractional parentage coefficients. In the limit of the electronic state $|\alpha\rangle$ being composed of a single configuration, ionization will access the continuum corresponding to the ion state α^+ that has the same core electronic configuration. In the single active electron approximation, for a single configuration, the electronic transition dipole matrix element in Eq. (34) may be rewritten as [51, 78]

$$d^{\alpha\alpha_+}_{\Gamma\mu hl}(q;\boldsymbol{R}) = \int \psi^*_{\Gamma\mu hl}(\boldsymbol{r}';\epsilon,\boldsymbol{R}) d_q \phi_i(\boldsymbol{r}';\boldsymbol{R})\,\mathrm{d}\boldsymbol{r}' \tag{37}$$

where $\phi_i(\boldsymbol{r}';\boldsymbol{R})$ is the initial bound molecular orbital from which photoionization takes place. In order for Eq. (37) to be nonzero, the following condition must be met [51]:

$$\Gamma \otimes \Gamma_q \otimes \Gamma_i \subset A_1 \tag{38}$$

Within the independent electron and single active electron approximations, the symmetries of the contributing photoelectron partial waves will be determined by the symmetry of the orbital(s) from which ionization occurs, and so the PAD will directly reflect the evolution of the molecular orbital configuration. Example calculations demonstrating this are shown in Fig. 3 for a model C_{3v} molecule, where a clear difference in the PAD is observed according to whether ionization occurs from an a_1 or an a_2 symmetry orbital [55] (discussed in more detail below).

We return now to considering the detailed form of the PAD in time-resolved pump–probe PES experiments. It is convenient to describe the excited-state population dynamics in terms of the density matrix, defined by [40, 87]

$$\rho(n_\alpha, n'_{\alpha'};t)_{M_\alpha M'_{\alpha'}} = C_{n_\alpha M_\alpha}(t) C^*_{n'_{\alpha'} M'_{\alpha'}}(t) \tag{39}$$

The diagonal elements of the density matrix contain the populations of each of the BO states, whereas off-diagonal elements contain the relative phases of the BO states. The components of the density matrix with $\alpha = \alpha'$ describe the vibrational and rotational dynamics in the electronic state α, while the rotational dynamics within a vibronic state are described by the density matrix elements with $\alpha = \alpha'$ and $v_\alpha = v'_{\alpha'}$. The density matrix components with $n_\alpha = n'_{\alpha'}$ describe the angular momentum polarization of the state J_α, often referred to as angular momentum orientation and alignment [40, 87–89]. The density matrix may be expanded in terms of multipole moments as:

$$\rho(n_\alpha, n'_{\alpha'};t)_{M_\alpha M'_{\alpha'}} = \sum_{KQ} (-1)^{J_\alpha - M_\alpha} [K]^{1/2} \begin{pmatrix} J_\alpha & J'_{\alpha'} & K \\ M_\alpha & -M'_{\alpha'} & -Q \end{pmatrix} \langle T(n_\alpha, n'_{\alpha'};t)^\dagger_{KQ} \rangle \tag{40}$$

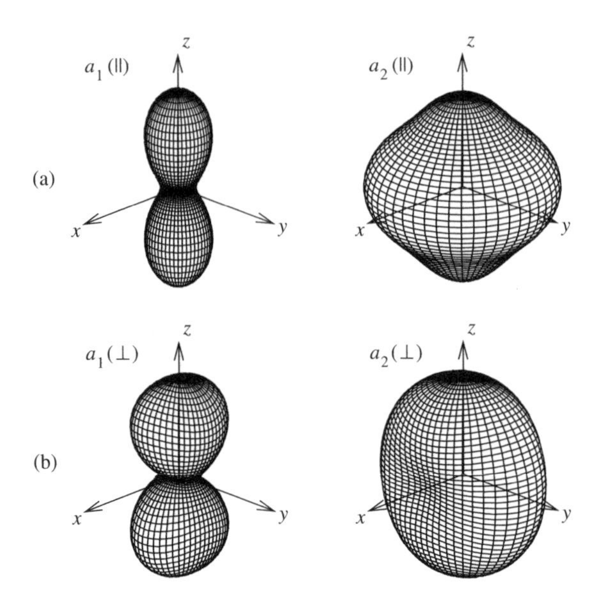

Figure 3. Calculated LF PADs for ionization of a model C_{3v} molecule. PADs are shown for ionization of a_1 and a_2 symmetry orbitals for the same set of dynamical parameters. The molecular axis distribution in these calculations was described as a $\cos^2\theta$ distribution, where θ is the angle between the direction of linear polarization of the pump laser and the principal molecular axis. The linear probe polarization is along the z axis. Panel (a) shows PADs for parallel pump and probe polarizations, while panel (b) shows PADs for perpendicular pump and probe polarizations. See Ref. [55] for the dynamical parameters used in these calculations.

where $\begin{pmatrix} \cdot & \cdot & \cdot \\ \cdot & \cdot & \cdot \end{pmatrix}$ is a Wigner $3j$ symbol. The multipole moments $\langle T(n_\alpha n'_{\alpha'};t)^\dagger_{KQ}\rangle$ are the expectation values of the irreducible spherical tensor operators $T(n_\alpha, n'_{\alpha'};t)^\dagger_{KQ,}$ which transform under rotation as the spherical harmonics [40, 87] and are termed state multipoles. From the properties of the Wigner $3j$ symbol, the possible range of K is given by $K = 0\ldots(J_\alpha + J'_{\alpha'})$, and $Q = -K\ldots K$. The multipole moments with $K = Q = 0$ contain the vibronic state populations (terms with $\alpha = \alpha'$ and $v_\alpha = v'_{\alpha'}$) and coherences (terms with $\alpha \neq \alpha'$ and/or $v_\alpha \neq v_{\alpha'}$). The multipole moments with $K = Q = 0$ also contain the populations of the rotational states, as well as the coherences between rotational states with the same values of J and M in different electronic states. Multipole moments with $K > 0$ describe the angular momentum polarization and the coherence among rotational states. In the perturbative limit, the maximum value of K is given by $2n$, where n is the number of photons involved in the pump step. For example, a single-photon pump step will prepare multipole moments with $K = 0, 2$.

The integral over the Euler angles in Eq. (32) is found analytically using the Clebsch–Gordan series [40, 64, 68]

$$\begin{aligned}
&\int D^{l}_{m\lambda}(\phi,\theta,\chi)D^{J_{\alpha_+}}_{M_{\alpha_+}K_{\alpha_+}}(\phi,\theta,\chi)D^{1*}_{pq}(\phi,\theta,\chi)D^{J_\alpha *}_{M_\alpha K_\alpha}(\phi,\theta,\chi)\mathrm{d}\boldsymbol{\Omega}\\
&\quad=(-1)^{M_{\alpha_+}-K_{\alpha_+}+p+q}\sum_{j_t}[j_t]\begin{pmatrix} J_{\alpha_+} & J_\alpha & j_t\\ -M_{\alpha_+} & M_\alpha & m_t\end{pmatrix}\begin{pmatrix} J_{\alpha_+} & J_\alpha & j_t\\ -K_{\alpha_+} & K_\alpha & k_t\end{pmatrix}\\
&\quad\times\begin{pmatrix} l & 1 & j_t\\ m & -p & m_t\end{pmatrix}\begin{pmatrix} l & 1 & j_t\\ \lambda & -q & k_t\end{pmatrix}
\end{aligned}\tag{41}$$

where j_t corresponds to the angular momentum transferred to the ion during the ionization process, with m_t and k_t denoting the projections of j_t on the LF and MF z axes, respectively.

Expanding Eq. (24) and substituting in Eqs (32), (39)–(41), and carrying out the summations over all LF projection quantum numbers (see Appendix A) gives an expression for the LF PAD as an expansion in spherical harmonics,

$$\sigma(\epsilon,\hat{\boldsymbol{k}}_{\mathrm{L}};t)=\frac{\sigma_{\mathrm{total}}(\epsilon;t)}{4\pi}\sum_{LM}\beta_{LM}(\epsilon;t)Y_{LM}(\hat{\boldsymbol{k}}_{\mathrm{L}})\tag{42}$$

where $\sigma_{\mathrm{total}}(\epsilon;t)$ is the total cross-section for producing electrons with an energy ϵ. The expansion coefficients $\beta_{LM}(\epsilon;t)$ are given by

$$\begin{aligned}
\beta_{LM}(\epsilon;t)=&-[L]^{1/2}\sum_{PR}(-1)^{P}[P]^{1/2}E_{PR}(\hat{\boldsymbol{e}})\sum_{KQ}(-1)^{K+Q}\begin{pmatrix} P & K & L\\ R & -Q & M\end{pmatrix}\\
&\times\sum_{n_\alpha n_{\alpha'}}(-1)^{J_\alpha}[J_\alpha,J'_{\alpha'}]^{1/2}\langle T(n_\alpha,n'_{\alpha'};t)^{\dagger}_{KQ}\rangle\sum_{ll'}(-1)^{l}[l,l']^{1/2}\begin{pmatrix} l & l' & L\\ 0 & 0 & 0\end{pmatrix}\\
&\times\sum_{j_tj'_t}(-1)^{j_t}[j_t,j'_t]\begin{Bmatrix} 1 & 1 & P\\ j_t & j'_t & K\\ l & l' & L\end{Bmatrix}\sum_{qq'}\sum_{\lambda\lambda'}\sum_{k_tk'_t}(-1)^{q+q'}\begin{pmatrix} l & 1 & j_t\\ \lambda & -q & k_t\end{pmatrix}\\
&\times\begin{pmatrix} l' & 1 & j'_t\\ \lambda' & -q' & k'_t\end{pmatrix}\sum_{J_{\alpha_+}\tau_{\alpha_+}}(-1)^{J_{\alpha_+}}[J_{\alpha_+}]\begin{Bmatrix} J_\alpha & j_t & J_{\alpha_+}\\ j'_t & J'_{\alpha'} & K\end{Bmatrix}\sum_{K_{\alpha_+}}\left|a^{J_{\alpha_+}\tau_{\alpha_+}}_{K_{\alpha_+}}\right|^2\\
&\times\sum_{K_\alpha K'_{\alpha'}}a^{J_\alpha\tau_\alpha}_{K_\alpha}a^{J'_{\alpha'}\tau'_{\alpha'}}_{K'_{\alpha'}}\begin{pmatrix} J_{\alpha_+} & J_\alpha & j_t\\ -K_{\alpha_+} & K_\alpha & k_t\end{pmatrix}\begin{pmatrix} J_{\alpha_+} & J'_{\alpha'} & j'_t\\ -K_{\alpha_+} & K'_{\alpha'} & k'_t\end{pmatrix}\\
&\times\sum_{v_{\alpha_+}\alpha_+}\mathcal{E}(n_{\alpha_+},n_\alpha,\epsilon)\mathcal{E}^*(n_{\alpha_+},n'_{\alpha'},\epsilon)\sum_{\Gamma\mu h}\sum_{\Gamma'\mu'h'}b^{\Gamma\mu}_{hl\lambda}b^{\Gamma'\mu'*}_{h'l'\lambda'}(-\mathrm{i})^{l-l'}\\
&\times\mathrm{e}^{\mathrm{i}(\sigma_l(\epsilon)-\sigma_{l'}(\epsilon))}D^{\alpha v_\alpha\alpha_+v_{\alpha_+}}_{\Gamma\mu hl}(q)D^{\alpha'v'_{\alpha'}\alpha_+v_{\alpha_+}*}_{\Gamma'\mu'h'l'}(q')
\end{aligned}\tag{43}$$

where $\begin{Bmatrix} \cdot & \cdot & \cdot \\ \cdot & \cdot & \cdot \end{Bmatrix}$ and $\begin{Bmatrix} \cdot & \cdot & \cdot \\ \cdot & \cdot & \cdot \\ \cdot & \cdot & \cdot \end{Bmatrix}$ are Wigner $6j$ and $9j$ coefficients, respectively [40].

The functions $E_{PR}(\hat{\boldsymbol{e}})$ describe the polarization of the probe laser pulse, and are given by

$$E_{PR}(\hat{\boldsymbol{e}}) = [e \otimes e]_R^P = [P]^{1/2} \sum_p (-1)^p \begin{pmatrix} 1 & 1 & P \\ p & -(R+p) & R \end{pmatrix} e_{-p} e^*_{-(R+p)} \tag{44}$$

From the properties of the Wigner $3j$ symbol, P can take the values $0, 1, 2$, and for linear polarization along the lab z axis, $P = 0, 2$ only. The Wigner $3j$ symbol also restricts the values of R to $-P \ldots P$.

If we make the assumption that the rotational states of the ion are not resolved and the Fourier transform of the probe laser pulse remains constant over the spectrum of transitions to ion rotational states, we can replace $\mathcal{E}(n_{\alpha_+}, n_\alpha, \epsilon)$ in Eq. (43) with an averaged Fourier transform at a frequency $2\pi(\bar{E}_{\alpha_+ v_{\alpha_+}} - \bar{E}_{\alpha v_\alpha} + \epsilon)/\hbar$, which we denote by $\mathcal{E}(\alpha, v_\alpha, \alpha_+, v_{\alpha_+}, \epsilon)$. This allows the summations over the ion rotational states and also j_t, k_t in Eq. (43) to be carried out analytically (see Appendix B), yielding a simplified expression for the coefficients

$$\begin{aligned} \beta_{LM}(\epsilon; t) = [L]^{1/2} & \sum_{\alpha v_\alpha} \sum_{\alpha' v'_{\alpha'}} \sum_{KQS} (-1)^{K+Q} A(\alpha, v_\alpha, \alpha', v'_{\alpha'}; t)^K_{QS} \\ & \times \sum_P (-1)^P [P]^{1/2} E_{PQ-M}(\hat{\boldsymbol{e}}) \begin{pmatrix} P & K & L \\ Q-M & -Q & M \end{pmatrix} \\ & \times \sum_{qq'} (-1)^q \begin{pmatrix} 1 & 1 & P \\ q & -q' & q'-q \end{pmatrix} \begin{pmatrix} P & K & L \\ q'-q & -S & S+q-q' \end{pmatrix} F_{LS}^{\alpha v_\alpha \alpha' v'_{\alpha'}}(q, q') \end{aligned} \tag{45}$$

where the dynamical parameters $F_{LS}^{\alpha v_\alpha \alpha' v'_{\alpha'}}(q, q')$ describe the ionization dynamics and are given by

$$\begin{aligned} F_{LS}^{\alpha v_\alpha \alpha' v'_{\alpha'}}(q,q') = & \sum_{ll'} [l,l']^{1/2} \begin{pmatrix} l & l' & L \\ 0 & 0 & 0 \end{pmatrix} \sum_{\lambda\lambda'} (-1)^{\lambda'} \begin{pmatrix} l & l' & L \\ -\lambda & \lambda' & S+q-q' \end{pmatrix} \\ & \times \sum_{v_{\alpha_+} \alpha_+} \mathcal{E}(\alpha, v_\alpha, \alpha_+, v_{\alpha_+}, \epsilon) \mathcal{E}^*(\alpha', v'_{\alpha'}, \alpha_+, v_{\alpha_+}, \epsilon) \\ & \times \sum_{\Gamma\mu h} \sum_{\Gamma'\mu' h'} b^{\Gamma\mu}_{hl\lambda} b^{\Gamma'\mu' *}_{h'l'\lambda'} (-\mathrm{i})^{l-l'} \mathrm{e}^{\mathrm{i}(\sigma_l(\epsilon) - \sigma_{l'}(\epsilon))} D^{\alpha v_\alpha \alpha_+ v_{\alpha_+}}_{\Gamma\mu hl}(q) D^{\alpha' v'_{\alpha'} \alpha_+ v_{\alpha_+} *}_{\Gamma'\mu' h' l'}(q') \end{aligned} \tag{46}$$

The parameters $A(\alpha, v_\alpha, \alpha', v'_{\alpha'}; t)^K_{QS}$ in Eq. (45) are defined as:

$$A(\alpha, v_\alpha, \alpha', v'_{\alpha'}; t)^K_{QS} = \frac{[K]^{1/2}}{8\pi^2} \sum_{J_\alpha, \tau_\alpha} \sum_{J'_{\alpha'}, \tau'_{\alpha'}} (-1)^{2J_\alpha + J'_{\alpha'}} [J_\alpha, J'_{\alpha'}]^{1/2}$$
$$\times \sum_{K_\alpha K'_{\alpha'}} (-1)^{-K_\alpha} a^{J_\alpha \tau_\alpha}_{K_\alpha} a^{J'_{\alpha'} \tau'_{\alpha'}}_{K'_{\alpha'}} \begin{pmatrix} J_\alpha & J'_{\alpha'} & K \\ -K_\alpha & K'_{\alpha'} & S \end{pmatrix} \langle T(n_\alpha, n'_{\alpha'}; t)^\dagger_{KQ} \rangle \tag{47}$$

The parameters in Eq. (47) have an immediate geometrical interpretation: they describe the LF distribution of molecular axes of the excited-state neutral molecules prior to ionization [87]. For this reason, we refer to them as the axis distribution moments (ADMs). The molecular axis distribution in a vibronic level may be expressed as an expansion of Wigner rotation matrices with the coefficients $A(\alpha, v_\alpha, \alpha, v_\alpha; t)^K_{QS}$,

$$P_{\alpha v_\alpha}(\phi, \theta, \chi; t) = \sum_{KQS} A(\alpha, v_\alpha, \alpha, v_\alpha; t)^K_{QS} D^{K*}_{QS}(\phi, \theta, \chi) \tag{48}$$

and the molecular axis distribution of the whole excited-state ensemble of molecules is given by

$$P(\phi, \theta, \chi; t) = \sum_{\alpha v_\alpha} \sum_{\alpha' v'_{\alpha'}} \sum_{KQS} A(\alpha, v_\alpha, \alpha', v'_{\alpha'}; t)^K_{QS} D^{K*}_{QS}(\phi, \theta, \chi) \tag{49}$$

The ADMs connect the multipole moments $\langle T(n_\alpha n'_{\alpha'}; t)^\dagger_{KQ} \rangle$, which characterize the angular momentum polarization and coherence, with the molecular axis distribution. Nonzero ADMs with even K characterize molecular axis alignment, whereas nonzero ADMs with odd K characterize molecular axis orientation. A cylindrically symmetric molecular axis distribution along the lab frame z axis will have nonzero ADMs with $Q = S = 0$ only. Linear and symmetric top molecules, for which only the two angles (θ, ϕ) are required to characterize the molecular orientation [40, 87], require only ADMs with $S = 0$ moments to fully characterize the molecular axis distribution. Asymmetric top molecules may have nonzero ADMs with both $Q \neq 0$ and $S \neq 0$ only when there is localization of all three Euler angles. An isotropic distribution of molecular axes has the only nonzero ADMs with $K = 0$.

Equation (45) explicitly expresses the sensitivity of the LF PAD to the molecular axis distribution. In fact, an equivalent expression is obtained by convolution of the molecular axis distribution with the vibronic transition dipole matrix elements without explicit consideration of molecular rotation [55]. The general form of Eq. (45) is obtained for other angular momentum coupling cases (e.g., in the presence of strong spin–orbit coupling): the lack of resolution of the ion rotational states essentially removes the details of the angular momentum coupling from the problem [64] (although the expression for the ADMs Eq. (47) may be different for other angular momentum coupling schemes).

From the properties of the Wigner $3j$ symbols, we see from Eq. (45) that the maximum value of L in the expansion in Eq. (42) is the smaller of $2l_{\max}$, where $l_{\max}$ is the largest value of l in the partial wave expansion Eq. (23), and $(K_{\max} + 2)$, where $K_{\max}$ is the maximum value of K in the axis distribution in Eq. (49). Each $\beta_{LM}(\epsilon; t)$ is sensitive to ADMs with values of K from $(L - 2)$ (or zero if $(L - 2)$ is negative) to $(L + 2)$ (since the maximum value of P is 2 and L, K, and P must satisfy the triangle condition for nonzero Wigner $3j$ coefficients). In other words, the more anisotropic the molecular axis distribution is, the higher the anisotropy of the LF PAD. The distribution of molecular axes geometrically determines the relative contributions of the molecular frame ionization transition dipole components that contribute to the LF PAD. The sensitivity of the LF PAD to the molecular orientation is determined by the relative magnitudes of the dynamical parameters $F_{LS}^{\alpha v_\alpha \alpha' v'_{\alpha'}}(q, q')$ that reflect the anisotropy of the ionization transition dipole and PAD in the MF. While the total cross-section for ionization (i.e., $\beta_{00}(\epsilon; t)$) is sensitive to ADMs with $K = 0 \ldots 2$, as is the total cross-section of all one-photon absorption processes, we see that measuring the PAD reveals information regarding ADMs with higher K values than single-photon absorption would normally. Note also that the PAD from aligned or oriented molecules yield far more information regarding the photoionization dynamics, and as such provides a route to performing "complete" photoionization experiments [90].

TRPES experiments frequently employ linearly polarized pump and probe pulses, with the excited-state rovibronic wave packet prepared via a single-photon resonant transition and for this reason we will briefly discuss this situation. The linearly polarized pump pulse excites molecules with their transition dipole moment aligned toward the direction of the laser polarization, due to the scalar product interaction $\boldsymbol{d} \cdot \hat{\boldsymbol{e}}_{\text{pump}}$ of the transition dipole moment with the polarization vector of the pump pulse. Since the transition dipole typically has a well-defined direction in the MF, this will create an ensemble of axis-aligned excited-state molecules. Since the pump pulse is linearly polarized, the excited-stated molecular axis distribution possesses cylindrical symmetry, and so is described by ADMs with $Q = S = 0$ in a LF whose z–axis is defined by the pump polarization. The single-photon nature of this pump step limits the values of K to 0 and 2. The fact that only even K moments are prepared, and the molecular axis distribution is aligned and not oriented, reflects the up–down symmetry of the pump interaction [87]. Since the maximum value of K is 2, the maximum value of L is 4 in Eq. (42). This means that the PAD contains information concerning the interference of partial waves with l differing by at most 4: The LF PAD will contain terms with at most $|l - l'| = 0, 2, 4$, and does not contain any information regarding the interference of odd and even partial waves. The rotational wave packet created by the pump pulse will subsequently evolve under the field-free Hamiltonian of the molecule, initially causing a reduction of the molecular axis alignment, and subsequently causing a realignment of the molecules when the rotational wave packet rephases [91–96]. If the probe pulse is timed to arrive when the molecules are strongly aligned there will be a strong dependence of the LF PAD upon the direction of the probe pulse polarization relative to the pump pulse polarization. If the pump and probe polarizations are parallel, then the LF PAD will maintain cylindrical symmetry, and ionization transition dipole moments along the molecular symmetry axis will be favored. Rotating the probe polarization away from that of the pump will remove the cylindrical symmetry of the PAD and increase the contributions of the ionization transition dipole moments perpendicular to the symmetry axis. An example of this effect can be seen in the model calculations shown in Fig. 3, where the shape of the PADs clearly depend on the molecular axis distribution in the frame defined by the ionizing laser polarization. An experimental example of this effect will also be discussed in Section V.B. Clearly, the LF PAD will be extremely sensitive to the rotational motion of the molecule, and is able to yield detailed information pertaining to molecular rotation and molecular axis alignment [53, 55, 57]. Furthermore, since the coupling of vibrational and rotational motion will cause changes in the evolution of the molecular axis alignment, the LF PAD can provide important information regarding such couplings [54, 56].

Note also that, in an experiment in which the species ionized by the probe laser is a product of photodissociation initiated by the pump, the PAD will be sensitive to the LF photofragment axis distribution, and as such will provide a probe of the photofragment angular momentum coherence and polarization. Furthermore, with a suitably designed experiment, PADs will allow measurement of product vector correlations, such as that between the photofragment velocity and angular momentum polarization. Such vector correlations in molecular photodissociation have long been studied in a non-time-resolved fashion [89, 97–99], and have provided detailed information concerning the photodissociation dynamics. Whereas these studies have focused upon the rotational state resolved angular momentum polarization, time-resolved measurements may yield information regarding the rotational coherence of the photofragments, as well as the angular momentum polarization.

In the preceding discussion, we discussed the form of the PAD as measured in the LF, that is, relative to the polarization direction of the ionizing laser polarization. However, by employing experimental techniques to measure the photoelectron in coincidence with a fragment ion following dissociative ionization, it is possible to measure the PAD referenced to the MF rather than the LF, removing all averaging over molecular orientation. These experimental techniques are described in Section IV.B, and examples of such measurements are subsequently given in Section VI.E. The form of the MF PAD can be expressed in a form similar to the LF PAD,

$$\sigma(\epsilon, \hat{\boldsymbol{k}}_{\mathrm{M}}; t) \propto \sum_{\alpha_+ v_{\alpha_+}} \left| \sum_{\alpha v_\alpha} C_{\alpha v_\alpha}(t) \langle \hat{\boldsymbol{k}}_{\mathrm{M}} \epsilon \alpha_+ v_{\alpha_+} | \boldsymbol{d} \cdot \hat{\boldsymbol{e}} | \alpha v_\alpha \rangle \mathcal{E}(\alpha, v_\alpha, \alpha_+, v_{\alpha_+}, \epsilon) \right|^2 \quad (50)$$

where $C_{\alpha v_\alpha}(t)$ are the time-dependent complex coefficients for each excited-state vibronic level. The scalar product of the transition dipole moment $\boldsymbol{d}$ with the polarization vector of the probe laser field $\hat{\boldsymbol{e}}$ in terms of the spherical tensor components in the MF as [40]

$$\boldsymbol{d} \cdot \hat{\boldsymbol{e}} = \sum_{q=-1}^{1} (-1)^q d_q e_{-q} \quad (51)$$

The electric field polarization is conveniently described in the LF. The MF spherical tensor components of the electric field polarization tensor are related to the components in the LF through a rotation

$$e_{-q} = \sum_{p=-1}^{1} D^1_{-p-q}(\phi, \theta, \chi) e_p \quad (52)$$

Substitution of Eqs. (21), (52), and (51) into Eq. (50) yields an expression similar to Eq. (42) for the MF PAD (see Appendix C) [51, 55, 100],

$$\sigma(\epsilon, \hat{\boldsymbol{k}}_{\mathrm{M}}; t) = \frac{\sigma_{\mathrm{total}}(\epsilon; t)}{4\pi} \sum_{LM} \beta^{\mathrm{M}}_{LM}(\epsilon; t) Y_{LM}(\hat{\boldsymbol{k}}_{\mathrm{M}}) \quad (53)$$

where the $\beta^{\mathrm{M}}_{LM}(\epsilon; t)$ coefficients are given by [51, 100]

$$\begin{aligned}
\beta^{\mathrm{M}}_{LM}(\epsilon;t) = {} & [L]^{1/2} \sum_{PR} \sum_{qq'} [P]^{1/2} (-1)^{q'} \begin{pmatrix} 1 & 1 & P \\ q & -q' & q'-q \end{pmatrix} D^P_{Rq'-q}(\phi,\theta,\chi) E_{PR}(\hat{\boldsymbol{e}}) \\
& \times \sum_{ll'} [l,l']^{1/2} \begin{pmatrix} l & l' & L \\ 0 & 0 & 0 \end{pmatrix} \sum_{\lambda\lambda'} (-1)^{\lambda} \begin{pmatrix} l & l' & L \\ \lambda & -\lambda' & M \end{pmatrix} (-\mathrm{i})^{l-l'} \mathrm{e}^{\mathrm{i}(\sigma_l(\epsilon) - \sigma_{l'}(\epsilon))} \\
& \times \sum_{\alpha v_\alpha} \sum_{\alpha' v'_{\alpha'}} C_{\alpha v_\alpha}(t) C^*_{\alpha' v'_{\alpha'}}(t) \sum_{\Gamma\mu h} \sum_{\Gamma'\mu' h'} b^{\Gamma\mu}_{hl\lambda} b^{\Gamma'\mu' *}_{h'l'\lambda'} \\
& \times \sum_{\alpha_+ v_{\alpha_+}} D^{\alpha v_\alpha \alpha_+ v_{\alpha_+}}_{\Gamma\mu hl}(q) D^{\alpha' v'_{\alpha'} \alpha_+ v_{\alpha_+} *}_{\Gamma'\mu' h' l'}(q') \mathcal{E}(\alpha, v_\alpha, \alpha_+, v_{\alpha_+}, \epsilon) \mathcal{E}^*(\alpha', v'_{\alpha'}, \alpha_+, v_{\alpha_+}, \epsilon)
\end{aligned} \quad (54)$$

Naturally, since this expression is for a property measured in the reference frame connected to the molecule, molecular rotations do not appear in this expression. The range of L in the summation Eq. (53) is $0...2l_{\max}$, and includes both odd and even values. In general, the MF PAD is far more anisotropic than the LF PAD, for which $L = 0, 2, 4$ in a two-photon linearly polarized pump–probe experiment in the perturbative limit. Clearly, the MF PAD contains far more detailed information than the LF PAD concerning the ionization dynamics of the molecule, as well as the structure and symmetry of the electronic state from which ionization occurs, since the partial waves that may interfere are no longer geometrically limited as they are for the LF PAD. The contributing MF ionization transition dipole components are determined by the laser polarization

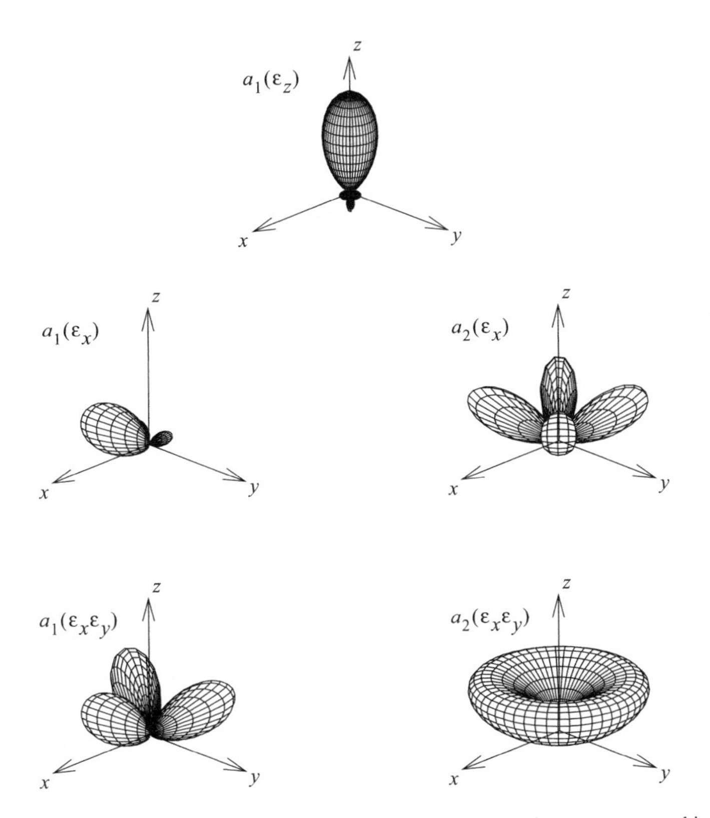

Figure 4. The MF PADs for single-photon ionization of a_1 and a_2 symmetry orbitals of a model C_{3v} molecule for light linearly polarized along different axes of the molecule (indicated in parentheses). Note that no photoionization can occur from the a_2 orbital for light polarized along the z axis (molecular symmetry axis). The same dynamical parameters as for the calculations of the LF PADs shown in Fig. 3 were used. For further details see Ref. [55].

and the Euler angles between the LF and MF. Example calculations for a model C_{3v} molecule are shown in Fig. 4 employing the same dynamical parameters as for the calculations of the LF PADs shown in Fig. 3 that demonstrate the much higher anisotropy of the MF PAD. These calculations demonstrate the dependence of the MF PAD upon the probe pulse MF polarization direction (or equivalently the MF ionization transition dipole direction). The LF PAD corresponds to a coherent summation over such MF PADs, weighted by the molecular axis distribution in the LF.

Finally, in closing, we note that significant advances have recently been made toward defining the direction of the molecular orientation in the LF using strong nonresonant laser fields [101–117]. This will provide us with the opportunity to make PAD measurements that approximate well the MF PAD without coincident detection of dissociative ionization. The alignment and orientation achievable with such techniques produces extremely high K valued ADMs due to the highly nonlinear (multiphoton) nature of the matter-laser interaction that produces the alignment and orientation. The prospects offered by these techniques are exciting and will herald a new generation of PAD measurements.

IV. EXPERIMENTAL TECHNIQUES

A. Photoelectron Spectrometers

In a PES measurement, the observables are the electron kinetic energy distribution, the PAD, the electron spin and the set of scalar and vector correlations between these electron distributions, as well as those of the ion. Spectrometers for femtosecond TRPES have modest energy resolution requirements as compared to modern standards for photoelectron spectrometers. For Gaussian optical pulses, the time-bandwidth product is $\Delta\nu\Delta t = 0.441$ and, therefore, the bandwidth (FWHM) of a Gaussian 100 fs pulse is $\sim 150\,\text{cm}^{-1}$. A pump–probe measurement typically involves the convolution of two such pulses, leading to an effective bandwidth of ~ 25 meV. This limits the energy resolution required in measuring the energy of the photoelectrons. We emphasize that in femtosecond pump–probe experiments, the laser intensity must be kept below multiphoton ionization thresholds. This simply requires a reduction of the laser intensity until one-photon processes dominate. At this level the ionization probabilities are small and, usually, single-particle counting techniques are required. Therefore, TRPES experiments are very data-intensive and require the collection of many photoelectron spectra. As a result, most neutral TRPES experiments performed to date make use of high efficiency electron energy analyzers in which a large fraction of the photoelectrons are collected.

A commonly used analyzer in TRPES experiments has been the "magnetic bottle" time-of-flight (TOF) spectrometer [118–120]. This technique uses a strong inhomogeneous magnetic field (1 T) to rapidly parallelize electron trajectories into a flight tube, followed by a constant magnetic field (10 G) to guide the electrons to the detector. The collection efficiency of magnetic bottle spectrometers can exceed 50%, while maintaining an energy resolution essentially equivalent to the femtosecond laser bandwidth. Highest resolution is obtained for electrons created within a small volume ($< 100\,\mu m$) at the very center of the interaction region. Magnetic bottle analyzers have been used in many neutral TRPES experiments. They are relatively simple, have high collection efficiency, and rapid data readout. Magnetic bottles have the general disadvantage that they can only be used to determine electron kinetic energy distributions; the complex electron trajectories in magnetic bottle analyzers make it impractical to extract angular information.

Time-resolved (2D) photoelectron imaging techniques, in which position-sensitive detection is used to measure the photoelectron kinetic energy and angular distributions simultaneously, is becoming increasingly popular due to its sensitivity and ease of implementation [28]. When used, as is common, with CCD camera systems for image collection, particle hits are usually averaged on the CCD chip because rapid CCD readout at kilohertz rates remains very challenging. The most straightforward 2D imaging technique is photoelectron velocity-map imaging (VMI) [121, 122], a variant of the photofragment imaging method [123]. Typically, a strong electric field images nascent charged particles onto a microchannel plate (MCP) detector. The ensuing electron avalanche falls onto a phosphor screen, which is imaged by the CCD camera. Analysis of the resultant image allows for the extraction of both energy- and angle-resolved information. In this case, a 2D projection of the three-dimensional (3D) distribution of recoil velocity vectors is measured; various image reconstruction techniques [124, 125] are then used to recover the full 3D distribution. Photoelectron VMI thus yields close to the theoretical limit for collection efficiency, along with simultaneous determination of the photoelectron energy and angular distributions. The 2D particle imaging approach may be used when the image is a projection of a cylindrically symmetric distribution whose symmetry axis lies parallel to the 2D detector surface. This requirement precludes the use of pump and probe laser polarization geometries other than parallel polarizations. It may therefore be preferable to adopt fully 3D imaging techniques based upon "time slicing" [126, 127] or full-time-and-position sensitive detection [128], where the full 3D distribution is obtained directly without mathematical reconstruction (see below). In femtosecond pump–probe experiments where the intensities must be kept below a certain limit (requiring single particle counting methods), time slicing may not be practical, leaving only full-time-and-position sensitive detection as the option.

Modern MCP detectors permit the direct 3D measurement of LF recoil momentum vectors by measuring both spatial position (x, y) on–and time of arrival (z) at–the detector face [128]. Importantly, this development does not require inverse transformation to reconstruct 3D distributions, and so is not restricted to experimental geometries with cylindrical symmetry, allowing any desired polarization geometry to be implemented. In 3D particle imaging, a weak electric field is used to extract nascent charged particles from the interaction region. Readout of the (x, y) position (i.e., the polar angle) yields information about the velocity distributions parallel to the detector face, equivalent to the information obtained from 2D detectors. However, the additional timing information allows measurement of the third (z) component (i.e., the azimuthal angle) of the LF velocity, via the "turn-around" time of the particle in the weak extraction field. Thus, these detectors allow for full 3D velocity vector measurements, with no restrictions on the symmetry of the distribution or any requirement for image reconstruction techniques. Very successful methods for full-time-and-position sensitive detection are based upon interpolation (rather than pixellation) using either charge-division (e.g., wedge-and- strip) [129] or crossed delay-line anode timing MCP detectors [130]. In the former case, the avalanche charge cloud is divided among three conductors: a wedge, a strip, and a zigzag. The (x, y) positions are obtained from the ratios of the wedge and strip charges to the zigzag (total) charge. Timing information can be obtained from a capacitive pick-off at the back of the last MCP plate. In the latter case, the anode is formed by a pair of crossed, impedance-matched delay-lines (i.e., x and y delay lines). The avalanche cloud that falls on a delay line propagates in both directions toward two outputs. Measurement of the timing difference of the output pulses on a given delay line yields the x (or y) positions on the anode. Measurement of the sum of the two output pulses (relative to, say, a pickoff signal from the ionization laser or the MCP plate itself) yields the particle arrival time at the detector face. Thus, direct anode timing yields a full 3D velocity vector measurement. An advantage of delay line anodes over charge division anodes is that the latter can tolerate only a single hit per laser shot, precluding the possibility of multiple coincidences, and additionally make the experiment sensitive to background scattered ultraviolet (UV) light.

B. Coincidence Techniques

Photoionization always produces two species available for analysis: the ion and the electron. By measuring both photoelectrons and photoions in coincidence, the kinetic electron may be assigned to its correlated parent ion partner, which may be identified by mass spectrometry. The extension of the photoelectron–photoion–coincidence (PEPICO) technique to the femtosecond time-resolved domain was shown to be very important for studies of dynamics in clusters [131, 132]. In these experiments, a simple yet efficient permanent magnet design "magnetic bottle" electron spectrometer was used for photoelectron

TOF measurements. A collinear TOF mass spectrometer was used to determine the mass of the parent ion. Using coincidence electronics, the electron TOF (yielding electron kinetic energy) is correlated with an ion TOF (yielding the ion mass). In this manner, TRPES experiments may be performed on neutral clusters, yielding time-resolved spectra for each parent cluster ion (assuming cluster fragmentation plays no significant role). Signal levels must be kept low (much less than one ionization event per laser shot) in order to minimize false coincidences. The reader is referred to a recent review for a detailed discussion on TRPEPICO methods [29].

Coincident detection of photoions and photoelectrons has long been recognized as a route to recoil or molecular frame photoelectron angular distributions in non-time-resolved studies [133–135]. For the case of nanosecond laser photodetachment, correlated photofragment and photoelectron velocities can provide a complete probe of the dissociation process [130, 136]. The photofragment recoil measurement defines the energetics of the dissociation process and the alignment of the recoil axis in the LF, the photoelectron energy provides spectroscopic identification of the products and the photoelectron angular distribution can be transformed to the recoil frame in order to make measurements approaching the MF PAD. Measuring the recoil frame PAD can also provide vector correlations, such as that between the photofragment angular momentum polarization and the recoil vector. Time- and angle-resolved PEPICO measurements showing the evolution of photoion and photoelectron kinetic energy and angular correlations will undoubtedly shed new light on the photodissociation dynamics of polyatomic molecules. The integration of photoion-photoelectron timing-imaging (energy and angular correlation) measurements with femtosecond time-resolved spectroscopy was first demonstrated, using wedge-and-strip anode detectors, in 1999 [129, 137]. This Time-Resolved Coincidence-Imaging Spectroscopy (TRCIS) method allows the time evolution of complex dissociation processes to be studied with unprecedented detail [138] and was first demonstrated for the case of the photodissociation dynamics of NO_2 [129] (discussed in more detail in Section VI.E).

TRCIS allows for kinematically complete energy- and angle-resolved detection of both electrons and ions in coincidence and as a function of time, representing the most differential TRPES measurements made to date. This time-resolved six-dimensional (6D) information can be projected, filtered, and/or averaged in many different ways, allowing for the determination of various time-resolved scalar and vector correlations in molecular photodissociation. For example, an interesting scalar correlation is the photoelectron kinetic energy plotted as a function of the coincident photofragment kinetic energy. This 2D correlation allows for the fragment kinetic energy distributions of specific channels to be extracted. For experimentalists, an important practical consequence of this is the ability to separate dissociative ionization (i.e., ionization followed by dissociation) of the parent molecule from photoionization of neutral fragments (i.e., dissociation followed by ionization). In both cases, the same ionic fragment may be produced and the separation of these very different processes may be challenging. TRCIS, via the 2D energy–energy correlation map, does this naturally. The coincident detection of the photoelectron separates these channels: in one case (dissociative ionization) the photoelectron comes from the parent molecule, whereas in the other case (neutral photodissociation) the photoelectron comes from the fragment. In most cases, these photoelectron spectra will be very different, allowing complete separation of the two processes.

A very interesting vector correlation is the recoil direction of the photoelectron as a function of the recoil direction of the coincident photofragment. Although for each dissociation event the fragment may recoil in a different laboratory direction, TRCIS determines this direction and, simultaneously, the direction of the coincident electron. Therefore, event-by-event detection via TRCIS allows the PAD to be measured in the fragment recoil frame rather than the usual LF. In other words, it is time-resolved dynamics from the molecule's point of view. This is important because the usual LF PADs are generally averaged over all molecular orientations, leading to a loss of information. Specifically, for a one-photon pump, one-photon probe TRPES experiment on a randomly aligned sample, conservation of angular momentum in the LF limits the PAD anisotropy, as discussed in Section III. In the recoil frame, these limitations are relaxed, and an unprecedentedly detailed view of the excited-state electronic dynamics obtains. Other types of correlations, such as the time evolution of photofragment angular momentum polarization, may also be constructed from the 6D data of TRCIS.

C. Femtosecond Laser Technology

Progress in femtosecond TRPES benefits from developments in femtosecond laser technology, since techniques for PES have been highly developed for some time. There are several general requirements for such a femtosecond laser system. Most of the processes of interest are initiated by absorption of a photon in the wavelength range ~200–350 nm, produced via nonlinear optical processes, such as harmonic generation, frequency mixing, and parametric generation. Thus the output pulse energy of the laser system must be high enough for efficient use of nonlinear optical techniques and ideally should be tunable over a wide wavelength range. Another important consideration in a femtosecond laser system for time-resolved PES is the repetition rate. To avoid domination of the signal by multiphoton processes, the laser pulse intensity must be limited, thus also limiting the available signal per laser pulse. As a result, for many experiments a high pulse repetition rate can be more beneficial than high energy per pulse. Finally, the signal level in PES is often low in any case and, for time-resolved experiments, spectra must be obtained at many time delays. This requires that any practical laser system must run very reliably for many hours at a time.

Modern Ti:Sapphire based femtosecond laser oscillators have been the most important technical advance for performing almost all types of femtosecond time-resolved measurements [139]. The Ti:Sapphire oscillators are tunable over a 725–1000 nm wavelength range, have an average output power of several hundred milliwatts or greater and can produce pulses as short as 8 fs, but more commonly 50–130 fs, at repetition rates of 80–100 MHz. Broadly tunable femtosecond pulses can be derived directly from amplification and frequency conversion of the fundamental laser frequency.

The development of chirped-pulse amplification and Ti:Sapphire regenerative amplifier technology now provides millijoule pulse energies at repetition rates of 1 kHz with <100 fs pulse widths [140]. Chirped pulse amplification typically uses a grating stretcher to dispersively stretch femtosecond pulses from a Ti:Sapphire oscillator to several hundred picoseconds. This longer pulse can now be efficiently amplified in a Ti:Sapphire amplifier to energies of several milliJoules while avoiding nonlinear propagation effects in the solid-state gain medium. The amplified pulse is typically recompressed in a grating compressor.

The most successful approach for generating tunable output is optical parametric amplification (OPA) of spontaneous parametric fluorescence or a white light continuum, using the Ti:Sapphire fundamental or second harmonic as a pump source. Typically, an 800 nm pumped femtosecond OPA can provide a continuous tuning range of 1200–2600 nm [141]. Noncollinear OPAs (NOPAs) [142] pumped at 400 nm provide μJ-level $\sim$10–20 fs pulses that are continuously tunable within a range of 480–750 nm, allowing for measurements with extremely high temporal resolution. A computer controlled stepper motor is normally used to control the time delay between the pump and probe laser systems. The development of femtosecond laser sources with photon energies in the vacuum ultraviolet (VUV, 100–200 nm), extreme ultraviolet (XUV, <100 nm), and beyond (soft X-ray) opens new possibilities for TRPES, including the preparation of high lying molecular states, the projection of excited states onto a broad set of cation electronic states and, in the soft X-ray regime, time-resolved inner-shell PES. High harmonic generation in rare gases is a well-established and important method for generating femtosecond VUV, XUV [143], and soft X-ray radiation [144–146]. Harmonics as high as the $\sim$300th order have been reported, corresponding to photon energies in excess of 500 eV. Both pulsed rare gas jets and hollow-core optical waveguides [145, 147] have been used for high harmonic generation. Lower harmonics of the Ti:sapphire laser have been used in TRPES experiments [148–151]. As these techniques become more commonplace, the range of applicability of TRPES will be increased significantly.

V. COMPARISON OF TIME-RESOLVED ION WITH TRPES MEASUREMENTS

A. Mass-Resolved Ion Yield Measurements

A powerful version of femtosecond pump–probe spectroscopy combines photoionization with mass-resolved ion detection (i.e., mass spectrometry). The mass of the parent ion directly identifies the species under interrogation and measurement of fragment ions can provide information on dissociation pathways in the excited molecule [12]. However, fragmentation may also be a consequence of the photoionization dynamics (i.e., dynamics in the ionic continuum upon photoionization). As photoionization dynamics are revealed by PES, it is worth comparing time-resolved mass spectrometry with TRPES in more detail. As a vehicle for this comparison, first we discuss the illustrative example of excited-state dynamics in linear polyenes. Nonadiabatic dynamics in linear polyenes generally leads to the fundamental process of cis–trans photoisomerization. All-*trans*-(2,4,6,8)-decatetraene (DT, $C_{10}H_{14}$) provides a classic example of internal conversion in a linear polyene [44, 47]. In DT, the lowest excited state is the one-photon forbidden S_1 2^1A_g state, whereas the second excited state is the one-photon allowed S_2 1^1B_u state (a classic $\pi \rightarrow \pi^*$ transition). When the energy gap between S_2 and S_1 is large, the density of S_1 vibronic levels can be very large compared to the reciprocal electronic energy spacing and the "dark" state forms an apparently smooth quasicontinuum (the statistical limit for the radiationless transition problem). The S_2–S_1 energy gap in DT is 5764 cm^{-1} (0.71 eV) placing this large molecule (with 66 vibrational modes) in this statistical limit.

In the following we consider a time-resolved photoionization experiment using mass-resolved ion detection, as illustrated in Fig. 5. All-*trans*-(2, 4, 6, 8)-decatetraene was excited to its S_2 origin and the ensuing dynamics followed by probing via single photon ionization. In DT, the S_2 electronic origin is at 4.32 eV (287 nm) and the ionization potential is 7.29 eV. Hence, all probe laser wavelengths <417 nm, permit single-photon ionization of the excited state. By using $\lambda_{\text{pump}} = 287$ nm and $\lambda_{\text{probe}} = 352$ nm, one could therefore perform a time-resolved experiment using mass-resolved ion detection, as shown in Fig. 6. The time-resolution (pump–probe cross-correlation) in these experiments was 80 fs. It can be seen in Fig. 6a that the parent ion $C_{10}H_{14}^+$ signal rises with the laser cross-correlation and then decays with a 0.4 ps time constant. This suggests that the S_2 state lifetime is 0.4 ps. The fate of the molecule following this ultrafast

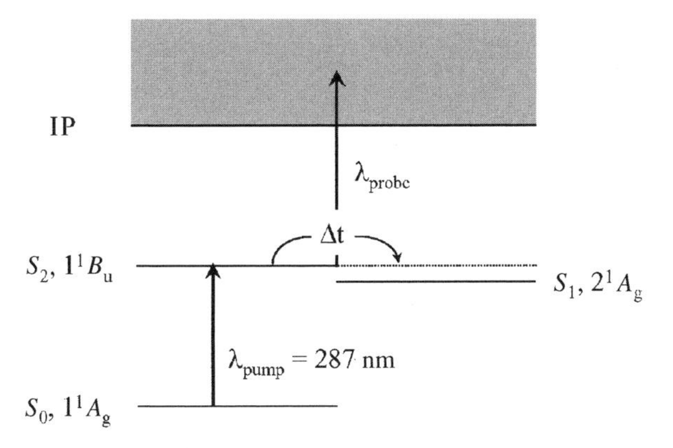

Figure 5. A femtosecond pump–probe photoionization scheme for studying excited-state dynamics in DT. The molecule is excited to its S_2 electronic origin with a pump pulse at 287 nm (4.32 eV). Due to nonadiabatic coupling, DT undergoes rapid internal conversion to the lower lying S_1 state (3.6 eV). The excited-state evolution is monitored via single-photon ionization. As the ionization potential is 7.29 eV, all probe wavelengths <417 nm permit single-photon ionization of the excited state.

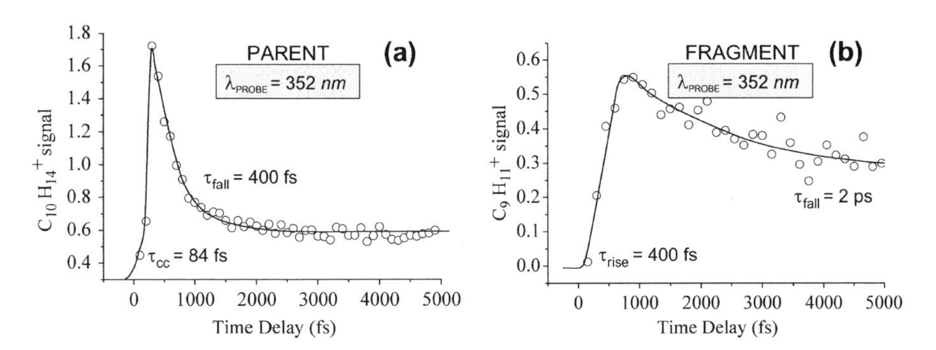

Figure 6. Time-resolved mass spectrometry. All-*trans*-(2, 4, 6, 8) decatetraene was excited to its S_2 electronic origin with a femtosecond pulse at $\lambda_{\mathrm{pump}} = 287$ nm. The excited-state evolution was probed via single-photon ionization using a femtosecond pulse at $\lambda_{\mathrm{probe}} = 352$ nm. The time resolution in these experiments was 80 fs (0.08 ps). (a) Time evolution of the parent ion $C_{10}H_{14}^+$ signal. The parent ion signal rises with the pump laser and then decays with a single exponential time constant of 0.4 ps, suggesting that this is the lifetime of the S_2 state. The fate of the molecule subsequent to this decay is unknown from these data. (b) Time evolution of the fragment ion $C_9H_{11}^+$ signal, corresponding to methyl loss. The rise time of this fragment signal is 0.4 ps, matching the decay time of the S_2 state. This might suggest that the methyl loss channel follows directly from internal conversion to S_1. The $C_9H_{11}^+$ signal subsequently decays with a time constant of 2 ps, suggesting some further step in the excited-state dynamics.

internal conversion, however, cannot be discerned from these data. As mass-resolved ion signals are the observable in this experiment, one could therefore look for the appearance of any potential reaction products (e.g., fragments) following the internal conversion.

In Fig. 6b, we present the time evolution of a fragment ion $C_9H_{11}^+$ that corresponds to the loss of a methyl group from the parent molecule. The rise time of this signal is 0.4 ps, matching the decay of the parent molecule. It might be concluded from these data that the 0.4 ps decay of the S_2 state leads directly to methyl loss on the S_1 manifold. The subsequent ~2 ps decay of this signal would then be the signature of some competing process in the S_1 state, perhaps internal conversion to the S_0 ground state. In the following, we will go on to show that this conclusion is, in fact, incorrect.

One might think that the specific wavelength of the photoionization laser is of little import as long as it sufficiently exceeds the ionization potential. In Fig. 7, the results of the same experiment, but repeated this time using a probe laser wavelength of 235 nm, are presented. As the pump laser remained invariant, the same excited-state wave packet was prepared in these two experiments. Contrasting with the 352 nm probe experiment, we see that the parent ion signal does not decay in 0.4 ps, but rather remains almost constant,

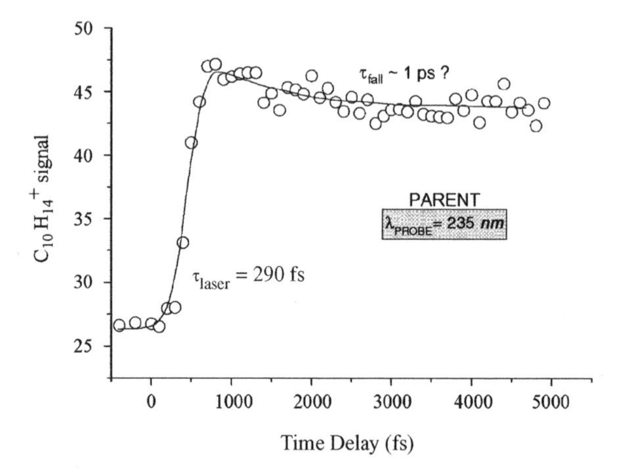

Figure 7. Time-resolved mass spectrometry. All-*trans*-(2, 4, 6, 8) decatetraene was excited to its S_2 electronic origin with a femtosecond pulse at $\lambda_{pump} = 287$ nm. The excited-state evolution was probed via single-photon ionization using a femtosecond pulse at $\lambda_{probe} = 235$ nm. The time resolution in these experiments was 290 fs (0.3 ps). The parent ion $C_{10}H_{14}^+$ signal rises with the pump laser, but then seems to stay almost constant with time. The modest decay observed can be fit with a single exponential time constant of ~1 ps. Note that this result is in apparent disagreement with the same experiment performed at $\lambda_{probe} = 352$ nm, which yields a lifetime of 0.4 ps for the S_2 state. The disagreement between these two results can be only reconciled by analyzing the time-resolved photoelectron spectrum.

perhaps decaying slightly with a ~1 ps time constant. Furthermore, no daughter ion fragments were observed. These very different results seem hard to reconcile. Which probe laser wavelength gives the right answer and why? As discussed below, the time evolution of mass-resolved ion signals can be misleading.

The solution to this apparent paradox lies in the photoionization dynamics. Clearly, the form of the parent ion signal depends strongly on the specific photoionization dynamics and, in order to avoid misleading conclusions, must be analyzed for each specific case. The (Koopmans) photoionization correlations of excited-state electronic configurations with those of the cation play a critical role.

B. TRPES: The Role of Electronic Continua

The above pump–probe experiments on DT were repeated using TRPES rather than mass-resolved ion detection. More detailed discussions of these experimental studies of Koopmans-type correlations with TRPES can be found in the literature [44, 47, 50]. The ultrafast internal conversion of DT provides an example of Type (I) Koopmans' correlations, and below we will also discuss two experimental TRPES studies of Type (II) correlations, namely, the internal conversion in the polyaromatic hydrocarbon phenanthrene, and in 1,4-diazabicyclo[2.2.2]octane (DABCO). As discussed in Section III, Type (I) ionization correlations are defined as being the case when the neutral excited states α and β in Fig. 2 correlate to different ion electronic continua, and are referred to as *complementary* ionization correlations. By contrast, Type (II) correlations are defined as being the case when the neutral excited states α and β correlate to the same ion electronic continua, a situation labeled *corresponding* ionization correlations. As detailed elsewhere [44, 47], the S_2 1^1B_u state of DT is a singly excited configuration and has Koopmans' correlations with the D_0 ^2B_g electronic ground state of the cation. The dipole forbidden $S_1 2^1A_g$ arises from configuration interaction between singly and doubly excited A_g configurations and has preferential Koopmans' correlations with the $D_1{}^2A_u$ first excited state of the cation. These Koopmans' correlations are illustrated in Fig. 8a. In Fig. 8b, we present femtosecond TRPES results on DT for 287 nm pump excitation followed by 235 nm probe laser ionization. The experimental photoelectron kinetic energy spectra reveal a rapid shift of electrons from an energetic peak ($\varepsilon_1 = 2.5$ eV) to a broad, structured low energy component (ε_2). The 2.5 eV band is due to ionization of S_2 into the D_0 ion state. The broad, low energy band arises from photoionization of S_1 that correlates with the D_1 ion state. Its appearance is due to population of the S_1 state by internal conversion. Integration of each photoelectron band directly reveals the S_2 to S_1 internal conversion time scale of 386 ± 65 fs. It is important to note that these results contain more information than the overall (integrated) internal conversion time. The vibrational structure in

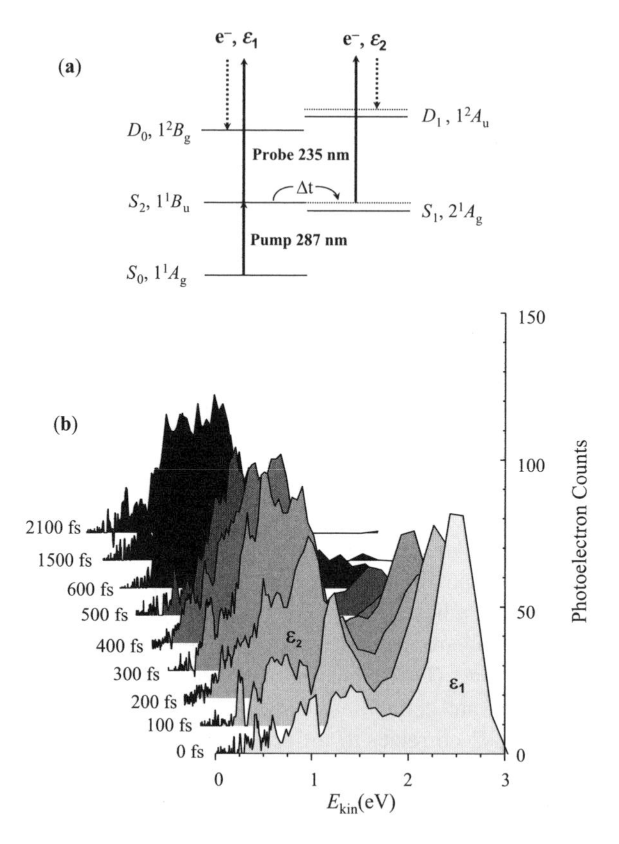

Figure 8. Time-resolved photoelectron spectra revealing vibrational and electronic dynamics during internal conversion in DT. (a) Level scheme in DT for one-photon probe ionization. The pump laser prepares the optically bright state S_2. Due to ultrafast internal conversion, this state converts to the lower lying state S_1 with 0.7 eV of vibrational energy. The expected ionization propensity rules are shown $S_2 \rightarrow D_0 + e^-(\varepsilon_1)$ and $S_1 \rightarrow D_1 + e^-(\varepsilon_2)$. (b) Femtosecond time-resolved photoelectron kinetic energy spectra of DT pumped at $\lambda_{pump} = 287\,\text{nm}$ and probed at $\lambda_{probe} = 235\,\text{nm}$. There is a rapid ($\sim 400\,\text{fs}$) shift in the distribution: From an energetic peak ε_1 at 2.5 eV due to photoionization of S_2 into the D_0 cation ground electronic state; to a broad, structured band ε_2 at lower energies due to photoionization of vibrationally hot S_1 into the D_1 cation first excited electronic state. These results show a disentangling of electronic population dynamics from vibrational dynamics. The structure in the low energy band reflects the vibrational dynamics in S_1.

each photoelectron band yields information about the vibrational dynamics, which promote and tune the electronic population transfer. In addition, it gives a direct view of the evolution of the ensuing intramolecular vibrational energy redistribution (IVR) in the "hot molecule" that occurs on the "dark" S_1 potential surface [47].

It is instructive to compare these TRPES results with the mass-resolved ion yield experiment at the same pump and probe laser wavelengths, discussed above (see Fig. 7). The parent ion signal would be the same as integrating the photoelectron spectra in Fig. 8b over all electron kinetic energies. In doing so, we would sum together a decaying photoelectron band ε_1 with a growing photoelectron band ε_2, leading to a signal that is more or less constant in time and provides little information about the decay dynamics. This is the reason why the parent ion signal in Fig. 7 does not show the 0.4 ps decay that corresponds to the lifetime of the S_2 state. It provides a clear example of how the parent ion signal as a function of time can be misleading.

Why then does the time-resolved parent ion signal at 352 nm probe give the correct 0.4 ps S_2 lifetime (Fig. 6a)? It turns out that 352 nm is just below the energy threshold for reaching the D_1 state of the cation. Therefore, upon internal conversion the formed S_1 state cannot be easily ionized via a single photon since, as discussed above, it does not have Koopmans' correlations with the D_0 ground state of the cation. Therefore, single photon ionization probes only the decaying S_2 state that does have Koopmans-allowed ionization into the D_0 ion ground state. Hence, the correct 0.4 ps decay is observed in the parent ion signal. The fragment ion signal has a 0.4 ps growth curve, indicating that it arises from photoionization of the S_1 state formed by internal conversion. Importantly, the observation of the fragment ion has nothing to do with a possible neutral channel dissociation in the S_1 excited state : there is none. Why then is a fragment ion observed? As both the D_0 and D_1 states of the ion are stable with respect to dissociation, it must be the case that a second probe photon is absorbed and higher lying (predissociative) states of the cation are accessed. A probe laser power study supported this point, yielding a quadratic dependence for the fragment ion and a linear dependence for the parent ion signal. But why, under invariant laser intensity, is a second probe photon absorbed only by the S_1 state and not the S_2 state? Again a consideration of the photoionization dynamics is required.

In Fig. 9 shows the time-resolved photoelectron spectrum for a 352 nm probe laser ionization. Initially, the spectrum is characterized by a low energy band, ε_1 at 0.56 eV, which decays with time. As indicated in Fig. 9, the ε_1 band is due to one-photon ionization of S_2 into D_0 and corresponds exactly with the 235 nm ε_1 band of Fig. 8, simply shifted to lower energy by the reduction in probe photon energy. This peak is also narrower due to the improved kinetic resolution at low energy. A broad energetic band, ε_2, ranging from 0.6 to 4 eV grows with time as the ε_1 band decays, and therefore arises from photoionization of the formed S_1 state. The ε_2 band must, via energy conservation, arise from two-photon probe ionization. As can be seen from Fig. 9a, due to the symmetry of the two-photon dipole operator, the ion continua accessed via two-photon ionization may also include D_0, D_3 and D_4. This explains the broad range and high kinetic energy of

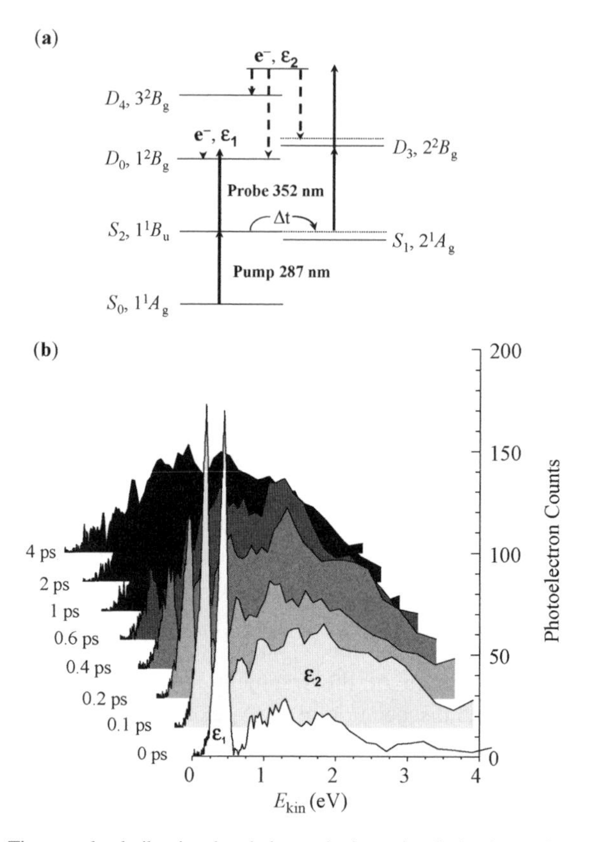

Figure 9. Time-resolved vibrational and electronic dynamics during internal conversion for DT pumped at $\lambda_{\text{pump}} = 287$ nm and probed at $\lambda_{\text{probe}} = 352$ nm. (a) Level scheme in DT for one- and two-photon probe ionization. The pump laser is identical to that in Fig. 8 and prepares the identical S_2 state wave packet. The expected ionization propensity rules are $S_2 \rightarrow D_0 + e^-(\epsilon_1)$ for 1-photon ($u \leftrightarrow g$) ionization and $S_1 \rightarrow D_0, D_3, D_4 + e^-(\varepsilon_2)$ for two-photon ($g \leftrightarrow g$) ionization. (b) Femtosecond time-resolved photoelectron kinetic energy spectra of DT pumped at 287 and probed at 352 nm, using both one- and two-photon probes. At 352 nm, the D_1 ion state is not energetically accessible from the S_1 state via a single-photon transition. Confirming the results of Fig. 8, there is a rapid shift (~400 fs) in the distribution: from ϵ_1 a peak at 0.4 eV due to one-photon ionization of S_2 into the D_0 cation ground electronic state; to ϵ_2 a broad, structured band at higher energies (1–3.5 eV) due to two-photon ionization of the vibrationally hot S_1 into the D_0 cation ground and excited electronic states. The photoionization channel switches from a one-photon to a two-photon process during the internal conversion indicating again that the electronic structure of the ionization continuum is selective of the evolving electronic symmetry in the neutral state.

the photoelectrons in the ε_2 band. Integration of the ε_1 and ε_2 bands provides yet another independent confirmation of the internal conversion time scale, 377 ± 47 fs, fully in agreement with the 235 nm probe results. It is interesting to consider why, at invariant probe laser intensity, the photoionization process switches from single-photon ionization of S_2 to two-photon ionization of S_1. Note that in both cases, the first probe photon is sufficient to ionize the excited state, and therefore the S_1 ionization is due to absorption of a second photon in the ionization continuum. This can be rationalized by consideration of the relative rates of two competing processes: second photon absorption versus autoionization. For the case of S_2, the photoionization correlation is with D_0, and therefore the ionization is direct. In other words, the "autoionization" is extremely rapid and second photon absorption cannot compete. For the case of S_1, the photoionization correlation is with D_1. The D_1 state, however, is energetically inaccessible, and therefore the transition is most likely into Rydberg series converging on the D_1 threshold. For these to emit an electron into the open D_0 continuum channel, there must be an electronic rearrangement, for which there is a finite autoionization rate. In this case, the absorption of a second photon competes effectively with autoionization. These two-photon experiments not only confirm the one photon results, but also demonstrate the symmetry selectivity of the photoionization process itself.

The other limiting Koopmans' case, Type (II), is where the one-electron correlations upon ionization correspond to the same cationic states. An example of Type (II) correlations is seen in the S_2–S_1 internal conversion in the polyaromatic hydrocarbon phenanthrene (PH), discussed in more detail elsewhere [50]. In the case of PH, both the S_2 and the S_1 states correlate similarly with the electronic ground state, as well as the first excited state of the cation. In this experiment, PH was excited from the $S_0{}^1A_1$ ground state to the origin of the $S_2{}^1B_2$ state with a 282 nm (4.37 eV) femtosecond pump pulse, and then ionized after a time delay Δt using a 250 nm (4.96 eV) probe photon. The $S_2{}^1B_2$ state rapidly internally converted to the lower lying $S_1{}^1A_1$ state at 3.63 eV, transforming electronic into vibrational energy. In PH, both the $S_2{}^1B_2$ and $S_1{}^1A_1$ states can correlate with the $D_0{}^2B_1$ ion ground state. The time-resolved photoelectron spectra for PH, shown in Fig. 10, revealed a rapidly decaying but energetically narrow peak at $\varepsilon_1 \sim 1.5$ eV due to photoionization of the vibrationless $S_2{}^1B_2$ state into the ionic ground state $D_0{}^2B_1$, resulting in a decay-time constant of 520 ± 8 fs. A broad photoelectron band, centered at ~ 0.7 eV, in these photoelectron spectra was due to ionization of vibrationally hot molecules in the S_1 state, formed by the S_2–S_1 internal conversion. At times $t > 1500$ fs or so (i.e., after internal conversion), the photoelectron spectrum is comprised exclusively of signals due to S_1 ionization. The S_1 state itself is long lived on the time scale of the experiment. Despite the fact that Type (II) molecules present an unfavorable case for disentangling electronic from vibrational dynamics, in PH a dramatic shift in the photoelectron spectrum was seen as a function of time. This is due to the fact that PH is a rigid molecule and the S_2, S_1, and D_0 states all have similar geometries. The photoionization probabilities are therefore dominated by small Δv transitions. Hence, the 0.74 eV vibrational energy in the populated S_1 state

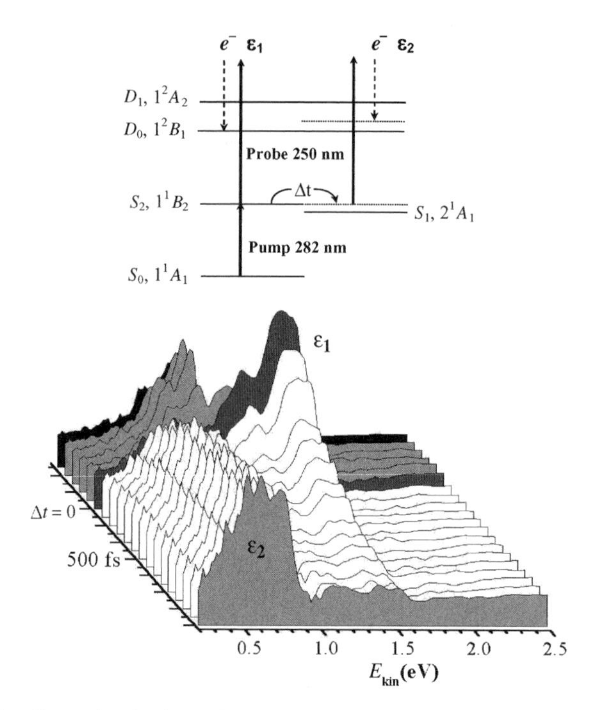

Figure 10. Energy level scheme for TRPES of PH, an example of a Type (II) ionization correlation. (a) The pump laser prepares the optically bright state S_2. Due to ultrafast internal conversion, this state converts to the lower lying state S_1 with ~0.74 eV of vibrational energy. The expected corresponding Type (II) Koopmans' correlations are shown $S_2 \to D_0 + e^-(\varepsilon_1)$ and $S_1 \to D_0 + e^-(\varepsilon_2)$. (b) The TRPES spectra of phenanthrene for a pump wavelength of $\lambda_{pump} = 282$ nm and a probe wavelength of $\lambda_{probe} = 250$ nm. The disappearance of the band ε_1 at ~1.5 eV and growth of the band at ε_2 at ~0.5 eV represents a direct measure of the S_2–S_1 internal conversion time (520 fs). Despite the unfavorable Type(II) ionization correlations, the rigidity of this molecule allows for direct observation of the internal conversion via vibrational propensities alone.

should be roughly conserved upon ionization into the D_0 ionic state. Small geometry changes favor conservation of vibrational energy upon ionization and thereby permit the observation of the excited-state electronic population dynamics via a photoelectron kinetic energy analysis alone. In general, however, significant geometry changes will lead to overlapping photoelectron bands, hindering the disentangling of vibrational from electronic dynamics.

As mentioned in Section III, where a molecule has Type (II) ionization correlations, it might be expected that the coupled electronic states of the neutral molecule would not be resolved in the photoelectron spectrum: the rigidity of PH and the large energy gap between the S_2 and S_1 origins allowed for the resolution of the excited state dynamics in the PES, but this is by no means a general situation. In such circumstances, the measurement of time-resolved PADs (TRPADS) offers an complementary approach to unraveling the dynamics. As discussed in Section III, the requirement that the direct product of the irreducible representations of the neutral electronic state, the transition dipole component, the ion electronic state, and the free electron contains the totally symmetric representation [see Eq. (35)] means that, if the coupled electronic states are of different symmetry, the PAD will differ for the two electronic states. The evolution of the PAD can therefore be expected to provide a mechanism for unraveling the electronic dynamics. A first demonstration of a TRPADs measurement of nonadiabatically coupled electronic states was provided by Hayden and co-workers who studied the molecule DABCO. In this experiment, 251 nm pump pulse excited the origin of the optically bright $S_2{}^1E$ electronic state, a $3p$ Rydberg state centred on the nitrogen atoms [152], via a single-photon excitation. The S_2 state is coupled to the lower lying optically dark $S_1{}^1A'_1$ state, a $3s$ Rydberg state centered on the nitrogen atoms, and internal conversion takes place on a $\sim$ 1 ps time scale. In the experiment, it was possible to resolve the S_1 and S_2 states directly in the PES. In Fig. 11, we show PADs measured for each electronic state at different time delays, measured with a photoelectron imaging apparatus.

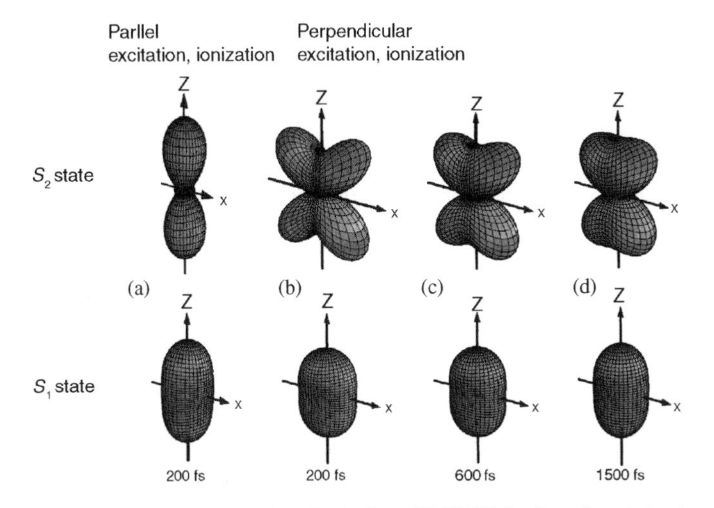

Figure 11. Time-resolved PADs from ionization of DABCO for linearly polarized pump and probe pulses. Here, the optically bright $S_2{}^1E$ state internally converts to the dark $S_1\ {}^1A\prime_1$ state on picosecond time scales. (a) PADs at 200 fs time delay for pump and probe polarization vector both parallel to the spectrometer axis. The difference in electronic symmetry between S_2 and S_1 leads to significant changes in the form of the PAD. (b) The PADs at 200 fs time delay for pump polarization parallel and probe polarization perpendicular to the spectrometer axis, showing the effects of lab frame molecular alignment. (c) and (d) The PADs evolve as a function of time due to molecular axis rotational wavepacket dynamics. Taken with permission from C .C. Hayden, unpublished.

Although the S_2 and S_1 states of DABCO are easily resolved in the PES, due again to the rigidity of the molecule, these measurements clearly demonstrate the sensitivity of the PAD to the electronic symmetry of the excited state. Furthermore, the data displayed in Fig. 11 demonstrates the sensitivity of the PAD to the molecular axis alignment. The single-photon pump step in this experiment produces a coherent superposition of rotational states of the molecules and an anisotropic distribution of molecular axes exhibiting alignment. By comparing Fig. 11 a and b, we see that the PAD changes dramatically when the probe polarization is changed from being parallel to being perpendicular to the pump polarization. When the pump and probe polarizations are parallel, the LF PAD possesses cylindrical symmetry. When the pump and probe polarizations are perpendicular, cylindrical symmetry is lost, although the LF PAD exhibits reflection symmetry in the plane containing the laser polarizations. As the pump–probe time delay increases, the rotational wave packet in the excited state dephases and the molecular axis alignment decreases accordingly. As the anisotropy of the distribution of molecular axes decreases, the PAD is also seen to become less anisotropic, and also less sensitive to the probe laser polarization direction, as discussed in Section III.

In closing this section, we note that although the Koopmans picture is a simplification of the ionization dynamics, it provides a very useful zeroth order picture from which to consider the TRPES results. Any potential failure of this independent electron picture can always be experimentally tested directly through variation of the photoionization laser frequency: resonance structures should lead to variations in the form of the spectra with electron kinetic energy, although the effect of resonances is more likely to be prominent in PAD measurements, and indeed an observation of a shape resonance in *p*-difluorobenzene has been reported [153, 154].

VI. APPLICATIONS

As discussed in the Introduction, a natural application of TRPES is to problems of excited-state nonadiabatic dynamics. Nonadiabatic dynamics involve a breakdown of the adiabatic (Born–Oppenheimer) approximation, which assumes that electrons instantaneously follow the nuclear dynamics. This approximation is exact provided that the nuclear kinetic energy is negligible and its breakdown is therefore uniquely due to the nuclear kinetic energy operator. Spin–orbit coupling, leading to intersystem crossing, is not a nonadiabatic process in this sense: the Born–Oppenheimer states could be chosen to be the fully relativistic eigenstates, and hence would be perfectly valid adiabatic states. Nevertheless, the description of intersystem crossing as a nonadiabatic process is seen in the literature, and therefore we include spin–orbit coupling problems in this section. Furthermore, in this section we have chosen to include examples from work that highlight some of the most recent advances in the use of TRPES for studying molecular dynamics. These include the application of TRCIS to the study of photodissociation dynamics, as well as the use of PADs to measure molecular axis distributions, as suggested by Eq. (45). Examples are also given of the utility of TRPES for the study of excited-state vibrational dynamics.

A. Internal Conversion: Electronic Relaxation in Substituted Benzenes

Internal conversion, also referred to as spin-conserving electronic relaxation or a radiationless transition, is one of the most important nonadiabatic processes in polyatomic molecules, and is often the trigger for any ensuing photochemistry [1–8]. As discussed in more detail in Section VI.D, in order to establish rules relating molecular structure to function—a concept central to the development of molecular scale electronics—it is first necessary to develop an understanding of the relationship between molecular structure and excited-state dynamics. The underlying "rules" governing these processes have yet to be fully established. A phenomenological approach to such rules involves the study of substituent effects in electronic relaxation (internal conversion) dynamics. For this reason, a series of monosubstituted benzenes was studied as model compounds [155]. The focus of this work was on the first and second $\pi\pi^*$ states of these aromatic systems and on substituents that were expected to affect the electronic structure and relaxation rates of the $\pi\pi^*$ states. The photophysics of benzenes is well understood, and therefore the major purpose of this study was to establish the quantitative accuracy of the internal conversion rates determined via TRPES.

As shown in Fig. 12, six benzene derivatives were studied: benzaldehyde (BZA), styrene (STY), indene (IND), acetophenone (ACP), α-methylstyrene (α-MeSTY), and phenylacetylene (ϕ-ACT). This choice of substituents addressed several points: (1) the effect of the substituent on the electronic states and couplings; (2) the effect of the substituent on the rigidity or floppiness of the MF; (3) a comparison of Type (I) with Type (II) Koopmans' systems; (4) to investigate the potential effects of autoionization resonances (i.e., non-Koopmans behavior) on the observed dynamics. Three electronically distinct substituents were chosen: C=O, C=C, and C≡C. For the C=C, potential off-axis conjugation effects with the ring in STY was contrasted with the lack of these in the C≡C of ϕ-ACT. For the heteroatomic substituent C=O, the influence of the additional $n\pi^*$ state on the $\pi\pi^*$ dynamics was investigated by comparing BZA with STY and ACP with α-MeSTY. The effects of vibrational dynamics and densities of states on the electronic relaxation rates were studied via both methyl (floppier) and alkyl ring (more rigid) substitution: STY was compared with both the floppier α-MeSTY and the much more rigid IND; BZA was compared with the floppier ACP. Both BZA and ACP have Type (I) Koopmans' ionization correlations, the rest have Type (II) correlations, allowing for another comparison of these two cases. In

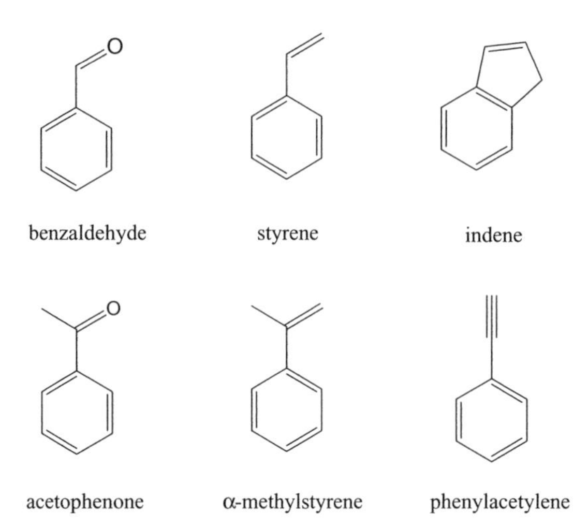

Figure 12. Molecular structures of some monosubstituted benzenes studied via TRPES in order to determine the quantitative accuracy of the extracted internal conversion rates. Three different electronic substituents were used, C=O, C=C, and C≡C, leading to different state interactions. The effects of vibrational dynamics were investigated via the use of methyl group (floppier), as in α-MeSTY and ACP, or a ring structure (more rigid), as in IND, side-group additions. Both BZA and ACP have favorable Type (I) ionization correlations, whereas STY, IND, α-MeSTY, and ACT have unfavorable Type (II) ionization correlations.

order to investigate the potential effects of autoionization resonances, the ionization probe photon energy was varied (~0.4 eV). In these systems, the form of the photoelectron spectra and the fits to the lifetime data were invariant with respect to probe laser frequency [155].

A sample TRPES spectrum, STY at $\lambda_{\text{pump}} = 254.3$ nm and $\lambda_{\text{probe}} = 218.5$ nm, is shown in Fig. 13. The $S_1(\pi\pi^*)$ component grows in rapidly, corresponding to the ultrafast internal conversion of the $S_2(\pi\pi^*)$ state. The $S_1(\pi\pi^*)$ component subsequently decays on a much longer picosecond time scale (not shown). It can be seen that despite STY being an unfavorable Type (II) case, the two bands are well enough resolved to allow for unambiguous separation of the two channels and determination of the sequential electronic relaxation time scales. Energy integration over each band allows for extraction of the electronic relaxation dynamics. The time-dependent $S_2(\pi\pi^*)$ photoelectron band integral yields for STY are also shown in Fig. 13. The open circles represent the pump–probe cross-correlation (i.e., the experimental time resolution) at these wavelengths. Integration over the $S_2(\pi\pi^*)$ photoelectron band is shown as the solid circles. The solid line is the best fit to the $S_2(\pi\pi^*)$ channel, yielding a lifetime of 52 ± 5 fs. In Fig. 13, the time-dependent $S_1(\pi\pi^*)$ photoelectron band integral yields for STY

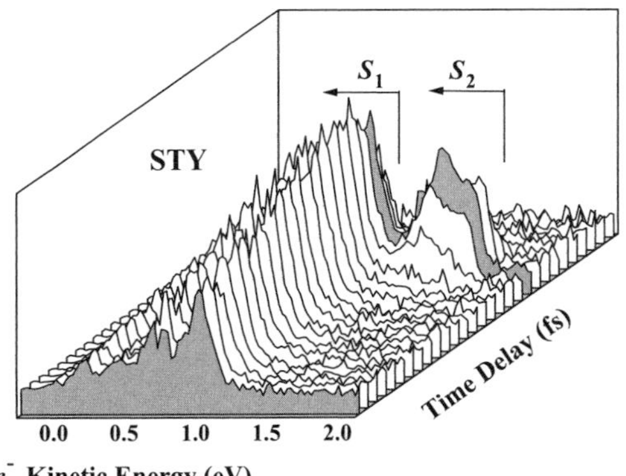

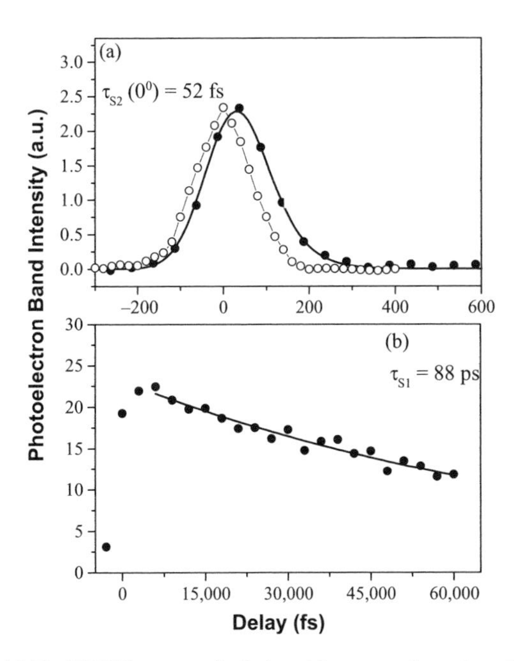

Figure 13. (a) The TRPES spectra of substituted benzenes, shown here for STY with $\lambda_{\text{pump}} = 254.3$ nm and photoionization $\lambda_{\text{probe}} = 218.5$ nm. The energetics and Koopmans' correlations allow for assignment of the photoelectron bands to ionization of $S_2(\pi\pi^*)$ and $S_1(\pi\pi^*)$, as indicated. The S_2 state decays on ultrafast time scales. The S_1 state decays on a much longer (ps) time scale. (Bottom) (a) Time-dependent $S_2(\pi\pi^*)$ 0^0 photoelectron band integral yields for STY, yielding a decay time constant of 52 fs. Open circles represent the laser cross-correlation at these wavelengths. (b) Time-dependent $S_1(\pi\pi^*)$ photoelectron band integral yields for STY obtained from a fit to the long time-delay part of the data (not shown).

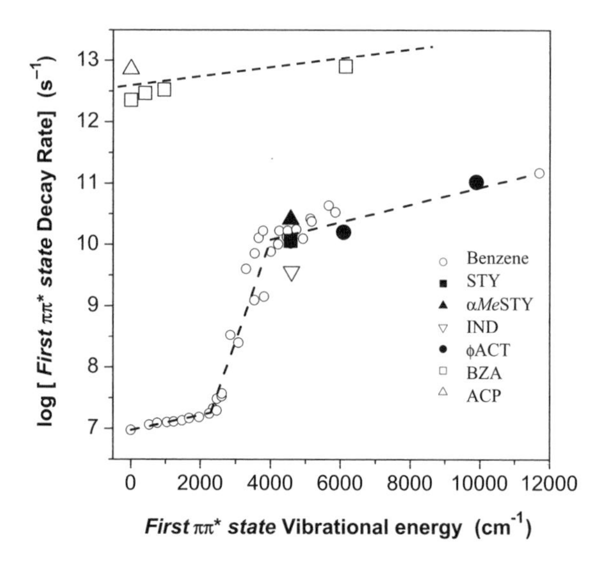

Figure 14. Excess vibrational energy dependence of the internal conversion rates of the first $\pi\pi^*$ state of benzene and its derivatives.

are shown, obtained from a fit to the long delay part of the data, yielding a lifetime of 88 ± 8 ps for the state $S_1(\pi\pi^*)$.

The excess vibrational energy dependence of the internal conversion rates of the first $\pi\pi^*$ state of benzene and its derivatives is shown in Fig. 14. These data suggest that the first $\pi\pi^*$ states of STY, α-MeSTY, IND, and ACT internally convert essentially via benzene ring dynamics. By contrast, the first $\pi\pi^*$ states of BZA and ACP internally convert orders of magnitude faster, indicating a completely different mechanism due to the presence of low lying $n\pi^*$ states in BZA and ACP, absent in the other systems, which lead to ultrafast intersystem crossing and the formation of triplet states. Overall, these results demonstrate that the TRPES method is well suited to the quantitative study of electronic relaxation processes, producing direct and accurate measurements of electronic relaxation rates that are in quantitative agreement with the currently accepted understanding of aromatic photophysics [155].

B. Excited State Nuclear Dynamics

As discussed in Section III, TRPES is sensitive to vibrational and rotational dynamics, as well as electronic dynamics. In this section, we give examples of the use of TRPES to the study of intramolecular vibrational energy redistribution (IVR), and the use of time-resolved PAD measurements as a probe of rotational dynamics.

A problem central to chemical reaction dynamics is that of IVR [156, 157], the flow of energy between zeroth-order vibrational modes. Indeed, IVR generally accompanies (and mediates) nonadiabatic dynamics, such as internal conversion and isomerization. The description of separated rigid rotational and normal-mode vibrational motions employed in Sections II.B and III provides an adequate description only in regions of low state density. As the state density increases, the vibrational dynamics become "dissipative" as normal-mode vibrations become mixed and energy flows between these zeroth-order states. Much work has been undertaken studying IVR using, for example, fluroscence techniques [156–158]. However, such techniques generally monitor flow of energy out of the initially excited vibrational states, but do not directly observe the optically dark "bath" vibrational modes into which vibrational energy flows. TRPES provides a window to these dark states, allowing for direct monitoring of IVR in molecules due to the Franck–Condon correlations described in Section III.

Reid and co-workers reported a picosecond TRPES study of IVR in the electronically excited S_1 state of *p*-fluorotoluene [159]. By selective excitation of specific vibrational modes in S_1 and measuring the evolution of the PES as a function of time delay, information regarding vibrational population dynamics was obtained. Analysis of this TRPES data also employed high resolution PES data and required a detailed understanding of the Franck–Condon factors for ionization. Reid and co-workers were able to measure the rates of IVR for the the initially prepared 7^1, 8^1, and 11^1 vibrational states (using Mulliken notation) of the first electronically excited state S_1 of *p*-fluorotoluene [159]. By selective population of vibrational states in S_1, it is possible to use TRPES to test for vibrational mode-specificity in IVR, as well as varying the excess vibrational energy. In Fig. 15, example TRPES data is shown for excitation of the 7^1 mode of *p*-fluorotoluene, corresponding to one quanta of excitation in the C—F stretching mode, and representing $1230\,cm^{-1}$ vibrational energy in S_1. The evolution of the PES shown in Fig. 15, and the disappearance of resolved structure at long time delays is a direct measure of the IVR of energy out of the C—F stretching mode and into other modes of the molecule. By analyzing spectra, such as those in Fig. 15 in terms of the population in the initially prepared state compared to the populations in all other "dark" modes, the rate of IVR may be extracted. An example of such an analysis is shown in Fig. 16.

In a second example of the utility of TRPES for the study of vibrational dynamics, Reid and co-workers studied the dynamics associated with a Fermi resonance between two near-degenerate vibrational modes in the S_1 state of toluene [160]. In this study, the pump pulse prepared a coherent superposition of the $6a^1$ state, corresponding to one quanta of vibrational excitation in the totally symmetric ring breathing mode, and the $10b^1 6b^1$ state, corresponding to one quanta in the CH_3 wagging mode (10*b*) and one quanta in the C—H out-of-plane bending mode (16*b*). The anharmonic coupling between these two states gives rise to an oscillation in the form of the PES, as shown in Fig. 17, the time scale of which corresponds to the energy separation of the two vibrational modes.

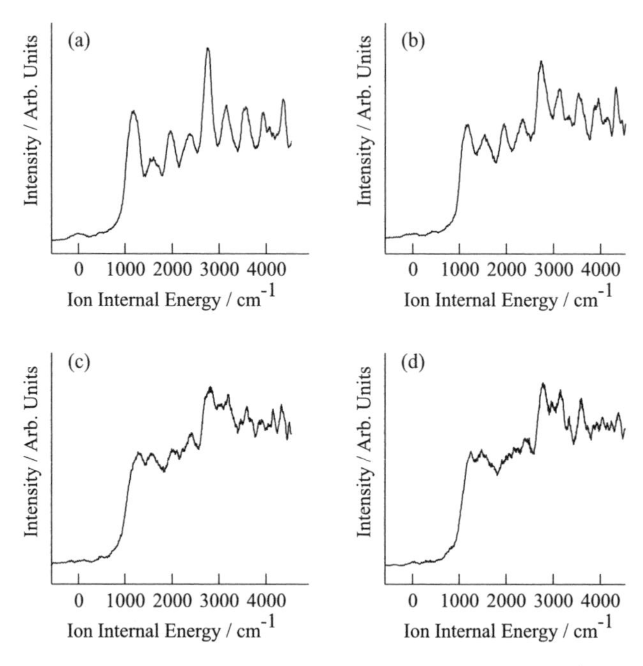

Figure 15. The TRPES of the S_1 state of *p*-fluortoluene prepared in the 7^1 state, that is, with one quanta of excitation in the C—F stretching mode. At early times, the PES contains a well-resolved Franck–Condon progression corresponding to ionization of the localized C—F stretching mode. At later times, the onset of IVR obscures the PES as many more vibrational modes become populated. Reproduced with permission from Ref. [159].

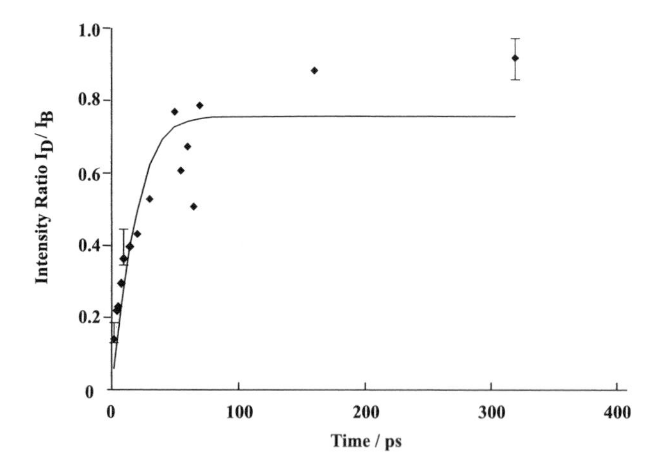

Figure 16. Ratio of the population of the initially prepared vibrational state to the population of the "dark" vibrational states populated through IVR for the data shown in Fig. 15. Taken with permission from Ref. [159].

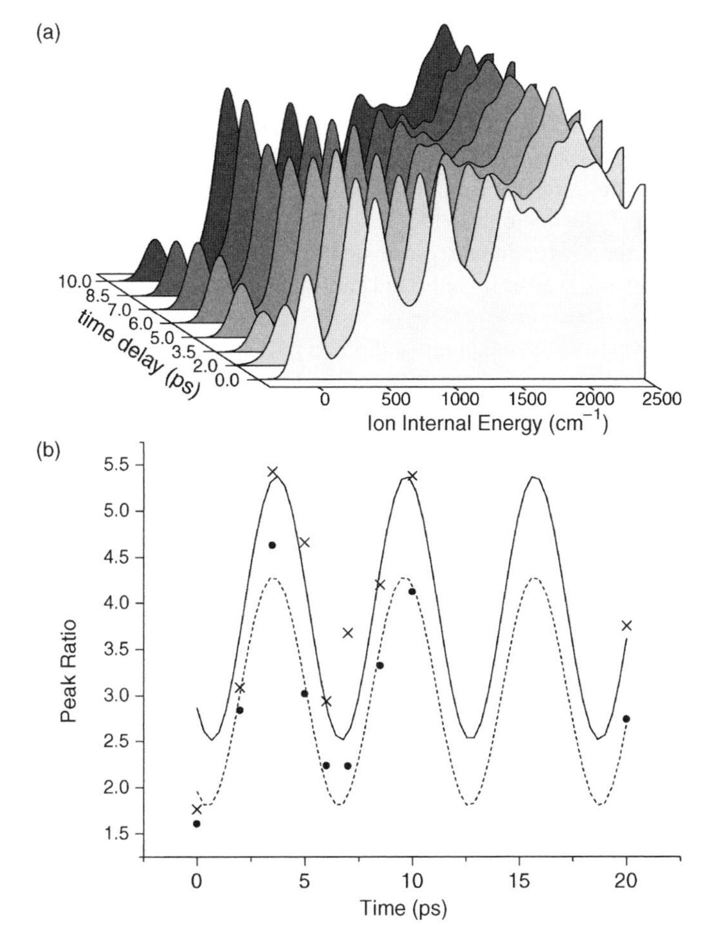

Figure 17. (a) Photoelectron spectra following preparation of the $6a^1 + 10b^1 6b^1$ Fermi resonance at $457\,\text{cm}^{-1}$ in the S_1 state of toluene with a 1-ps pump pulse. The PES are shown as a function of time delay between the pump pulse and a 1-ps probe pulse. (b) Fits to the oscillation of the PES according to two different models of the data shown in (a) clearly showing a period of oscillation of ~6 ps, corresponding to the energy separation of the two vibrational states comprising the wave packet. For more detail see Ref. [160]. Taken with permission from Ref. [160].

As discussed in Section III, TRPAD measurements are sensitive to molecular rotational motion by virtue of their geometric dependence on the molecular axis distribution in the LF. An elegant experimental demonstration of this has been performed by Suzuki and co-workers who measured the PAD temporal evolution from excited-state pyrazine [59]. In these experiments, the origin of the S_1 electronic state of pyrazine was excited by a pump pulse at 323 nm, and subsequently probed by a time-delayed probe pulse via a two photon ionization. In this energy region, the excited-state dynamics of pyrazine involve a well-studied intersystem crossing caused by strong spin–orbit coupling of the $S_1B_{3u}(n\pi^*)$ state with a manifold of triplet states, denoted T_1, resulting in a complex energy spectrum [161–164]. While earlier studies focused on the monitoring of fluorescence from the S_1 state, this work directly measured the S_1 decay and the T_1 formation via TRPES and TRPAD measurements. Two photon ionization in these experiments proceeded via $3s$ and $3p_z$ Rydberg states of the neutral, producing well-resolved bands attributable to the S_1 and T_1 states (see Fig. 18).

In these experiments, the pump and probe pulses were linearly polarized with their electric fields vectors mutually parallel. The pump pulse created an initially aligned $\cos^2\theta$ distribution of the principal molecular axes, with θ the polar angle between the principal molecular axis and the laser field polarization. This initially prepared rotational wave packet subsequently evolved with time, and the ionization yield (shown in Fig. 18b) from the singlet-state directly reflected this wave packet evolution, exhibiting the expected rotational recurrence behavior of a near oblate symmetric top [91, 93–95, 156, 165]. The sensitivity of the photoelectron yield to the molecular axis alignment arises due to the well-defined molecular frame direction of the transition dipole momement for excitation of the intermediate Rydberg states in the probe step – the $3p_z \leftarrow S_1$ and the $3s \leftarrow S_1$ transitions being perpendicular and parallel to the principal molecular axis, respectively, resulting in the opposite (out of phase) behaviors in the black and red lines shown in Fig. 18b. Interestingly, the rotational coherence is also directly observed in the signal representing the formation of the T_1 manifold (blue line in Fig. 18), demonstrating that rotational coherence is (perhaps partially) preserved upon intersystem crossing [28, 166]. Additionally, the PAD also reflected the wave packet evolution with the time dependence of the value of β_{20}/β_{00} mapping the rotational recurrence behavior, as shown in Fig. 18c. In this case, the different dependence on molecular axis alignment of the LF PAD for ionization via the $3s$ versus the $3p_z$ Rydberg states reflects the different MF PADs for ionization of these two Rydberg states. These measurements demonstrate the utility of TRPAD measurements as a probe of rotational dynamics. Such measurements are sensitive to vibration–rotation coupling [54, 56–58].

C. Excited-State Intramolecular Proton Transfer

Excited-state intramolecular proton transfer (ESIPT) processes are important for both practical and fundamental reasons. *o*-Hydroxybenzaldehyde (OHBA) is the simplest aromatic molecule displaying ESIPT and serves as a model system for comparison with theory. TRPES was used to study ESIPT in OHBA, monodeuterated ODBA and an analogous two-ring system hydroxyacetonaphtone (HAN) as a function of pump laser wavelength, tuning over the entire enol

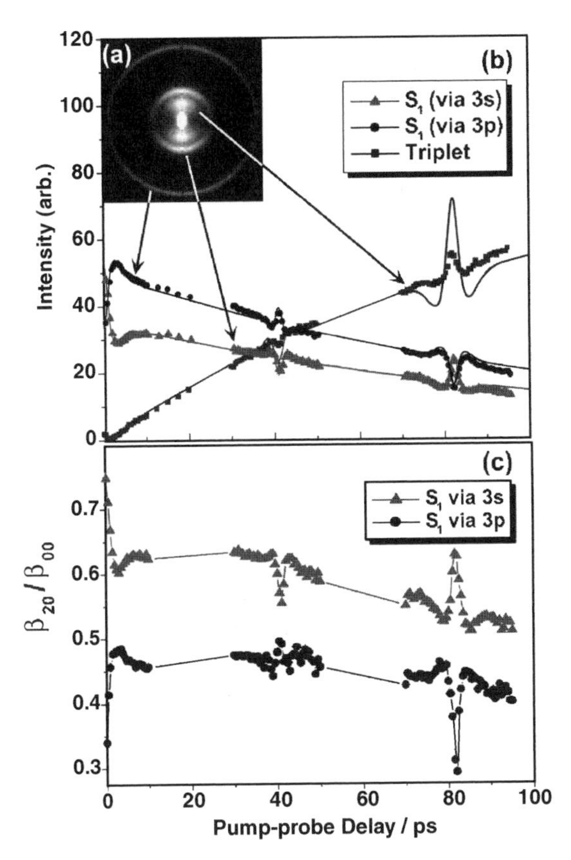

Figure 18. (a) Inverse Abel transformed photoelectron image showing the lab frame PAD for ionization of pyrazine with a pump pulse at 323 nm and a probe pulse at 401 nm. The laser pulses had parallel linear polarizations and a temporal separation of 30 ps. The outer two rings correspond to two photon ionization of the S_1 electronic state via 3*s* and 3*p* Rydberg states, and the inner-ring corresponds to two-photon ionization of the triplet-state manifold T_1 in the neutral formed by intersystem crossing from the S_1 state. (b) Time-dependence of the angle-integrated signals in (a). (c) Time-dependence of the PAD anisotropy for the three signals in (b) as monitored by the ratio of the PAD parameters β_{20}/β_{00} from a fit to an expansion in spherical harmonics, Eq. (42). See color insert.

$S_1(\pi\pi^*)$ absorption band of these molecules [167, 168]. The experimental scheme is depicted in Fig. 19, showing energetics for the case of OHBA. Excitation with a tuneable pump laser $h\nu_{\text{pump}}$ forms the enol tautomer in the $S_1(\pi\pi^*)$ state. The ESIPT leads to ultrafast population transfer from the S_1 enol to the S_1 keto tautomer. On a longer time scale, the S_1 keto population decays via internal conversion to the ground state. Both the enol and keto excited-state

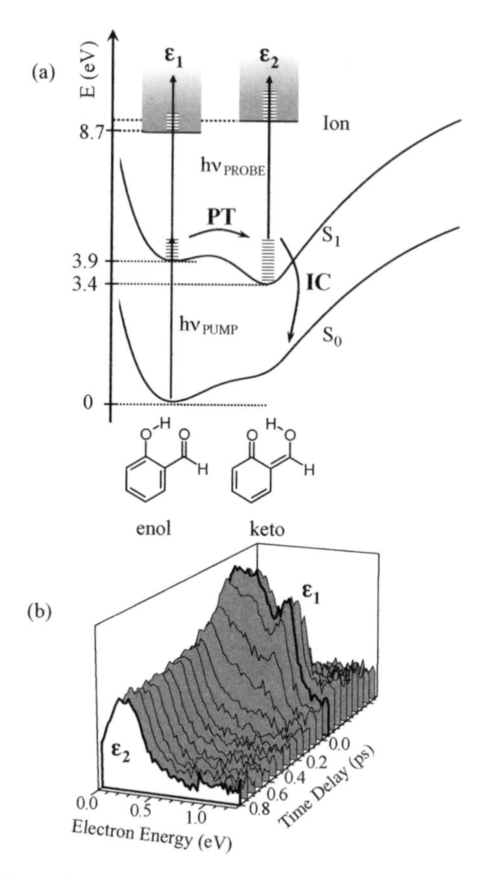

Figure 19. (a) Energetics for ESIPT in OHBA, showing the enol and keto forms. Excitation with a pump laser forms the enol tautomer in the $S_1(\pi\pi^*)$ state. The ESIPT leads to ultrafast population transfer from the S_1 enol to the S_1 keto tautomer. On a longer time scale, the keto S_1 population decays via internal conversion to the keto ground state. Both the enol and keto excited state populations are probed via TRPES, producing the two photoelectron bands ε_1 and ε_2. (b) The TRPES spectra of OHBA at an excitation wavelength of 326 nm and a probe wavelength of 207 nm. Two photoelectron bands were observed: ε_1 due to ionization of the S_1 enol, and ε_2 due to ionization of the S_1 keto. Band ε_1 was observed only when the pump and probe laser beams overlapped in time, indicating a sub-50-fs time scale for the proton transfer. Band ε_2 displayed a pump wavelength dependent lifetime in the picosecond range corresponding to the energy dependent internal conversion rate of the dark S_1 keto state formed by the proton transfer.

populations are probed by photoionization with a probe laser $h\nu_{\text{probe}}$, producing the two photoelectron bands ε_1 and ε_2.

Figure 19 shows TRPES spectra of OHBA at an excitation wavelength of 326 nm. Two photoelectron bands ε_1 and ε_2 with distinct dynamics were observed. Band ε_1 is due to photoionization of the initially populated S_1 enol tautomer, and band ε_2 is due to the photoionization of the S_1 keto tautomer. The decay of band ε_1 yields an estimated upper limit of 50 fs for the lifetime of the S_1 enol tautomer. Proton transfer reactions often proceed via tunneling of the proton through a barrier. Deuteration of the transferred proton should then significantly prolong the lifetime of the S_1 enol tautomer. In experiments with ODBA, an isotope effect was not observed (i.e., the ESIPT reaction was again complete within the laser cross-correlation).

Figure 20 shows examples of fits to OHBA at 326 nm, ODBA at 316 nm, and HAN at 363 nm. The proton transfer rates for all three molecules were sub-50 fs over their entire S_1 enol absorption bands. It was concluded that the barrier in the OH stretch coordinate must be very small or nonexistent. This interpretation

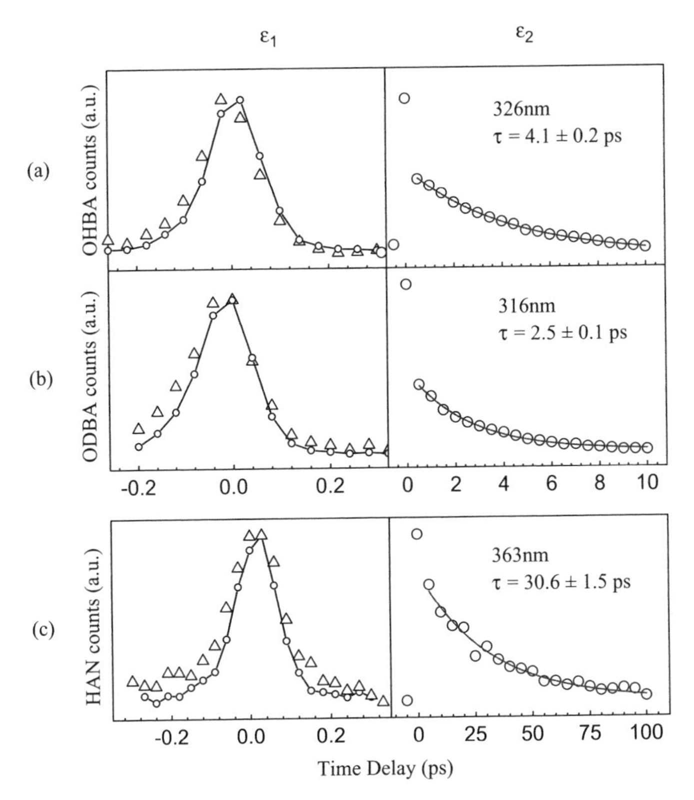

Figure 20. Integrated signals ε_1 and ε_2 for OHBA (a), ODBA (b), and HAN (c) plotted as a function of the time delay at the indicated excitation wavelength. Note the change in ordinate time scales. Signal ε_1 always followed the laser cross-correlation, indicating a rapid proton transfer reaction. The decay of signal ε_2 was fitted via single exponential decay, yielding the time constant for internal conversion of the S_1 keto state in each molecule. See color insert.

is consistent with *ab initio* calculations that predict no barrier for the proton transfer [169, 170]. An estimate of the corresponding reaction rate using an instanton calculation, which takes into account the multimode character of proton transfer, resulted in S_1 enol lifetimes of $\sim$ 20 fs for the transfer of a proton and <50 fs for the transfer of a deuteron when the barrier was lowered to 2.4 kcal/mol^{-1} [167, 168]. This value was considered to be an upper limit for the proton transfer barrier.

As is common in TRPES, these spectra also give insights into the dynamics on the "dark" S_1 keto state. The picosecond decay of band ε_2 corresponds to S_1 keto internal conversion to the ground state. The wavelength-dependent S_1 keto internal conversion rates for OHBA and ODBA shown in Fig. 21 revealed no significant isotope effect. Interestingly, the measured internal conversion rates for OHBA/ODBA are very fast (1.6–6 ps over the range 286–346 nm) considering the large energy gap of 3.2 eV between the ground and excited state. One possibility is that fast internal conversion in such systems is due to an efficient conical intersection involving a $\pi\pi^*$ state with a $\pi\sigma^*$ via large amplitude hydroxy H-atom motion [169, 170]. However, the observed absence of an isotope effect on S_1 keto internal conversion rates in ODBA does not support this mechanism. A clue is found in the comparison with internal conversion rates of OHBA/ODBA with the larger HAN, shown in Fig. 20. The HAN has both a smaller S_1–S_0 energy gap and a higher density of states, leading to the expectation that its internal conversion rate should be faster than that of

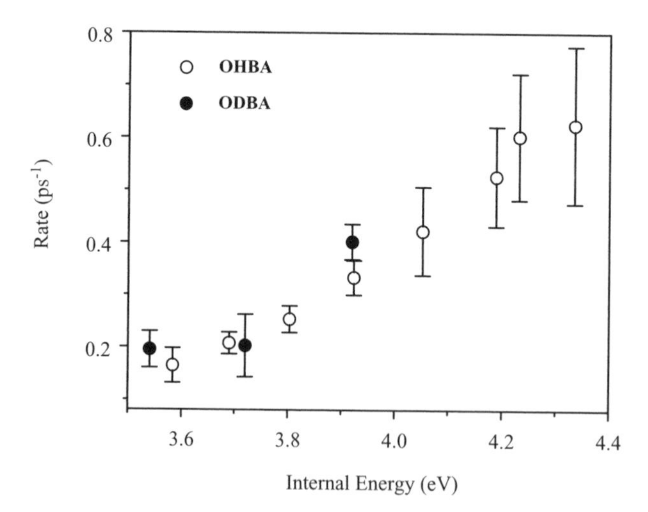

Figure 21. Internal conversion rates of the S_1 keto state of OHBA (open circles) and ODBA (filled circles), determined by single exponential fits to the ε_2 band decay. Both show a monotonic increase in rate as a function of the excitation energy, but without a significant isotope effect.

OHBA. Surprisingly, it is ~10 times slower, indicating that some other effect must be operative. A major difference between the two molecules is the position of a $n\pi^*$ state, which is almost isoenergetic with the $\pi\pi^*$ state in OHBA, but 0.5 eV higher in HAN. The coupling of the $\pi\pi^*$ and $n\pi^*$ states, mediated by out-of-plane vibrations, greatly increases the internal conversion rate in OHBA. The local mode character of the OH out-of-plane bending vibration makes this mode inefficient for the coupling of the $n\pi^*$ and $\pi\pi^*$ states. As a result, the bending modes of the aromatic ring dominate this interaction, which explains the absence of an isotope effect [167, 168]. This example serves to illustrate how TRPES can be used to study the dynamics of biologically relevant processes, such as ESIPT and that it reveals details of both the proton transfer step and the subsequent dynamics in the "dark" state formed after the proton transfer.

D. Dynamics of Molecular Electronic Switches

The burgeoning area of active molecular electronics involves the use of molecules or molecular assemblies acting as switches, transistors, or modulators. A central theme is that structural rearrangement processes, such as isomerization, should lead to changes in either optical or electrical properties, generating the desired effect. It is often proposed that these structural rearrangements be induced via electronic excitation. The rational design of active molecular electronic devices must include a detailed consideration of the dynamics of the "switching" process for several reasons. Foremost is that activation of the device (e.g., by a photon) must indeed lead to the desired change in optical or electrical properties and therefore this basic mechanism must be present. Two other issues, however, are of great practical significance. The efficiency of the molecular electronic process is a critical element because excited organic molecules often have a variety of complex decay paths that compete with the desired process. The efficiency of a device can be defined simply as the rate of the desired process divided by the sum of the rates of all competing processes. As certain of these competing processes can occur on ultrafast time scales (e.g., dissipation, dissociation), the rate of the desired process must be very fast indeed, even if the required overall response is slow. A directly related issue is that of stability. A molecular modulator that operates at 1 GHz and lasts for 3 years must "switch" $\sim 10^{17}$ times without malfunction. The quantum yields of any "harmful" processes must therefore be exceedingly small. Unfortunately, excited organic molecules have a number of destructive decay pathways, such as photodissociation and triplet formation (often leading to reaction). The relative rates and quantum yields of these processes, as well as their dependence on substituent and environmental effects, will be critical elements in the design of efficient, stable active molecular devices. *trans*-Azobenzene is often considered the canonical molecular switch and its photoisomerization is the basis for numerous functional materials [171]. Azobenzene provides an important

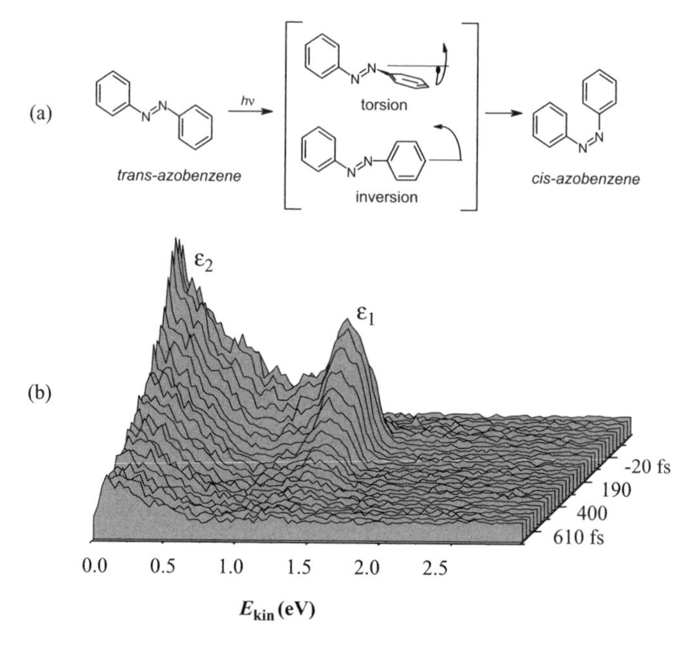

Figure 22. (a) Photoisomerization dynamics of *trans*- to *cis*-azobenzene, indicating torsional and inversion pathways. (b) The TRPES spectra of *trans*-azobenzene excited at 330 nm and probed at 207 nm. Two photoelectron bands ε_1 and ε_2 were observed, having identical laser-limited rise times, but differing decay rates ($\tau_1 = 130\,\text{fs}$, $\tau_2 = 410\,\text{fs}$) and, importantly differing Koopmans' ionization correlations. These results indicate that there is a previously unrecognized $\pi\pi^*$ state, S_3 (centered on the aromatic rings), involved in the dynamics.

example for the study of the dynamics of Molecular Electronic switches via TRPES [172].

Despite great interest in azobenzene photophysics, the basic photoisomerization mechanism remains disputed [173]: in contrast to the expectations of Kasha's rule, the isomerization quantum yield decreases rather than increases with increasing photon energy. In Fig. 22, the two possible isomerization channels, proceeding via either a planar pathway (inversion) or a nonplanar, twisted pathway (torsion) are shown. Previous studies determined that isomerization in the first excited state S_1 state proceeds along the inversion coordinate [171]. The second excited state $S_2(\pi\pi^*_{\mathrm{N=N}})$ is generally thought to be the N=N analogue of the C=C $\pi\pi^*$-state in stilbene and that, somehow, motion along the torsional coordinate in $S_2(\pi\pi^*_{\mathrm{N=N}})$ is responsible for the observed reduction in isomerization yield. Time-resolved studies suggested that photoisomerization proceeds via the inversion coordinate in S_1 [171]. The role of the torsional isomerization pathway remains controversial. Theoretical studies have supported both torsion and inversion pathways, but disagreed on the states involved in the excited-state relaxation. Any successful model considering $\pi\pi^*$ state relaxation in AZ must address three puzzling features [172]: (1) the violation of Kasha's rule, that is, Φ_{isom} ~25% for $S_1(n\pi^*)$, but drops to Φ_{isom} ~12% for the higher lying $\pi\pi^*$ state(s); (2) inhibition of the torsional coordinate in sterically restrained AZ increases Φ_{isom} of the $\pi\pi^*$ states to a level identical to that observed for S_1 photoexcitation; (3) the observation of efficient relaxation of $S_2(\pi\pi^*)$ to the S_1 state via planar geometries.

In Fig. 22, a time-resolved photoelectron spectrum for excitation of AZ to the origin of its $S_2(\pi\pi^*_{N=N})$ state is shown. Two photoelectron bands ε_1 and ε_2 with differing lifetimes and differing Koopmans' correlations were observed. Due to these two differences, the ε_1 and ε_2 bands must be understood as arising from the ionization of two different electronic states. Furthermore, as both bands rise within the laser cross-correlation, they are due to direct photoexcitation from S_0 and not to secondary processes. Therefore, in order to account for different lifetimes, different Koopmans' correlations and simultaneous excitation from S_0, the existence of an additional state, labeled $S_3(\pi\pi^*_\phi)$, which overlaps spectroscopically with $S_2(\pi\pi^*_{N=N})$ must be invoked. According to the Koopmans analysis (based upon assignment of the photoelectron bands) and to high level, large active space CASSCF calculations, this new state $S_3(\pi\pi^*_\phi)$ corresponds to $\pi\pi^*$ excitation of the phenyl rings [172], as opposed to the $S_2(\pi\pi^*_{N=N})$ state where excitation is localized on the N=N bond. Therefore, $\pi\pi^*$ excitation in the phenyl rings does not directly "break" the N=N bond and leads to reduced isomerization quantum yields.

A new model for AZ photophysics was proposed as a result of these TRPES studies. The $S_2(\pi\pi^*_{N=N})$ state internally converts to S_1 in a planar geometry, explaining puzzle (3) above. The subsequent relaxation of S_1 does indeed follow Kasha's rule and yields Φ_{isom} ~25% for the population originating from $S_2(\pi\pi^*_{N=N})$. Different dynamics are observed in the TRPES experiments for the $S_3(\pi\pi^*_\phi)$ state, indicating a different relaxation pathway. To explain puzzle (1), relaxation of $S_3(\pi\pi^*_\phi)$ with reduced isomerization must be assumed: The ring-localized character of $S_3(\pi\pi^*_\phi)$ suggests a relaxation pathway involving phenyl ring dynamics. This could involve torsion and lead directly to the *trans*-AZ ground state—explaining both puzzles (1) and (2). *Ab initio* Molecular Dynamics (AIMD) simulations [172] starting from the Franck–Condon geometry in $S_2(\pi\pi^*_{N=N})$ agree with result (3) and predict that the molecule quickly (< 50 fs) samples geometries near conical intersections while still in a planar geometry, with no evidence for torsion or inversion. For S_1, the AIMD simulations predict that a conical intersection involving inversion is approached within 50 fs [172]. This mechanism differs greatly from that of all earlier models in that those always assumed that only a single bright state, $S_2(\pi\pi^*_{N=N})$, exists in this wavelength region. This example shows how TRPES can be used to study competing electronic relaxation pathways in a model molecular switch, revealing hidden yet important electronic states that can be very hard to discern via conventional means.

E. Photodissociation Dynamics

From the point of view of chemical reaction dynamics, the most interesting case is that of unbound excited states or excited states coupled to a dissociative continuum: that is, photodissociation dynamics. The dissociative electronically excited states of polyatomic molecules can exhibit very complex dynamics, usually involving nonadiabatic processes. The TRPES and TRCIS may be used to study the complex dissociation dynamics of neutral polyatomic molecules, and below we will give two examples of dissociative molecular systems that have been studied by these approaches, NO_2 and $(NO)_2$.

TRCIS was first applied to dissociative multiphoton ionization of NO_2 at 375.3 nm [129]. This was identified as a three-photon transition to a repulsive surface correlating with $NO(C^2\Pi) + O(^3P)$ fragments. The NO(C) was subsequently ionizing by a single photon, yielding $NO^+(X^1\Sigma^+) + e^-$.

As an illustration of the multiply differential information obtained via TRCIS, energy–energy correlations plotting photoelectron kinetic energy versus NO(C) photofragment kinetic energy, as a function of time, are shown in Fig. 23. At early times, 0 and 350 fs, there is a negative correlation between electron and fragment recoil energy. This form is expected for a molecule in the process of dissociating where there is a trade-off between ionization energy and fragment recoil energy. At longer time delays, 500 fs and 10 ps, the NO(C) fragment is no longer recoiling from the O atom—it is a free particle—and the photoelectron spectrum obtained is simply that of free NO(C). Hence, the negative correlation vanishes [129]. By measuring the angle of recoil of both photoelectron and photofragment in coincidence, the PAD may be transformed into the RF at each time delay [137]. In Fig. 24, the time-resolved RF PADs are shown for the case of photofragments ejected parallel to the laser polarization axis. It can be seen that at early times, 0 and 350 fs, the PAD is highly asymmetric. The breaking of forward–backward symmetry in the RF originates from NO(C) polarization due to the presence of the O atom from which it is recoiling. At longer times, 1 and 10 ps, this forward–backward asymmetry vanishes, as the NO(C) becomes a free particle. This once again shows the power of TRCIS in obtaining highly detailed information about molecules in the process of dissociating.

A second illustrative example of the utility of TRPES and TRCIS for studying complex molecular photodissociation dynamics that involve multiple electronic state is the case of the weakly bound cis-planar C_{2v} nitric oxide dimer [174]. The weak ($D_0 = 710\,\text{cm}^{-1}$) $^1A_1$ ground-state covalent bond is formed by the pairing of two singly occupied π^* orbitals, one from each $NO(X^2\Pi)$ monomer. The very intense UV absorption spectrum of the NO dimer appears

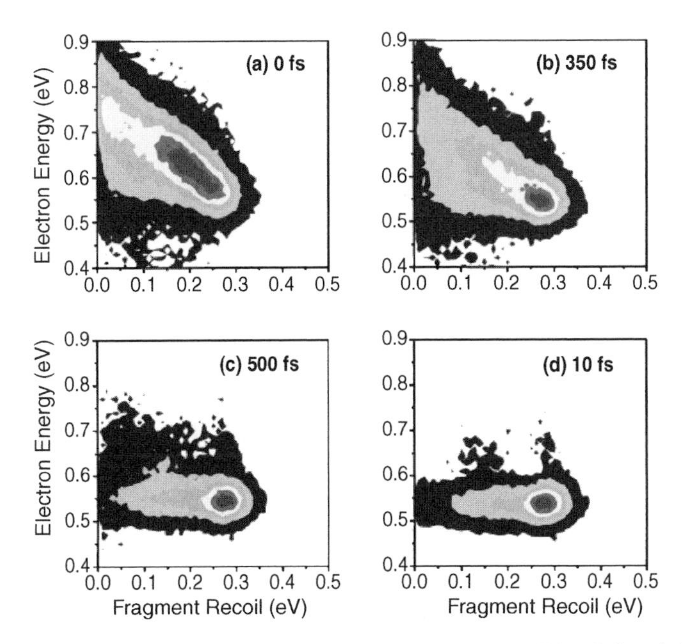

Figure 23. Time-resolved coincidence-imaging spectroscopy (TRCIS) of dissociative multi-photon ionization processes in NO_2 using 100-fs laser pulses at 375.3 nm, using energy–energy correlations. The 2D maps show, at time delays of 0 fs, 350 fs, 500 fs, and 10 ps, the correlation between the photoelectron kinetic energy (abscissa) and NO photofragment recoil energy (ordinate). The intensity distributions change from a negative correlation at early times to uncorrelated at later times, yielding information about the molecule as it dissociates. Taken with permission from Ref. [129]

broad and featureless and spans a 190–240 nm range, with a maximum at ~205 nm. This transition was assigned as ${}^1B_2 \leftarrow {}^1A_1$, and therefore has a transition dipole along the N—N bond direction (with B_2 symmetry). Recent *ab initio* studies of the excited electronic states of the dimer revealed a complex set of interactions between two very strongly absorbing states of mixed-valence–Rydberg character that play a central role in the photodissociation dynamics [175]. As we will see from the following measurements, these "diabatic" states are roughly comprised of a diffuse $3p_y$ Rydberg function (the y axis is along the N—N bond) and a localized valence function that has charge-transfer character and therefore carries most of the oscillator strength in the Franck–Condon region, as the oscillator strengths are much too high for a pure Rydberg state [175]. At 210 nm excitation one product channel is dominant:

$$(\mathrm{NO})_2^* \rightarrow \mathrm{NO}(\mathrm{A}^2\Sigma^+, v, J) + \mathrm{NO}(\mathrm{X}^2\Pi, v', J') \tag{55}$$

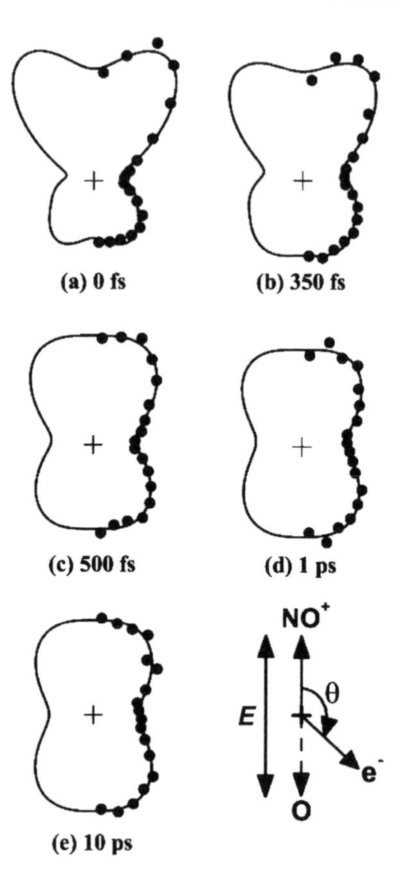

Figure 24. Coincidence-imaging spectroscopy of dissociative multiphoton ionization processes in NO_2 with ~100-fs laser pulses at 375.3 nm, using angle–angle correlations. The polar plots show, at time delays of 0 fs, 350 fs, 500 fs, 1 ps, and 10 ps, the angular correlation between the ejected electron and NO photofragment when the latter is ejected parallel to the laser field polarization vector. The intensity distributions change from a forward–backward asymmetric distribution at early times to a symmetric angular distribution at later times, yielding detailed information about the molecule as it dissociates. Taken with permission from Ref. [137]

The fragment excited-state NO($A^2\Sigma^+$) is a molecular 3*s* Rydberg state, and we shall refer to this as NO(A, 3*s*). The observed NO(A, 3*s*) product state distributions supported the notion of a planar dissociation involving restricted intramolecular vibrational energy redistribution (IVR) [176]. A scheme for studying NO dimer photodissociation dynamics via TRPES is depicted in Fig. 25. The NO(A, 3*s*) + NO(X) product elimination channel, its scalar and vector properties, and its evolution on the femtosecond time scale have been discussed in a number of recent publications (see Ref. [175] and references cited therein).

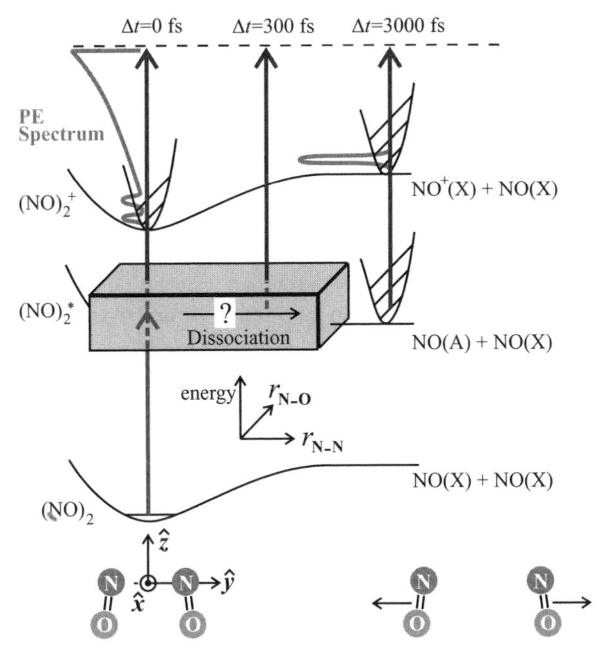

Figure 25. A femtosecond TRPES scheme for studying NO dimer photodissociation. A UV pump pulse creates the excited state $(NO)_2^*$. Its subsequent evolution is monitored all the way from initial excitation to final product emission via a UV probe pulse, projecting the wave packet onto the ionization continuum. The resulting photoelectron spectrum, reflecting vibrational and electronic changes during dissociation, is depicted in green. See color insert.

The first TRPES study of NO dimer photodissociation at 210 nm excitation (and 287 nm probe) showed that the decaying $(NO)_2^+$ parent ion signal disappeared more rapidly (when fit to a single exponential decay of 0.3 ps) than the NO(A, 3*s*) state product signal appeared to rise (when fit to a single exponential growth of 0.7 ps) [174]. This result shows once again that the time dependence of the parent ion signal alone can be misleading. Due to its Rydberg character, the NO(A 3*s*, *v*, *J*) products produced a single sharp peak in the photoelectron spectrum, due to the well-known $NO(A^2\Sigma^+, v, J) \rightarrow NO^+ (X, {}^1\Sigma^+, v)$ ionizing transition that has predominantly $\Delta v = 0$. The dissociation dynamics was interpreted in terms of a two-step sequential process involving an unknown intermediate configuration. Subsequent femtosecond time-resolved ion and photoelectron imaging studies further considered the dissociation dynamics of the NO dimer [177–179]. These reported the observation that both the decaying NO dimer cation signal and the rising NO(A) photoelectron signal could be fit using single exponential functions. Furthermore, the emerging NO(A, 3*s*) photoelectron peak changed shape and shifted in energy (by 15–20 meV) at early times. This was taken as evidence for formation of a dimer 3*s* Rydberg state that was expected to correlate directly to NO(A, 3*s*) + NO(X) products. It was argued that when the shifting of this peak is taken into consideration, the decay of the parent signal and the rise of the product signal could be fit with the same single exponential time constant, suggesting no need for an intermediate configuration.

Recently, the photodissociation dynamics of the NO dimer was reinvestigated using a high sensitivity magnetic bottle technique combined with TRCIS (discussed below) [138]. In Figure 26 shows a magnetic bottle TRPES spectrum of $(NO)_2$ photodissociation. At $\Delta t = 0$, a broad spectrum due to photoionization of $(NO)_2^*$ shows two resolved vibrational peaks assigned to 0 and 1 quanta of the cation N=O stretch mode (ν_1). The $\nu_1 = 2$ peaks merges with a broad, intense Franck–Condon dissociative continuum. At long times ($\Delta t = 3500$ fs), a sharp photoelectron spectrum of the free NO(A, 3*s*) product is seen. The 10.08 eV band shows the decay of the $(NO)_2^*$ excited state. The 9.66 eV band shows both the decay of $(NO)_2^*$ and the growth of free NO(A, 3*s*) product. It is not possible to fit these via single exponential kinetics. However, these 2D data are fit very accurately at all photoelectron energies and all time delays simultaneously by a two-step sequential model, implying that an initial bright state $(NO)_2^*$ evolves to an intermediate configuration $(NO)_2^{*\dagger}$, which itself subsequently decays to yield free NO(A, 3*s*) products [138]

$$(NO)_2^* \rightarrow (NO)_2^{*\dagger} \rightarrow NO(A, 3s) + NO(X) \qquad (56)$$

The requirement for a sequential model is seen in the 9.66 eV photoelectron band, showing NO(A, 3*s*) product growth. The delayed rise of the free NO(A, 3*s*) signal simply cannot be fit by a single exponential decay followed by single exponential growth with the same time constant. The 10.08 eV dissociative ionization band, dominant at early times, is revealing of $(NO)_2^*$ configurations preceding dissociation. Its time evolution, which also cannot be fit by single exponential decay, provides another clear view of the intermediate step. The decay time constant of the initial $(NO)_2^*$ state is 140 ± 30 fs, which matches the rise time of the intermediate $(NO)_2^{*\dagger}$ configuration. This intermediate configuration has a subsequent decay time of 590 ± 20 fs. These two time constants result in a maximum for $(NO)_2^{*\dagger}$ at $\Delta t \sim 330$ fs delay. The two components can be seen as the dashed lines in the fits to the 10.08 eV data (along with a small instrumental response signal). In the 9.66 eV band, the dashed lines from the fits show that the rise of the NO(A, 3*s*) product channel is first delayed by 140 ± 30 fs but then grows with a 590 ± 20 fs time constant. Although only two cuts are shown, the data are fit at all time delays and photoelectron energies simultaneously. These results show that the decay of the parent molecule does not match the rise of the free products and, therefore, an intermediate

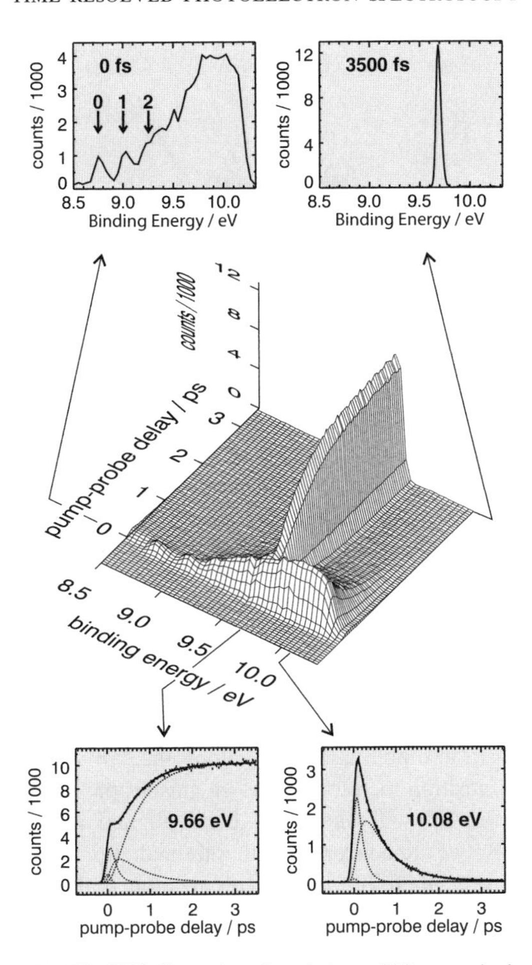

Figure 26. The TRPES of NO dimer photodissociation at 210-nm excitation [138]. The broad, decaying spectrum apparent at early times is due to photoionization of the dissociating excited parent molecule. The sharp peak emerging with time is due to growth of the free NO(A, 3*s*) products. These 2D data are globally fit at all energies and time delays simultaneously. The green inserts (top) are 1D cuts along the energy axis, showing photoelectron spectra at two selected time delays. The blue inserts (bottom) are 1D cuts along the time axis, showing the evolution of the photoelectron intensity at two selected binding energies. The solid lines in the blue graphs are from the 2D fits to the sequential two-step dissociation model discussed in the text. The dashed lines are the respective initial-, intermediate-, and final-state signal components. See color insert.

configuration that has differing ionization dynamics is required to model the data. The nature of this $(NO)_2^{*\dagger}$ configuration cannot be discerned from TRPES data alone. In order to uncover its character, this system was also studied using the TRCIS technique [138].

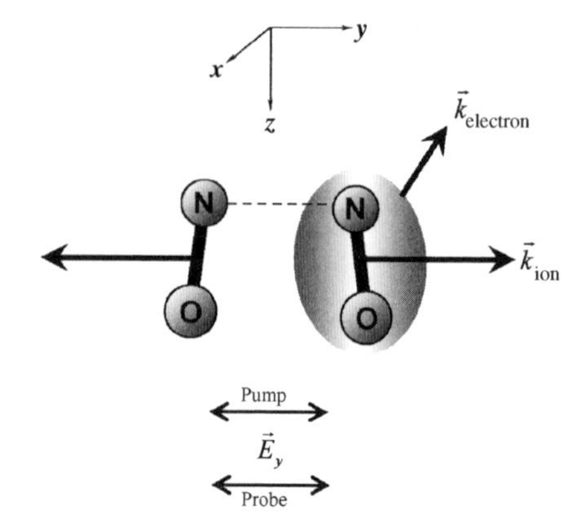

Figure 27. Molecular frame axis convention for the C_{2v} NO dimer. The y axis is along the N—N bond. Both pump and probe laser polarizations are parallel to the y axis.

The 6D fully correlated TRCIS data set may be cut, projected, or filtered to reveal both scalar and vector correlations as a function of time. We restrict our discussion here to angular correlations. The molecular frame axis convention for the NO dimer is shown in Fig. 27. Note that the pump and probe laser polarizations were parallel to each other in these experiments.

The pump transition dipole is directed along the MF y axis (the N—N bond axis). The pump transition therefore forms an anisotropic distribution of excited $(NO)_2^*$ states in the LF with the N—N bond aligned along the pump laser polarization axis. As we are concerned with intermediate configurations in $(NO)_2^*$ evolution, we consider therefore the photoionization probing of $(NO)_2^*$, which predominately leads to dissociative ionization, as shown in Fig. 26. The dissociative ionization of $(NO)_2^*$ produces NO^+ fragments strongly directed along the laser polarization axis. The NO^+ fragment recoil direction therefore indicates the lab frame direction of the N—N bond (MF y axis) prior to ionization. Rotating the electron momentum vector into the fragment recoil frame (RF) on an event-by-event basis allows for reconstruction of the $(NO)_2^*$ photoelectron angular distribution in the RF, rather than the LF. Here the RF coincides with the MF, differing only by azimuthal averaging about the N—N bond direction. Out of all fragment recoil events, only those directed (up or down) along the parallel pump and probe laser polarization axis were selected. Importantly, by choosing events from this selected set, the data is restricted to the excited-state ionization events arising from interactions with the MF y component of the ionization transition dipole only. As discussed below, this restriction greatly limits the allowed partial waves for the emitted electron, especially in the present case where only a single cation electronic continuum is accessed [138].

In Fig. 28, time-resolved lab and RF PADs arising from photoionization of $(NO)_2^*$ in the 9.9–10.3 eV band of Fig. 26 are presented. This dissociative ionization region contains significant contributions from the intermediate $(NO)_2^{*\dagger}$ configuration. In general, the time dependence of PADs relates to the evolution of the excited state electronic structure, as discussed in Section III. Here, the LF PADs have a largely isotropic character that show no discernible change with time, obscuring information about excited-state dynamics. By contrast, the RF PADs show a highly anisotropic character and a variation with time delay. The solid lines in the polar plots of Fig. 28 are fits to an expansion in Legendre polynomials $P_L(\cos\theta)$,

$$I(\theta) = \sum_L \mathcal{B}_L P_L(\cos\theta) \tag{57}$$

For the RF PADs only even L terms were nonzero with $L \leq 8$ in this fit. Increasing the maximum value of L did not improve the fit to the data, and odd L coefficients were found to converge to zero in the fits, in agreement with the up–down symmetry of the RF PADs.

Interestingly, the RF PADs have dominant intensity perpendicular to the laser polarization axis. An A_1 Rydberg $3s$ intermediate state would most likely yield maximum intensity parallel to the laser polarization axis, contrary to what is observed, since a $3s$ Rydberg state would ionize to primarily form a p-wave ($l = 1$ electrons). As can be seen from visual inspection of the data, the ratio of perpendicular-to-parallel photoelectron intensity varies with time, going through a maximum at ~0.3 ps before decaying again to smaller values. This "model-free" result rules out the A_1 Rydberg $3s$ state as the intermediate configuration. Corroborated by *ab initio* calculations [175], the RF PADs were modeled using states of B_2 symmetry. It was also assumed that the molecule largely retains C_{2v} symmetry, supported by the retention of planarity during dissociation [138, 175] as deduced from vector correlation measurements.

To proceed further, detailed analysis of the RF PADs is required. The outgoing free electron partial waves are decomposed into symmetry-adapted spherical harmonics [51, 55], as given by Eq. (3.4). For C_{2v} , these harmonics are described by their C_{2v} symmetry and by $l_{|\lambda|}$, where l, $|\lambda|$, are the orbital angular momentum and projection quantum numbers, respectively. Values of $l = 0, 1, 2\ldots$ are labeled s, p, d ... whereas values of $|\lambda| = 0, 1, 2\ldots$ are labeled $\sigma, \pi, \delta\ldots$. For the case of the NO dimer, ionization of a B_2 electronic state to an A_1 cation state via a y-polarized transition (also of B_2 symmetry) means that the free electron must have A_1 symmetry in order to satisfy the requirement in Eq. (35). This significantly restricts the allowed free electron states. Since the fit to Legendre polynomials required $L \leq 8$, partial waves with $l = 0\ldots4$ are required to model the data. The A_1 symmetry partial waves with $l \leq 4$ are

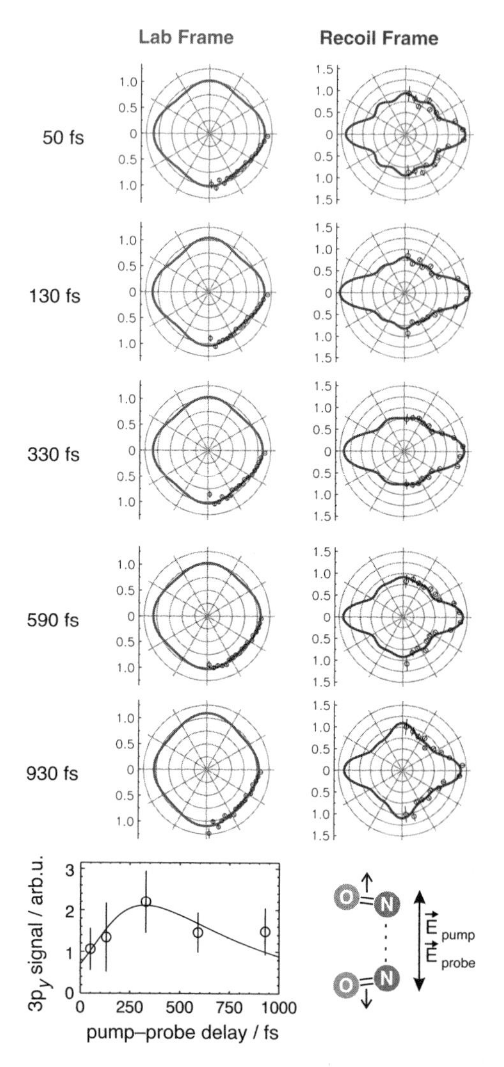

Figure 28. Coincidence-imaging spectroscopy of $(NO)_2$ photodissociation at 210 nm showing LF (left) and RF (right) photoelectron angular distributions (PADs) from the 9.9–10.3eV dissociative ionization region of Fig. 26. The laser polarizations and RF axes are along the y direction, as shown (bottom right). The LF PADs show featureless and almost invariant behavior. The RF PADs show strong anisotropies that vary with time. The fit curves (solid lines) include even-order Legendre polynomials $P_L(\cos\theta)$ up to $L = 4$ for the LF and up to $L = 8$ for the RF. The average partial wave contribution expected from Rydberg $3p_y$ ionization is plotted as a function of time (bottom left). The time dependence of the intermediate configuration extracted from the TRPES data of Fig. 26 and is plotted here as the solid line, agreeing well with the time dependence of the $3p_y$ ionization contribution. This substantiates the intermediate configuration as being of Rydberg $3p_y$ character. For details see the text. See color insert.

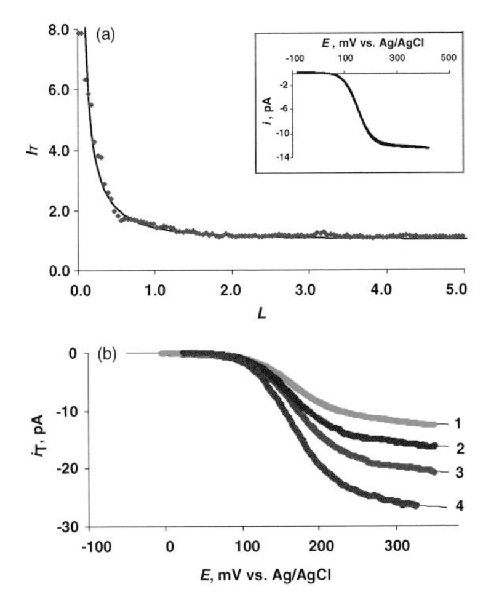

Figure 16. (a) The SECM current versus distance curve and a steady-state voltammogram (inset) obtained with a 46-nm radius polished Pt electrode. Aqueous solution contained 1 m*M* $FcCH_2OH$ and 0.2 *M* NaCl. (a) Theoretical approach curve (solid line) for diffusion-controlled positive feedback was calculated from Eq. (19). Symbols are experimental data. The tip approached the unbiased Au film substrate with a 5-nm s^{-1} speed. (b) Experimental (symbols) and theoretical (sold lines) steady-state voltammograms of 1 m*M* ferrocenemethanol obtained at different separation distances between a 36-nm Pt tip and a Au substrate. $d = \infty$ (1), 54 nm (2), 29 nm (3), and 18 nm (4). $v = 50\,\text{mV}\,\text{s}^{-1}$. Theoretical curves were calculated from Eq. (22). Adapted with permission from Ref. [51]. Copyright 2006, American Chemical Society.

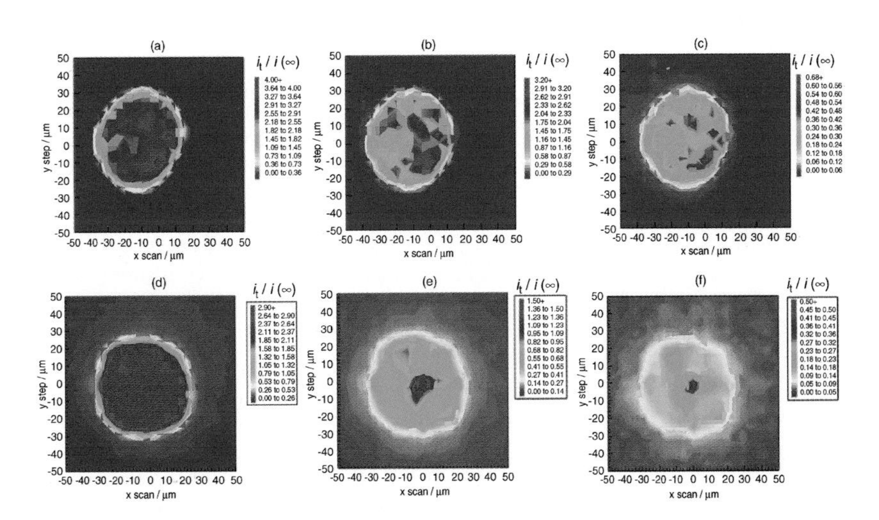

Figure 21. The SG–TC 100× 100-µm images of a two different boron doped regions (a, b, and c - region 1; d, e and f - region 2) in an intrinsic diamond disk. The substrate was biased at (a, d) −0.4 V, (b, e) − 0.3 V and (c, f) −0.2 V. The 5-µm radius Pt UME tip was kept 1 µm above region 1 and 0.6 µm above region 2. The substrate potential was 0 V versus Ag/AgCl. Adapted with permission from Ref. [104]. Copyright 2006, American Chemical Society.

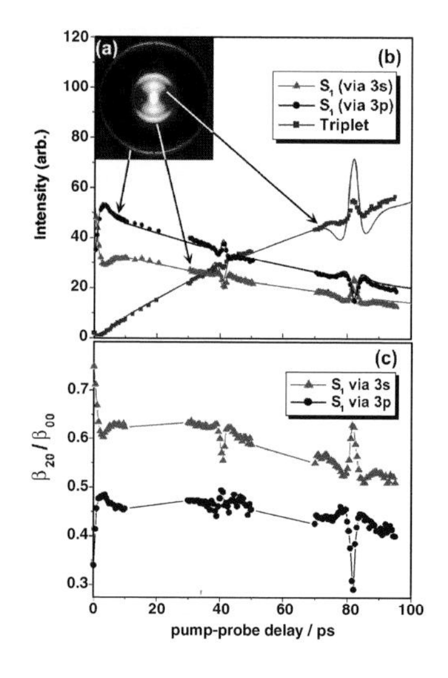

Figure 18. (a) Inverse Abel transformed photoelectron image showing the lab frame PAD for ionization of pyrazine with a pump pulse at 323 nm and a probe pulse at 401 nm. The laser pulses had parallel linear polarizations and a temporal separation of 30 ps. The outer two rings correspond to two photon ionization of the S_1 electronic state via $3s$ and $3p$ Rydberg states, and the inner-ring corresponds to two-photon ionization of the triplet-state manifold T_1 in the neutral formed by intersystem crossing from the S_1 state. (b) Time-dependence of the angle-integrated signals in (a). (c) Time-dependence of the PAD anisotropy for the three signals in (b) as monitored by the ratio of the PAD parameters β_{20}/β_{00} from a fit to an expansion in spherical harmonics, Eq. (42).

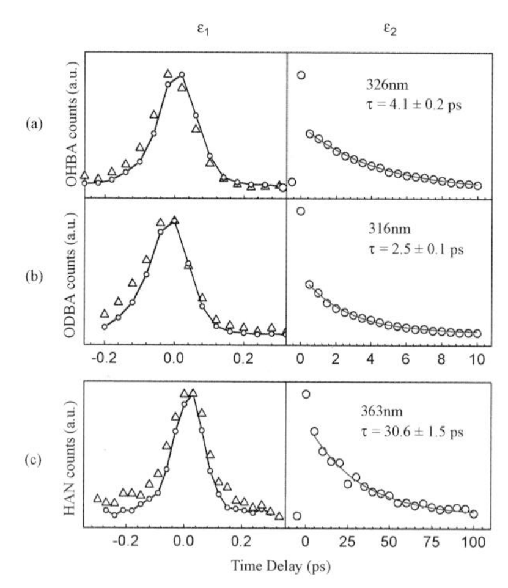

Figure 20. Integrated signals ε_1 and ε_2 for OHBA (a), ODBA (b), and HAN (c) plotted as a function of the time delay at the indicated excitation wavelength. Note the change in ordinate time scales. Signal ε_1 always followed the laser cross-correlation, indicating a rapid proton transfer reaction. The decay of signal ε_2 was fitted via single exponential decay, yielding the time constant for internal conversion of the S_1 keto state in each molecule.

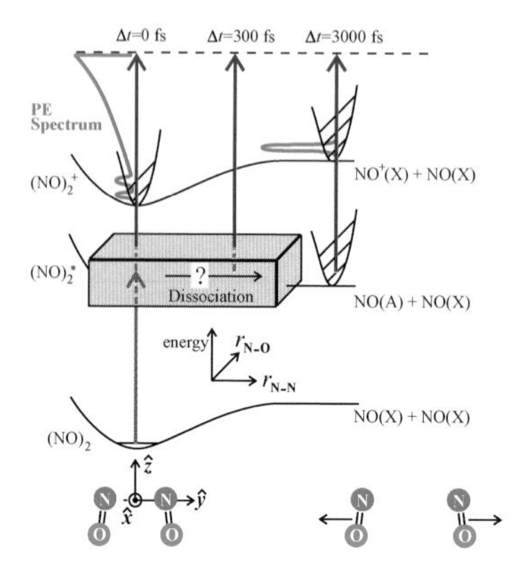

Figure 25. A femtosecond TRPES scheme for studying NO dimer photodissociation. A UV pump pulse creates the excited state $(NO)_2^*$. Its subsequent evolution is monitored all the way from initial excitation to final product emission via a UV probe pulse, projecting the wave packet onto the ionization continuum. The resulting photoelectron spectrum, reflecting vibrational and electronic changes during dissociation, is depicted in green.

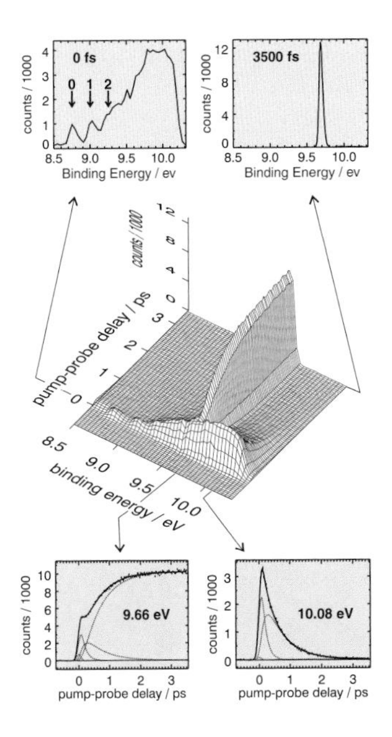

Figure 26. The TRPES of NO dimer photodissociation at 210 nm excitation [138]. The broad, decaying spectrum apparent at early times is due to photoionization of the dissociating excited parent molecule. The sharp peak emerging with time is due to growth of the free NO(A, 3*s*) products. These 2D data are globally fit at all energies and time delays simultaneously. The green inserts (top) are 1D cuts along the energy axis, showing photoelectron spectra at two selected time delays. The blue inserts (bottom) are 1D cuts along the time axis, showing the evolution of the photoelectron intensity at two selected binding energies. The solid lines in the blue graphs are from the 2D fits to the sequential two-step dissociation model discussed in the text. The dashed lines are the respective initial-, intermediate-, and final-state signal components.

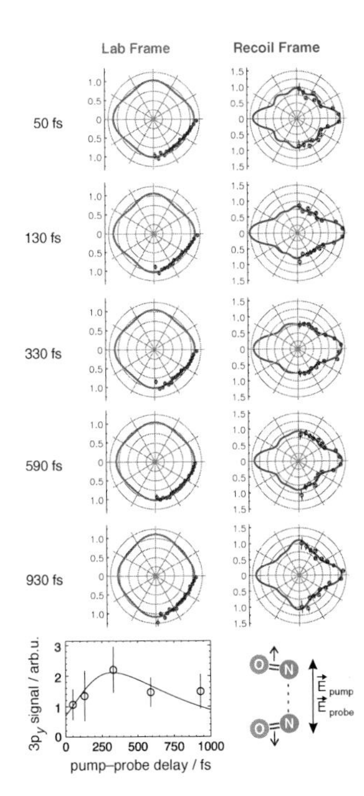

Figure 28. Coincidence-imaging spectroscopy of $(NO)_2$ photodissociation at 210 nm showing LF (left) and RF (right) photoelectron angular distributions (PADs) from the 9.9–10.3-eV dissociative ionization region of Fig. 26. The laser polarizations and RF axes are along the *y* direction, as shown (bottom right). The LF PADs show featureless and almost invariant behavior. The RF PADs show strong anisotropies that vary with time. The fit curves (solid lines) include even-order Legendre polynomials $P_L(\cos\theta)$ up to $L = 4$ for the LF and up to $L = 8$ for the RF. The average partial wave contribution expected from Rydberg $3p_y$ ionization is plotted as a function of time (bottom left). The time dependence of the intermediate configuration extracted from the TRPES data of Fig. 26 and is plotted here as the solid line, agreeing well with the time dependence of the $3p_y$ ionization contribution. This substantiates the intermediate configuration as being of Rydberg $3p_y$ character. For details see the text.

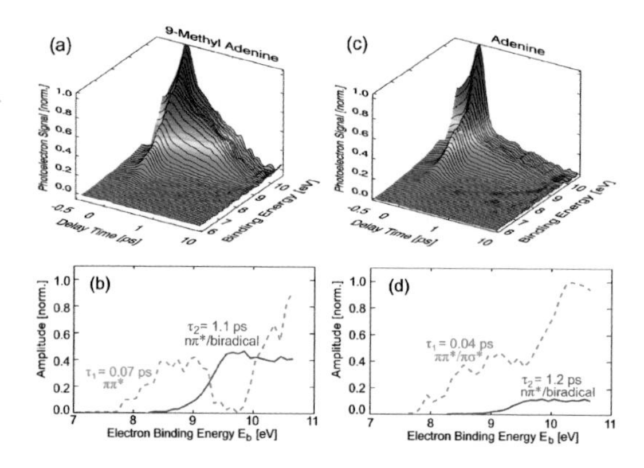

Figure 29. The TRPES spectra for adenine (left) and 9-methyl adenine (right), pumped at $\lambda_{\text{pump}} = 267\,\text{nm}$ and and probed at $\lambda_{\text{probe}} = 200\,\text{nm}$. The time dependence is plotted using a linear/logarithmic scale with a linear scale in the region -0.4–$1.0\,\text{ps}$ and a logarithmic scale for delay times 1.0–10.0 ps.

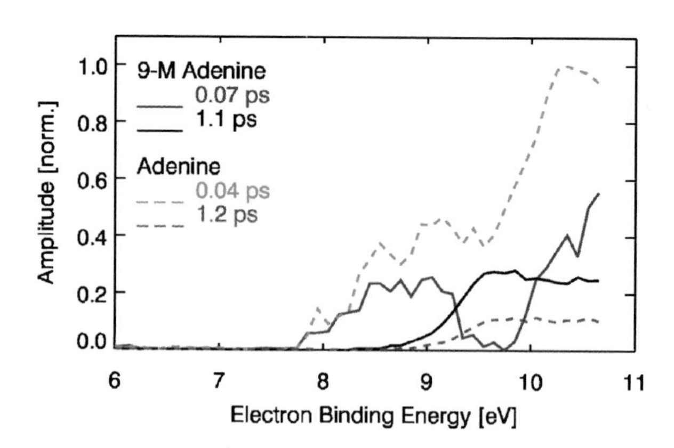

Figure 30. Decay associated spectra for adenine (dashed lines) and 9-methyl adenine (solid lines), extracted from the 2D TRPES spectra using global fitting procedures. Both molecules were fit by the same two time constants: $\tau_1 \leq 0.1\,\text{ps}$ and $\tau_2 \sim 1.1\,\text{ps}$, agreeing quantitatively with previous results. The spectra, however, are very different for adenine as compared to 9-methyl adenine. For details, see the text.

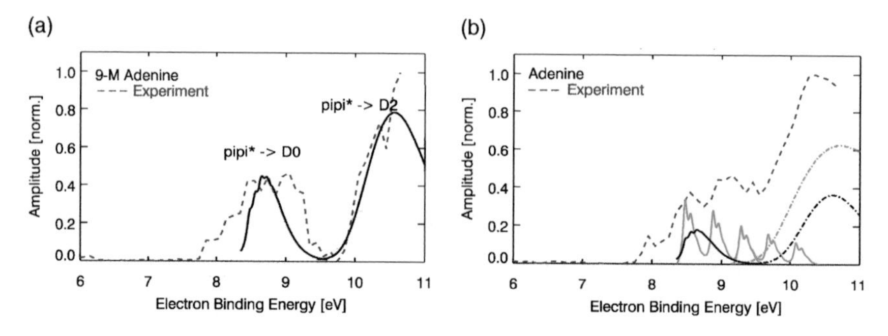

Figure 31. Decay associated spectra of the short-lived state compared with calculated FC spectra for 9-methyl adenine (a) and adenine (b). In 9-methyl adenine, the $\pi\pi^* \rightarrow D_0(\pi^{-1}), D_2(\pi^{-1})$ transitions leave a FC gap. In adenine, this gap is filled by the $\pi\sigma^*$ ionizing transitions.

s_σ, p_σ, d_σ, d_δ, f_σ, f_δ, g_σ, g_δ, and g_γ. In general, the s, p, and d waves were dominant. Modeling of the data would therefore require nine partial wave amplitudes and eight relative phases, and so clearly a unique fit to the data was not possible. However, it was possible to determine the range of partial wave amplitudes that could reproduce the shape of the RF PADs using the following method to systematically vary the model parameters. From a starting set of initial partial wave parameters (amplitudes and phases), the downhill simplex method [180] was employed to adjust the sum of differences between the model and experimental $\mathcal{B}_L$ coefficients. This optimization process adjusted the parameters such that the agreement between model and experimental B_L coefficients was better than the experimental uncertainty. This optimization process was carried out in three stages: (1) only s_σ, p_σ, d_σ, and d_δ amplitudes and phases were optimized with all other parameters held constant; (2) s_σ, p_σ, d_σ, and d_δ amplitudes and phases were held constant at the optimized values found in the previous step; (3) all parameters were optimized, starting with the values found in the two previous steps. This process was carried out for 32 different sets of starting parameters using the same set of initial parameters for the five time delays.

In order to calculate the RF PAD for a set of partial wave amplitudes and phases we use Eq. (54) to first calculate the MF PAD. The MF is defined with the z axis along the C_{2v} symmetry axis, the y-axis along the N—N bond, and the x axis perpendicular to the molecular plane. The RF plane is defined with the z axis along the N—N bond direction. In order to calculate the RF PAD from the MF PAD, a rotation is applied to bring the MF z axis to the RF z axis. The resulting PAD is then azimuthally averaged about the z axis (the N—N direction),

$$I(\theta) = \int d\phi \sum_{LMM'} \beta^{M}_{LM} D^{L}_{MM'}(\pi/2, \pi/2, 0) Y_{LM'}(\theta, \phi) \tag{58}$$

Performing the integration over ϕ analytically yields

$$\mathcal{B}_L = 2\pi \sum_{M} \beta^{M}_{LM} Y^{*}_{LM}(\pi/2, 0) \tag{59}$$

In order to obviate the dependence of our conclusions upon any specific partial wave amplitude, the amplitudes were contracted into two sets: those expected from $3p_y$ ionization and those not. Ionization of a dimer $3p_y$ Rydberg state via a y-polarized transition would, in an "atomic" $\Delta l = \pm 1$ picture of Rydberg orbital ionization, produce only electrons with s_σ, d_σ, d_δ character. Therefore, the ratio of $[s_\sigma + d_\sigma + d_\delta]$ to the sum of all other contributions Σ_{pfg} is a measure of $3p_y$ Rydberg character in the $(NO)_2^*$ excited electronic states. In Fig. 28 (bottom), we plot the time dependence of this ratio, labeled the "$3p_y$ signal", showing that dimer $3p_y$ Rydberg character rises from early times, peaks at 330 fs, and then subsequently falls. The solid curve is the time dependence of the intermediate configuration extracted from Fig. 26, showing that the $3p_y$ character follows the time behavior of the intermediate $(NO)_2^{*\dagger}$ configuration. The agreement substantiates $(NO)_2^{*\dagger}$ as being of $3p_y$ character.

Ab initio studies fully support this picture [175]. Briefly, a very bright diabatic charge-transfer (valence) state carries the transition oscillator strength. At 210 nm, a vibrationally excited (roughly estimated, $v_1 \sim 4$) adiabatic $(NO)_2^*$ state of mixed charge-transfer/Rydberg character is populated. This quickly evolves, via N=O stretch dynamics, toward increasing $3p_y$ Rydberg character, forming the $(NO)_2^{*\dagger}$ state. The 140 fs initial decay constant is the time scale for the initial valence state to develop intermediate $3p_y$ character and explains the emergence of $3p_y$ ionization dynamics seen in Fig. 28 at intermediate time scales ($(NO)_2^{*\dagger}$). The 590 fs sequential time constant is the time scale for evolution of the dimer $3p_y$ configuration to free products via IVR, coupling the N=O stretch to the low frequency N–N stretch and other modes. Due to photofragment indistinguishability, the dimer $3p_y$ state correlates adiabatically to free NO(A, 3*s*) + NO(X) products without any curve crossings. With respect to the 3*s* Rydberg state, a dimer A_1 Rydberg 3*s* state was indeed found, but at lower energy than the bright valence state and does not cross the latter in the FC region [175]. It is therefore likely that the dimer 3*s* state does not participate in the dissociation dynamics except perhaps far out in the exit valley where the dimer 3*s* and 3*p* states become degenerate and strongly mix.

F. Photostability of the DNA Bases

The UV photostability of biomolecules is determined by the competition between ultrafast excited-state electronic relaxation processes. Some of these, such as excited-state reaction, photodissociation, or triplet formation, can be destructive to the molecule. In order to protect against these, nature designed mechanisms that convert dangerous electronic energy to less dangerous vibrational energy. However, in order to have nonzero efficiency, any such protective mechanisms must operate on ultrafast time scales in order to dominate over competing photochemical mechanisms that potentially lead to destruction of the biomolecule. In DNA, the nucleic bases are not only the building blocks of genetic material, but are also the UV chromophores of the double helix. It was suggested that DNA must have photoprotective mechanisms that rapidly convert dangerous electronic energy into heat [181].

The purine bases adenine and guanine and the pyrimidine bases cytosine, thymine, and uracil are all heterocycles. They typically have strong $\pi\pi^*$ UV absorption bands and, due to the lone electron pairs on the heteroatoms, have additional low lying $n\pi^*$ transitions. Furthermore, for some bases $\pi\sigma^*$ states are also in a similarly low energy range. This can lead to rather complex photophysical properties. Of all the bases, adenine has been most extensively studied [181]. In the gas phase, the 9H tautomer of isolated adenine is the lowest energy and most abundant form. Two competing models were proposed to explain the photophysics of isolated adenine involving these low lying states. One predicted internal conversion from the initially excited $\pi\pi^*$ state to the lower $n\pi^*$ state along a coordinate involving six-membered ring puckering [182]. This would be followed by further out-of-plane distortion, initiating relaxation back to the S_0 ground state. An alternate model suggested that along the 9-N H-stretch coordinate a two-step relaxation pathway evolves via conical intersections of the $\pi\pi^*$ state with a repulsive $\pi\sigma^*$ state, followed by decay back to the S_0 ground state [183]. Due to the repulsive character of the $\pi\sigma^*$ state, this mechanism was suggested to be highly efficient. More recently, various other possible relaxation pathways have been suggested [184–186]. The relative importance of the electronic relaxation channels in adenine has been a matter of some debate.

A time-resolved ion yield study of the adenine excited-state dynamics yielded an excited-state lifetime of $\sim$1 ps and seemed to support the model of internal conversion via the $n\pi^*$ state along a coordinate involving six-membered ring puckering [187]. In order to determine the global importance of the $\pi\sigma^*$ channel, a comparison of the primary photophysics of adenine with 9-methyl adenine will be useful, as the latter lacks a $\pi\sigma^*$ channel at the excitation energies of concern here. The first study of this type revealed no apparent changes in excited-state lifetime upon methylation at the N9 position [188]: a lifetime of $\sim$1 ps was observed for both adenine and 9-methyl adenine. This was interpreted as evidence that the $\pi\sigma^*$ is not involved in adenine electronic relaxation.

By contrast, the first TRPES studies compared adenine electronic relaxation dynamics at two different wavelengths, 266 versus 250 nm, and concluded that the $\pi\sigma^*$ state may indeed be important [189, 190]. Additional evidence of $\pi\sigma^*$ state participation obtained from H-atom loss experiments [191, 192]. Hydrogen atom detection is highly sensitive and can reveal even minor H-atom loss channels. The observation of fast hydrogen atoms following UV excitation of adenine is a compelling argument for the $\pi\sigma^*$ state: fast H atoms result from an excited-state potential that is repulsive in the N9H coordinate. Although this shows that a $\pi\sigma^*$ channel exists, it might play only a minor role since the H-atom quantum yield remains unknown. A more detailed time-resolved ion yield study comparing adenine with 9-methyl adenine photophysics revealed further insights [193]. The excited-state decay dynamics of adenine at 266 nm excitation required a biexponential fit using two time constants: a fast component decaying in 0.1 ps followed by a slower component with a 1.1 ps lifetime. Interestingly, 9-methyl adenine also exhibited that same two time

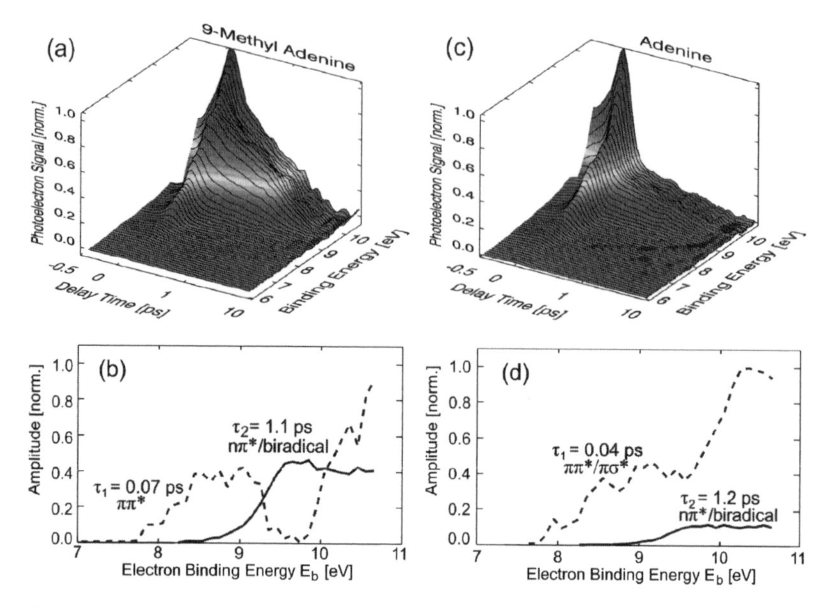

Figure 29. The TRPES spectra for adenine (left) and 9-methyl adenine (right), pumped at $\lambda_{\text{pump}} = 267\,\text{nm}$ and and probed at $\lambda_{\text{probe}} = 200\,\text{nm}$. The time dependence is plotted using a linear/logarithmic scale with a linear scale in the region -0.4–$1.0\,\text{ps}$ and a logarithmic scale for delay times 1.0–10.0 ps. See color insert.

constants of 0.1 and 1.1 ps. This again led to the suggestion that the $\pi\sigma^*$ state was not strongly involved in the dynamics, supporting the earlier ion yield experiments, but contradicting the TRPES results.

More recently, a new TRPES study compared adenine with 9-methyl adenine [194], as shown in Fig. [29]. The behavior of the two molecules appears quite similar, but there are important differences, as discussed below. Both molecules exhibit a broad spectral feature that covers the 7.5–10.8 eV electron-binding energy (E_b) range. This feature, peaking toward 10.8 eV, decays quickly and, beyond 500 fs, transforms into a second spectral feature spanning the 8.5–10.8 eV (E_b) range. This second spectrum grows smoothly between 8.5–9.6 eV and is flat between 9.6–10.8 eV. This feature decays more slowly, in 3 ps. Beyond $\sim$ 6 ps, no remaining photoelectron signal was observed. Global 2D nonlinear fitting algorithms determined that two exponential time constants were needed to fit these data. For 9-methyl adenine, these were $\tau_1 = 70 \pm 25\,\text{fs}$ and $\tau_2 = 1.1 \pm 0.1\,\text{ps}$. For adenine, these were $\tau_1 = 40 \pm 20\,\text{fs}$ and $\tau_2 = 1.2 \pm 0.2\,\text{ps}$. Note that the two time constants for these molecules are the same within errors and agree quantitatively with the two time constants previously reported in the ion yield experiments [193].

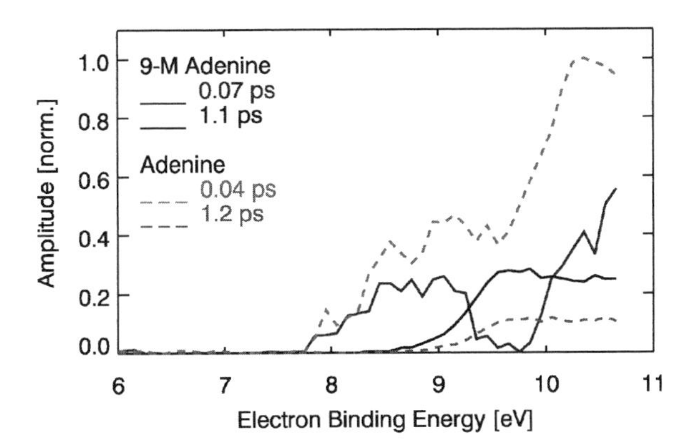

Figure 30. Decay associated spectra for adenine (dashed lines) and 9-methyl adenine (solid lines), extracted from the 2D TRPES spectra using global fitting procedures. Both molecules were fit by the same two time constants: $\tau_1 \leq 0.1$ and $\tau_2 \sim 1.1$ ps, agreeing quantitatively with previous results. The spectra, however, are very different for adenine as compared to 9-methyl adenine. For details, see the text. See color insert.

Although the time constants for adenine and 9-methyl adenine are very similar, the associated photoelectron spectra reveal important differences that are obscured in ion-yield measurements. The decay associated spectra obtained from the fitting algorithm are shown in Fig. 30. The spectra of the fast (<0.1 ps) components are shown for adenine (dashed green) and 9-methyl adenine (solid blue). Likewise, the spectra of the 1.1 ps components for adenine (dashed red) and 9-methyl adenine (solid black) are given. The electronic states of the cations are $D_0(\pi^{-1})$, $D_1(n^{-1})$, and D_2 (π^{-1}). The expected Koopmans' correlations would therefore be: $\pi\pi^* \rightarrow D_0(\pi^{-1})$, $D_2(\pi^{-1})$, and $n\pi^* \rightarrow D_1(n^{-1})$. As detailed elsewhere [194], the spectra of the 1.1 ps component correspond to the $n\pi^* \rightarrow D_1$ $(n^{-1}) + e^-$ ionizing transitions. Although the form of the $n\pi^*$ spectra are similar, the yield (amplitude) of $n\pi^*$ state is considerably reduced in adenine as compare to 9-methyl adenine. The most significant difference lies in the form of the spectra of the short-lived 0.1 ps component: the spectrum of 9-methyl adenine (blue solid) appears as two lobes with a gap in between whereas the spectrum of adenine (dashed green) appears as a broad spectrum without a gap.

In Figure 31, we compare the associated spectrum of the fast component in 9-methyl adenine with calculated [194] Franck–Condon structures for the $\pi\pi^* \rightarrow D_0(\pi^{-1}) + e^-$ (solid line) and $\pi\pi^* \rightarrow D_2(\pi^{-1}) + e^-$ (dash–dotted line) ionizing transitions. The two separated peaks agree well with the FC calculations, strongly suggesting that the short-lived state in 9-methyl adenine is the $\pi\pi^*$ state. By contrast, adenine contains an additional contribution that

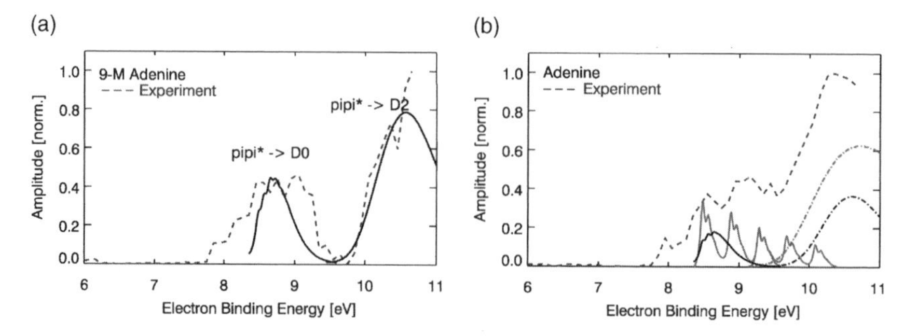

Figure 31. Decay associated spectra of the short-lived state compared with calculated FC spectra for 9-methyl adenine (a) and adenine (b). In 9-methyl adenine, the $\pi\pi^* \rightarrow D_0(\pi^{-1}), D_2(\pi^{-1})$ transitions leave a FC gap. In adenine, this gap is filled by the $\pi\sigma^*$ ionizing transitions. See color insert.

fills in the gap between the $\pi\pi^* \rightarrow D_0(\pi^{-1}) + e^-$ (black solid) and $\pi\pi^* \rightarrow D_2(\pi^{-1}) + e^-$ (black dotted) transitions. This gap is filled from the left by transitions due to $\pi\sigma^* \rightarrow D_0(\pi^{-1}) + e^-$ (green solid) and from the right by $\pi\sigma^* \rightarrow D_2(\pi^{-1}) + e^-$ (green dashed) ionizing transitions. These calculated FC structures provide strong evidence that the $\pi\sigma^*$ is present in adenine, but absent in 9-methyl adenine [194]. Adenine has two fast relaxation channels from the $\pi\pi^*$ state, whereas 9-methyl adenine has only one. This also explains why the yield (amplitude) of $n\pi^*$ state is reduced in adenine as compare to 9-methyl adenine. The fact that the two fast relaxation channels in adenine have very similar time constants is the reason why the ion yield experiments showed no apparent difference in lifetimes between adenine and 9-methyl adenine. Once again, the importance of measuring (dispersed) photoelectron spectra as opposed to (integrated) ion yield spectra is apparent.

VII. CONCLUSION

Our goals were to elucidate important physical concepts in energy-angle resolved TRPES and to illustrate the range of its applicability to problems in molecular dynamics. We discussed general aspects of femtosecond pump–probe experiments from both the wave packet and the frequency domain point of view. Experimentalists are, in principle, free to choose a final state in which to observe the wave packet dynamics of interest. We emphasized the critical role of the choice of the final state in determining both the experimental technique (e.g., collection of photons or particles) and the information content of an experiment (averaged or state-resolved). The molecular ionization continuum has a rich structure that can act as a template onto which multidimensional wave packet dynamics may be projected. The set of electronic states of the cation are sensitive to both the electronic population dynamics and the vibrational dynamics in the excited state, whereas the free electron continua are sensitive to the electronic population dynamics and the molecular frame alignment dynamics. In sum, TRPES and its variants are well suited to the study of excited-state polyatomic dynamics because of their sensitivity to both electronic configurations and vibrational dynamics, the universal nature of photoionization as a probe, and the dispersed (energy- and angle-resolved) nature of the measurement.

A powerful variant, TRCIS, measures energy-resolved and 3D angle-resolved photoions and photoelectrons in coincidence, yielding unprecedented details about complex molecular photodissociations. However, TRCIS has potential beyond the ability to observe time-resolved molecular frame excited-state dynamics. For example, in even more complex dissociation problems, it may be very difficult to "follow" the excited-state dynamics all the way from initial excitation to final product emission. In such cases, one is tempted to resort to statistical models of the dynamics, such as phase-space theory. The TRCIS provides a new opportunity to follow the time evolution of the product states distributions. For example, product attributes, such as photofragment kinetic energy and angular distributions, photofragment angular momentum polarization, and $\mu - v - J$ correlations may all now be measured as a function of time. We expect that the time evolution of these will be related to the divergence of phase space flux during dissociation and may well provide new insights into the time scales for the onset of and the extent of statisticality in energized molecules.

Future applications of TRPES and its variants will undoubtedly benefit from ongoing developments in detector technologies, femtosecond, and attosecond laser sources, nonlinear optical frequency conversion schemes, and developments in free electron lasers and fourth generation synchrotron light sources. TRPES research will include molecular-frame measurements, photofragment–photoelectron scalar and vector correlations, extreme time scales, and inner-shell dynamics. The use of shaped, intense nonresonant laser fields to create field-free alignment in polyatomic systems [101–117] will combine with TRPES and TRCIS to help probe molecular-frame dynamics. Further development of the multiply differential photoelectron–photofragment coincidence and coincidence-imaging methods will permit highly detailed investigation of statistical and nonstatistical photoinduced charge and energy flow, an area of fundamental dynamical interest and of interest in applications to the gas-phase photophysics of biomolecules. The development of high average power femtosecond VUV–XUV sources and the dawn of attosecond science present the possibility of probing highly excited states, core dynamics, and electron correlation in real time. Equally important are ongoing theoretical developments in *ab initio* molecular dynamics methods for studying nonadiabatic processes in polyatomic molecules (see, e.g., Ref. [195] and references cited therein). New methods for calculating photoionization differential cross-sections (see, e.g., Ref. [26] and references cited therein) will play an increasingly important role in the future of TRPES. These experimental and theoretical challenges will, we expect, be met by many researchers, surely leading to exciting new developments in the dynamics of polyatomic molecules.

Acknowledgments

We thank our co-workers and collaborators who have contributed both materially and intellectually to the work described here: C. Bisgaard, V. Blanchet, A. Boguslavskiy, A.L.L. East, N. Gador, O. Gessner, C.C. Hayden, A. Krylov, A.M.D. Lee, S. Lochbrunner, T.J. Martinez, K.L. Reid, H. Reisler, H. Satzger, M. Schmitt, T. Schultz, T. Seideman, J.P. Shaffer, D. Townsend, S. Ullrich and M.Z. Zgierski. We thank T. Suzuki and K.L Reid for permission to use figure 18 and figures 15,16,17, respectively.

APPENDIX A DERIVATION OF EQ. (43)

Expanding Eq. (25) and substituting in Eqs. (32), (39)–(41) yields

$$
\begin{aligned}
\sigma(\varepsilon,\hat{\boldsymbol{k}}_{\mathrm{L}};t) \propto & \frac{1}{64\pi^2}\sum_{n_\alpha n'_{\alpha'}}\sum_{n_{\alpha_+}}\sum_{K_{\alpha_+}M_{\alpha_+}}\sum_{lm}\sum_{l'm'}\sum_{\lambda\lambda'}\sum_{K_\alpha M_\alpha}\sum_{K'_{\alpha'}M'_{\alpha'}}\sum_{KQ}\sum_{pp'}\sum_{qq'}\sum_{j_t j'_t}\sum_{k_t k'_t}\sum_{m_t m'_t} \\
& \times(-1)^{J_\alpha+q+q'+M_\alpha-2K_\alpha}[j_t,j'_t,J_{\alpha_+}][K,J_\alpha,J'_{\alpha'}]^{1/2} \\
& \times\begin{pmatrix} J_{\alpha_+} & J_\alpha & j_t \\ -M_{\alpha_+} & M_\alpha & m_t \end{pmatrix}\begin{pmatrix} J_{\alpha_+} & J'_{\alpha'} & j'_t \\ -M_{\alpha_+} & M'_{\alpha'} & m'_t \end{pmatrix}\begin{pmatrix} l & 1 & j_t \\ m & -p & m_t \end{pmatrix}\begin{pmatrix} l' & 1 & j'_t \\ m' & -p' & m'_t \end{pmatrix} \\
& \times\begin{pmatrix} J_{\alpha_+} & J_\alpha & j_t \\ -K_{\alpha_+} & K_\alpha & k_t \end{pmatrix}\begin{pmatrix} J_{\alpha_+} & J'_{\alpha'} & j'_t \\ -K_{\alpha_+} & K'_\alpha & k'_t \end{pmatrix}\begin{pmatrix} l & 1 & j_t \\ \lambda & -q & k_t \end{pmatrix}\begin{pmatrix} l' & 1 & j'_t \\ \lambda' & -q' & k'_t \end{pmatrix} \\
& \times\begin{pmatrix} J_\alpha & J'_{\alpha'} & K \\ M_\alpha & -M'_{\alpha'} & -Q \end{pmatrix} Y_{lm}(\hat{\boldsymbol{k}})Y^*_{l'm'}(\hat{\boldsymbol{k}})\langle T(n_\alpha,n'_{\alpha'};t)^\dagger_{KQ}\rangle e_{-p}e^*_{-p'} \\
& \times a^{J_\alpha\tau_\alpha}_{K_\alpha}a^{J'_{\alpha'}\tau'_{\alpha'}}_{K'_{\alpha'}}\left|a^{J_{\alpha_+}\tau_{\alpha_+}}_{K_{\alpha_+}}\right|^2\mathcal{E}(n_{\alpha_+},n_\alpha,\epsilon)\mathcal{E}^*(n_{a_+},n'_{\alpha'},\epsilon) \\
& \times\sum_{\Gamma\mu h}\sum_{\Gamma'\mu'h'} b^{\Gamma\mu}_{hl\lambda}b^{\Gamma'\mu'*}_{h'l'\lambda'}(-\mathrm{i})^{l-l'}e^{\mathrm{i}(\sigma_l(\epsilon)-\sigma_{l'}(\epsilon))}D^{\alpha v_\alpha\alpha_+v_{\alpha_+}}_{\Gamma\mu hl}(q)D^{\alpha'v'_{\alpha'}\alpha_+v_{\alpha_+}*}_{\Gamma'\mu'h'l'}(q')
\end{aligned}
\tag{A.1}
$$

The various angular momentum algebraic manipulations outlined below draw on the text by Zare [40]. The two spherical harmonics in Eq. (A.1) may be combined using the Clebsch–Gordan series,

$$
Y_{lm}(\hat{\boldsymbol{k}}_{\mathrm{L}})Y^*_{l'm'}(\hat{\boldsymbol{k}}_{\mathrm{L}}) = \sqrt{\frac{[l,l']}{4\pi}}(-1)^m\sum_L[L]^{1/2}\begin{pmatrix} l & l' & L \\ -m & m' & M \end{pmatrix}\begin{pmatrix} l & l' & L \\ 0 & 0 & 0 \end{pmatrix}Y_{LM}(\hat{\boldsymbol{k}}_{\mathrm{L}}) \tag{A.2}
$$

Equation (4.16) of Zare [40] is used to perform the following manipulations,

$$\begin{pmatrix} J_{\alpha_+} & J_\alpha & j_t \\ -M_{\alpha_+} & M_\alpha & m_t \end{pmatrix}\begin{pmatrix} J_{\alpha_+} & J'_{\alpha'} & j'_t \\ -M_{\alpha_+} & M'_{\alpha'} & m'_t \end{pmatrix} = \sum_X [X](-1)^{J+j_t-J_{\alpha_+}+j'_t+J'_{\alpha'}+X-M_\alpha+m'_t} \\ \times \begin{Bmatrix} J_\alpha & j_t & J_{\alpha_+} \\ j'_t & J'_{\alpha'} & X \end{Bmatrix}\begin{pmatrix} J'_{\alpha'} & J_\alpha & X \\ -M'_{\alpha'} & M_\alpha & x \end{pmatrix}\begin{pmatrix} j_t & j'_t & X \\ m_t & -m'_t & -x \end{pmatrix} \tag{A.3}$$

$$\begin{pmatrix} l & 1 & j_t \\ m & -p & m_t \end{pmatrix}\begin{pmatrix} j_t & j'_t & X \\ m_t & -m'_t & -x \end{pmatrix} = \sum_Y [Y](-1)^{l+1+2j'_t+2X+Y-m-m'_t} \\ \times \begin{Bmatrix} l & 1 & j_t \\ j'_t & X & Y \end{Bmatrix}\begin{pmatrix} X & l & Y \\ x & m & y \end{pmatrix}\begin{pmatrix} 1 & j'_t & Y \\ -p & m'_t & -y \end{pmatrix} \tag{A.4}$$

$$\begin{pmatrix} l' & 1 & j'_t \\ m' & -p' & m'_t \end{pmatrix}\begin{pmatrix} 1 & j'_t & Y \\ -p & -m'_t & -y \end{pmatrix} = \sum_P [P](-1)^{l'-j'_t+Y+P-m'-p} \\ \times \begin{Bmatrix} l' & 1 & j'_t \\ 1 & Y & P \end{Bmatrix}\begin{pmatrix} Y & l' & P \\ y & m' & p-p' \end{pmatrix}\begin{pmatrix} 1 & 1 & P \\ -p' & p & p'-p \end{pmatrix} \tag{A.5}$$

$$\begin{pmatrix} Y & l' & P \\ y & m' & p-p' \end{pmatrix}\begin{pmatrix} X & l & Y \\ x & m & y \end{pmatrix} = \sum_G [G](-1)^{X+l+G+m'-x} \\ \times \begin{Bmatrix} l' & P & Y \\ X & l & G \end{Bmatrix}\begin{pmatrix} l & l' & G \\ m & -m' & g \end{pmatrix}\begin{pmatrix} P & X & G \\ p'-p & x & -g \end{pmatrix} \tag{A.6}$$

The orthogonality of the Wigner $3j$ symbols is then used to perform the summations over m, m', M_α, and $M'_{\alpha'}$,

$$\sum_{mm'}\begin{pmatrix} l & l' & G \\ m & -m' & g \end{pmatrix}\begin{pmatrix} l & l' & L \\ -m & m' & M \end{pmatrix} = (-1)^{l+l'+L}[L]^{-1}\delta_{LG}\delta_{-Mg}, \tag{A.7}$$

$$\sum_{M_\alpha M'_{\alpha'}}\begin{pmatrix} J'_{\alpha'} & J_\alpha & X \\ -M'_{\alpha'} & M_\alpha & x \end{pmatrix}\begin{pmatrix} J_\alpha & J'_{\alpha'} & K \\ M_\alpha & -M'_{\alpha'} & -Q \end{pmatrix} = (-1)^{J_\alpha+J'_{\alpha'}+K}[K]^{-1}\delta_{KX}\delta_{-Qx} \tag{A.8}$$

The sum over Y is carried out analytically using the following identity relating the Wigner $9j$ symbol to Wigner $6j$ symbols:

$$\sum_Y (-1)^{2Y}[Y]\begin{Bmatrix} l' & 1 & j'_t \\ 1 & Y & P \end{Bmatrix}\begin{Bmatrix} l' & P & Y \\ K & l & L \end{Bmatrix}\begin{Bmatrix} l & 1 & j_t \\ j'_t & K & Y \end{Bmatrix} = \begin{Bmatrix} 1 & 1 & P \\ j_t & j'_t & K \\ l & l' & L \end{Bmatrix} \tag{A.9}$$

APPENDIX B DERIVATION OF EQ. (45)

Equation (4.16) of Zare [40] is used to perform the following manipulations,

$$\begin{pmatrix} J_{\alpha_+} & J_\alpha & j_t \\ -K_{\alpha_+} & K_\alpha & k_t \end{pmatrix}\begin{pmatrix} J_{\alpha_+} & J'_{\alpha'} & j'_t \\ -K_{\alpha_+} & K'_{\alpha'} & k'_t \end{pmatrix} = \sum_R [R](-1)^{J_\alpha + j_t - J'_{\alpha'} + j'_t + J'_{\alpha'} + R - K_\alpha + k'_t} \times \begin{Bmatrix} J_\alpha & j_t & J_{\alpha_+} \\ j'_t & J'_{\alpha'} & R \end{Bmatrix}\begin{pmatrix} J'_{\alpha'} & J_\alpha & R \\ -K'_{\alpha'} & K_\alpha & r \end{pmatrix}\begin{pmatrix} j_t & j'_t & R \\ k_t & -k'_t & -r \end{pmatrix} \tag{B.1}$$

$$\begin{pmatrix} l & 1 & j_t \\ \lambda & -q & k_t \end{pmatrix}\begin{pmatrix} j_t & j'_t & R \\ k_t & -k'_t & -r \end{pmatrix} = \sum_S [S](-1)^{l+1-j_t+R+j'_t+S-\lambda-r} \times \begin{Bmatrix} l & 1 & j_t \\ R & j'_t & S \end{Bmatrix}\begin{pmatrix} j'_t & l & S \\ k'_t & \lambda & s \end{pmatrix}\begin{pmatrix} 1 & R & S \\ -q & r & -s \end{pmatrix} \tag{B.2}$$

$$\begin{pmatrix} l' & 1 & j'_t \\ \lambda' & -q' & k'_t \end{pmatrix}\begin{pmatrix} j'_t & l & S \\ k'_t & \lambda & s \end{pmatrix} = \sum_T [T](-1)^{l'+1-j'_t+S+l+T-\lambda'+s} \times \begin{Bmatrix} l' & 1 & j'_t \\ S & l & T \end{Bmatrix}\begin{pmatrix} l & l' & T \\ -\lambda & \lambda' & t \end{pmatrix}\begin{pmatrix} 1 & S & T \\ -q' & -s & -t \end{pmatrix} \tag{B.3}$$

$$\begin{pmatrix} 1 & R & S \\ -q & r & -s \end{pmatrix}\begin{pmatrix} 1 & S & T \\ -q' & -s & -t \end{pmatrix} = \sum_U [U](-1)^{R+2T+1+U+q-t} \times \begin{Bmatrix} 1 & R & S \\ T & 1 & U \end{Bmatrix}\begin{pmatrix} 1 & 1 & U \\ q' & -q & q-q' \end{pmatrix}\begin{pmatrix} R & T & U \\ r & t & q'-q \end{pmatrix} \tag{B.4}$$

The sum over S can be carried out analytically by relating the Wigner $6j$ symbols to the Wigner $9j$ symbol,

$$\sum_S (-1)^{2S}[S]\begin{Bmatrix} l & 1 & j_t \\ R & j'_t & S \end{Bmatrix}\begin{Bmatrix} l' & 1 & j'_t \\ S & l & T \end{Bmatrix}\begin{Bmatrix} 1 & R & S \\ T & 1 & U \end{Bmatrix} = \begin{Bmatrix} 1 & 1 & U \\ l & l' & \mathrm{T} \\ j_t & j'_t & R \end{Bmatrix} \tag{B.5}$$

The summation over J_{α_+} is then completed using the orthogonality of the Wigner $6j$ symbols,

$$\sum_{J_{\alpha_+}} [J_{\alpha_+}, R]\begin{Bmatrix} J_\alpha & j_t & j_{\alpha_+} \\ j'_t & J'_{\alpha'} & R \end{Bmatrix}\begin{Bmatrix} J_\alpha & j_t & J_{\alpha_+} \\ j'_t & J'_{\alpha'} & K \end{Bmatrix} = \delta_{RK} \tag{B.6}$$

together with the fact that $\sum_{J_{\alpha_+}} |a_{K_{\alpha_+}}^{J_{\alpha_+}\tau_{\alpha_+}}|^2 = 1$. Rearranging the Wigner 9j symbol in Eq. (42),

$$\begin{Bmatrix} 1 & 1 & P \\ j_t & j_t' & K \\ l & l' & L \end{Bmatrix} = (-1)^{l+l'+L+P+j_t+j_t'+K} \begin{Bmatrix} 1 & 1 & P \\ l & l' & L \\ j_t & j_t' & K \end{Bmatrix} \tag{B.7}$$

allows the use of the orthogonality of the Wigner 9j symbols to remove the summation over j_t and j_t',

$$\sum_{j_t j_t'} [j_t, j_t', L, P] \begin{Bmatrix} 1 & 1 & P \\ l & l' & L \\ j_t & j_t' & K \end{Bmatrix} \begin{Bmatrix} 1 & 1 & U \\ l & l' & T \\ j_t & j_t' & K \end{Bmatrix} = \delta_{PU}\delta_{LT} \tag{B.8}$$

APPENDIX C DERIVATION OF EQ. (53)

Substitution of Eqs. (21), (52), and (51) into Eq. (50) yields

$$\begin{aligned}\sigma(\epsilon, \hat{\boldsymbol{k}}_M; t) \propto & \sum_{pp'}\sum_{qq'}\sum_{ll'}\sum_{\lambda\lambda'} (-1)^{q+q'} Y_{l\lambda}(\hat{\boldsymbol{k}}) Y^*_{l'\lambda'}(\hat{\boldsymbol{k}}) D^1_{-p-q}(\phi,\theta,\chi) D^{1*}_{-p'-q'}(\phi,\theta,\chi) \\ & \times e_{-p} e^*_{-p'} (-\mathrm{i})^{l-l'} e^{\mathrm{i}(\sigma_l(\epsilon)-\sigma_{l'}(\epsilon))} \sum_{\alpha v_\alpha}\sum_{\alpha' v'_{\alpha'}} C_{\alpha v_\alpha}(t) C^*_{\alpha' v'_{\alpha'}}(t) \\ & \times \sum_{\Gamma\mu h}\sum_{\Gamma'\mu' h'} b^{\Gamma\mu}_{hl\lambda} b^{\Gamma'\mu' *}_{h'l'\lambda'} \sum_{\alpha_+ v_{\alpha_+}} D^{\alpha v_\alpha \alpha_+ v_{\alpha_+}}_{\Gamma\mu hl}(q) D^{\alpha' v'_{\alpha'} \alpha_+ v_{\alpha_+} *}_{\Gamma'\mu' h' l'}(q') \\ & \times \mathcal{E}(\alpha, v_\alpha, \alpha_+, v_{\alpha_+}, \epsilon)\mathcal{E}^*(\alpha', v'_{\alpha'}, \alpha_+, v_{\alpha_+}, \epsilon)\end{aligned} \tag{C.1}$$

This equation may be simplified using the Clebsch–Gordan series,

$$Y_{l\lambda}(\hat{\boldsymbol{k}}_M) Y^*_{l'\lambda'}(\hat{\boldsymbol{k}}_M) = \sqrt{\frac{[l,l']}{4\pi}} (-1)^{\lambda} \sum_L [L]^{1/2} \begin{pmatrix} l & l' & L \\ -\lambda & \lambda' & M \end{pmatrix} \begin{pmatrix} l & l' & L \\ 0 & 0 & 0 \end{pmatrix} Y_{LM}(\hat{\boldsymbol{k}}_M) \tag{C.2}$$

$$\begin{aligned}D^1_{-p-q}(\phi,\theta,\chi) D^{1*}_{-p'-q'}(\phi,\theta,\chi) = & (-1)^{p+q} \sum_P [P] \begin{pmatrix} 1 & 1 & P \\ p & -p' & p'-p \end{pmatrix} \\ & \times \begin{pmatrix} 1 & 1 & P \\ q & -q' & q'-q \end{pmatrix} D^P_{p'-p, q'-q}(\phi,\theta,\chi)\end{aligned} \tag{C.3}$$

References

1. M. Bixon and J. Jortner, *J. Chem. Phys.* **48**, 715 (1968).
2. J. Jortner, S. A. Rice, and R. M. Hochstrasser, *Adv. Photochem.* **7**, 149 (1969).
3. S. R. Henry and W. Siebrand, in J. B. Birks (ed.), *Organic Molecular Photophysics*, John Wiley & Sons, Inc., London, 1973, vol. 1, p. 152.
4. K. F. Freed, in F. K. Fong (ed.), *Radiationless Processes in Molecules and Condensed Phases*, Springer-Verlag, Berlin, 1976, p. 23.
5. G. Stock and W. Domcke, *Adv. Chem. Phys.* **100**, 1 (1997).
6. G. A. Worth and L. S. Cederbaum, *Annu. Rev. Phys. Chem.* **55**, 127 (2004).
7. M. Klessinger and J. Michl, *Excited States and Photochemistry of Organic Molecules*, VCH, New York, 1994.
8. H. Köppel, W. Domcke, and L. S. Cederbaum, *Adv. Chem. Phys.* **57**, 59 (1984).
9. J. Michl and V. Bonacic-Koutecky, *Electronic Aspects of Organic Photochemistry*, John Wiley & Sons, Inc., New York, 1990.
10. R. W. Schoenlein, L. A. Peteanu, R. A. Mathies, and C. V. Shank, *Science* **254**, 412 (1991).
11. J. Jortner and M. A. Ratner, *Molecular Electronics*, IUPAC, Blackwell, Oxford, 1997.
12. A. H. Zewail, *J. Phys. Chem. A* **104**, 5660 (2000).
13. J. H. D. Eland, *Photoelectron Spectroscopy*, 2nd ed., Butterworths, London, 1984.
14. I. Fischer, D. M. Villeneuve, M. J. J. Vrakking, and A. Stolow, *J. Chem. Phys.* **102**, 5566 (1995).
15. I. Fischer, M. Vrakking, D. Villeneuve, and A. Stolow, in M. Chergui (ed.), *Femtosecond Chemistry*, World Scientific, Singapore, 1996.
16. A. Stolow, *Philos. Trans. R. Soc. (London) A* **356**, 345 (1998).
17. C. C. Hayden and A. Stolow, in C.-Y. Ng (ed.), *Advanced Physical Chemistry*, World Scientific, Singapore, 2000, vol. 10.
18. W. Radloff, in C.-Y. Ng (ed.), *Advanced Physical Chemistry*, World Scientific, Singapore, 2000, vol. 10.
19. K. Takatsuka, Y. Arasaki, K. Wang, and V. McKoy, *Faraday Discus.* 1 (2000).
20. D. M. Neumark, *Annu. Rev. Phys. Chem.* **52**, 255 (2001).
21. T. Suzuki and B. J. Whitaker, *Int. Rev. Phys. Chem.* **20**, 313 (2001).
22. T. Seideman, *Annu. Rev. Phys. Chem.* **53**, 41 (2002).
23. K. L. Reid, *Annu. Rev. Phys. Chem.* **54**, 397 (2003).
24. A. Stolow, *Annu. Rev. Phys. Chem.* **54**, 89 (2003).
25. A. Stolow, *Inter. Rev. Phys. Chem.* **22**, 377 (2003).
26. T. Suzuki, Y. Seideman, and M. Stener, *J. Chem. Phys.* **120**, 1172 (2004).
27. V. Wollenhaupt, M. Engel, and T. Baumert, *Annu. Rev. Phys. Chem.* **56**, 25 (2005).
28. T. Suzuki, *Annu. Rev. Phys. Chem.* **57**, 555 (2006).
29. I. V. Hertel and W. Radloff, *Rep. Prog. Phys.* **69**, 1897 (2006).
30. A. Stolow, A. Bragg, and D. Neumark, *Chem. Rev.* **104**, 1719 (2004).
31. M. Born and K. Huang, *Dynamical Theory of Crystal Lattices*, Oxford University Press, London, 1954.
32. A. Raab, G. Worth, H. D. Meyer, and L. S. Cederbaum, *J. Chem. Phys.* **110**, 936 (1999).
33. C. Woywod, W. Domcke, A. L. Sobolewski, and H. J. Werner, *J. Chem. Phys.* **100**, 1400 (1994).

34. L. S. Cederbaum, W. Domcke, and H. Köppel, *Int. J. Quant. Chem.* **15**, 251 (1981).
35. L. S. Cederbaum, W. Domcke, H. Köppel, and W. von Niessen, *Chem. Phys.* **26**, 169 (1977).
36. E. B. Wilson, J. C. Decius, and P. C. Cross, *Molecular Vibrations. The Theory of Infrared and Raman Vibrational Spectra*, McGraw-Hill Book Company Inc., New York, 1955.
37. B. H. Bransden and C. J. Joachain, *Physics of Atoms and Molecules*, 2nd ed., Prentice Hall, Harlow, England, 2003.
38. P. R. Bunker and P. Jensen, *Molecular Symmetry and Spectroscopy*, 2nd ed. NRC Research Press, Ottawa, Canada, 1998.
39. H. W. Kroto, *Molecular Rotation Spectra*, John Wiley & Sons, Inc. New York, 1975.
40. R. N. Zare, *Angular Momentum: Understanding Spatial Aspects in Chemistry and Physics*, John Wiley & Sons Inc, New York, 1988.
41. T. Koopmans, *Physica* **1**, 104 (1933).
42. A. M. Ellis, M. Feher, and T. G. Wright, *Electronic and Photoelectron Spectroscopy. Fundamentals and Case Studies*, Cambridge University Press, Cambridge, UK, 2005.
43. V. Blanchet and A. Stolow, in T. Elsaesser, J. G. Fujimoto, D. A. Wiersma, and W. Zinth (eds.), *Ultrafast Phenomena XI*, Springer-Verlag, Berlin, 1998, vol. 63 of *Springer Series in Chemical Physics*, pp. 456–458.
44. V. Blanchet, M. Z. Zgierski, T. Seideman, and A. Stolow, *Nature London* **401**, 52 (1999).
45. V. Blanchet, S. Lochbrunner, M. Schmitt, J. P. Shaffer, J. J. Larsen, M. Z. Zgierski, T. Seideman, and A. Stolow, *Faraday Discuss.* 33 (2000).
46. K. Resch, V. Blanchet, A. Stolow, and T. Seideman, *J. Phys. Chem. A* **105**, 2756 (2001).
47. V. Blanchet, M. Z. Zgierski, and A. Stolow, *J. Chem. Phys.* **114**, 1194 (2001).
48. M. Seel and W. Domcke, *J. Chem. Phys.* **95**, 7806 (1991).
49. M. Seel and W. Domcke, *Chem. Phys.* **151**, 59 (1991).
50. M. Schmitt, S. Lochbrunner, J. P. Shaffer, J. J. Larsen, M. Z. Zgierski, and A. Stolow, *J. Chem. Phys.* **114**, 1206 (2001).
51. N. Chandra, *J. Phys. B* **20**, 3405 (1987).
52. P. G. Burke, N. Chandra, and F. A. Gianturco, *J. Phys. B.* **5**, 2212 (1972).
53. K. L. Reid and J. G. Underwood, *J. Chem. Phys.* **112**, 3643 (2000).
54. K. L. Reid, T. A. Field, M. Towrie, and P. Matousek, *J. Chem. Phys.* **111**, 1438 (1999).
55. J. G. Underwood and K. L. Reid, *J. Chem. Phys.* **113**, 1067 (2000).
56. S. C. Althorpe and T. Seideman, *J. Chem. Phys.* **113**, 7901 (2000).
57. T. Seideman and S. C. Althorpe, *J. Elec. Spec. Relat. Phenom.* **108**, 99 (2000).
58. S. C. Althorpe and T. Seideman, *J. Chem. Phys.* **110**, 147 (1999).
59. M. Tsubouchi, B. J. Whitaker, L. Wang, H. Kohguchi, and T. Suzuki, *Phys. Rev. Lett.* **86**, 4500 (2001).
60. Y. Suzuki, M. Stener, and T. Seideman, *Phys. Rev. Lett.* **89**, 233002 (2002).
61. T. Seideman, *Phys. Rev. A* **6404**, 042504 (2001).
62. T. Seideman, *J. Chem. Phys.* **107**, 7859 (1997).
63. J. Cooper and R. N. Zare, *J. Chem. Phys.* **48**, 942 (1968).
64. A. D. Buckingham, B. J. Orr, and J. M. Sichel, *Philos. Trans. R. Soc. London A.* **268**, 147 (1970).
65. S. J. Smith and G. Leuchs, *Adv. At. Mol. Phys.* **24**, 157 (1988).
66. H. Park and R. N. Zare, *J. Chem. Phys.* **104**, 4554 (1996).

67. U. Fano and A. R. P. Rau, *Atomic Collisions and Spectra*, Academic Press, Orlando, USA, 1986.
68. S. N. Dixit and V. McKoy, *J. Chem. Phys.* **82**, 3546 (1985).
69. F. C. V. D. Lage and H. A. Bethe, *Phys. Rev.* **71**, 612 (1947).
70. N. Chandra, *J. Chem. Phys.* **89**, 5987 (1988).
71. N. Chandra and M. Chakraborty, *J. Chem. Phys.* **95**, 6382 (1991).
72. N. Chandra and M. Chakraborty, *Eur. Phys. J. D* **2**, 253 (1998).
73. R. Conte, J. Raynal, and E. Soulié, *J. Math. Phys.* **25**, 1176 (1984), see also, references cited therein.
74. B. S. Tsukerblat, *Group Theory in Chemistry and Spectroscopy: A simple guide to advanced useage*, Academic Press, New York, 1994.
75. R. McWeeny, *Symmetry: An introduction to group theory and its applications*, Pergamon Press, 1963.
76. C. D. H. Chrishol, *Group Theoretical Techniques in Quantum Chemistry*, Academic Press, 1976.
77. R. Signorell and F. Merkt, *Mol. Phys.* **92**, 793 (1997).
78. N. Chandra, *J. Phys. B* **20**, 3417 (1987).
79. M. I. Al-Joboury and D. W. Turner, *J. Chem. Soc.* 5141 (1963).
80. D. C. Frost, C. A. McDowell, and D. A. Vroom, *Phys. Rev. Lett.* **15**, 612 (1965).
81. D. C. Frost, C. A. McDowell, and D. A. Vroom, *Proc. R. Soc. London A* **296**, 566 (1967).
82. D. W. Turner, *Nature (London)* **213**, 795 (1967).
83. G. R. Branton, D. C. Frost, T. Makita, C. A. McDowell, and I. A. Stenhouse, *Philas. Trans. R. Soc. London A* **268**, 77 (1970).
84. A. J. Blake, J. L. Bahr, J. H. Carver, and V. Kumar, *Philos. Trans. R. Soc. London A* **268**, 159 (1970).
85. D. W. Turner, *Philos. Trans. R. Soc. London. A* **268**, 7 (1970).
86. R. S. Berry, *J. Chem. Phys.* **45**, 1228 (1966).
87. K. Blum, *Density Matrix Theory and Applications (Physics of Atoms and Molecules)*, 2nd ed. Springer, 1996.
88. C. H. Greene and R. N. Zare, *Annu. Rev. Phys. Chem.* **33**, 119 (1982).
89. A. J. Orr-Ewing and R. N. Zare, *Annu. Rev. Phys. Chem.* **45**, 315 (1994).
90. D. J. Leahy, K. L. Reid and R. N. Zare, *Phys. Rev. Lett.* **68**, 3527 (1992).
91. S. Baskin, P. M. Felker, and A. H. Zewail, *J. Chem. Phys.* **84**, 4708 (1986).
92. P. M. Felker and A. H. Zewail, *J. Chem. Phys.* **86**, 2460 (1987).
93. J. S. Baskin, P. M. Felker, and A. H. Zewail, *J. Chem. Phys.* **86**, 2483 (1987).
94. P. W. Joireman, L. L. Connell, S. M. Ohline, and P. M. Felker, *J. Chem. Phys.* **96**, 4118 (1992).
95. P. M. Felker and A. H. Zewail, in J. Manz and L. Wöste (eds.), *Femtosecond Chemistry*, VCH Publishers Inc., 1994).
96. C. Riehn, *Chem. Phys.* **283**, 297 (2002).
97. R. N. Dixon, *J. Chem. Phys.* **85**, 1866 (1986).
98. P. L. Houston, *J. Phys. Chem.* **91**, 5388 (1987).
99. G. E. Hall and P. L. Houston, *Annu. Rev. Phys. Chem.* **40**, 375 (1989).
100. D. Dill, *J. Chem. Phys.* **65**, 1130 (1976).
101. K. F. Lee, D. M. Villeneuve, P. B. Corkum, A. Stolow, and J. G. Underwood, *Phys. Rev. Lett.* **97**, 173001 (2006).

102. B. Friedrich and D. Herschbach, *Phys. Rev. Lett.* **74**, 4623 (1995).
103. T. Seideman, *J. Chem. Phys.* **103**, 7887 (1995).
104. E. Peronne, M. D. Poulson, C. Z. Bisgaard, H. Stapelfeldt, and T. Seideman, *Phys. Rev. Lett.* **91**, 043003 (2003).
105. K. F. Lee, I. V. Litvinyuk, P. W. Dooley, M. Spanner, D. M. Villeneuve, and P. B. Corkum, *J. Phys. B* **37**, L43 (2004).
106. C. Z. Bisgaard, M. D. Poulsen, E. Péronne, S. S. Viftrup, and H. Stapelfeldt, *Phys. Rev. Lett.* **92**, 173004 (2004).
107. A. Matos-Abiague and J. Berakdar, *Phys. Rev. A* **68**, 063411 (2003).
108. P. W. Dooley, I. V. Litvinyuk, K. F. Lee, D. M. Rayner, M. Spanner, D. M. Villeneuve, and P. B. Corkum, *Phys. Rev. A* **68**, 023406 (2003).
109. J. J. Larsen, K. Hald, N. Bjerre, H. Stapelfeldt, and T. Seideman, *Phys. Rev. Lett.* **85**, 2470 (2000).
110. M. Machholm and N. E. Henriksen, *Phys. Rev. Lett.* **87**, 193001 (2001).
111. H. Stapelfeldt and T. Seideman, *Rev. Mod. Phys.* **75**, 543 (2003).
112. D. Sugny, A. Keller, O. Atabek, D. Daems, C. M. Dion, S. Gurin, and H. R. Jauslin, *Phys. Rev. A.* **69**, 033402 (2004).
113. J. G. Underwood, M. Spanner, M. Y. Ivanov, J. Mottershead, B. J. Sussman, and A. Stolow, *Phys. Rev. Lett.* **90**, 223001 (2003).
114. J. G. Underwood, B. J. Sussman, and A. Stolow, *Phys. Rev. Lett.* **94**, 143002 (2005).
115. D. Daems, S. Guérin, D. Sugny, and H. R. Jauslin, *Phys. Rev. Lett.* **94**, 153003 (2005).
116. H. Tanji, S. Minemoto, and H. Sakai, *Phys. Rev. A* **72**, 063401 (2005).
117. H. Sakai, S. Minemoto, H. Nanjo, H. Tanji, and T. Suzuki, *Phys. Rev. Lett.* 90, 083001 (2003).
118. P. Kruit and F. H. Read, *J. Phys. E* **16**, 313 (1983).
119. C. A. de Lange, in I. Powis, T. Baer, and C. Y. Ng (eds.), *High Resolution Laser Photoionization and Photoelectron Studies*, John Wiley & Sons, Inc., 1995, p. 195.
120. S. Lochbrunner, J. J. Larsen, J. P. Shaffer, M. Schmitt, T. Schultz, J. G. Underwood, and A. Stolow, *J. Elec. Spec. Relat. Phenom.* **112**, 183 (2000).
121. A. T. J. B. Eppink and D. H. Parker, *Rev. Sci. Instr.* **68**, 3477 (1997).
122. D. H. Parker and A. T. J. B. Eppink, *J. Chem. Phys.* **107**, 2357 (1997).
123. A. J. R. Heck and D. W. Chandler, *Annu. Rev. Phys. Chem.* **46**, 335 (1995).
124. B. J. Whitaker, in A. G. Suits and R. E. Continetti (eds.), *Imaging in Chemical Dynamics*, American Chemical Society, Washington, D.C., 2000, pp. 68–86.
125. V. Dribinski, A. Ossadtchi, V. A. Mandelshtam, and H. Reisler, *Rev. Sci. Instr.* **73**, 2634 (2002).
126. C. R. Gebhardt, T. P. Rakitzis, P. C. Samartzis, V. Ladopoulos, and T. N. Kitsopoulos, *Rev. Sci. Instr.* **72**, 3848 (2001).
127. D. Townsend, M. P. Minitti, and A. G. Suits, *Rev. Sci. Instr.* **74**, 2530 (2003).
128. R. E. Continetti and C. C. Hayden, in K. Liu and X. Yang (eds.), *Advanced Series in Physical Chemistry: Modern Trends in Chemical Reaction Dynamics*, World Scientific, Singapore, 2003.
129. J. A. Davies, J. E. LeClaire, R. E. Continetti, and C. C. Hayden, *J. Chem. Phys.* **111**, 1 (1999).
130. K. A. Hanold, M. C. Garner, and R. E. Continetti, *Phys. Rev. Lett.* **77**, 3335 (1996).
131. V. Stert, W. Radloff, T. Freudenberg, F. Noack, I. V. Hertel, C. Jouvet, C. Dedonder-Lardeux, and D. Solgadi, *Europhys. Lett.* **40**, 515 (1997).
132. V. Stert, W. Radloff, C. P. Schulz, and I. V. Hertel, *Eur. Phys. J. D* **5**, 97 (1999).

133. J. H. D. Eland, *J. Chem. Phys.* **70**, 2926 (1979).
134. K. Low, P. Hampton, and I. Powis, *Chem. Phys.* **100**, 401 (1985).
135. I. Powis, *Chem. Phys. Lett.* **189**, 473 (1992).
136. M. C. Garner, K. A. Hanold, M. S. Resat, and R. E. Continetti, *J. Phys. Chem.* **101**, 6577 (1997).
137. J. A. Davies, R. E. Continetti, D. W. Chandler, and C. C. Hayden, *Phys. Rev. Lett.* **84**, 5983 (2000).
138. O. Gessner, et al., *Science* **311**, 219 (2006).
139. F. Krausz, M. E. Fermann, T. Brabec, P. F. Curley, M. Hofer, M. H. Ober, C. Spielmann, E. Wintner, and A. J. Schmidt, *IEEE J. Quant. Elec.* **28**, 2097 (1992).
140. J. Squier, F. Salin, G. Mourou, and D. Harter, *Opt. Lett.* **16**, 324 (1991).
141. M. Nisoli, S. Desilvestri, V. Magni, O. Svelto, R. Danielius, A. Piskarskas, G. Valiulis, and A. Varanavicius, *Opt. Lett.* **19**, 1973 (1994).
142. T. Wilhelm, J. Piel, and E. Riedle, *Opt. Lett.* **22**, 1494 (1997).
143. A. L'Huillier, *J. Nonlinear Opt. Phys. Mater.* **4**, 647 (1995).
144. C. Spielmann, N. H. Burnett, S. Sartania, R. Koppitsch, M. Schnurer, C. Kan, M. Lenzner, P. Wobrauschek, and F. Krausz, *Science* **278**, 661 (1997).
145. A. Rundquist, C. G. Durfee, Z. H. Chang, C. Herne, S. Backus, M. M. Murnane, and H. C. Kapteyn, *Science* **280**, 1412 (1998).
146. R. Bartels, S. Backus, E. Zeek, L. Misoguti, G. Vdovin, I. P. Christov, M. M. Murnane, and H. C. Kapteyn, *Nature (London)* **406**, 164 (2000).
147. C. G. Durfee, A. R. Rundquist, S. Backus, C. Herne, M. M. Murnane, and H. C. Kapteyn, *Phys. Rev. Lett.* **83**, 2187 (1999).
148. S. L. Sorensen et al., *J. Chem. Phys.* **112**, 8038 (2000).
149. L. Nugent-Glandorf, M. Scheer, D. A. Samuels, V. Bierbaum, and S. R. Leone, *Rev. Sci. Instr.* **73**, 1875 (2002).
150. L. Nugent-Glandorf, M. Scheer, D. A. Samuels, A. M. Mulhisen, E. R. Grant, X. M. Yang, V. M. Bierbaum, and S. R. Leone, *Phys. Rev. Lett.* **87**, 193002 (2001).
151. L. Nugent-Glandorf, M. Scheer, D. A. Samuels, V. M. Bierbaum, and S. R. Leone, *J. Chem. Phys.* **117**, 6108 (2002).
152. D. H. Parker and P. Avouris, *J. Chem. Phys.* **71**, 1241 (1979).
153. S. M. Bellm and K. L. Reid, *Phys. Rev. Lett.* **91**, 263002 (2003).
154. S. M. Bellm, J. A. Davies, P. T. Whiteside, J. Guo, I. Powis, and K. L. Reid, *J. Chem. Phys.* **122**, 224306 (2005).
155. S. H. Lee, K. C. Tang, I. C. Chen, M. Schmitt, J. P. Shaffer, T. Schultz, J. G. Underwood, M. Z. Zgierski, and A. Stolow, *J. Phys. Chem. A* **106**, 8979 (2002).
156. P. M. Felker and A. H. Zewail, in M. Hollas and D. Phillips (eds.), *Jet Spectroscopy and Molecular Dynamics*, Chapman and Hall, Blackie Academic, New York, 1995.
157. J. C. Keske and B. H. Pate, *Annu. Rev. Phys. Chem.* **51**, 323 (2000).
158. D. B. Moss and C. S. Parmenter, *J. Chem. Phys.* **98**, 6897 (1993).
159. A. K. King, S. M. Bellm, C. J. Hammond, K. L. Reid, M. Towrie, and P. Matousek, *Mol. Phys.* **103**, 1821 (2005).
160. C. J. Hammond, K. L. Reid, and K. L. Ronayne, *J. Chem. Phys.* **124**, 201102 (2006).
161. D. B. McDonald, G. R. Fleming, and S. A. Rice, *Chem. Phys.* **60**, 335 (1981).
162. J. L. Knee, F. E. Doany, and A. H. Zewail, *J. Chem. Phys.* **82**, 1042 (1985).

163. A. Lorincz, D. D. Smith, F. Novak, R. Kosloff, D. J. Tannor, and S. A. Rice, *J. Chem. Phys.* **82**, 1067 (1985).

164. P. M. Felker and A. H. Zewail, *Chem. Phys. Lett.* **128**, 221 (1986).

165. P. M. Felker, *J. Phys. Chem.* **96**, 7844 (1992).

166. M. Tsubouchi, B. J. Whitaker, and T. Suzuki, *J. Phys. Chem. A* **108**, 6823 (2004).

167. S. Lochbrunner, M. Schmitt, J. P. Shaffer, T. Schultz, and A. Stolow, in T. Elsaesser, S. Mukamel, M. Murnane, and N. Scherer (eds.), *Ultrafast Phenomena XII*, (Springer-Verlag, Berlin, 2000), vol. 66 of *Springer Series in Chemical Physics*, p. 642.

168. S. Lochbrunner, T. Schultz, M. Schmitt, J. P. Shaffer, M. Z. Zgierski, and A. Stolow, *J. Chem. Phys.* **114**, 2519 (2001).

169. A. L. Sobolewski and W. Domcke, *Chem. Phys.* **184**, 115 (1994).

170. A. L. Sobolewski and W. Domcke, *Phys. Chem. Chem. Phys.* **1**, 3065 (1999).

171. H. Rau, in H. Durr and H. Buas-Laurent (eds.), *Photochromism, Molecules and Systems*, Elsevier, Amsterdam, The Netherlands, 1990, p. 165.

172. T. Schultz, J. Quenneville, B. Levine, A. Toniolo, S. Lochbrunner, M. Schmitt, J. P. Shaffer, M. Z. Zgierski, and A. Stolow, *J. Am. Chem. Soc.* **125**, 8098 (2003).

173. N. Tamai and H. Miyasaka, *Chem. Rev.* **100**, 1875 (2000).

174. V. Blanchet and A. Stolow, *J. Chem. Phys.* **108**, 4371 (1998).

175. S. Levchenko, H. Reisler, A. Krylov, O. Gessner, A. Stolow, H. Shi, and A. East, *J. Chem. Phys.* **125**, 084301 (2006).

176. A. V. Demyanenko, A. B. Potter, V. Dribinski, and H. Reisler, *J. Chem. Phys.* **117**, 2568 (2002).

177. M. Tsubouchi, C. A. de Lange, and T. Suzuki, *J. Chem. Phys.* **119**, 11728 (2003).

178. M. Tsubouchi and T. Suzuki, *Chem. Phys. Lett.* **382**, 418 (2003).

179. M. Tsubouchi, C. de Lange, and T. Suzuki, *J. Elec. Spec. Relat. Phenom.* **142**, 193 (2005).

180. J. A. Nelder and R. Mead, *Comp. J.* **7**, 308 (1965).

181. C. E. Crespo-Hernandez, B. Cohen, P. M. Hare, and B. Kohler, *Chem. Rev.* **104**, 1977 (2004).

182. A. Broo, J. Phys. *Chem. A* **102**, 526 (1998).

183. A. Sobelewski and W. Domcke, *Eur. Phys.* J. D **20**, 369 (2002).

184. A. Perun, S. Sobelewski and W. Domcke, *Chem. Phys.* **313**, 107 (2005).

185. C. Marian, J. *Chem. Phys.* **122**, 104314 (2005).

186. L. Blancafort, *J. Am. Chem. Soc.* **128**, 210 (2006).

187. H. Kang, K. Lee, B. Jung, Y. Ko, and S. Kim, *J. Am. Chem. Soc.* **124**, 12958 (2002).

188. B. Kang, H. Jung and S. Kim, *J. Chem. Phys.* **118**, 6717 (2003).

189. S. Ullrich, T. Schultz, M. Zgierski, and A. Stolow, *J. Am. Chem. Soc.* **126**, 2262 (2004).

190. S. Ullrich, T. Schultz, M. Zgierski, and A. Stolow, *Phys. Chem. Chem. Phys.* **6**, 2796 (2004).

191. I. Hünig, C. Plützer, K. A. Seefeld, D. Löwenich, M. Nispel, and K. Kleinermanns, *Chem Phys Chem* **5**, 1427 (2004).

192. M. Zierhut, W. Roth, and I. Fischer, *Phys. Chem. Chem. Phys.* **6**, 5168 (2004).

193. C. Canuel, M. Mons, F. Piuzzi, B. Tardivel, I. Dimicoli, and M. Elhanine, *J. Chem. Phys.* **122**, 074316 (2005).

194. H. Satzger, D. Townsend, M. Zgierski, S. Patchkovskii, S. Ullrich, and A. Stolow, *PNAS* **103**, 10196 (2006).

195. T. J. Martinez, *Acc. Chem. Res.* **39**, 119 (2006).

AUTHOR INDEX

Numbers in parentheses are reference numbers and indicate that the author's work is referred to although his name is not mentioned in the text. Numbers in *italic* show the page on which the complete references are listed.

Advances in Chemical Physics, Volume 139, edited by Stuart A. Rice

SUBJECT INDEX

Advances in Chemical Physics, Volume 139, edited by Stuart A. Rice